ELECTRONIC FILTER DESIGN HANDBOOK

Arthur B. Williams
Fred J. Taylor

Fourth Edition

McGRAW-HILL

New York Chicago San Francisco Lisbon London Madrid
Mexico City Milan New Delhi San Juan Seoul
Singapore Sydney Toronto

The McGraw-Hill Companies

1 2 3 4 5 6 7 8 9 0 DOC/DOC 0 1 9 8 7 6

ISBN 0-07-147171-5

The sponsoring editor for this book was Wendy Rinaldi, the editing supervisor was Jody McKenzie, and the production supervisor was Jean Bodeaux. It was set in Times Roman by International Typesetting and Composition. The art director for the cover was Handel Low.

Printed and bound by RR Donnelley.

From Arthur B. Williams
To my Family
Ellen, Howard, Bonnie, Robin, Mitchell
and grandchildren Leviah and Ilona

From Dr. Fred J. Taylor
To my grandchildren Schuyler, Bennett,
and Graysen and their
devoted grandmother Lori

ABOUT THE AUTHORS

Arthur Williams is the Chief Scientist at Telebyte Inc., a developer and manufacturer of broadband test equipment and data communication products. Previously, he was Senior Staff Engineer and Manager of Engineering for Tellabs Inc. Author of five books and holder of eleven patents, Mr. Williams has served as a consultant to the industry and is currently the Chairman of the IEEE Circuits and Systems (CAS) chapter on Long Island.

Fred J. Taylor is co-founder and Chairman of the Board of the Athena Group, a DSP semiconductor silicon intellectual property company and education technology innovator. He is also professor of electrical and computer engineering, and computer and information science engineering at the University of Florida. Author of ten textbooks and holder of three patents, Dr. Taylor serves as a consultant to Fortune 500 companies and government agencies.

CONTENTS

Chapter 9. Design and Selection of Inductors for *LC* Filters 361

Chapter 10. Component Selection for *LC* and Active Filters 393

Chapter 11. Normalized Filter Design Tables 425

Chapter 12. Introduction to Digital Filters 497

Appendix A. Discrete Systems Mathematics 733

Appendix B. Software Summary 767

Index 769

PREFACE

This is the fourth edition of the *Electronic Filter Design Handbook,* which was first published in 1981. It was expanded in 1988 to include five additional chapters on digital filters and then updated in 1995. This revised edition contains new material on both analog and digital filters. A CD-ROM has been included containing a number of programs which allow the rapid design of analog filters for input requirements without the tedious mathematical computations normally encountered. The digital filter chapters are all integrated with a profusion of MATLAB examples.

Prior to the introduction of this book in 1981, the design of *LC* and active filters had been reserved for specialists. The *Electronic Filter Design Handbook* treated the design of these filters in a practical manner and provided extensive tabulated data so that the average engineer who had no previous experience could design passive or active filters. This philosophy was expanded to include digital filters in the second edition in 1988, where, for the first time in any book, the design of all three classes of filters was covered in a practical easy-to-follow style. The book was then further updated in its third edition in 1995 to include a number of new technologies and design methods. In this fourth edition, the book now contains additional material and chapters on analog filters, and has been updated on available components to include surface mount technology. A number of powerful design programs have been included on the CD-ROM. Some of the previously included tables on normalized elliptic-function low-pass filters have been replaced by a powerful program, *Filter Solutions* (from Nuhertz Technologies®), which can directly design these filters and create a schematic without tedious calculations. An EXCEL spreadsheet contains formulas from the individual chapters keyed to the text so that the tedious calculations required in the past are no longer necessary.

The digital filter chapters have been completely revised and the vast majority of the material is new. This new edition contains all the topics and studies found in the previous version of the *Electronic Filter Design Handbook,* plus many more that cover the full range of modern fixed-coefficients digital filter design. The coverage provides the reader with both a conceptual understanding of digital filters and the ability to design digital filters for use in a number of application domains. The presentation keeps in mind that traditional digital filters, used a decade ago, remain in popular use today. It's also recognized that non-traditional and multirate filters are of growing importance. Thus, basic digital filter design methods are presented, along with an enriched treatment on multirate solutions. Strong emphasis is placed on achieving filter-technology synergy, which considers issues related to the physical implementation of a digital filter in either hardware or software. This fourth edition also presumes that contemporary engineers will increasingly turn to the computer software to support filter design and analysis activities. These activities are motivated and supported with a profusion of computer-generated examples using the ubiquitous MATLAB software package.

Chapter 1 introduces the concept of modern network theory and discusses the trade-off between active and passive filter implementations.

The mathematical properties of standard filter response types are covered in Chapter 2, including Butterworth, Chebyshev, Bessel, linear phase with equiripple error, transitional, synchronously tuned, and constant delay with Chebyshev stopband. Extensive normalized curves for both frequency and time-domain parameters of these standard polynomial transfer functions are provided. The highly efficient elliptic-function filter response is also discussed in this chapter and emphasized throughout the handbook. Two programs contained on the CD-ROM, *Filter Solutions (book version)* and *ELI 1.0* (for the design of elliptic-function filters), are introduced.

In Chapter 3 the design of both passive and active low-pass filters is covered using normalized tables. Specialized passive low-pass filter design techniques are illustrated, such as designing for unequal terminations and compensating for the effects of component dissipation (low Q). Various active low-pass filter structures are covered for both all-pole and elliptic-function types.

High-pass filters for both passive and active implementations are discussed in Chapter 4.

Chapter 5 covers bandpass filters. Various passive filter transformations, approximations, and identities are illustrated to ensure practical element values even for extreme conditions of center frequency, bandwidth, or impedance level. Some active bandpass implementations are offered that exhibit low sensitivity at frequencies previously considered too high for active filters.

Techniques for band-reject filter design are presented in Chapter 6, where passive and active types are covered.

Chapter 7 covers the design of networks having properties best described in the time domain. All-pass delay and amplitude equalizers are discussed in detail. Methods are shown for the design of *LC*, as well as active delay lines and wideband 90° phase shift networks.

Refinements in *LC* filter design are covered in Chapter 8. Special techniques are presented to manipulate element values so that practical values can always be obtained. Here, measurement techniques are shown, and the design of various forms of resistive attenuators for attenuation and impedance matching is covered. The theory and design of power splitting networks is also explained.

The successful operation of *LC* filter design is highly dependent upon the proper selection and manufacture of inductors. The design of magnetic components is presented in Chapter 9. The entire process—ranging from the selection of various magnetic material types and shapes to coil-winding methods to achieve the optimum characteristics over the operating frequency range—is explained in detail. New magnetic materials and shapes are covered. Q curves are also provided, as well as those for MPP torroidal cores, ferrite RM and potcores, and surface mount RF inductors.

The component selection for *LC* and active filters is discussed in Chapter 10. Coverage includes capacitor characteristics and the selection of fixed and variable resistor types. Johnson (Thermal) noise is also discussed. Operational amplifier theory is reviewed both from a theoretical and a practical standpoint, and expanded and updated device selection charts are provided to enable the rapid choice of the appropriate operational amplifier for a given filter configuration and required operating frequency range. Surface-mount (SMD) components are also emphasized, as well as the manufacturing considerations using this technology.

Chapter 11 contains normalized tables for the rapid design of both passive and active filters. In addition to the standard polynomial types, tables are provided for the unique constant-delay low-pass filters with Chebyshev stopband characteristics.

Chapter 12 introduces digital filters. The presentation begins with a differentiation of analog and digital systems, representations, and design strategies. The sampling theory is established as being a core element in the understanding and realization of digital filters, and issues such as quantifying sampling modalities and aliasing are discussed. Data conversion principles and mechanics (to and from the analog domain) are developed, and the chapter closes with a discussion of computer arithmetic and spectral analysis.

Finite impulse-response filters (FIR) are discussed in Chapter 13. The chapter begins with a general characterization of FIR filters, including stability. Both linear and non-linear phase FIR are studied in detail, and FIR design procedures based on the window, least-squares, and equiripple methods are presented, as well as the extension of their use as Hilbert and differentiating FIRs. FIR special cases, such as comb, moving average, L-band, mirror, complement, and frequency sampling forms are also described and illustrated. The FIR design strategies are reinforced using MATLAB FIR architectures, including direct, transpose, symmetric, lattice, distribute, and canonic digit form are developed in detail. The chapter concludes with a study of finite word length effects.

Chapter 14 covers the infinite impulse-response (IIR) filters. The presentation begins with a general characterization of FIR filters, including stability. A review of classic analog filters, Butterworth, Chebyshev I and II, and elliptic (Cauer) are presented and used to define classic digital IIR filters. The conversion of analog filters to digital IIR filters are examined in terms of impulse invariant and bilinear z-transforms. The design of an IIR, based upon measured input-output responses, is also presented in terms of auto-regressive models. The IIR filter design process is illustrated using MATLAB IIR filters are described using a natural state variable analysis framework, which is interpreted in terms of a Direct I and II, Cascade, Parallel, Normal, and Lattice architectures. The state variable models are next reinforced using MATLAB. The chapter concludes with an exposition of the analysis procedures required to insure a successful fixed-point IIR implementation and how to minimize run-time finite wordlength (fixed-point) errors.

Multirate filters are discussed in Chapter 15. The presentation begins with a general characterization of multirate systems and their use. Following this, the process of decimation and interpolation are developed, along with sample rate conversion. Multirate systems are then presented using a polyphase framework that is extended to the design filter banks and DFT filter banks. The chapter concludes with the development of high decimation rate filters and frequency masked filters.

Chapter 16 covers the field of digital filter technology. The presentation begins with an overview of technology types including general purpose μps, DSP μps, application-specific integrated circuits (ASIC), and field programmable gate arrays (FPGA). Various processor organizations are examined. Processor architectural variations and their impact on design choices are developed and compared. A discussion of analog to digital (ADC) architectures is presented and used to define and compare various types of converter architectures. Next, software issues and opportunities are developed and analyzed. The chapter concludes with a filter implementation case study based on Texas Instruments DSP μp processor architecture.

Chapter 17 covers switched-capacitor filters. The underlying theory behind this technology is presented, and some design examples are shown using standard building-block ICs. A survey and a convenient selection guide (updated) is included, as well as a program called FilterCAD on the CD-ROM. The latter program (from Linear Technology) can help to quickly design a switched-capacitor filter from a set of input parameters. It then provides a schematic of the filter along with both the predicted frequency response and time response.

Chapter 18 is an introduction to microwave filters. It discusses Kuroda's Identities and Richards' Transformation and illustrates some design examples of microstrip filters.

Appendix A provides a review of DSP mathematics. The presentation begins with a comprehensive study of the z-transform and inverse z-transform. Transforms are then investigated using MATLAB. The chapter concludes with a discussion of the discrete Fourier transform (DFT).

The authors would like to thank Leo Moodenbaugh of Telebyte Inc., and Michael Christensen of the University of Florida for their assistance.

ARTHUR B. WILLIAMS
FRED J. TAYLOR

CHAPTER 1
INTRODUCTION TO MODERN NETWORK THEORY

1.1 MODERN NETWORK THEORY

A generalized filter is shown in Figure 1-1. The filter block may consist of inductors, capacitors, resistors, and possibly active elements such as operational amplifiers and transistors. The terminations shown are a voltage source E_s, a source resistance R_s, and a load resistor R_L.

The circuit equations for the network of Figure 1-1 can be written by using circuit-analysis techniques. Modern network theory solves these equations to determine the network values for optimum performance in some respect.

The Pole-Zero Concept

The frequency response of the generalized filter can be expressed as a ratio of two polynomials in s where $s = j\omega$ ($j = \sqrt{-1}$, and ω, the frequency in radians per second, is $2\pi f$) and is referred to as a transfer function. This can be stated mathematically as

$$T(s) = \frac{E_L}{E_s} = \frac{N(s)}{D(s)} \tag{1-1}$$

The roots of the denominator polynomial $D(s)$ are called poles and the roots of the numerator polynomial $N(s)$ are referred to as zeros.

Deriving a network's transfer function could become quite tedious and is beyond the scope of this book. The following discussion explores the evaluation and representation of a relatively simple transfer function.

Analysis of the low-pass filter of Figure 1-2a results in the following transfer function:

$$T(s) = \frac{1}{s^3 + 2s^2 + 2s + 1} \tag{1-2}$$

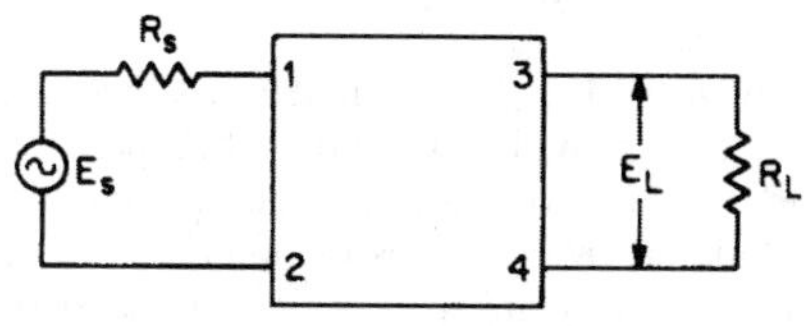

FIGURE 1-1 A generalized filter.

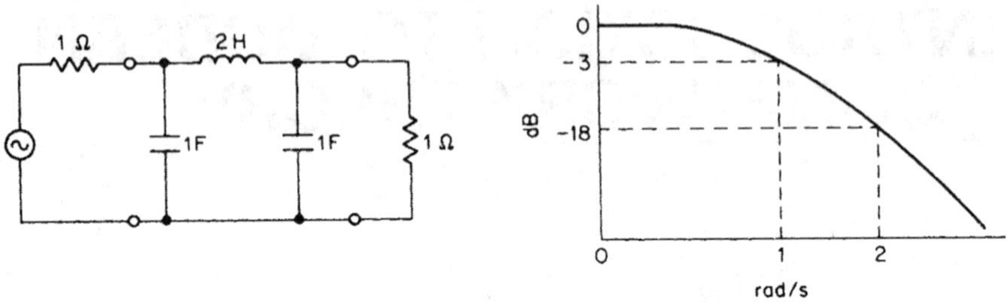

(a) (b)

FIGURE 1-2 An all-pole $n = 3$ low-pass filter: (*a*) a filter circuit; and (*b*) a frequency response.

Let us now evaluate this expression at different frequencies after substituting $j\omega$ for s. The result will be expressed as the absolute magnitude of $T(j\omega)$ and the relative attention in decibels with respect to the response at DC.

$$T(j\omega) = \frac{1}{1 - 2\omega^2 + j(2\omega - \omega^3)} \tag{1-3}$$

ω	$\lvert T(j\omega)\rvert$	$20 \log \lvert T(j\omega)\rvert$
0	1	0 dB
1	0.707	−3 dB
2	0.124	−18 dB
3	0.0370	−29 dB
4	0.0156	−36 dB

The frequency-response curve is plotted in Figure 1-2*b*.

Analysis of Equation (1-2) indicates that the denominator of the transfer function has three roots or poles and the numerator has none. The filter is therefore called an all-pole type. Since the denominator is a third-order polynomial, the filter is also said to have an $n = 3$ complexity. The denominator poles are $s = -1$, $s = -0.500 + j0.866$, and $s = -0.500 - j0.866$.

These complex numbers can be represented as symbols on a complex-number plane. The abscissa is α, the real component of the root, and the ordinate is β, the imaginary part. Each pole is represented as the symbol *X*, and a zero is represented as 0. Figure 1-3 illustrates the complex-number plane representation for the roots of Equation (1-2).

Certain mathematical restrictions must be applied regarding the location of poles and zeros in order for the filter to be realizable. They must occur in pairs which are conjugates of each other, except for real-axis poles and zeros, which may occur singly. Poles must also be restricted to the left plane (in other words, the real coordinate of the pole must be negative), while zeros may occur in either plane.

Synthesis of Filters from Polynomials. Modern network theory has produced families of standard transfer functions that provide optimum filter performance in some desired respect. Synthesis is the process of deriving circuit component values from these transfer functions. Chapter 11 contains extensive tables of transfer functions and their associated component values so that design by synthesis is not required. Also, computer programs on the CD-ROM simplify the design process. However, in order to gain some understanding

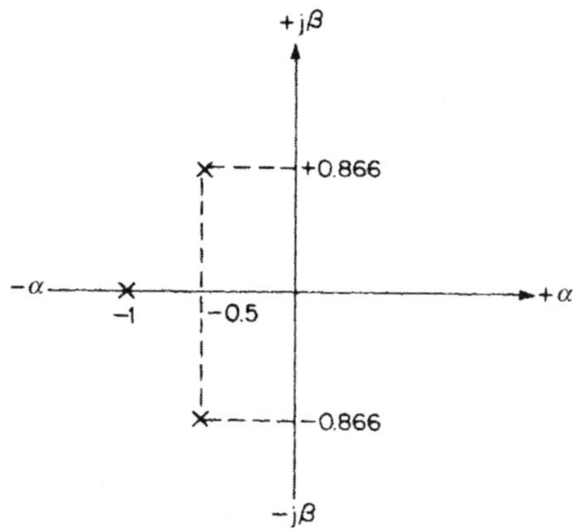

FIGURE 1-3 A complex-frequency plane representation of Equation (1-2).

as to how these values have been determined, we will now discuss a few methods of filter synthesis.

Synthesis by Expansion of Driving-Point Impedance. The input impedance to the generalized filter of Figure 1-1 is the impedance seen looking into terminals 1 and 2 with terminals 3 and 4 terminated, and is referred to as the driving-point impedance or Z_{11} of the network. If an expression for Z_{11} could be determined from the given transfer function, this expression could then be expanded to define the filter.

A family of transfer functions describing the flattest possible shape and a monotonically increasing attenuation in the stopband is known as the *Butterworth low-pass response*. These all-pole transfer functions have denominator polynomial roots, which fall on a circle having a radius of unity from the origin of the $j\omega$ axis. The attenuation for this family is 3 dB at 1 rad/s.

The transfer function of Equation (1-2) satisfies this criterion. It is evident from Figure 1-3 that if a circle were drawn having a radius of 1, with the origin as the center, it would intersect the real root and both complex roots.

If R_s in the generalized filter of Figure 1-1 is set to 1 Ω, a driving-point impedance expression can be derived in terms of the Butterworth transfer function as

$$Z_{11} = \frac{D(s) - s^n}{D(s) + s^n} \tag{1-4}$$

where $D(s)$ is the denominator polynomial of the transfer function and n is the order of the polynomial.

After $D(s)$ is substituted into Equation (1-4), Z_{11} is expanded using the continued fraction expansion. This expansion involves successive division and inversion of a ratio of two polynomials. The final form contains a sequence of terms, each alternately representing a capacitor and an inductor and finally the resistive termination. This procedure is demonstrated by the following example.

Example 1-1 Synthesis of $N = 3$ Butterworth Low-Pass Filter by Continued Fraction Expansion

Required:

A low-pass LC filter having a Butterworth $n = 3$ response.

Result:

(*a*) Use the Butterworth transfer function:

$$T(s) = \frac{1}{s^3 + 2s^2 + 2s + 1} \tag{1-2}$$

(*b*) Substitute $D(s) = s^3 + 2s^2 + 2s + 1$ and $s^n = s^3$ into Equation (1-4), which results in

$$Z_{11} = \frac{2s^2 + 2s + 1}{2s^3 + 2s^2 + 2s + 1} \tag{1-4}$$

(*c*) Express Z_{11} so that the denominator is a ratio of the higher-order to the lower-order polynomial:

$$Z_{11} = \frac{1}{\dfrac{2s^3 + 2s^2 + 2s + 1}{2s^2 + 2s + 1}}$$

(*d*) Dividing the denominator and inverting the remainder results in

$$Z_{11} = \frac{1}{s + \dfrac{1}{\dfrac{2s^2 + 2s + 1}{s + 1}}}$$

(*e*) After further division and inversion, we get as our final expression:

$$Z_{11} = \frac{1}{s + \dfrac{1}{2s + \dfrac{1}{s + 1}}} \tag{1-5}$$

The circuit configuration of Figure 1-4 is called a ladder network, since it consists of alternating series and shunt branches. The input impedance can be expressed as the following continued fraction:

$$Z_{11} = \frac{1}{Y_1 + \dfrac{1}{Z_2 + \dfrac{1}{Y_3 + \cdots \dfrac{1}{Z_{n-1} + \dfrac{1}{Y_n}}}}} \tag{1-6}$$

where $Y = sC$ and $Z = sL$ for the low-pass all-pole ladder except for a resistive termination where $Y_n = sC + 1/R_L$.

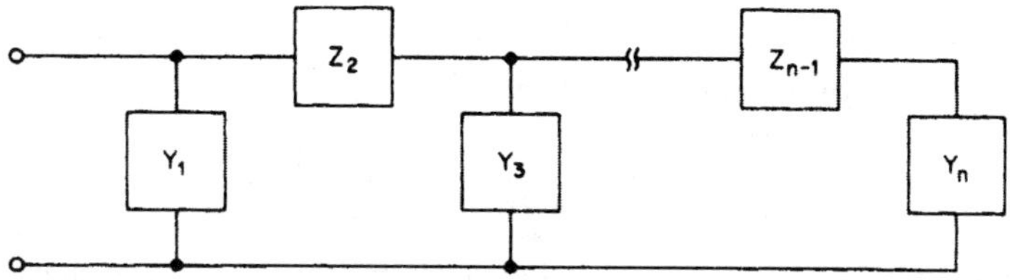

FIGURE 1-4 A general ladder network.

Figure 1-5 can then be derived from Equation (1-5) and (1-6) by inspection. This can be proved by reversing the process of expanding Z_{11}. By alternately adding admittances and impedances while working toward the input, Z_{11} is verified as being equal to Equation (1-5).

Synthesis for Unequal Terminations. If the source resistor is set equal to 1 Ω and the load resistor is desired to be infinite (unterminated), the impedance looking into terminals 1 and 2 of the generalized filter of Figure 1-1 can be expressed as

$$Z_{11} \frac{D(s \text{ even})}{D(s \text{ odd})} \tag{1-7}$$

$D(s$ even) contains all the even-power s terms of the denominator polynomial and $D(s$ odd) consist of all the odd-power s terms of any realizable all-pole low-pass transfer function. Z_{11} is expanded into a continued fraction, as in Example 1-1, to define the circuit.

Example 1-2 Synthesis of $N = 3$ Butterworth Low-Pass Filter for an Infinite Termination

Required:
Low-pass filter having a Butterworth $n = 3$ response with a source resistance of 1 Ω and an infinite termination.

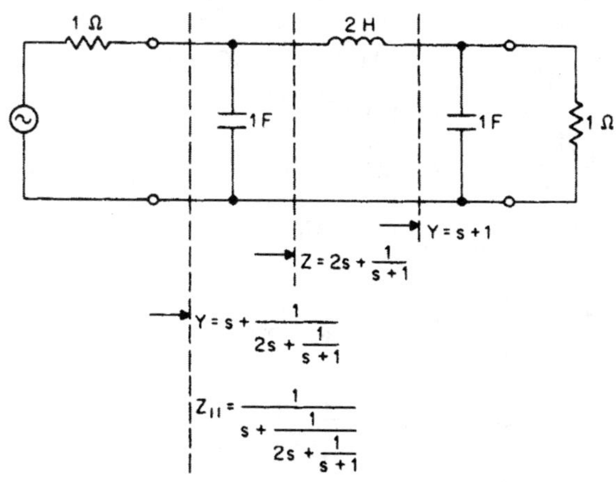

FIGURE 1-5 The low-pass filter for Equation (1-5).

Result:

(*a*) Use the Butterworth transfer function:

$$T(s) = \frac{1}{s^3 + 2s^2 + 2s + 1} \tag{1-2}$$

(*b*) Substitute $D(s \text{ even}) = 2s^2 + 1$ and $D(s \text{ odd}) = s^3 + 2s$ into Equation (1-7):

$$Z_{11} = \frac{2s^2 + 1}{s^3 + 2s} \tag{1-7}$$

(*c*) Express Z_{11} so that the denominator is a ratio of the higher- to the lower-order polynomial:

$$Z_{11} = \frac{1}{\dfrac{s^3 + 2s}{2s^2 + 1}}$$

(*d*) Dividing the denominator and inverting the remainder results in

$$Z_{11} = \frac{1}{0.5s + \dfrac{1}{\dfrac{2s^2 + 1}{1.5s}}}$$

(*e*) Dividing and further inverting results in the final continued fraction:

$$Z_{11} = \frac{1}{0.5s + \dfrac{1}{1.333s + \dfrac{1}{1.5s}}} \tag{1-8}$$

The circuit is shown in Figure 1-6.

Synthesis by Equating Coefficients. An active three-pole low-pass filter is shown in Figure 1-7. Its transfer function is given by

$$T(s) = \frac{1}{s^3A + s^2B + sC + 1} \tag{1-9}$$

where

$$A = C_1C_2C_3 \tag{1-10}$$

$$B = 2C_3(C_1 + C_2) \tag{1-11}$$

and

$$C = C_2 + 3C_3 \tag{1-12}$$

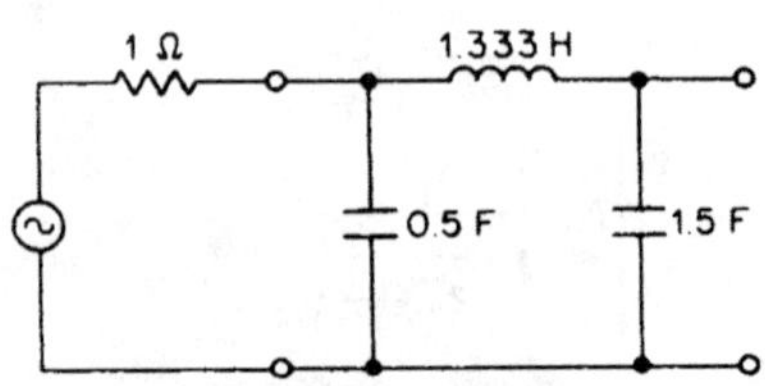

FIGURE 1-6 The low-pass filter of Example 1-2.

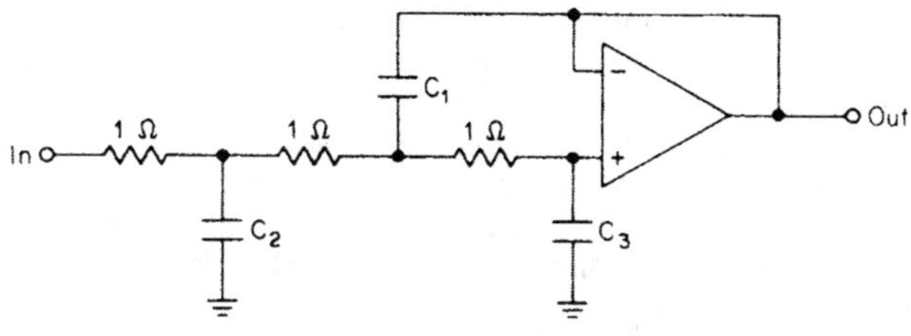

FIGURE 1-7 The general $n = 3$ active low-pass filter.

If a Butterworth transfer function is desired, we can set Equation (1-9) equal to Equation (1-2).

$$T(s) = \frac{1}{s^3A + s^2B + sC + 1} = \frac{1}{s^3 + 2s^2 + 2s + 1} \tag{1-13}$$

By equating coefficients, we obtain

$$A = 1$$
$$B = 2$$
$$C = 2$$

Substituting these coefficients in Equation (1-10) through (1-12) and solving for C_1, C_2, and C_3 results in the circuit of Figure 1-8.

Synthesis of filters directly from polynomials offers an elegant solution to filter design. However, it also may involve laborious computations to determine circuit element values. Design methods have been greatly simplified by the curves, tables, computer programs, and step-by-step procedures provided in this handbook, so design by synthesis can be left to the advanced specialist.

Active vs. Passive Filters. The LC filters of Figures 1-5 and 1-6 and the active filter of Figure 1-8 all satisfy an $n = 3$ Butterworth low-pass transfer function. The filter designer is frequently faced with the sometimes difficult decision of choosing whether to use an active or LC design. A number of factors must be considered. Some of the limitations and considerations for each filter type will now be discussed.

Frequency Limitations. At subaudio frequencies, LC filter designs require high values of inductance and capacitance along with their associated bulk. Active filters are more practical because they can be designed at higher impedance levels so that capacitor magnitudes are reduced.

Above 20 MHz or so, most commercial-grade operational amplifiers have insufficient open-loop gain for the average active filter requirement. However, amplifiers are available

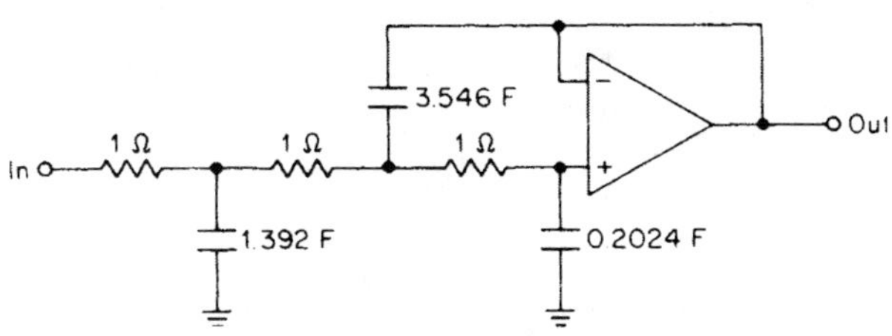

FIGURE 1-8 A Butterworth $n = 3$ active low-pass filter.

with extended bandwidth at an increased cost so that active filters at frequencies up to 100 MHz are possible. LC filters, on the other hand, are practical at frequencies up to a few hundred megahertz. Beyond this range, filters become impractical to build in lumped form, and so distributed parameter techniques are used, such as stripline or microstrip, where a PC board functions as a distributed transmission line.

Size Considerations. Active filters are generally smaller than their LC counterparts since inductors are not required. Further reduction in size is possible with microelectronic technology. Surface mount components for the most part have replaced Hybrid technology, whereas in the past Hybrids were the only way to reduce the size of active filters.

Economics and Ease of Manufacture. LC filters generally cost more than active filters because they use inductors. High-quality coils require efficient magnetic cores. Sometimes, special coil-winding methods are needed as well. These factors lead to the increased cost of LC filters.

Active filters have the distinct advantage that they can be easily assembled using standard off-the-shelf components. LC filters require coil-winding and coil-assembly skills. In addition, eliminating inductors prevents magnetic emissions, which can be troublesome.

Ease of Adjustment. In critical LC filters, tuned circuits require adjustment to specific resonances. Capacitors cannot be made variable unless they are below a few hundred picofarads. Inductors, however, can easily be adjusted, since most coil structures provide a means for tuning, such as an adjustment slug for a Ferrite potcore.

Many active filter circuits are not easily adjustable, however. They may contain RC sections where two or more resistors in each section have to be varied in order to control resonance. These types of circuit configurations are avoided. The active filter design techniques presented in this handbook include convenient methods for adjusting resonances where required, such as for narrowband bandpass filters.

BIBLIOGRAPHY

Guillemin, E. A. (1957). *Introduction to Circuit Theory.* New York: John Wiley and Sons.

Stewart, J. L. (1956). *Circuit Theory and Design.* New York: John Wiley and Sons.

White Electromagnetics. (1963). *A Handbook on Electrical Filters.* White Electromagnetics, Inc.

CHAPTER 2
SELECTING THE RESPONSE CHARACTERISTIC

2.1 FREQUENCY-RESPONSE NORMALIZATION

Several parameters are used to characterize a filter's performance. The most commonly specified requirement is frequency response. When given a frequency-response specification, the engineer must select a filter design that meets these requirements. This is accomplished by transforming the required response to a normalized low-pass specification having a cutoff of 1 rad/s. This normalized response is compared with curves of normalized low-pass filters which also have a 1-rad/s cutoff. After a satisfactory low-pass filter is determined from the curves, the tabulated normalized element values of the chosen filter are transformed or denormalized to the final design.

Modern network theory has provided us with many different shapes of amplitude versus frequency which have been analytically derived by placing various restrictions on transfer functions. The major categories of these low-pass responses are

- Butterworth
- Chebyshev
- Linear Phase
- Transitional
- Synchronously tuned
- Elliptic-function

With the exception of the elliptic-function family, these responses are all normalized to a 3-dB cutoff of 1 rad/s.

Frequency and Impedance Scaling

The basis for normalization of filters is the fact that a given filter's response can be scaled (shifted) to a different frequency range by dividing the reactive elements by a frequency-scaling factor (FSF). The FSF is the ratio of a reference frequency of the desired response to the corresponding reference frequency of the given filter. Usually 3-dB points are selected as reference frequencies of low-pass and high-pass filters, and the center frequency is chosen as the reference for bandpass filters. The FSF can be expressed as

$$\text{FSF} = \frac{\text{desired reference frequency}}{\text{existing reference frequency}} \tag{2-1}$$

The FSF must be a dimensionless number; so both the numerator and denominator of Equation (2-1) must be expressed in the same units, usually radians per second. The following example demonstrates the computation of the FSF and frequency scaling of filters.

Example 2-1 Frequency Scaling of a Low-Pass Filter

Required:
A low-pass filter, either *LC* or active, with an $n = 3$ Butterworth transfer function having a 3-dB cutoff at 1000 Hz.

Result:
Figure 2-1 illustrates the *LC* and active $n = 3$ Butterworth low-pass filters discussed in Chapter 1 and their response.

(*a*) Compute FSF.

$$\text{FSF} = \frac{2\pi 1000 \text{ rad/s}}{1 \text{ rad/s}} = 6280 \tag{2-1}$$

(*b*) Dividing all the reactive elements by the FSF results in the filters of Figure 2-2*a* and *b* and the response of Figure 2-2*c*.

Note that all points on the frequency axis of the normalized response have been multiplied by the FSF. Also, since the normalized filter has its cutoff at 1 rad/s, the FSF can be directly expressed by $2\pi f_c$, where f_c is the desired low-pass cutoff frequency in hertz.

Frequency scaling a filter has the effect of multiplying all points on the frequency axis of the response curve by the FSF. Therefore, a normalized response curve can be directly used to predict the attenuation of the denormalized filter.

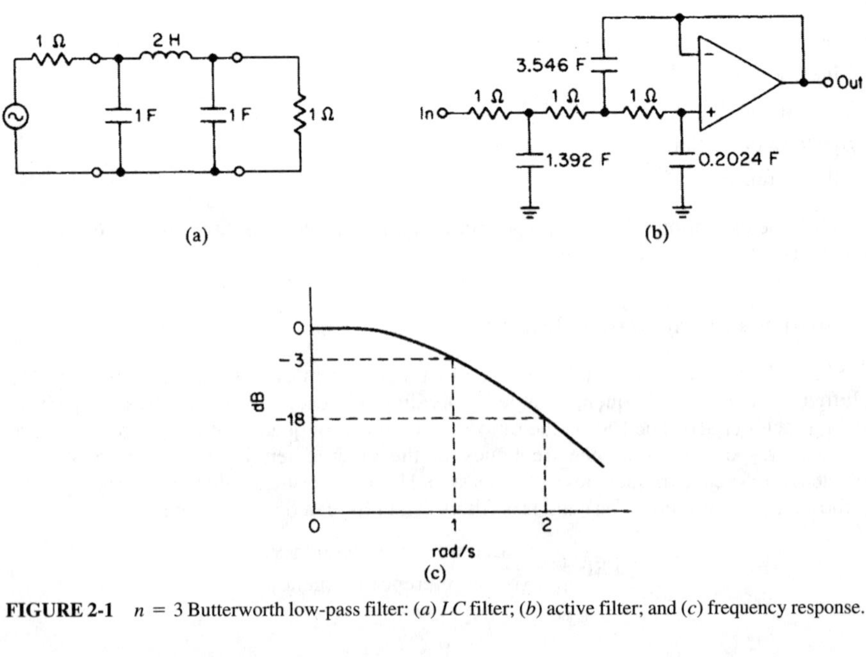

FIGURE 2-1 $n = 3$ Butterworth low-pass filter: (*a*) *LC* filter; (*b*) active filter; and (*c*) frequency response.

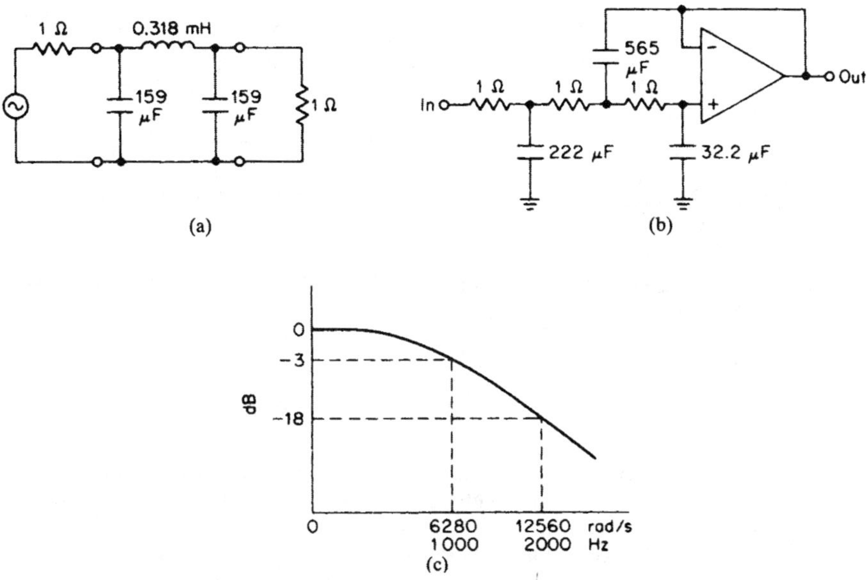

FIGURE 2-2 The denormalized low-pass filter of Example 2-1: (*a*) *LC* filter; (*b*) active filter; and (*c*) frequency response.

When the filters of Figure 2-1 were denormalized to those of Figure 2-2, the transfer function changed as well. The denormalized transfer function became

$$T(s) = \frac{1}{4.03 \times 10^{-12}s^3 + 5.08 \times 10^{-9}s^2 + 3.18 \times 10^{-4}s + 1} \tag{2-2}$$

The denominator has roots:

$$s = -6280, s = -3140 + j5438, \text{and } s = -3140 - j5438.$$

These roots can be obtained directly from the normalized roots by multiplying the normalized root coordinates by the FSF. Frequency scaling a filter also scales the poles and zeros (if any) by the same factor.

The component values of the filters in Figure 2-2 are not very practical. The capacitor values are much too large and the 1-Ω resistor values are not very desirable. This situation can be resolved by impedance scaling. Any linear active or passive network maintains its transfer function if all resistor and inductor values are multiplied by an impedance-scaling factor *Z*, and all capacitors are divided by the same factor *Z*. This occurs because the *Z*s cancel in the transfer function. To prove this, let's investigate the transfer function of the simple two-pole low-pass filter of Figure 2-3*a*, which is

$$T(s) = \frac{1}{s^2LC + sCR + 1} \tag{2-3}$$

Impedance scaling can be mathematically expressed as

$$R' = ZR \tag{2-4}$$

$$L' = ZL \tag{2-5}$$

$$C' = \frac{C}{Z} \tag{2-6}$$

where the primes denote the values after impedance scaling.

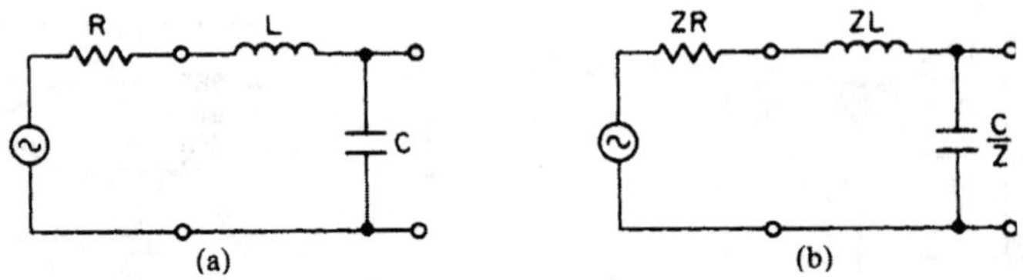

FIGURE 2-3 A two-pole low-pass *LC* filter: (*a*) a basic filter; and (*b*) an impedance-scaled filter.

If we impedance-scale the filter, we obtain the circuit of Figure 2-3*b*. The new transfer function then becomes

$$T(s) = \frac{1}{s^2 ZL\dfrac{C}{Z} + s\dfrac{C}{Z}ZR + 1} \tag{2-7}$$

Clearly, the *Z*s cancel, so both transfer functions are equivalent.

We can now use impedance scaling to make the values in the filters of Figure 2-2 more practical. If we use impedance scaling with a *Z* of 1000, we obtain the filters of Figure 2-4. The values are certainly more suitable.

Frequency and impedance scaling are normally combined into one step rather than performed sequentially. The denormalized values are then given by

$$R' = R \times Z \tag{2-8}$$

$$L' = \frac{L \times Z}{\text{FSF}} \tag{2-9}$$

$$C' = \frac{C}{\text{FSF} \times Z} \tag{2-10}$$

where the primed values are both frequency- and impedance-scaled.

Low-Pass Normalization. In order to use normalized low-pass filter curves and tables, a given low-pass filter requirement must first be converted into a normalized requirement. The curves can now be entered to find a satisfactory normalized filter which is then scaled to the desired cutoff.

The first step in selecting a normalized design is to convert the requirement into a steepness factor A_s, which can be defined as

$$A_s = \frac{f_s}{f_c} \tag{2-11}$$

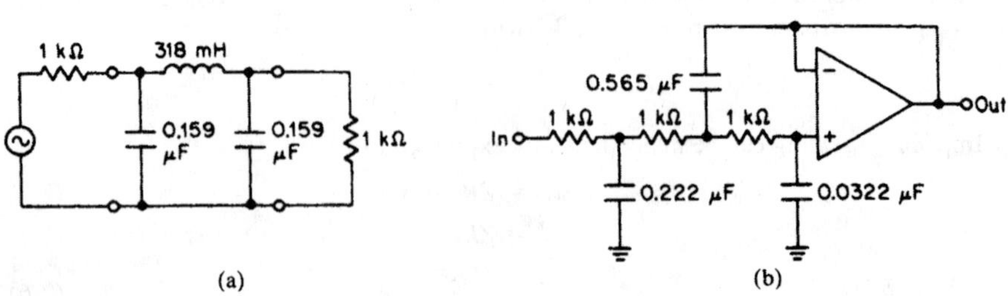

FIGURE 2-4 The impedance-scaled filters of Example 2-1: (*a*) *LC* filter; and (*b*) active filter.

where f_s is the frequency having the minimum required stopband attenuation and f_c is the limiting frequency or cutoff of the passband, usually the 3-dB point. The normalized curves are compared with A_s, and a design is selected that meets or exceeds the requirement. The design is often frequency scaled so that the selected passband limit of the normalized design occurs at f_c.

If the required passband limit f_c is defined as the 3-dB cutoff, the steepness factor A_s can be directly looked up in radians per second on the frequency axis of the normalized curves.

Suppose that we required a low-pass filter that has a 3-dB point at 100 Hz and more than 30-dB attenuation at 400 Hz. A normalized low-pass filter that has its 3-dB point at 1 rad/s and over 30-dB attenuation at 4 rad/s would meet the requirement if the filter were frequency-scaled so that the 3-dB point occurred at 100 Hz. Then there would be over 30-dB attenuation at 400 Hz, or four times the cutoff, because a response shape is retained when a filter is frequency scaled.

The following example demonstrates normalizing a simple low-pass requirement.

Example 2-2 Normalizing a Low-Pass Specification for a 3-dB cutoff

Required:

Normalize the following specification:

A low-pass filter

3 dB at 200 Hz

30-dB minimum at 800 Hz

Result:

(*a*) Compute A_s.

$$A_s = \frac{f_s}{f_c} = \frac{800 \text{ Hz}}{200 \text{ Hz}} = 4 \tag{2-11}$$

(*b*) Normalized requirement:

3 dB at 1 rad/s

30-dB minimum at 4 rad/s

In the event f_c does not correspond to the 3-dB cutoff, A_s can still be computed and a normalized design found that will meet the specifications. This is illustrated in the following example.

Example 2-3 Normalizing a Low-Pass Specification for a 1-dB cutoff

Required:

Normalize the following specification:

A low-pass filter

1 dB at 200 Hz

30-dB minimum at 800 Hz

Result:

(*a*) Compute A_s.

$$A_s = \frac{f_s}{f_c} = \frac{800 \text{ Hz}}{200 \text{ Hz}} = 4 \tag{2-11}$$

(*b*) Normalized requirement:

1 dB at K rad/s
30-dB minimum at 4 K rad/s
(where K is arbitrary)

A possible solution to Example 2-3 would be a normalized filter which has a 1-dB point at 0.8 rad/s and over 30 dB attenuation at 3.2 rad/s. The fundamental requirement is that the normalized filter makes the transition between the passband and stopband limits within a frequency ratio A_s.

High-Pass Normalization. A normalized $n = 3$ low-pass Butterworth transfer function was given in section 1.1 as

$$T(s) = \frac{1}{s^3 + 2s^2 + 2s + 1} \tag{1-2}$$

and the results of evaluating this transfer function at various frequencies were

ω	$\lvert T(j\omega)\rvert$	$20 \log \lvert T(j\omega)\rvert$
0	1	0 dB
1	0.707	−3 dB
2	0.124	−18 dB
3	0.0370	−29 dB
4	0.0156	−36 dB

Let's now perform a high-pass transformation by substituting $1/s$ for s in Equation (1-2). After some algebraic manipulations, the resulting transfer function becomes

$$T(s) = \frac{s^3}{s^3 + 2s^2 + 2s + 1} \tag{2-12}$$

If we evaluate this expression at specific frequencies, we can generate the following table:

ω	$\lvert T(j\omega)\rvert$	$20 \log \lvert T(j\omega)\rvert$
0.25	0.0156	−36 dB
0.333	0.0370	−29 dB
0.500	0.124	−18 dB
1	0.707	−3 dB
∞	1	0 dB

The response is clearly that of a high-pass filter. It is also apparent that the low-pass attenuation values now occur at high-pass frequencies that are exactly the reciprocals of the corresponding low-pass frequencies. A high-pass transformation of a normalized low-pass filter transposes the low-pass attenuation values to reciprocal frequencies and retains the 3-dB cutoff at 1 rad/s. This relationship is evident in Figure 2-5, where both filter responses are compared.

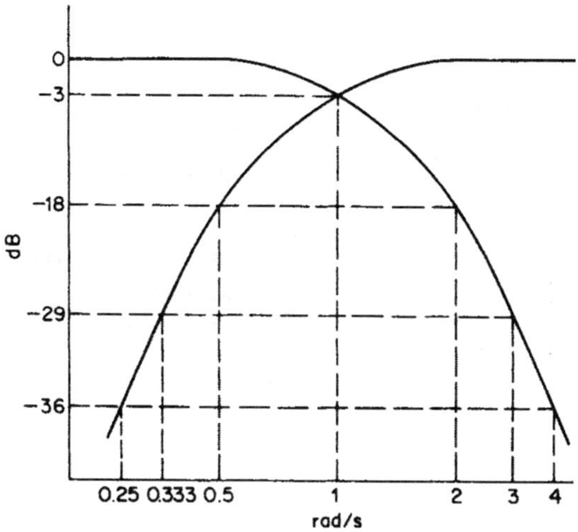

FIGURE 2-5 A normalized low-pass high-pass relationship.

The normalized low-pass curves could be interpreted as normalized high-pass curves by reading the attenuation as indicated and taking the reciprocals of the frequencies. However, it is much easier to convert a high-pass specification into a normalized low-pass requirement and use the curves directly.

To normalize a high-pass filter specification, calculate A_s, which in the case of high-pass filters is given by

$$A_s = \frac{f_c}{f_s} \tag{2-13}$$

Since the A_s, for high-pass filters is defined as the reciprocal of the A_s for low-pass filters, Equation (2-13) can be directly interpreted as a low-pass requirement. A normalized low-pass filter can then be selected from the curves. A high-pass transformation is performed on the corresponding low-pass filter, and the resulting high-pass filter is scaled to the desired cutoff frequency.

The following example shows the normalization of a high-pass filter requirement.

Example 2-4 Normalizing a High-Pass Specification

Required:

Normalize the following requirement:

A high-pass filter
3 dB at 200 Hz
30-dB minimum at 50 Hz

Result:

(*a*) Compute A_s.

$$A_s = \frac{f_c}{f_s} = \frac{200 \text{ Hz}}{50 \text{ Hz}} = 4 \tag{2-13}$$

(*b*) Normalized equivalent low-pass requirement:

3 dB at 1 rad/s
30-dB minimum at 4 rad/s

Bandpass Normalization. Bandpass filters fall into two categories: narrowband and wideband. If the ratio of the upper cutoff frequency to the lower cutoff frequency is over 2 (an octave), the filter is considered a wideband type.

Wideband Bandpass Filters. Wideband filter specifications can be separated into individual low-pass and high-pass requirements which are treated independently. The resulting low-pass and high-pass filters are then cascaded to meet the composite response.

Example 2-5 Normalizing a Wideband Bandpass Filter

Required:
Normalize the following specification:

bandpass filter
3 dB at 500 and 1000 Hz
40-dB minimum at 200 and 2000 Hz

Result:

(*a*) Determine the ratio of upper cutoff to lower cutoff.

$$\frac{1000 \text{ Hz}}{500 \text{ Hz}} = 2$$

wideband type

(*b*) Separate requirement into individual specifications.

High-pass filter:	**Low-pass filter:**
3 dB at 500 Hz	3 dB at 1000 Hz
40-dB minimum at 200 Hz	40-dB minimum at 2000 Hz
$A_s = 2.5$ (2-13)	$A_s = 2.0$ (2-11)

(*c*) Normalized high-pass and low-pass filters are now selected, scaled to the required cutoff frequencies, and cascaded to meet the composite requirements. Figure 2-6 shows the resulting circuit and response.

Narrowband Bandpass Filters. Narrowband bandpass filters have a ratio of upper cutoff frequency to lower cutoff frequency of approximately 2 or less and cannot be designed as separate low-pass and high-pass filters. The major reason for this is evident from Figure 2-7. As the ratio of upper cutoff to lower cutoff decreases, the loss at the center frequency will increase, and it may become prohibitive for ratios near unity.

If we substitute $s + 1/s$ for s in a low-pass transfer function, a bandpass filter results. The center frequency occurs at 1 rad/s, and the frequency response of the low-pass filter is directly transformed into the bandwidth of the bandpass filter at points of equivalent attenuation. In other words, the attenuation bandwidth ratios remain unchanged. This is shown in Figure 2-8, which shows the relationship between a low-pass filter and its transformed bandpass equivalent. Each pole and zero of the low-pass filter is transformed into a *pair* of poles and zeros in the bandpass filter.

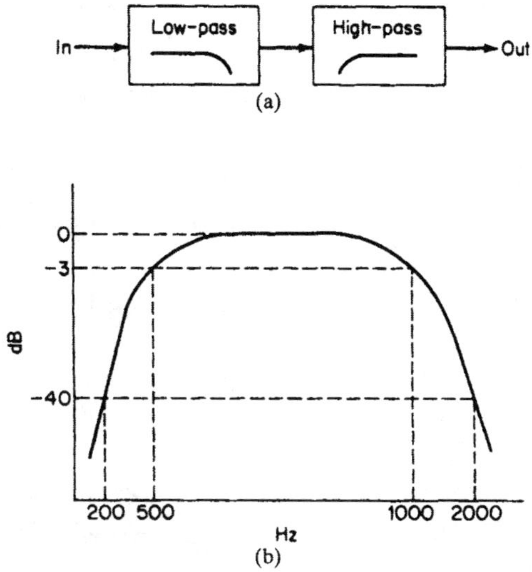

FIGURE 2-6 The results of Example 2-5: *(a)* cascade of low-pass and high-pass filters; and *(b)* frequency response.

In order to design a bandpass filter, the following sequence of steps is involved.

1. Convert the given bandpass filter requirement into a normalized low-pass specification.
2. Select a satisfactory low-pass filter from the normalized frequency-response curves.
3. Transform the normalized low-pass parameters into the required bandpass filter.

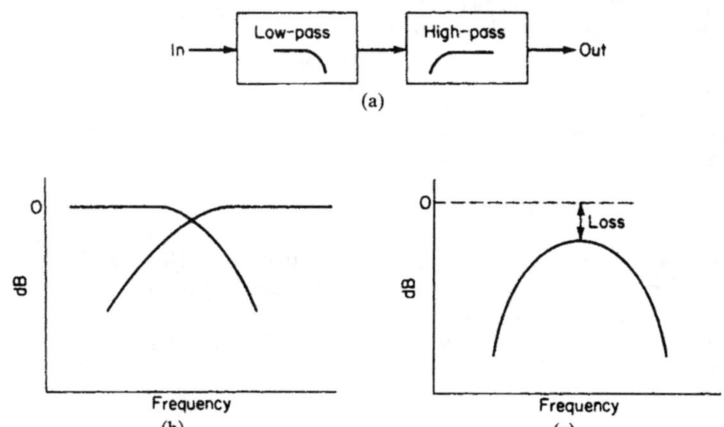

FIGURE 2-7 Limitations of the wideband approach for narrowband filters: (*a*) a cascade of low-pass and high-pass filters; (*b*) a composite response; and (*c*) algebraic sum of attenuation.

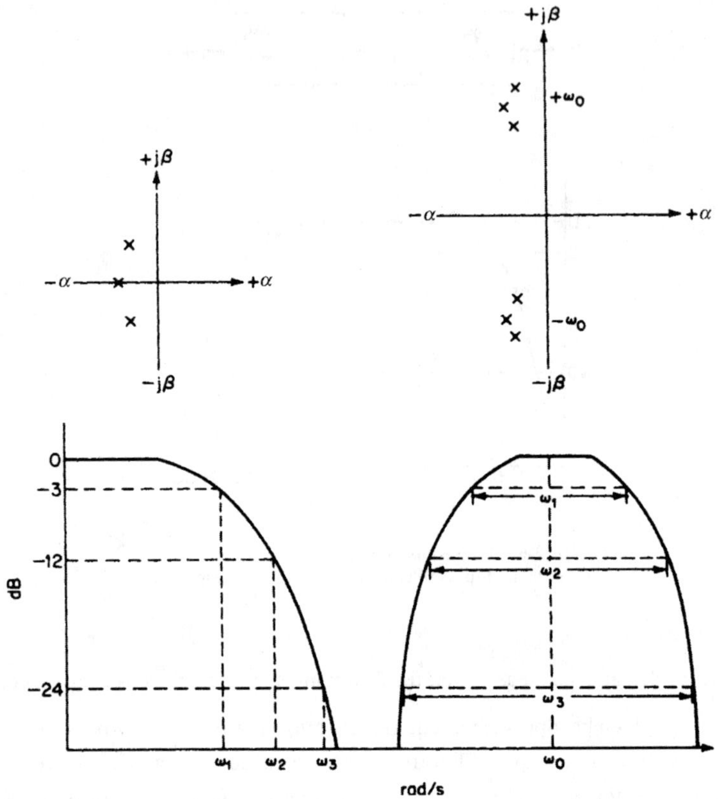

FIGURE 2-8 A low-pass to bandpass transformation.

The response shape of a bandpass filter is shown in Figure 2-9, along with some basic terminology. The center frequency is defined as

$$f_0 = \sqrt{f_L f_u} \tag{2-14}$$

where f_L is the lower passband limit and f_u is the upper passband limit, usually the 3-dB attenuation frequencies. For the more general case

$$f_0 = \sqrt{f_1 f_2} \tag{2-15}$$

where f_1 and f_2 are any two frequencies having equal attenuation. These relationships imply geometric symmetry; that is, the entire curve below f_0 is the mirror image of the curve above f_0 when plotted on a *logarithmic* frequency axis.

An important parameter of bandpass filters is the filter selectivity factor or Q, which is defined as

$$Q = \frac{f_0}{\text{BW}} \tag{2-16}$$

where BW is the passband bandwidth or $f_u - f_L$.

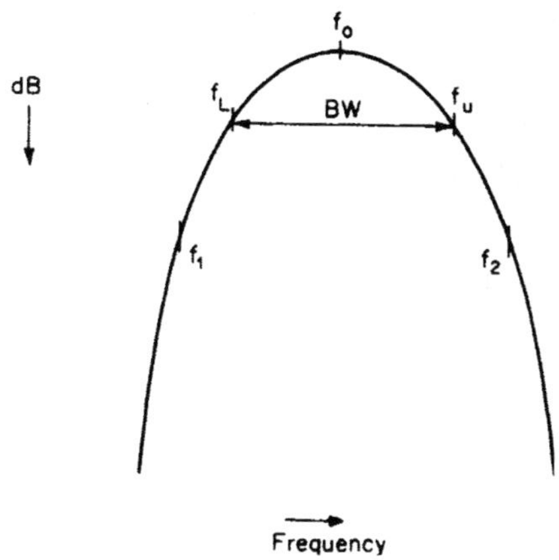

FIGURE 2-9 A general bandpass filter response shape.

As the filter Q increases, the response shape near the passband approaches the arithmetically symmetrical condition which is mirror-image symmetry near the center frequency, when plotted using a *linear* frequency axis. For Qs of 10 or more, the center frequency can be redefined as the arithmetic mean of the passband limits, so we can replace Equation (2-14) with

$$f_0 = \frac{f_L + f_u}{2} \tag{2-17}$$

In order to utilize the normalized low-pass filter frequency-response curves, a given narrowband bandpass filter specification must be transformed into a normalized low-pass requirement. This is accomplished by first manipulating the specification to make it geometrically symmetrical. At equivalent attenuation points, corresponding frequencies above and below f_0 must satisfy

$$f_1 f_2 = f_0^2 \tag{2-18}$$

which is an alternate form of Equation (2-15) for geometric symmetry. The given specification is modified by calculating the corresponding opposite geometric frequency for each stopband frequency specified. Each pair of stopband frequencies will result in two new frequency pairs. The pair having the lesser separation is retained, since it represents the more severe requirement.

A bandpass filter steepness factor can now be defined as

$$A_s = \frac{\text{stopband bandwidth}}{\text{passband bandwidth}} \tag{2-19}$$

This steepness factor is used to select a normalized low-pass filter from the frequency-response curves that makes the passband to stopband transition within a frequency ratio of A_s.

The following example shows the normalization of a bandpass filter requirement.

Example 2-6 Normalizing a Bandpass Filter Requirement

Required:
Normalize the following bandpass filter requirement:

A bandpass filter
A center frequency of 100 Hz
3 dB at ±15 Hz (85 Hz, 115 Hz)
40 dB at ±30 Hz (70 Hz, 130 Hz)

Result:

(*a*) First, compute the center frequency f_0.

$$f_0 = \sqrt{f_L f_u} = \sqrt{85 \times 115} = 98.9 \text{ Hz} \tag{2-14}$$

(*b*) Compute two geometrically related stopband frequency pairs for each pair of stopband frequencies given.

Let $f_1 = 70$ Hz.

$$f_2 = \frac{f_0^2}{f_1} = \frac{(98.9)^2}{70} = 139.7 \text{ Hz} \tag{2-18}$$

Let $f_2 = 130$ Hz.

$$f_1 = \frac{f_0^2}{f_2} = \frac{(98.9)^2}{130} = 75.2 \text{ Hz} \tag{2-18}$$

The two pairs are

$$f_1 = 70 \text{ Hz}, f_2 = 139.7 \text{ Hz } (f_2 - f_1 = 69.7 \text{ Hz})$$

and

$$f_1 = 75.2 \text{ Hz}, f_2 = 130 \text{ Hz } (f_2 - f_1 = 54.8 \text{ Hz})$$

Retain the second frequency pair, since it has the lesser separation. Figure 2-10 compares the specified filter requirement and the geometrically symmetrical equivalent.

(*c*) Calculate A_s.

$$A_s = \frac{\text{stopband bandwidth}}{\text{passband bandwidth}} = \frac{54.8 \text{ Hz}}{30 \text{ Hz}} = 1.83 \tag{2-19}$$

(*d*) A normalized low-pass filter can now be selected from the normalized curves. Since the passband limit is the 3-dB point, the normalized filter is required to have over 40 dB of rejection at 1.83 rad/s or 1.83 times the 1-rad/s cutoff.

The results of Example 2-6 indicate that when frequencies are specified in an arithmetically symmetrical manner, the narrower stopband bandwidth can be directly computed by

$$\text{BW}_{\text{stopband}} = f_2 - \frac{f_0^2}{f_2} \tag{2-20}$$

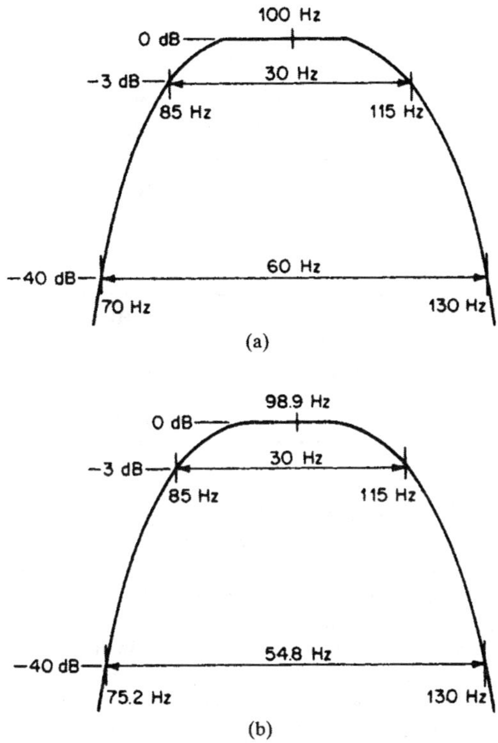

FIGURE 2-10 The frequency-response requirements of Example 2-6: (*a*) a given filter requirement; and (*b*) a geometrically symmetrical requirement.

The narrower stopband bandwidth corresponds to the more stringent value of A_s, the steepness factor.

It is sometimes desirable to compute two geometrically related frequencies that correspond to a given bandwidth. Upon being given the center frequency f_0 and the bandwidth BW, the lower and upper frequencies are respectively computed by

$$f_1 = \sqrt{\left(\frac{\text{BW}}{2}\right)^2 + f_0^2} - \frac{\text{BW}}{2} \tag{2-21}$$

$$f_2 = \sqrt{\left(\frac{\text{BW}}{2}\right)^2 + f_0^2} + \frac{\text{BW}}{2} \tag{2-22}$$

Use of these formulas is illustrated in the following example.

Example 2-7 Determining Bandpass Filter Bandwidths at Equal Attenuation Points

Required:

For a bandpass filter having a center frequency of 10 kHz, determine the frequencies corresponding to bandwidths of 100 Hz, 500 Hz, and 2000 Hz.

Result:
Compute f_1 and f_2 for each bandwidth, using

$$f_1 = \sqrt{\left(\frac{\text{BW}}{2}\right)^2 + f_0^2} - \frac{\text{BW}}{2} \tag{2-21}$$

$$f_2 = \sqrt{\left(\frac{\text{BW}}{2}\right)^2 + f_0^2} + \frac{\text{BW}}{2} \tag{2-22}$$

BW, Hz	f_1, Hz	F_2, Hz
100	9950	10,050
500	9753	10,253
2000	9050	11,050

The results of Example 2-7 indicate that for narrow percentage bandwidths (1 percent) f_1 and f_2 are arithmetically spaced about f_0. For the wider cases, the arithmetic center of f_1 and f_2 would be slightly above the actual geometric center frequency f_0. Another and more meaningful way of stating the converse is that for a given pair of frequencies, the geometric mean is below the arithmetic mean.

Bandpass filter requirements are not always specified in an arithmetically symmetrical manner as in the previous examples. Multiple stopband attenuation requirements may also exist. The design engineer is still faced with the basic problem of converting the given parameters into geometrically symmetrical characteristics so that a steepness factor (or factors) can be determined. The following example demonstrates the conversion of a specification somewhat more complicated than the previous example.

Example 2-8 Normalizing a Non-Symmetrical Bandpass Filter Requirement

Required:
Normalize the following bandpass filter specification:

bandpass filter
1-dB passband limits of 12 kHz and 14 kHz
20-dB minimum at 6 kHz
30-dB minimum at 4 kHz
40-dB minimum at 56 kHz

Result:

(*a*) First, compute the center frequency, using

$$f_L = 12 \text{ kHz} \qquad f_u = 14 \text{ kHz}$$

$$f_0 = 12.96 \text{ kHz} \tag{2-14}$$

(*b*) Compute the corresponding geometric frequency for each stopband frequency given, using Equation (2-18).

$$f_1 f_2 = f_0^2 \tag{2-18}$$

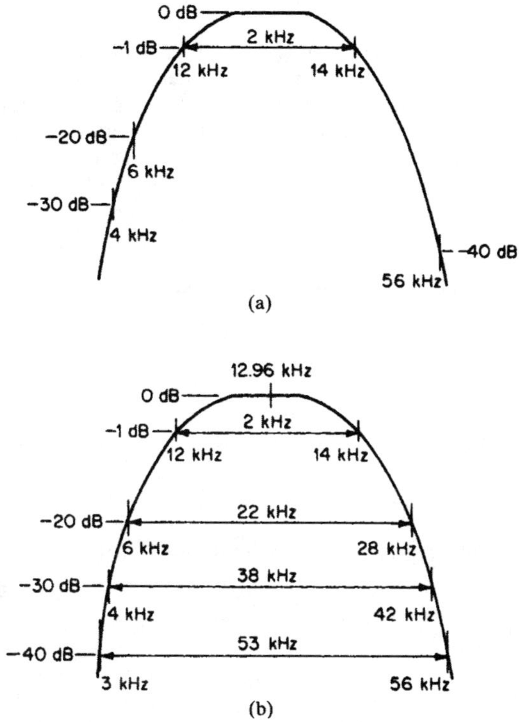

FIGURE 2-11 The given and transformed responses of Example 2-7: (*a*) a given requirement; and (*b*) geometrically symmetrical response.

Figure 2-11 illustrates the comparison between the given requirement and the corresponding geometrically symmetrical equivalent response.

f_1	f_2
6 kHz	28 kHz
4 kHz	42 kHz
3 kHz	56 kHz

(*c*) Calculate the steepness factor for each stopband bandwidth in Figure 2-11*b*.

20 dB: $$A_s = \frac{22\text{ kHz}}{2\text{ kHz}} = 11 \qquad (2\text{-}19)$$

30 dB: $$A_s = \frac{38\text{ kHz}}{2\text{ kHz}} = 19$$

40 dB: $$A_s = \frac{53\text{ kHz}}{2\text{ kHz}} = 26.5$$

(*d*) Select a low-pass filter from the normalized tables. A filter is required that has over 20, 30, and 40 dB of rejection at, respectively, 11, 19, and 26.5 times its 1-dB cutoff.

Band-Reject Normalization

Wideband Band-Reject Filters. Normalizing a band-reject filter requirement proceeds along the same lines as for a bandpass filter. If the ratio of the upper cutoff frequency to the lower cutoff frequency is an octave or more, a band-reject filter requirement can be classified as wideband and separated into individual low-pass and high-pass specifications. The resulting filters are paralleled at the input and combined at the output. The following example demonstrates normalization of a wideband band-reject filter requirement.

Example 2-9 Normalizing a Wideband Band-Reject Filter

Required:

A band-reject filter

3 dB at 200 and 800 Hz

40-dB minimum at 300 and 500 Hz

Result:

(*a*) Determine the ratio of upper cutoff to lower cutoff, using

$$\frac{800 \text{ Hz}}{200 \text{ Hz}} = 4$$

wideband type

(*b*) Separate requirements into individual low-pass and high-pass specifications.

Low-pass filter:	**High-pass filter:**
3 dB at 200 Hz	3 dB at 800 Hz
40-dB minimum at 300 Hz	40-dB minimum at 500 Hz
$A_s = 1.5$ (2-11)	$A_s = 1.6$ (2-13)

(*c*) Select appropriate filters from the normalized curves and scale the normalized low-pass and high-pass filters to cutoffs of 200 Hz and 800 Hz, respectively. Figure 2-12 shows the resulting circuit and response.

The basic assumption of the previous example is that when the filter outputs are combined, the resulting response is the superimposed individual response of both filters. This is a valid assumption if each filter has sufficient rejection in the band of the other filter so that there is no interaction when the outputs are combined. Figure 2-13 shows the case where inadequate separation exists.

The requirement for a minimum separation between cutoffs of an octave or more is by no means rigid. Sharper filters can have their cutoffs placed closer together with minimal interaction.

Narrowband Band-Reject Filters. The normalized transformation described for bandpass filters where $s + 1/s$ is substituted into a low-pass transfer function can instead be applied to a high-pass transfer function to obtain a band-reject filter. Figure 2-14 shows the direct equivalence between a high-pass filter's frequency response and the transformed band-reject filter's bandwidth.

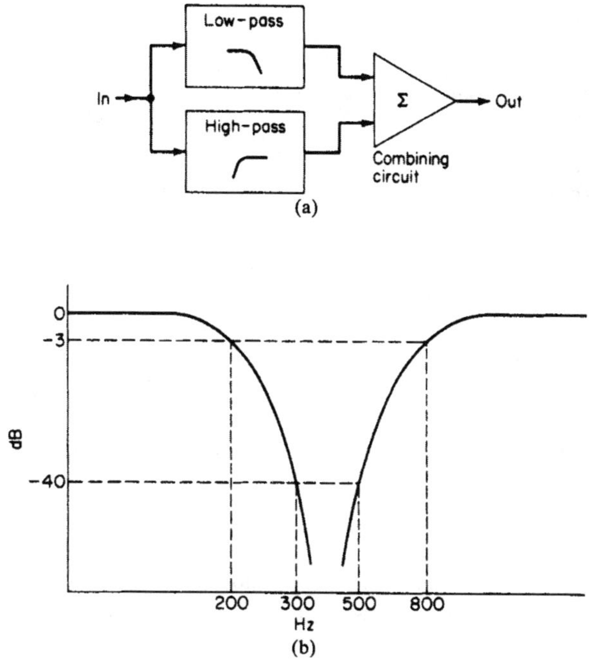

FIGURE 2-12 The results of Example 2-9: (*a*) combined low-pass and high-pass filters; and (*b*) a frequency response.

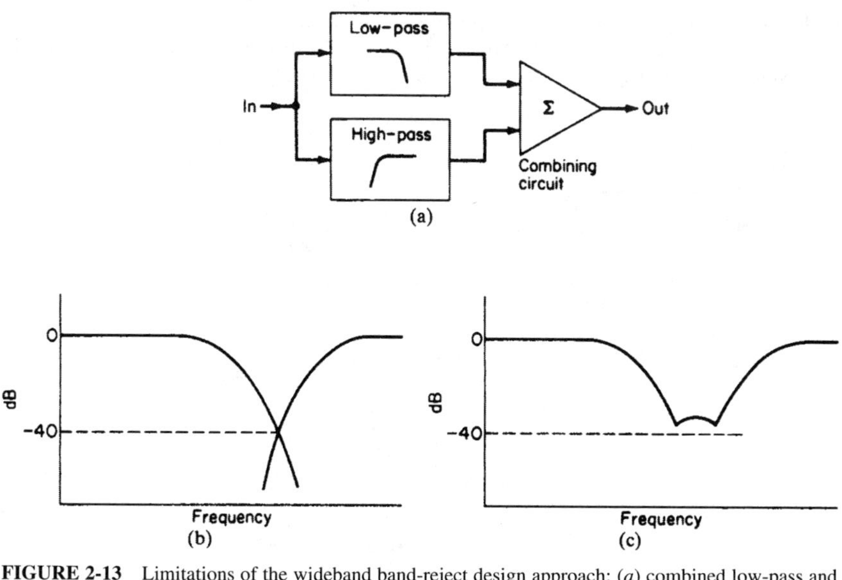

FIGURE 2-13 Limitations of the wideband band-reject design approach: (*a*) combined low-pass and high-pass filters; (*b*) composite response; and (*c*) combined response by the summation of outputs.

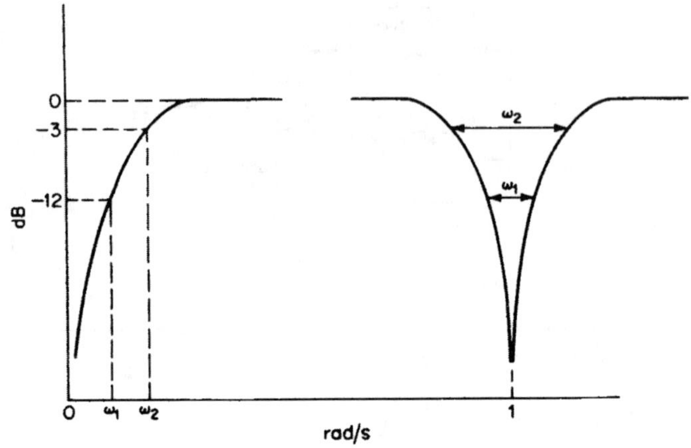

FIGURE 2-14 The relationship between band-reject and high-pass filters.

The design method for narrowband band-reject filters can be defined as follows:

1. Convert the band-reject requirement directly into a normalized low-pass specification.
2. Select a low-pass filter (from the normalized curves) that meets the normalized requirements.
3. Transform the normalized low-pass parameters into the required band-reject filter. This may involve designing the intermediate high-pass filter, or the transformation may be direct.

The band-reject response has geometric symmetry just as bandpass filters have. Figure 2-15 defines this response shape. The parameters shown have the same relationship to each other as they do for bandpass filters. The attenuation at the center frequency is theoretically infinite since the response of a high-pass filter at DC has been transformed to the center frequency.

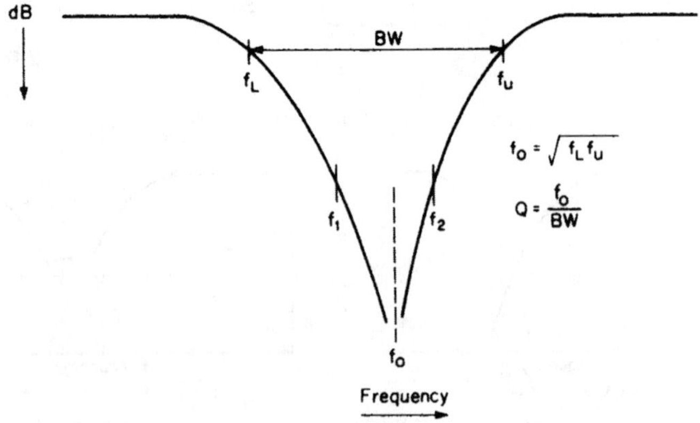

FIGURE 2-15 The band-reject response.

The geometric center frequency can be defined as

$$f_0 = \sqrt{f_L f_u} \tag{2-14}$$

where f_L and f_u are usually the 3-dB frequencies, or for the more general case:

$$f_0 = \sqrt{f_1 f_2} \tag{2-15}$$

The selectivity factor Q is defined as

$$Q = \frac{f_0}{\text{BW}} \tag{2-16}$$

where BW is $f_u - f_L$. For Qs of 10 or more, the response near the center frequency approaches the arithmetically symmetrical condition, so we can then state

$$f_0 = \frac{f_L + f_u}{2} \tag{2-17}$$

To use the normalized curves for the design of a band-reject filter, the response requirement must be converted to a normalized low-pass filter specification. In order to accomplish this, the band-reject specification should first be made geometrically symmetrical—that is, each pair of frequencies having equal attenuation should satisfy

$$f_1 f_2 = f_0^2 \tag{2-18}$$

which is an alternate form of Equation (2-15). When two frequencies are specified at a particular attenuation level, two frequency pairs will result from calculating the corresponding opposite geometric frequency for each frequency specified. Retain the pair having the wider separation since it represents the more severe requirement. In the bandpass case, the pair having the lesser separation represented the more difficult requirement.

The band-reject filter steepness factor is defined by

$$A_s = \frac{\text{passband bandwidth}}{\text{stopband bandwidth}} \tag{2-23}$$

A normalized low-pass filter can now be selected that makes the transition from the passband attenuation limit to the minimum required stopband attenuation within a frequency ratio A_s.

The following example demonstrates the normalization procedure for a band-reject filter.

Example 2-10 Normalizing a Narrowband Band-Reject Filter

Required:

band-reject filter

center frequency of 1000 Hz

3 dB at 300 Hz (700 Hz, 1300 Hz)

40 dB at 200 Hz (800 Hz, 1200 Hz)

Result:

(*a*) First, compute the center frequency f_0.

$$f_0 = \sqrt{f_L f_u} = \sqrt{700 \times 1300} = 954 \text{ Hz} \tag{2-14}$$

(*b*) Compute two geometrically related stopband frequency pairs for each pair of stopband frequencies given:

Let $f_1 = 800$ Hz

$$f_2 = \frac{f_0^2}{f_1} = \frac{(954)^2}{800} = 1138 \text{ Hz} \qquad (2\text{-}18)$$

Let $f_2 = 1200$ Hz

$$f_1 = \frac{f_0^2}{f_2} = \frac{(954)^2}{1200} = 758 \text{ Hz} \qquad (2\text{-}18)$$

The two pairs are

$$f_1 = 800 \text{ Hz}, f_2 = 1138 \text{ Hz } (f_2 - f_1 = 338 \text{ Hz})$$

and

$$f_1 = 758 \text{ Hz}, f_2 = 1200 \text{ Hz } (f_2 - f_1 = 442 \text{ Hz})$$

Retain the second pair since it has the *wider* separation and represents the more severe requirement. The given response requirement and the geometrically symmetrical equivalent are compared in Figure 2-16

(*c*) Calculate A_s.

$$A_s = \frac{\text{passband bandwidth}}{\text{stopband bandwidth}} = \frac{600 \text{ Hz}}{442 \text{ Hz}} = 1.36 \qquad (2\text{-}23)$$

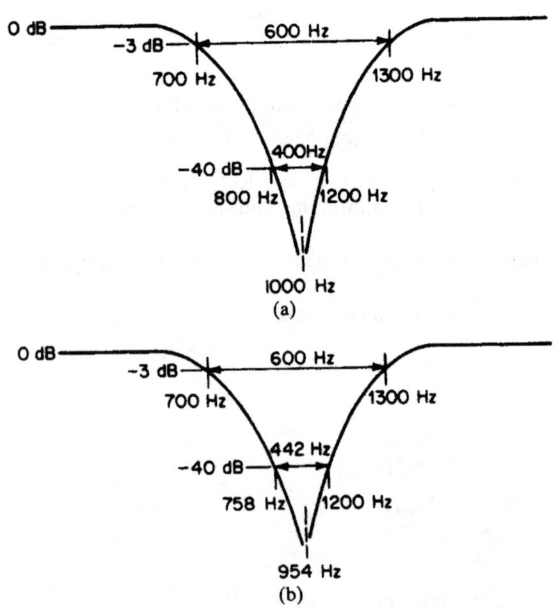

FIGURE 2-16 The response of Example 2-10: (*a*) given requirement; and (*b*) geometrically symmetrical response.

(*d*) Select a normalized low-pass filter from the normalized curves that makes the transition from the 3-dB point to the 40-dB point within a frequency ratio of 1.36. Since these curves are all normalized to 3 dB, a filter is required with over 40 dB of rejection at 1.36 rad/s.

2.2 TRANSIENT RESPONSE

In our previous discussions of filters, we have restricted our interest to frequency-domain parameters such as frequency response. The input forcing function was a sine wave. In real-world applications of filters, input signals consist of a variety of complex waveforms. The response of filters to these nonsinusoidal inputs is called *transient response*.

A filter's transient response is best evaluated in the time domain since we are usually dealing with input signals which are functions of time, such as pulses or amplitude steps. The frequency- and time-domain parameters of a filter are directly related through the Fourier or Laplace transforms.

The Effect of Nonuniform Time Delay

Evaluating a transfer function as a function of frequency results in both a magnitude and phase characteristic. Figure 2-17 shows the amplitude and phase response of a normalized $n = 3$ Butterworth low-pass filter. Butterworth low-pass filters have a phase shift of exactly n times $-45°$ at the 3-dB frequency. The phase shift continuously increases as the transition is made into the stopband and eventually approaches n times $-90°$ at frequencies far removed from the passband. Since the filter described by Figure 2-17 has a complexity of $n = 3$, the phase shift is $-135°$ at the 3-dB cutoff and approaches $-270°$ in the stopband. Frequency scaling will transpose the phase characteristics to a new frequency range as determined by the FSF.

It is well known that a square wave can be represented by a Fourier series of odd harmonic components, as indicated in Figure 2-18. Since the amplitude of each harmonic is reduced as the harmonic order increases, only the first few harmonics are of significance. If a square wave is applied to a filter, the fundamental and its significant harmonics must have a proper relative amplitude relationship at the filter's output in order to retain the square waveshape. In addition, these components must not be displaced in time with respect to each other. Let's now consider the effect of a low-pass filter's phase shift on a square wave.

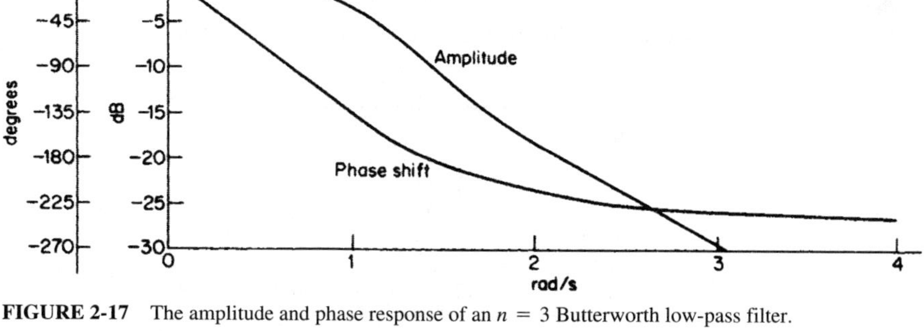

FIGURE 2-17 The amplitude and phase response of an $n = 3$ Butterworth low-pass filter.

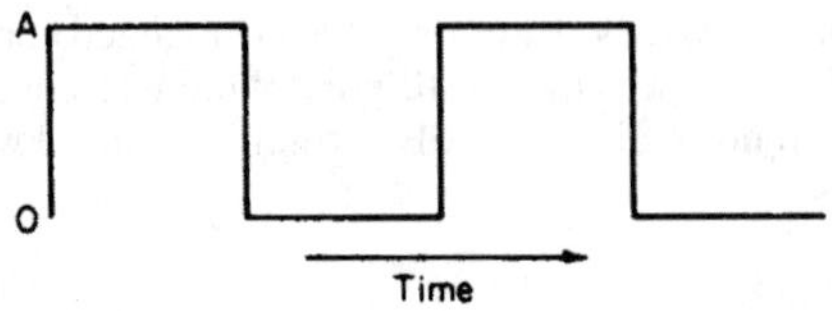

$$A(t) = A\left(\frac{1}{2} + \frac{2}{\pi}\cos\omega_1\tau - \frac{2}{3\pi}\cos 3\omega_1\tau + \frac{2}{5\pi}\cos 5\omega_1\tau + \ldots\right)$$

FIGURE 2-18 The frequency analysis of a square wave.

If we assume that a low-pass filter has a linear phase shift between 0° at DC and n times $-45°$ at the cutoff, we can express the phase shift in the passband as

$$\phi = -\frac{45nf_x}{f_c} \tag{2-24}$$

where f_x is any frequency in the passband, and f_c is the 3-dB cutoff frequency.

A phase-shifted sine wave appears displaced in time from the input waveform. This displacement is called *phase delay* and can be computed by determining the time interval represented by the phase shift, using the fact that a full period contains 360°. Phase delay can then be computed by

$$T_{pd} = \frac{\phi}{360}\frac{1}{f_x} \tag{2-25}$$

or, as an alternate form,

$$T_{pd} = -\frac{\beta}{\omega} \tag{2-26}$$

where β is the phase shift in radians (1 rad $= 360/2\pi$ or 57.3°) and ω is the input frequency expressed in radians per second ($\omega = 2\pi f_x$).

Example 2-11 Effect of Nonlinear Phase on a Square Wave

Required:

Compute the phase delay of the fundamental and the third, fifth, seventh, and ninth harmonics of a 1 kHz square wave applied to an $n = 3$ Butterworth low-pass filter having a 3-dB cutoff of 10 kHz. Assume a linear phase shift with frequency in the passband.

Result:

Using Equations (2-24) and (2-25), the following table can be computed:

Frequency	ϕ	T_{pd}
1 kHz	−13.5°	37.5 μs
3 kHz	−40.5°	37.5 μs
5 kHz	−67.5°	37.5 μs
7 kHz	−94.5°	37.5 μs
9 kHz	−121.5°	37.5 μs

The phase delays of the fundamental and each of the significant harmonics in Example 2-11 are identical. The output waveform would then appear nearly equivalent to the input except for a delay of 37.5 μs. If the phase shift is not linear with frequency, the ratio ϕ/f_x in Equation (2-25) is not constant, so each significant component of the input square wave would undergo a different delay. This displacement in time of the spectral components, with respect to each other, introduces a distortion of the output waveform. Figure 2-19 shows some typical effects of a nonlinear phase shift upon a square wave. Most filters have nonlinear phase versus frequency characteristics, so some waveform distortion will usually occur for complex input signals.

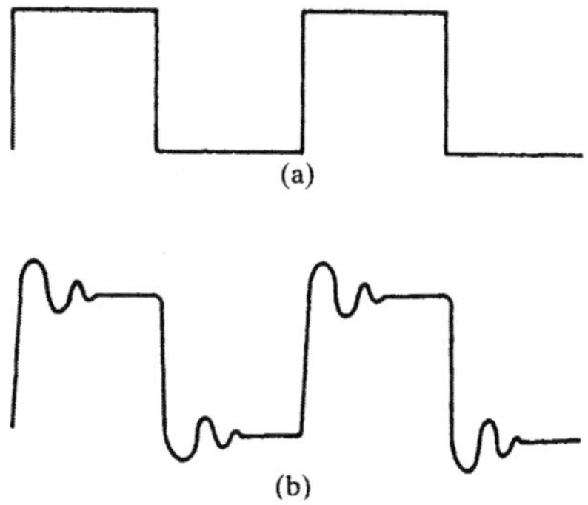

FIGURE 2-19 The effect of a nonlinear phase: (*a*) an ideal square wave; and (*b*) a distorted square wave.

Not all complex waveforms have harmonically related spectral components. An amplitude-modulated signal, for example, consists of a carrier and two sidebands, each sideband separated from the carrier by a modulating frequency. If a filter's phase characteristic is linear with frequency and intersects zero phase shift at zero frequency (DC), both the carrier and the two sidebands will have the same delay in passing through the filter—thus, the output will be a delayed replica of the input. If these conditions are not satisfied, the carrier and both sidebands will be delayed by different amounts. The carrier delay will be in accordance with the equation for phase delay:

$$T_{pd} = -\frac{\beta}{\omega} \tag{2-26}$$

(The terms *carrier delay* and *phase delay* are used interchangeably.)

A new definition is required for the delay of the sidebands. This delay is commonly called *group delay* and is defined as the derivative of phase versus frequency, which can be expressed as

$$T_{gd} = -\frac{d\beta}{d\omega} \tag{2-27}$$

Linear phase shift results in constant group delay since the derivative of a linear function is a constant. Figure 2-20 illustrates a low-pass filter phase shift which is non-linear in the vicinity of a carrier ω_c and the two sidebands: $\omega_c - \omega_m$ and $\omega_c + \omega_m$. The phase delay at ω_c is the negative slope of a line drawn from the origin to the phase shift corresponding to ω_c, which is in agreement with Equation (2-26). The group delay at ω_c is shown as the negative slope of a line which is tangent to the phase response at ω_c. This can be mathematically expressed as

$$T_{gd} = -\left.\frac{d\beta}{d\omega}\right|_{\omega=\omega_c}$$

If the two sidebands are restricted to a region surrounding ω_c and having a constant group delay, the envelope of the modulated signal will be delayed by T_{gd}. Figure 2-21 compares the input and output waveforms of an amplitude-modulated signal applied to the filter depicted by Figure 2-20. Note that the carrier is delayed by the phase delay, while the envelope is delayed by the group delay. For this reason, group delay is sometimes called *envelope delay*.

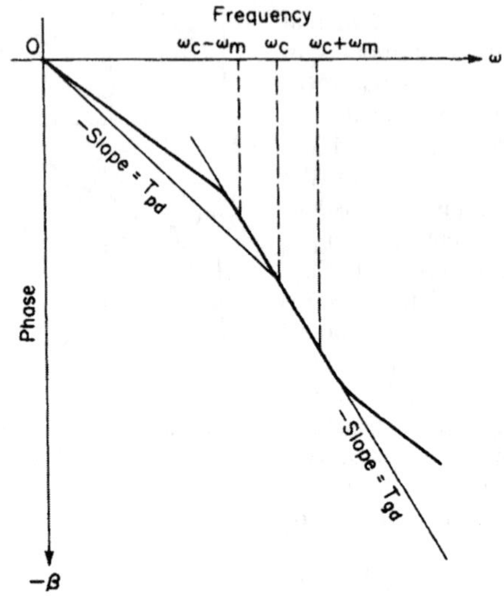

FIGURE 2-20 The nonlinear phase shift of a low-pass filter.

If the group delay is not constant over the bandwidth of the modulated signal, waveform distortion will occur. Narrow-bandwidth signals are more likely to encounter constant group delay than signals having a wider spectrum. It is common practice to use a group-delay variation as a criterion to evaluate phase nonlinearity and subsequent waveform distortion. The absolute magnitude of the nominal delay is usually of little consequence.

Step Response of Networks. If we were to define a hypothetical ideal low-pass filter, it would have the response shown in Figure 2-22. The amplitude response is unity from DC

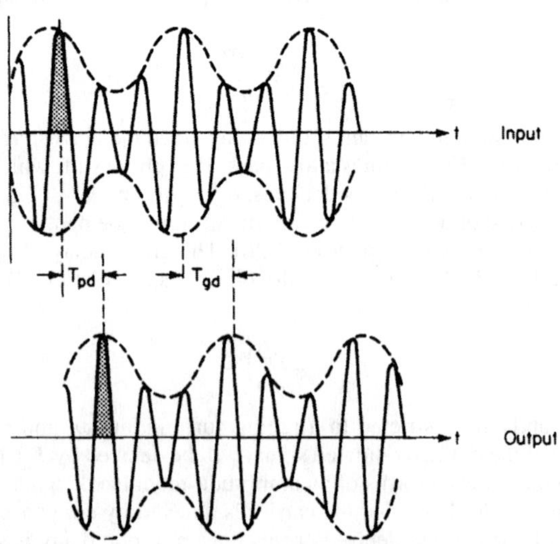

FIGURE 2-21 The effect of nonlinear phase on an AM signal.

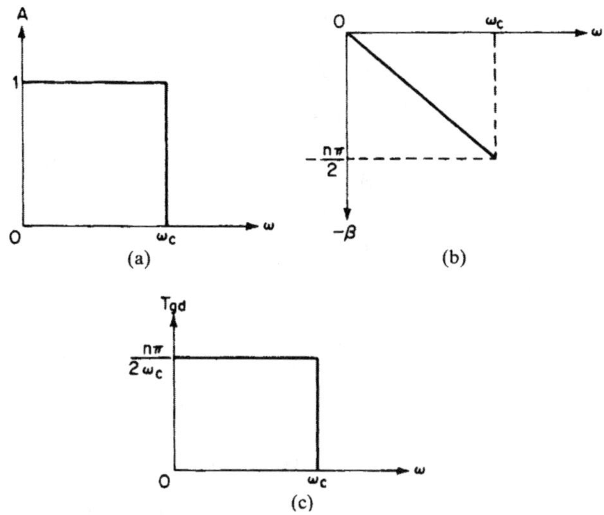

FIGURE 2-22 An ideal low-pass filter: (*a*) frequency response; (*b*) phase shift; and (*c*) group delay.

to the cutoff frequency ω_c, and zero beyond the cutoff. The phase shift is a linearly increasing function in the passband, where n is the order of the ideal filter. The group delay is constant in the passband and zero in the stopband. If a unity amplitude step were applied to this ideal filter at $t = 0$, the output would be in accordance with Figure 2-23. The delay of the half-amplitude point would be $n\pi/2\omega_c$, and the rise time, which is defined as the interval required to go from zero amplitude to unity amplitude with a slope equal to that at the half-amplitude point, would be equal to π/ω_c. Since rise time is inversely proportional to ω_c, a wider filter results in reduced rise time. This proportionality is in agreement with a fundamental rule of thumb relating rise time to bandwidth, which is

$$T_r \approx \frac{0.35}{f_c} \tag{2-28}$$

where T_r is the rise time in seconds and f_c is the 3-dB cutoff in hertz.

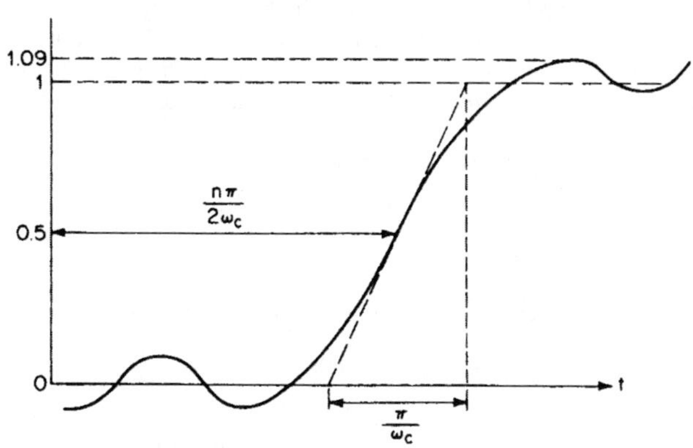

FIGURE 2-23 The step response of an ideal low-pass filter.

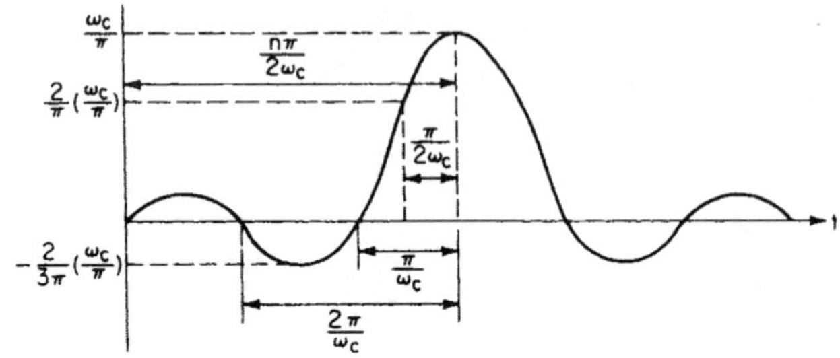

FIGURE 2-24 The impulse response of an ideal low-pass filter.

A 9-percent overshoot exists on the leading edge. Also, a sustained oscillation occurs having a period of $2\pi/\omega_c$, which eventually decays, and then unity amplitude is established. This oscillation is called *ringing*. Overshoot and ringing occur in an ideal low-pass filter, even though we have linear phase. This is because of the abrupt amplitude roll-off at cutoff. Therefore, both linear phase and a prescribed roll-off are required for minimum transient distortion.

Overshoot and prolonged ringing are both very undesirable if the filter is required to pass pulses with minimum waveform distortion. The step-response curves provided for the different families of normalized low-pass filters can be very useful for evaluating the transient properties of these filters.

Impulse Response. A unit impulse is defined as a pulse which is infinitely high and infinitesimally narrow, and has an area of unity. The response of the ideal filter of Figure 2-22 to a unit impulse is shown in Figure 2-24. The peak output amplitude is ω_c/π, which is proportional to the filter's bandwidth. The pulse width, $2\pi/\omega_c$, is inversely proportional to the bandwidth.

An input signal having the form of a unit impulse is physically impossible. However, a narrow pulse of finite amplitude will represent a reasonable approximation, so the impulse response of normalized low-pass filters can be useful in estimating the filter's response to a relatively narrow pulse.

Estimating Transient Characteristics. Group-delay, step-response, and impulse-response curves are given for the normalized low-pass filters discussed in the latter section of this chapter. These curves are useful for estimating filter responses to nonsinusoidal signals. If the input waveforms are steps or pulses, the curves may be used directly. For more complex inputs, we can use the method of superposition, which permits the representation of a complex signal as the sum of individual components. If we find the filter's output for each individual input signal, we can combine these responses to obtain the composite output.

Group Delay of Low-Pass Filters. When a normalized low-pass filter is frequency-scaled, the delay characteristics are frequency-scaled as well. The following rules can be applied to derive the resulting delay curve from the normalized response:

1. Divide the delay axis by $2\pi f_c$, where f_c is the filter's 3-dB cutoff.
2. Multiply all points on the frequency axis by f_c.

The following example demonstrates the denormalization of a low-pass curve.

Example 2-12 Frequency Scaling the Delay of a Low-Pass Filter

Required:

Using the normalized delay curve of an $n = 3$ Butterworth low-pass filter given in Figure 2-25*a*, compute the delay at DC and the delay variation in the passband if the filter is frequency-scaled to a 3-dB cutoff of 100 Hz.

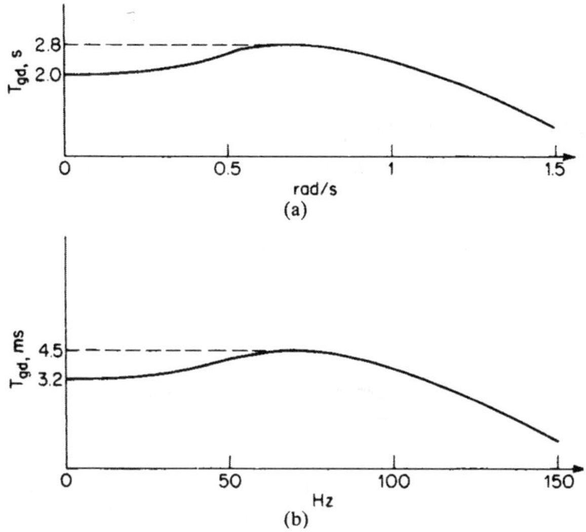

FIGURE 2-25 The delay of an $n = 3$ Butterworth low-pass filter: (*a*) normalized delay; and (*b*) delay with $f_c = 100$ Hz.

Result:

To denormalize the curve, divide the delay axis by $2\pi f_c$ and multiply the frequency axis by f_c, where f_c is 100 Hz. The resulting curve is shown in Figure 2-25*b*. The delay at DC is 3.2 ms, and the delay variation in the passband is 1.3 ms.

The nominal delay of a low-pass filter at frequencies well below the cutoff can be estimated by the following formula:

$$T \approx \frac{125n}{f_c} \tag{2-29}$$

where T is the delay in milliseconds, n is the order of the filter, and f_c is the 3-dB cutoff in hertz. Equation (2-29) is an approximation which usually is accurate to within 25 percent.

Group Delay of Bandpass Filters. When a low-pass filter is transformed to a narrowband bandpass filter, the delay is transformed to a nearly symmetrical curve mirrored about the center frequency. As the bandwidth increases from the narrow-bandwidth case, the symmetry of the delay curve is distorted approximately in proportion to the filter's bandwidth.

For the narrowband condition, the bandpass delay curve can be approximated by implementing the following rules:

1. Divide the delay axis of the normalized delay curve by πBW, where BW is the 3-dB bandwidth in hertz.
2. Multiply the frequency axis by BW/2.
3. A delay characteristic symmetrical around the center frequency can now be formed by generating the mirror image of the curve obtained by implementing steps 1 and 2. The total 3-dB bandwidth thus becomes BW.

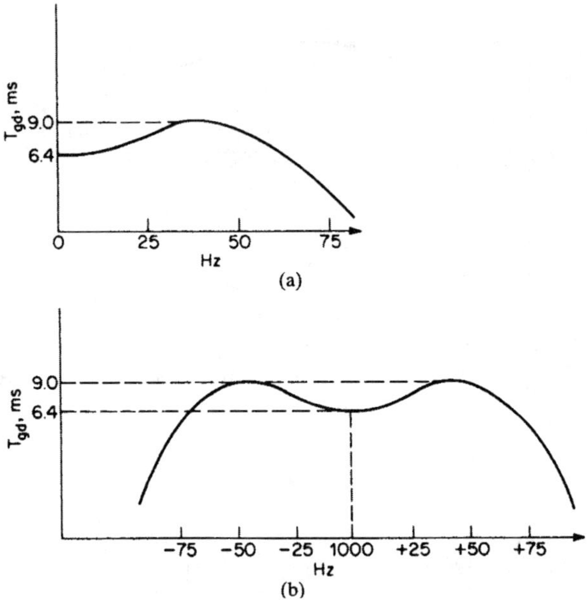

FIGURE 2-26 The delay of a narrow-band bandpass filter: (*a*) a low-pass delay; and (*b*) a bandpass delay.

The following example demonstrates the approximation of a narrowband bandpass filter's delay curve.

Example 2-13 Estimate the Delay of a Bandpass Filter

Required:

Estimate the group delay at the center frequency and the delay variation over the passband of a bandpass filter having a center frequency of 1000 Hz and a 3-dB bandwidth of 100 Hz. The bandpass filter is derived from a normalized $n = 3$ Butterworth low-pass filter.

Result:

The delay of the normalized filter is shown in Figure 2-25*a*. If we divide the delay axis by πBW and multiply the frequency axis by BW/2, where BW = 100 Hz, we obtain the delay curve of Figure 2-26*a*. We can now reflect this delay curve on both sides of the center frequency of 1000 Hz to obtain Figure 2-26*b*. The delay at the center frequency is 6.4 ms, while the delay variation over the passband is 2.6 ms.

The technique used in Example 2-13 to approximate a bandpass delay curve is valid for bandpass filter Qs of 10 or more (f_0/BW $\geq$ 10). As the fractional bandwidth increases, the delay becomes less symmetrical and peaks toward the low side of the center frequency, as shown in Figure 2-27.

The delay at the center frequency of a bandpass filter can be estimated by

$$T \approx \frac{250n}{\text{BW}} \tag{2-30}$$

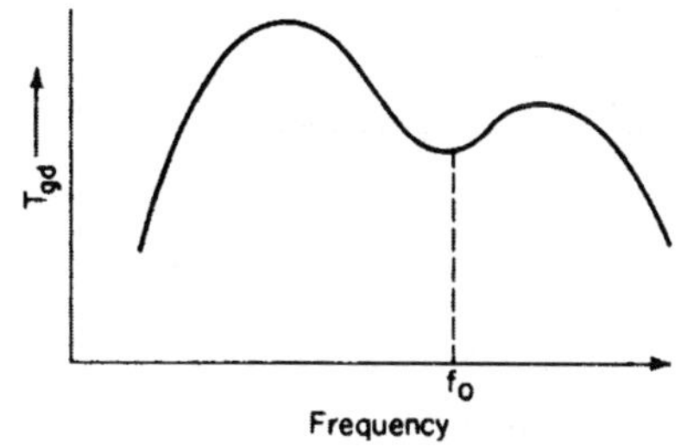

FIGURE 2-27 The delay of a wideband bandpass filter.

where T is the delay in milliseconds. This approximation is usually accurate to within 25 percent.

A comparison of Figures 2-25*b* and 2-26*b* indicates that a bandpass filter has twice the delay of the equivalent low-pass filter of the same bandwidth. This results from the low-pass to bandpass transformation where a low-pass filter transfer function of order n always results in a bandpass filter transfer function with an order $2n$. However, a bandpass filter is conventionally referred to as having the same order n as the low-pass filter it was derived from.

Step Response of Low-Pass Filters. Delay distortion usually cannot be directly used to determine the extent of the distortion of a modulated signal. A more direct parameter would be the step response, especially where the modulation consists of an amplitude step or pulse.

The two essential parameters of a filter's step response are overshoot and ringing. Overshoot should be minimized for accurate pulse reproduction. Ringing should decay as rapidly as possible to prevent interference with subsequent pulses. Rise time and delay are usually less important considerations.

Step-response curves for standard normalized low-pass filters are provided in the latter part of this chapter. These responses can be denormalized by dividing the time axis by $2\pi f_c$, where f_c is the 3-dB cutoff of the filter. Denormalization of the step response is shown in the following example.

Example 2-14 Determining the Overshoot of a Low-Pass Filter

Required:

Determine the amount of overshoot of an $n = 3$ Butterworth low-pass filter having a 3-dB cutoff of 100 Hz. Also determine the approximate time required for the ringing to decay substantially—for instance, the settling time.

Result:

The step response of the normalized low-pass filter is shown in Figure 2-28*a*. If the time axis is divided by $2\pi f_c$, where $f_c = 100$ Hz, the step response or Figure 2-28*b* is obtained. The overshoot is slightly under 10 percent. After 25 ms, the amplitude will have almost completely settled.

If the input signal to a filter is a pulse rather than a step, the step-response curves can still be used to estimate the transient response, provided that the pulse width is greater than the settling time.

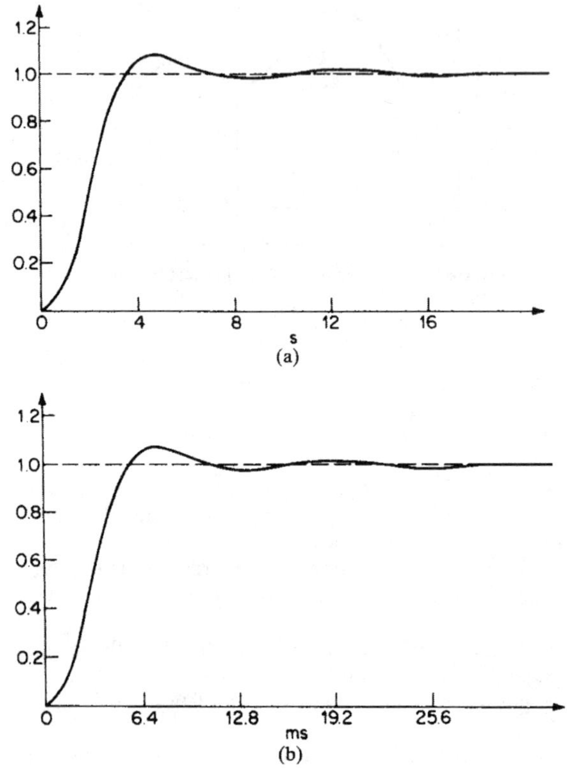

FIGURE 2-28 The step response of Example 2-14: (*a*) normalized step response; and (*b*) denormalized step response.

Example 2-15 Determining the Pulse Response of a Low-Pass Filter

Required:

Estimate the output waveform of the filter of Example 2-14 if the input is the pulse of Figure 2-29*a*.

Result:

Since the pulse width is in excess of the settling time, the step response can be used to estimate the transient response. The leading edge is determined by the shape of the denormalized step response of Figure 2-28*b*. The trailing edge can be derived by inverting the denormalized step response. The resulting waveform is shown in Figure 2-29*b*.

The Step Response of Bandpass Filters. The envelope of the response of a narrow bandpass filter to a step of the center frequency is almost identical to the step response of the equivalent low-pass filter having half the bandwidth. To determine this envelope shape, denormalize the low-pass step response by dividing the time axis by πBW, where BW is the 3-dB bandwidth of the bandpass filter. The previous discussions of overshoot, ringing, and so on, can be applied to the carrier envelope.

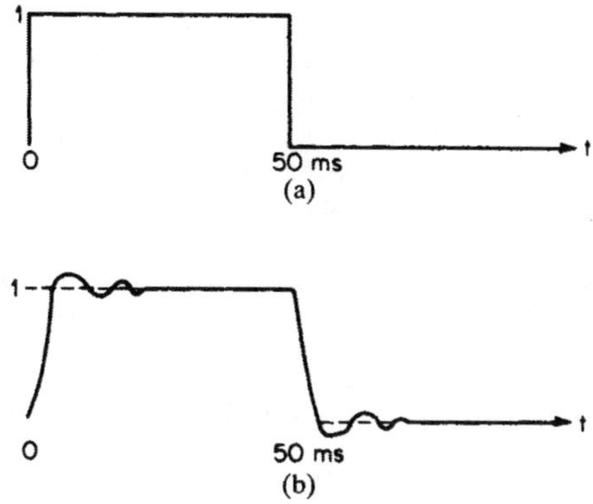

FIGURE 2-29 The pulse response of Example 2-15: (*a*) input pulse; and (*b*) output pulse.

Example 2-16 Determining the Step Response of a Bandpass Filter

Required:

Determine the envelope of the response to a 1000 Hz step for an $n = 3$ Butterworth bandpass filter having a center frequency of 1000 Hz and a 3-dB bandwidth of 100 Hz.

Result:

Using the normalized step response of Figure 2-28*a*, divide the time axis by πBW, where BW = 100 Hz. The results are shown in Figure 2-30.

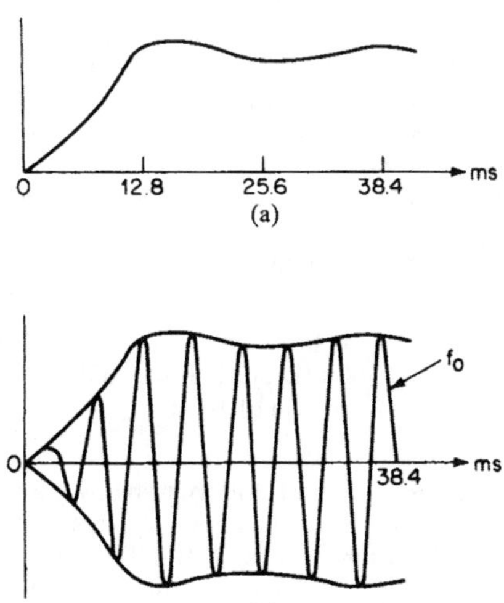

FIGURE 2-30 The bandpass response to a center frequency step: (*a*) denormalized low-pass step response; and (*b*) bandpass envelope response.

The Impulse Response of Low-Pass Filters. If the duration of a pulse applied to a low-pass filter is much less than the rise time of the filter's step response, the filter's impulse response will provide a reasonable approximation to the shape of the output waveform.

Impulse-response curves are provided for the different families of low-pass filters. These curves are all normalized to correspond to a filter having a 3-dB cutoff of 1 rad/s, and have an area of unity. To denormalize the curve, multiply the amplitude by the FSF and divide the time axis by the same factor.

It is desirable to select a normalized low-pass filter having an impulse response whose peak is as high as possible. The ringing, which occurs after the trailing edge, should also decay rapidly to avoid interference with subsequent pulses.

Example 2-17 Determining the Impulse Response of a Low-Pass Filter

Required:

Determine the approximate output waveform if a 100-μs pulse is applied to an $n = 3$ Butterworth low-pass filter having a 3-dB cutoff of 100 Hz.

Result:

The denormalized step response of the filter is given in Figure 2-28*b*. The rise time is well in excess of the given pulse width of 100 μs, so the impulse response curve should be used to approximate the output waveform.

The impulse response of a normalized $n = 3$ Butterworth low-pass filter is shown in Figure 2-31*a*. If the time axis is divided by the FSF and the amplitude is multiplied by this same factor, the curve of Figure 2-31*b* results.

Since the input pulse amplitude of Example 2-17 is certainly not infinite, the amplitude axis is in error. However, the pulse shape is retained at a lower amplitude. As the input pulse width is reduced in relation to the filter rise time, the output amplitude decreases and eventually the output pulse vanishes.

The Impulse Response of Bandpass Filters. The envelope of the response of a narrow-band bandpass filter to a short tone burst of center frequency can be found by denormalizing the low-pass impulse response. This approximation is valid if the burst width is much less than the rise time of the denormalized step response of the bandpass filter. Also, the center frequency should be high enough so that many cycles occur during the burst interval.

To transform the impulse-response curve, multiply the amplitude axis by πBW and divide the time axis by the same factor, where BW is the 3-dB bandwidth of the bandpass filter. The resulting curve defines the shape of the envelope of the filter's response to the tone burst.

Example 2-18 Determining the Impulse Response of a Bandpass Filter

Required:

Determine the approximate shape of the response of an $n = 3$ Butterworth bandpass filter having a center frequency of 1000 Hz and a 3-dB bandwidth of 10 Hz to a tone burst of the center frequency having a duration of 10 ms.

Result:

The step response of a normalized $n = 3$ Butterworth low-pass filter is shown in Figure 2-28*a*. To determine the rise time of the bandpass step response, divide the normalized low-pass rise time by πBW, where BW is 10 Hz. The resulting rise time is approximately 120 ms, which well exceeds the burst duration. Also, 10 cycles of the center frequency occur during the burst interval, so the impulse response can be used to approximate the output envelope. To denormalize the impulse response, multiply the amplitude axis by πBW and divide the time axis by the same factor. The results are shown in Figure 2-32.

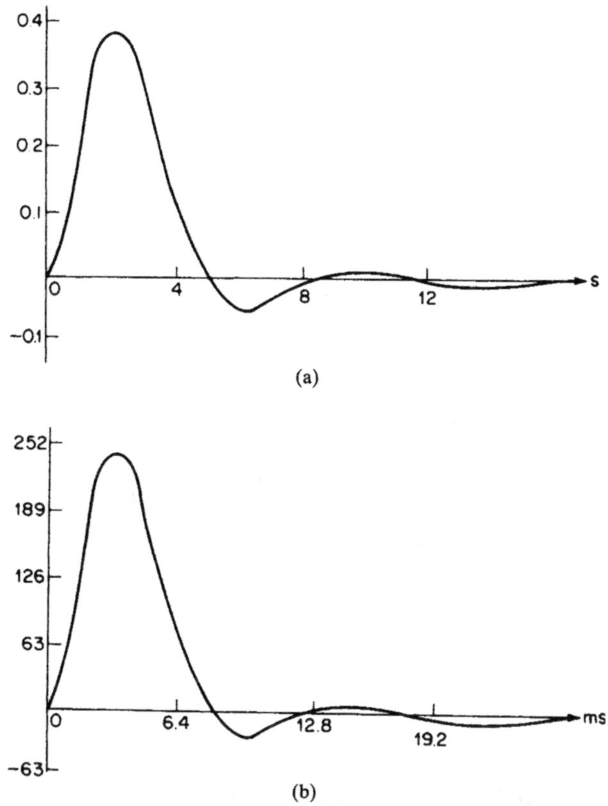

FIGURE 2-31 The impulse response for Example 2-17: (*a*) normalized response; and (*b*) denormalized response.

Effective Use of the Group-Delay, Step-Response, and Impulse-Response Curves. Many signals consist of complex forms of modulation rather than pulses or steps, so the transient response curves cannot be directly used to estimate the amount of distortion introduced by the filters. However, the curves are useful as a figure of merit, since networks having desirable step- or impulse-response behavior introduce minimal distortion to most forms of modulation.

Examination of the step- and impulse-response curves in conjunction with group delay indicates that a necessary condition for good pulse transmission is a flat group delay. A gradual transition from the passband to the stopband is also required for low transient distortion but is highly undesirable from a frequency-attenuation point of view.

In order to obtain a rapid pulse rise time, the higher-frequency spectral components should not be delayed with respect to the lower frequencies. The curves indicate that low-pass filters which do have a sharply increasing delay at higher frequencies have an impulse response which comes to a peak at a later time.

When a low-pass filter is transformed to a high-pass, a band-reject, or a wideband bandpass filter, the transient properties are not preserved. Lindquist and Zverev (see Bibliography) provide computational methods for the calculation of these responses.

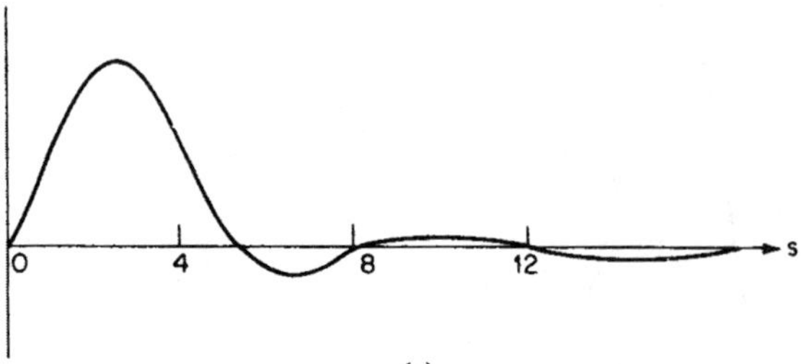

(a)

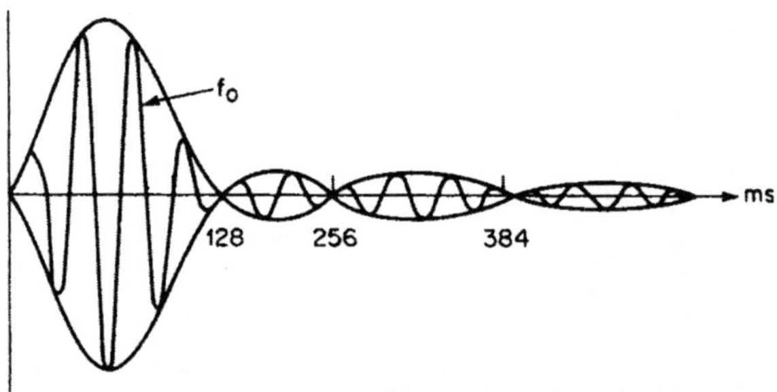

(b)

FIGURE 2-32 The results of Example 2-18: (*a*) normalized low-pass impulse response; and (*b*) impulse response of bandpass filter.

2.3 BUTTERWORTH MAXIMALLY FLAT AMPLITUDE

The Butterworth approximation to an ideal low-pass filter is based on the assumption that a flat response at zero frequency is more important than the response at other frequencies. A normalized transfer function is an all-pole type having roots which all fall on a unit circle. The attenuation is 3 dB at 1 rad/s.

The attenuation of a Butterworth low-pass filter can be expressed by

$$A_{dB} = 10 \log\left[1 + \left(\frac{\omega_x}{\omega_c}\right)^{2n}\right] \tag{2.31}$$

where ω_x/ω_c is the ratio of the given frequency ω_x to the 3-dB cutoff frequency ω_c, and n is the order of the filter.

For the more general case,

$$A_{dB} = 10 \log(1 + \Omega^{2n}) \tag{2-32}$$

where Ω is defined by the following table.

The value Ω is a dimensionless ratio of frequencies or a normalized frequency. $BW_{3\,dB}$ is the 3-dB bandwidth, and BW_x is the bandwidth of interest. At high values of Ω, the attenuation increases at a rate of $6n$ dB per octave, where an octave is defined as a frequency ratio of 2 for the low-pass and high-pass cases, and a *bandwidth* ratio of 2 for bandpass and band-reject filters.

Filter Type	Ω
Low-pass	ω_x/ω_c
High-pass	ω_c/ω_x
Bandpass	$BW_x/BW_{3\ dB}$
Band-reject	$BW_{3\ dB}/BW_x$

The pole positions of the normalized filter all lie on a unit circle and can be computed by

$$-\sin\frac{(2K-1)\pi}{2n} + j\cos\frac{(2K-1)\pi}{2n}, \qquad K = 1, 2, \ldots, n \tag{2-33}$$

and the element values for an *LC* normalized low-pass filter operating between equal $1\text{-}\Omega$ terminations can be calculated by

$$L_K \text{ or } C_K = 2\sin\frac{(2K-1)\pi}{2n}, \qquad K = 1, 2, \ldots, n \tag{2-34}$$

where $(2K - 1)\pi/2n$ is in radians.

Equation (2-34) is exactly equal to twice the real part of the pole position of Equation (2-33), except that the sign is positive.

Example 2-19 Calculating the Frequency Response, Pole Locations, and *LC* Element Values of a Butterworth Low-Pass Filter

Required:

Calculate the frequency response at 1, 2, and 4 rad/s, the pole positions, and the *LC* element values of a normalized $n = 5$ Butterworth low-pass filter.

Result:

(*a*) Using Equation (2-32) with $n = 5$, the following frequency-response table can be derived:

Ω	Attenuation
1	3 dB
2	30 dB
4	60 dB

(*b*) The pole positions are computed using Equation (2-33) as follows:

K	$-\sin\frac{(2K-1)\pi}{2n}$	$j\cos\frac{(2K-1)\pi}{2n}$
1	−0.309	$+j\,0.951$
2	−0.809	$+j\,0.588$
3	−1	
4	−0.809	$-j\,0.588$
5	−0.309	$-j\,0.951$

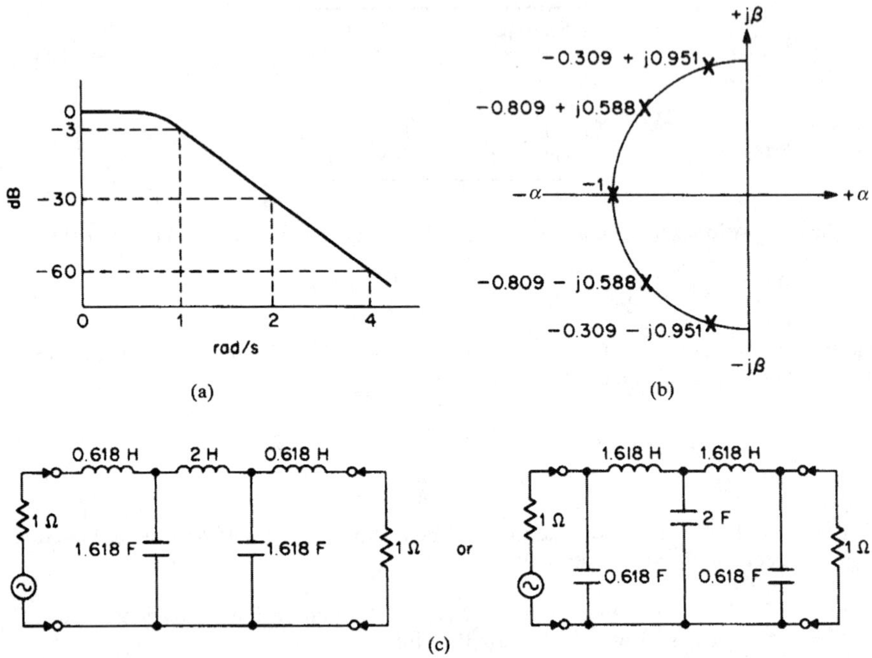

FIGURE 2-33 The Butterworth low-pass filter of Example 2-19: (*a*) frequency response; (*b*) pole locations; and (*c*) circuit configuration.

(*c*) The element values can be computed by Equation (2-34) and have the following values:

$L_1 = 0.618$ H		$C_1 = 0.618$ F
$C_2 = 1.618$ F		$L_2 = 1.618$ H
$L_3 = 2$ H	or	$C_3 = 2$ F
$C_4 = 1.618$ F		$L_4 = 1.618$ H
$L_5 = 0.618$ H		$C_5 = 0.618$ F

The results of Example 2-19 are shown in Figure 2-33.

Chapter 11 provides pole locations and element values for both *LC* and active Butterworth low-pass filters having complexities up to $n = 10$.

The Butterworth approximation results in a class of filters which have moderate attenuation steepness and acceptable transient characteristics. Their element values are more practical and less critical than those of most other filter types. The rounding of the frequency response in the vicinity of cutoff may make these filters undesirable where a sharp cutoff is required; nevertheless, they should be used wherever possible because of their favorable characteristics.

Figures 2-34 through 2-37 indicate the frequency response, group delay, impulse response, and step response for the Butterworth family of low-pass filters normalized to a 3-dB cutoff of 1 rad/s.

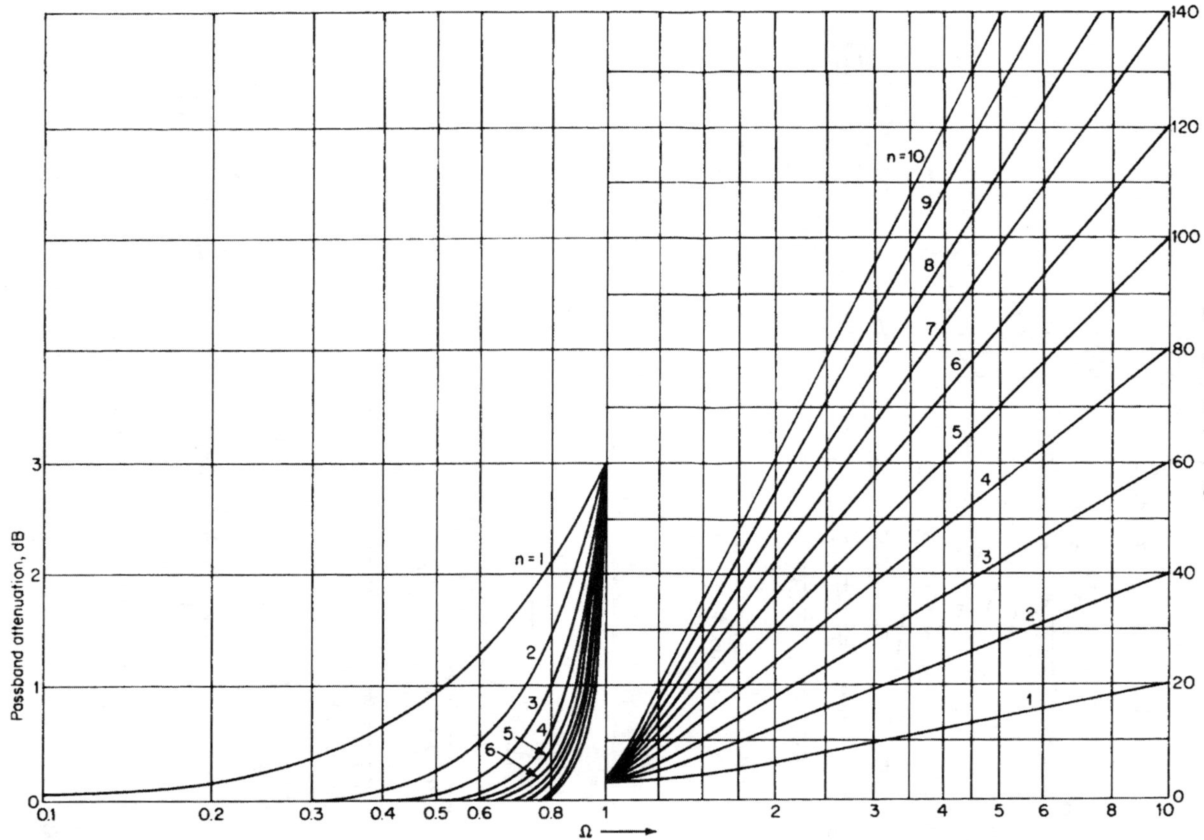

FIGURE 2-34 Attenuation characteristics for Butterworth filters. (From A. I. Zverev, *Handbook of Filter Synthesis* [New York: John Wiley and Sons, 1967.] By permission of the publishers.)

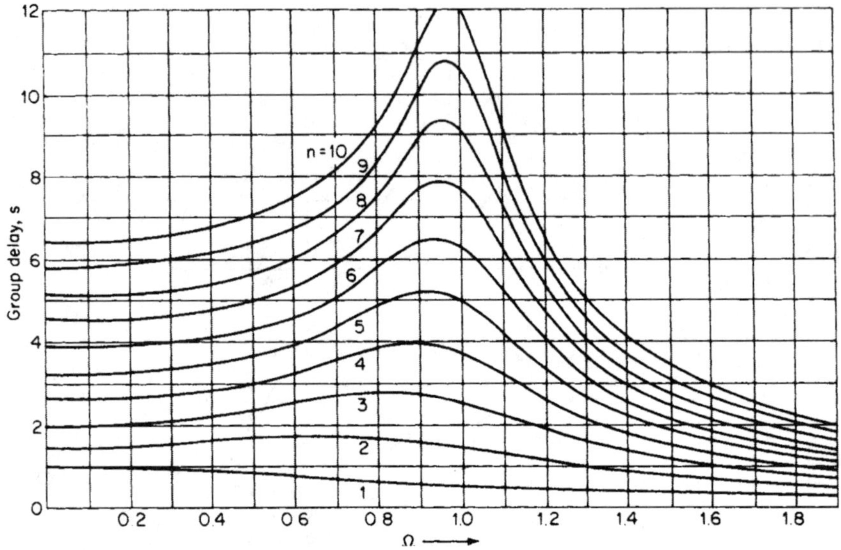

FIGURE 2-35 Group-delay characteristics for Butterworth filters. (From A. I. Zverev, *Handbook of Filter Synthesis* [New York: John Wiley and Sons, 1967.] By permission of the publishers.)

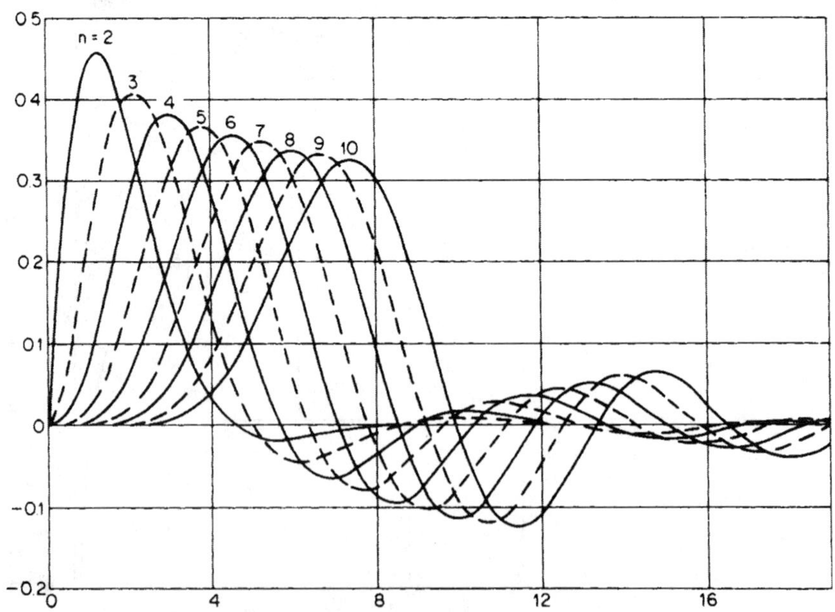

FIGURE 2-36 Impulse response for Butterworth filters. (From A. I. Zverev, *Handbook of Filter Synthesis* [New York: John Wiley and Sons, 1967.] By permission of the publishers.)

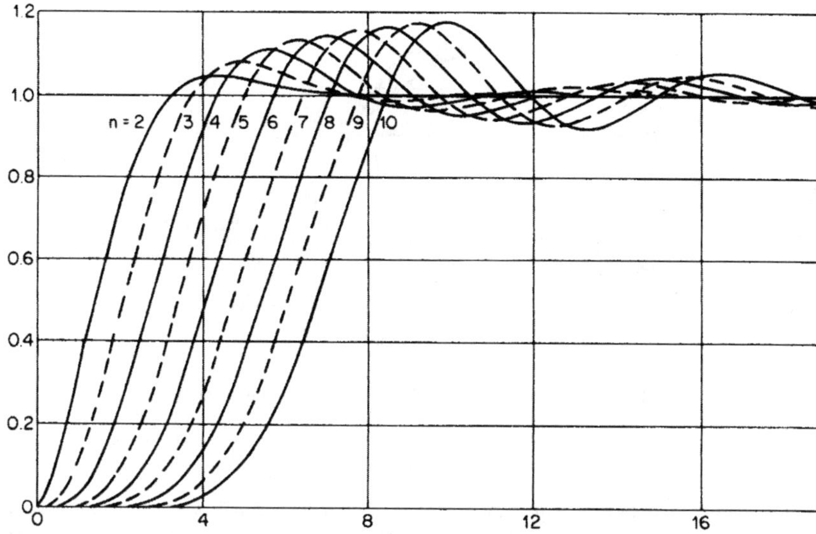

FIGURE 2-37 Step response for Butterworth filters. (From A. I. Zverev, *Handbook of Filter Synthesis* [New York: John Wiley and Sons, 1967.] By permission of the publishers.)

2.4 CHEBYSHEV RESPONSE

If the poles of the normalized Butterworth low-pass transfer function were moved to right by multiplying the real parts of the pole position by a constant k_r, and the imaginary parts by a constant k_j, where both k's are <1, the poles would now lie on an ellipse instead of a unit circle. The frequency response would ripple evenly and have an attenuation at 1 rad/s equal to the ripple. The resulting response is called the Chebyshev or equiripple function.

The Chebyshev approximation to an ideal filter has a much more rectangular frequency response in the region near cutoff than the Butterworth family of filters. This is accomplished at the expense of allowing ripples in the passband.

The factors k_r and k_j are computed by

$$k_r = \sinh \mathrm{A} \tag{2-35a}$$

$$k_j = \cosh \mathrm{A} \tag{2-35b}$$

The parameter A is given by

$$A = \frac{1}{n} \sinh^{-1} \frac{1}{\varepsilon} \tag{2-36}$$

where

$$\varepsilon = \sqrt{10^{R_{\mathrm{dB}}/10} - 1} \tag{2-37}$$

and R_{dB} is the ripple in decibels.

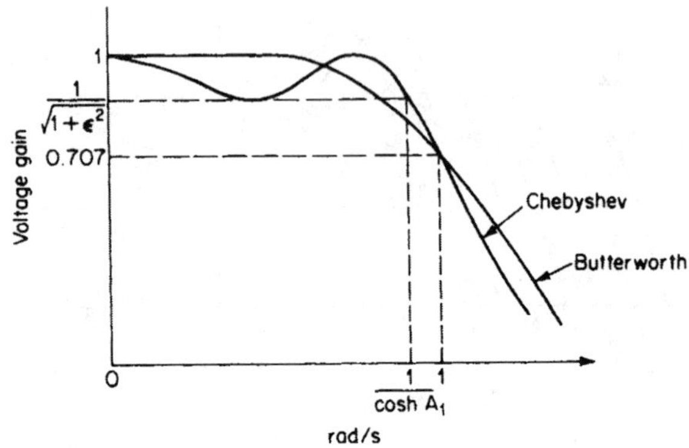

FIGURE 2-38 A comparison of Butterworth and Chebyshev low-pass filters.

Figure 2-38 compares the frequency response of an $n = 3$ Butterworth normalized low-pass filter and the Chebyshev filter generated by applying Equations 2-35a and 2-35b. The Chebyshev filter response has also been normalized so that the attenuation is 3 dB at 1 rad/s. The actual 3-dB bandwidth of a Chebyshev filter computed using Equations 2-35a and 2-35b is cosh A_1, where A_1 is given by

$$A_1 = \frac{1}{n}\cosh^{-1}\left(\frac{1}{\varepsilon}\right) \tag{2-37a}$$

The attenuation of Chebyshev filters can be expressed as

$$A_{\text{dB}} = 10 \log[1 + \varepsilon^2 C_n^2(\Omega)] \tag{2-38}$$

where $C_n(\Omega)$ is a Chebyshev polynomial whose magnitude oscillates between ± 1 for $\Omega \leq 1$. Table 2-1 lists the Chebyshev polynomials up to order $n = 10$.

At $\Omega = 1$, Chebyshev polynomials have a value of unity, so the attenuation defined by Equation (2-38) would be equal to the ripple. The 3-dB cutoff is slightly above $\Omega = 1$ and is equal to cosh A_1. In order to normalize the response equation so that 3 dB

TABLE 2-1 Chebyshev Polynomials

1. Ω
2. $2\Omega^2 - 1$
3. $4\Omega^3 - 3\Omega$
4. $8\Omega^4 - 8\Omega^2 + 1$
5. $16\Omega^5 - 20\Omega^3 + 5\Omega$
6. $32\Omega^6 - 48\Omega^4 + 18\Omega^2 - 1$
7. $64\Omega^7 - 112\Omega^5 + 56\Omega^3 - 7\Omega$
8. $128\Omega^8 - 256\Omega^6 + 160\Omega^4 - 32\Omega^2 + 1$
9. $256\Omega^9 - 576\Omega^7 + 432\Omega^5 - 120\Omega^3 + 9\Omega$
10. $512\Omega^{10} - 1280\Omega^8 + 1120\Omega^6 - 400\Omega^4 + 50\Omega^2 - 1$

of attenuation occurs at $\Omega = 1$, the Ω of Equation (2-38) is computed by using the following table:

Filter Type	Ω
Low-pass	$(\cosh A_1)\ \omega_x/\omega_c$
High-pass	$(\cosh A_1)\ \omega_c/\omega_x$
Bandpass	$(\cosh A_1)\ BW_x/BW_{3\ dB}$
Band-reject	$(\cosh A_1)\ BW_{3\ dB}/BW_x$

Figure 2-39 compares the ratios of 3-dB bandwidth to ripple bandwidth ($\cosh A_1$) for Chebyshev low-pass filters ranging from $n = 2$ to $n = 10$.

n	0.001 dB	0.005 dB	0.01 dB	0.05 dB
2	5.7834930	3.9027831	3.3036192	2.2685899
3	2.6427081	2.0740079	1.8771819	1.5120983
4	1.8416695	1.5656920	1.4669048	1.2783955
5	1.5155888	1.3510908	1.2912179	1.1753684
6	1.3495755	1.2397596	1.1994127	1.1207360
7	1.2531352	1.1743735	1.1452685	1.0882424
8	1.1919877	1.1326279	1.1106090	1.0673321
9	1.1507149	1.1043196	1.0870644	1.0530771
10	1.1215143	1.0842257	1.0703312	1.0429210

n	0.10 dB	0.25 dB	0.50 dB	1.00 dB
2	1.9432194	1.5981413	1.3897437	1.2176261
3	1.3889948	1.2528880	1.1674852	1.0948680
4	1.2130992	1.1397678	1.0931019	1.0530019
5	1.1347180	1.0887238	1.0592591	1.0338146
6	1.0929306	1.0613406	1.0410296	1.0234422
7	1.0680005	1.0449460	1.0300900	1.0172051
8	1.0519266	1.0343519	1.0230107	1.0131638
9	1.0409547	1.0271099	1.0181668	1.0103963
10	1.0331307	1.0219402	1.0147066	1.0084182

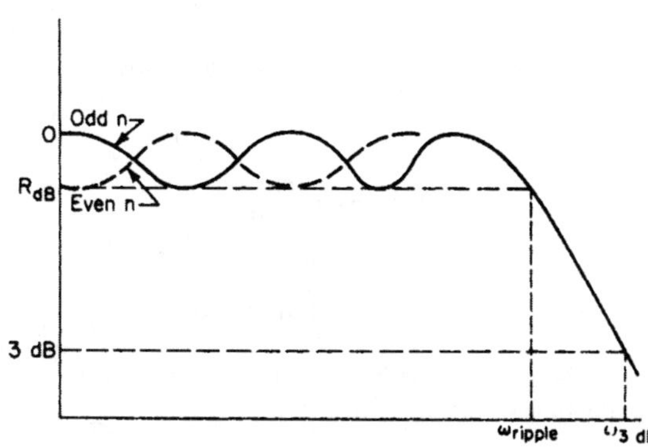

FIGURE 2-39 The ratio of 3-dB bandwidth to ripple bandwidth.

Odd-order Chebyshev LC filters have zero relative attenuation at DC. Even-order filters, however, have a loss at DC equal to the passband ripple. As a result, the even-order networks must operate between unequal source and load resistances, whereas for odd n's, the source and the load may be equal.

The element values for an LC normalized low-pass filter operating between equal 1-Ω terminations and having an odd n can be calculated from the following series of relations.

$$G_1 = \frac{2A_1 \cosh A}{Y} \tag{2-39}$$

$$G_k = \frac{4A_{k-1}A_k \cosh^2 A}{B_{k-1}G_{k-1}} \qquad k = 2, 3, 4, \ldots, n \tag{2-40}$$

where

$$Y = \sinh \frac{\beta}{2n} \tag{2-41}$$

$$\beta = \ln\left(\coth \frac{R_{\mathrm{dB}}}{17.37}\right) \tag{2-42}$$

$$A_k = \sin \frac{(2k-1)\pi}{2n} \qquad k = 1, 2, 3, \ldots, n \tag{2-43}$$

$$B_k = Y^2 + \sin^2\left(\frac{k\pi}{n}\right) \qquad k = 1, 2, 3, \ldots, n \tag{2-44}$$

Coefficients G_1 through G_n are the element values.

An alternate form of determining LC element values can be done by synthesizing the driving-point impedance directly from the transfer function. Closed form formulas are given in Matthaei (see Bibliography). These methods include both odd- and even-order n's.

Example 2-20 Calculating the Pole Locations, Frequency Response, and LC Element Values of a Chebyshev Low-Pass Filter

Required:

Compute the pole positions, the frequency response at 1, 2, and 4 rad/s, and the element values of a normalized $n = 5$ Chebyshev low-pass filter having a ripple of 0.5 dB.

Result:

(*a*) To compute the pole positions, first solve for k_c as follows:

$$\varepsilon = \sqrt{10^{R_{\mathrm{dB}}/10} - 1} = 0.349 \tag{2-37}$$

$$A = \frac{1}{n}\sinh^{-1}\frac{1}{\varepsilon} = 0.355 \tag{2-36}$$

$$k_r = \sinh A = 0.3625 \tag{2-35a}$$

$$k_j = \cosh A = 1.0637 \tag{2-35b}$$

Multiplication of the real parts of the normalized Butterworth poles of Example 2-19 by k_r and the imaginary parts by k_j results in

$$-0.1120 \pm j1.0116; -0.2933 \pm j0.6255; -0.3625$$

To denormalize these coordinates for 3 dB at 1 rad/s, divide all values by cosh A_1, where A_1 is given by

$$A_1 = 0.3428 \quad \text{(2-37a)}$$

so cosh $A_1 = 1.0593$. The resulting pole positions are

$$-0.1057 \pm j0.9549; \; -0.2769 \pm j0.5905; \; -0.3422$$

(*b*) To calculate the frequency response, substitute a fifth-order Chebyshev polynomial and $\varepsilon = 0.349$ into Equation (2-38). The following results are obtained:

Ω	A_{dB}
1.0	3 dB
2.0	45 dB
4.0	77 dB

(*c*) The element values are computed as follows:

$$A_1 = 0.309 \quad \text{(2-43)}$$

$$\beta = 3.55 \quad \text{(2-42)}$$

$$Y = 0.363 \quad \text{(2-41)}$$

$$G_1 = 1.81 \quad \text{(2-39)}$$

$$G_2 = 1.30 \quad \text{(2-40)}$$

$$G_3 = 2.69 \quad \text{(2-40)}$$

$$G_4 = 1.30 \quad \text{(2-40)}$$

$$G_5 = 1.81 \quad \text{(2-40)}$$

Coefficients G_1 through G_5 represent the element values of a normalized Chebyshev low-pass filter having a 0.5-dB ripple and a 3-dB cutoff of 1 rad/s.

Figure 2-40 shows the results of this example.

Chebyshev filters have a narrower transition region between the passband and stopband than Butterworth filters but have more delay variation in their passband. As the passband ripple is made larger, the rate of roll-off increases, but the transient properties rapidly deteriorate. If no ripples are permitted, the Chebyshev filter degenerates to a Butterworth.

The Chebyshev function is useful where frequency response is a major consideration. It provides the maximum theoretical rate of roll-off of any all-pole transfer function for a given order. It does not have the mathematical simplicity of the Butterworth family, which should be evident from comparing Examples 2-20 and 2-19. Fortunately, the computation of poles and element values is not required since this information is provided in Chapter 11.

Figures 2-41 through 2-54 show the frequency and time-domain parameters of Chebyshev low-pass filters for ripples of 0.01, 0.1, 0.25, 0.5, and 1 dB, all normalized for a 3-dB cutoff of 1 rad/s.

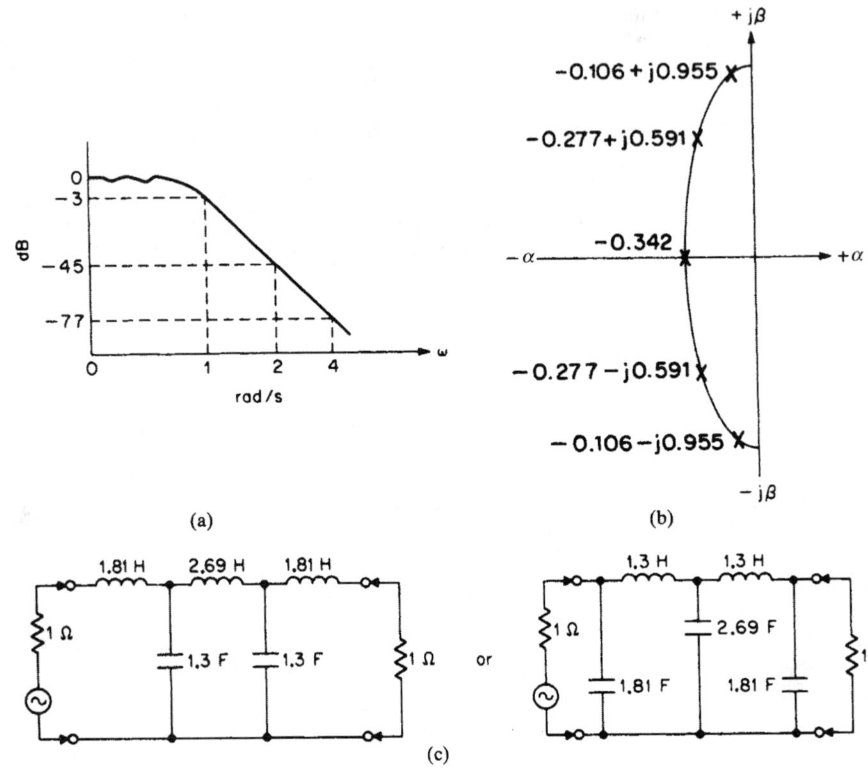

FIGURE 2-40 The Chebyshev low-pass filter of Example 2-20: (*a*) frequency response; (*b*) pole locations; and (*c*) circuit configuration.

2.5 BESSEL MAXIMALLY FLAT DELAY

Butterworth filters have fairly good amplitude and transient characteristics. The Chebyshev family of filters offers increased selectivity but poor transient behavior. Neither approximation to an ideal filter is directed toward obtaining a constant delay in the passband.

The Bessel transfer function has been optimized to obtain a linear phase—in other words, a maximally flat delay. The step response has essentially no overshoot or ringing, and the impulse response lacks oscillatory behavior. However, the frequency response is much less selective than in the other filter types.

The low-pass approximation to a constant delay can be expressed as the following general transfer function:

$$T(s) = \frac{1}{\sinh s + \cosh s} \tag{2-45}$$

If a continued-fraction expansion is used to approximate the hyperbolic functions and the expansion is truncated at different lengths, the Bessel family of transfer functions will result.

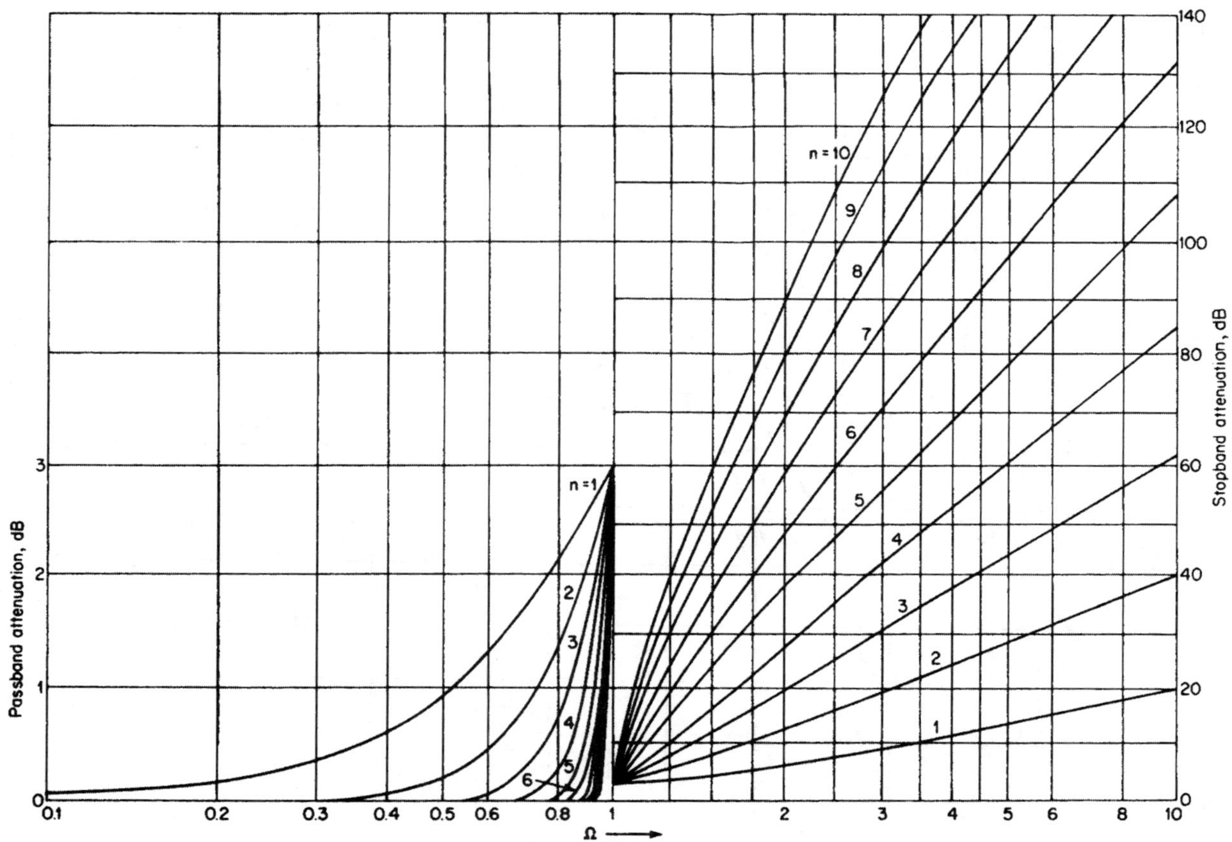

FIGURE 2-41 Attenuation characteristics for Chebyshev filters with 0.01-dB ripple. (From A. I. Zverev, *Handbook of Filter Synthesis* [New York: John Wiley and Sons, 1967.] By permission of the publishers.)

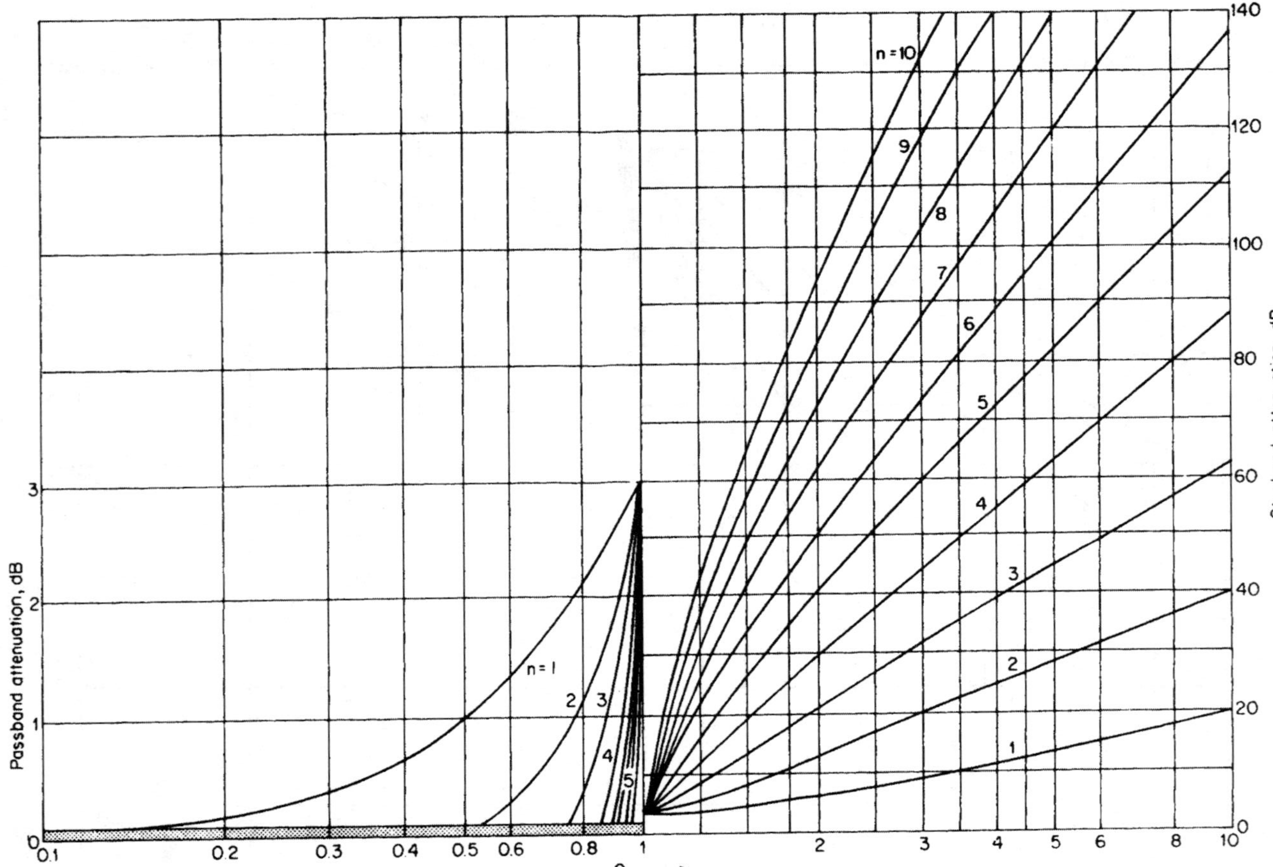

FIGURE 2-42 Attenuation characteristics for Chebyshev filters with 0.1-dB ripple. (From A. I. Zverev, *Handbook of Filter Synthesis* [New York: John Wiley and Sons, 1967.] By permission of the publishers.)

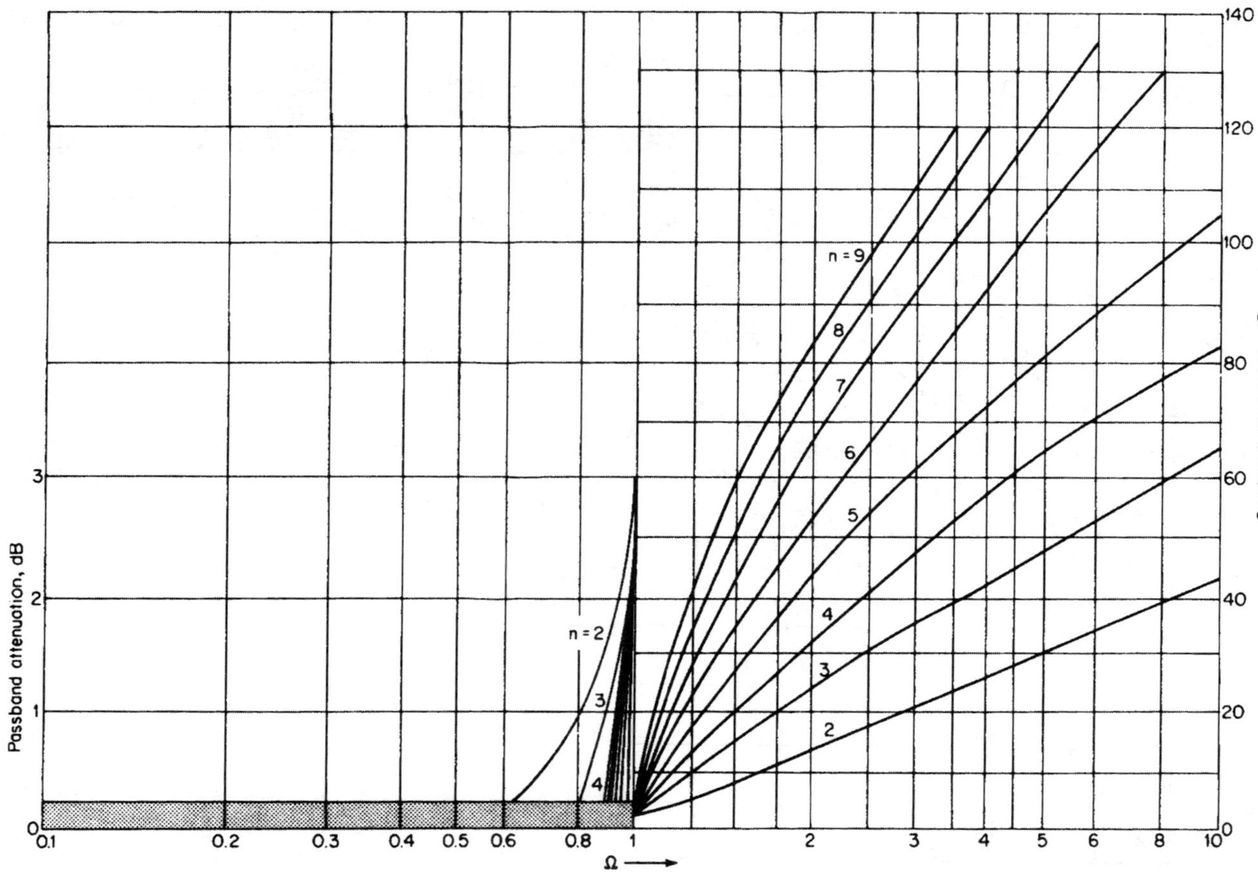

FIGURE 2-43 Attenuation characteristics for Chebyshev filters with 0.25-dB ripple.

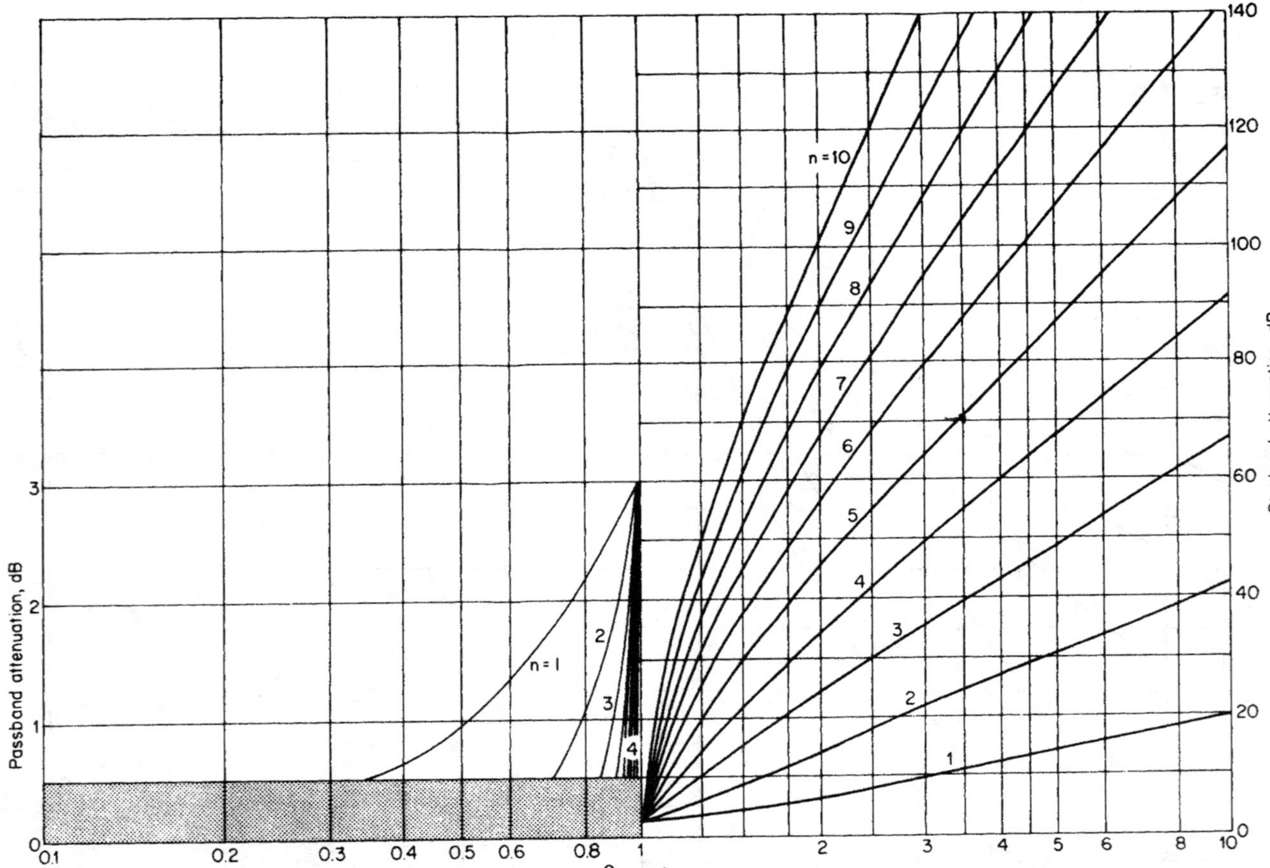

FIGURE 2-44 Attenuation characteristics for Chebyshev filters with 0.5-dB ripple. (From A. I. Zverev, *Handbook of Filter Synthesis* [New York: John Wiley and Sons, 1967.] By permission of the publishers.)

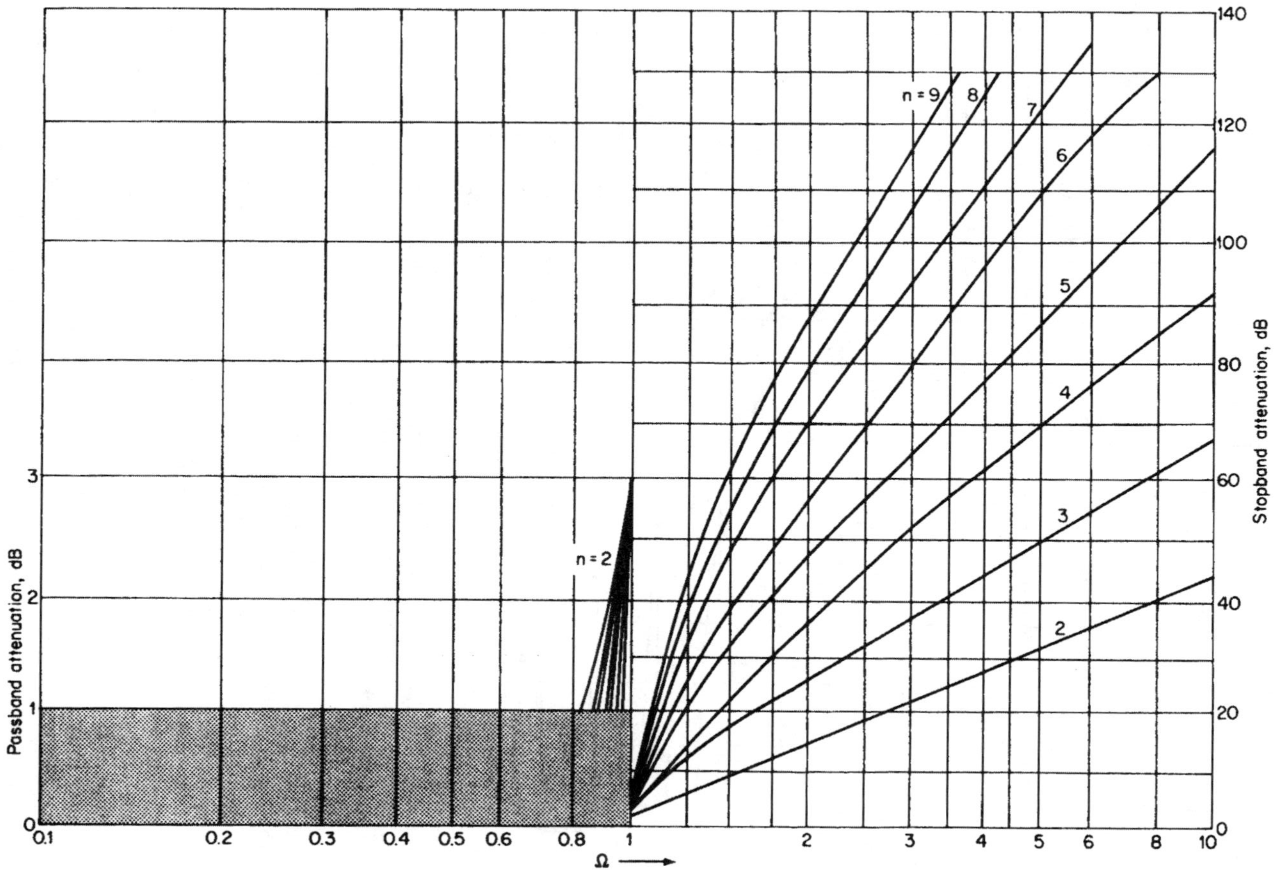

FIGURE 2-45 Attenuation characteristics for Chebyshev filters with 1-dB ripple.

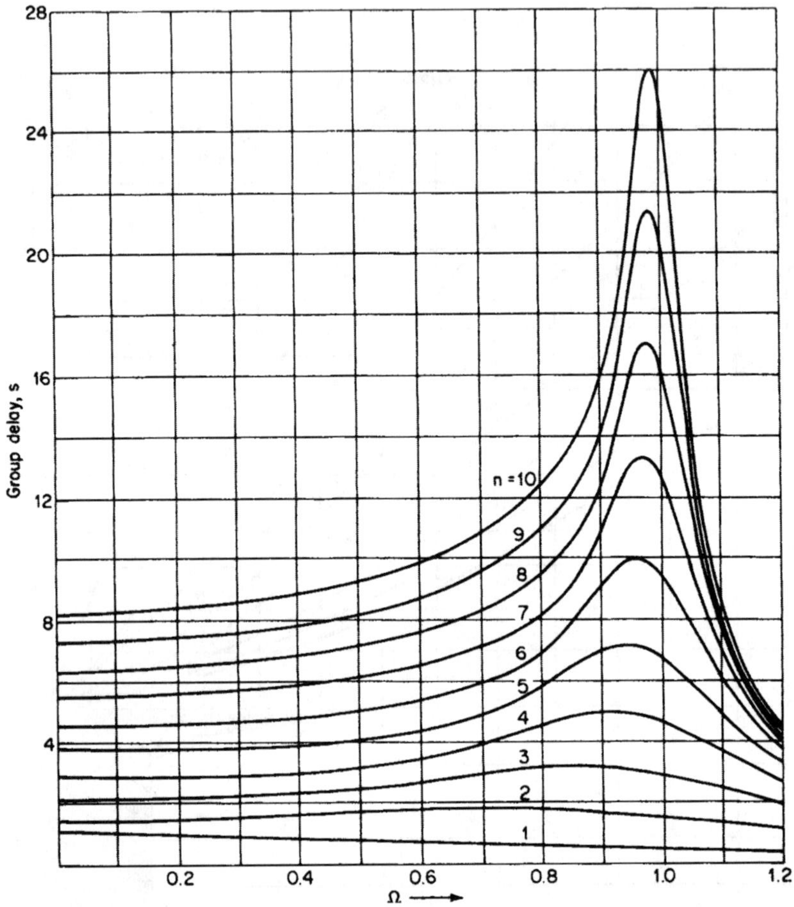

FIGURE 2-46 Group-delay characteristics for Chebyshev filters with 0.01-dB ripple. (From A. I. Zverev, *Handbook of Filter Synthesis* [New York: John Wiley and Sons, 1967.] By permission of the publishers.)

A crude approximation to the pole locations can be found by locating all the poles on a circle and separating their imaginary parts by $2/n$, as shown in Figure 2-55. The vertical spacing between poles is equal, whereas in the Butterworth case the angles were equal.

The relative attenuation of a Bessel low-pass filter can be approximated by

$$A_{\mathrm{dB}} = 3\left(\frac{\omega_x}{\omega_c}\right)^2 \tag{2-46}$$

This expression is reasonably accurate for ω_x/ω_c ranging between 0 and 2.

Figures 2-56 through 2-59 indicate that as the order n is increased, the region of flat delay is extended farther into the stopband. However, the steepness of roll-off in the transition region does not improve significantly. This restricts the use of Bessel filters to applications where the transient properties are the major consideration.

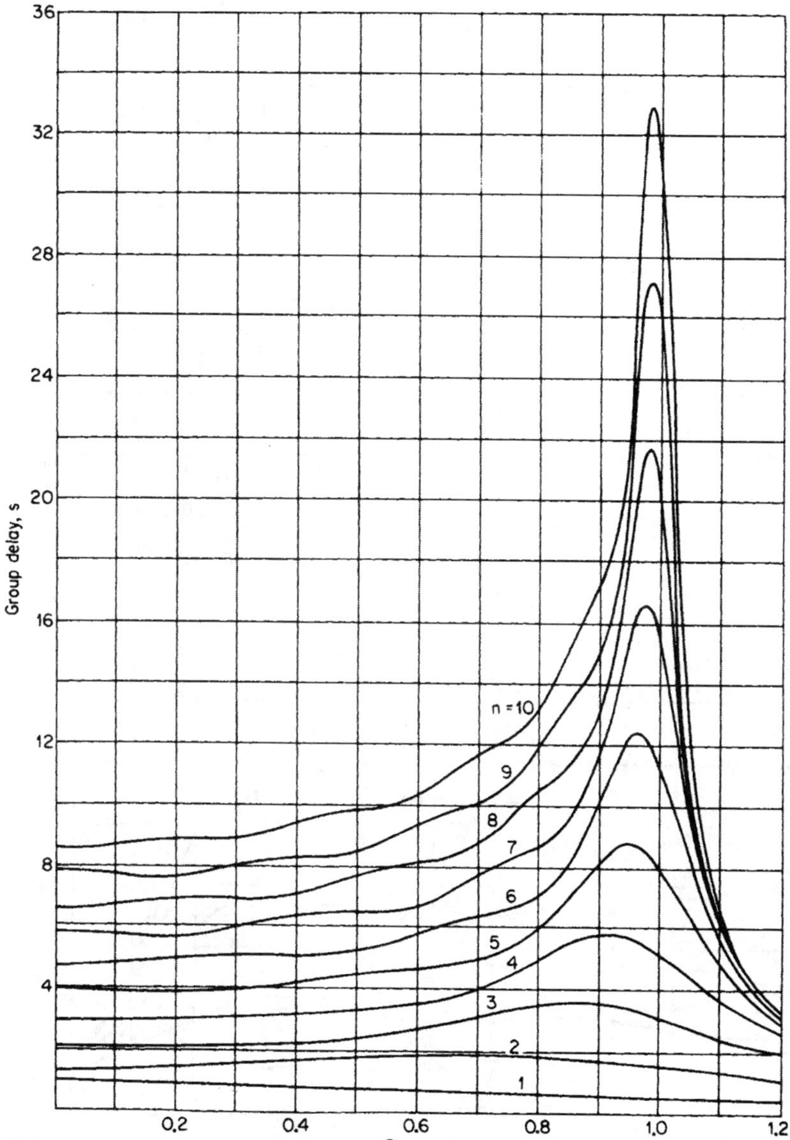

FIGURE 2-47 Group-delay characteristics for Chebyshev filters with 0.1-dB ripple. (From A. I. Zverev, *Handbook of Filter Synthesis* [New York: John Wiley and Sons, 1967.] By permission of the publishers.)

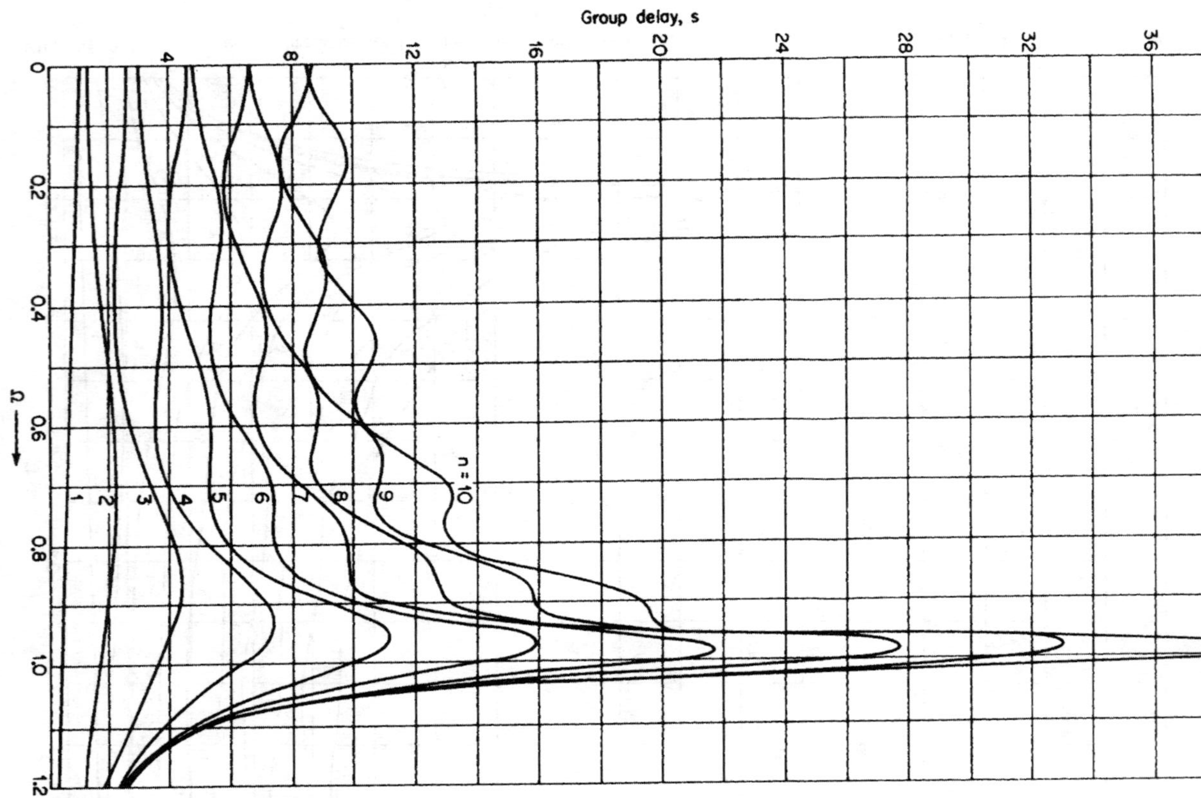

FIGURE 2-48 Group-delay characteristics for Chebyshev filters with 0.5-dB ripple. (From A. I. Zverev, *Handbook of Filter Synthesis* [New York: John Wiley and Sons, 1967.] By permission of the publishers.)

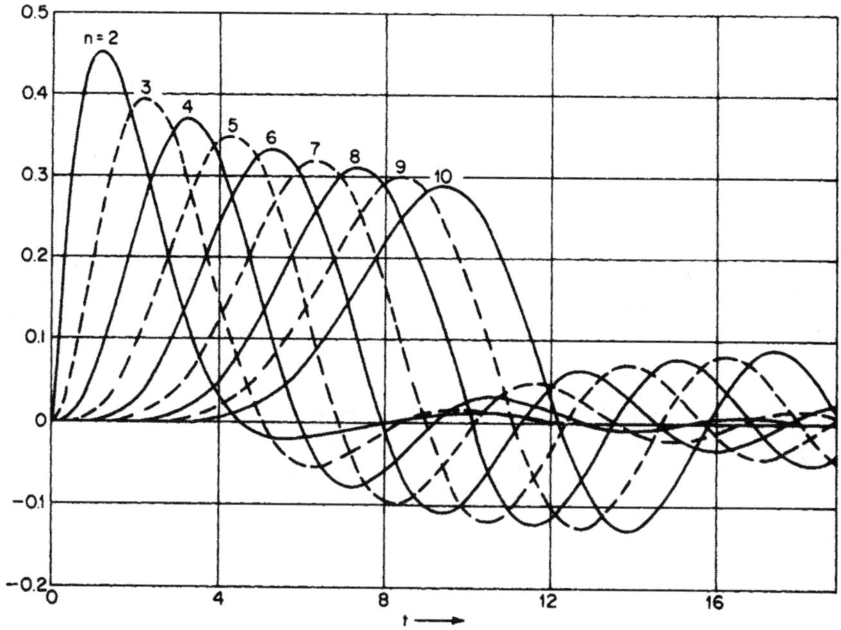

FIGURE 2-49 Impulse response for Chebyshev filters with 0.01-dB ripple. (From A. I. Zverev, *Handbook of Filter Synthesis* [New York: John Wiley and Sons, 1967.] By permission of the publishers.)

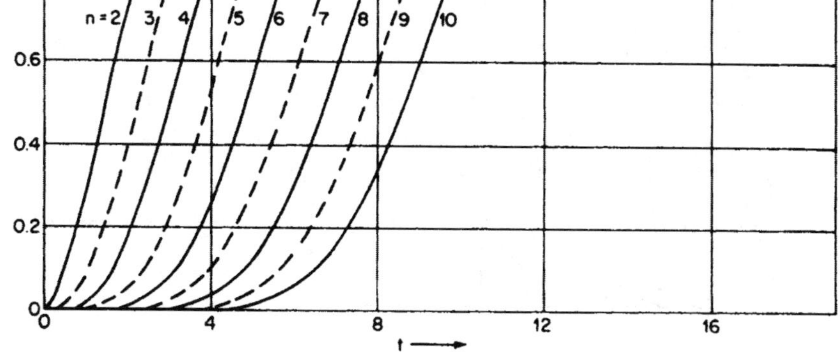

FIGURE 2-50 Step response for Chebyshev filters with 0.01-dB ripple. (From A. I. Zverev, *Handbook of Filter Synthesis* [New York: John Wiley and Sons, 1967.] By permission of the publishers.)

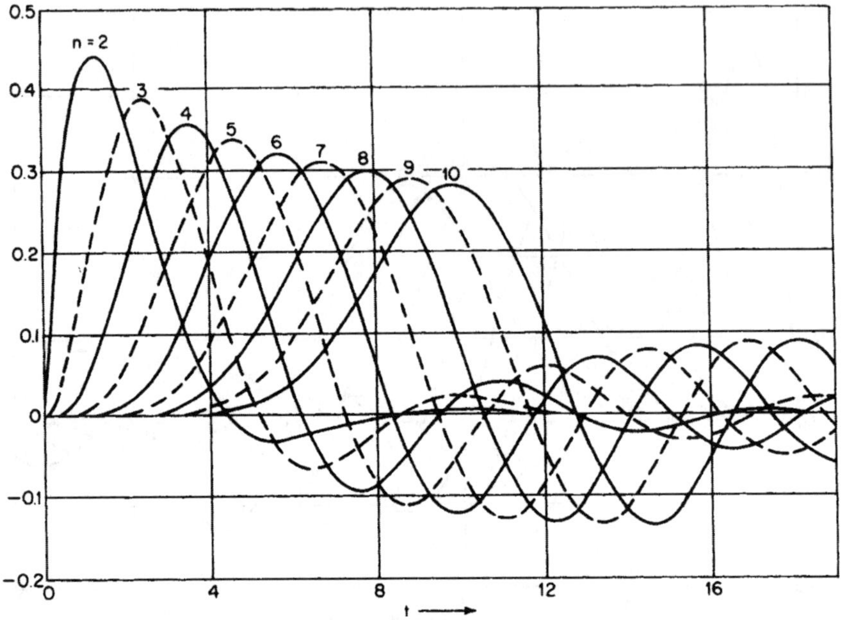

FIGURE 2-51 Impulse response for Chebyshev filters with 0.1-dB ripple. (From A. I. Zverev, *Handbook of Filter Synthesis* [New York: John Wiley and Sons, 1967.] By permission of the publishers.)

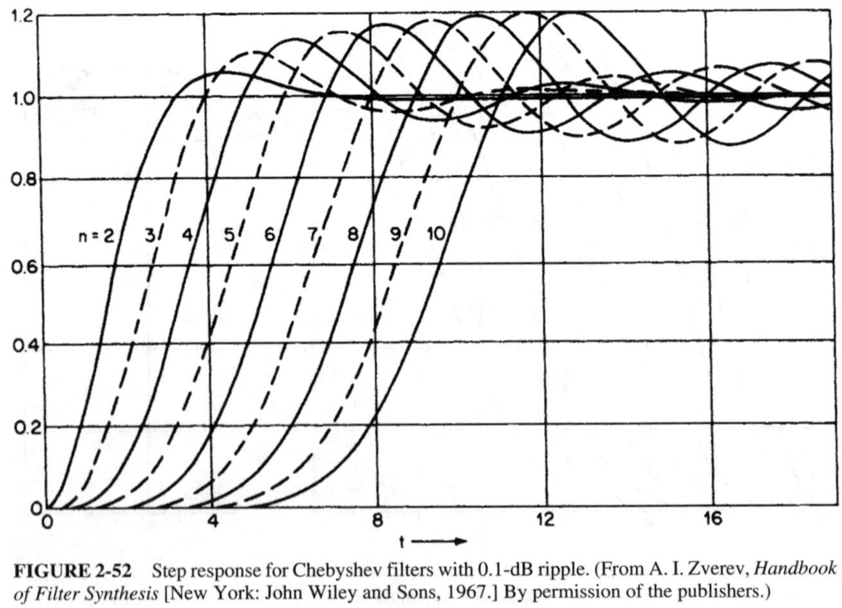

FIGURE 2-52 Step response for Chebyshev filters with 0.1-dB ripple. (From A. I. Zverev, *Handbook of Filter Synthesis* [New York: John Wiley and Sons, 1967.] By permission of the publishers.)

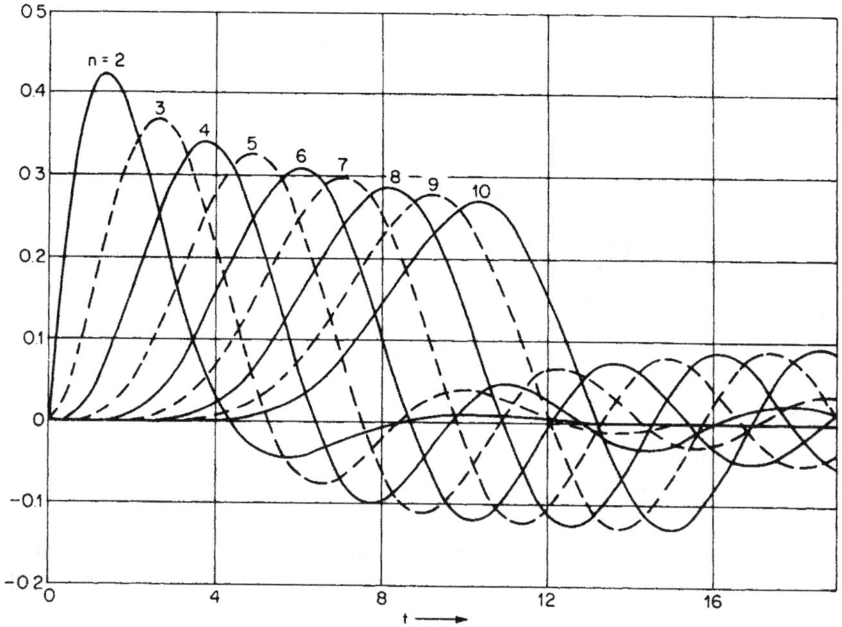

FIGURE 2-53 Impulse response for Chebyshev filters with 0.5-dB ripple. (From A. I. Zverev, *Handbook of Filter Synthesis* [New York: John Wiley and Sons, 1967.] By permission of the publishers.)

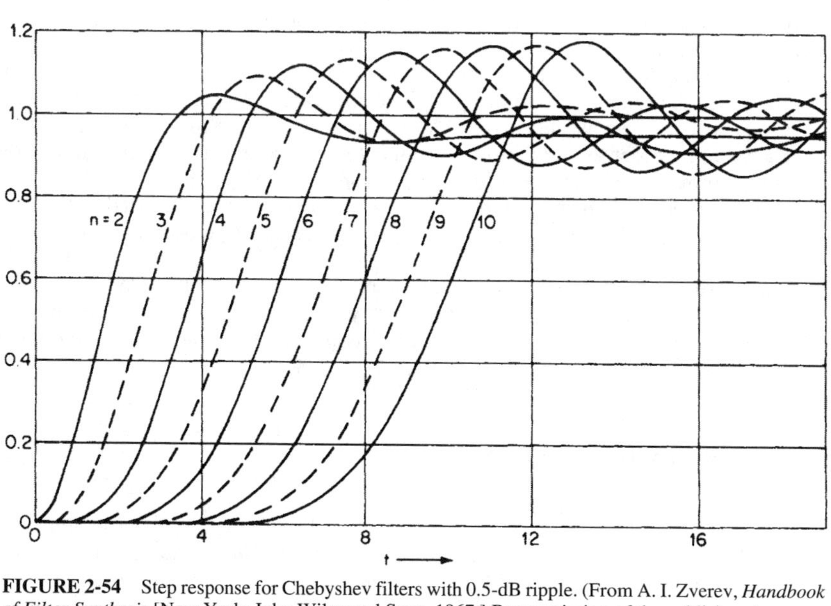

FIGURE 2-54 Step response for Chebyshev filters with 0.5-dB ripple. (From A. I. Zverev, *Handbook of Filter Synthesis* [New York: John Wiley and Sons, 1967.] By permission of the publishers.)

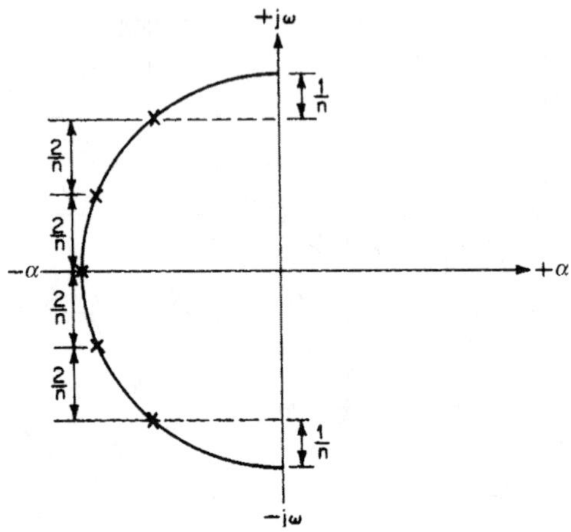

FIGURE 2-55 Approximate Bessel pole locations.

A similar family of filters is the gaussian type. However, the gaussian phase response is not as linear as the Bessel for the same number of poles, and the selectivity is not as sharp.

2.6 LINEAR PHASE WITH EQUIRIPPLE ERROR

The Chebyshev (equiripple amplitude) function is a better approximation of an ideal amplitude curve than the Butterworth. Therefore, it stands to reason that an equiripple approximation of a linear phase will be more efficient than the Bessel family of filters.

Figure 2-60 illustrates how a linear phase can be approximated to within a given ripple of ε degrees. For the same n, the equiripple-phase approximation results in a linear phase and, consequently, a constant delay over a larger interval than the Bessel approximation. Also the amplitude response is superior far from cutoff. In the transition region and below cutoff, both approximations have nearly identical responses.

As the phase ripple ε is increased, the region of constant delay is extended farther into the stopband. However, the delay develops ripples. The step response has slightly more overshoot than Bessel filters.

A closed-form method for computation of the pole positions is not available. The pole locations tabulated in Chapter 11 were developed by iterative techniques. Values are provided for phase ripples of 0.05° and 0.5°, and the associated frequency and time-domain parameters are given in Figures 2-61 through 2-68.

2.7 TRANSITIONAL FILTERS

The Bessel filters discussed in Section 2.5 have excellent transient properties but poor selectivity. Chebyshev filters, on the other hand, have steep roll-off characteristics but poor time-domain behavior. A transitional filter offers a compromise between a gaussian filter, which is similar to the Bessel family, and Chebyshev filters.

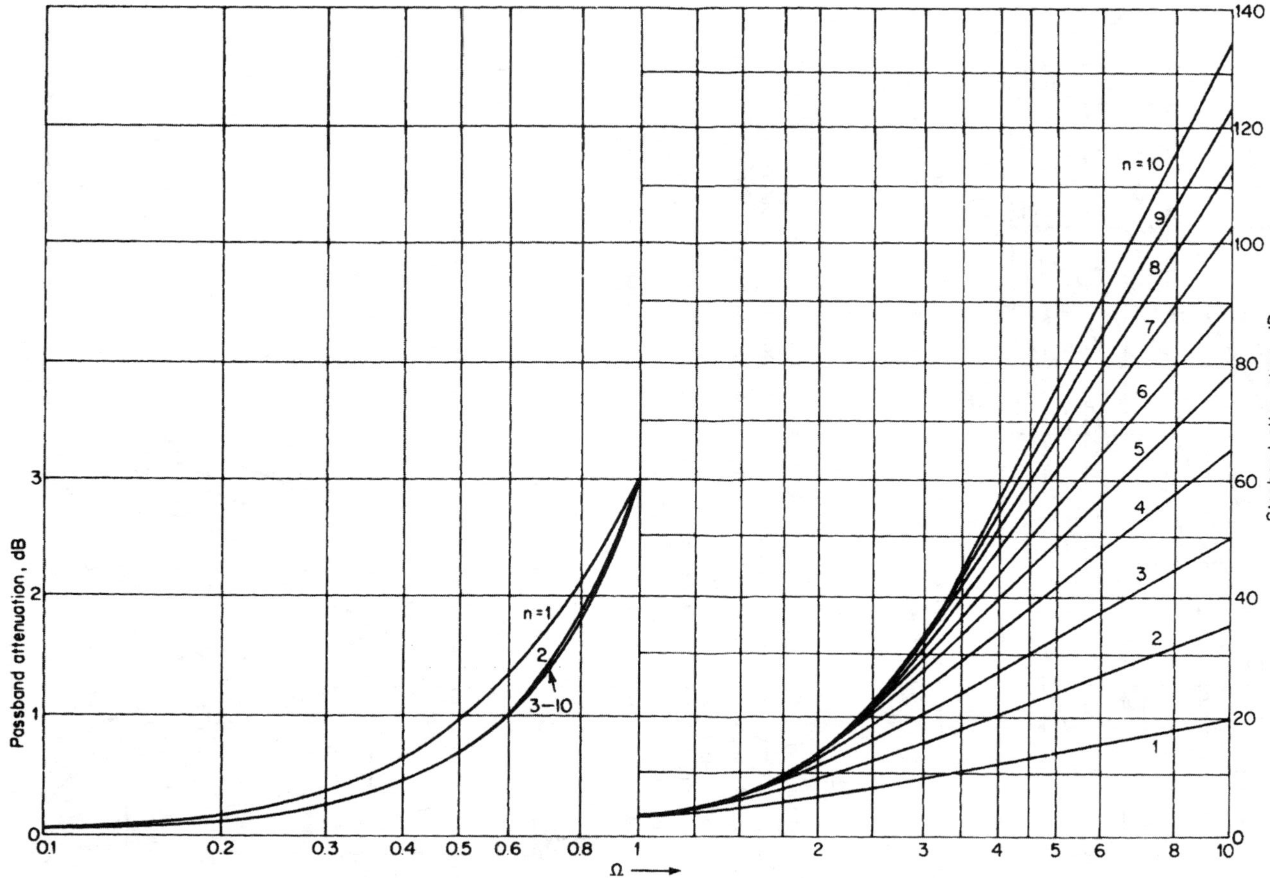

FIGURE 2-56 Attenuation characteristics for maximally flat delay (Bessel) filters. (From A. I. Zverev, *Handbook of Filter Synthesis* [New York: John Wiley and Sons, 1967.] By permission of the publishers.)

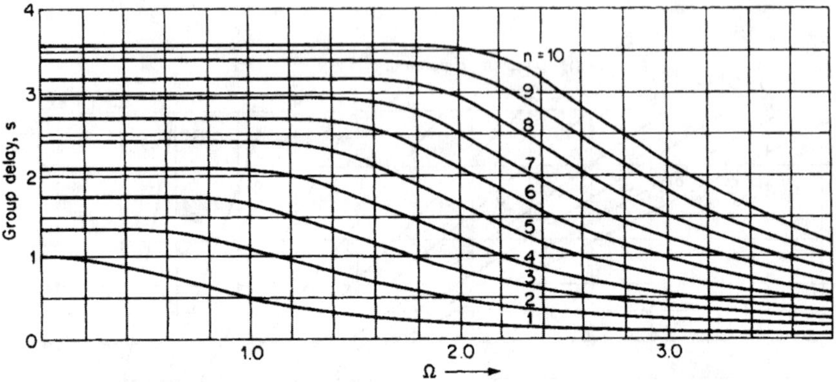

FIGURE 2-57 Group-delay characteristics for maximally flat delay (Bessel) filters. (From A. I. Zverev, *Handbook of Filter Synthesis* [New York: John Wiley and Sons, 1967.] By permission of the publishers.)

Transitional filters have a near linear phase shift and smooth amplitude roll-off in the passband. Outside the passband, a sharp break in the amplitude characteristics occurs. Beyond this breakpoint, the attenuation increases quite abruptly in comparison with Bessel filters, especially for the higher *n*'s.

In the tables in Chapter 11, transnational filters are listed which have gaussian characteristics to both 6 dB and 12 dB. The transient properties of the gaussian to 6-dB filters are somewhat superior to those of the Butterworth family. Beyond the 6-dB point, which

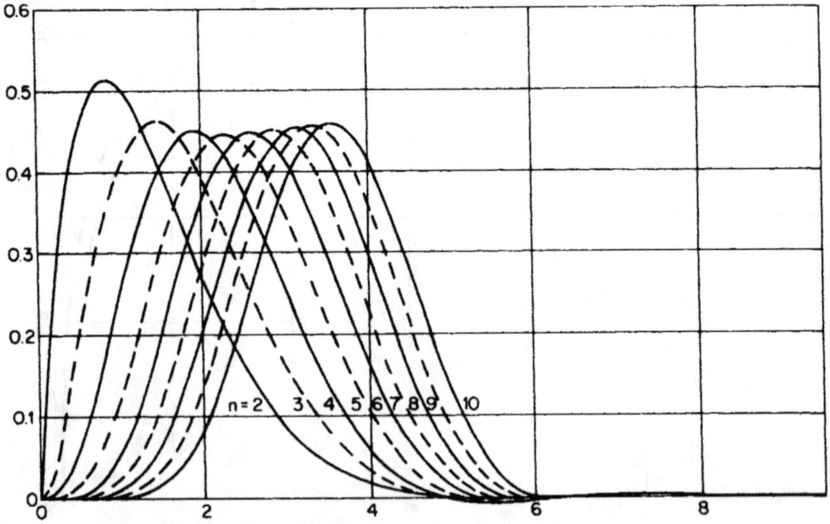

FIGURE 2-58 Impulse response for maximally flat delay (Bessel) filters. (From A. I. Zverev, *Handbook of Filter Synthesis* [New York: John Wiley and Sons, 1967.] By permission of the publishers.)

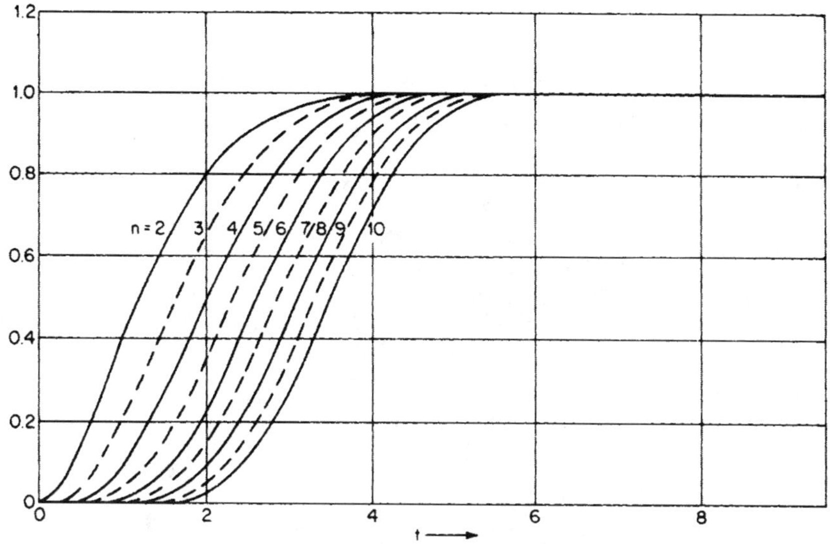

FIGURE 2-59 Step response for maximally flat delay (Bessel) filters. (From A. I. Zverev, *Handbook of Filter Synthesis* [New York: John Wiley and Sons, 1967.] By permission of the publishers.)

occurs at approximately 1.5 rad/s, the attenuation characteristics are nearly comparable with Butterworth filters. The gaussian to 12-dB filters have time-domain parameters far superior to those of Butterworth filters. However, the 12-dB breakpoint occurs at 2 rad/s, and the attenuation characteristics beyond this point are inferior to those of Butterworth filters.

The transnational filters tabulated in Chapter 11 were generated using mathematical techniques which involve interpolation of pole locations. Figures 2-69 through 2-76 indicate the frequency and time-domain properties of both the gaussian to 6-dB and gaussian to 12-dB transitional filters.

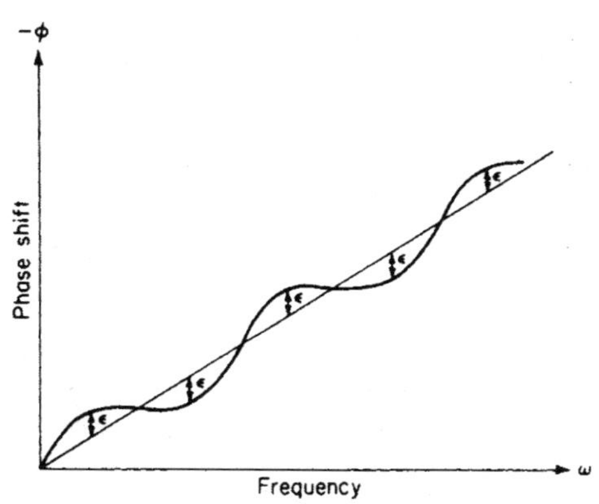

FIGURE 2-60 An equiripple linear-phase approximation.

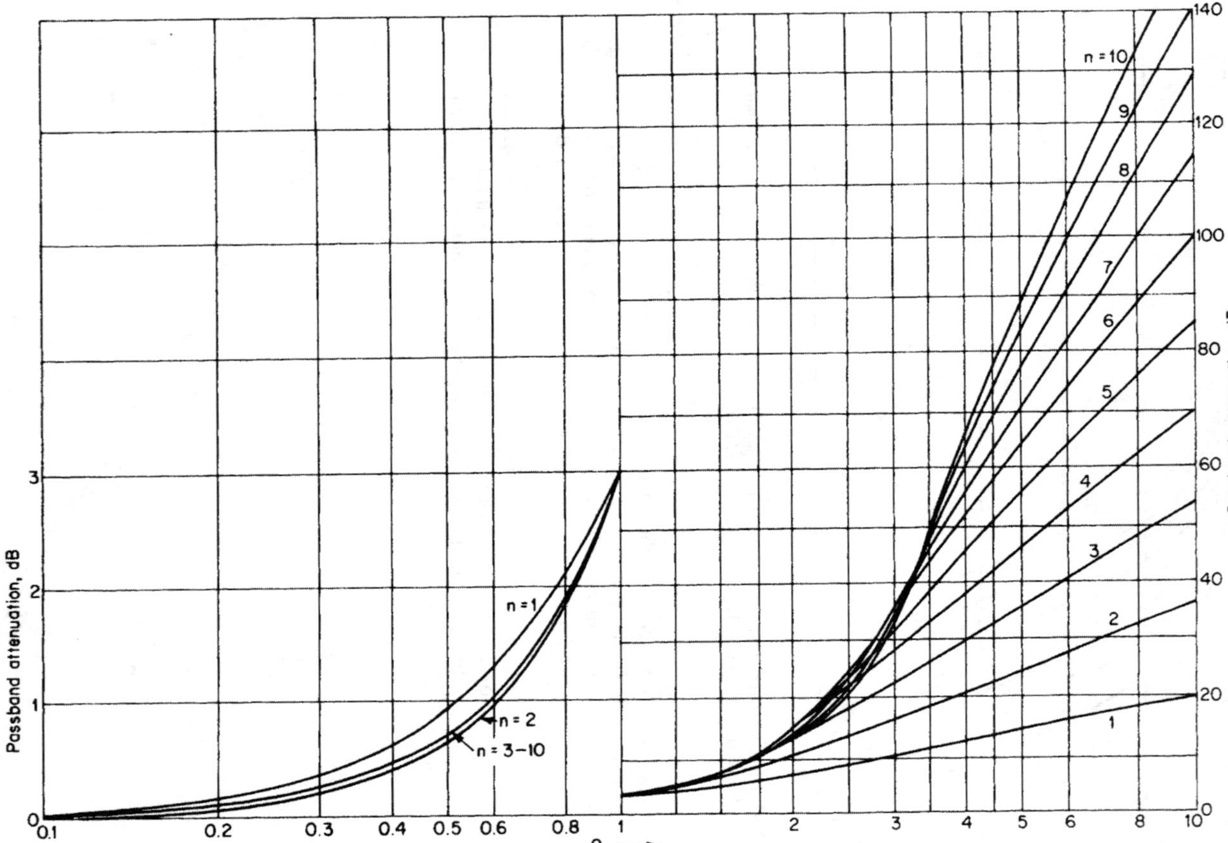

FIGURE 2-61 Attenuation characteristics for linear phase with equiripple error filters (phase error =0.05°). (From A. I. Zverev, *Handbook of Filter Synthesis* [New York: John Wiley and Sons, 1967.] By permission of the publishers.)

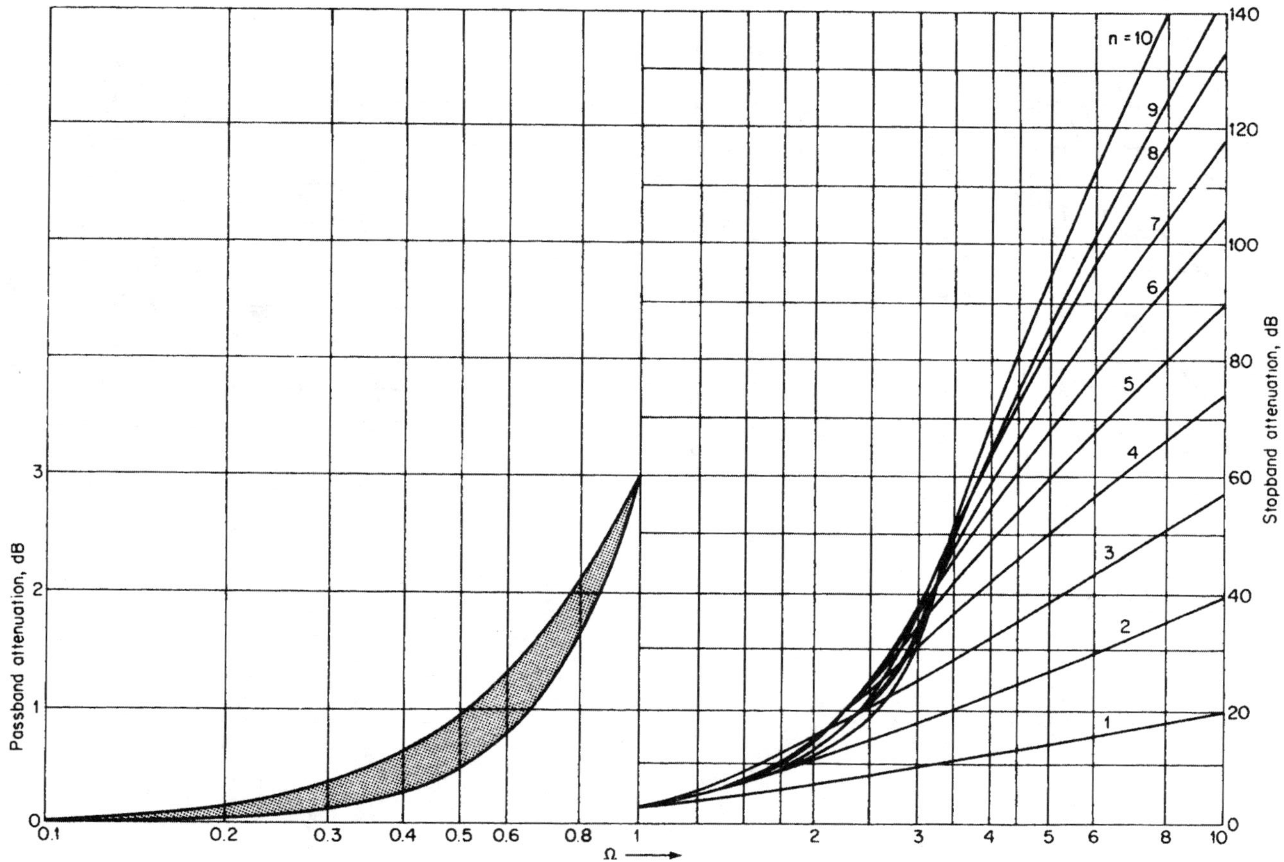

FIGURE 2-62 Attenuation characteristics for linear phase with equiripple error filters (phase error =0.5°). (From A. I. Zverev, *Handbook of Filter Synthesis* [New York: John Wiley and Sons, 1967.] By permission of the publishers.)

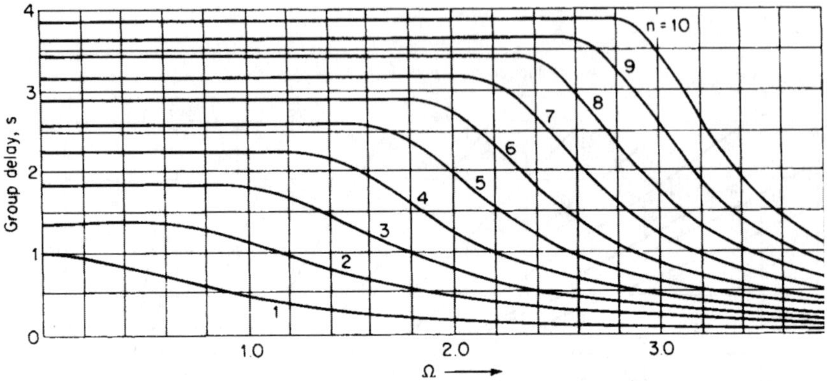

FIGURE 2-63 Group-delay characteristics for linear phase with equiripple error filters (phase error = 0.05°). (From A. I. Zverev, *Handbook of Filter Synthesis* [New York: John Wiley and Sons, 1967.] By permission of the publishers.)

2.8 SYNCHRONOUSLY TUNED FILTERS

Synchronously tuned filters are the most basic filter type and are the easiest to construct and align. They consist of identical multiple poles. A typical application is in the case of a bandpass amplifier, where a number of stages are cascaded, with each stage having the same center frequency and Q.

The attenuation of a synchronously tuned filter can be expressed as

$$A_{dB} = 10n \log [1 + (2^{1/n} - 1)\Omega^2] \tag{2-47}$$

Equation (2-47) is normalized so that 3 dB of attenuation occurs at $\Omega = 1$.

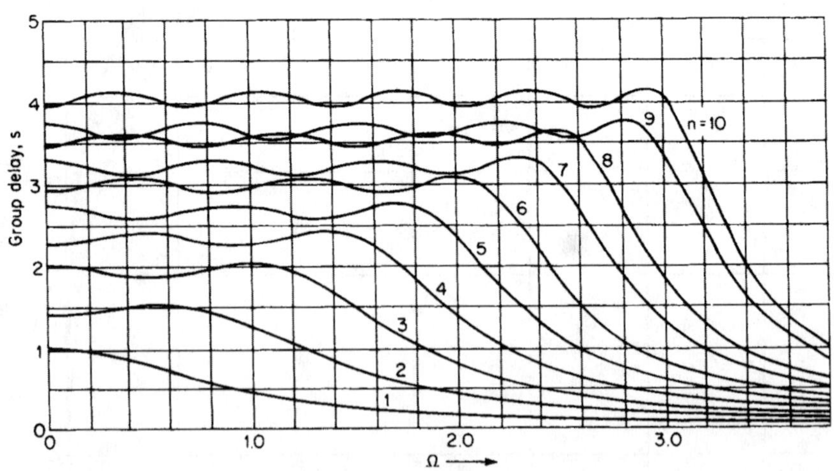

FIGURE 2-64 Group-delay characteristics for linear phase with equiripple error filters (phase error = 0.5°). (From A. I. Zverev, *Handbook of Filter Synthesis* [New York: John Wiley and Sons, 1967.] By permission of the publishers.)

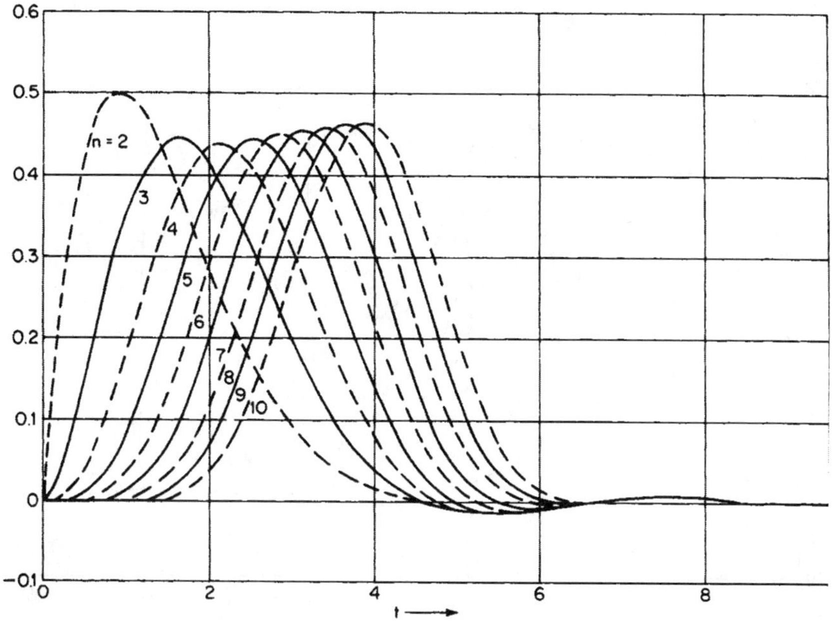

FIGURE 2-65 Impulse response for linear phase with equiripple error filters (phase error = 0.05°). (From A. I. Zverev, *Handbook of Filter Synthesis* [New York: John Wiley and Sons, 1967.] By permission of the publishers.)

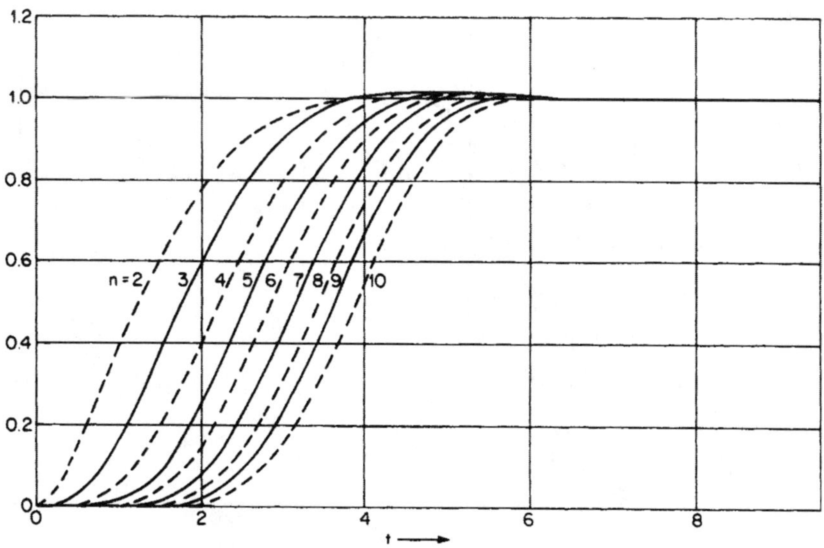

FIGURE 2-66 Step response for linear phase with equiripple error filters (phase error = 0.05°). (From A. I. Zverev, *Handbook of Filter Synthesis* [New York: John Wiley and Sons, 1967.] By permission of the publishers.)

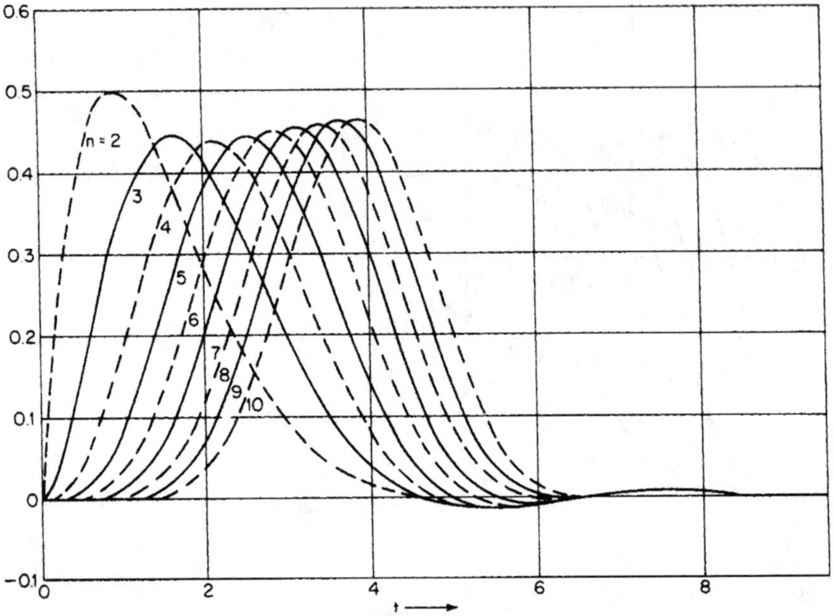

FIGURE 2-67 Impulse response for linear phase with equiripple error filters (phase error = 0.5°). (From A. I. Zverev, *Handbook of Filter Synthesis* [New York: John Wiley and Sons, 1967.] By permission of the publishers.)

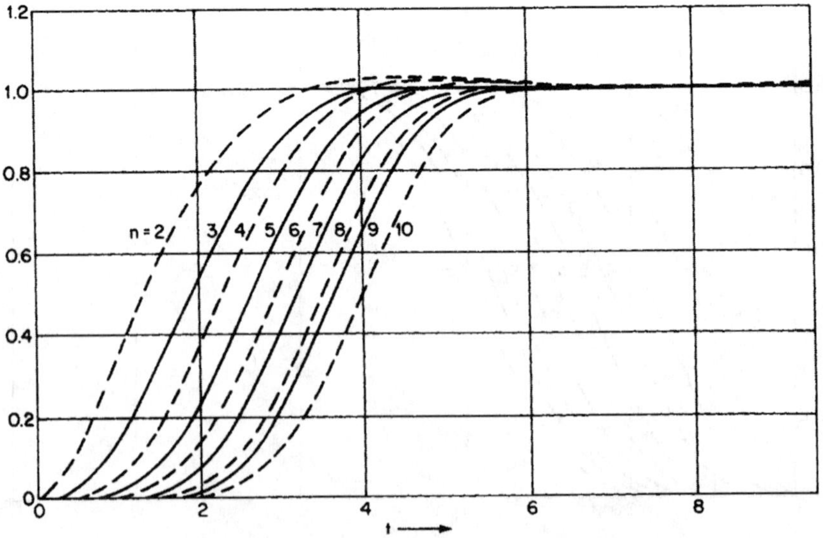

FIGURE 2-68 Step response for linear phase with equiripple error filters (phase error = 0.05°). (From A. I. Zverev, *Handbook of Filter Synthesis* [New York: John Wiley and Sons, 1967.] By permission of the publishers.)

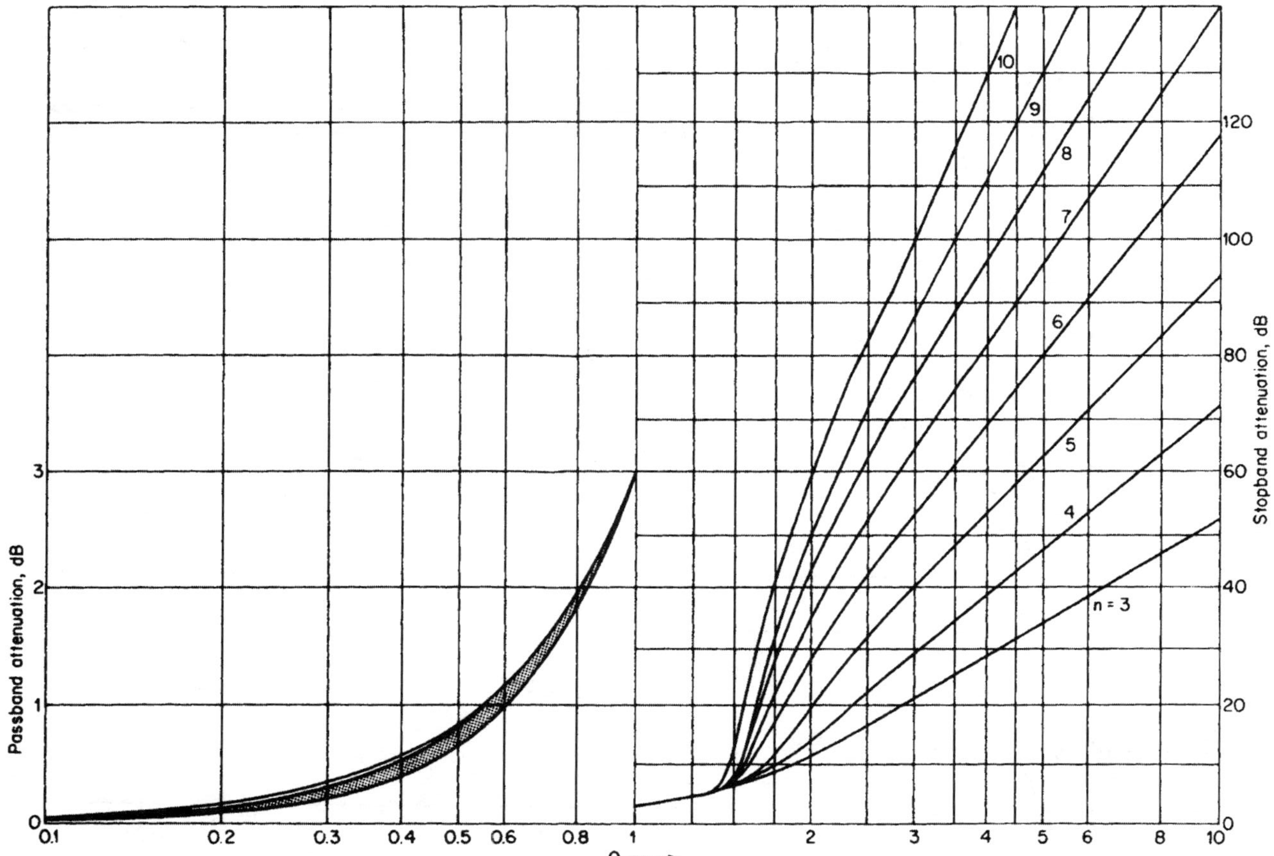

FIGURE 2-69 Attenuation characteristics for transitional filters (gaussian to 6 dB). (From A. I. Zverev, *Handbook of Filter Synthesis* [New York: John Wiley and Sons, 1967.] By permission of the publishers.)

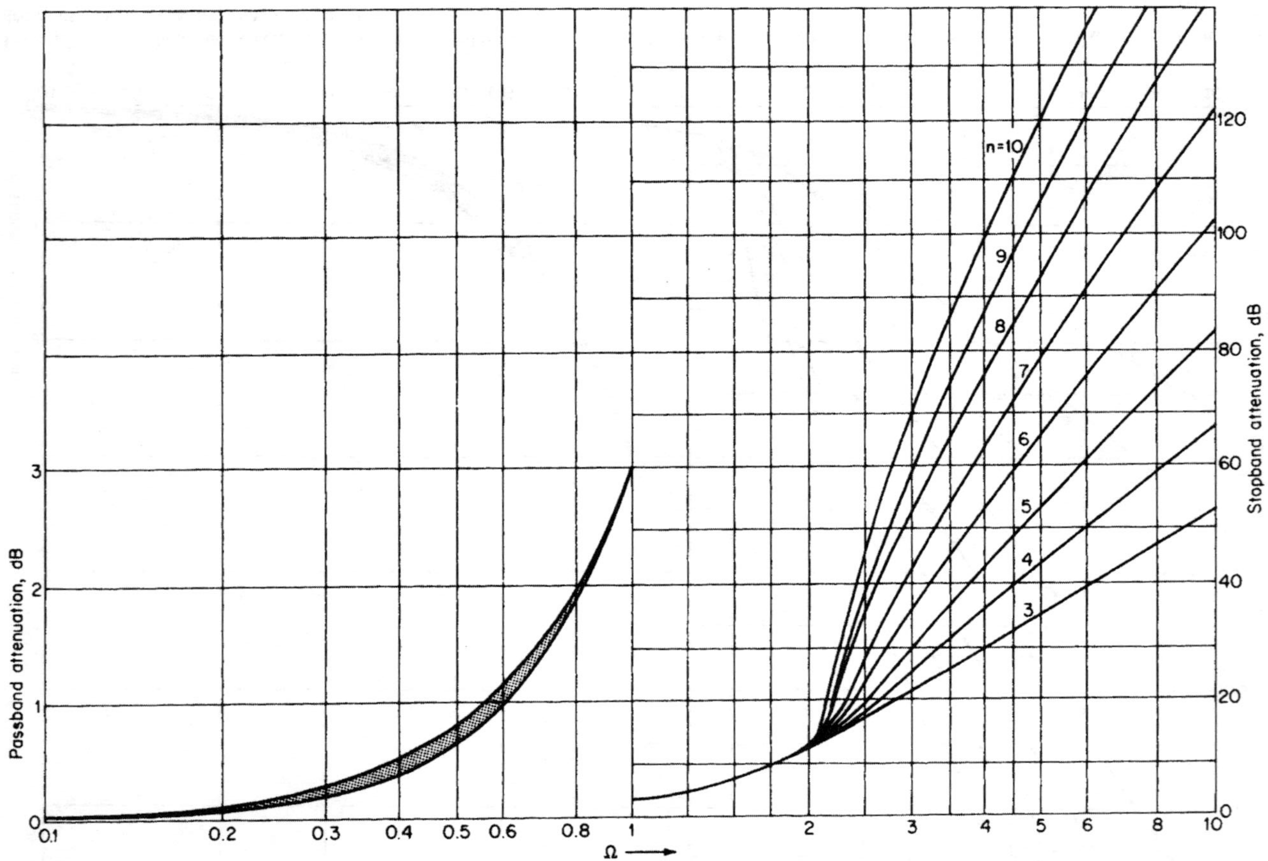

FIGURE 2-70 Attenuation characteristics for transitional filters (gaussian to 12 dB). (From A. I. Zverev, *Handbook of Filter Synthesis* [New York: John Wiley and Sons, 1967.] By permission of the publishers.)

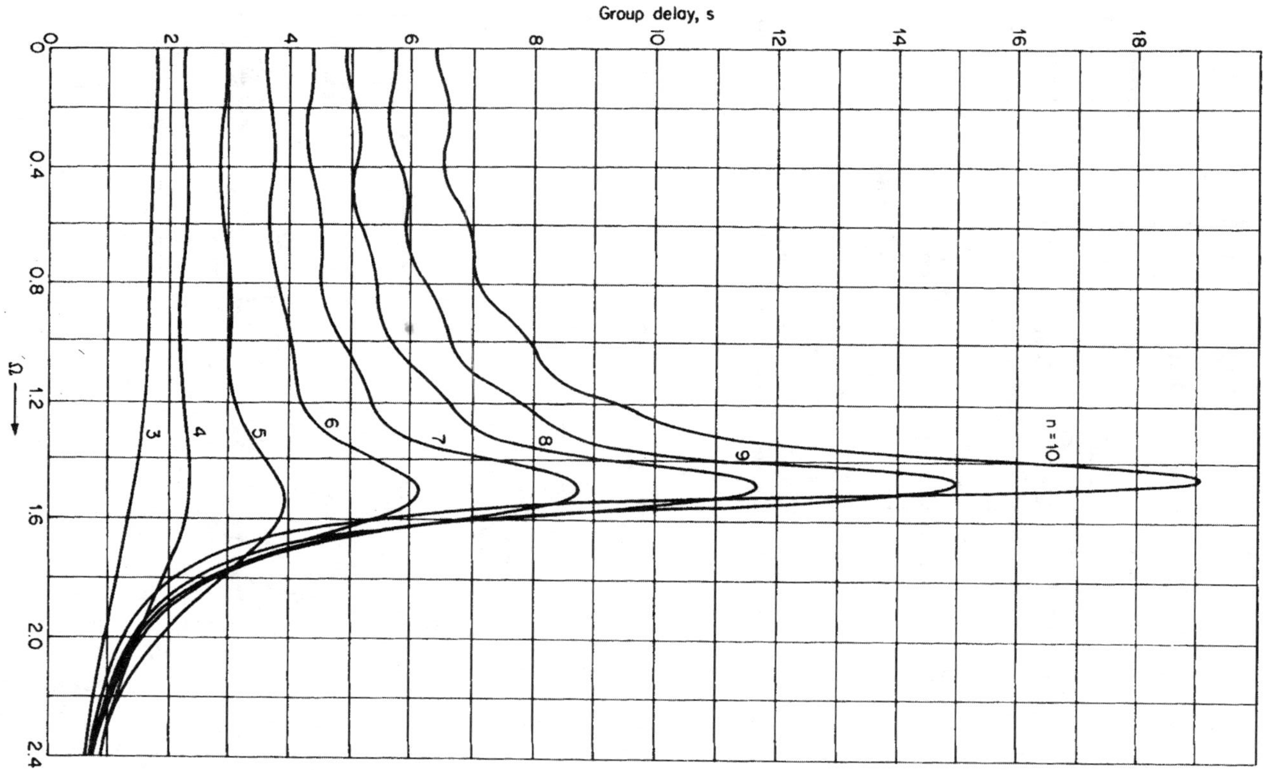

FIGURE 2-71 Group-delay characteristics for transitional filters (gaussian to 6 dB). (From A. I. Zverev, *Handbook of Filter Synthesis* [New York: John Wiley and Sons, 1967.] By permission of the publishers.)

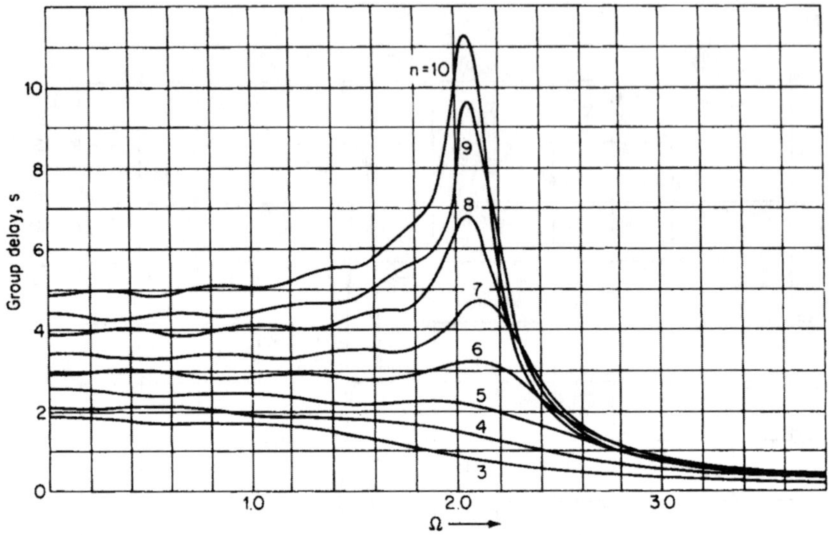

FIGURE 2-72 Group-delay characteristics for transitional filters (gaussian to 12 dB). (From A. I. Zverev, *Handbook of Filter Synthesis* [New York: John Wiley and Sons, 1967.] By permission of the publishers.)

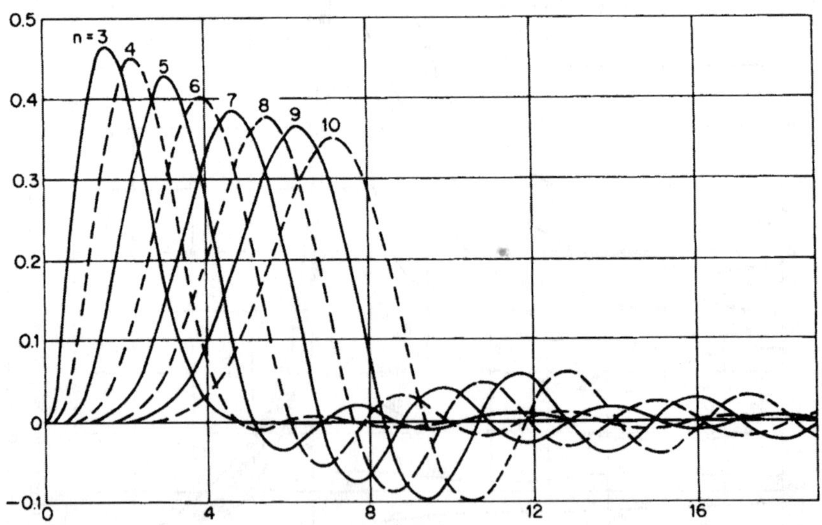

FIGURE 2-73 Impulse response for transitional filters (gaussian to 6 dB). (From A. I. Zverev, *Handbook of Filter Synthesis* [New York: John Wiley and Sons, 1967.] By permission of the publishers.)

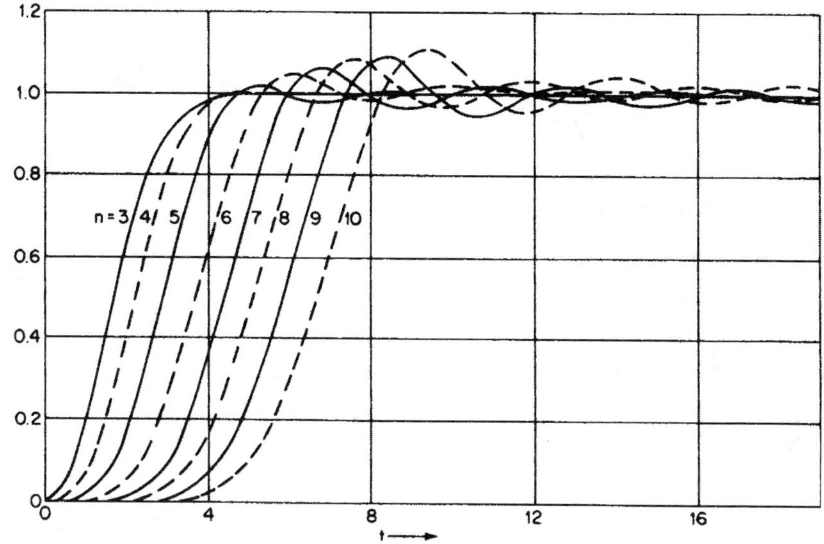

FIGURE 2-74 Step response for transitional filters (gaussian to 6 dB). (From A. I. Zverev, *Handbook of Filter Synthesis* [New York: John Wiley and Sons, 1967.] By permission of the publishers.)

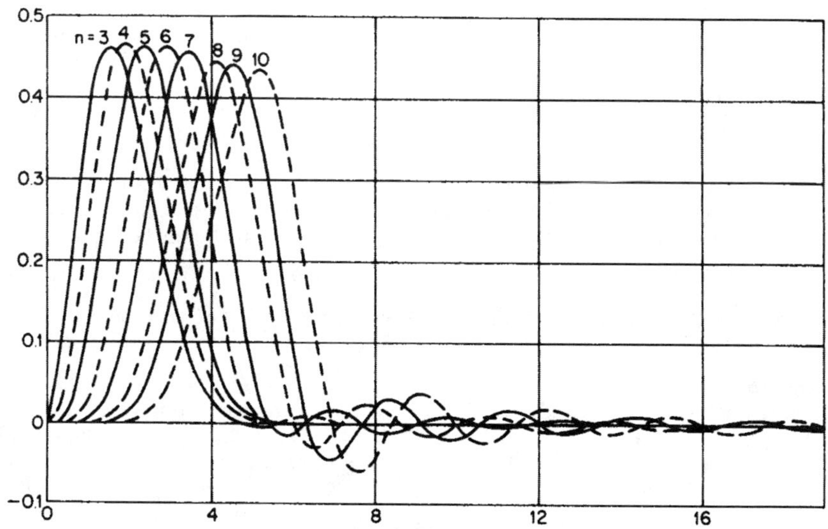

FIGURE 2-75 Impulse response for transitional filters (gaussian to 12 dB). (From A. I. Zverev, *Handbook of Filter Synthesis* [New York: John Wiley and Sons, 1967.] By permission of the publishers.)

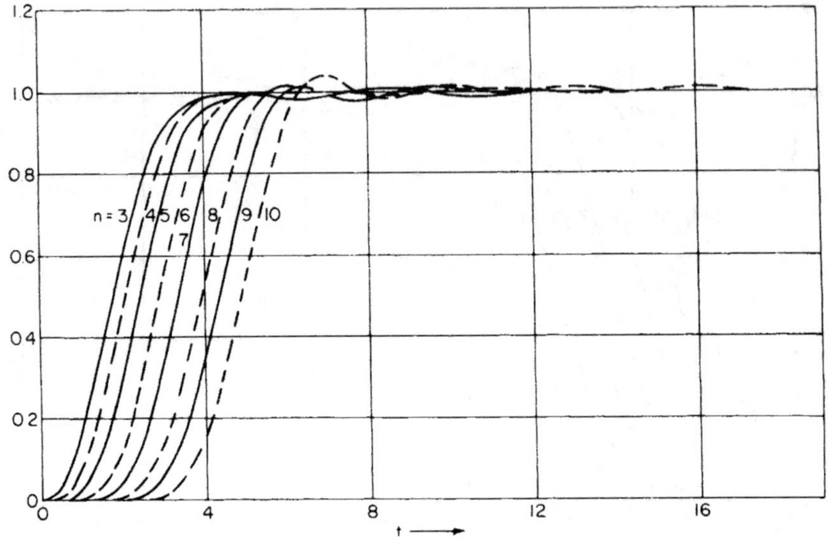

FIGURE 2-76 Step response for transitional filters (gaussian to 12 dB). (From A. I. Zverev, *Handbook of Filter Synthesis* [New York: John Wiley and Sons, 1967.] By permission of the publishers.)

The individual section Q can be defined in terms of the composite circuit Q requirement using the following relationship:

$$Q_{\text{section}} = Q_{\text{overall}} \sqrt{2^{1/n} - 1} \tag{2-48}$$

Alternatively, we can state that the 3-dB bandwidth of the individual sections is reduced by the shrinkage factor $(2^{1/n} - 1)^{1/2}$. The individual section Q is less than the overall Q, whereas in the case of nonsynchronously tuned filters the section Qs may be required to be much higher than the composite Q.

Example 2-21 Calculate the Attenuation and Section Q's of a Synchronously Tuned Bandpass Filter

Required:

A three-section synchronously tuned bandpass filter is required to have a center frequency of 10 kHz and a 3-dB bandwidth of 100 Hz. Determine the attenuation corresponding to a bandwidth of 300 Hz, and calculate the Q of each section.

Result:

(*a*) The attenuation at the 300Hz bandwidth can be computed as

$$A_{\text{dB}} = 10n \log [1 + (2^{1/n} - 1)\Omega^2] = 15.7 \text{ dB} \tag{2-47}$$

where $n = 3$ and Ω, the bandwidth ratio, is 300 Hz/100 Hz, or 3. (Since the filter is a narrowband type, conversion to a geometrically symmetrical response requirement was not necessary.)

(*b*) The Q of each section is

$$Q_{\text{section}} = Q_{\text{overall}}\sqrt{2^{1/n} - 1} = 51 \tag{2-48}$$

where Q_{overall} is 10 kHz/100 Hz, or 100.

The synchronously tuned filter of Example 2-21 has only 15.7 dB of attenuation at a normalized frequency ratio of 3, and for $n = 3$. Even the gradual roll-off characteristics of the Bessel family provide better selectivity than synchronously tuned filters for equivalent complexities.

The transient properties, however, are near optimum. The step response exhibits no overshoot at all and the impulse response lacks oscillatory behavior.

The poor selectivity of synchronously tuned filters limits their application to circuits requiring modest attenuation steepness and simplicity of alignment. The frequency and time-domain characteristics are illustrated in Figures 2-77 through 2-80.

2.9 ELLIPTIC-FUNCTION FILTERS

All the filter types previously discussed are all-pole networks. They exhibit infinite rejection only at the extremes of the stopband. Elliptic-function filters have zeros as well as poles at finite frequencies. The location of the poles and zeros creates equiripple behavior in the passband similar to Chebyshev filters. Finite transmission zeros in the stopband reduce the transition region so that extremely sharp roll-off characteristics can be obtained. The introduction of these transmission zeros allows the steepest rate of descent theoretically possible for a given number of poles.

Figure 2-81 compares a five-pole Butterworth, a 0.1-dB Chebyshev, and a 0.1-dB elliptic-function filter having two transmission zeros. Clearly, the elliptic-function filter has a much more rapid rate of descent in the transition region than the other filter types.

Improved performance is obtained at the expense of return lobes in the stopband. Elliptic-function filters are also more complex than all-pole networks. Return lobes usually are acceptable to the user, since a minimum stopband attenuation is required and the chosen filter will have return lobes that meet this requirement. Also, even though each filter section is more complex than all-pole filters, fewer sections are required.

The following definitions apply to normalized elliptic-function low-pass filters and are illustrated in Figure 2-82:

R_{dB} = the passband ripple

$A_{\min}$ = the minimum stopband attenuation in decibels

Ω_s = the lowest stopband frequency at which $A_{\min}$ occurs

The response in the passband is similar to that of Chebyshev filters except that the attenuation at 1 rad/s is equal to the passband ripple instead of 3 dB. The stopband has transmission zeros, with the first zero occurring slightly beyond Ω_s. All returns (comebacks) in the stopband are equal to $A_{\min}$.

The attenuation of elliptic filters can be expressed as

$$A_{\text{dB}} = 10 \log[1 + \varepsilon^2 Z_n^2(\Omega)] \tag{2-49}$$

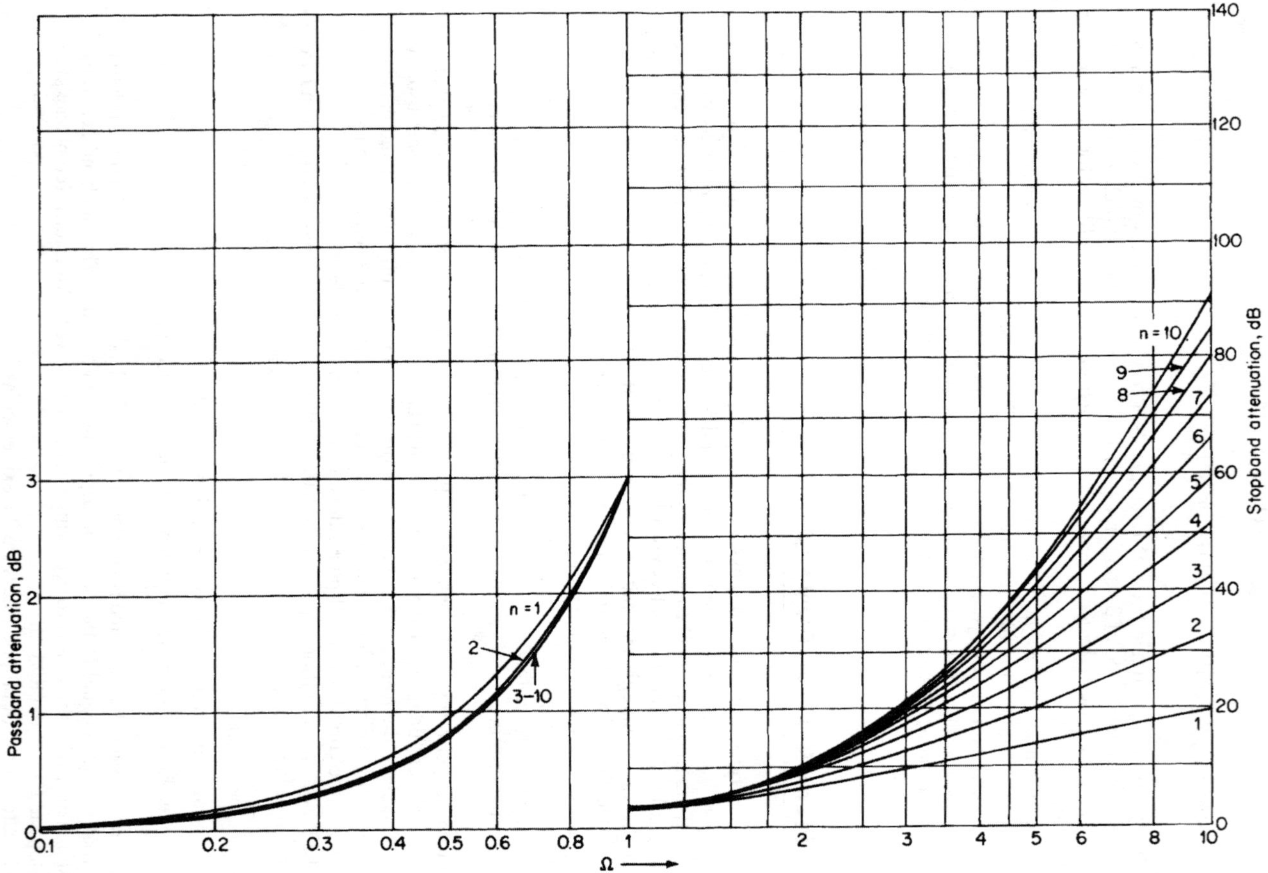

FIGURE 2-77 Attenuation characteristics for synchronously tuned filters. (From A. I. Zverev, *Handbook of Filter Synthesis* [New York: John Wiley and Sons, 1967.] By permission of the publishers.)

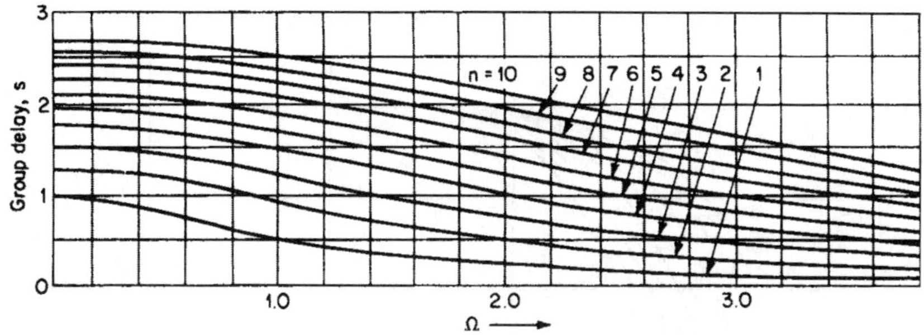

FIGURE 2-78 Group-delay characteristics for synchronously tuned filters. (From A. I. Zverev, *Handbook of Filter Synthesis* [New York: John Wiley and Sons, 1967.] By permission of the publishers.)

where ε is determined by the ripple (Equation 2-37) and $Z_n(\Omega)$ is an elliptic function of the nth order. Elliptic functions have both poles and zeros and can be expressed as

$$Z_n(\Omega) = \frac{\Omega(a_2^2 - \Omega^2)(a_4^2 - \Omega^2)\ldots(a_m^2 - \Omega^2)}{(1 - a_2^2\Omega^2)(1 - a_4^2\Omega^2)\ldots(1 - a_m^2\Omega^2)} \tag{2-50}$$

where n is odd and $m = (n - 1)/2$, or

$$Z_n(\Omega) = \frac{(a_2^2 - \Omega^2)(a_4^2 - \Omega^2)\ldots(a_m^2 - \Omega^2)}{(1 - a_2^2\Omega^2)(1 - a_4^2\Omega^2)\ldots(1 - a_m^2\Omega^2)} \tag{2-51}$$

where n is even and $m = n/2$.

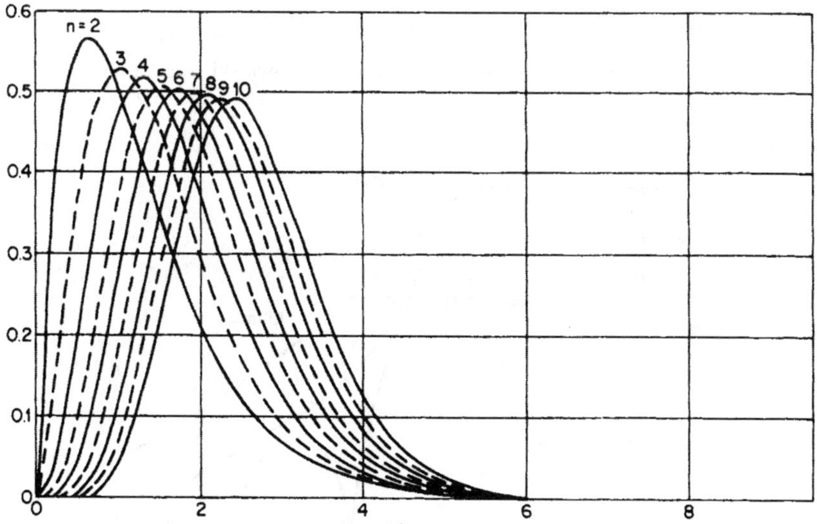

FIGURE 2-79 Impulse response for synchronously tuned filters. (From A. I. Zverev, *Handbook of Filter Synthesis* [New York: John Wiley and Sons, 1967.] By permission of the publishers.)

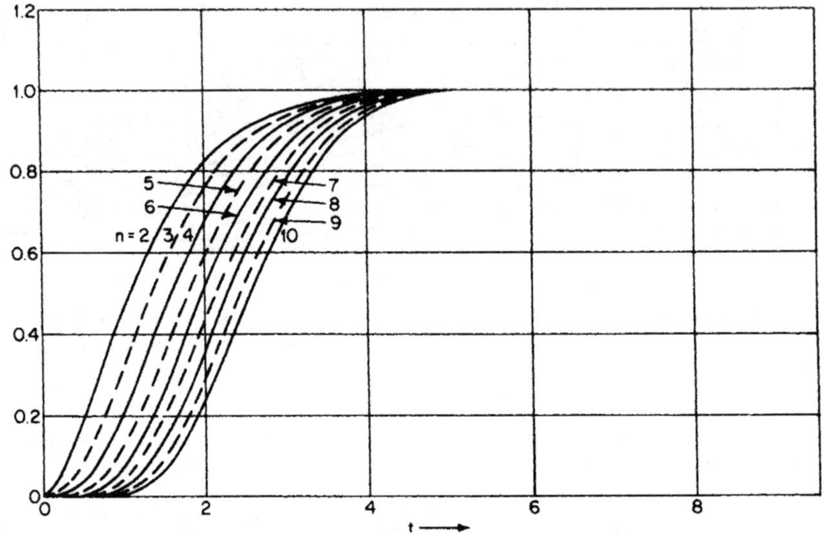

FIGURE 2-80 Step response for synchronously tuned filters. (From A. I. Zverev, *Handbook of Filter Synthesis* [New York: John Wiley and Sons, 1967.] By permission of the publishers.)

The zeros of Z_n are $a_2, a_4, \ldots, a_m$, whereas the poles are $1/a_2, 1/a_4, \ldots, 1/a_m$. The reciprocal relationship between the poles and zeros of Z_n results in equiripple behavior in both the stopband and the passband.

The values for a_2 through a_m are derived from the elliptic integral, which is defined as

$$K_e = \int_0^{\pi/2} \frac{d\theta}{\sqrt{1 - k^2 \sin^2\theta}} \tag{2-52}$$

Numerical evaluation may be somewhat difficult. Glowatski (see Bibliography) contains tables specifically intended for determining the poles and zeros of $Z_n(\Omega)$.

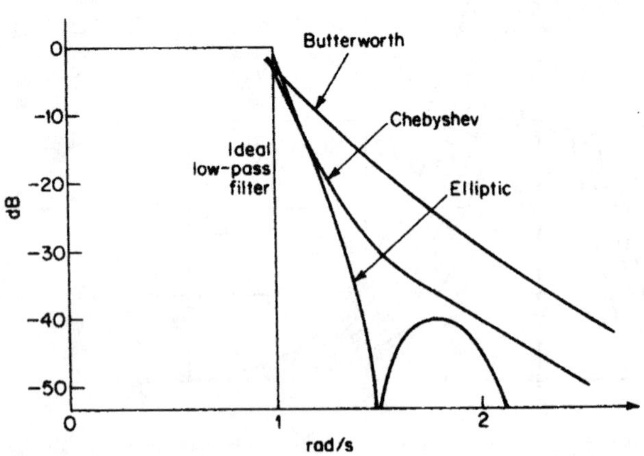

FIGURE 2-81 A comparison of $n = 5$ Butterworth, Chebyshev, and elliptic-function filters.

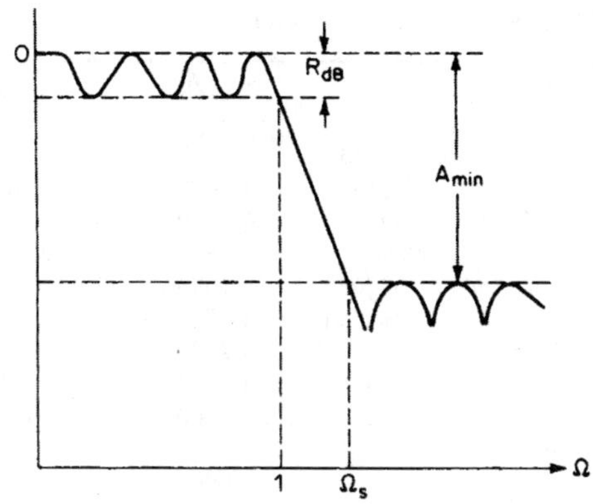

FIGURE 2-82 Normalized elliptic-function low-pass filter response.

Elliptic-function filters have been extensively tabulated by Saal and Zverev (see Bibliography). The basis for these tabulations was the order n and the parameters θ (degrees) and reflection coefficient ρ (percent).

Elliptic-function filters are sometimes called *Cauer filters* in honor of network theorist Professor Wilhelm Cauer. They were tabulated using the following convention

$$C\,n\,\rho\,\theta$$

where C represents Cauer, n is the filter order, ρ is the reflection coefficient, and θ is the modular angle. A fifth-order filter having a ρ of 15 percent and a θ of 29° would be described as CO5 15 $\theta = 29°$.

The angle θ determines the steepness of the filter and is defined as

$$\theta = \sin^{-1}\frac{1}{\Omega_s} \tag{2-53}$$

or, alternatively, we can state

$$\Omega_s = \frac{1}{\sin\theta} \tag{2-54}$$

Table 2-2 gives some representative value of θ and Ω_s.

TABLE 2-2 Ω_s vs. θ

θ, degrees	Ω_s
0	∞
10	5.759
20	2.924
30	2.000
40	1.556
50	1.305
60	1.155
70	1.064
80	1.015
90	1.000

TABLE 2-3 ρ vs. R_{dB} VSWR, and ε

ρ, %	R_{dB}	VSWR	ε (ripple factor)
1	0.0004343	1.0202	0.0100
2	0.001738	1.0408	0.0200
3	0.003910	1.0619	0.0300
4	0.006954	1.0833	0.0400
5	0.01087	1.1053	0.0501
8	0.02788	1.1739	0.0803
10	0.04365	1.2222	0.1005
15	0.09883	1.3529	0.1517
20	0.1773	1.5000	0.2041
25	0.2803	1.6667	0.2582
50	1.249	3.0000	0.5774

The parameter ρ, the reflection coefficient, can be derived from

$$\rho = \frac{\text{VSWR} - 1}{\text{VSWR} + 1} = \sqrt{\frac{\varepsilon^2}{1 + \varepsilon^2}} \tag{2-55}$$

where VSWR is the standing-wave ratio and ε is the ripple factor (see Section 2.4 on the Chebyshev response). The passband ripple and reflection coefficient are related by

$$R_{dB} = -10 \log(1 - \rho^2) \tag{2-56}$$

Table 2-3 interrelates these parameters for some typical values of the reflection coefficient, where ρ is expressed as a percentage.

As the parameter θ approaches 90°, the edge of the stopband Ω_s approaches unity. For θs near 90°, extremely sharp roll-offs are obtained. However, for a fixed n, the stopband attenuation A_{mni} is reduced as the steepness increases. Figure 2-83 shows the frequency response of an $n = 3$ elliptic filter for a fixed ripple of 1 dB ($\rho = 50$ percent) and different values of θ.

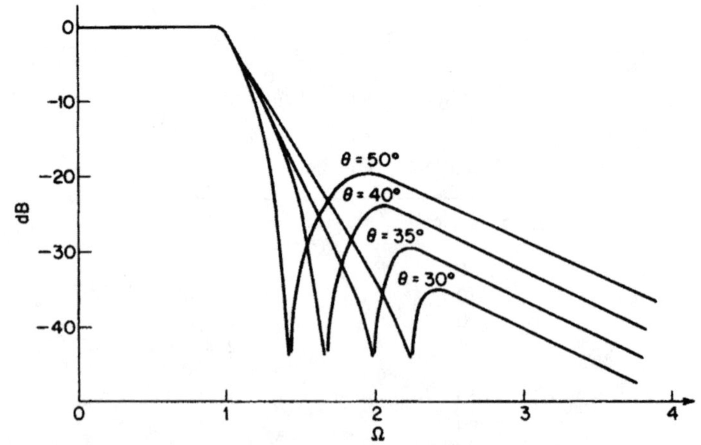

FIGURE 2-83 The elliptic-function low-pass filter response for $n = 3$ and $R_{dB} = 1$ dB.

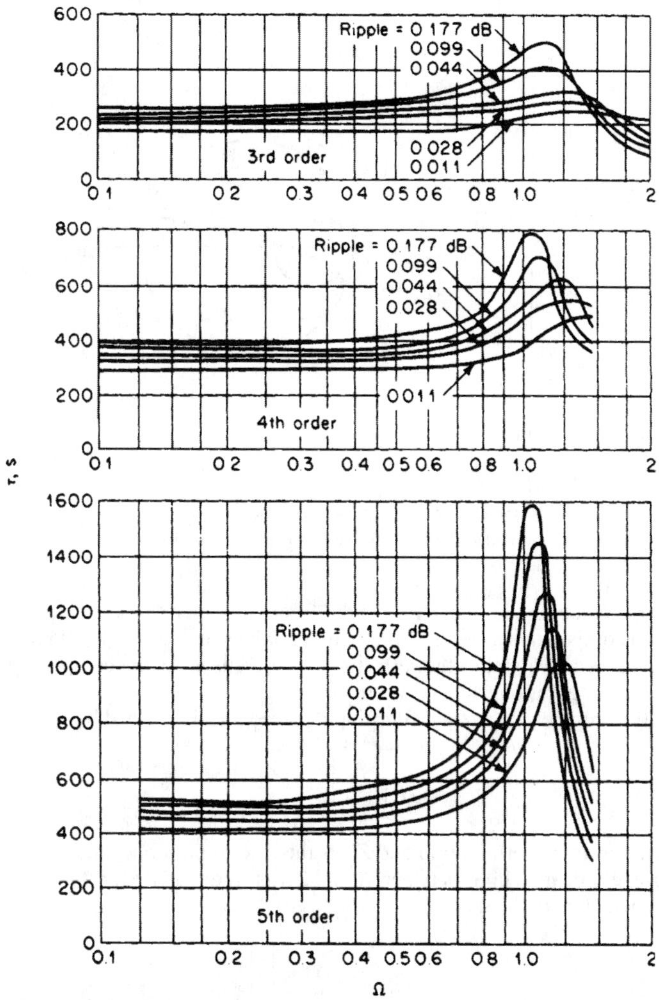

FIGURE 2-84 Delay characteristics of elliptic-function filters $n = 3, 4, 5$, and with an A_{min} of 60 dB. (From Lindquist, C. S. (1977). *Active Network Design*. California: Steward and Sons.)

For a given θ and order n, the stopband attenuation parameter A_{min} increases as the ripple is made larger. Since the poles of elliptic-function filters are approximately located on an ellipse, the delay curves behave in a similar manner to those of the Chebyshev family. Figure 2-84 compares the delay characteristics of $n = 3, 4$, and 5 elliptic filters, all having an A_{min} of 60 dB. The delay variation tends to increase sharply with increasing ripple and filter order n.

The factor ρ determines the input impedance variation with frequency of *LC* elliptic filters, as well as the passband ripple. As ρ is reduced, a better match is achieved between the resistive terminations and the filter impedance. Figure 2-85 illustrates the input impedance variation with frequency of a normalized $n = 5$ elliptic-function low-pass filter. At DC,

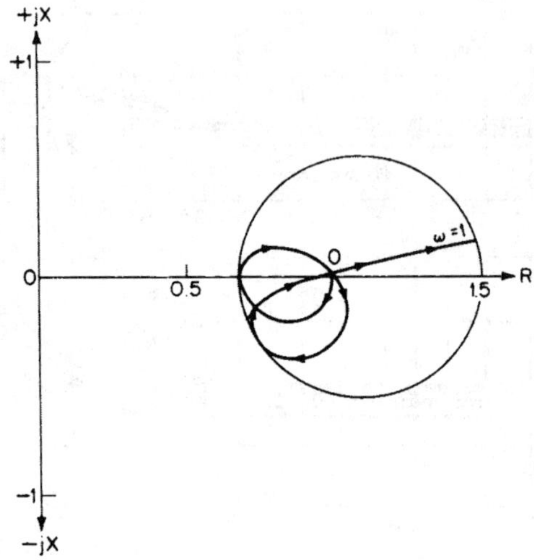

FIGURE 2-85 Impedance variation in the passband of a normalized $n = 5$ elliptic-function low-pass filter.

the input impedance is 1 Ω resistive. As the frequency increases, both positive and negative reactive components appear. All maximum values are within the diameter of a circle whose radius is proportional to the reflection coefficient ρ. As the complexity of the filter is increased, more gyrations occur within the circle.

The relationship between ρ and filter input impedance is defined by

$$|\rho|^2 = \left|\frac{R - Z_{11}}{R + Z_{11}}\right|^2 \tag{2-57}$$

where R is the resistive termination and Z_{11} is the filter input impedance.

The closeness of matching between R and Z_{11} is frequently expressed in decibels as a returns loss, which is defined as

$$A_\rho = 20 \log \left|\frac{1}{\rho}\right| \tag{2-58}$$

Using Filter Solutions (Book Version) Software for Design of Elliptic Function Low-Pass Filters. Previous editions of this book have contained extensive numerical tables of normalized values which have to be scaled to the operating frequencies and impedance levels during the design process. This is no longer the case.

A program called *Filter Solutions* is included on the CD-ROM. This program is limited to Elliptic Function *LC* filters (up to $n = 10$) and is a subset of the complete program which is available from Nuhertz Technologies® (www.nuhertz.com). The reader is encouraged to obtain the full version, which in addition to passive implementations covers many filter polynomial types, and includes transmission line, active, switched capacitor, and digital along with many very powerful features. It has also been integrated into Applied Wave Research's (AWR) popular Microwave Office software.

The program is quite intuitive and self-explanatory; thus, the reader is encouraged to explore its many features on his/her own. Nevertheless, all design examples using this program will elaborate on its usage and provide helpful hints.

You can extract and install this program by running FSBook.exe, which is contained on the CD-ROM. All examples in the book using *Filter Solutions* are based on starting with the program default settings. To restore these settings, click the *Initialize* button, then *Default*, and then *Save*.

Example 2-22 Determining the Order of an Elliptic Function Filter using *Filter Solutions*

Required:
Determine the order of an elliptic-function filter having a passband ripple less than 0.2 dB up to 1000 Hz, and a minimum rejection of 60 dB at 1300 Hz and above. Use *Filter Solutions.*

Result:

(*a*) Open ***Filter Solutions***.
Check the ***Stop Band Freq*** box.
Enter **0.2** in the ***Pass Band Ripple(dB)*** box.
Enter **1000** in the ***Pass Band Freq*** box.
Enter **1300** in the ***Stop Band Freq*** box.
Check the ***Frequency Scale Hertz*** box.

(*b*) Click the ***Set Order*** control button to open the second panel.
Enter **60** for the ***Stop band Attenuation (dB)***.
Click the ***Set Minimum Order*** button and then click ***Close***.
7 Order is displayed on the main control panel.

(*c*) The result is that a 7th order elliptic-function low-pass filter provides the required attenuation. By comparison, a 27th-order Butterworth low-pass filter would be needed to meet the requirements of Example 2-22, so the elliptic-function family is a *must* for steep filter requirements.

Using the ELI 1.0 Program for the Design of Odd-Order Elliptic-Function Low-Pass Filters up to the 31st Order. This program allows the design of odd-order elliptic function *LC* low-pass filters up to a complexity of 15 nulls (transmission zeros), or the 31st order. It is based on an algorithm developed by Amstutz. (See Bibliography)

The program inputs are passband edge (Hz), stopband edge (Hz), number of nulls (up to 15), stopband rejection in dB, and source and load terminations (which are always equal). The output parameters are critical Q (theoretical minimum Q), passband ripple (dB), nominal 3-dB cutoff and a list of component values along with resonant null frequencies.

To install the program, first copy **ELI1.zip** from the CD-ROM to the desktop and then double-click it to extract it to the C:\ root directory. A folder "eli1" will be created in the C:\ root directory, and a desktop shortcut "eli1.bat" will be created on the desktop. (If not, go to the C:\eli1 folder and create a shortcut on the desktop from "eli1.bat".

To run the program, double-click the "eli1.bat" shortcut and enter inputs as requested. Upon completing the execution, a dataout.text file will open using Notepad and containing the resulting circuit description.

If the number of nulls is excessive for the response requirements (indicated by zero passband ripple) the final capacitor may have a negative value as a result of the algorithm. Reduce the number of nulls, increase the required attention, define a steeper filter, or do a combination of these.

2.10 MAXIMALLY FLAT DELAY WITH CHEBYSHEV STOPBAND

The Bessel, linear phase with equiripple error, and transitional filter families all exhibit either maximally flat or equiripple-delay characteristics over most of the passband and, except for the transitional type, even into the stopband. However, the amplitude versus frequency response is far from ideal. The passband region in the vicinity of the cutoff is very rounded, while the stopband attenuation in the first few octaves is poor.

Elliptic-function filters have an extremely steep rate of descent into the stopband because of transmission zeros. However, the delay variation in the passband is unacceptable when the transient behavior is significant.

The maximally flat delay with Chebyshev stopband filters is derived by introducing transmission zeros into a Bessel-type transfer function. The constant delay properties in the passband are retained. However, the stopband rejection is significantly improved because of the effectiveness of the transmission zeros.

The step response exhibits no overshoot or ringing, and the impulse response has essentially no oscillatory behavior. Constant delay properties extend well into the stopband for higher-order networks.

Normalized tables of element values for the maximally flat delay with the Chebyshev stopband family of filters are provided in Table 11-56. These tables are normalized so that the 3-dB response occurs at 1 rad/s. The tables also provide the delay at DC and the normalized frequencies corresponding to a 1-percent and 10-percent deviation from the delay at DC. The amplitude response below the 3-dB point is identical to the attenuation characteristics of the Bessel filters shown in Figure 2-56.

BIBLIOGRAPHY

Amstutz, P. "Elliptic Approximation and Elliptic Filter Design on Small Computers." *IEEE Transactions on Circuits and Systems* CAS-25, No.12 (December, 1978).

Feistel, V. K., and R. Unbehauen. "Tiefpasse mit Tschebyscheff—Charakter der Betriebsdampfung im Sperrbereich und Maximal geebneter Laufzeit." *Frequenz* 8 (1965).

Glowatski, E. "Sechsstellige Tafel der Cauer-Parameter." *Verlag der Bayr*, Akademie der Wissenchaften (1955).

Lindquist, C. S. *Active Network Design*. California: Steward and Sons, 1977.

Matthaei, G. L., Young, L., and E. M. T. Jones. "*Microwave Filters, Impedance-Matching Networks, and Coupling Structures*." Massachusetts: Artech House, 1980.

Saal, R. "Der Entwurf von Filtern mit Hilfe des Kataloges Normierter Tiefpasse." *Telefunken GMBH* (1963).

White Electromagnetics. *A Handbook on Electrical Filters*. White Electromagnetics Inc., 1963.

Zverev, A. I. *Handbook of Filter Synthesis*. New York: John Wiley and Sons, 1967.

CHAPTER 3
LOW-PASS FILTER DESIGN

3.1 LC LOW-PASS FILTERS

All-Pole Filters

LC low-pass filters can be designed from the tables provided in Chapter 11 or the software available on the CD-ROM. A suitable filter must first be selected using the guidelines established in Chapter 2, however. The chosen design is then frequency- and impedance-scaled to the desired cutoff and impedance level when using the tables, or directly designed when using the software.

Example 3-1 Design of an *LC* Low-Pass Filter from the Tables

Required:

An *LC* low-pass filter
3 dB at 1000 Hz
20-dB minimum at 2000 Hz
$R_s = R_L = 600\ \Omega$

Result:

(*a*) To normalize the low-pass requirement, compute A_s.

$$A_s = \frac{f_s}{f_c} = \frac{2000\text{ Hz}}{1000\text{ Hz}} = 2 \qquad (2\text{-}11)$$

(*b*) Choose a normalized low-pass filter from the curves of Chapter 2 having at least 20 dB of attenuation at 2 rad/s.

Examination of the curves indicates that an $n = 4$ Butterworth or third-order 0.1-dB Chebyshev satisfies this requirement. Let us select the latter, since fewer elements are required.

(*c*) Table 11-28 contains element values for normalized 0.1-dB Chebyshev *LC* filters ranging from $n = 2$ through $n = 10$. The circuit corresponding to $n = 3$ and equal source and load resistors ($R_s = 1\ \Omega$) is shown in Figure 3-1*a*.

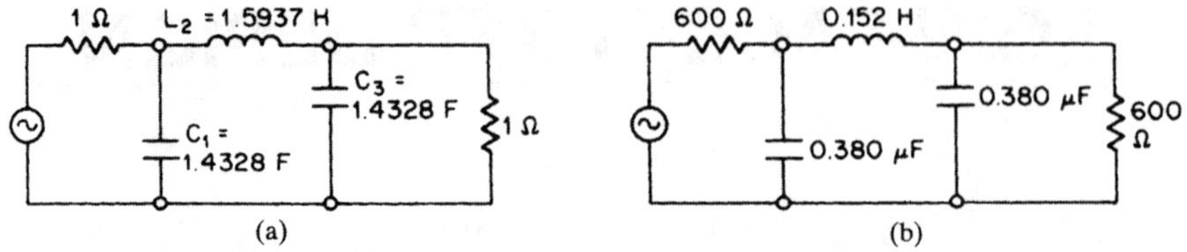

FIGURE 3-1 The results of Example 3-1: (*a*) normalized filter from Table 11-28; (*b*) frequency- and impedance-scaled filter.

(*d*) Denormalize the filter using a Z of 600 and a frequency-scaling factor (FSF) of $2\pi f_c$ or 6280.

$$R'_s = R'_L = 600\ \Omega \tag{2-8}$$

$$L'_2 = \frac{L \times Z}{\text{FSF}} = \frac{1.5937 \times 600}{6280} = 0.152\ \text{H} \tag{2-9}$$

$$C'_1 = C'_3 = \frac{C}{\text{FSF} \times Z} = \frac{1.4328}{6280 \times 600} = 0.380\ \mu\text{F} \tag{2-10}$$

The resulting filter is shown in Figure 3-1*b*.

The normalized filter used in Example 3-1 is shown in Table 11-28 (in Chapter 11) as having a current source input with a parallel resistor of 1 Ω. The reader will recall that Thévenin's theorems permit the replacement of this circuit with a voltage source having an equivalent series source resistance.

Elliptic-Function Filters

Elliptic Function Low-Pass Filters Using the Filter Solutions Program. The following example illustrates the design of an elliptic-function low-pass filter using the *Filter Solutions* program introduced in Section 2.9.

Example 3-2 Design of an Elliptic Function Low-Pass Filter using *Filter Solutions* Program

Required:

LC low-pass filter

0.25-dB maximum ripple DC to 100 Hz

60-dB minimum at 132 Hz

$R_s = R_L = 900\ \Omega$

Result:

(*a*) Open ***Filter Solutions***.

Check the ***Stop Band Freq*** box.

Enter **0.18** in the ***Pass Band Ripple(dB)*** box.

Enter **100** in the ***Pass Band Freq*** box.

Enter **132** in the ***Stop Band Freq*** box.

Check the ***Frequency Scale Hertz*** box.

Enter **900** for ***Source Res*** and ***Load Res***.

(*b*) Click the ***Set Order*** control button to open the second panel.

Enter **60** for ***Stop band Attenuation (dB)***.

Click the ***Set Minimum Order*** button and then click ***Close***.

7 Order is displayed on the main control panel.

(*c*) Click the ***Circuits*** button.

Two schematics are presented and shown in Figure 3-2. The circuit of Figure 3-2*a* has a shunt capacitor as its first element, and the circuit of Figure 3-2*b* has a series inductor as its first element. Normally, one would select the configuration having less inductors, which is the first circuit.

Note: All examples in the book using *Filter Solutions* are based on starting with program default settings. To restore these settings click the **Initialize** button, then **Default**, and then click **Save**.

Using the "ELI 1.0" Program for Designing Odd-Order Elliptic Function Low-Pass Filters up to the 31st Order. The following example illustrates the design of an elliptic-function low-pass filter using the **ELI1.0** program first introduced in Section 2.9. This program allows the design of odd-order elliptic function *LC* low-pass filters up to a complexity

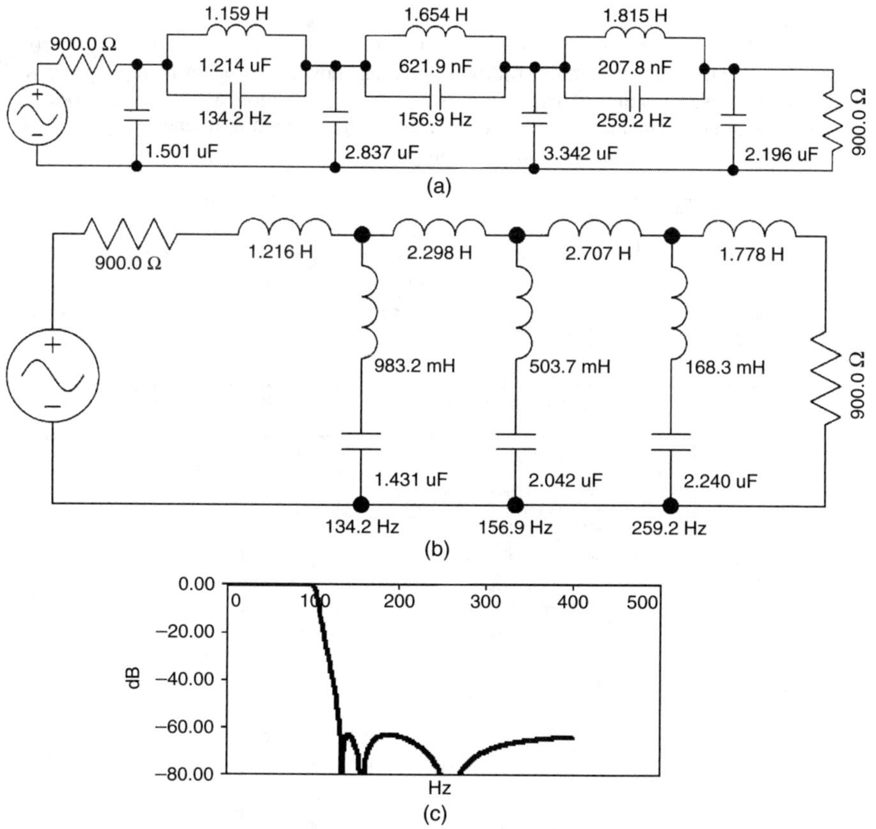

FIGURE 3-2 Filters of Example 3-2: (*a*) first element shunt capacitor; (*b*) first element series inductor; and (*c*) a frequency response.

of 15 nulls (transmission zeros) or the 31st order. It is based on an algorithm developed by Amstutz (see Bibliography).

The program inputs are passband edge (Hz), stopband edge (Hz), number of nulls (up to 15), stopband rejection in dB, and source and load terminations (which are always equal). The output parameters are critical *Q* (theoretical minimum *Q*), passband ripple (dB), nominal 3dB cutoff, and a list of component values along with resonant null frequencies.

If the number of nulls is excessive for the response requirements (indicated by zero passband ripple), the final capacitor may have a negative value as a result of the algorithm. Reduce the number of nulls, increase the required attention, or define a steeper filter—or use a combination of these.

Example 3-3 Design of an Elliptic Function Low-Pass Filter using *ELI 1.0* Program

Required:

An *LC* low-pass filter

0.25-dB maximum ripple DC to 100 Hz

35-dB minimum at 105 Hz

$R_s = R_L = 10\ \text{k}\Omega$

Result:

To run, double-click the "eli1.bat" shortcut and enter inputs as requested. Upon completing execution, a dataout.text file will open (as shown next using Notepad) and will contain the resulting circuit description. Note that the capacitors are all listed in one column, the inductors in another, and the corresponding resonant frequencies in a third column lined up with the parallel tuned circuits.

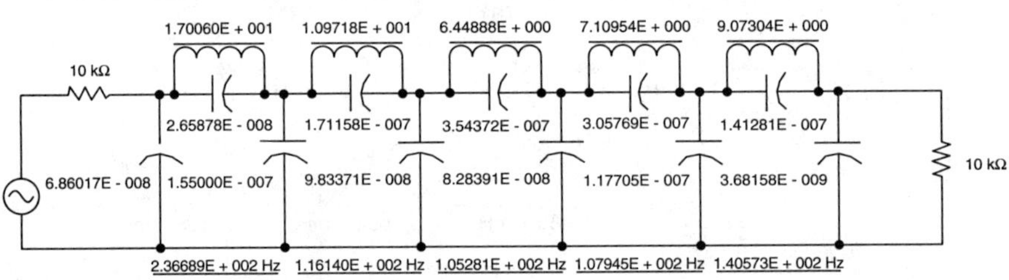


```
*****Odd-Order Elliptical Filter Synthesis*****

Passband edge (Hz) = 1.00000E + 002
Stopband edge (Hz) = 1.05000E + 002
Number of nulls (1–15) = 5
Critical Q = 144.56
Stopband rejection (dB) = 40.00
Passband Ripple (dB) = 0.000395
Nominal 3dB Cutoff (Hz) = 1.02487E + 002
Termination Impedance (Ohms) = 10000.00

                  ** Low-Pass Filter **

  Hz                Farads            Henries

6.86017E - 008
2.36689E + 002    2.65878E - 008    1.70060E + 001
1.55000E - 007
1.16140E + 002    1.71158E - 007    1.09718E + 001
9.83371E - 008
1.05281E + 002    3.54372E - 007    6.44888E + 000
8.28391E - 008
1.07945E + 002    3.05769E - 007    7.10954E + 000
1.17705E - 007
1.40573E + 002    1.41281E - 007    9.07304E + 000
3.68158E - 009
```

Duality and Reciprocity. A network and its dual have identical response characteristics. Each all-pole *LC* filter tabulated in Chapter 11 has an equivalent dual network. The circuit configuration shown at the bottom of each table, and the bottom set of nomenclature, corresponds to the dual of the upper filter. For elliptic filters using *Filter Solutions*, a checkmark in **1st Ele Shunt** and **1st Ele Series** will give you dual networks in the normalized case of equal 1 ohm source and load terminations.

Any ladder-type network can be transformed into its dual by implementing the following rules:

1. Convert every series branch into a shunt branch and every shunt branch into a series branch.
2. Convert circuit branch elements in series to elements in parallel, and vice versa.
3. Transform each inductor into a capacitor, and vice versa. The values remain unchanged—for instance, 4 H becomes 4 F.
4. Replace each resistance with a conductance—for example, 3 Ω becomes 3 mhos or $^1\!/_3$ Ω.
5. Change a voltage source into a current source, and vice versa.

Figure 3-3 shows a network and its dual.

The theorem of reciprocity states that if a voltage located at one point of a linear network produces a current at any other point, the same voltage acting at the second point results in the same current at the first point. Alternatively, if a current source at one point of a linear network results in a voltage measured at a different point, the same current source at the second point produces the same voltage at the first point. As a result, the response of an *LC* filter is the same regardless of which direction the signal flows in, except for a constant multiplier. It is perfectly permissible to turn a filter schematic completely around with regard to its driving source, provided that the source- and load-resistive terminations are also interchanged.

The laws of duality and reciprocity are used to manipulate a filter to satisfy termination requirements or to force a desired configuration.

Designing for Unequal Terminations. Tables of all-pole filter *LC* element values are provided in Chapter 11 for both equally terminated and unequally terminated networks. A number of different ratios of source-to-load resistance are tabulated, including the impedance extremes of infinity and zero.

To design an unequally terminated filter, first determine the desired ratio of R_s/R_L. Select a normalized filter from the table that satisfies this ratio. The reciprocity theorem can be applied to turn a network around end for end and the source and load resistors can be

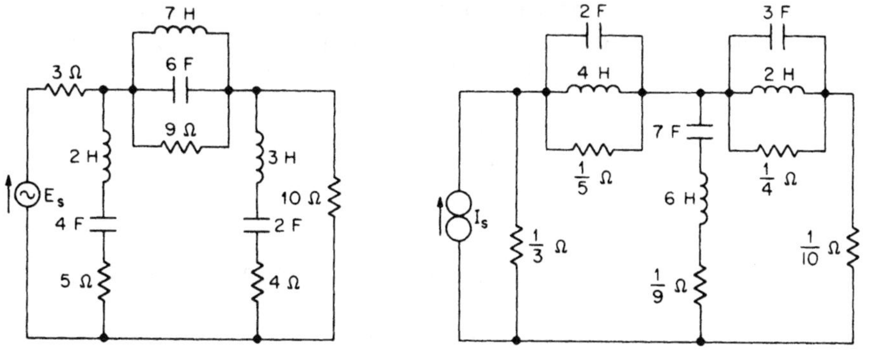

FIGURE 3-3 An example of dual networks.

interchanged. The tabulated impedance ratio is inverted if the dual network given by the lower schematic is used. The chosen filter is then frequency- and impedance-scaled.

For unequally terminated elliptic filters, you can enter the required source and load terminations in the **Source Res** and **Load Res** boxes of *Filter Solutions* before clicking the ***Circuits*** button.

Example 3-4 Design of an *LC* Low-Pass Filter for Unequal Terminations

Required:

An *LC* low-pass filter

1 dB at 900 Hz

20-dB minimum at 2700 Hz

$R_s = 1\ \text{k}\Omega$

$R_L = 5\ \text{k}\Omega$

Result:

(*a*) Compute A_s.

$$A_s = \frac{f_s}{f_c} = \frac{2700\ \text{Hz}}{900\ \text{Hz}} = 3 \tag{2-11}$$

(*b*) Normalized requirement:

1 dB at X rad/s

20-dB minimum at $3X$ rad/s

(where X is arbitrary)

(*c*) Select a normalized low-pass filter that makes the transition from 1 dB to at least 20 dB over a frequency ratio of 3:1. A Butterworth $n = 3$ design satisfies these requirements since Figure 2-34 indicates that the 1-dB point occurs at 0.8 rad/s and that more than 20 dB of attenuation is obtained at 2.4 rad/s. Table 11-2 provides element values for normalized Butterworth low-pass filters for a variety of impedance ratios. Since the ratio of R_s/R_L is 1:5, we will select a design for $n = 3$, corresponding to $R_s = 0.2\ \Omega$, and use the upper schematic. (Alternatively, we could have selected the lower schematic corresponding to $R_s = 5\ \Omega$ and turned the network end for end, but an additional inductor would have been required.)

(*d*) The normalized filter from Table 11-2 is shown in Figure 3-4*a*. Since the 1-dB point is required to be 900 Hz, the FSF is calculated by

$$\text{FSF} = \frac{\text{desired reference frequency}}{\text{existing reference frequency}}$$

$$= \frac{2\pi 900\ \text{rad/s}}{0.8\ \text{rad/s}} = 7069 \tag{2-1}$$

Using a Z of 5000 and an FSF of 7069, the denormalized component values are

$$R'_s = R \times Z = 1\ \text{k}\Omega \tag{2-8}$$

$$R'_L = 5\ \text{k}\Omega$$

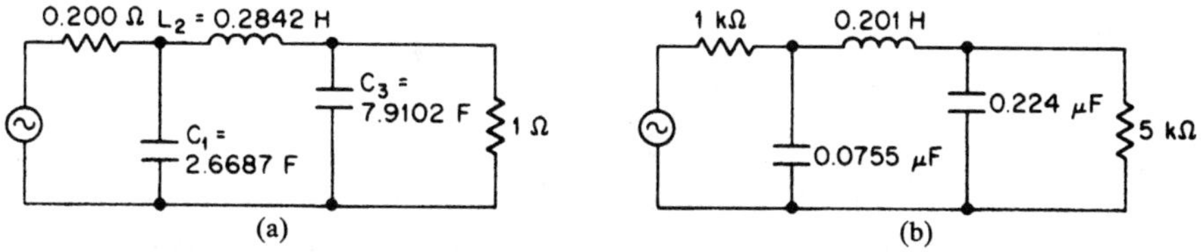

FIGURE 3-4 low-pass filter with unequal terminations: (*a*) normalized low-pass filter; (*b*) frequency- and impedance-scaled filter.

$$C_1' = \frac{C}{\text{FSF} \times Z} = \frac{2.6687}{7069 \times 5000} = 0.0755\ \mu\text{F} \tag{2-10}$$

$$C_3' = 0.22\ \mu\text{F}$$

$$L_2 = \frac{L \times Z}{\text{FSF}} = \frac{0.2842 \times 5000}{7069} = 0.201\ \text{H} \tag{2-9}$$

The scaled filter is shown in Figure 3-4*b*.

If an infinite termination is required, a design having an R_s of infinity is selected. When the input is a current source, the configuration is used as given. For an infinite load impedance, the entire network is turned end for end.

If the design requires a source impedance of 0 Ω, the dual network is used corresponding to $1/R_s$ of infinity or $R_s = 0\ \Omega$.

In practice, impedance extremes of near zero or infinity are not always possible. However, for an impedance ratio of 20 or more, the load can be considered infinite in comparison with the source, and the design for an infinite termination is used. Alternatively, the source may be considered zero with respect to the load and the dual filter corresponding to $R_s = 0\ \Omega$ may be used. When n is odd, the configuration having the infinite termination has one less inductor than its dual.

An alternate method of designing filters to operate between unequal terminations involves partitioning the source or load resistor between the filter and the termination. For example, a filter designed for a 1-kΩ source impedance could operate from a 250-Ω source if a 750-Ω resistor were placed within the filter network in series with the source. However, this approach would result in a higher insertion loss.

Bartlett's Bisection Theorem. A filter network designed to operate between equal terminations can be modified for unequal source and load resistors if the circuit is symmetrical. Bartlett's bisection theorem states that if a symmetrical network is bisected and one half is impedance-scaled, including the termination, the response shape will not change. All tabulated odd-order Butterworth and Chebyshev filters having equal terminations satisfy the symmetry requirement.

Example 3-5 Design of an *LC* Low-Pass Filter for Unequal Terminations using Bartletts' Bisection Theorem

Required:

An *LC* low-pass filter

3 dB at 200 Hz

15-dB minimum at 400 Hz

$R_s = 1\ \text{k}\Omega$

$R_L = 1.5\ \text{k}\Omega$

Result:

(*a*) Compute A_s.

$$A_s = \frac{f_s}{f_c} = \frac{400}{200} = 2 \tag{2-11}$$

(*b*) Figure 2-34 indicates that an $n = 3$ Butterworth low-pass filter provides 18-dB rejection at 2 rad/s. Normalized *LC* values for Butterworth low-pass filters are given in Table 11-2. The circuit corresponding to $n = 3$ and equal terminations is shown in Figure 3-5*a*.

(*c*) Since the circuit of Figure 3-5*a* is symmetrical, it can be bisected into two equal halves, as shown in Figure 3-5*b*. The requirement specifies a ratio of load-to-source resistance of 1.5 (1.5 kΩ/1 kΩ), so we must impedance-scale the right half of the circuit by a factor of 1.5. The circuit of Figure 3-5*c* is thus obtained.

(*d*) The recombined filter of Figure 3-5*d* can now be frequency- and impedance-scaled using an FSF of $2\pi 200$ or 1256 and a *Z* of 1000.

$$R_s' = 1\ \text{k}\Omega$$

$$R_L' = 1.5\ \text{k}\Omega$$

$$C_1' = \frac{C}{\text{FSF} \times Z} = \frac{1}{1256 \times 1000} = 0.796\ \mu\text{F} \tag{2-10}$$

$$C_3' = 0.530\ \mu\text{F}$$

$$L_2' = \frac{L \times Z}{\text{FSF}} = \frac{2.5 \times 1000}{1256} = 1.99\ \text{H} \tag{2-9}$$

The final filter is shown in Figure 3-5*e*.

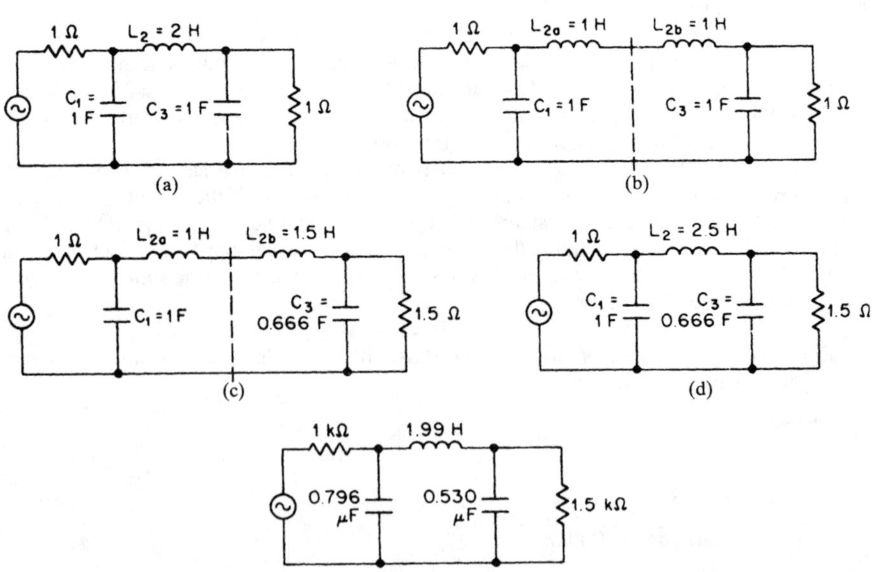

FIGURE 3-5 An example of Bartlett's bisection theorem: (*a*) normalized filter having equal terminations; (*b*) bisected filter; (*c*) impedance-scaled right half section; (*d*) recombined filter; and (*e*) final scaled network.

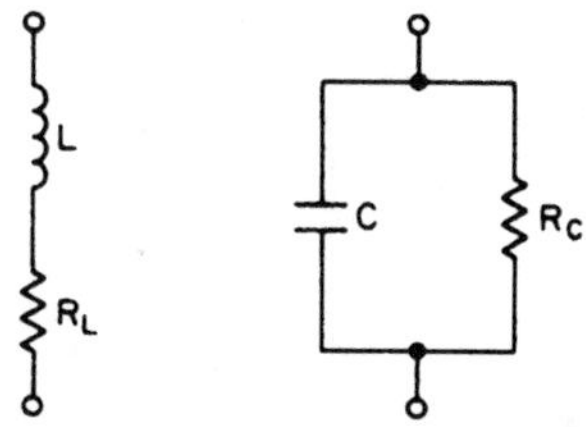

FIGURE 3-6 Low-frequency equivalent circuits of practical inductors and capacitors.

Effects of Dissipation. Filters designed using the tables of *LC* element values in Chapter 11 require lossless coils and capacitors to obtain the theoretical response predicted in Chapter 2. In the practical world, capacitors are usually obtainable that have low losses, but inductors are generally lossy, especially at low frequencies. Losses can be defined in terms of Q, the figure of merit or quality factor of a reactive component.

If a lossy coil or capacitor is resonated in parallel with a lossless reactance, the ratio of resonant frequency to 3-dB bandwidth of the resonant circuit's impedance (in other words, the band over which the magnitude of the impedance remains within 0.707 of the resonant value) is given by

$$Q = \frac{f_0}{\text{BW}_{3\text{ dB}}} \tag{3-1}$$

Figure 3-6 gives the low-frequency equivalent circuits for practical inductors and capacitors. Their Qs can be calculated by

Inductors: $$Q = \frac{\omega L}{R_L} \tag{3-2}$$

Capacitor: $$Q = \omega C R_c \tag{3-3}$$

where ω is the frequency of interest, in radians per second.

Using elements having a finite Q in a design intended for lossless reactances has the following mostly undesirable effects:

- At the passband edge, the response shape becomes more rounded. Within the passband, the ripples are diminished and may completely vanish.
- The insertion loss of the filter is increased. The loss in the stopband is maintained (except in the vicinity of transmission zeros), so the relative attenuation between the passband and the stopband is reduced.

Figure 3-7 shows some typical examples of these effects on all-pole and elliptic function low-pass filters.

The most critical problem caused by the finite element Q is the effect on the response shape near cutoff. Estimating the extent of this effect is somewhat difficult without extensive empirical data. The following variations in the filter design parameters will cause *increased* rounding of the frequency response near cutoff for a fixed Q:

- Going to a larger passband ripple
- Increasing the filter order n
- Decreasing the transition region of elliptic-function filters

Changing these parameters in the opposite direction, of course, reduces the effects of dissipation.

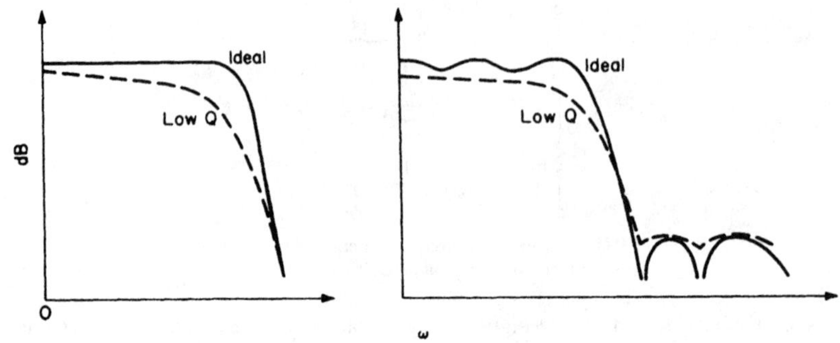

FIGURE 3-7 The effects of finite Q.

Filters can be designed to have the responses predicted by modern network theory using finite element Qs. Figure 3-8 shows the minimum Qs required at the cutoff for different low-pass responses. If elements used are having Qs slightly above the minimum values given in Figure 3-8, the desired response can be obtained provided that certain predistorted element values are used. However, the insertion loss will be prohibitive. It is therefore highly desirable that element Qs be several times higher than the values indicated.

The effect of low Q on the response near cutoff can usually be compensated for by going to a higher-order network or a steeper filter and using a larger design bandwidth to allow for rounding. However, this design approach does not always result in satisfactory results, since the Q requirement may also increase. A method of compensating for low Q by using amplitude equalization is discussed in Section 8.4.

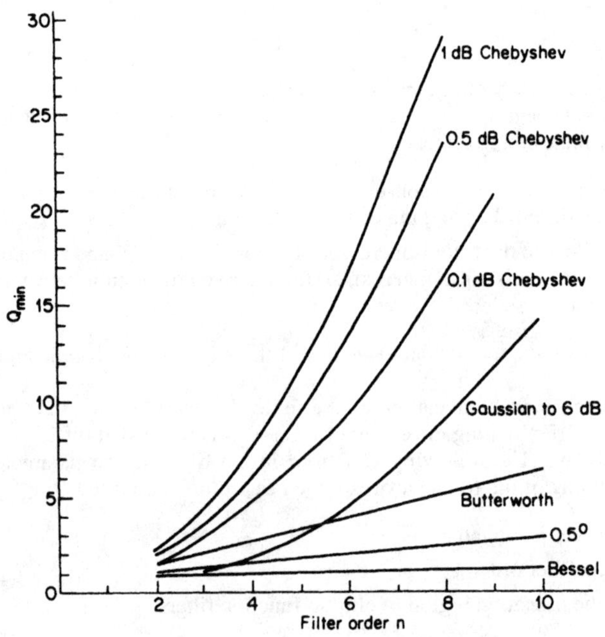

FIGURE 3-8 Minimum Q requirements for low-pass filters.

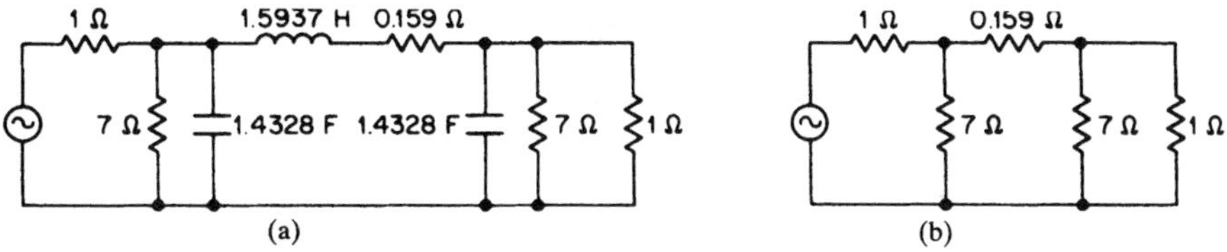

FIGURE 3-9 Calculation of insertion loss: (*a*) third-order 0.1-dB Chebyshev low-pass filter with $Q = 10$; (*b*) equivalent circuit at DC.

The insertion loss of low-pass filters can be computed by replacing the reactive elements with resistances corresponding to their *Q*s since at DC the inductors become short circuits and capacitors become open, which leaves the resistive elements only.

Figure 3-9*a* shows a normalized third-order 0.1-dB Chebyshev low-pass filter where each reactive element has a *Q* of 10 at the 1-rad/s cutoff. The series and shunt resistors for the coil and capacitors are calculated using Equations (3-2) and (3-3), respectively. At 1 rad/s these equations can be simplified and reexpressed as

$$R_L = \frac{L}{Q} \tag{3-4}$$

$$R_c = \frac{Q}{C} \tag{3-5}$$

The equivalent circuit at DC is shown in Figure 3-9*b*. The insertion loss is 1.9 dB. The actual loss calculated was 7.9 dB, but the 6-dB loss due to the source and load terminations is normally not considered a part of the filter's insertion loss since it would also occur in the event that the filter was completely lossless.

Using Predistorted Designs. The effect of finite element *Q* on an *LC* filter transfer function is to increase the real components of the pole positions by an amount equal to the dissipation factor *d*, where

$$d = \frac{1}{Q} \tag{3-6}$$

Figure 3-10 shows this effect. All poles are displaced to the left by an equal amount.

If the desired poles were first shifted to the right by an amount equal to *d*, the introduction of the appropriate losses into the corresponding *LC* filter would move the poles back to the desired locations. This technique is called *predistortion*. Predistorted filters are obtained by predistorting the required transfer function for a desired *Q* and then synthesizing an *LC* filter from the resulting transfer function. When the reactive elements of the filter have the required losses added, the response shape will correspond to the original transfer function.

The maximum amount that a group of poles can be displaced to the right in the process of predistortion is equal to the smallest real part among the poles given that further movement corresponds to locating a pole in the right half plane, which is an unstable condition. The minimum *Q* therefore is determined by the highest *Q* pole (in other words, the pole having the smallest real component). The *Q*s shown in Figure 3-8 correspond to $1/d$, where *d* is the real component of the highest *Q* pole.

Tables are provided in Chapter 11 for all-pole predistorted low-pass filters. These designs are all singly terminated with a source resistor of 1 Ω and an infinite termination. Their duals turned end for end can be used with a voltage source input and a 1-Ω termination.

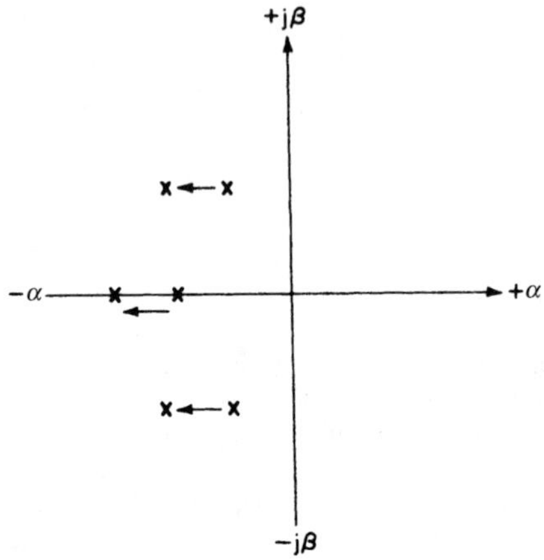

FIGURE 3-10 The effects of dissipation on the pole pattern.

Two types of predistorted filters are tabulated for various *d*s. The uniform dissipation networks require uniform losses in both the coils and the capacitors. The second type are the Butterworth lossy-*L* filters, where only the inductors have losses, which closely agrees with practical components. It is important for both types that the element *Q*s are closely equal to 1/*d* at the cutoff frequency. In the case of the uniform dissipation networks, losses must usually be added to the capacitors.

Example 3-6 Design of a Predistorted Lossy-L *LC* Low-Pass filter

Required:

An *LC* low-pass filter
3 dB at 500 Hz
24-dB minimum at 1200 Hz
$R_s = 600\ \Omega$
$R_L = 100\ \text{k}\Omega$ minimum
Inductor *Q*s of 5 at 500 Hz
Lossless capacitors

Result:

(*a*) Compute A_s.

$$A_s = \frac{1200}{500} = 2.4 \tag{2-11}$$

(*b*) The curves of Figure 2-34 indicate that an $n = 4$ Butterworth low-pass filter has over 24 dB of rejection at 2.4 rad/s. Table 11-14 contains the element values for Butterworth lossy-*L* network where $n = 4$. The circuit corresponding to $d = 0.2$ ($d = 1/Q$) is shown in Figure 3-11*a*.

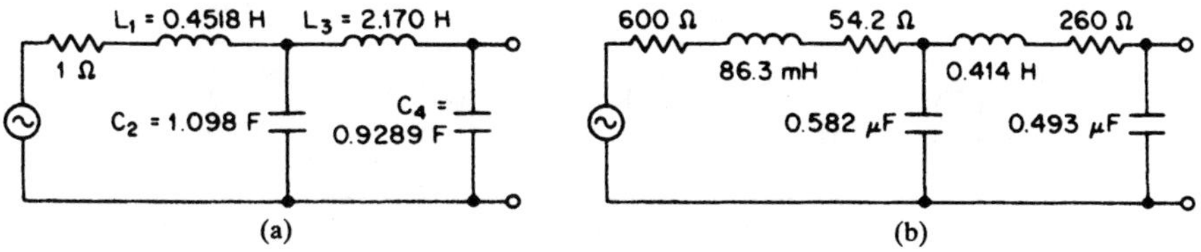

FIGURE 3-11 The lossy-*L* low-pass filter of Example 3-6: (*a*) normalized filter; and (*b*) scaled filter.

(*c*) The normalized filter can now be frequency- and impedance-scaled using an FSF of $2\pi 500 = 3142$ and a *Z* of 600.

$$R_s' = 600\ \Omega$$

$$L_1' = \frac{L \times Z}{\text{FSF}} = \frac{0.4518 \times 600}{3142} = 86.3\ \text{mH} \qquad (2\text{-}9)$$

$$L_3' = 0.414\text{H}$$

$$C_2' = \frac{C}{\text{FSF} \times Z} = \frac{1.098}{3142 \times 600} = 0.582\ \mu\text{F} \qquad (2\text{-}10)$$

$$C_4' = 0.493\ \mu\text{F}$$

(*d*) The resistive coil losses are

$$R_1 = \frac{\omega L}{Q} = 54.2\ \Omega \qquad (3\text{-}2)$$

and

$$R_3 = 260\ \Omega$$

where

$$\omega = 2\pi f_c = 3142$$

The final filter is given in Figure 3-11*b*.

Example 3-7 Design of a Uniform Distortion *LC* Low-Pass filter

Required:

An *LC* low-pass filter
3 dB at 100 Hz
58-dB minimum at 300 Hz
$R_s = 1\ \text{k}\Omega$
$R_L = 100\ \text{k}\Omega$ minimum
Inductor *Q*s of 11 at 100 Hz
Lossless capacitors

Result:

(*a*) Compute A_s.

$$A_s = \frac{300}{100} = 3 \qquad (2\text{-}11)$$

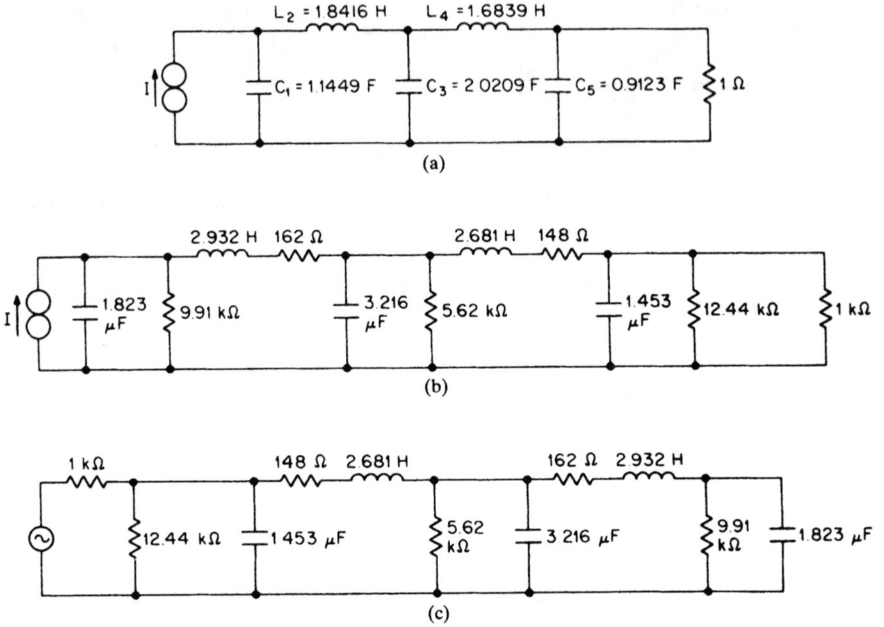

FIGURE 3-12 The design of the uniform dissipation network from Example 3-7: (*a*) normalized fifth-order 0.1-dB Chebyshev with $d = 0.0881$; (*b*) frequency- and impedance-scaled filter including losses; and (*c*) final network.

(*b*) Figure 2-42 indicates that a fifth-order 0.1-dB Chebyshev has about 60 dB of rejection at 3 rad/s. Table 11-32 provides *LC* element values for 0.1-dB Chebyshev uniform dissipation networks. The available inductor Q of 11 corresponds to a d of 0.091 ($d = 1/Q$). Values are tabulated for an $n = 5$ network having a d of 0.0881, which is sufficiently close to the requirement. The corresponding circuit is shown in Figure 3-12*a*.

(*c*) The normalized filter is frequency-and impedance-scaled using an FSF of $2\pi 100 = 628$ and a Z of 1000.

$$C_1' = \frac{C}{\text{FSF} \times Z} = \frac{1.1449}{628 \times 1000} = 1.823\ \mu\text{F} \tag{2-10}$$

$$C_3' = 3.216\ \mu\text{F}$$

$$C_5' = 1.453\ \mu\text{F}$$

$$L_2' = \frac{L \times Z}{\text{FSF}} = \frac{1.8416 \times 1000}{628} = 2.932\ \text{H} \tag{2-9}$$

$$L_4' = 2.681\ \text{H}$$

(*d*) The shunt resistive losses for capacitors C_1', C_3', and C_5' are

$$R_1 = \frac{Q}{\omega C} = 9.91\,\text{k}\Omega \tag{3-3}$$

$$R_3' = 5.62\ \text{k}\Omega$$

$$R_5' = 12.44\ \text{k}\Omega$$

The series resistive inductor losses are

$$R_1 = \frac{\omega L}{Q} = 162\ \Omega \tag{3-2}$$

$$R_4' = 148\ \Omega$$

where
$$Q = \frac{1}{d} = \frac{1}{0.0881} = 11.35$$

and
$$\omega = 2\pi f_c = 2\pi 100 = 628$$

The resulting circuit, including all losses, is shown in Figure 3-12*b*. This circuit can be turned end for end so that the requirement for a 1-kΩ source resistance is met. The final filter is given in Figure 3-12*c*.

It is important to remember that uniform dissipation networks require the presence of losses in both the coils and capacitors, thus resistors must usually be added. Component *Q*s within 20 percent of 1/*d* are usually sufficient for satisfactory results.

Resistors can sometimes be combined to eliminate components. In the circuit of Figure 3-12*c*, the 1-kΩ source and the 12.44-kΩ resistor can be combined, which results in a 926-Ω equivalent source resistance. The network can then be impedance scaled to restore a 1-kΩ source.

3.2 ACTIVE LOW-PASS FILTERS

Active low-pass filters are designed using a sequence of operations similar to the design of *LC* filters. The specified low-pass requirement is first normalized and a particular filter type of the required complexity is selected using the response characteristics given in Chapter 2. Normalized tables of active filter component values are provided in Chapter 11 for each associated transfer function. The corresponding filter is denormalized by frequency and impedance scaling.

Active filters can also be designed directly from the poles and zeros. This approach sometimes offers some additional degrees of freedom and will also be covered.

All-Pole Filters

The transfer function of a passive *RC* network has poles that lie only on the negative real axis of the complex frequency plane. In order to obtain the complex poles required by the all-pole transfer functions of Chapter 2, active elements must be introduced. Integrated circuit operational amplifiers are readily available that have nearly ideal properties, such as high gain. However, these properties are limited to frequencies below a few MHz, so active filters beyond this range are difficult.

Unity-Gain Single-Feedback Realization. Figure 3-13 shows two active low-pass filter configurations. The two-pole section provides a pair of complex conjugate poles, whereas the three-pole section produces a pair of complex conjugate poles and a single real-axis pole. The operational amplifier is configured in the voltage-follower configuration, which has a closed-loop gain of unity, very high-input impedance, and nearly zero output impedance.

The two-pole section has the transfer function

$$T(s) = \frac{1}{C_1 C_2 s^2 + 2C_2 s + 1} \tag{3-7}$$

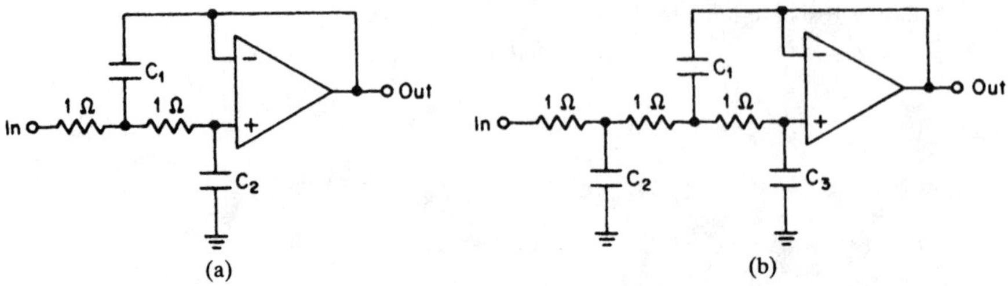

FIGURE 3-13 Unity-gain active low-pass configurations: (*a*) two-pole section; and (*b*) three-pole section.

A second-order low-pass transfer function can be expressed in terms of the pole locations as

$$T(s) = \frac{1}{\frac{1}{\alpha^2 + \beta^2} s^2 + \frac{2\alpha}{\alpha^2 + \beta^2} s + 1} \tag{3-8}$$

Equating coefficients and solving for the capacitors results in

$$C_1 = \frac{1}{\alpha} \tag{3-9}$$

$$C_2 = \frac{\alpha}{\alpha^2 + \beta^2} \tag{3-10}$$

where α and β are the real and imaginary coordinates of the pole pair.

The transfer function of the normalized three-pole section was discussed in Section 1.2 and given by

$$T(S) = \frac{1}{s^3 A + s^2 B + sC + 1} \tag{1-17}$$

where

$$A = C_1 C_2 C_3 \tag{1-18}$$

$$B = 2C_3(C_1 + C_2) \tag{1-19}$$

and

$$C = C_2 + 3C_3 \tag{1-20}$$

The solution of these equations to find the values of C_1, C_2, and C_3 in terms of the poles is somewhat laborious and is best accomplished with a digital computer.

If the filter order n is an even order, $n/2$ two-pole filter sections are required. Where n is odd, $(n - 3)/2$, two-pole sections and a single three-pole section are necessary. This occurs because even-order filters have complex poles only, whereas an odd-order transfer function has a single real pole in addition to the complex poles.

At DC, the capacitors become open circuits; so the circuit gain becomes equal to that of the amplifier, which is unity. This can also be determined analytically from the transfer functions given by Equations (3-7) and (1-17). At DC, $s = 0$ and $T(s)$ reduces to 1. Within the passband of a low-pass filter, the response of individual sections may have sharp peaks and some corresponding gain.

All resistors are 1 Ω in the two normalized filter circuits of Figure 3-13. Capacitors C_1, C_2, and C_3 are tabulated in Chapter 11. These values result in the normalized all-pole transfer functions of Chapter 2 where the 3-dB cutoff occurs at 1 rad/s.

To design a low-pass filter, a filter type is first selected from Chapter 2. The corresponding active low-pass filter values are then obtained from chapter 11. The normalized filter is denormalized by dividing all the capacitor values by FSF × Z, which is identical to the denormalization formula for *LC* filters, as shown in the following

$$C' = \frac{C}{\text{FSF} \times Z} \tag{2-10}$$

where FSF is the frequency-scaling factor $2\pi f_c$ and Z is the impedance-scaling factor. The resistors are multiplied by Z, which results in equal resistors throughout, having a value of $Z\ \Omega$.

The factor Z does not have to be the same for each filter section, since the individual circuits are isolated by the operational amplifiers. The value of Z can be independently chosen for each section so that practical capacitor values occur, but the FSF must be the same for all sections. The sequence of the sections can be rearranged if desired.

The frequency response obtained from active filters is usually very close to theoretical predictions, provided that the component tolerances are small and that the amplifier has satisfactory properties. The effects of low Q, which occurs in *LC* filters, do not apply, so the filters have no insertion loss and the passband ripples are well-defined.

Example 3-8 Design of an Active All-pole Low-Pass filter

Required:

An active low-pass filter
3 dB at 100 Hz
70-dB minimum at 350 Hz

Result:

(*a*) Compute the low-pass steepness factor A_s.

$$A_s = \frac{f_s}{f_c} = \frac{350}{100} = 3.5 \tag{2-11}$$

(*b*) The response curve of Figure 2-44 indicates that a fifth-order 0.5-dB Chebyshev low-pass filter meets the 70-dB requirement at 3.5 rad/s.

(*c*) The normalized values can be found in Table 11-39. The circuit consists of a three-pole section followed by a two-pole section and is shown in Figure 3-14*a*.

(*d*) Let us arbitrarily select an impedance-scaling factor of 5×10^4. Using an FSF of $2\pi f_c$ or 628, the resulting new values are

Three-pole section:

$$C_1' = \frac{C}{\text{FSF} \times Z} = \frac{6.842}{628 \times 5 \times 10^4} = 0.218\ \mu\text{F} \tag{2-10}$$

$$C_2' = 0.106\ \mu\text{F}$$

$$C_3' = 0.00966\ \mu\text{F}$$

Two-pole section:

$$C_1' = \frac{C}{\text{FSF} \times Z} = \frac{9.462}{628 \times 5 \times 10^4} = 0.301\ \mu\text{F}$$

$$C_2' = 0.00364\ \mu\text{F}$$

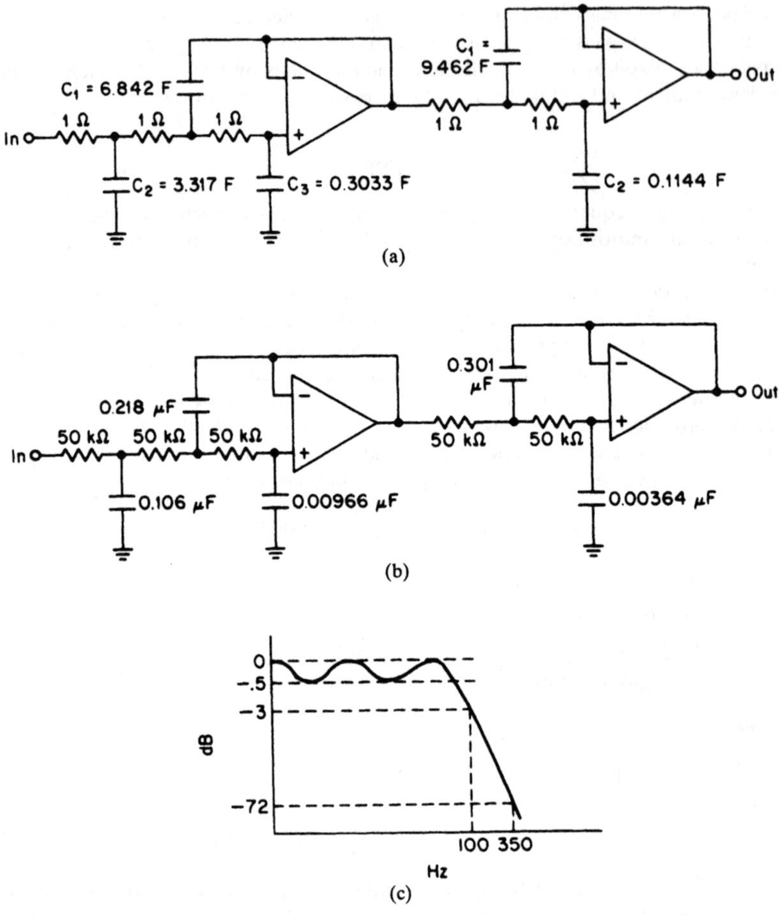

FIGURE 3-14 The low-pass filter of Example 3-8: (*a*) normalized fifth-order 0.5-dB Chebyshev low-pass filter; (*b*) denormalized filter; and (*c*) frequency response.

The resistors in both sections are multiplied by Z, resulting in equal resistors throughout of 50-kΩ. The denormalized circuit is given in Figure 3-14*b*, and having the frequency response of Figure 3-14*c*.

The first section of the filter should be driven by a voltage source having a source impedance much less than the first resistor of the section. The input must have a DC return to ground if a blocking capacitor is present. Since the filter's output impedance is low, the frequency response is independent of the terminating load, provided that the operational amplifier has sufficient driving capability.

Real-Pole Configurations. All odd-order low-pass transfer functions have a single real-axis pole. This pole is realized as part of the $n = 3$ section of Figure 3-13*b* when the tables of active all-pole low-pass values in Chapter 11 are used. If an odd-order filter is designed directly from the tabulated poles, the normalized real-axis pole can be generated using one of the configurations given in Figure 3-15.

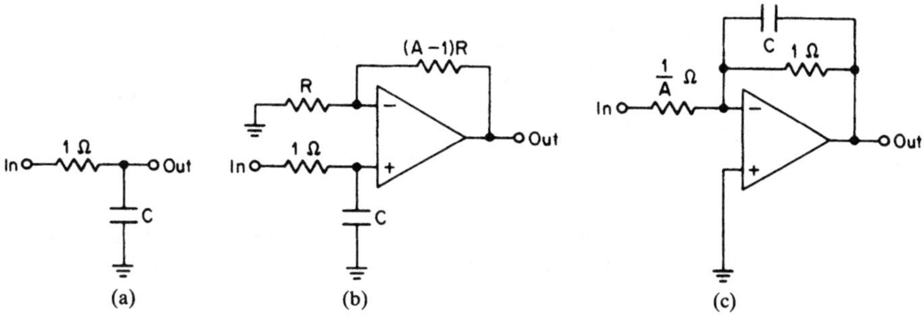

FIGURE 3-15 The first-order pole configurations: (*a*) a basic *RC* section; (*b*) a noninverting gain configuration; and (*c*) an inverting gain circuit.

The most basic form of a real pole is the circuit of Figure 3-15*a*. The capacitor *C* is defined by

$$C = \frac{1}{\alpha_0} \tag{3-11}$$

where α_0 is the normalized real-axis pole. The circuit gain is unity with a high-impedance termination.

If gain is desirable, the circuit of Figure 3-15*a* can be followed by a noninverting amplifier, as shown in Figure 3-15*b*, where *A* is the required gain. When the gain must be inverting, the circuit of Figure 3-15*c* is used.

The chosen circuit is frequency- and impedance-scaled in a manner similar to the rest of the filter. The value *R* in Figure 3-15*b* is arbitrary since only the ratio of the two feedback resistors determines the gain of the amplifier.

Example 3-9 Design of an Active All-pole Low-Pass filter with a Separate Real-Pole section

Required:

An active low-pass filter

3 dB at 75 Hz

15-dB minimum at 150 Hz

A gain of 40 dB ($A = 100$)

Result:

(*a*) Compute the steepness factor.

$$A_s = \frac{f_s}{f_c} = \frac{150}{75} = 2 \tag{2-11}$$

(*b*) Figure 2-34 indicates that an $n = 2$ Butterworth low-pass response satisfies the attenuation requirement. Since a gain of 100 is required, we will use the $n = 2$ section of Figure 3-13*a*, followed by the $n = 1$ section of Figure 3-15*b*, which provides the gain. The circuit configuration is shown in Figure 3-16*a*.

(*c*) The following pole locations of a normalized $n = 3$ Butterworth low-pass filter are obtained from Table 11-1:

Complex pole $\alpha = 0.5000$ $\beta = 0.8660$

Real pole $\alpha_0 = 1.0000$

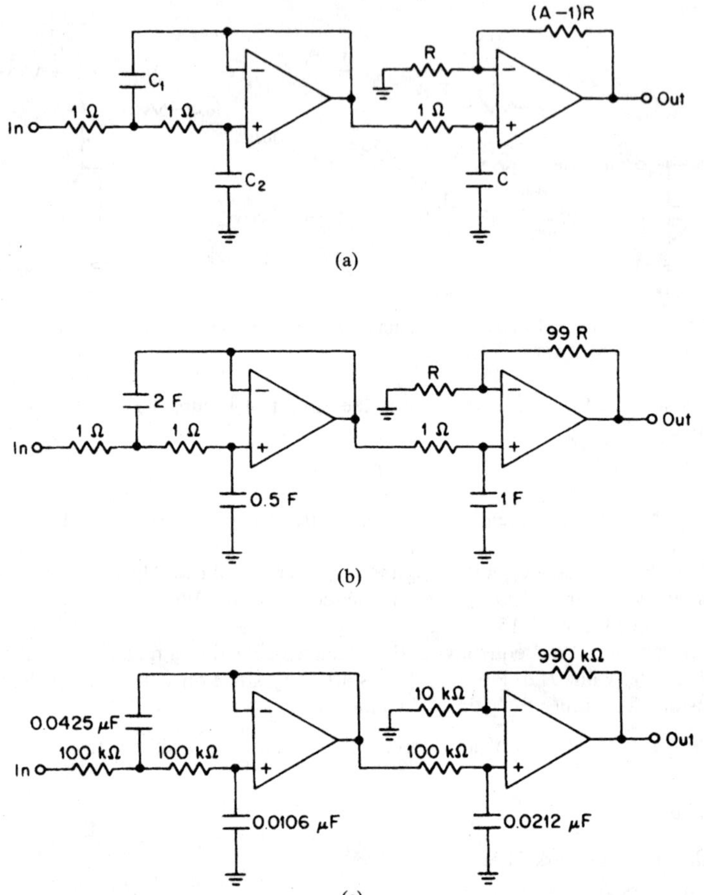

FIGURE 3-16 The low-pass filter of Example 3-9: (*a*) circuit configuration; (*b*) normalized circuit; and (*c*) scaled filter.

The component values for the $n = 2$ section are

$$C_1 = \frac{1}{\alpha} = \frac{1}{0.5} = 2\text{ F} \tag{3-9}$$

$$C_2 = \frac{\alpha}{\alpha^2 + \beta^2} = \frac{0.5}{0.5^2 + 0.866^2} = 0.5\text{ F} \tag{3-10}$$

The capacitor in the $n = 1$ circuit is computed by

$$C = \frac{1}{\alpha_0} = \frac{1}{1.0} = 1\text{ F} \tag{3-11}$$

Since $A = 100$, the feedback resistor is $99R$ in the normalized circuit shown in Figure 3-16*b*.

(*d*) Using an FSF of $2\pi f_c$ or 471 and selecting an impedance-scaling factor of 10^5, the denormalized capacitor values are

***n* = 2 section:**

$$C_1' = \frac{C}{\text{FSF} \times Z} = \frac{2}{471 \times 10^5} = 0.0425\ \mu\text{F} \tag{2-10}$$

$$C_2' = 0.0106\ \mu\text{F}$$

***n* = 1 section:**

$$C' = 0.0212\ \mu\text{F}$$

The value R for the $n = 1$ section is arbitrarily selected at 10 kΩ. The final circuit is given in Figure 3-16*c*.

Although these real-pole sections are intended to be part of odd-order low-pass filters, they can be independently used as an $n = 1$ low-pass filter, and have the transfer function

$$T(s) = K\frac{1}{sC + 1} \tag{3-12}$$

where $K = 1$ for Figure 3-15*a*, $K = A$ for Figure 3-15*b*, and $K = -A$ for Figure 3-15*c*. If $C = 1$ F, the 3-dB cutoff occurs at 1 rad/s.

The attenuation of a first-order filter can be expressed as

$$A_{\text{dB}} = 10 \log\left[1 + \left(\frac{\omega_x}{\omega_c}\right)^2\right] \tag{3-13}$$

where ω_x/ω_c is the ratio of a given frequency to the cutoff frequency. The normalized frequency response corresponds to the $n = 1$ curve of the Butterworth low-pass filter response curves of Figure 2-34. The step response has no overshoot and the impulse response does not have any oscillatory behavior.

Example 3-10 Design of an Active All-pole Low-Pass filter with a Gain of 10 dB

Required:

An active low-pass filter

3 dB at 60 Hz

12-dB minimum attenuation at 250 Hz

A gain of 20 dB with inversion

Result:

(*a*) Compute A_s.

$$A_s = \frac{f_s}{f_c} = \frac{250}{60} = 4.17 \tag{2-11}$$

(*b*) Figure 2-34 indicates that an $n = 1$ filter provides over 12 dB attenuation at 4.17 rad/s. Since an inverting gain of 20 dB is required, the configuration of Figure 3-15*c* will be used. The normalized circuit is shown in Figure 3-17*a*, where $C = 1$ F and $A = 10$, corresponding to a gain of 20 dB.

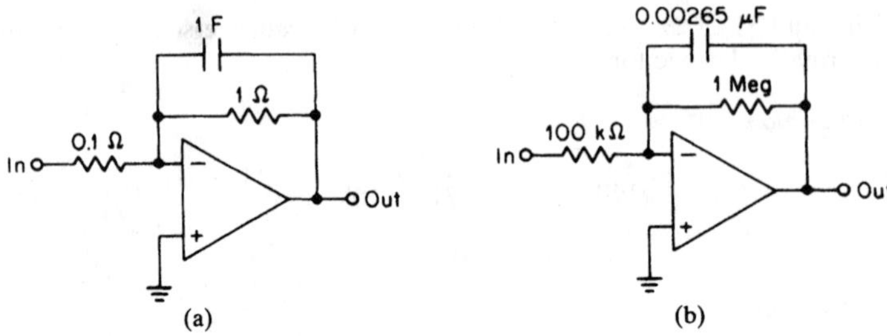

FIGURE 3-17 The $n = 1$ low-pass filter of Example 3-10: (*a*) normalized filter; and (*b*) frequency- and impedance-scaled filter.

(*c*) Using an FSF of $2\pi 60$ or 377 and an impedance-scaling factor of 10^6, the denormalized capacitor is

$$C' = \frac{C}{\text{FSF} \times Z} = \frac{1}{377 \times 10^6} = 0.00265\ \mu\text{F} \tag{2-10}$$

The input and output feedback resistors are 100 kΩ and 1 MΩ, respectively. The final circuit is shown in Figure 3-17*b*.

Second-Order Section with Gain. If an active low-pass filter is required to have a gain higher than unity and the order is even, the $n = 1$ sections of Figure 3-15 cannot be used since a real pole is not contained in the transfer function.

The circuit of Figure 3-18 realizes a pair of complex poles and provides a gain of $-A$. The element values are computed using the following formulas:

$$C_1 = (A + 1)\left(1 + \frac{\beta^2}{\alpha^2}\right) \tag{3-14}$$

$$R_1 = \frac{\alpha}{A(\alpha^2 + \beta^2)} \tag{3-15}$$

$$R_2 = \frac{AR_1}{A + 1} \tag{3-16}$$

$$R_3 = AR_1 \tag{3-17}$$

This section is used in conjunction with the $n = 2$ section of Figure 3-13*a* to realize even-order low-pass filters with gain. This is shown in the following example:

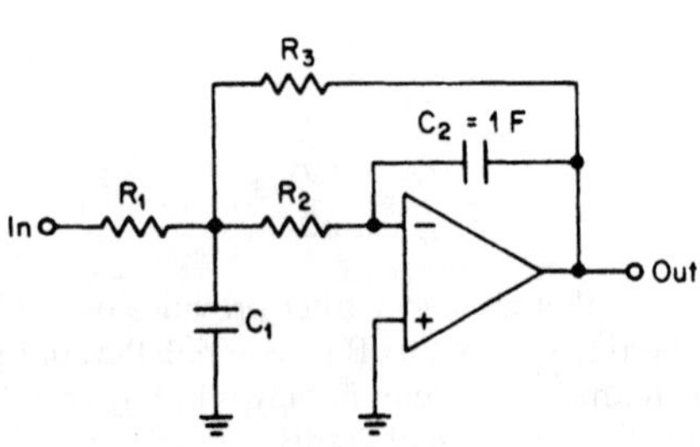

FIGURE 3-18 Second-order section with gain.

Example 3-11 Design of an Active All-pole Low-Pass filter with a Gain of 2

Required:

An active low-pass filter

3 dB at 200 Hz

30-dB minimum at 800 Hz

No step-response overshoot

A gain of 6 dB with inversion ($A = 2$)

Result:

(*a*) Compute A_s.

$$A_s = \frac{f_s}{f_c} = \frac{800}{200} = 4 \tag{2-11}$$

(*b*) Since no overshoot is permitted, a Bessel filter type will be used. Figure 2-56 indicates that a fourth-order network provides over 30 dB of rejection at 4 rad/s. Since an inverting gain of 2 is required and $n = 4$, the circuit of Figure 3-18 will be used, followed by the two-pole section of Figure 3-13*a*. The basic circuit configuration is given in Figure 3-19*a*.

(*c*) The following pole locations of a normalized $n = 4$ Bessel low-pass filter are obtained from Table 11-41:

$$\alpha = 1.3596 \quad \beta = 0.4071$$

and

$$\alpha = 0.9877 \quad \beta = 1.2476$$

The normalized component values for the first section are determined by the following formulas, where $\alpha = 1.3596$, $\beta = 0.4071$, and $A = 2$:

$$C_1 = (A + 1)\left(1 + \frac{\beta^2}{\alpha^2}\right) = 3\left(1 + \frac{0.4071^2}{1.3596^2}\right) = 3.27 \text{ F} \tag{3-14}$$

$$R_1 = \frac{\alpha}{A(\alpha^2 + \beta^2)} = \frac{1.3596}{2(1.3596^2 + 0.4071^2)} = 0.3375\ \Omega \tag{3-15}$$

$$R_2 = \frac{AR_1}{A + 1} = \frac{2 \times 0.3375}{3} = 0.225\ \Omega \tag{3-16}$$

$$R_3 = AR_1 = 2 \times 0.3375 = 0.675\ \Omega \tag{3-17}$$

The remaining pole pair of $\alpha = 0.9877$ and $\beta = 1.2476$ is used to compute the component values of the second section.

$$C_1 = \frac{1}{\alpha} = \frac{1}{0.9877} = 1.012 \text{ F} \tag{3-9}$$

$$C_2 = \frac{\alpha}{\alpha^2 + \beta^2} = \frac{0.9877}{0.9877^2 + 1.2476^2} = 0.39 \text{ F} \tag{3-10}$$

The normalized low-pass filter is shown in Figure 3-19*b*.

(*d*) Using an FSF of $2\pi f_c$ or 1256 and an impedance-scaling factor of 10^4 for both sections, the denormalized values are

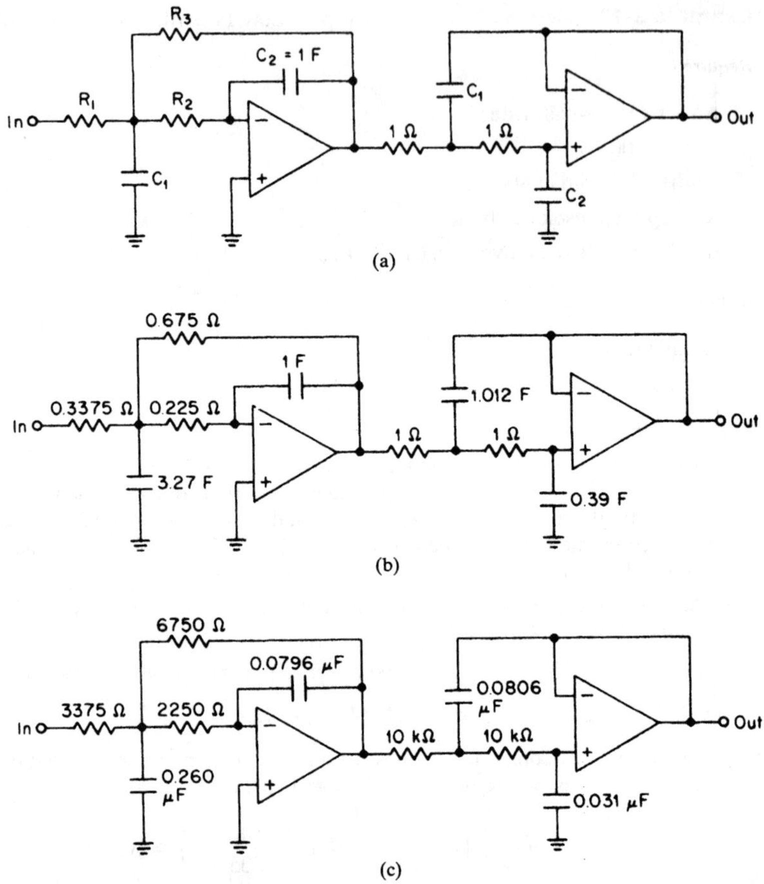

FIGURE 3-19 The $n = 4$ Bessel low-pass filter of Example 3-11: (*a*) circuit configuration; (*b*) normalized filter; and (*c*) frequency- and impedance-scaled filter.

***n* = 2 section with *A* = 2:**

$$R_1' = R \times Z = 0.3375 \times 10^4 = 3375\ \Omega \tag{2-8}$$

$$R_2' = 2250\ \Omega$$

$$R_3' = 6750\ \Omega$$

$$C_1' = \frac{C}{\text{FSF} \times Z} = \frac{3.27}{1256 \times 10^4} = 0.260\ \mu\text{F} \tag{2-10}$$

$$C_2' = 0.0796\ \mu\text{F}$$

***n* = 2 section having unity gain:**

$$R' = 10\ \text{k}\Omega$$

$$C_1' = \frac{C}{\text{FSF} \times Z} = \frac{1.012}{1256 \times 10^4} = 0.0806\ \mu\text{F} \tag{2-10}$$

$$C_2' = 0.0310\ \mu\text{F}$$

The final circuit is shown in Figure 3-19*c*.

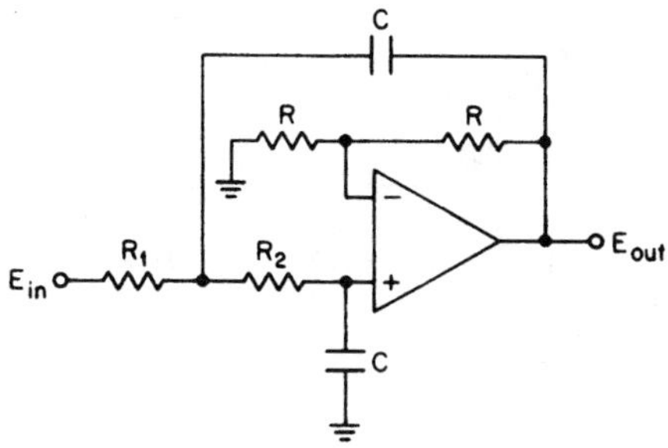

FIGURE 3-20 An all-pole configuration.

VCVS Uniform Capacitor Structure. The unity-gain $n = 2$ all-pole configuration of Figure 3-13*a* requires unequal capacitor values and noninteger capacitor ratios. The inconvenience usually results in either the use of nonstandard capacitor values or the paralleling of two or more standard values.

An alternate configuration is given in this section. This structure features equal capacitors. However, the circuit sensitivities are somewhat higher than the previously discussed configuration. Nevertheless, the more convenient capacitor values may justify its use in many instances where higher sensitivities are tolerable.

The $n = 2$ low-pass circuit of Figure 3-20 features equal capacitors and a gain of 2. The element values are computed as follows:

Select C.

Then

$$R_1 = \frac{1}{2\alpha' C} \tag{3-18}$$

and

$$R_2 = \frac{2\alpha'}{C(\alpha'^2 + \beta'^2)} \tag{3-19}$$

where α' and β' are the denormalized real and imaginary pole coordinates. R may be conveniently chosen.

Example 3-12 Design of an Active All-pole Low-Pass filter Using Uniform Capacitor Values

Required:

Design a fourth-order 0.1-dB Chebyshev active low-pass filter for a 3-dB cutoff of 100 Hz using 0.01 μF capacitors throughout.

Result:

(*a*) The pole locations for a normalized 0.1-dB Chebyshev low-pass filter are obtained from Table 11-23 and are as follows:

$$\alpha = 0.2177 \quad \beta = 0.9254$$

and

$$\alpha = 0.5257 \quad \beta = 0.3833$$

(*b*) Two sections of the filter of Figure 3-20 will be cascaded. The value of C is 0.01 μF, and R is chosen at 10 kΩ.

Section 1:

$$\alpha = 0.2177 \quad \alpha' = \alpha \times \text{FSF} = 136.8$$
$$\beta = 0.9254 \quad \beta' = \beta \times \text{FSF} = 581.4$$

where FSF $2\pi f_c = 628.3$

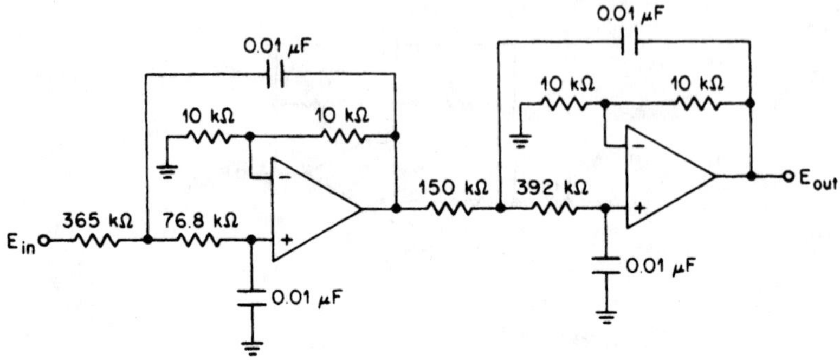

FIGURE 3-21 The equal capacitor circuit of Example 3-12.

$$R_1 = \frac{1}{2\alpha' C} = 365.5 \text{ k}\Omega \tag{3-18}$$

$$R_2 = \frac{2\alpha'}{C(\alpha'^2 + \beta'^2)} = 76.7 \text{ k}\Omega \tag{3-19}$$

Section 2:

$$\alpha = 0.5257 \quad \alpha' = 330.0$$
$$\beta = 0.3833 \quad \beta' = 240.8$$

$$R_1 = 151.4 \text{ k}\Omega \tag{3-18}$$

$$R_2 = 395.4 \text{ k}\Omega \tag{3-19}$$

The final filter is shown in Figure 3-21. The gain is 2^2, or 4.

The Low-Sensitivity Second-Order Section. The low-pass filter section of Figure 3-22 realizes a second-order transfer function which can be expressed as

$$T(s) = \frac{1}{\tau_1 \tau_2 s^2 + \tau_2 s + 1} \tag{3-20}$$

where $\tau_1 = R_1C_1$ and $\tau_2 = R_2C_2$.

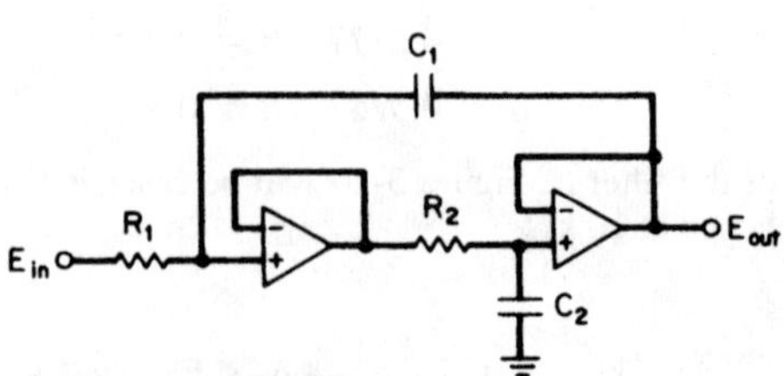

FIGURE 3-22 The low-sensitivity second-order section.

If we first equate Equation (3-20) with Equation (3-7), the general form for a second-order transfer function, and then solve for R_1 and R_2, we obtain

$$R_1 = \frac{1}{2\alpha C_1} \tag{3-21}$$

and

$$R_2 = \frac{2\alpha}{(\alpha^2 + \beta^2)C_2} \tag{3-22}$$

Two important observations can be made from Figure 3-22 and the associated design equations. Since both operational amplifiers are configured as voltage followers, the circuit sensitivity to amplifier open-loop gain is not as severe as for the previous circuit which requires a gain of 2. Secondly, both the transfer function and design equations clearly indicate that the circuit operation is dictated by two time constants: R_1C_1 and R_2C_2. Thus, C_1 and C_2 can be independently selected for convenient values or made equal, as desired.

Example 3-13 illustrates the application of this configuration.

Example 3-13 Design of an Active All-pole Low-Pass filter Using Low-sensitivity Second-Order Sections

Required:

Design a fourth-order 0.1-dB Chebyshev low-pass filter for a 3-dB cutoff frequency of 10 kHz using the low-sensitivity second-order section.

Result:

(*a*) The pole locations for a normalized 0.1-dB Chebyshev low-pass filter (given in Table 11-23) are as follows:

$$\alpha = 0.2177 \quad \beta = 0.9254$$

and

$$\alpha = 0.5257 \quad \beta = 0.3833$$

(*b*) Denormalizing the pole locations (multiply α and β by the FSF):

$$\alpha' = 13{,}678 \quad \beta' = 58{,}145$$

and

$$\alpha' = 33{,}031 \quad \beta' = 24{,}083$$

(*c*) Compute the component values as follows:

Section 1:

$$\alpha' = 13{,}678 \quad \beta' = 58{,}145$$

Let

$$C_1 = C_2 = 0.001\ \mu\text{F}$$

$$R_1 = \frac{1}{2\alpha' C} = 36.56\ \text{k}\Omega \tag{3-21}$$

$$R_2 = \frac{2\alpha'}{(\alpha'^2 + \beta'^2)C_2} = 7.667\ \text{k}\Omega \tag{3-22}$$

Section 2:

$$\alpha' = 33{,}031 \quad \beta' = 24{,}083$$

$$R_1 = 15.14\ \text{k}\Omega \tag{3-21}$$

$$R_2 = 39.53\ \text{k}\Omega \tag{3-22}$$

The resulting circuit is shown in Figure 3-23. The overall gain is unity.

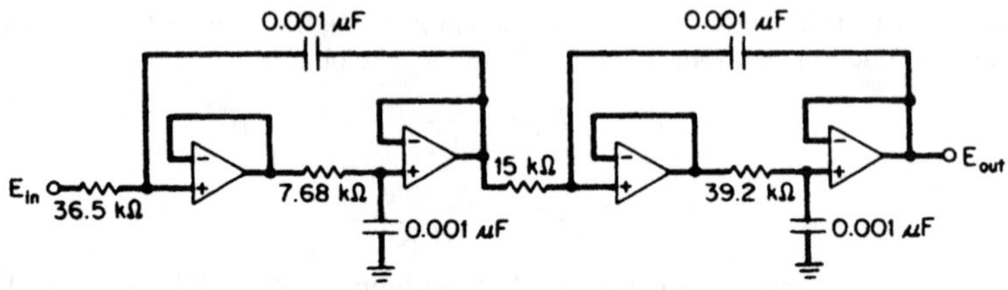

FIGURE 3-23 The circuit of Example 3-13.

Elliptic-Function VCVS Filters. Elliptic-function filters were first discussed in section 2.9. They contain zeros as well as poles. The zeros begin just outside the passband and force the response to decrease rapidly as Ω_s is approached. (Refer to Figure 2-82 for frequency-response definitions.)

Because of these finite zeros, the active filter circuit configurations of the previous section cannot be used since they are restrained to the realization of poles only.

The schematic of an elliptic-function low-pass filter section is shown in Figure 3-24*a*. This section provides a pair of complex conjugate poles and a pair of imaginary zeros, as shown in Figure 3-24*b*. The complex pole pair has a real component of α and an imaginary coordinate of β. The zeros are located at $\pm j\omega_\infty$. The *RC* section consisting of R_5 and C_5 introduces a real pole at α_0.

The configuration contains a voltage-controlled voltage source (VCVS) as the active element and is frequently referred to as a VCVS realization. Although this structure requires additional elements when compared with other VCVS configurations, it has been found to yield more reliable results and has lower sensitivity factors.

The normalized element values are determined by the following relations:

First calculate

$$a = \frac{2\alpha'}{\sqrt{\alpha'^2 + \beta'^2}} = \frac{1}{Q} \tag{3-23}$$

$$b = \frac{\omega'^2_\infty}{\alpha'^2 + \beta'^2} = \frac{\omega^2_\infty}{\omega^2_0} \tag{3-24}$$

$$c = \sqrt{\alpha'^2 + \beta'^2} = \omega_0 \tag{3-25}$$

where α', β', and ω'_∞ are the denormalized pole-zero coordinates. The second forms of Equations (3-23) through (3-25) involving Q and ω_0 are used when these parameters are directly provided already denormalized by the *Filter Solutions* program.

A controlled amplification of K is required between the noninverting amplifier input and the section output. Since the gain of a noninverting operational amplifier is the ratio of the feedback resistors plus 1, R_6 and R_7 are R and $(K - 1)R$, respectively, where R can be any convenient value.

In the event that K is less than 1, the amplifier is reconfigured as a voltage follower and R_4 is split into resistors R_{4a} and R_{4b} where

$$R_{4a} = (1 - K)R_4 \tag{3-26}$$

$$R_{4b} = KR_4 \tag{3-27}$$

The modified circuit is shown in Figure 3-25.

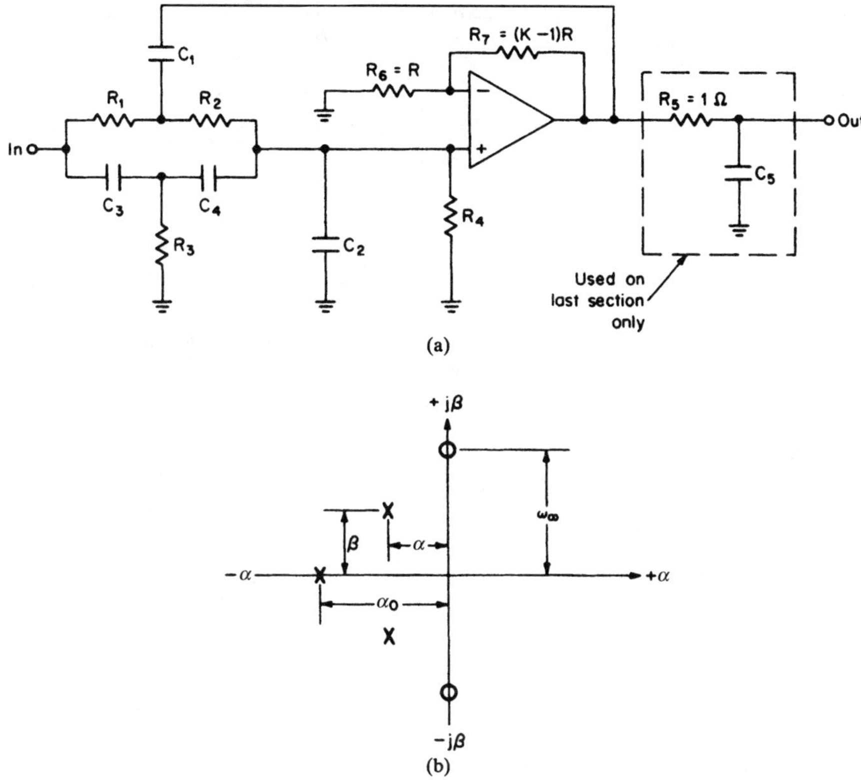

FIGURE 3-24 The elliptic-function low-pass filter section: (*a*) VCVS circuit configuration for $K > 1$; and (*b*) pole-zero pattern.

The design of active elliptic-function filters utilizes the *Filter Solutions* program provided on the CD-ROM for obtaining pole-zero locations which are already denormalized. The design method proceeds as if a passive elliptic low-pass filter is being designed. However, once a design is completed, the ***Transfer Function*** button is depressed and then

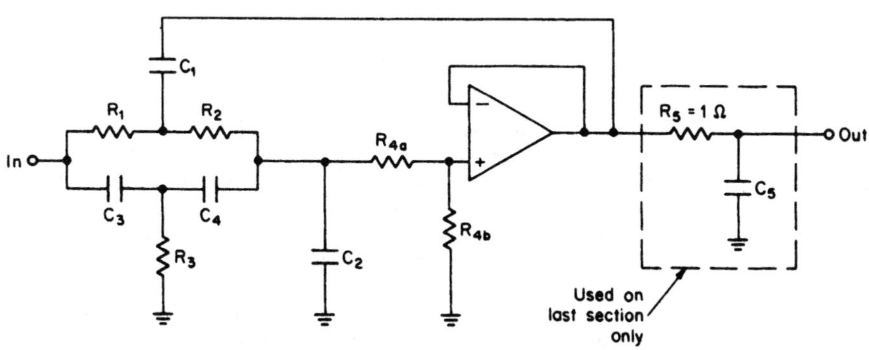

FIGURE 3-25 The elliptic-function VCVS low-pass filter section for $K < 1$.

the ***Casc*** box is checked. The poles and zeros are displayed in cascaded form (rather than rectangular form), which must now be utilized to compute **a**, **b**, and **c** of Equations (3-23), (3-24), and (3-25), as follows:

$$a = \frac{2\alpha'}{\sqrt{\alpha'^2 + \beta'^2}} = \frac{1}{Q} \tag{3-23}$$

$$b = \frac{\omega'^2_\infty}{\alpha'^2 + \beta'^2} = \frac{\omega^2_\infty}{\omega^2_0} \tag{3-24}$$

$$c = \sqrt{\alpha'^2 + \beta'^2} = \omega_0 \tag{3-25}$$

For odd-order filters, the real pole α_0 is presented as $(S + \alpha_0)$ in the denominator. The element values are computed as follows:

Select C

Then
$$C_1 = C \tag{3-28}$$

$$C_3 = C_4 = \frac{C_1}{2} \tag{3-29}$$

let
$$C_2 \geq \frac{C_1(b - 1)}{4} \tag{3-30}$$

$$R_3 = \frac{1}{cC_1\sqrt{b}} \tag{3-31}$$

$$R_1 = R_2 = 2R_3 \tag{3-32}$$

$$R_4 = \frac{4\sqrt{b}}{cC_1(1 - b) + 4cC_2} \tag{3-33}$$

$$K = 2 + \frac{2C_2}{C_1} - \frac{a}{2\sqrt{b}} + \frac{2}{C_1\sqrt{b}}\left(\frac{1}{cR_4} - aC_2\right) \tag{3-34}$$

$$\text{Section gain} = \frac{bKC_1}{4C_2 + C_1} \tag{3-35}$$

Capacitor C_5 is determined from the denormalized real pole by

$$C_5 = \frac{1}{R_5\alpha'_0} \tag{3-36}$$

where both R and R_5 can be arbitrarily chosen and α'_0 is $\alpha_0 \times$ FSF.

Odd-order elliptic-function filters are more efficient than even-order since maximum utilization is made of the number of component elements used. Since the circuit of Figure 3-24 provides a single pole pair (along with a pair of zeros), the total number of sections required for an odd-order filter is determined by $(n - 1)/2$, where n is the order of the filter. Because an odd-order transfer function has a single real pole, R_5 and C_5 appear on the output section only.

In the absence of a detailed analysis, it is a good rule of thumb to pair poles with their nearest zeros when allocating poles and zeros to each active section. This applies to highpass, bandpass, and band-reject filters as well.

Example 3-14 Design of an Active Elliptic Function Low-Pass Filter using the VCVS Structure

Required:
Design an active elliptic-function low-pass filter corresponding to a 0.177-dB ripple, a cutoff of 100 Hz and a minimum attenuation of 37 dB at 292.4 Hz using the VCVS structure of Figure 3-24.

Result:

(*a*) Open ***Filter Solutions***.

Check the ***Stop Band Freq*** box.
Enter **.177** in the ***Pass Band Ripple (dB)*** box.
Enter **100** in the ***Pass Band Freq*** box.
Enter **292.4** in the ***Stop Band Freq*** box.
Check the ***Frequency Scale Hertz*** box.

(b) Click the Set Order control button to open the second panel.

Enter **37** for ***Stop band Attenuation (dB)***.
Click the ***Set Minimum Order*** button and then click ***Close***.
3 Order is displayed on the main control panel.

(*c*) Click the ***Transfer Function*** button.
Check the ***Casc*** box.

The following is displayed:

Continuous Transfer Function

Wn = 2105

$$\frac{71.83 \quad (S^2 + 4.432e{+}06)}{(S^2 + 485.5{*}S + 5.713e{+}05) \quad (S + 557.4)}$$

Wo = 755.8
Q = 1.557

3rd Order Low Pass Elliptic

Pass Band Frequency = 100.0 Hz
Pass Band Ripple = 177.0 mdB
Stop Band Ratio = 2.924
Stop Band Frequency = 292.4 Hz
Stop Band Attenuation = 37.43 dB

(*d*) The design parameters are summarized as follows:

$$\text{Section } Q = 1.557$$

$$\text{Section } \omega_0 = 755.8$$

$$\text{Section } \omega_\infty = 2105$$

$$\alpha_0 = 557.4 \text{ (from the denominator)}$$

(*e*) The element values are computed as follows:

$$a = 0.6423 \quad (3\text{-}23)$$

$$b = 7.7569 \quad (3\text{-}24)$$

$$c = 755.8 \quad (3\text{-}25)$$

Select $$C_1 = C = 0.1\ \mu\text{F} \quad (3\text{-}28)$$

$$C_3 = C_4 = 0.05\ \mu\text{F} \quad (3\text{-}29)$$

$$C_2 \geq 0.169\ \mu\text{F} \quad (3\text{-}30)$$

Let $$C_2 = 0.22\ \mu\text{F}$$

$$R_3 = 4751\ \Omega \quad (3\text{-}31)$$

$$R_1 = R_2 = 9502\ \Omega \quad (3\text{-}32)$$

$$R_4 = 72.2\ \text{k}\Omega \quad (3\text{-}33)$$

$$K = 5.402 \quad (3\text{-}34)$$

Let $$R = R_5 = 10\ \text{k}\Omega \quad (3\text{-}35)$$

then $$C_5 = 0.180\ \mu\text{F} \quad (3\text{-}36)$$

The resulting circuit is shown in Figure 3-26 using standard values.

State-Variable Low-Pass Filters. The poles and zeros of the previously discussed active filter configurations cannot be easily adjusted because of the interaction of circuit elements. For most industrial requirements, sufficient accuracy is obtained by specifying 1-percent resistors and 1- or 2-percent capacitors. In the event that greater precision is required, the state-variable approach features independent adjustment of the pole and zero coordinates. Also, the state-variable configuration has a lower sensitivity to many of the inadequacies of operational amplifiers such as finite bandwidth and gain.

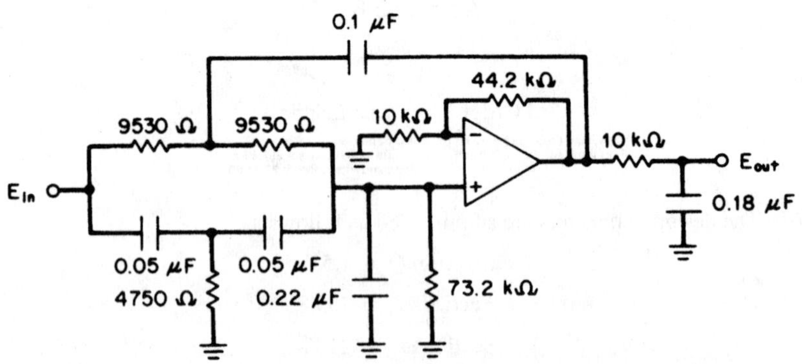

FIGURE 3-26 The circuit for the active elliptic-function low-pass filter.

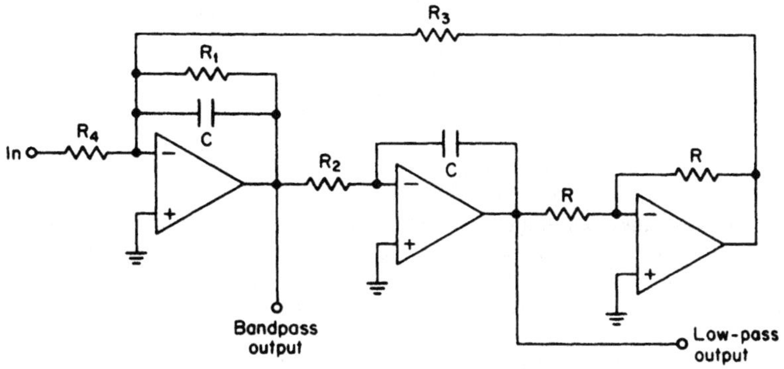

FIGURE 3-27 The state-variable all-pole low-pass configuration.

All-Pole Configuration. The circuit of Figure 3-27 realizes a single pair of complex poles. The low-pass transfer function is given by

$$T(s) = \frac{1}{R_2R_4C^2} \frac{1}{s^2 + \dfrac{1}{R_1C}s + \dfrac{1}{R_2R_3C^2}} \tag{3-37}$$

If we equate Equation (3-37) to the second-order low-pass transfer function expressed by Equation (3-8) and solve for the element values, after some algebraic manipulation we obtain the following design equations:

$$R_1 = \frac{1}{2\alpha C} \tag{3-38}$$

$$R_2 = R_3 = R_4 = \frac{1}{C\sqrt{\alpha^2 + \beta^2}} \tag{3-39}$$

where α and β are the real and imaginary components, respectively, of the pole locations and C is arbitrary. The value of R in Figure 3-27 is also optional.

The element values computed by Equations (3-38) and (3-39) result in a DC gain of unity. If a gain of $-A$ is desired, R_4 can instead be defined by

$$R_4 = \frac{1}{AC\sqrt{\alpha^2 + \beta^2}} \tag{3-40}$$

Sometimes it is desirable to design a filter directly at its cutoff frequency instead of calculating the normalized values and then frequency- and impedance-scaling the normalized network. Equations (3-38) and (3-39) result in the denormalized values if α and β are first denormalized by the frequency-scaling factor FSF as follows:

$$\alpha' = \alpha \times \text{FSF} \tag{3-41}$$

$$\beta' = \beta \times \text{FSF} \tag{3-42}$$

Direct design of the denormalized filter is especially advantageous when the design formulas permit the arbitrary selection of capacitors and all network capacitors are equal. A standard capacitance value can then be chosen.

Figure 3-27 indicates that a bandpass output is also provided. Although a discussion of bandpass filters will be deferred until Chapter 5, this output is useful for tuning of the low-pass filter. To adjust the low-pass real and imaginary pole coordinates, first compute the bandpass resonant frequency:

$$f_0 = \frac{\sqrt{(\alpha')^2 + (\beta')^2}}{2\pi} \tag{3-43}$$

Trim the value of R_3 until resonant conditions occur at the bandpass output with f_0 applied. Resonance can be determined by exactly 180° of phase shift between input and output or by peak output amplitude. The 180° phase shift method normally results in more accuracy and resolution. By connecting the vertical channel of an oscilloscope to the section input and the horizontal channel to the bandpass output, a *Lissajous pattern* is obtained. This pattern is an ellipse that will collapse to a straight line (at a 135° angle) when the phase shift is 180°.

For the final adjustment, trim R_1 for a bandpass Q (for example, f_0/3-dB bandwidth) equal to

$$Q = \frac{\pi f_0}{\alpha'} \tag{3-44}$$

Resistor R_1 can be adjusted for a measured bandpass output gain at f_0 equal to the computed ratio of R_1/R_4. Amplifier phase shift creates a Q-enhancement effect where the Q of the section is increased. This effect also increases the gain at the bandpass output, so adjustment of R_1 for the calculated gain will usually restore the desired Q. Alternatively, the 3-dB bandwidth can be measured and the Q computed. Although the Q measurement approach is the more accurate method, it certainly is slower than a simple gain adjustment.

Example 3-15 Design of an Active State-Variable All-Pole Low-Pass Filter

Required:

An active low-pass filter
3 dB ± 0.25 dB at 500 Hz
40-dB minimum at 1375 Hz

Result:

(*a*) Compute A_s.

$$A_s = \frac{1375}{500} = 2.75 \tag{2-11}$$

(*b*) Figure 2-42 indicates that a fourth-order 0.1-dB Chebyshev low-pass filter has over 40 dB of rejection at 2.75 rad/s. Since a precise cutoff is required, we will use the state-variable approach so the filter parameters can be adjusted if necessary.

The pole locations for a normalized fourth-order 0.1-dB Chebyshev low-pass filter are obtained from Table 11-23 and are as follows:

$$\alpha = 0.2183 \quad \beta = 0.9262$$

and

$$\alpha = 0.5271 \quad \beta = 0.3836$$

(*c*) Two sections of the circuit of Figure 3-27 will be cascaded. The denormalized filter will be designed directly. The capacitor value C is chosen to be 0.01 μF, and R is arbitrarily selected to be 10 kΩ.

Section 1:

$$\alpha = 0.2183 \quad \alpha' = \alpha \times \text{FSF} = 685.5 \tag{3-41}$$

$$\beta = 0.9262 \quad \beta' = \beta \times \text{FSF} = 2908 \tag{3-42}$$

where FSF $= 2\pi f_c = 2\pi 500 = 3140$

$$R_1 = \frac{1}{2\alpha' C} = 72.94 \text{ k}\Omega \tag{3-38}$$

$$R_2 = R_3 = R_4 = \frac{1}{C\sqrt{(\alpha')^2 + (\beta')^2}} = 33.47 \text{ k}\Omega \tag{3-39}$$

Section 2:

$$\alpha = 0.5271 \quad \alpha' = \alpha \times \text{FSF} = 1655 \tag{3-41}$$

$$\beta = 0.3836 \quad \beta' = \beta \times \text{FSF} = 1205 \tag{3-42}$$

where FSF $= 3140$

$$R_1 = \frac{1}{2\alpha' C} = 30.21 \text{ k}\Omega \tag{3-38}$$

$$R_2 = R_3 = R_4 = \frac{1}{C\sqrt{(\alpha')^2 + (\beta')^2}} = 48.85 \text{ k}\Omega \tag{3-39}$$

(*d*) The resulting filter is shown in Figure 3-28. The resistor values have been modified so that standard 1-percent resistors are used and adjustment capability is provided.

The bandpass resonant frequency and Q are

Section 1:

$$f_0 = \frac{\sqrt{(\alpha')^2 + (\beta')^2}}{2\pi} = 476 \text{ Hz} \tag{3-43}$$

$$Q = \frac{\pi f_0}{\alpha'} = 2.18 \tag{3-44}$$

Section 2:

$$f_0 = 326 \text{ Hz} \tag{3-43}$$

$$Q = 0.619 \tag{3-44}$$

Elliptic-Function Configuration. When precise control of the parameters of elliptic-function filters is required, a state-variable elliptic-function approach is necessary. This is especially true in the case of very sharp filters where the location of the poles and zeros is highly critical.

The circuit in Figure 3-29 has the transfer function

$$T(s) = -\frac{R_6}{R}\left[\frac{s^2 + \dfrac{1}{R_2R_3C^2}\left(1 + \dfrac{R_3R}{R_4R_5}\right)}{s^2 + \dfrac{1}{R_1C}s + \dfrac{1}{R_2R_3C^2}}\right] \tag{3-45}$$

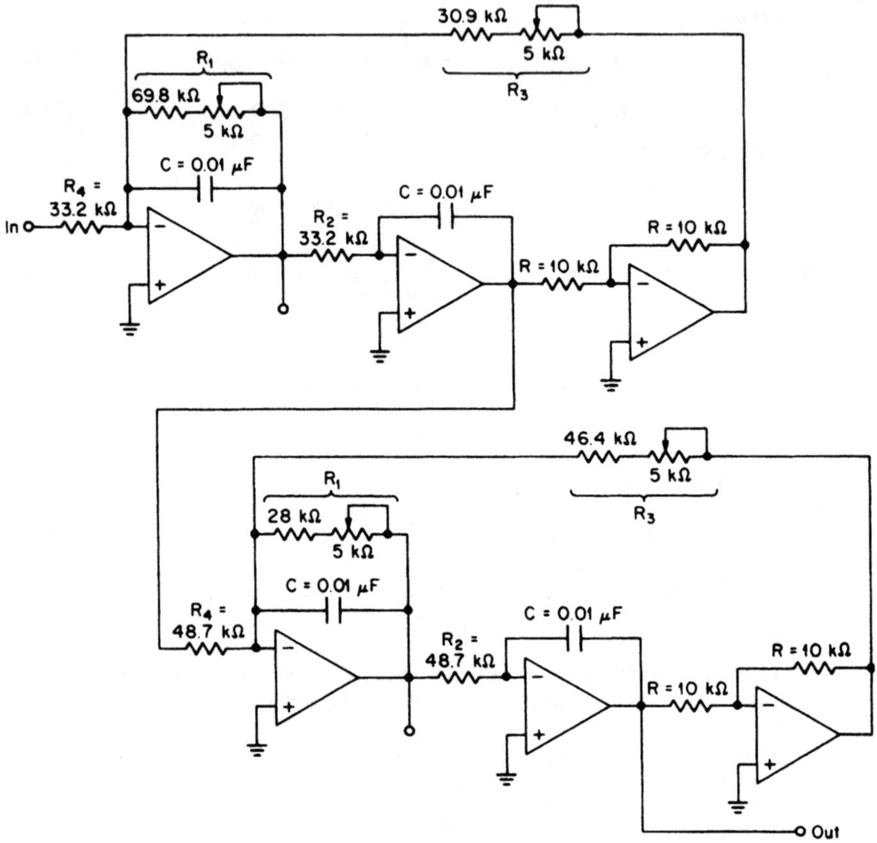

FIGURE 3-28 The state-variable low-pass filter of Example 3-15.

where $R_1 = R_4$ and $R_2 = R_3$. The numerator roots result in a pair of imaginary zeros, and the denominator roots determine a pair of complex poles. Since both the numerator and denominator are second-order, this transfer function form is frequently referred to as *biquadratic*, while the circuit is called a *biquad*. The zeros are restricted to frequencies beyond the pole locations—for example, the stopband of elliptic-function low-pass filters. If R_5 in Figure 3-29 were connected to node 2 instead of node 1, the zeros would occur below the poles as in high-pass elliptic-function filters.

The design of active elliptic-function filters using biquads utilizes the *Filter Solutions* program provided on the CD-ROM for obtaining pole-zero locations, which are already denormalized. The design method proceeds as if a passive elliptic low-pass filter is being designed. However once a design is completed the ***Transfer Function*** button is depressed and then the ***Casc*** box is checked. The poles and zeros are displayed in cascaded form (rather than rectangular form), which can then be used in the design equations.

The parameters obtained from *Filter Solutions* are ω_∞, ω_0, Q, and α_0, which are *already denormalized.*

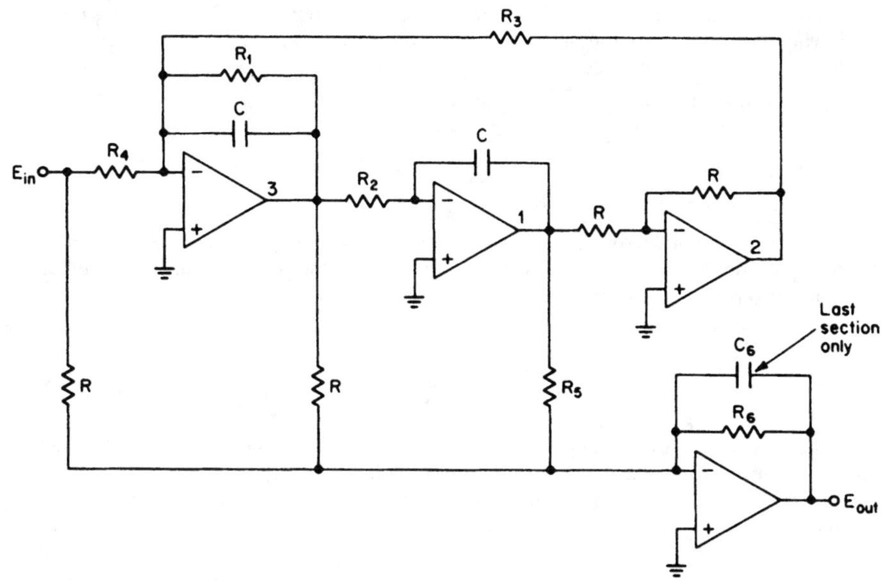

FIGURE 3-29 The state-variable (biquad) configuration for elliptic-function low-pass filters.

First, compute:

$$\alpha = \frac{\omega_0}{2Q} \tag{3-46}$$

The component values are

$$R_1 = R_4 = \frac{1}{2\alpha C} \tag{3-47}$$

$$R_2 = R_3 = \frac{1}{\omega_0 C} \tag{3-48}$$

$$R_5 = \frac{2\alpha\omega_0 R}{(\omega_\infty)^2 - (\omega_0)^2} \tag{3-49}$$

$$R_6 = \left(\frac{\omega_0}{\omega_\infty}\right)^2 AR \tag{3-50}$$

where C and R are arbitrary and A is the desired low-pass gain at DC.

Since odd-order elliptic-function filters contain a real pole, the last section of a cascade of biquads should contain capacitor C_6 in parallel with R_6. To compute C_6

$$C_6 = \frac{1}{\alpha_0 R_6} \tag{3-51}$$

The poles and zeros of the biquad configuration of Figure 3-29 can be adjusted by implementing the following sequence of steps:

1. *Resonant frequency:* The bandpass resonant frequency is defined by

$$f_0 = \frac{\omega_0}{2\pi} \tag{3-52}$$

If R_3 is made adjustable, the section's resonant frequency can be tuned to f_0 by monitoring the bandpass output at node 3. The 180° phase shift method is preferred for the determination of resonance.

2. *Q adjustment:* The bandpass Q is given by

$$Q = \frac{\pi f_0}{\alpha} \tag{3-53}$$

Adjustment of R_1 for unity gain at f_0 measured between the section input and the bandpass output at node 3 will usually compensate for any Q enhancement resulting from amplifier phase shift.

3. *Notch frequency:* The notch frequency is given by

$$f_\infty = \omega_\infty/2\pi \tag{3-54}$$

Adjustment of f_∞ usually is not required if the circuit is first tuned to f_0, since f_∞ will then fall in. However, if independent tuning of the notch frequency is desired, R_5 should be made adjustable. The notch frequency is measured by determining the input frequency where E_{out} is nulled.

Example 3-16 Design of an Active Elliptic-Function Low-Pass Filter using the State-Variable Configuration

Required:
Design an active elliptic-function low-pass filter corresponding to a 0.18-dB ripple, a cutoff of 1000 Hz and a minimum attenuation of 18 dB at 1556 Hz using the state-variable (biquad) configuration for elliptic-function low-pass filters of Figure 3-29.

Result:

(*a*) Open ***Filter Solutions***.

Check the ***Stop Band Freq*** box.
Enter **.18** in the ***Pass Band Ripple (dB)*** box.
Enter **1000** in the ***Pass Band Freq*** box.
Enter **1556** in the ***Stop Band Freq*** box.
Check the ***Frequency Scale Hertz*** box.

(*b*) Click the ***Set Order*** control button to open the second panel.

Enter **18** for ***Stop band Attenuation (dB)***.
Click the ***Set Minimum Order*** button and then click ***Close***.
3 Order is displayed on the main control panel.

(*c*) Click the ***Transfer Function*** button.
Check the ***Casc*** box.

The following is displayed:

Continuous Transfer Function

$$W_n = 1.095e{+}04$$

$$\frac{3040 \quad (S^2 + 1.199e{+}08)}{(S^2 + 3573{*}S + 5.511e{+}07) \quad (S + 6613)}$$

$$W_o = 7423$$
$$Q = 2.077$$

3rd Order Low Pass Elliptic

Pass Band Frequency = 1.000 KHz
Pass Band Ripple = 180.0 mdB
Stop Band Ratio = 1.556
Stop Band Frequency = 1.556 KHz
Stop Band Attenuation = 18.63 dB

(*d*) The design parameters are summarized as follows:

$$\text{Section } Q = 2.077$$

$$\text{Section } \omega_0 = 7{,}423$$

$$\text{Section } \omega_\infty = 10{,}950$$

$$\alpha_0 = 6613 \quad \text{(from the denominator)}$$

(*e*) A single section is required. Let $R = 100\ \text{k}\Omega$ and $C = 0.1\ \mu\text{F}$, and let the gain equal unity ($A = 1$). The values are computed as follows:

$$\alpha = \frac{\omega_0}{2Q} = 1787 \tag{3-46}$$

$$R_1 = R_4 = \frac{1}{2\alpha C} = 2798\ \Omega \tag{3-47}$$

$$R_2 = R_3 = \frac{1}{\omega_0 C} = 1347\ \Omega \tag{3-48}$$

$$R_5 = \frac{2\alpha\omega_0 R}{(\omega_\infty)^2 - (\omega_0)^2} = 40.94\ \text{k}\Omega \tag{3-49}$$

$$R_6 = \left(\frac{\omega_0}{\omega_\infty}\right)^2 AR = 45.96\ \text{k}\Omega \tag{3-50}$$

Since a real pole is required, C_6 is introduced in parallel with R_6 and is calculated by

$$C_6 = \frac{1}{\alpha_0 R_6} = 3260\ \text{pF} \tag{3-51}$$

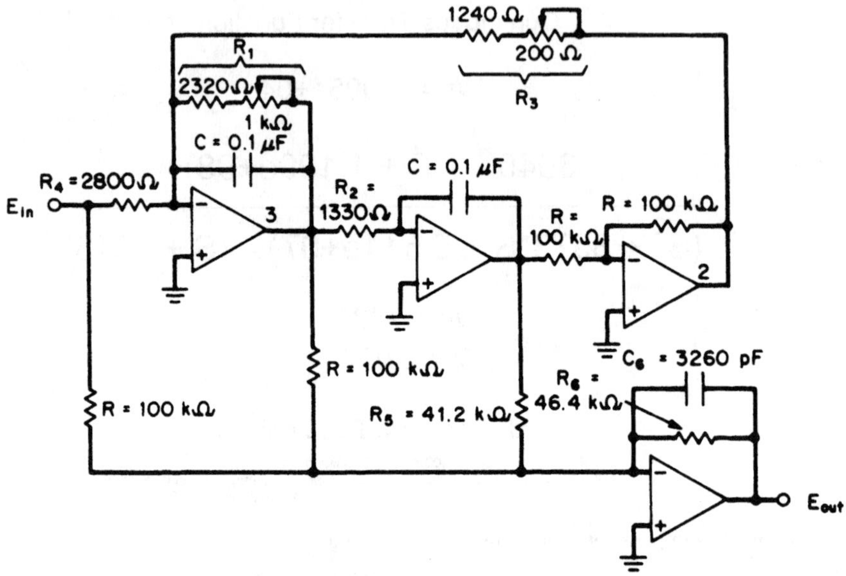

FIGURE 3-30 The elliptic-function low-pass filter of Example 3-16.

(*f*) The resulting filter is shown in Figure 3-30. The resistor values are modified so that standard 1-percent values are used and the circuit is adjustable. The sections f_0 and Q are computed by

$$f_0 = \frac{\omega_0}{2\pi} = 1181 \text{ Hz} \tag{3-52}$$

$$Q = \frac{\pi f_0}{\alpha} = 2.076 \tag{3-53}$$

The frequency of infinite attenuation is given by

$$f_\infty = \omega_\infty/2\pi = 1743 \text{ Hz} \tag{3-54}$$

Generalized Impedance Converters. The circuit of Figure 3-31 is known as a generalized impedance converter (GIC). The driving-point impedance can be expressed as

$$Z_{11} = \frac{Z_1 Z_3 Z_5}{Z_2 Z_4} \tag{3-55}$$

By substituting RC combinations of up to two capacitors for Z_1 through Z_5, a variety of impedances can be simulated. If, for instance, Z_4 consists of a capacitor having an impedance $1/sC$, where $s = j\omega$ and all other elements are resistors, the driving-point impedance is given by

$$Z_{11} = \frac{sCR_1R_3R_5}{R_2} \tag{3-56}$$

The impedance is proportional to frequency and is therefore identical to an inductor, having a value of

$$L = \frac{CR_1R_3R_5}{R_2} \tag{3-57}$$

as shown in Figure 3-32.

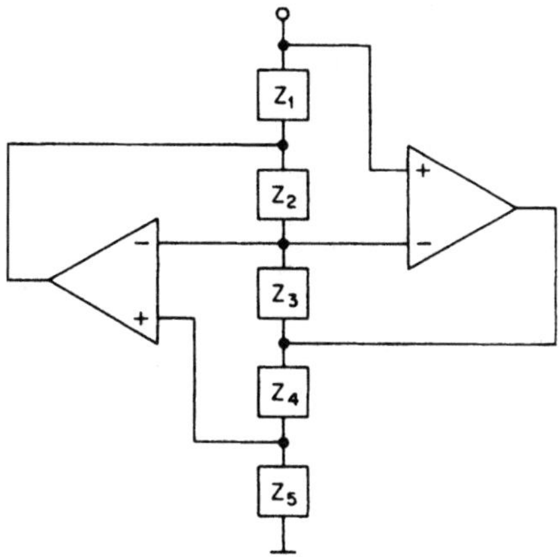

FIGURE 3-31 A generalized impedance converter (GIC).

If two capacitors are introduced for Z_1 and Z_3, and Z_2, Z_4, and Z_5 are resistors, the resulting driving-point impedance expression can be expressed in the form of

$$Z_{11} = \frac{R_5}{s^2C^2R_2R_4} \tag{3-58}$$

An impedance proportional to $1/s^2$ is called a D element, whose driving point impedance is given by

$$Z_{11} = \frac{1}{s^2D} \tag{3-59}$$

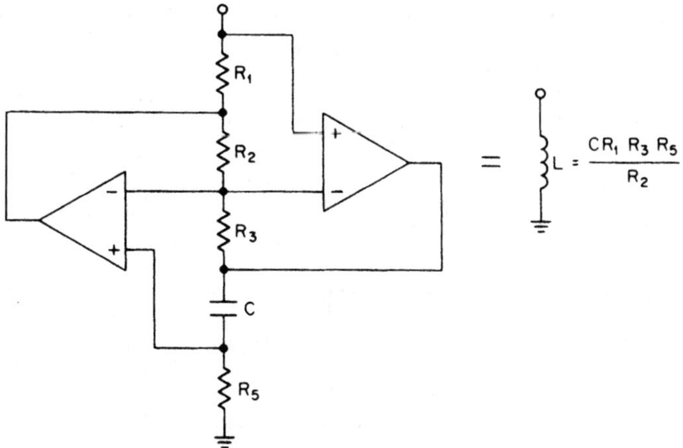

FIGURE 3-32 A GIC inductor simulation.

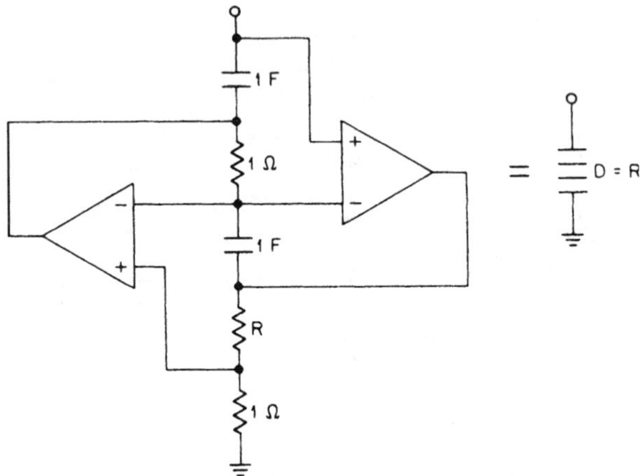

FIGURE 3-33 A GIC realization of a normalized D element.

Equation (3-58) therefore defines a D element having the value

$$D = \frac{C^2 R_2 R_4}{R_5} \tag{3-60}$$

If we let $C = 1$ F, $R_2 = R_5 = 1\ \Omega$, and $R_4 = R$, Equation (3-64) simplifies to $D = R$.

In order to gain some insight into the nature of this element, let us substitute $s = j\omega$ into Equation (3-58). The resulting expression is

$$Z_{11} = -\frac{R_5}{\omega^2 C^2 R_2 R_4} \tag{3-61}$$

Equation (3-61) corresponds to a frequency-dependent negative resistor (FDNR).

A GIC in the form of a normalized D element and its schematic designation are shown in Figure 3-33. Bruton (see Bibliography) has shown how the FDNR or D element can be used to generate active filters directly from the LC normalized low-pass prototype values. This technique will now be described.

The 1/s Transformation. If all the impedances of an LC filter network are multiplied by $1/s$, the transfer function remains unchanged. This operation is equivalent to impedance-scaling a filter by the factor $1/s$ and should not be confused with the high-pass transformation which involves the substitution of $1/s$ for s. Section 2.1 under "Frequency and Impedance Scaling" demonstrated that the impedance scaling of a network by any factor Z does not change the frequency response, since the Zs cancel in the transfer function, so the validity of this transformation should be apparent.

When the elements of a network are impedance-scaled by $1/s$, they undergo a change in form. Inductors are transformed into resistors, resistors into capacitors, and capacitors into D elements, which are summarized in Table 3-1. Clearly, this design technique is extremely powerful. It enables us to design active filters directly from passive LC circuits. Knowledge of the pole and zero locations is unnecessary.

The design method proceeds by first selecting a normalized low-pass LC filter. All capacitors must be restricted to the shunt arms only, since they will be transformed into D elements which are connected to ground. The dual LC filter (defined by the lower schematic

TABLE 3-1 The 1/*s* Impedance Transformation

Element	Impedance	Transformed Element	Transformed Impedance
L	sL	L	L
C	$\frac{1}{sC}$	C	$\frac{1}{s^2C}$
R	R	$\frac{1}{R}$	$\frac{R}{s}$

in the tables of Chapter 11 for the all-pole case) is usually chosen to be transformed to minimize the number of *D* elements. The circuit elements are modified by the 1/*s* transformation, and the *D* elements are realized using the normalized GIC circuit of Figure 3-33. The transformed filter is then frequency- and impedance-scaled in the conventional manner. The following example demonstrates the design of an all-pole active low-pass filter using the 1/*s* impedance transformation and the GIC.

Example 3-17 Design of an Active All-Pole Low-Pass filter using a *D* Element

Required:

An active low-pass filter
3 dB to 400 Hz
20-dB minimum at 1200 Hz
Minimal ringing and overshoot

Result:

(*a*) Compute the steepness factor.

$$A_s = \frac{f_s}{f_c} = \frac{1200 \text{ Hz}}{400 \text{ Hz}} = 3 \qquad (2\text{-}11)$$

(*b*) Since low transient distortion is desired, a linear phase filter with a phase error of 0.5° will be selected. The curves of Figure 2-62 indicate that a filter complexity of $n = 3$ provides over 20 dB of attenuation at 3 rad/s.
The 1/*s* transformation and a GIC realization will be used.

(*c*) The normalized *LC* low-pass filter from Table 11-47 corresponding to $n = 3$ is shown in Figure 3-34*a*. The dual circuit has been selected so that only a single *D* element will be required.

(*d*) The normalized filter is transformed in accordance with Table 3-1, resulting in the circuit of Figure 3-34*b*. The *D* element is realized using the normalized GIC configuration of Figure 3-33, as shown in Figure 3-34*c*.

(*e*) Since all normalized capacitors are 1 F, it would be desirable if they were all denormalized to a standard value such as 0.01 μF. Using an FSF of $2\pi f_c$ or 2513 and a

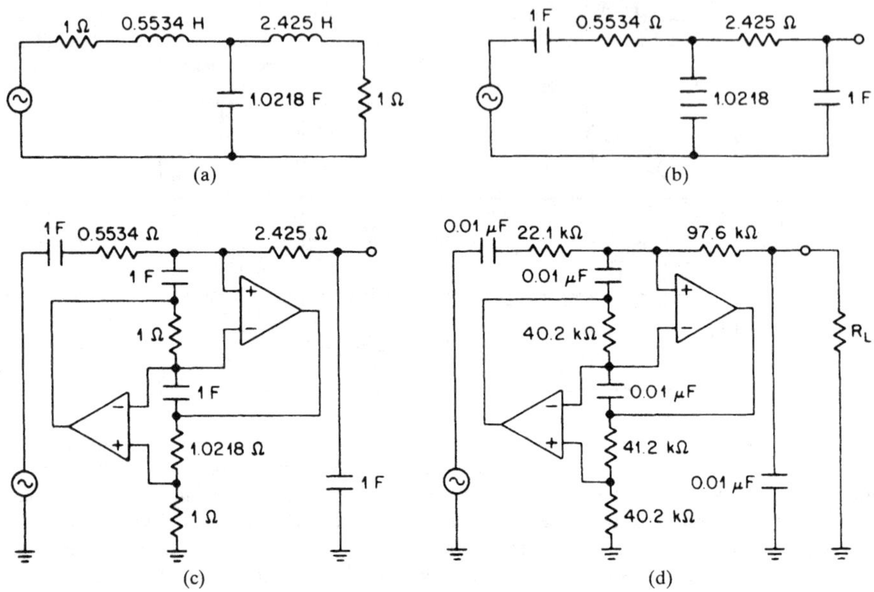

FIGURE 3-34 The network of Example 3-17: (*a*) normalized low-pass prototype; (*b*) normalized circuit after 1/*s* transformation; (*c*) realization of *D* element; and (*d*) final circuit.

C' of 0.01 μF, the required impedance-scaling factor can be found by solving Equation (2-10) for Z as follows:

$$Z = \frac{C}{\text{FSF} \times C'} = \frac{1}{2513 \times 0.01 \times 10^{-6}} = 39{,}800 \tag{2-10}$$

Using an FSF of $2\pi f_c$ or 2513 and an impedance-scaling factor Z of 39.8×10^3, the normalized filter is scaled by dividing all capacitors by $Z \times$ FSF and multiplying all resistors by Z. The final circuit is given in Figure 3-34*d*. The resistor values were modified for standard 1-percent values. The filter loss is 6 dB, corresponding to the loss due to the resistive source and load terminations of the *LC* filter.

The *D* elements are usually realized with dual operational amplifiers which are available as a matched pair in a single package. In order to provide a bias current for the noninverting input of the upper amplifier, a resistive termination to ground must be provided. This resistor will cause a low-frequency roll-off, so true DC-coupled operation will not be possible. However, if the termination is made much larger than the nominal resistor values of the circuit, the low-frequency roll-off can be made to occur well below the frequency range of interest. If a low output impedance is required, the filter can be followed by a voltage follower or an amplifier if gain is also desired. The filter input should be driven by a source impedance much less than the input resistor of the filter. A voltage follower or amplifier could be used for input isolation.

Elliptic-Function Low-Pass Filters Using the GIC. The 1/*s* transformation and GIC realization are particularly suited for the realization of active high-order elliptic-function low-pass filters. These circuits exhibit low sensitivity to component tolerances and amplifier characteristics. They can be made tunable, and are less complex than the state-variable configurations. The following example illustrates the design of an elliptic-function low-pass filter using the GIC as a *D* element.

Example 3-18 Design of an Active Elliptic-Function Low-Pass Filter using *D* Elements

Required:

An active low-pass filter

0.18-dB ripple at 260 Hz

45-dB minimum at 270 Hz

Result:

Note: The passive elliptic low-pass filter will be designed for a 1 *rad/sec cutoff and* 1 Ω *terminations to obtain the initial low-pass filter prototype.*

(*a*) Compute the steepness factor.

$$A_s = \frac{f_s}{f_c} = \frac{270\text{ Hz}}{260\text{ Hz}} = 1.0385 \qquad (2\text{-}11)$$

(*b*) Open ***Filter Solutions***.

Check the ***Stop Band Freq*** box.

Enter **0.18** in the ***Pass Band Ripple (dB)*** box.

Enter **1** in the ***Pass Band Freq*** box.

Enter **1.0385** in the ***Stop Band Freq*** box.

The ***Frequency Scale Rad/Sec*** box should be checked.

Enter **1** for ***Source Res*** and ***Load Res***.

(*c*) Click the ***Set Order*** control button to open the second panel.

Enter **45** for ***Stop band Attenuation (dB)***.

Click the ***Set Minimum Order*** button and then click ***Close***.

9 Order is displayed on the main control panel.

(*d*) Click the ***Circuits*** button.

Two schematics are presented by *Filter Solutions*. Select the one representing the dual (Passive Filter 2), which is shown in Figure 3-35*a*.

(*e*) The 1/*s* impedance transformation modifies the elements in accordance with Table 3-1, resulting in the circuit of Figure 3-35*b*. The *D* elements are realized using the GIC of Figure 3-33, as shown in Figure 3-35*c*.

(*f*) The normalized circuit can now be frequency- and impedance-scaled. Since all normalized capacitors are equal, it would be desirable if they could all be scaled to a standard value such as 0.1 μF. The required impedance-scaling factor can be determined from Equation (2-10) by using an FSF of $2\pi f_c$ or 1634, corresponding to a cutoff of 260 Hz. Therefore,

$$Z = \frac{C}{\text{FSF} \times C'} = \frac{1}{1634 \times 0.1 \times 10^{-6}} = 6120 \qquad (2\text{-}10)$$

Frequency and impedance scaling by dividing all capacitors by $Z \times$ FSF and multiplying all resistors by Z results in the final filter circuit of Figure 3-35*d* having the frequency response of Figure 3-35*e*. The resistor values have been modified so that 1-percent values are used and the transmission zeros can be adjusted. The frequency of each zero was computed by multiplying each zero in rad/sec of Figure 3-35*a* by f_c. A resistive termination is provided so that bias current can be supplied to the amplifiers.

The transmission zeros generated by the circuit of Figure 3-35*d* occur because at specific frequencies the value of each FDNR is equal to the positive resistor in series, therefore creating cancellations or null in the shunt branches. By adjusting each *D* element, these nulls can be tuned to the required frequencies. These adjustments are usually sufficient to obtain satisfactory results. State-variable filters permit the adjustment of the poles and zeros directly for greater accuracy. However, the realization is more complex—for instance,

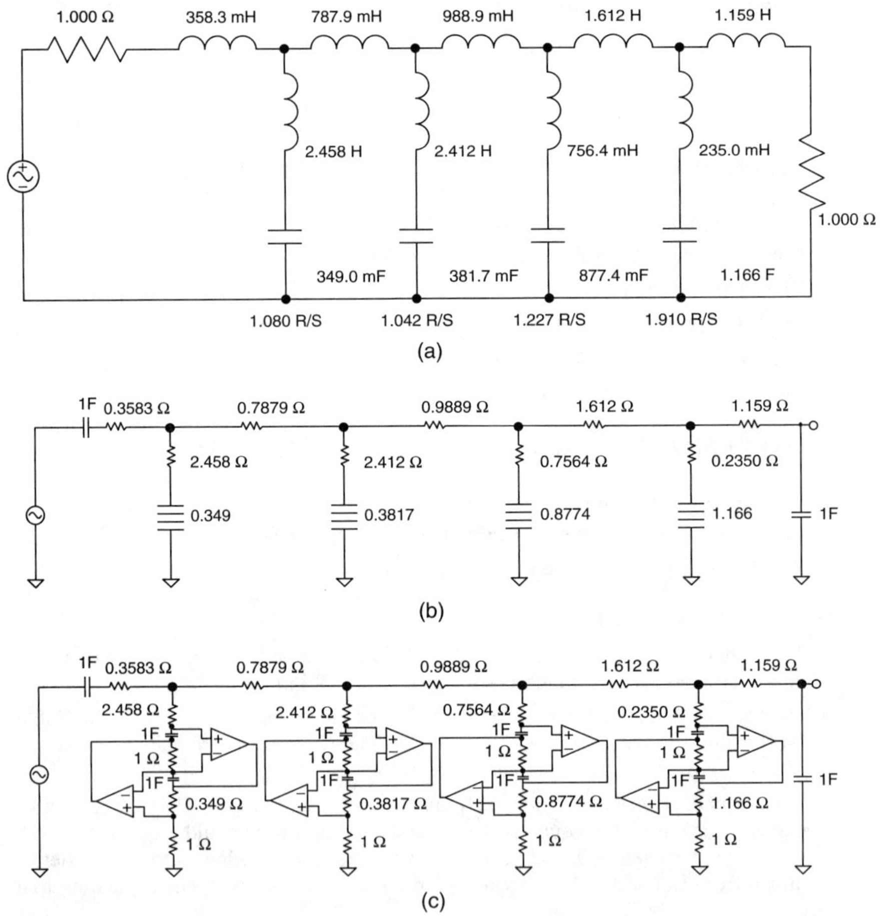

FIGURE 3-35 The filter of Example 3-18: (*a*) normalized low-pass filter; (*b*) circuit after 1/*s* transformation; (*c*) normalized configuration using GICs for *D* elements; (*d*) denormalized filer; and (*e*) frequency response.

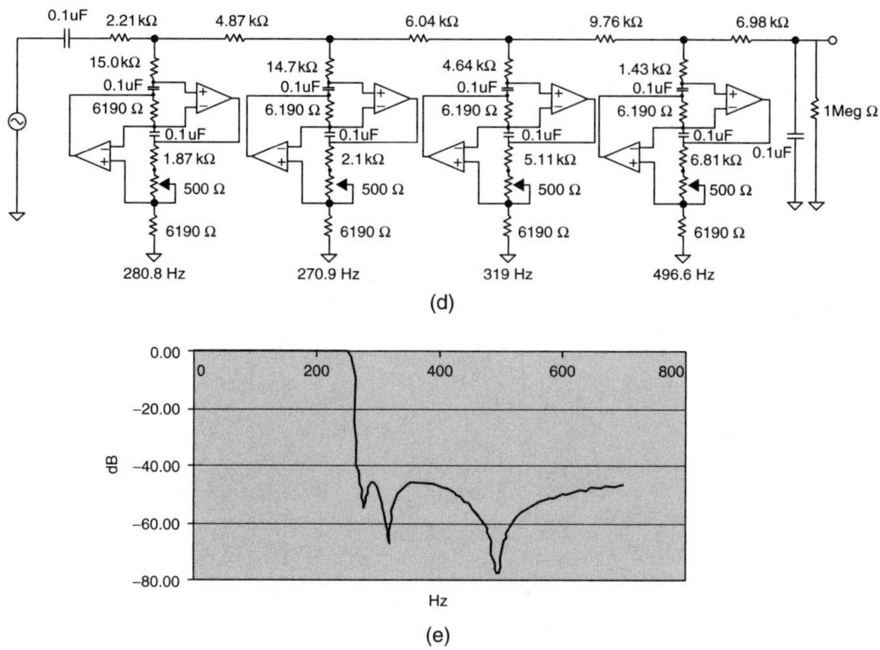

FIGURE 3-35 *(Continued)*

the filter of Example 3-18 would require twice as many amplifiers, resistors, and potentiometers if the state-variable approach were used.

BIBLIOGRAPHY

Amstutz, P. "Elliptic Approximation and Elliptic Filter Design on Small Computers." *IEEE Transactions on Circuits and Systems* CAS-25, No.12 (December, 1978).

Bruton, L. T. "Active Filter Design Using Generalized Impedance Converters." *EDN* (February, 1973).

———. "Network Transfer Functions Using the Concept of Frequency-Dependent Negative Resistance." *IEEE Transactions on Circuit Theory* CT-16 (August, 1969): 406–408.

Christian, E., and E, Eisenmann. *Filter Design Tables and Graphs*. New York: John Wiley Sons, 1966.

Geffe, P. *Simplified Modern Filter Design*. New York: John F. Rider, 1963.

Huelsman, L. P. *Theory and Design of Active RC Circuits*. McGraw-Hill, 1968.

Saal, R., and E. Ulbrich. "On the Design of Filters by Synthesis." *IRE Transactions on Circuit Theory* (December, 1958).

Shepard, B. R. "Active Filters Part 12." *Electronics* (August 18, 1969): 82–91.

Thomas, L. C. "The Biquad: Part I—Some Practical Design Considerations." *IEEE Transactions on Circuit Theory* CT-18 (May, 1971): 350–357.

Tow, J. "A Step-by-Step Active Filter Design." *IEEE Spectrum* vol. 6 (December, 1969) 64–68.

Williams, A. B. "Design Active Elliptic Filters Easily from Tables." *Electronic Design* 19, no. 21 (October 14, 1971): 76–79.

———. *Active Filter Design*. Dedham, Massachusetts: Artech House, 1975.

Zverev, A. I. *Handbook of Filter Synthesis*. New York: John Wiley and Sons, 1967.

CHAPTER 4
HIGH-PASS FILTER DESIGN

4.1 LC *HIGH-PASS FILTERS*

LC high-pass filters can be directly designed by mapping the values of a normalized *LC* low-pass filter into a high-pass filter. This allows use of existing tables of normalized low-pass values to create high-pass filters.

The Low-Pass to High-Pass Transformation

If $1/s$ is substituted for s in a normalized low-pass transfer function, a high-pass response is obtained. The low-pass attenuation values will now occur at high-pass frequencies, which are the reciprocal of the corresponding low-pass frequencies. This was demonstrated in Section 2.1.

A normalized *LC* low-pass filter can be transformed into the corresponding high-pass filter by simply replacing each coil with a capacitor and vice versa, using reciprocal element values. This can be expressed as

$$C_{\text{hp}} = \frac{1}{L_{\text{Lp}}} \tag{4-1}$$

and

$$L_{\text{hp}} = \frac{1}{C_{\text{Lp}}} \tag{4-2}$$

The source and load resistive terminations are unaffected.

The transmission zeros of a normalized elliptic-function low-pass filter are also reciprocated when the high-pass transformation occurs. Therefore,

$$\omega_{\infty}(\text{hp}) = \frac{1}{\omega_{\infty}(\text{Lp})} \tag{4-3}$$

To minimize the number of inductors in the high-pass filter, the dual low-pass circuit defined by the lower schematic in the tables of Chapter 11 is usually chosen to be transformed except for even-order all-pole filters, where either circuit may be used. For elliptic function high-pass filters, the *Filter Solutions* program is used to obtain a low-pass filter prototype normalized to 1 radian/sec and 1-Ω. Thus, the circuit representing the dual (Passive Filter 2) is used.

The objective is to start with a low-pass prototype containing more inductors than capacitors since after the low-pass to high-pass transformation the result will contain more capacitors than inductors.

After the low-pass to high-pass transformation, the normalized high-pass filter is frequency- and impedance-scaled to the required cutoff frequency. The following two examples demonstrate the design of high-pass filters.

Example 4-1 Design of an All-Pole *LC* High-Pass Filter from a Normalized Low-Pass Filter

Required:

An LC high-pass filter
3 dB at 1 MHz
28-dB minimum at 500 kHz
$R_s = R_L = 300\ \Omega$

Result:

(*a*) To normalize the requirement, compute the high-pass steepness factor A_s.

$$A_s = \frac{f_c}{f_s} = \frac{1 \text{ MHz}}{500 \text{ kHz}} = 2 \tag{2-13}$$

(*b*) Select a normalized low-pass filter that offers over 28 dB of attenuation at 2 rad/s.

Inspection of the curves of Chapter 2 indicates that a normalized $n = 5$ Butterworth low-pass filter provides the required attenuation. Table 11-2 contains element values for the corresponding network. The normalized low-pass filter for $n = 5$ and equal terminations is shown in Figure 4-1*a*. The dual circuit as defined by the lower schematic of Table 11-2 was chosen.

(*c*) To transform the normalized low-pass circuit to a high-pass configuration, replace each coil with a capacitor and vice versa, using reciprocal element values, as shown in Figure 4-1*b*.

(*d*) Denormalize the high-pass filter using a *Z* of 300 and a frequency-scaling factor (FSF) of $2\pi f_c$ of 6.28×10^6.

$$C_1' = \frac{C}{\text{FSF} \times Z} = \frac{\frac{1}{0.618}}{6.28 \times 10^6 \times 300} = 858 \text{ pF} \tag{2-10}$$

$$C_3' = 265 \text{ pF}$$

$$C_5' = 858 \text{ pF}$$

$$L_2' = \frac{L \times Z}{\text{FSF}} = \frac{\frac{1}{1.618} \times 300}{6.28 \times 10^6} = 29.5\ \mu\text{H} \tag{2-9}$$

$$L_4' = 29.5\ \mu\text{H}$$

The final filter is given in Figure 4-1*c*, having the frequency response shown in Figure 4-1*d*.

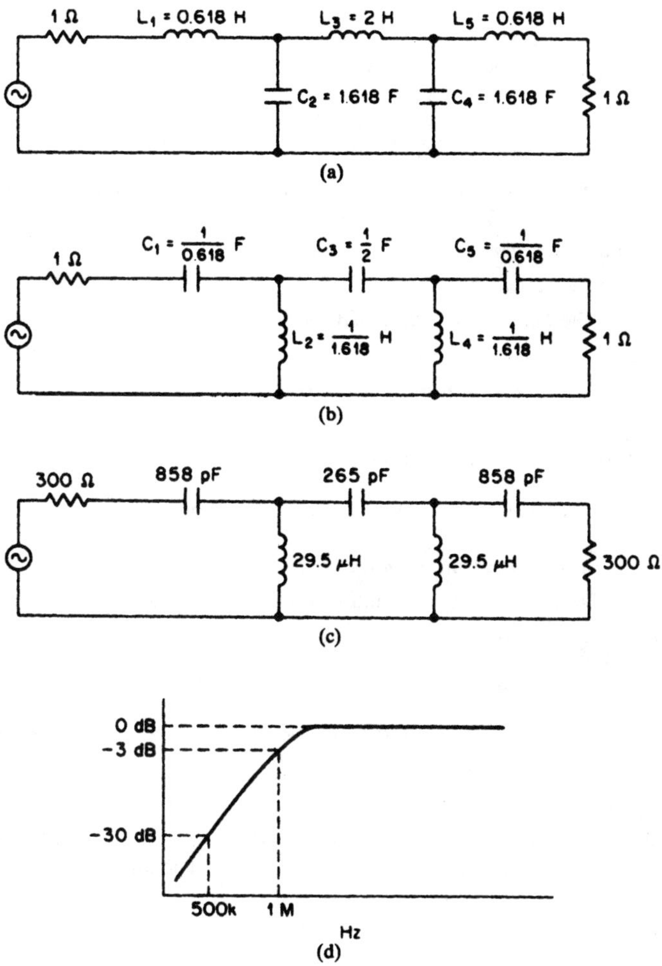

FIGURE 4-1 The high-pass filter of Example 4-1: (*a*) normalized low-pass filter; (*b*) high-pass transformation; (*c*) frequency- and impedance-scaled filter; and (*d*) frequency response.

Example 4-2 Design of an Elliptic-Function *LC* High-Pass Filter using *Filter Solutions*

Required:

An LC high-pass filter

2-dB maximum at 3220 Hz

52-dB minimum at 3020 Hz

$R_s = R_L = 300\ \Omega$

Result:

(*a*) Compute the high-pass steepness factor A_s.

$$A_s = \frac{f_c}{f_s} = \frac{3220\text{ Hz}}{3020\text{ Hz}} = 1.0662 \tag{2-13}$$

(*b*) Since the filter requirement is very steep, an elliptic-function will be selected.

Open ***Filter Solutions***.
Check the ***Stop Band Freq*** box.
Enter **.2** in the ***Pass Band Ripple (dB)*** box.
Enter **1** in the ***Pass Band Freq*** box.
Enter **1.0662** in the ***Stop Band Freq*** box.
The ***Frequency Scale Rad/Sec*** box should be checked.
Enter **1** for the ***Source Res*** and ***Load Res***.

(*c*) Click the ***Set Order*** control button to open the second panel.

Enter **52** for ***Stop band Attenuation (dB)***.
Click the ***Set Minimum Order*** button and then click ***Close***.
9 Order is displayed on the main control panel

(*d*) Click the ***Circuits*** button.

Two schematics are presented by ***Filter Solutions***. Select the one representing the dual (Passive Filter 2), which is shown in Figure 4-2*a*.

(*e*) To transform the normalized low-pass circuit into a high-pass configuration, convert inductors into capacitors and vice versa, using reciprocal values. The transformed high-pass filter is illustrated in Figure 4-2*b*. The transmission zeros are also transformed by conversion to reciprocal values.

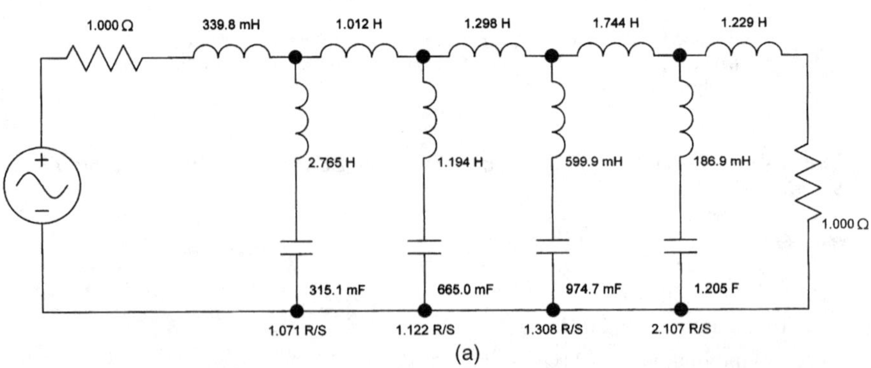

FIGURE 4-2 The high-pass filter of Example 4-2: (*a*) normalized low-pass filter from *Filter Solutions*; (*b*) transformed high-pass filter; (*c*) frequency- and impedance-scaled high-pass filter; and (*d*) frequency response.

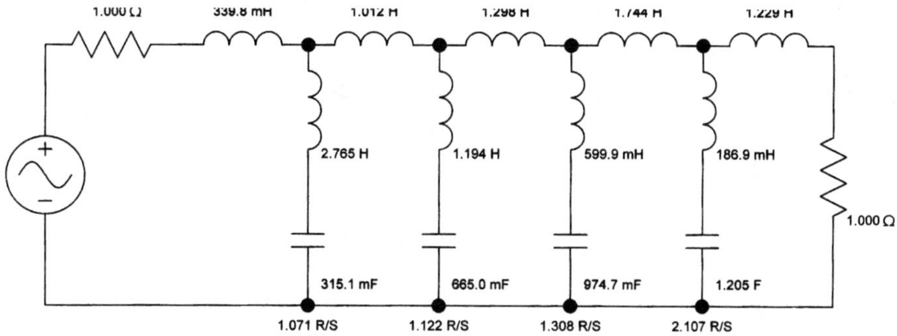

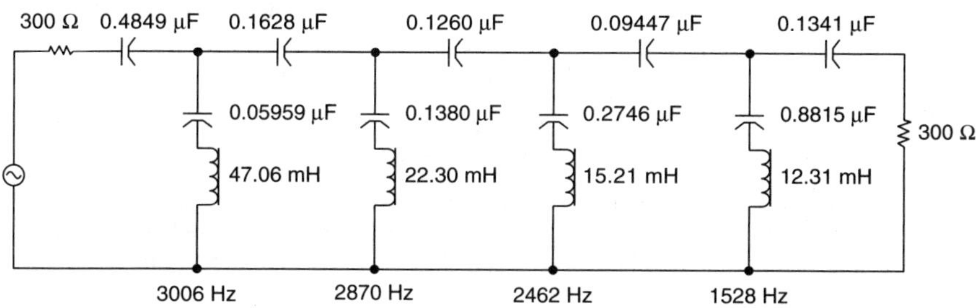

(c)

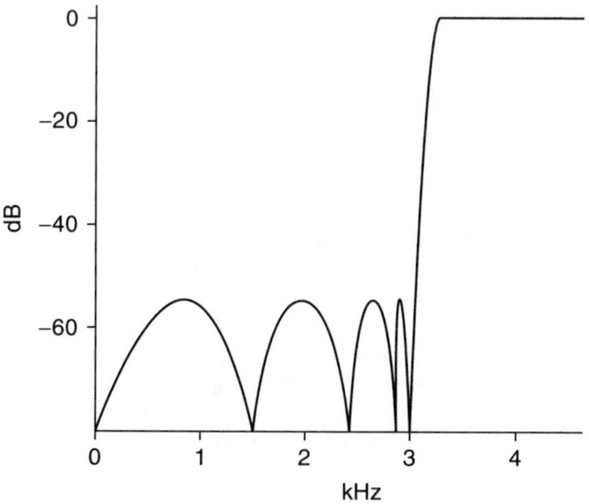

(d)

FIGURE 4-2 (*Continued*)

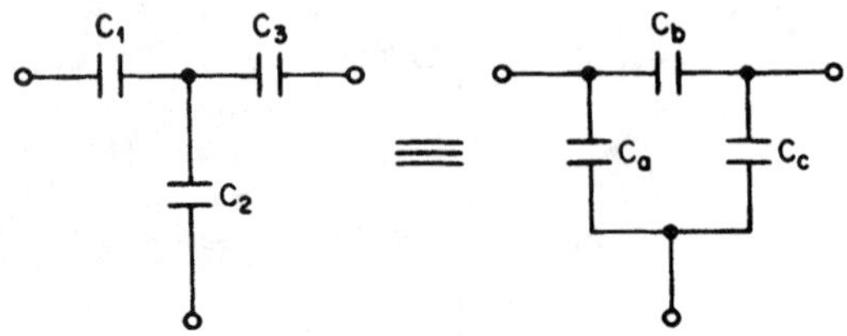

FIGURE 4-3 The T-to-pi capacitance transformation.

(*f*) Denormalize the high-pass filter using a Z of 300 and a frequency-scaling factor of $2\pi f_c$ or 20,232. The denormalized elements are computed by

$$L' = \frac{L \times Z}{\text{FSF}} \tag{2-9}$$

and

$$C' = \frac{C}{\text{FSF} \times Z} \tag{2-10}$$

The resulting denormalized high-pass filter is illustrated in Figure 4-2*c*. The frequency of each zero was computed by multiplying each zero in rad/sec of Figure 4-2*b* by the design cutoff frequency of $f_c = 3220$ Hz. The frequency response is given in Figure 4-2*d*.

The T-to-Pi Capacitance Conversion. When the elliptic-function high-pass filters are designed for audio frequencies and at low impedance levels, the capacitor values tend to be large. The T-to-pi capacitance conversion will usually restore practical capacitor values.

The two circuits of Figure 4-3 have identical terminal behavior and are therefore equivalent if

$$C_a = \frac{C_1C_2}{\Sigma C} \tag{4-4}$$

$$C_b = \frac{C_1C_3}{\Sigma C} \tag{4-5}$$

$$C_c = \frac{C_2C_3}{\Sigma C} \tag{4-6}$$

where $\Sigma C = C_1 + C_2 + C_3$. The following example demonstrates the effectiveness of this transformation in reducing large capacitances.

Example 4-3 The T- to PI-Capacitance Transformation to Reduce Capacitor Values

Required:

The high-pass filter of Figure 4-2*c* contains a 0.8815-μF capacitor in the last shunt branch. Use the T-to-pi transformation to provide some relief.

Result:

The circuit of Figure 4-2*c* is repeated in Figure 4-4*a* showing a "T" of capacitors including the undesirable 0.8815 μF capacitor. The T-to-pi transformation results in

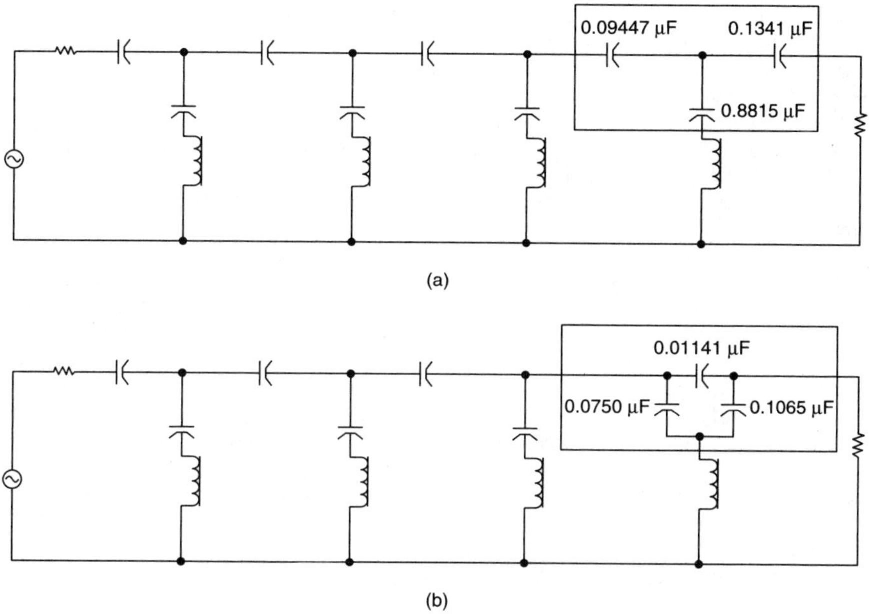

FIGURE 4-4 The T-to-pi transformation of Example 4-3: (*a*) the high-pass filter of Example 4-2; and (*b*) the modified configuration.

$$C_a = \frac{C_1 C_2}{\Sigma C} = 0.0750\ \mu\text{F} \tag{4-4}$$

$$C_b = \frac{C_1 C_3}{\Sigma C} = 0.01141\ \mu\text{F} \tag{4-5}$$

and

$$C_c = \frac{C_2 C_3}{\Sigma C} = 0.1065\ \mu\text{F} \tag{4-6}$$

where $C_1 = 0.09447\ \mu\text{F}$, $C_2 = 0.8815\ \mu\text{F}$, and $C_3 = 0.1341\ \mu\text{F}$. The transformed circuit is given in Figure 4-4*b*, where the maximum capacitor value has undergone more than an 8:1 reduction.

4.2 ACTIVE HIGH-PASS FILTERS

The Low-Pass to High-Pass Transformation

Active high-pass filters can be derived directly from the normalized low-pass configurations by a suitable transformation in a similar manner to *LC* high-pass filters. To make the conversion, replace each resistor by a capacitor having the reciprocal value and vice versa, as follows:

$$C_{\text{hp}} = \frac{1}{R_{\text{Lp}}} \tag{4-7}$$

$$R_{\text{hp}} = \frac{1}{C_{\text{Lp}}} \tag{4-8}$$

It is important to recognize that only the resistors that are a part of the low-pass *RC* networks are transformed into capacitors by Equation (4-7). Feedback resistors that strictly determine operational amplifier gain, such as R_6 and R_7 in Figure 3-20*a*, are omitted from the transformation.

After the normalized low-pass configuration is transformed into a high-pass filter, the circuit is frequency- and impedance-scaled in the same manner as in the design of low-pass filters. The capacitors are divided by Z × FSF and the resistors are multiplied by Z. A different Z can be used for each section, but the FSF must be uniform throughout the filter.

All-Pole High-Pass Filters. Active two-pole and three-pole low-pass filter sections were shown in Figure 3-13 and correspond to the normalized active low-pass values obtained from relevant tables in Chapter 11. These circuits can be directly transformed into high-pass filters by replacing the resistors with capacitors and vice versa using reciprocal element values and then frequency- and impedance-scaling the filter network. The filter gain is unity at frequencies well into the passband corresponding to unity gain at DC for low-pass filters. The source impedance of the driving source should be much less than the reactance of the capacitors of the first filter section at the highest passband frequency of interest. The following example demonstrates the design of an all-pole high-pass filter.

Example 4-4 Design of an Active All-Pole High-Pass Filter

Required:

An active high-pass filter
3 dB at 100 Hz
75-dB minimum at 25 Hz

Result:

(*a*) Compute the high-pass steepness factor.

$$A_s = \frac{f_c}{f_s} = \frac{100}{25} = 4 \qquad (2\text{-}13)$$

(*b*) A normalized low-pass filter must first be selected that makes the transition from 3 to 75 dB within a frequency ratio of 4:1. The curves of Figure 2-44 indicate that a fifth-order 0.5-dB Chebyshev filter is satisfactory. The corresponding active filter consists of a three-pole section and a two-pole section whose values are obtained from Table 11-39 and are shown in Figure 4-5*a*.

(*c*) To transform the normalized low-pass filter into a high-pass filter, replace each resistor with a capacitor, and vice versa, using reciprocal element values. The normalized high-pass filter is given in Figure 4-5*b*.

(*d*) Since all normalized capacitors are equal, the impedance-scaling factor Z will be computed so that all capacitors become 0.015 μF after denormalization. Since the cutoff frequency is 100 Hz, the FSF is $2\pi f_c$ or 628, so that

$$Z = \frac{C}{\text{FSF} \times C'} = \frac{1}{628 \times 0.015 \times 10^{-6}} = 106.1 \times 10^3 \qquad (2\text{-}10)$$

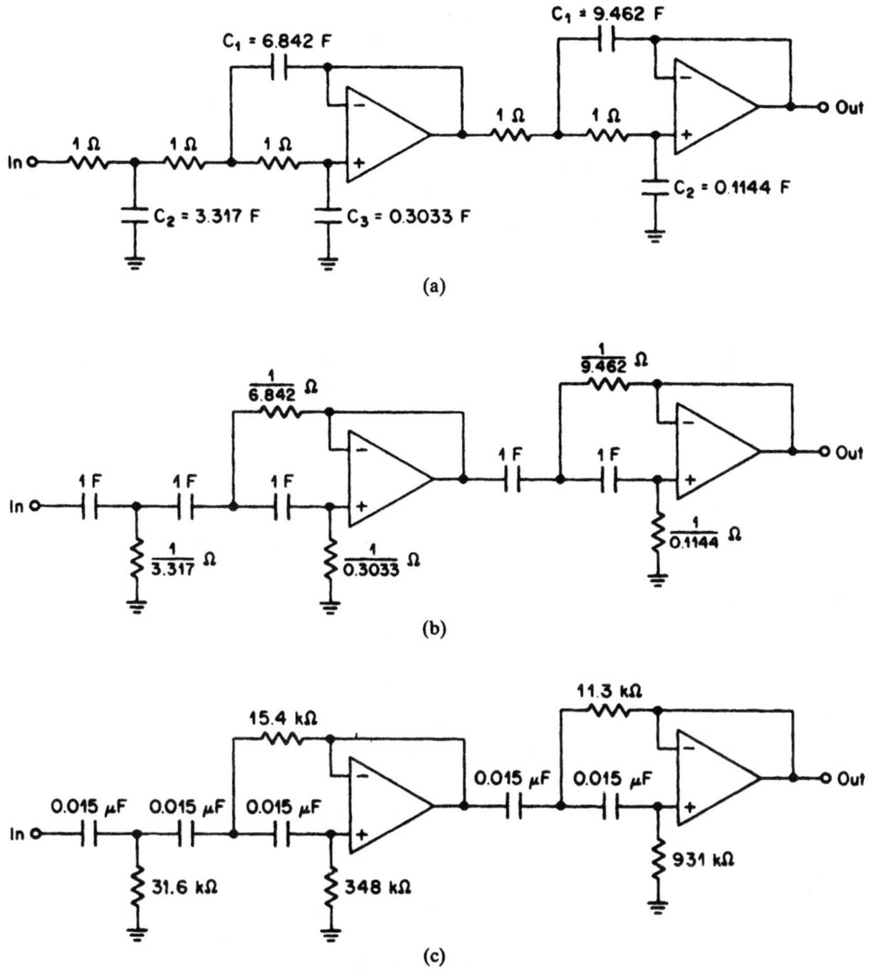

FIGURE 4-5 The all-pole high-pass filter of Example 4-4: (*a*) normalized low-pass filter; (*b*) high-pass transformation; and (*c*) frequency- and impedance-scaled high-pass filter.

If we frequency- and impedance-scale the normalized high-pass filter by dividing all capacitors by $Z \times \text{FSF}$ and multiplying all resistors by Z, the circuit of Figure 4-5*c* is obtained. The resistors were rounded off to standard 1-percent values.

Elliptic-Function High-Pass Filters. High-pass elliptic-function filters can be designed using the elliptic-function VCVS configuration discussed in Section 3.2. This structure can introduce transmission zeros either above or below the pole locations and is therefore suitable for elliptic-function high-pass requirements.

The normalized low-pass poles and zeros must first be transformed to the high-pass form. Each complex low-pass pole pair consisting of a real part α and imaginary part β is transformed into a normalized high-pass pole pair as follows:

$$\alpha_{hp} = \frac{\alpha}{\alpha^2 + \beta^2} = \frac{1}{2Q\omega_0} \tag{4-9}$$

$$\beta_{hp} = \frac{\beta}{\alpha^2 + \beta^2} = \frac{1}{\omega_o}\sqrt{1 - \frac{1}{4Q^2}} \tag{4-10}$$

The second forms of Equations 4-9 and 4-10 involving Q and ω_0 are used when these parameters are provided by the *Filter Solutions* program.

The transformed high-pass pole pair can be denormalized by

$$\alpha'_{hp} = \alpha_{hp} \times \text{FSF} \tag{4-11}$$

$$\beta'_{hp} = \beta_{hp} \times \text{FSF} \tag{4-12}$$

where FSF is the frequency scaling factor $2\pi f_c$. If the pole is real, the normalized pole is transformed by

$$\alpha_{0,hp} = \frac{1}{\alpha_0} \tag{4-13}$$

The denormalized real pole is obtained from

$$\alpha'_{0,hp} = \alpha_{0,hp} \times \text{FSF} \tag{4-14}$$

To transform zeros, we first compute

$$\omega_\infty(\text{hp}) = \frac{1}{\omega_\infty(\text{Lp})} \tag{4-3}$$

Denormalization occurs by

$$\omega'_\infty(\text{hp}) = \omega_\infty(\text{hp}) \times \text{FSF} \tag{4-15}$$

The elliptic-function VCVS circuit of Section 3.2 is repeated in Figure 4-6. The elements are then computed as follows:

First, calculate

$$a = \frac{2\alpha'_{hp}}{\sqrt{(\alpha'_{hp})^2 + (\beta'_{hp})^2}} \tag{4-16}$$

$$b = \frac{[\omega'_\infty(\text{hp})]^2}{(\alpha'_{hp})^2 + (\beta'_{hp})^2} \tag{4-17}$$

$$c = \sqrt{(\alpha'_{hp})^2 + (\beta'_{hp})^2} \tag{4-18}$$

where α'_{hp}, β'_{hp}, and $\omega'_\infty(\text{hp})$ are the denormalized high-pass pole-zero coordinates.

Select C

$$C_1 = C \tag{4-19}$$

Then

$$C_3 = C_4 = \frac{C_1}{2} \tag{4-20}$$

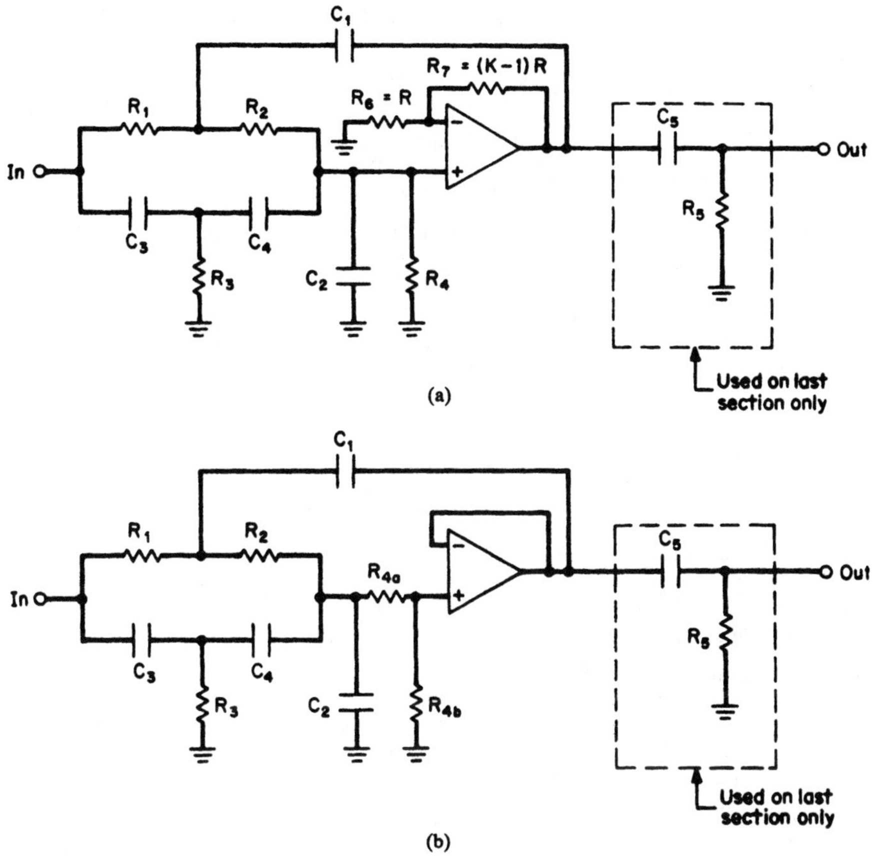

FIGURE 4-6 VCVS elliptic-function high-pass section: (*a*) circuit for $K > 1$; and (*b*) circuit for $K < 1$.

Let
$$C_2 \geq \frac{C_1(b-1)}{4} \tag{4-21}$$

$$R_3 = \frac{1}{cC_1\sqrt{b}} \tag{4-22}$$

$$R_1 = R_2 = 2R_3 \tag{4-23}$$

$$R_4 = \frac{4\sqrt{b}}{cC_1(1-b) + 4cC_2} \tag{4-24}$$

$$K = 2 + \frac{2C_2}{C_1} - \frac{a}{2\sqrt{b}} + \frac{2}{C_1\sqrt{b}}\left(\frac{1}{cR_4} - aC_2\right) \tag{4-25}$$

$$R_6 = R \tag{4-26}$$

$$R_7 = (K-1)R \tag{4-27}$$

where R can be arbitrarily chosen. If K is less than 1, the circuit of Figure 4-6b is used. Then

$$R_{4a} = (1 - K)R_4 \tag{4-28}$$

$$R_{4b} = KR_4 \tag{4-29}$$

$$\text{Section gain} = \frac{bKC_1}{4C_2 + C_1} \tag{4-30}$$

R_5 is determined from the denormalized real pole by

$$R_5 = \frac{1}{C_5\alpha'_{0,\text{hp}}} \tag{4-31}$$

where C_5 can be arbitrarily chosen and $\alpha'_{0,\text{hp}}$ is $\alpha'_{0,\text{hp}} \times$ FSF. The design procedure is illustrated in Example 4-5.

Example 4-5 Design of an Active Elliptic-Function High-Pass Filter using the VCVS Configuration

Required:

An active high-pass filter
0.2-dB maximum at 3000 Hz
35-dB minimum rejection at 1026 Hz

Result:

(a) Compute the high-pass steepness factor.

$$A_s = \frac{f_c}{f_s} = \frac{3000}{1026} = 2.924 \tag{2-13}$$

(b) Open ***Filter Solutions***.

Check the ***Stop Band Freq*** box.
Enter **.177** in the ***Pass Band Ripple (dB)*** box.
Enter **1** in the ***Pass Band Freq*** box.
Enter **2.924** in the ***Stop Band Freq*** box.
Check the ***Frequency Scale Radians*** *box.*

(c) Click the ***Set Order*** control button to open the second panel.

Enter **35** for ***Stop band Attenuation (dB)***.
Click the ***Set Minimum Order*** button and then click ***Close***.
3 Order is displayed on the main control panel.

(d) Click the ***Transfer Function*** button.

Check the ***Casc*** box.

The following is displayed:

Continuous Transfer Function

Wn = 3.351

$$\frac{.1143 \quad (S^2 + 11.23)}{(S^2 + .7727{*}S + 1.447) \quad (S + .8871)}$$

Wo = 1.203
Q = 1.557

3rd Order Low Pass Elliptic

Pass Band Frequency = 1.000 Rad/Sec
Pass Band Ripple = 177.0 mdB
Stop Band Ratio = 2.924
Stop Band Frequency = 2.924 Rad/Sec
Stop Band Attenuation = 37.43 dB

(*e*) The design parameters are summarized as follows:

$$\text{Section } Q = 1.557$$
$$\text{Section } \omega_0 = 1.203$$
$$\text{Section } \omega_\infty = 3.351$$
$$\alpha_0 = 0.8871 \text{ (from the denominator)}$$

(*f*) To compute the element values, first transform the low-pass poles and zeros to the high-pass form.

Complex pole:

$$\alpha_{\text{hp}} = \frac{1}{2Q\omega_0} = 0.2670 \tag{4-9}$$

$$\beta_{\text{hp}} = \frac{1}{\omega_0}\sqrt{1 - \frac{1}{4Q^2}} = 0.7872 \tag{4-10}$$

The pole-pair is denormalized by

$$\alpha'_{hp} = 5033 \qquad (4\text{-}11)$$

$$\beta'_{hp} = 14{,}838 \qquad (4\text{-}12)$$

where FSF $= 2\pi \times 3000$.

Real pole:

$$\alpha_0 = 0.8871$$

$$\alpha_{0,hp} = 1.127 \qquad (4\text{-}13)$$

$$\alpha'_{0,hp} = 21{,}248 \qquad (4\text{-}14)$$

Zero:

$$\omega_\infty = 3.351$$

$$\omega_\infty(\text{hp}) = 0.2984 \qquad (4\text{-}3)$$

$$\omega'_\infty(\text{hp}) = 5625 \qquad (4\text{-}15)$$

(*g*) The results of (*c*) are summarized as follows:

$$\alpha'_{hp} = 5033$$

$$\beta'_{hp} = 14{,}838$$

$$\alpha'_{0,hp} = 21{,}248$$

$$\omega'_\infty(\text{hp}) = 5625$$

The element values can now be computed:

$$a = 0.6424 \qquad (4\text{-}16)$$

$$b = 0.1289 \qquad (4\text{-}17)$$

$$c = 15{,}668 \qquad (4\text{-}18)$$

Let $C = 0.02\ \mu\text{F}$

then $C_1 = 0.02\ \mu\text{F}$ (4-19)

$$C_3 = C_4 = 0.01\ \mu\text{F} \qquad (4\text{-}20)$$

$$C_2 \geq -0.00436\ \mu\text{F} \qquad (4\text{-}21)$$

Let $C_2 = 0$

$$R_3 = 8.89\ \text{k}\Omega \qquad (4\text{-}22)$$

$$R_1 = R_2 = 17.8\ \text{k}\Omega \qquad (4\text{-}23)$$

$$R_4 = 5.26\ \text{k}\Omega \qquad (4\text{-}24)$$

$$K = 4.48 \qquad (4\text{-}25)$$

Let $R_6 = R = 10\ \text{k}\Omega$ (4-26)

$$R_7 = 34.8\ \text{k}\Omega \qquad (4\text{-}27)$$

Let $C_5 = 0.01\ \mu\text{F}$

$$R_5 = 4.70\ \text{k}\Omega \qquad (4\text{-}31)$$

The resulting circuit is illustrated in Figure 4-7.

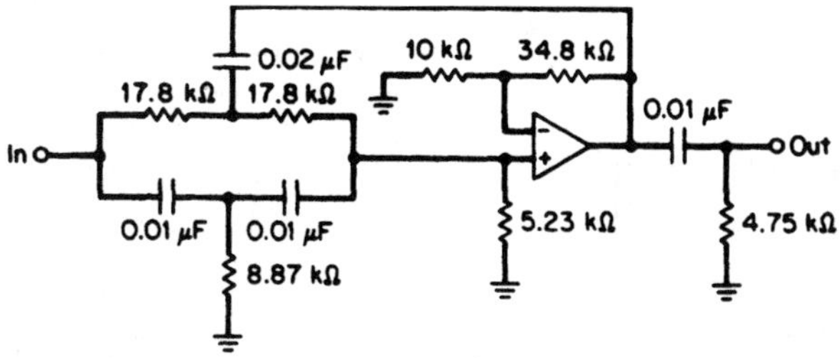

FIGURE 4-7 The circuit of Example 4-5.

State-Variable High-Pass Filters. The all-pole and elliptic-function active high-pass filters previously discussed cannot be easily adjusted. If the required degree of accuracy results in unreasonable component tolerances, the state-variable or biquad approach will permit independent adjustment of the filter's pole and zero coordinates. Another feature of this circuit is the reduced sensitivity of the response to many of the amplifier limitations such as finite bandwidth and gain.

All-Pole Configuration. In order to design a state-variable all-pole high-pass filter, the normalized low-pass poles must first undergo a low-pass to high-pass transformation. Each low-pass pole pair consisting of a real part α and imaginary part β is transformed into a normalized high-pass pole pair as follows:

$$\alpha_{hp} = \frac{\alpha}{\alpha^2 + \beta^2} \tag{4-9}$$

$$\beta_{hp} = \frac{\beta}{\alpha^2 + \beta^2} \tag{4-10}$$

The transformed high-pass pole pair can now be denormalized by

$$\alpha'_{hp} = \alpha_{hp} \times \text{FSF} \tag{4-11}$$

$$\beta'_{hp} = \beta_{hp} \times \text{FSF} \tag{4-12}$$

where FSF is the frequency-scaling factor $2\pi f_x$.

The circuit of Figure 4-8 realizes a high-pass second-order biquadratic transfer function. The element values for the all-pole case can be computed in terms of the high-pass pole coordinates as follows:

First, compute

$$\omega'_0 = \sqrt{(\alpha'_{hp})^2 + (\beta'_{hp})^2} \tag{4-32}$$

The component values are

$$R_1 = R_4 = \frac{1}{2\alpha'_{hp}C} \tag{4-33}$$

$$R_2 = R_3 = \frac{1}{\omega'_0 C} \tag{4-34}$$

$$R_5 = \frac{2\alpha'_{hp}}{\omega'_0} R \tag{4-35}$$

$$R_6 = AR \tag{4-36}$$

where C and R are arbitrary and A is the section gain.

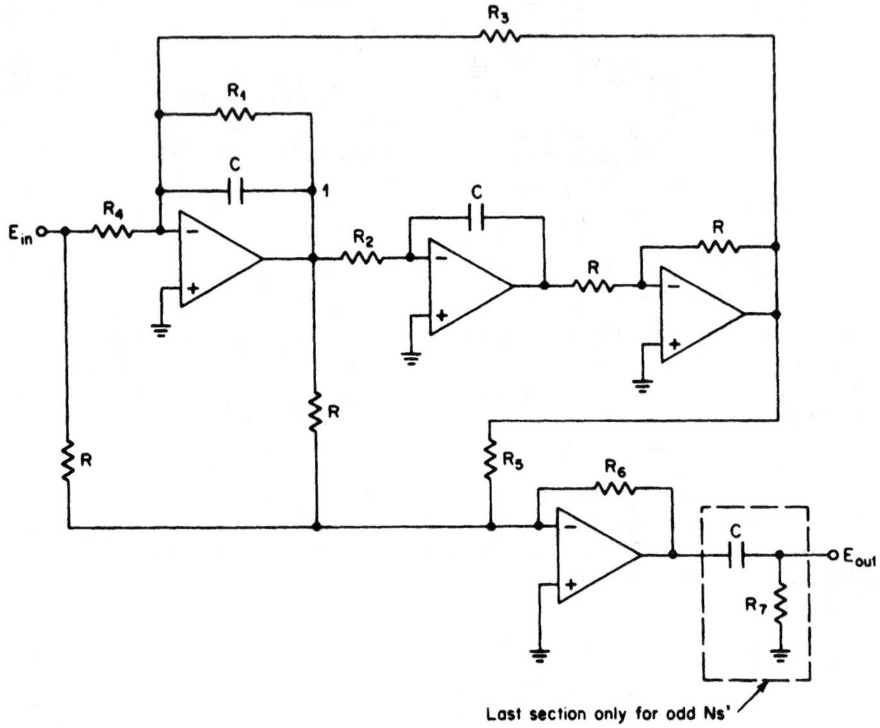

FIGURE 4-8 The biquadratic high-pass configuration.

If the transfer function is of an odd order, a real pole must be realized. To transform the normalized low-pass real pole α_0, compute

$$\alpha_{0,\text{hp}} = \frac{1}{\alpha_0} \tag{4-13}$$

Then denormalize the high-pass real pole by

$$\alpha'_{0,\text{hp}} = \alpha_{0,\text{hp}} \times \text{FSF} \tag{4-14}$$

The last section of the filter is followed by an *RC* network, as shown in Figure 4-8. The value of R_7 is computed by

$$R_7 = \frac{1}{\alpha'_{0,\text{hp}} C} \tag{4-37}$$

where *C* is arbitrary.

A bandpass output is provided at node 1 for tuning purposes. The bandpass resonant frequency is given by

$$f_0 = \frac{\omega'_0}{2\pi} \tag{4-38}$$

R_3 can be made adjustable and the circuit tuned to resonance by monitoring the phase shift between E_{in} and node 1 and adjusting R_3 for 180° of phase shift at f_0 using a Lissajous pattern.

The bandpass Q can then be monitored at node 1 and is given by

$$Q = \frac{\pi f_0}{\alpha'_{hp}} \tag{4-39}$$

R_1 controls the Q and can be adjusted until either the computed Q is obtained or, more conveniently, the gain is unity between E_{in} and node 1 with f_0 applied.

The following example demonstrates the design of an all-pole high-pass filter using the biquad configuration.

Example 4-6 Design of an Active All-Pole High-Pass Filter Using a State-Variable Approach

Required:

An active high-pass filter

3 ± 0.1 dB at 300 Hz

30-dB minimum at 120 Hz

A gain of 2

Result:

(*a*) Compute the high-pass steepness factor.

$$A_s = \frac{f_c}{f_s} = \frac{300}{120} = 2.5 \tag{2-13}$$

(*b*) The curves of Figure 2-45 indicate that a normalized third-order 1-dB Chebyshev low-pass filter has over 30 dB of attenuation at 2.5 rad/s. Since an accuracy of 0.1 dB is required at the cutoff frequency, a state-variable approach will be used so that an adjustment capability is provided.

(*c*) The low-pass pole locations are found in Table 11-26 and are as follows:

Complex pole	$\alpha = 0.2257$	$\beta = 0.8822$
Real pole	$\alpha_0 = 0.4513$	

Complex pole-pair realization:

The complex low-pass pole pair is transformed to a high-pass pole pair as follows:

$$\alpha_{hp} = \frac{\alpha}{\alpha^2 + \beta^2} = \frac{0.2257}{0.2257^2 + 0.8822^2} = 0.2722 \tag{4-9}$$

and

$$\beta_{hp} = \frac{\beta}{\alpha^2 + \beta^2} = \frac{0.8822}{0.2257^2 + 0.8822^2} = 1.0639 \tag{4-10}$$

The transformed pole pair is then denormalized by

$$\alpha'_{hp} = \alpha_{hp} \times \text{FSF} = 513 \tag{4-11}$$

$$\beta'_{hp} = \beta_{hp} \times \text{FSF} = 2005 \tag{4-12}$$

where FSF is $2\pi f_c$ or 1885 since $f_c = 300$ Hz. If we choose $R = 10$ kΩ and $C = 0.01$ μF, the component values are calculated by

$$\omega_0' = \sqrt{(\alpha_{hp}')^2 + (\beta_{hp}')^2} = 2070 \tag{4-32}$$

then
$$R_1 = R_4 = \frac{1}{2\alpha_{hp}' C} = 97.47 \text{ k}\Omega \tag{4-33}$$

$$R_2 = R_3 = \frac{1}{\omega_0' C} = 48.31 \text{ k}\Omega \tag{4-34}$$

$$R_5 = \frac{2\alpha_{hp}'}{\omega_0'} R = 4957 \ \Omega \tag{4-35}$$

$$R_6 = AR = 20 \text{ k}\Omega \tag{4-36}$$

where
$$A = 2$$

The bandpass resonant frequency and Q are

$$f_0 = \frac{\omega_0'}{2\pi} = \frac{2070}{2\pi} = 329 \text{ Hz} \tag{4-38}$$

$$Q = \frac{\pi f_0}{\alpha_{hp}'} = \frac{\pi 329}{513} = 2.015 \tag{4-39}$$

Real-pole realization:
Transform the real pole:

$$\alpha_{0,hp} = \frac{1}{\alpha_0} = \frac{1}{0.4513} = 2.216 \tag{4-13}$$

To denormalize the transformed pole, compute

$$\alpha_{0,hp}' = \alpha_{0,hp} \times \text{FSF} = 4177 \tag{4-14}$$

Using $C = 0.01$ μF the real pole section resistor is given by

$$R_7 = \frac{1}{\alpha_{0,hp}'} = 23.94 \text{ k}\Omega \tag{4-37}$$

The final filter configuration is shown in Figure 4-9. The resistors were rounded off to standard 1-percent values, and R_1 and R_3 were made adjustable.

Elliptic-Function Configuration. The biquadratic configuration of Figure 4-8 can also be applied to the design of elliptic-function high-pass filters. The design of active high-pass elliptic-function filters using biquads utilizes the *Filter Solutions* program provided on the CD-ROM for obtaining normalized low-pass pole-zero locations which are then converted into high-pass pole-zero locations, are denormalized, and then used to compute the component values. The parameters obtained from *Filter Solutions* are ω_∞, ω_0, Q, and α_0.

The normalized low-pass poles and zeros must first be transformed to the high-pass form. Each complex low-pass pole pair consisting of a real part α and imaginary part β is transformed into a normalized high-pass pole pair as follows:

$$\alpha_{hp} = \frac{\alpha}{\alpha^2 + \beta^2} = \frac{1}{2Q\omega_0} \tag{4-9}$$

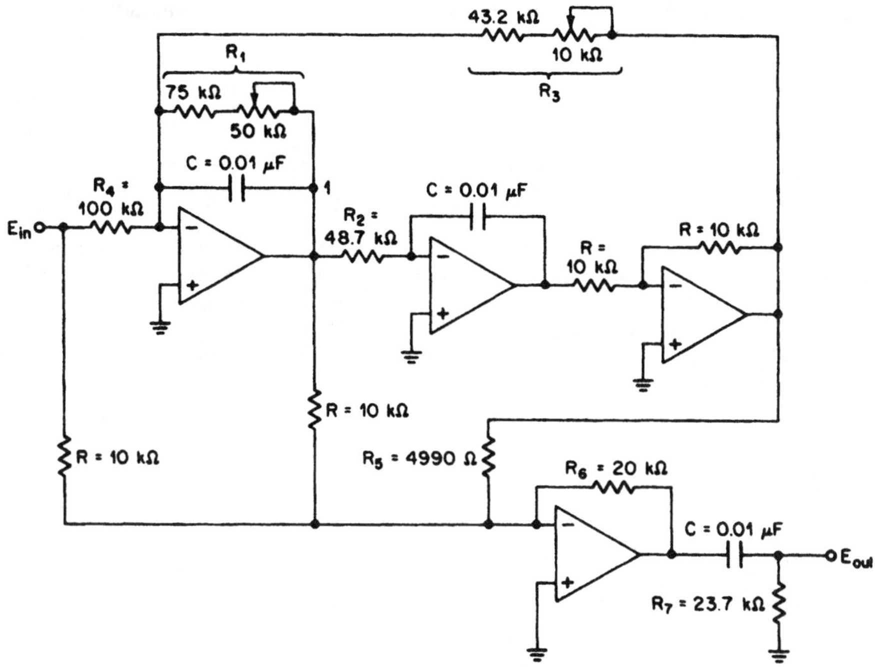

FIGURE 4-9 The all-pole high-pass filter of Example 4-6.

$$\beta_{hp} = \frac{\beta}{\alpha^2 + \beta^2} = \frac{1}{\omega_0}\sqrt{1 - \frac{1}{4Q^2}} \tag{4-10}$$

The second forms of Equations 4-9 and 4-10 involving Q and ω_0 are used when these parameters are provided by the *Filter Solutions* program.

The transformed high-pass pole pair can be denormalized by

$$\alpha'_{hp} = \alpha_{hp} \times \text{FSF} \tag{4-11}$$

$$\beta'_{hp} = \beta_{hp} \times \text{FSF} \tag{4-12}$$

where FSF is the frequency scaling factor $2\pi f_c$. If the pole is real, the normalized pole is transformed by

$$\alpha_{0,hp} = \frac{1}{\alpha_0} \tag{4-13}$$

The denormalized real pole is obtained from

$$\alpha'_{0,hp} = \alpha_{0,hp} \times \text{FSF} \tag{4-14}$$

To transform zeros, we compute

$$\omega_\infty(\text{hp}) = \frac{1}{\omega_\infty(\text{Lp})} \tag{4-3}$$

The component values are computed by using the same formulas as for the all-pole case, except for R_5, which is given by

$$R_5 = \frac{2\alpha'_{hp}\omega'_0}{(\omega'_0)^2 - [\omega'_\infty(\text{hp})]^2} R \qquad (4\text{-}40)$$

where $\omega'_\infty(\text{hp})$ is the denormalized high-pass transmission zero which is obtained from

$$\omega'_\infty(\text{hp}) = \omega_\infty(\text{hp}) \times \text{FSF} \qquad (4\text{-}41)$$

As in the all-pole circuit, the bandpass resonant frequency f_0 is controlled by R_3 and the bandpass Q is determined by R_1. In addition, the section notch can be adjusted if R_5 is made variable. However, this adjustment is usually not required if the circuit is first tuned to f_0, since the notch will then usually fall in.

The following example illustrates the design of an elliptic-function high-pass filter using the biquad configuration of Figure 4-8.

Example 4-7 Design of an Active Elliptic-Function High-Pass Filter Using a State-Variable Approach

Required:

An active high-pass filter
0.3-dB maximum ripple above 1000 Hz
18-dB minimum at 643 Hz

Result:

(*a*) Compute the high-pass steepness factor.

$$A_s = \frac{f_c}{f_s} = \frac{1000}{643} = 1.556 \qquad (2\text{-}13)$$

(*b*) Open ***Filter Solutions***.

Check the ***Stop Band Freq*** box.
Enter **.18** in the ***Pass Band Ripple (dB)*** box.
Enter **1** in the ***Pass Band Freq*** box.
Enter **1.556** in the ***Stop Band Freq*** box.
Check the ***Frequency Scale Radians*** box.

(*c*) Click the ***Set Order*** control button to open the second panel.

Enter **18** for ***Stop band Attenuation (dB)***.
Click the ***Set Minimum Order*** button and then click ***Close***.
3 Order is displayed on the main control panel.

(*d*) Click the ***Transfer Function*** button.

Check the ***Casc*** box.

The following is displayed:

Continuous Transfer Function

Wn = 1.743

$$\frac{.4838 \quad (S^2 + 3.037)}{(S + 1.053) \quad (S^2 + .5687*S + 1.396)}$$

Wo = 1.181

Q = 2.077

3rd Order Low Pass Elliptic

Pass Band Frequency = 1.000 Rad/Sec
Pass Band Ripple = 180.0 mdB
Stop Band Ratio = 1.556
Stop Band Frequency = 1.556 Rad/Sec
Stop Band Attenuation = 18.63 dB

The design parameters are summarized as follows:

$$\text{Section } Q = 2.077$$

$$\text{Section } \omega_0 = 1.181$$

$$\text{Section } \omega_\infty = 1.743$$

$$\alpha_0 = 1.053 \text{ (from the denominator)}$$

(*e*) To compute the element values, first transform the normalized low-pass poles and zeros to the high-pass form.

Complex pole:

$$\alpha_{\text{hp}} = \frac{1}{2Q\omega_0} = 0.2038 \tag{4-9}$$

$$\beta_{\text{hp}} = \frac{1}{\omega_0}\sqrt{1 - \frac{1}{4Q^2}} = 0.8218 \tag{4-10}$$

Zero:

$$\omega_\infty(\text{hp}) = \frac{1}{\omega_\infty(\text{Lp})} = \frac{1}{1.743} = 0.5737 \tag{4-3}$$

(*f*) The poles and zeros are denormalized as follows:

$$\alpha'_{\text{hp}} = \alpha_{\text{hp}} \times \text{FSF} = 1280 \tag{4-11}$$

$$\beta'_{\text{hp}} = \beta_{\text{hp}} \times \text{FSF} = 5163 \tag{4-12}$$

and

$$\omega'_\infty(\text{hp}) = \omega_\infty(\text{hp}) \times \text{FSF} = 3605 \tag{4-41}$$

where FSF $= 2\pi f_c$ or 6283.

(*g*) If we arbitrarily choose $C = 0.01\ \mu\text{F}$ and $R = 100\ \text{k}\Omega$, the component values can be obtained by

$$\omega'_0 = \sqrt{(\alpha'_{\text{hp}})^2 + (\beta'_{\text{hp}})^2} = 5319 \tag{4-32}$$

then

$$R_1 = R_4 = \frac{1}{2\alpha'_{\text{hp}}C} = 39.1\ \text{k}\Omega \tag{4-33}$$

$$R_2 = R_3 = \frac{1}{\omega'_0 C} = 18.8\ \text{k}\Omega \tag{4-34}$$

$$R_5 = \frac{2\alpha'_{\text{hp}}\omega'_0}{(\omega'_0)^2 - [\omega'_\infty(\text{hp})]^2} R = 89.0\ \text{k}\Omega \tag{4-40}$$

and

$$R_6 = AR = 100\ \text{k}\Omega \tag{4-36}$$

where the gain A is unity.

The bandpass resonant frequency and Q are determined from

$$f_0 = \frac{\omega'_0}{2\pi} = 847\ \text{Hz} \tag{4-38}$$

and

$$Q = \frac{\pi f_0}{\alpha'_{\text{hp}}} = 2.077 \tag{4-39}$$

The notch frequency occurs at $\omega_\infty(\text{hp}) \times f_c$ or 574 Hz.

(*d*) The normalized real low-pass pole is transformed to a high-pass pole:

$$\alpha_{0,\text{hp}} = \frac{1}{\alpha_0} = 0.950 \tag{4-13}$$

and is then denormalized by

$$\alpha'_{0,\text{hp}} = \alpha_{0,\text{hp}} \times \text{FSF} = 5967 \tag{4-14}$$

Resistor R_7 is found by

$$R_7 = \frac{1}{\alpha'_{0,\text{hp}}C} = 16.8\ \text{k}\Omega \tag{4-37}$$

where $C = 0.01\ \mu\text{F}$.

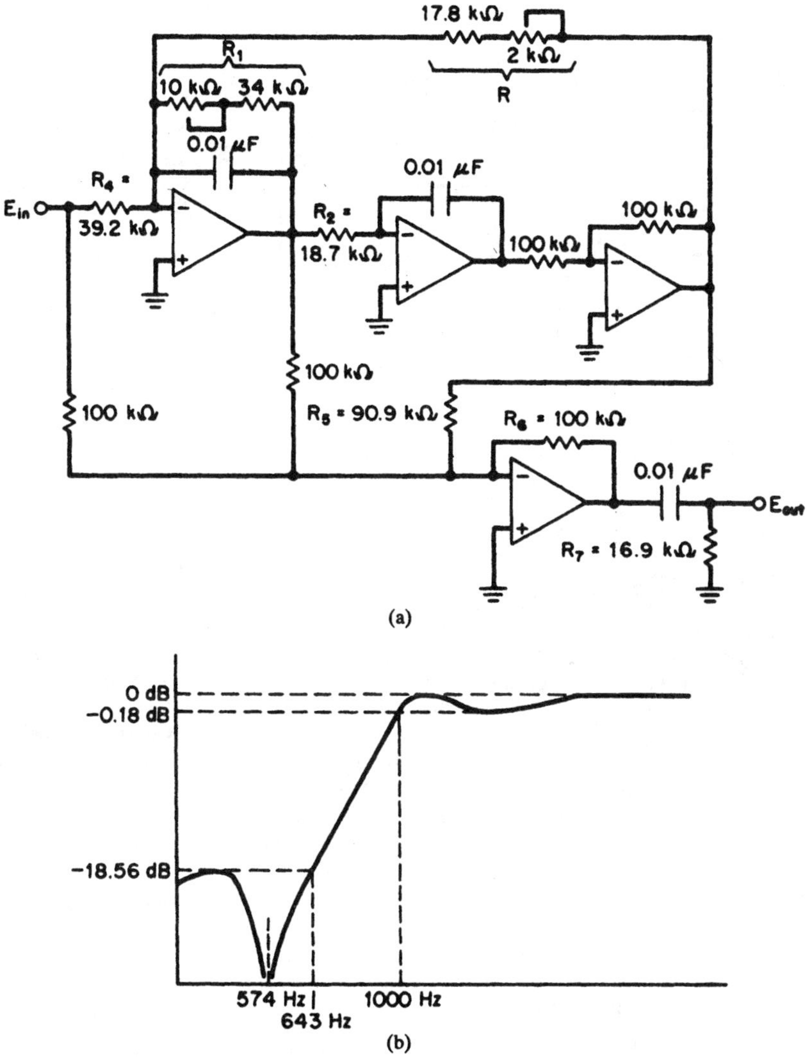

FIGURE 4-10 The elliptic-function high-pass filter of Example 4-7: (*a*) filter using the biquad configuration; and (*b*) frequency response.

The final circuit is given in Figure 4-10*a* using standard 1-percent values with R_1 and R_3 made adjustable. The frequency response is illustrated in Figure 4-10*b*.

High-Pass Filters Using the GIC. The generalized impedance converter (GIC) was first introduced in Section 3.2. This versatile device is capable of simulating a variety of different impedance functions. The circuit of Figure 3-28 simulated an inductor whose magnitude was given by

$$L = \frac{CR_1R_3R_5}{R_2} \tag{3-61}$$

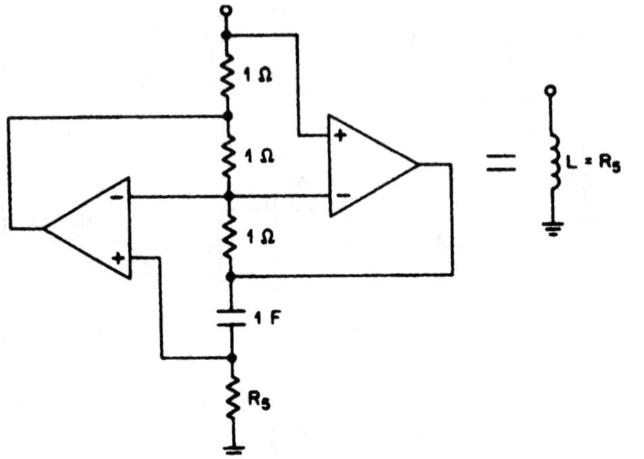

FIGURE 4-11 A normalized inductor using the GIC.

If we set R_1 through R_3 equal to 1 Ω and $C = 1$ F, a normalized inductor is obtained where $L = R_5$. This circuit is shown in Figure 4-11.

An active realization of a grounded inductor is particularly suited for the design of active high-pass filters. If a passive *LC* low-pass configuration is transformed into a high-pass filter, shunt inductors to ground are obtained which can be implemented using the GIC. The resulting normalized filter can then be frequency- and impedance-scaled. If R_5 is made variable, the equivalent inductance can be adjusted. This feature is especially desirable in the case of steep elliptic-function high-pass filters since the inductors directly control the location of the critical transmission zeros in the stopband.

The following example illustrates the design of an active all-pole high-pass filter directly from the *LC* element values using the GIC as a simulated inductance.

Example 4-8 Design of an Active All-Pole High-Pass Filter Using a GIC Approach

Required:

An active high-pass filter
3 dB at 1200 Hz
35-dB minimum at 375 Hz

Result:

(*a*) Compute the high-pass steepness factor.

$$A_s = \frac{f_c}{f_s} = \frac{1200}{375} = 3.2 \tag{2-13}$$

The curves of Figure 2-45 indicate that a third-order 1-dB Chebyshev low-pass filter provides over 35 dB of attenuation at 3.2 rad/s. For this example, we will use a GIC to simulate the inductor of an $n = 3$ *LC* high-pass configuration.

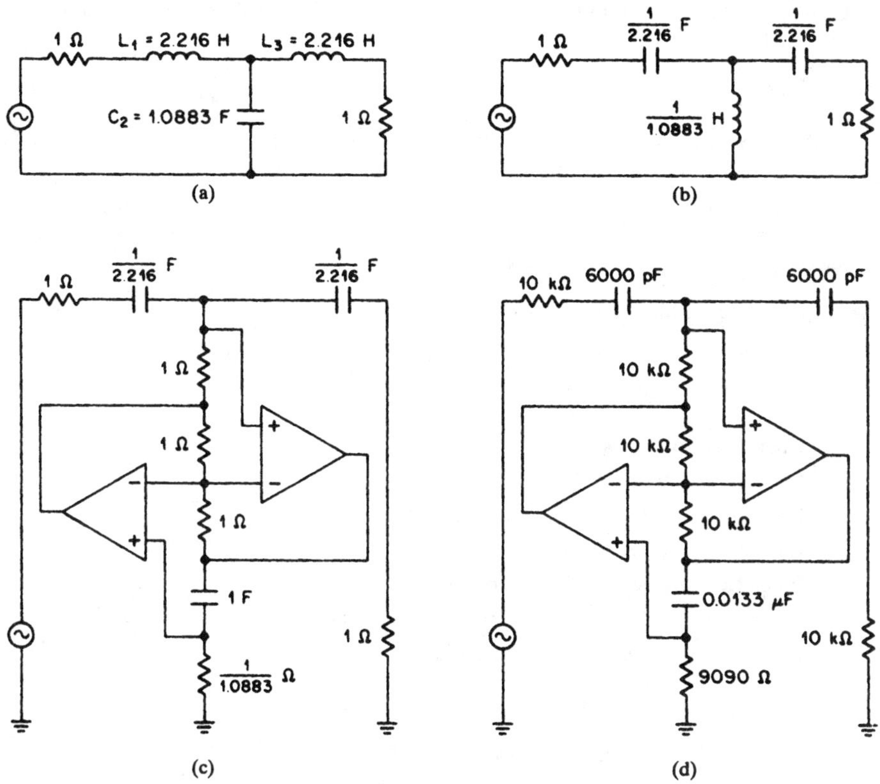

FIGURE 4-12 The all-pole high-pass filter of Example 4-8 using the GIC: (*a*) normalized low-pass filter; (*b*) transformed high-pass filter; (*c*) active inductor realization; and (*d*) the final network after scaling.

(*b*) The normalized low-pass filter is obtained from Table 11-31 and is shown in Figure 4-12*a*. The dual filter configuration is used to minimize the number of inductors in the high-pass filter.

(*c*) To transform the normalized low-pass filter into a high-pass configuration, replace the inductors with capacitors, and vice versa, using reciprocal element values. The normalized high-pass filter is shown in Figure 4-12*b*. The inductor can now be replaced by the GIC of Figure 4-11, resulting in the high-pass filter of Figure 4-12*c*.

(*d*) The filter is frequency- and impedance-scaled. Using an FSF of $2\pi f_c$ or 7540 and a Z of 10^4, divide all capacitors by $Z \times$ FSF and multiply all resistors by Z. The final configuration is shown in Figure 4-12*d* using standard 1-percent resistor values.

Active Elliptic-Function High-Pass Filters Using the GIC. Active elliptic-function high-pass filters can also be designed directly from normalized elliptic function low-pass filters using the GIC. This approach is much less complex than a high-pass configuration involving biquads, and still permits adjustment of the transmission zeros. Design of an elliptic-function high-pass filter using the GIC and the *Filter Solutions* program is demonstrated in the following example.

Example 4-9 Design of an Active Elliptic-Function High-Pass Filter Using a GIC Approach

Required:

An active high-pass filter
0.5-dB maximum at 2500 Hz
60-dB minimum at 1523 Hz

Result:

(*a*) Compute the high-pass steepness factor.

$$A_s = \frac{f_c}{f_s} = \frac{2500}{1523} = 1.641 \qquad (2\text{-}13)$$

Open ***Filter Solutions***.
Check the ***Stop Band Freq*** box.
Enter **.18** in the ***Pass Band Ripple (dB)*** box.
Enter **1** in the ***Pass Band Freq*** box.
Enter **1.641** in the ***Stop Band Freq*** box.
The ***Frequency Scale Rad/Sec*** box should be checked.
Enter **1** for the ***Source Res*** and ***Load Res***.

(*b*) Click the ***Set Order*** control button to open the second panel.

Enter **60** for the ***Stopband Attenuation (dB)***.
Click the ***Set Minimum Order*** button and then click ***Close***.
6 Order is displayed on the main control panel.
Check the ***Even Order Mod*** box.

(*c*) Click the ***Circuits*** button.

Two schematics are presented by *Filter Solutions*. Select the one representing the dual (Passive Filter 2), which is shown in Figure 4-13*a*.

(*d*) To transform the network into a normalized high-pass filter, replace each inductor with a capacitor having a reciprocal value, and vice versa. The zeros are also reciprocated. The normalized high-pass filter is given in Figure 4-13*b*.

(*e*) The inductors can be replaced using the GIC inductor simulation of Figure 4-11, resulting in the circuit of Figure 4-13*c*.

To scale the network, divide all capacitors by $Z \times$ FSF and multiply all resistors by Z, which is arbitrarily chosen at 10^4 and FSF is $2\pi f_c$ or 15,708. The final filter is shown in Figure 4-13*d*. The stopband peaks were computed by multiplying each normalized high-pass transmission zero by $f_c = 2500$ Hz, resulting in the frequencies indicated.

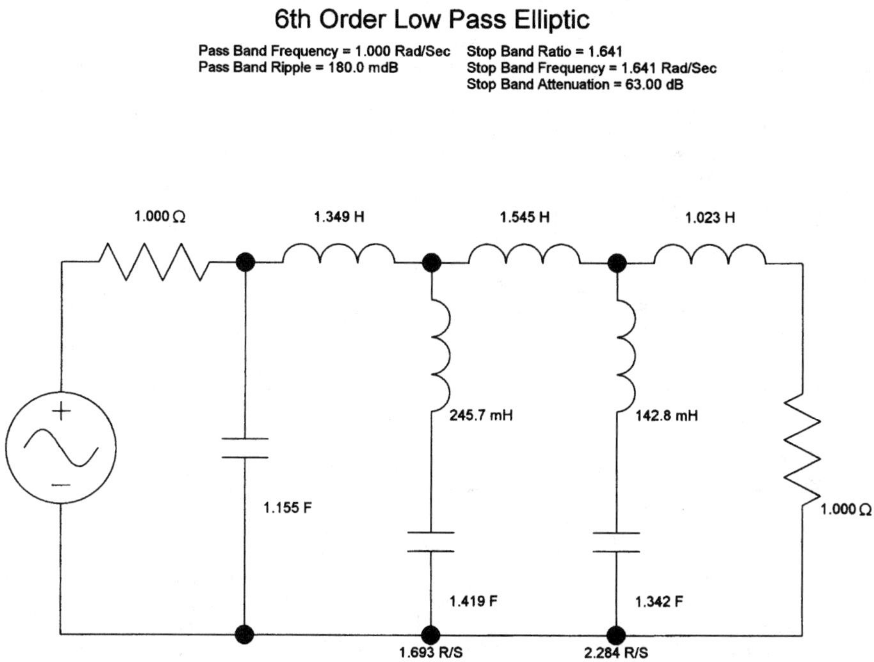

(a)

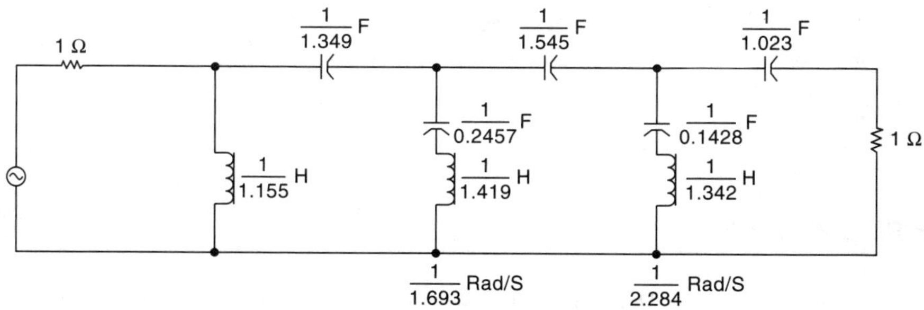

(b)

FIGURE 4-13 The elliptic-function high-pass filter of Example 4-9: (*a*) normalized low-pass filter; (*b*) transformed high-pass filter; (*c*) high-pass filter using the GIC; and (*d*) frequency- and impedance-scaled network.

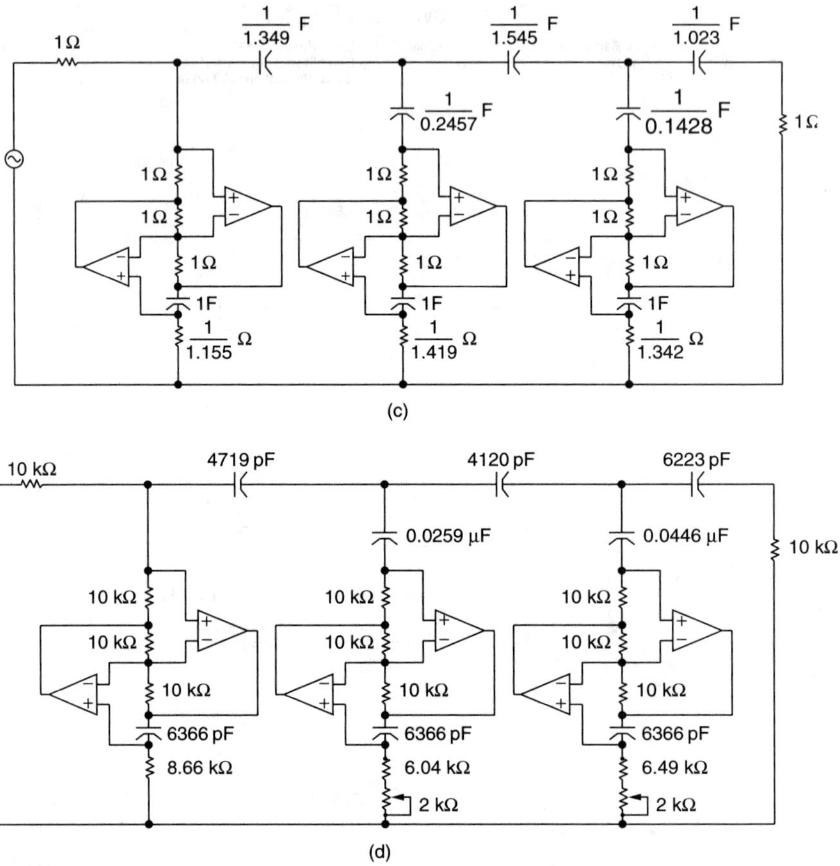

FIGURE 4-13 (*Continued*)

BIBLIOGRAPHY

Bruton, L. T. "Active Filter Design Using Generalized Impedance Converters." *EDN* (February, 1973).

Geffe, P. *Simplified Modern Filter Design.* New York: John F. Rider, 1963.

Williams, A. B. *Active Filter Design.* Dedham, Massachusetts: Artech House, 1975.

——————. "Design Active Elliptic Filters Easily from Tables." *Electronic Design* 19, no. 21 (October 14, 1971): 76–79.

CHAPTER 5
BANDPASS FILTERS

5.1 *LC BANDPASS FILTERS*

Bandpass filters were classified in Section 2.1 as either narrowband or wideband. If the ratio of upper cutoff frequency to lower cutoff frequency is over an octave, the filter is considered a wideband type. The specification is separated into individual low-pass and high-pass requirements and is simply treated as a cascade of low-pass and high-pass filters.

The design of narrow-band filters becomes somewhat more difficult. The circuit configuration must be appropriately chosen and suitable transformations may have to be applied to avoid impractical element values. In addition, as the filter becomes narrower, the element Q requirements increase and component tolerances and stability become more critical.

Wideband Filters

Wideband bandpass filters are obtained by cascading a low-pass filter and a high-pass filter. The validity of this approach is based on the assumption that the filters maintain their individual responses even though they are cascaded.

The impedance observed at the input or output terminals of an *LC* low-pass or high-pass filter approaches the resistive termination at the other end at frequencies well in the passband. This is apparent from the equivalent circuit of the low-pass filter at DC and the high-pass filter at infinite frequency. At DC, the inductors become short circuits and capacitors become open circuits, and at infinite frequency the opposite conditions occur. If a low-pass and high-pass filter are cascaded and both filters are designed to have equal source and load terminations and identical impedances, the filters will each be properly terminated in their passbands if the cutoff frequencies are separated by at least one or two octaves.

If the separation between passbands is insufficient, the filters will interact because of impedance variations. This effect can be minimized by isolating the two filters through an attenuator. Usually 3 dB of loss is sufficient. Further attenuation provides increased isolation. Table 5-1 contains values for T and π attenuators ranging from 1 to 10 dB at an impedance level of 500 Ω. These networks can be impedance-scaled to the filter impedance level R if each resistor value is multiplied by $R/500$.

Example 5-1 Design of a Wideband *LC* Bandpass Filter

Required:

An *LC* bandpass filter

3 dB at 500 and 2000 Hz

40-dB minimum at 100 and 4000 Hz

$R_s = R_L = 600\ \Omega$

TABLE 5-1 *T* and π Attenuators

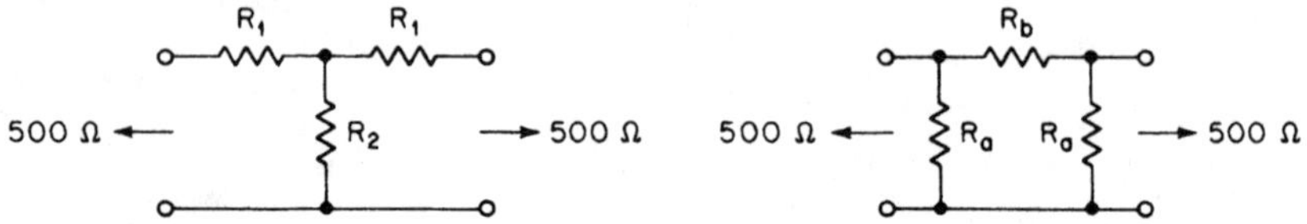

dB	R_1	R_2	R_a	R_b
1	28.8	4330	8700	57.7
2	57.3	2152	4362	116
3	85.5	1419	2924	176
4	113	1048	2210	239
5	140	822	1785	304
6	166	669	1505	374
7	191	558	1307	448
8	215	473	1161	528
9	238	406	1050	616
10	260	351	963	712

Result:

(*a*) Since the ratio of upper cutoff frequency to lower cutoff frequency is 4:1, a wide-band approach will be used. The requirement is first separated into individual low-pass and high-pass specifications:

High-pass filter:	*Low-pass filter:*
3 dB at 500 Hz	3 dB at 2000 Hz
40-dB minimum at 100 Hz	40-dB minimum at 4000 Hz

(*b*) The low-pass and high-pass filters are designed independently, using the design methods outlined in Sections 3.1 and 4.1 as follows:

Low-pass filter:

Compute the low-pass steepness factor.

$$A_s = \frac{f_s}{f_c} = \frac{4000 \text{ Hz}}{2000 \text{ Hz}} = 2 \tag{2-11}$$

Figure 2-43 indicates that a fifth-order 0.25-dB Chebyshev normalized low-pass filter provides over 40 dB of attenuation at 2 rad/s. The normalized low-pass filter is obtained from Table 11-29 and is shown in Figure 5-1*a*. The filter is frequency- and impedance-scaled by multiplying all inductors by *Z*/FSF and dividing all capacitors by Z × FSF, where *Z* is 600 and the frequency-scaling factor FSF is $2\pi f_c$ or 12,560. The denormalized low-pass filter is shown in Figure 5-1*b*.

High-pass filter:

Compute the high-pass steepness factor:

$$A_s = \frac{f_s}{f_c} = \frac{500 \text{ Hz}}{100 \text{ Hz}} = 5 \tag{2-13}$$

Using Figure 2-34, an $n = 3$ Butterworth normalized low-pass filter is selected to meet the attenuation requirement. The normalized filter values are found in Table 11-2 and

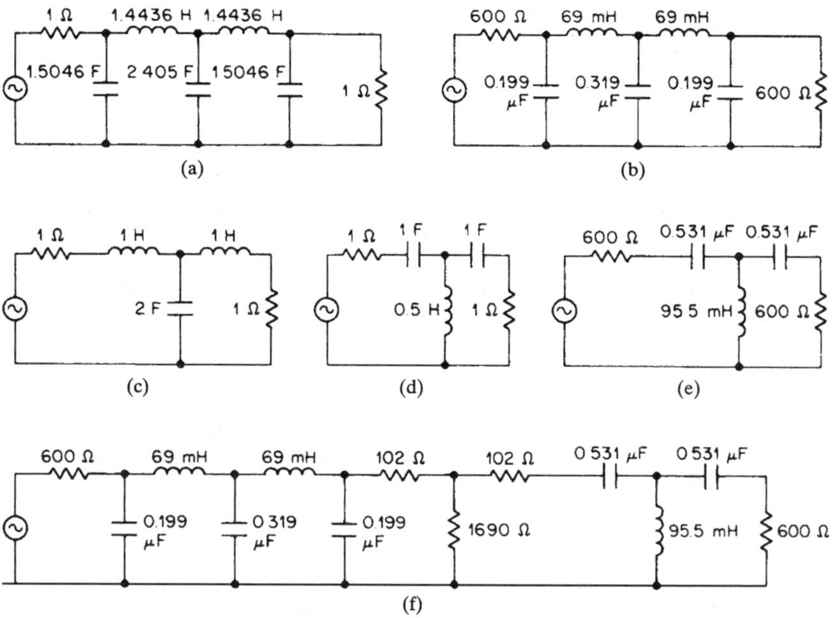

FIGURE 5-1 The *LC* wideband bandpass filter of Example 5-1: (*a*) normalized low-pass filter; (*b*) scaled low-pass filter; (*c*) normalized low-pass filter for high-pass requirement; (*d*) transformed high-pass filter; (*e*) scaled high-pass filter; and (*f*) combined network.

are shown in Figure 5-1*c*. Since the low-pass filter is to be transformed into a high-pass filter, the dual configuration was selected. By reciprocating element values and replacing inductors with capacitors and vice versa, the normalized high-pass filter of Figure 5-1*d* is obtained. The network is then denormalized by multiplying all inductors by *Z*/FSF and dividing all capacitors by $Z \times$ FSF, where *Z* is 600 Ω and FSF is 3140. The denormalized high-pass filter is illustrated in Figure 5-1*e*.

(*c*) The low-pass and high-pass filters can now be combined. A 3-dB T pad will be used to provide some isolation between filters since the separation of cutoffs is only two octaves. The pad values are obtained by multiplying the resistances of Table 5-1, corresponding to 3 dB by 600 Ω/500 Ω or 1.2, and rounding off to standard 1-percent values. The final circuit is shown in Figure 5-*f*.

Narrowband Filters. Narrowband bandpass filter terminology was introduced in Section 2.1 using the concept of bandpass *Q*, which was defined by

$$Q_{\text{bp}} = \frac{f_0}{\text{BW}_{3\text{ dB}}} \tag{2-16}$$

where f_0 is the geometric center frequency and BW is the 3-dB bandwidth. The geometric center frequency was given by

$$f_0 = \sqrt{f_L f_u} \tag{2-14}$$

where f_L and f_u are the lower and upper 3-dB limits.

Bandpass filters obtained by transformation from a low-pass filter exhibit geometric symmetry, that is,

$$f_0 = \sqrt{f_1 f_2} \tag{2-15}$$

where f_1 and f_2 are any two frequencies having equal attenuation. Geometric symmetry must be considered when normalizing a bandpass specification. For each stopband frequency specified, the corresponding geometric frequency is calculated and a steepness factor is computed based on the more severe requirement.

For bandpass Qs of 10 or more, the passband response approaches arithmetic symmetry. The center frequency then becomes the average of the 3-dB points, for instance

$$f_0 = \frac{f_L + f_u}{2} \tag{2-17}$$

The stopband will also become arithmetically symmetrical as the Q increases even further.

The Low-Pass to Bandpass Transformation. A bandpass transfer function can be obtained from a low-pass transfer function by replacing the frequency variable by a new variable, which is given by

$$f_{\text{bp}} = f_0\left(\frac{f}{f_0} - \frac{f_0}{f}\right) \tag{5-1}$$

When f is equal to f_0, the bandpass center frequency, the response corresponds to that at DC for the low-pass filter.

If the low-pass filter has a 3-dB cutoff of f_c, the corresponding bandpass frequency f can be found by solving

$$\pm f_c = f_0\left(\frac{f}{f_0} - \frac{f_0}{f}\right) \tag{5-2}$$

The $\pm$ signs occur because a low-pass filter has a mirrored response at negative frequencies in addition to the normal response. Solving Equation (5-2) for f, we obtain

$$f = \pm\frac{f_c}{2} \pm \sqrt{\left(\frac{f_c}{2}\right)^2 + f_0^2} \tag{5-3}$$

Equation (5-3) implies that the bandpass response has two positive frequencies corresponding to the low-pass response at $\pm f_c$, as well as two negative frequencies with identical responses. These frequencies can be obtained from Equation (5-3) and are given by

$$f_L = f_0\left[\sqrt{1 + \left(\frac{f_c}{2f_0}\right)^2} - \frac{f_c}{2f_0}\right] \tag{5-4}$$

and

$$f_u = f_0\left[\sqrt{1 + \left(\frac{f_c}{2f_0}\right)^2} + \frac{f_c}{2f_0}\right] \tag{5-5}$$

The bandpass 3-dB bandwidth is

$$\text{BW}_{3\text{ dB}} = f_u - f_L = f_c \tag{5-6}$$

The correspondence between a low-pass filter and the transformed bandpass filter is shown in Figure 5-2. The response of a low-pass filter to positive frequencies is transformed

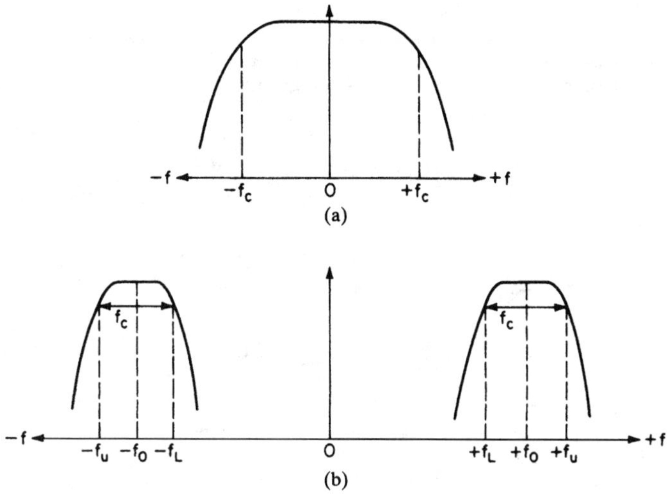

FIGURE 5-2 The low-pass to bandpass transformation: (*a*) low-pass filter; and (*b*) transformed bandpass filter.

into the response of the bandpass filter at an equivalent bandwidth. Therefore, a bandpass filter can be obtained by first designing a low-pass filter that has the required response corresponding to the desired bandwidth characteristics of the bandpass filter. The low-pass filter is then transformed into the bandpass filter.

The reactance of a capacitor in the low-pass filter is given by

$$X_c = \frac{1}{j\omega C} \tag{5-7}$$

where $\omega = 2\pi f$. If we replace the frequency variable f by the expression of Equation (5-1), the impedance expression becomes

$$Z = \frac{1}{j\omega C + \dfrac{1}{\dfrac{j\omega}{\omega_0^2 C}}} \tag{5-8}$$

where $\omega = 2\pi f_0$. This is the impedance of a parallel resonant LC circuit where the capacitance is still C and the inductance is $1/\omega_0^2 C$. The resonant frequency is ω_0.

The reactance of an inductor in the low-pass filter is

$$X_L = j\omega L \tag{5-9}$$

If we again replace the frequency variable using Equation (5-1), the resulting impedance expression becomes

$$Z = j\omega L + \frac{1}{\dfrac{j\omega}{\omega_0^2 L}} \tag{5-10}$$

TABLE 5-2 The Low-Pass to Bandpass Transformation

	Low-Pass Branch	Bandpass Configuration	Circuit Values
Type I	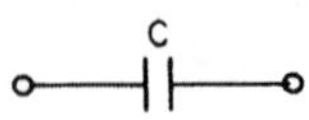	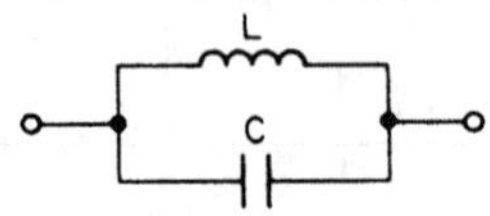	$L=\frac{1}{\omega_0^2 C}$ (5-11)
Type II	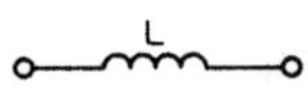	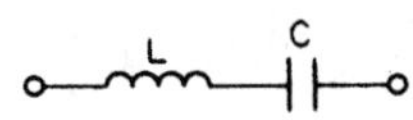	$C=\frac{1}{\omega_0^2 L}$ (5-12)
Type III	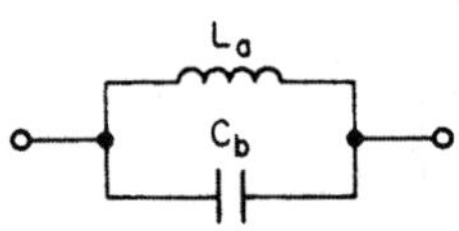	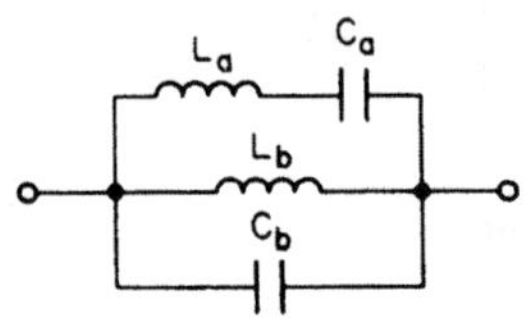	$C_a=\frac{1}{\omega_0^2 L_a}$ (5-13) $L_b=\frac{1}{\omega_0^2 C_b}$ (5-14)
Type IV	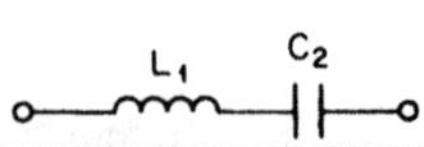	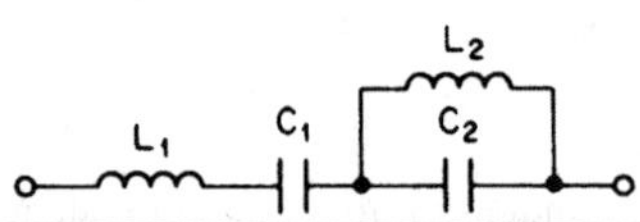	$C_1=\frac{1}{\omega_0^2 L_1}$ (5-15) $L_2=\frac{1}{\omega_0^2 C_2}$ (5-16)

This corresponds to a series resonant LC circuit where the inductance L is unchanged and C is $1/\omega_0^2 L$. The resonant frequency is ω_0.

We can summarize these results by stating that an LC low-pass filter can be transformed into a bandpass filter having the equivalent bandwidth by resonating each capacitor with a parallel inductor and each inductor with a series capacitor. The resonant frequency is f_0, the bandpass filter center frequency. Table 5-2 shows the circuits which result from the low-pass to bandpass transformation.

The Transformation of All-Pole Low-Pass Filters. The LC bandpass filters discussed in this section are probably the most important type of filter. These networks are directly obtained by the bandpass transformation from the LC low-pass values tabulated in Chapter 11. Each normalized low-pass filter defines an infinitely large family of bandpass filters having a geometrically symmetrical response predetermined by the low-pass characteristics.

A low-pass transfer function can be transformed to a bandpass type by substitution of the frequency variable using Equation (5-1). This transformation can also be made directly to the circuit elements by first scaling the low-pass filter to the required bandwidth and impedance level. Each coil is then resonated with a series capacitor to the center frequency f_0, and an inductor is introduced across each capacitor to form a parallel tuned circuit that's also resonant at f_0. Every low-pass branch is replaced by the associated bandpass branch, as illustrated by Table 5-2.

Some of the effects of dissipation in low-pass filters were discussed in Section 3.1. These effects are even more severe in bandpass filters. The minimum Q retirement for the low-pass elements can be obtained from Figure 3-8 for a variety of filter types. These minimum values are based on the assumption that the filter elements are predistorted so that the theoretical response is obtained. Since this is not always the case, the branch Qs should be several times higher than the values indicated. When the network undergoes a low-pass-to-bandpass transformation, the Q requirement is increased by the bandpass Q of the filter. This can be stated as

$$Q_{\text{min}}\,(\text{bandpass}) = Q_{\text{min}}\,(\text{low-pass}) \times Q_{\text{bp}} \tag{5-17}$$

where $Q_{\text{bp}} = f_0/\text{BW}_{3\text{ dB}}$. As in the low-pass case, the branch Qs should be several times higher than Q_{min}. Since capacitor losses are usually negligible, the branch Q is determined strictly by the inductor losses.

The spread of values in bandpass filters is usually wider than with low-pass filters. For some combinations of impedance and bandwidth, the element values may be impossible or impractical to realize because of their magnitude or the effects of parasitics. When this situation occurs, the designer can use a variety of circuit transformations to obtain a more practical circuit. These techniques are covered in Chapter 8.

The design method can be summarized as follows:

1. Convert the response requirement into a geometrically symmetrical specification.
2. Compute the bandpass steepness factor A_s. Select a normalized low-pass filter from the frequency-response curves of Chapter 2 that makes the passband to stopband transition within a frequency ratio of A_s.
3. Scale the corresponding normalized low-pass filter from the tables of Chapter 11 to the required bandwidth and impedance level of the bandpass filter.
4. Resonate each L and C to f_0 in accordance with Table 5-2.
5. The final design may require manipulation by various transformations so that the values are more practical. In addition, the branch Qs must be well in excess of Q_{min} (bandpass) as given by Equation (5-17) to obtain near theoretical results.

Example 5-2 Design of an All-Pole *LC* Bandpass Filter

Required:

Bandpass filter

A center frequency of 1000 Hz

3-dB points at 950 and 1050 Hz

25-dB minimum at 800 Hz and 1150 Hz

$R_s = R_L = 600\ \Omega$

An available inductor Q of 100

Result:

(*a*) Convert to a geometrically symmetrical bandpass requirement:

First, calculate the geometric center frequency.

$$f_0 = \sqrt{f_L f_u} = \sqrt{950 \times 1050} = 998.8\text{ Hz} \tag{2-14}$$

Compute the corresponding geometric frequency for each stopband frequency given, using Equation (2-18).

$$f_1 f_2 = f_0^2 \tag{2-18}$$

f_1	f_2	$f_2 - f_1$
800 Hz	1247 Hz	447 Hz
867 Hz	1150 Hz	283 Hz

The second pair of frequencies will be retained since they represent the more severe requirement. The resulting geometrically symmetrical requirement can be summarized as

$$f_0 = 998.8 \text{ Hz}$$

$$\text{BW}_{3\text{ dB}} = 100 \text{ Hz}$$

$$\text{BW}_{25\text{ dB}} = 283 \text{ Hz}$$

(*b*) Compute the bandpass steepness factor.

$$A_s = \frac{\text{stopband bandwidth}}{\text{passband bandwidth}} = \frac{283 \text{ Hz}}{100 \text{ Hz}} = 2.83 \qquad (2\text{-}19)$$

(*c*) Select a normalized low-pass filter that makes the transition from 3 dB to more than 25 dB within a frequency ratio of 2.83:1. Figure 2-34 indicates that an $n = 3$ Butterworth type will satisfy the response requirement. The normalized low-pass filter is found in Table 11-2 and is shown in Figure 5-3*a*.

(*d*) Denormalize the low-pass filter using a Z of 600 and a frequency-scaling factor (FSF) of $2\pi f_c$ or 628, where $f_c = 100$ Hz.

$$C'_1 = C'_3 = \frac{C}{\text{FSF} \times Z} = \frac{1}{628 \times 600} = 2.653\ \mu\text{F} \qquad (2\text{-}10)$$

$$L'_2 = \frac{L \times Z}{\text{FSF}} = \frac{2 \times 600}{628} = 1.91 \text{ H} \qquad (2\text{-}9)$$

The denormalized low-pass filter is illustrated in Figure 5-3*b*.

(*e*) To make the low-pass to bandpass transformation, resonate each capacitor with a parallel inductor and each inductor with a series capacitor using a resonance frequency of $f_0 = 998.8$ Hz.

$$L'_1 = \frac{1}{\omega_0^2 C'_1} = \frac{1}{(6275)^2 \times 2.653 \times 10^{-6}} = 9.573 \text{ mH} \qquad (5\text{-}11)$$

$$L'_3 = L'_1 = 9.573 \text{ mH}$$

$$C'_2 = \frac{1}{\omega_0^2 L'_2} = \frac{1}{(6275)^2 \times 1.91} = 0.01329\ \mu\text{F} \qquad (5\text{-}12)$$

where $\omega_0 = 2\pi f_0$. The resulting bandpass filter is given in Figure 5-3*c*.

(*f*) Estimate if the available inductor Q of 100 is sufficient.

$$Q_{\min}\text{ (bandpass)} = Q_{\min}\text{ (low-pass)} \times Q_{\text{bp}} = 2 \times 10 = 20 \qquad (5\text{-}17)$$

where $Q_{\min}$ (low-pass) was obtained from Figure 3-8 and Q_{bp} is $f_0/\text{BW}_{3\text{ dB}}$. Since the available Q is well in excess of $Q_{\min}$ (bandpass), the filter response will closely agree with the theoretical predictions.

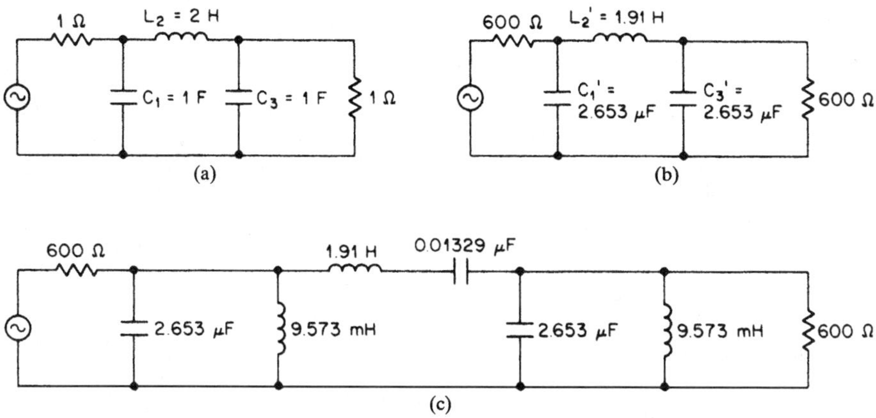

FIGURE 5-3 The bandpass filter of Example 5-2: (*a*) normalized $n = 3$ Butterworth low-pass filter; (*b*) low-pass filter scaled to 600 Ω and an f_c of 100 Hz; and (*c*) transformed bandpass filter.

The response requirement of Example 5-2 was converted to a geometrically symmetrical specification by calculating the corresponding frequency for each stopband frequency specified at a particular attenuation level using the relationship $f_1 f_2 = f_0^2$. The pair of frequencies having the lesser separation was chosen, since this would represent the steeper filter requirement. This technique represents a general method for obtaining the geometrically related frequencies that determine the response requirements of the normalized low-pass filter.

Stopband requirements are frequently specified in an arithmetically symmetrical manner where the deviation on both sides of the center frequency is the same for a given attenuation. Because of the geometric symmetry of bandpass filters, the attenuation for a particular deviation below the center frequency will be greater than for the same deviation above the center frequency. The response curve would then appear compressed on the low side of the passband if plotted on a linear frequency axis. On a logarithmic scale, the curve would be symmetrical.

When the specification is stated in arithmetic terms, the stopband bandwidth on a geometric basis can be computed directly by

$$\text{BW} = f_2 - \frac{f_0^2}{f_2} \tag{5-18}$$

where f_2 is the upper stopband frequency and f_0 is the geometric center frequency as determined from the passband limits. This approach is demonstrated in the following example.

Example 5-3 Design of All-Pole *LC* Bandpass Filter from Arithmetically Symmetrical Requirement

Required:

Bandpass filter

A center frequency of 50 kHz

3-dB points at ±3 kHz (47 kHz, 53 kHz)

30-dB minimum at ±7.5 kHz (42.5 kHz, 57.5 kHz)

40-dB minimum at ±10.5 kHz (39.5 kHz, 60.5 kHz)

$R_s = 150\ \Omega \qquad R_L = 300\ \Omega$

Result:

(*a*) Convert to the geometrically symmetrical bandpass requirement.

$$f_0 = \sqrt{f_L f_u} = \sqrt{47 \times 53 \times 10^6} = 49.91 \text{ kHz} \tag{2-14}$$

Since the stopband requirement is arithmetically symmetrical, compute the stopband bandwidth using Equation (5-18).

$$\text{BW}_{30 \text{ dB}} = f_2 - \frac{f_0^2}{f_2} = 57.5 \times 10^3 - \frac{(49.91 \times 10^3)^2}{57.5 \times 10^3} = 14.18 \text{ kHz}$$

$$\text{BW}_{40 \text{ dB}} = 19.33 \text{ kHz}$$

Requirement:

$$f_0 = 49.91 \text{ kHz}$$
$$\text{BW}_{3 \text{ dB}} = 6 \text{ kHz}$$
$$\text{BW}_{30 \text{ dB}} = 14.18 \text{ kHz}$$
$$\text{BW}_{40 \text{ dB}} = 19.33 \text{ kHz}$$

(*b*) Since two stopband bandwidth requirements are given, they must both be converted into bandpass steepness factors.

$$A_s(30 \text{ dB}) = \frac{\text{stopband bandwidth}}{\text{passband bandwidth}} = \frac{14.18 \text{ kHz}}{6 \text{ kHz}} = 2.36 \tag{2-19}$$

$$A_s(40 \text{ dB}) = 3.22$$

(*c*) A normalized low-pass filter must be chosen that provides over 30 dB of rejection at 2.36 rad/s and more than 40 dB at 3.22 rad/s. Figure 2-41 indicates that a fourth-order 0.01-dB Chebyshev filter will meet this requirement. The corresponding low-pass filter can be found in Table 11-27. Since a 2:1 ratio of R_L to R_s is required, the design for a normalized R_s of 2 Ω is chosen and is turned end for end. The circuit is shown in Figure 5-4*a*.

(*d*) The circuit is now scaled to an impedance level *Z* of 150 and a cutoff of $f_c = 6$ kHz. All inductors are multiplied by *Z*/FSF, and the capacitors are divided by *Z* × FSF, where FSF is $2\pi f_c$. The 1-Ω source and 2-Ω load become 150 and 300 Ω, respectively. The denormalized network is illustrated in Figure 5-4*b*.

(*e*) The scaled low-pass filter is transformed to a bandpass filter at $f_0 = 49.91$ kHz by resonating each capacitor with a parallel inductor and each inductor with a series capacitor using the general relationship $\omega_0^2 LC = 1$. The resulting bandpass filter is shown in Figure 5-4*c*.

The Design of Parallel Tuned Circuits. The simple *RC* low-pass circuit of Figure 5-5*a* has a 3-dB cutoff corresponding to

$$f_c = \frac{1}{2\pi RC} \tag{5-19}$$

If a bandpass transformation is performed, the circuit of Figure 5-5*b* results, where

$$L = \frac{1}{\omega_0^2 C} \tag{5-11}$$

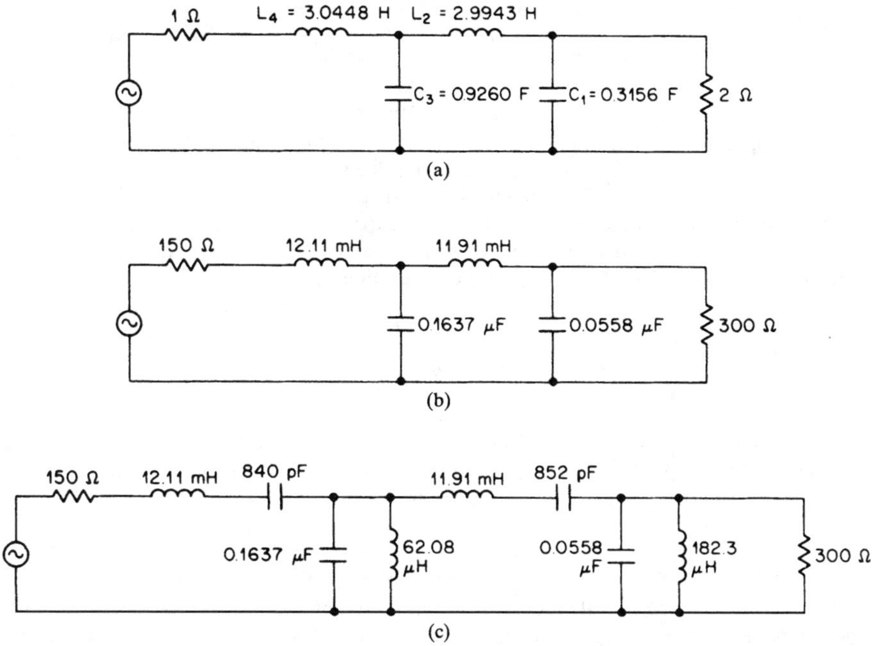

FIGURE 5-4 The bandpass filter of Example 5-3: (*a*) normalized low-pass filter; (*b*) scaled low-pass filter; and (*c*) transformed bandpass filter.

The center frequency is f_0 and the 3-dB bandwidth is equal to f_c. The bandpass Q is given by

$$Q_{\text{bp}} = \frac{f_0}{\text{BW}_{3\text{ dB}}} = \frac{f_0}{f_c} = \omega_0 RC \tag{5-20}$$

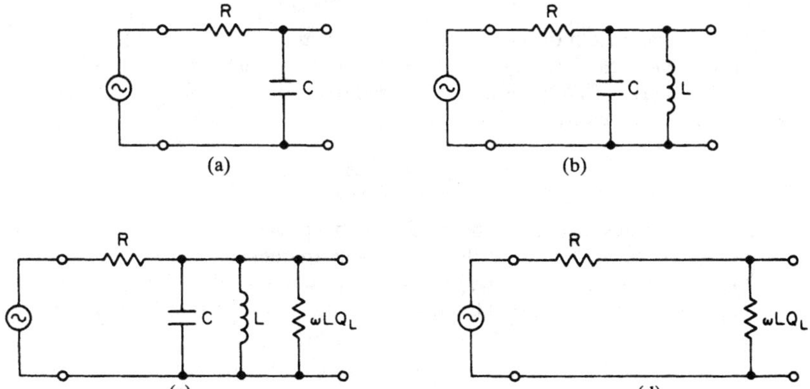

FIGURE 5-5 The single tuned circuit: (*a*) *RC* low-pass circuit; (*b*) result of bandpass transformation; (*c*) representation of coil losses; (*d*) equivalent circuit at resonance.

Since the magnitudes of the capacitive and inductive susceptances are equal at resonance by definition, we can substitute $1/\omega_0 L$ for $\omega_0 C$ in Equation (5-20) and obtain

$$Q_{\text{bp}} = \frac{R}{\omega_0 L} \tag{5-21}$$

The element R may be a single resistor as in Figure 5-5*b* or the parallel combination of both the input and output terminations if an output load resistor is also present.

The circuit of Figure 5-5*b* is somewhat ideal, since inductor losses are usually unavoidable. (The slight losses usually associated with the capacitor will be neglected.) If the inductor Q is given as Q_L, the inductor losses can be represented as a parallel resistor of $\omega L Q_L$, as shown in Figure 5-5*c*. The effective Q of the circuit thus becomes

$$Q_{\text{eff}} = \frac{\dfrac{R}{\omega_0 L} Q_L}{\dfrac{R}{\omega_0 L} + Q_L} \tag{5-22}$$

As a result, the effective circuit Q is somewhat less than the values computed by Equations (5-20) or (5-21). To compensate for the effect of finite inductor Q, the design Q should be somewhat higher. This value can be found from

$$Q_d = \frac{Q_{\text{eff}} Q_L}{Q_L - Q_{\text{eff}}} \tag{5-23}$$

At a resonance, the equivalent circuit is represented by the resistive voltage divider of Figure 5-5*d*, since the reactive elements cancel. The insertion loss at f_0 can be determined by the expression

$$IL_{\text{dB}} = 20 \log\left(1 + \frac{1}{k - 1}\right) \tag{5-24}$$

where $k = Q_L/Q_{\text{eff}}$, and can be obtained directly from the curve of Figure 5-6. Clearly the insertion loss increases dramatically as the inductor Q approaches the required effective Q of the circuit.

The frequency response of a single tuned circuit is expressed by

$$A_{\text{dB}} = 10 \log\left[1 + \left(\frac{\text{BW}_x}{\text{BW}_{3\text{ dB}}}\right)^2\right] \tag{5-25}$$

where BW_x is the bandwidth of interest, and $\text{BW}_{3\text{ dB}}$ is the 3-dB bandwidth. The response characteristics are identical to an $n = 1$ Butterworth, so the attenuation curves of Figure 2-34 can be applied using $\text{BW}_x/\text{BW}_{3\text{ dB}}$ as the normalized frequency in radians per second.

The phase shift is given by

$$\theta = \tan^{-1}\left(\frac{2\Delta f}{\text{BW}_{3\text{ dB}}}\right) \tag{5-26}$$

where Δf is the frequency deviation from f_0. The output phase shift lags by 45° at the upper 3-dB frequency, and leads by 45° at the lower 3-dB frequency. At DC and infinity, the phase shift reaches +90° and −90°, respectively. Equation (5-26) is plotted in Figure 5-7.

The group delay can be estimated by the slope of the phase shift at f_0 and results in the approximation

$$T_{gd} = \frac{318}{\text{BW}_{3\text{ dB}}} \tag{5-27}$$

where $\text{BW}_{3\text{ dB}}$ is the 3-dB bandwidth in hertz and T_{gd} is the resulting group delay in milliseconds.

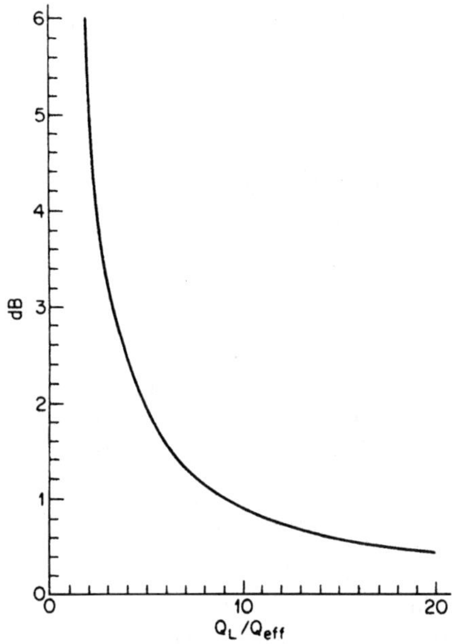

FIGURE 5-6 Insertion loss versus Q_L/Q_{eff}.

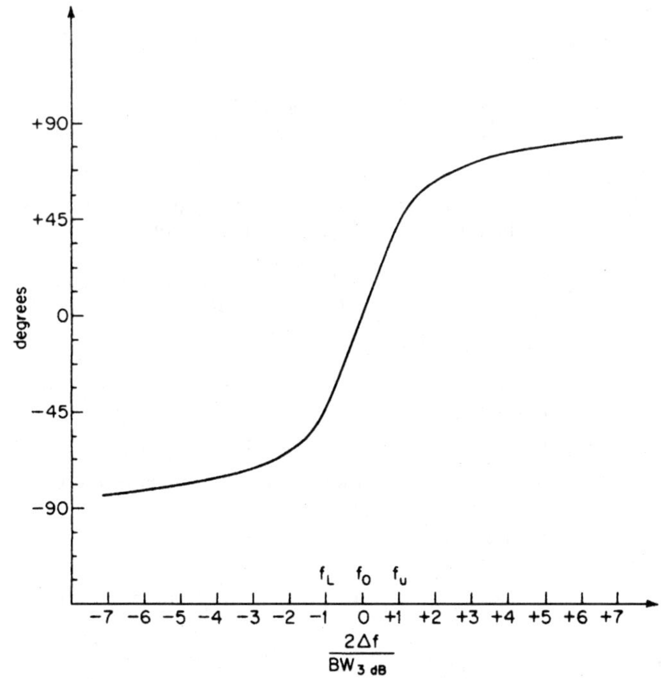

FIGURE 5-7 Phase shift versus frequency.

Example 5-4 Design of a Bandpass Filter Using a Parallel Tuned Circuit

Required:

An LC bandpass filter

A center frequency of 10 kHz

3 dB at ±100 Hz (9.9 kHz, 10.1 kHz)

15-dB minimum at ±1 kHz (9 kHz, 11 kHz)

Inductor $Q_L = 200$

$R_s = R_L = 6\ \text{k}\Omega$

Result:

(*a*) Convert to the geometrically symmetrical bandpass specification. Since the bandpass Q is much greater than 10, the specified arithmetically symmetrical frequencies are used to determine the following design requirements:

$$f_0 = 10\ \text{kHz}$$
$$\text{BW}_{3\ \text{dB}} = 200\ \text{Hz}$$
$$\text{BW}_{15\ \text{dB}} = 2000\ \text{Hz}$$

(*b*) Compute the bandpass steepness factor.

$$A_s = \frac{\text{stopband bandwidth}}{\text{passband bandwidth}} = \frac{2000}{200} = 10 \tag{2-19}$$

Figure 2-34 indicates that a single tuned circuit ($n = 1$) provides more than 15 dB of attenuation within a bandwidth ratio of 10:1.

(*c*) Calculate the design Q to obtain a Q_{eff} equal to $f_0/\text{BW}_{3\ \text{dB}} = 50$, considering the inductor Q_L of 200.

$$Q_d = \frac{Q_{\text{eff}} Q_L}{Q_L - Q_{\text{eff}}} = \frac{50 \times 200}{200 - 50} = 66.7 \tag{5-23}$$

(*d*) Since the source and load are both of 6 kΩ, the total resistive loading on the tuned circuit is the parallel combination of both terminations—thus, $R = 3\ \text{k}\Omega$. The design Q can now be used in Equation (5-20) to compute C.

$$C = \frac{Q_{\text{bp}}}{\omega_0 R} = \frac{66.7}{6.28 \times 10 \times 10^3 \times 3000} = 0.354\ \mu\text{F} \tag{5-20}$$

The inductance is given by Equation (5-11).

$$L = \frac{1}{\omega_0^2 C} = \frac{1}{(2\pi \times 10\ \text{kHz})^2 \times 3.54 \times 10^{-7}} = 716\ \mu\text{H} \tag{5-11}$$

The resulting circuit is shown in Figure 5-8*a*, which has the frequency response of Figure 5-8*b*. See Section 8.1 for a more practical implementation using a tapped inductor.

(*e*) The circuit insertion loss can be calculated from

$$IL_{\text{dB}} = 20 \log\left(1 + \frac{1}{k - 1}\right) = 20 \log 1.333 = 2.5\ \text{dB} \tag{5-24}$$

where
$$k = \frac{Q_L}{Q_{\text{eff}}} = \frac{200}{50} = 4$$

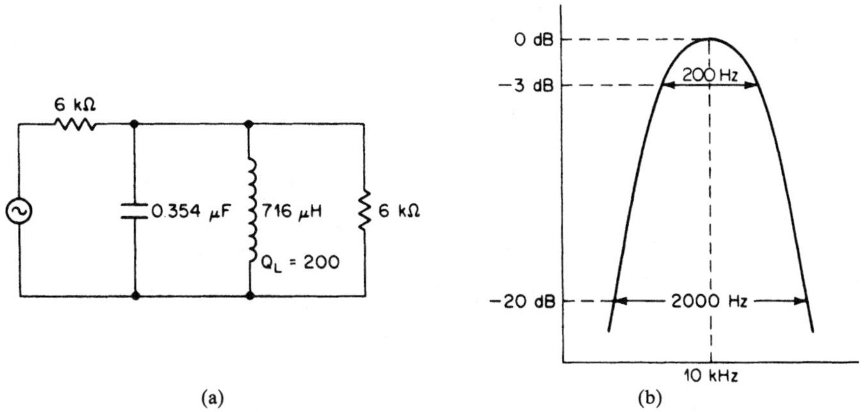

(a) (b)

FIGURE 5-8 The tuned circuit of Example 5-4: (*a*) circuit; and (*b*) frequency response.

The low-pass to bandpass transformation illustrated in Figure 5-5 can also be examined from a pole-zero perspective. The *RC* low-pass filter has a single real pole at $1/RC$, as shown in Figure 5-9*a*, and a zero at infinity. The bandpass transformation results in a pair of complex poles and zeros at the origin and infinity, as illustrated in Figure 5-9*b*. The radial distance from the origin to the pole is $1/(LC)^{1/2}$, corresponding to ω_0, the resonant frequency. The Q can be expressed by

$$Q = \frac{\omega_0}{2\alpha} \tag{5-28}$$

where α, the real part, is $1/2RC$. The transfer function of the circuit of Figure 5-9*b* becomes

$$T(s) = \frac{s}{s^2 + \frac{\omega_0}{Q}s + \omega_0^2} \tag{5-29}$$

At ω_0, the impedance of the parallel resonant circuit is a maximum and is purely resistive, resulting in zero phase shift. If the Q is much less than 10, these effects do not both occur at precisely the same frequency. Series losses of the inductor will also displace the zero from the origin onto the negative real axis.

The Series Tuned Circuit. The losses of an inductor can be conveniently represented by a series resistor determined by

$$R_{\text{coil}} = \frac{\omega L}{Q_L} \tag{5-30}$$

If we form a series resonant circuit and include the source and load resistors, we obtain the circuit of Figure 5-10. Equations (5-24) through (5-27) for insertion loss, frequency response, phase shift, and the group delay of the parallel tuned circuit apply since the two circuits are duals of each other. The inductance is calculated from

$$L = \frac{R_s + R_L}{\omega_0\left(\frac{1}{Q_{\text{bp}}} - \frac{1}{Q_L}\right)} \tag{5-31}$$

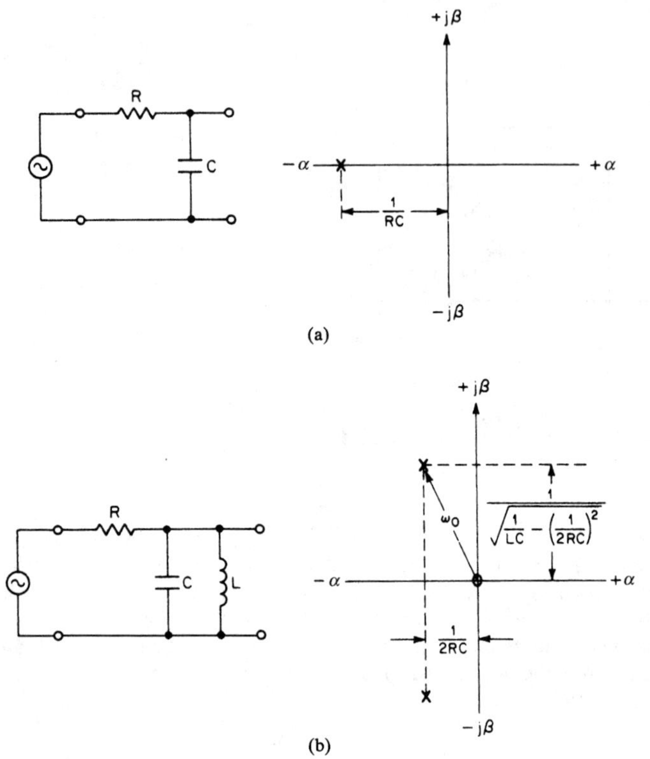

FIGURE 5-9 The bandpass transformation: (*a*) the low-pass circuit; and (*b*) the bandpass circuit.

where Q_{bp} is the required Q, and Q_L is the inductor Q. The capacitance is given by

$$C = \frac{1}{\omega_0^2 L} \tag{5-12}$$

Example 5-5 Design of a Bandpass Filter using a Series Tuned Circuit

Required:

A series tuned circuit

A center frequency of 100 kHz

A 3-dB bandwidth of 2 kHz

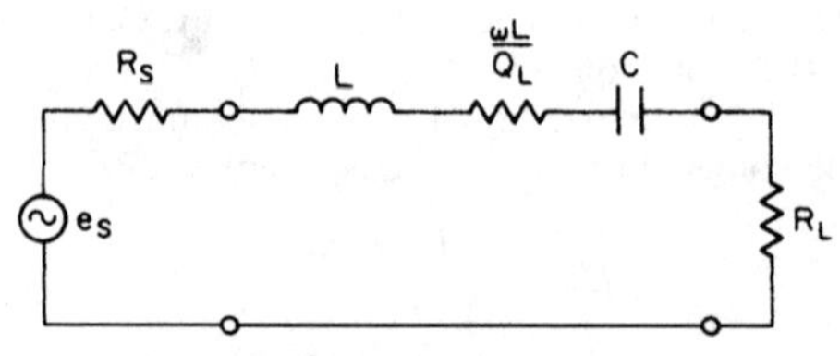

FIGURE 5-10 The series resonant circuit.

$R_s = R_L = 100\ \Omega$
Inductor Q of 400

Result:

(a) Compute the bandpass Q.

$$Q_{\text{bp}} = \frac{f_0}{\text{BW}_{3\text{ dB}}} = \frac{100\text{ kHz}}{2\text{ kHz}} = 50 \tag{2-16}$$

(b) Calculate the element values, using

$$L = \frac{R_s + R_L}{\omega_0\left(\frac{1}{Q_{\text{bp}}} - \frac{1}{Q_L}\right)} = \frac{200}{2\pi \times 10^5\left(\frac{1}{50} - \frac{1}{400}\right)} = 18.2\text{ mH} \tag{5-31}$$

$$C = \frac{1}{\omega_0^2 L} = 139\text{ pF} \tag{5-12}$$

The circuit is shown in Figure 5-11.

The reader may recall from AC circuit theory that one of the effects of series resonance is a buildup of voltage across both reactive elements. The voltage across either reactive element at resonance is equal to Q times the input voltage and may be excessively high, causing inductor saturation or capacitor breakdown. In addition, the L/C ratio becomes large as the bandwidth is reduced and will result in impractical element values where high Qs are required. As a result, series resonant circuits are less desirable than parallel tuned circuits.

Synchronously Tuned Filters. Tuned circuits can be cascaded to obtain bandpass filters of a higher complexity. Each stage must be isolated from the previous section. If all circuits are tuned to the same frequency, a synchronously tuned filter is obtained. The characteristics of synchronously tuned bandpass filters are discussed in Section 2.8, and the normalized frequency response is illustrated by the curves of Figure 2-77. The design Q of each section was given by

$$Q_{\text{section}} = Q_{\text{overall}}\sqrt{2^{1/n} - 1} \tag{2-45}$$

where Q_{overall} is defined by the ratio $f_0/\text{BW}_{3\text{ dB}}$ of the composite filter. The individual circuits may be of either the series or the parallel resonant type.

Synchronously tuned filters are the simplest approximation to a bandpass response. Since all stages are identical and tuned to the same frequency, they are simple to construct and easy to align. The Q requirement of each individual section is less than the overall Q, whereas the opposite is true for conventional bandpass filters. The transient behavior exhibits no overshoot or ringing. On the other hand, the selectivity is extremely poor. To obtain a particular

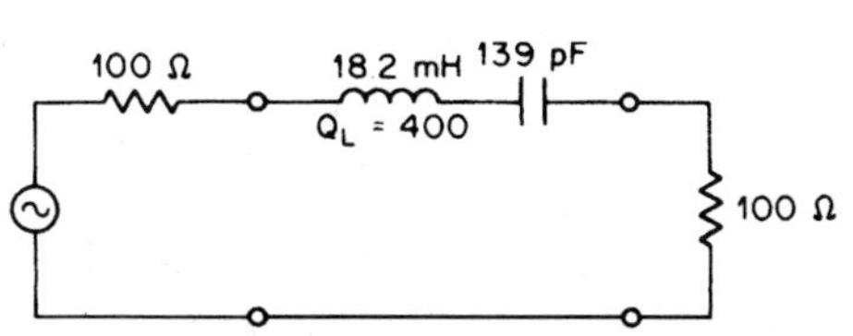

FIGURE 5-11 The series tuned circuit of Example 5-5.

attenuation for a given steepness factor A_s, many more stages are required than for the other filter types. In addition, each section must be isolated from the previous section, so interstage amplifiers are required. The disadvantages generally outweigh the advantages, thus synchronously tuned filters are usually restricted to special applications such as IF and RF amplifiers.

Example 5-6 Design of a Synchronously Tuned Bandpass Filter

Required:

A synchronously tuned bandpass filter
A center frequency of 455 kHz
3 dB at ±5 kHz
30-dB minimum at ±35 kHz
An inductor Q of 400

Result:

(*a*) Compute the bandpass steepness factor.

$$A_s = \frac{\text{stopband bandwidth}}{\text{passband bandwidth}} = \frac{70 \text{ kHz}}{10 \text{ kHz}} = 7 \qquad (2\text{-}19)$$

The curves of Figure 2-77 indicate that a third-order ($n = 3$) synchronously tuned filter satisfies the attenuation requirement.

(*b*) Three sections are required which are all tuned to 455 kHz and have identical Q. To compute the Q of the individual sections, first calculate the overall Q, which is given by

$$Q_{\text{bp}} = \frac{f_0}{\text{BW}_{3\text{ dB}}} = \frac{455 \text{ kHz}}{10 \text{ kHz}} = 45.5 \qquad (2\text{-}16)$$

The section Qs can be found from

$$Q_{\text{section}} = Q_{\text{overall}}\sqrt{2^{1/n} - 1} = 45.5\sqrt{2^{1/3} - 1} = 23.2 \qquad (2\text{-}48)$$

(*c*) The tuned circuits can now be designed using either a series or parallel realization. Let us choose a parallel tuned circuit configuration using a single-source resistor of 10 kΩ and a high-impedance termination. Since an effective circuit Q of 23.2 is desired and the inductor Q is 400, the design Q is calculated from

$$Q_d = \frac{Q_{\text{eff}}Q_L}{Q_L - Q_{\text{eff}}} = \frac{23.2 \times 400}{400 - 23.2} = 24.6 \qquad (5\text{-}23)$$

The inductance is then given by

$$L = \frac{R}{\omega_0 Q_{bp}} = \frac{10 \times 10^3}{2\pi 455 \times 10^3 \times 24.6} = 142 \ \mu\text{H} \qquad (5\text{-}21)$$

The resonating capacitor can be obtained from

$$C = \frac{1}{\omega_0^2 L} = 862 \text{ pF} \qquad (5\text{-}12)$$

The final circuit is shown in Figure 5-12 utilizing buffer amplifiers to isolate the three sections.

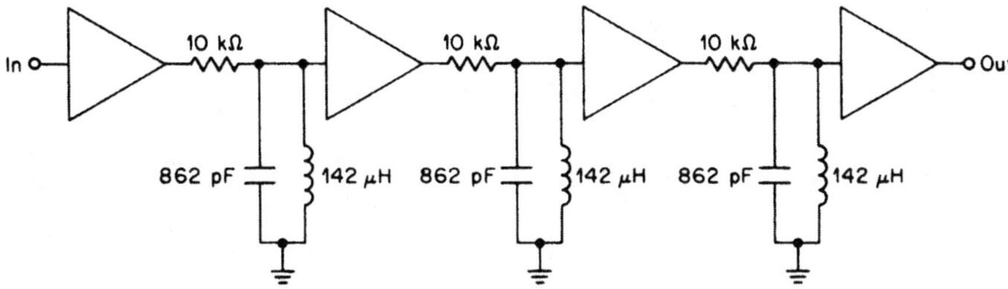

FIGURE 5-12 The synchronously tuned filter of Example 5-6.

Narrowband Coupled Resonators. Narrowband bandpass filters can be designed by using coupling techniques where parallel tuned circuits are interconnected by coupling elements such as inductors or capacitors. Figure 5-13 illustrates some typical configurations.

Coupled resonator configurations are desirable for narrowband filters having bandpass Qs of 10 or more. The values are generally more practical than the elements obtained by the low-pass to bandpass transformation, especially for very high Qs. The tuning is also simpler since it turns out that all nodes are resonated to the same frequency. Of the three configurations shown in Figure 5-13, the capacitive coupled configuration is the most desirable from the standpoint of economy and ease or manufacture.

The theoretical justification for the design method is based on the assumption that the coupling elements have a constant impedance with frequency. This assumption is approximately accurate over narrow bandwidths. At DC, the coupling capacitors will introduce

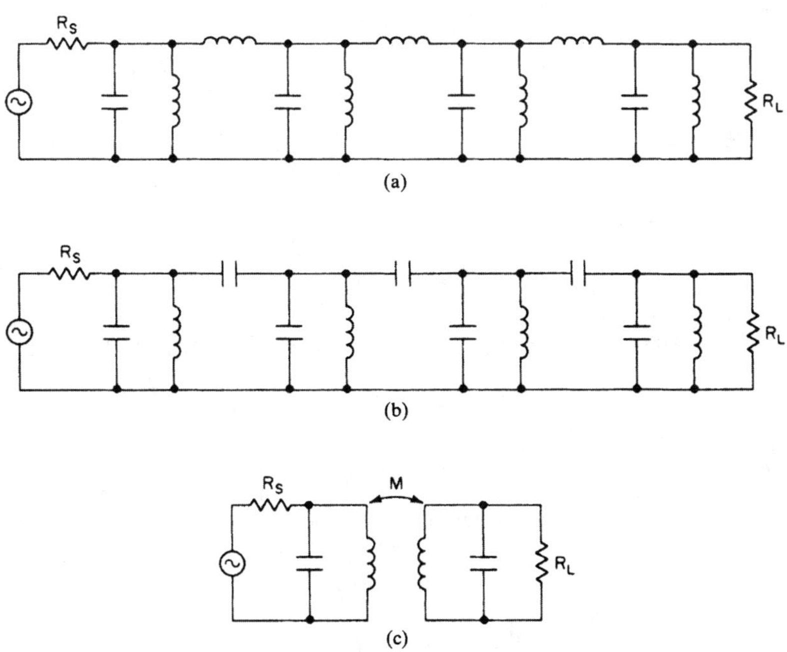

FIGURE 5-13 Coupled resonators: (*a*) inductive coupling; (*b*) capacitive coupling; and (*c*) magnetic coupling.

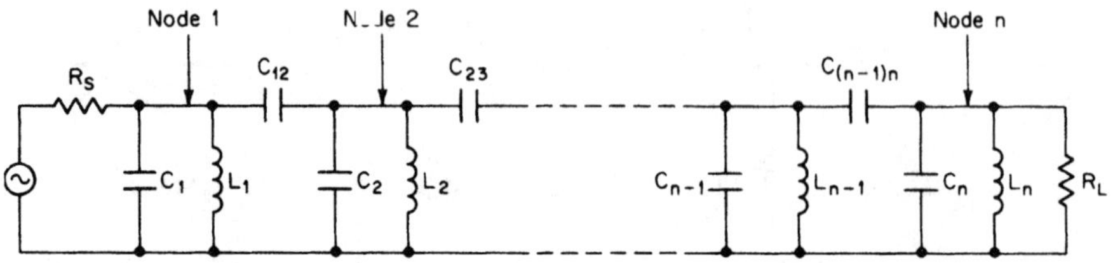

FIGURE 5-14 A general form of a capacitive coupled resonator filter.

additional response zeros. This causes the frequency response to be increasingly unsymmetrical both geometrically and arithmetically as we deviate from the center frequency. The response shape will be somewhat steeper on the low-frequency side of the passband, however.

The general form of a capacitive coupled resonator filter is shown in Figure 5-14. An nth-order filter requires n parallel tuned circuits and contains n nodes. Tables 5-3 through 5-12 present in tabular form q and k parameters for all-pole filters. These parameters are used to generate the component values for filters having the form shown in Figure 5-14. For each network, a q_1 and q_n is given that corresponds to the first and last resonant circuit. The k parameters are given in terms of k_{12}, k_{23}, and so on, and are related to the coupling capacitor shown in Figure 5-14. The design method proceeds as follows:

1. Compute the desired filter's passband Q, which was given by

$$Q_{\text{bp}} = \frac{f_0}{\text{BW}_{3\text{ dB}}} \tag{2-16}$$

2. Determine the q's and k's from the tables corresponding to the chosen filter type and the order of complexity n. Denormalize these coefficients as follows:

$$Q_1 = Q_{\text{bp}} \times q_1 \tag{5-32}$$

$$Q_n = Q_{\text{bp}} \times q_n \tag{5-33}$$

$$K_{xy} = \frac{k_{xy}}{Q_{\text{bp}}} \tag{5-34}$$

3. Choose a convenient inductance value L. The source and load terminations are found from

$$R_s = \omega_0 L Q_1 \tag{5-35}$$

and

$$R_L = \omega_0 L Q_n \tag{5-36}$$

4. The total nodal capacitance is determined by

$$C_{\text{node}} = \frac{1}{\omega_0^2 L} \tag{5-37}$$

The coupling capacitors are then computed from

$$C_{xy} = K_{xy}\, C_{\text{node}} \tag{5-38}$$

TABLE 5-3 Butterworth Capacitive Coupled Resonators

n	q_1	q_n	k_{12}	k_{23}	k_{34}	k_{45}	k_{56}	k_{67}	k_{78}
2	1.414	1.414	0.707						
3	1.000	1.000	0.707	0.707					
4	0.765	0.765	0.841	0.541	0.841				
5	0.618	0.618	1.000	0.556	0.556	1.000			
6	0.518	0.518	1.169	0.605	0.518	0.605	1.169		
7	0.445	0.445	1.342	0.667	0.527	0.527	0.667	1.342	
8	0.390	0.390	1.519	0.736	0.554	0.510	0.554	0.736	1.519

TABLE 5-4 0.01-dB Chebyshev Capacitive Coupled Resonators

n	q_1	q_n	k_{12}	k_{23}	k_{34}	k_{45}	k_{56}	k_{67}	k_{78}
2	1.483	1.483	0.708						
3	1.181	1.181	0.682	0.682					
4	1.046	1.046	0.737	0.541	0.737				
5	0.977	0.977	0.780	0.540	0.540	0.780			
6	0.937	0.937	0.809	0.550	0.518	0.550	0.809		
7	0.913	0.913	0.829	0.560	0.517	0.517	0.560	0.829	
8	0.897	0.897	0.843	0.567	0.520	0.510	0.520	0.567	0.843

TABLE 5-5 0.1-dB Chebyshev Capacitive Coupled Resonators

n	q_1	q_n	k_{12}	k_{23}	k_{34}	k_{45}	k_{56}	k_{67}	k_{78}
2	1.638	1.638	0.711						
3	1.433	1.433	0.662	0.662					
4	1.345	1.345	0.685	0.542	0.685				
5	1.301	1.301	0.703	0.536	0.536	0.703			
6	1.277	1.277	0.715	0.539	0.518	0.539	0.715		
7	1.262	1.262	0.722	0.542	0.516	0.516	0.542	0.722	
8	1.251	1.251	0.728	0.545	0.516	0.510	0.516	0.545	0.728

TABLE 5-6 0.5-dB Chebyshev Capacitive Coupled Resonators

n	q_1	q_n	k_{12}	k_{23}	k_{34}	k_{45}	k_{56}	k_{67}	k_{78}
2	1.950	1.950	0.723						
3	1.864	1.864	0.647	0.647					
4	1.826	1.826	0.648	0.545	0.648				
5	1.807	1.807	0.652	0.534	0.534	0.652			
6	1.796	1.796	0.655	0.533	0.519	0.533	0.655		
7	1.790	1.790	0.657	0.533	0.516	0.516	0.533	0.657	
8	1.785	1.785	0.658	0.533	0.515	0.511	0.515	0.533	0.658

TABLE 5-7 1-dB Chebyshev Capacitive Coupled Resonators

n	q_1	q_n	k_{12}	k_{23}	k_{34}	k_{45}	k_{56}	k_{67}
2	2.210	2.210	0.739					
3	2.210	2.210	0.645	0.645				
4	2.210	2.210	0.638	0.546	0.638			
5	2.210	2.210	0.633	0.535	0.538	0.633		
6	2.250	2.250	0.631	0.531	0.510	0.531	0.531	
7	2.250	2.250	0.631	0.530	0.517	0.517	0.530	0.631

TABLE 5-8 Bessel Capacitive Coupled Resonators

n	q_1	q_n	k_{12}	k_{23}	k_{34}	k_{45}	k_{56}	k_{67}	k_{78}
2	0.5755	0.148	0.900						
3	0.337	2.203	1.748	0.684					
4	0.233	2.240	2.530	1.175	0.644				
5	0.394	0.275	1.910	0.750	0.650	1.987			
6	0.415	0.187	2.000	0.811	0.601	1.253	3.038		
7	0.187	0.242	3.325	1.660	1.293	0.695	0.674	2.203	
8	0.139	0.242	4.284	2.079	1.484	1.246	0.678	0.697	2.286

TABLE 5-9 Linear Phase with Equiripple Error of 0.05° Capacitive Coupled Resonators

n	q_1	q_n	k_{12}	k_{23}	k_{34}	k_{45}	k_{56}	k_{67}	k_{78}
2	0.648	2.109	0.856						
3	0.433	2.254	1.489	0.652					
4	0.493	0.718	1.632	0.718	0.739				
5	0.547	0.446	1.800	0.848	0.584	1.372			
6	0.397	0.468	1.993	1.379	0.683	0.661	1.553		
7	0.316	0.484	2.490	1.442	1.446	0.927	0.579	1.260	
8	0.335	0.363	2.585	1.484	1.602	1.160	0.596	0.868	1.733

TABLE 5-10 Linear Phase with Equiripple Error of 0.5° Capacitive Coupled Resonators

n	q_1	q_n	k_{12}	k_{23}	k_{34}	k_{45}	k_{56}	k_{67}	k_{78}
2	0.825	1.980	0.783						
3	0.553	2.425	1.330	0.635					
4	0.581	1.026	1.575	0.797	0.656				
5	0.664	0.611	1.779	0.919	0.576	1.162			
6	0.552	0.586	1.874	1.355	0.641	0.721	1.429		
7	0.401	0.688	2.324	1.394	1.500	1.079	0.590	1.045	
8	0.415	0.563	2.410	1.470	1.527	1.409	0.659	0.755	1.335

TABLE 5-11 Transitional Gaussian to 6-dB Capacitive Coupled Resonators

n	q_1	q_n	k_{12}	k_{23}	k_{34}	k_{45}	k_{56}	k_{67}	k_{78}
3	0.404	2.338	1.662	0.691					
4	0.570	0.914	1.623	0.798	0.682				
5	0.891	0.670	1.418	0.864	0.553	1.046			
6	0.883	0.752	1.172	1.029	0.595	0.605	1.094		
7	0.736	0.930	1.130	0.955	0.884	0.534	0.633	1.104	
8	0.738	0.948	1.124	0.866	0.922	0.708	0.501	0.752	1.089

TABLE 5-12 Transitional Gaussian to 12-dB Capacitive Coupled Resonators

n	q_1	q_n	k_{12}	k_{23}	k_{34}	k_{45}	k_{56}	k_{67}	k_{78}
3	0.415	2.345	1.631	0.686					
4	0.419	0.766	1.989	0.833	0.740				
5	0.534	0.503	2.085	0.976	0.605	1.333			
6	0.543	0.558	1.839	1.442	0.686	0.707	1.468		
7	0.492	0.665	1.708	1.440	1.181	0.611	0.781	1.541	
8	0.549	0.640	1.586	1.262	1.296	0.808	0.569	1.023	1.504

5. The total capacity connected to each node must be equal to C_{node}. Therefore, the shunt capacitors of the parallel tuned circuits are equal to the total nodal capacitance C_{node}, minus the values of the coupling capacitors connected to that node. For example

$$C_1 = C_{node} - C_{12}$$

$$C_2 = C_{node} - C_{12} - C_{23}$$

$$C_7 = C_{node} - C_{67} - C_{78}$$

Each node is tuned to f_0 with the adjacent nodes shorted to ground so that the coupling capacitors connected to that node are placed in parallel across the tuned circuit.

The completed filter may require impedance scaling so that the source and load terminating requirements are met. In addition, some of the impedance transformations discussed in Chapter 8 may have to be applied.

The k and q values tabulated in Tables 5-3 through 5-12 are based on infinite inductor Q. In reality, satisfactory results will be obtained for inductor Qs several times higher than Q_{min} (bandpass), determined by Equation (5-17) in conjunction with Figure 3-8 which shows the minimum theoretical low-pass Qs.

Example 5-7 Design of a Capacitive Coupled Resonator Bandpass Filter

Required:

A bandpass filter

Center frequency of 100 kHz

3 dB at ±2.5 kHz

35-dB minimum at ±12.5 kHz

Constant delay over the passband

Result:

(*a*) Since a constant delay is required, a Bessel filter type will be chosen. The low-pass constant delay properties will undergo a minimum of distortion for the bandpass case since the bandwidth is relatively narrow—that is, the bandpass Q is high. Because the bandwidth is narrow, we can treat the requirements on an arithmetically symmetrical basis.

The bandpass steepness factor is given by

$$A_s = \frac{\text{stopband bandwidth}}{\text{passband bandwidth}} = \frac{25\text{ kHz}}{5\text{ kHz}} = 5 \qquad (2\text{-}19)$$

The frequency-response curves of Figure 2-56 indicate that an $n = 4$ Bessel filter provides over 35 dB of attenuation at 5 rad/s. A capacitive coupled resonator configuration will be used for the implementation.

(*b*) The q and k parameters for a Bessel filter corresponding to $n = 4$ are found in Table 5-8 and are as follows:

$$q_1 = 0.233$$

$$q_4 = 2.240$$

$$k_{12} = 2.530$$

$$k_{23} = 1.175$$

$$k_{34} = 0.644$$

To denormalize these values, divide each k by the bandpass Q and multiply each q by the same factor as follows:

$$Q_{\text{bp}} = \frac{f_0}{\text{BW}_{3\text{ dB}}} = \frac{100\text{ kHz}}{5\text{ kHz}} = 20 \qquad (2\text{-}16)$$

The resulting values are

$$Q_1 = Q_{\text{bp}} \times q_1 = 20 \times 0.233 = 4.66 \qquad (5\text{-}32)$$

$$Q_4 = 44.8$$

$$K_{12} = \frac{k_{12}}{Q_{\text{bp}}} = \frac{2.530}{20} = 0.1265 \qquad (5\text{-}34)$$

$$K_{23} = 0.05875$$

$$K_{34} = 0.0322$$

(*c*) Let's choose an inductance of $L = 2.5$ mH. The source and load terminations are

$$R_s = \omega_0 L Q_1 = 6.28 \times 10^5 \times 2.5 \times 10^{-3} \times 4.66 = 7.32\text{ k}\Omega \qquad (5\text{-}35)$$

and $$R_L = \omega_0 L Q_4 = 70.37\text{ k}\Omega \qquad (5\text{-}36)$$

where $$\omega_0 = 2\pi f_0$$

(*d*) The total nodal capacitance is determined by

$$C_{\text{node}} = \frac{1}{\omega_0^2 L} = 1013 \text{ pF} \tag{5-37}$$

The coupling capacitors can now be calculated, using

$$C_{12} = K_{12}\, C_{\text{node}} = 0.1265 \times 1.013 \times 10^{-9} = 128.1 \text{ pF} \tag{5-38}$$

$$C_{23} = K_{23}\, C_{\text{node}} = 59.5 \text{ pF}$$

$$C_{34} = K_{34}\, C_{\text{node}} = 32.6 \text{ pF}$$

The shunt capacitors are determined from

$$C_1 = C_{\text{node}} - C_{12} = 884.9 \text{ pF}$$

$$C_2 = C_{\text{node}} - C_{12} - C_{23} = 825.4 \text{ pF}$$

$$C_3 = C_{\text{node}} - C_{23} - C_{34} = 920.9 \text{ pF}$$

$$C_4 = C_{\text{node}} - C_{34} = 980.4 \text{ pF}$$

The final circuit is shown in Figure 5-15.

Predistorted Bandpass Filters. The inductor Q requirements of the bandpass filters are higher than those of low-pass filters since the minimum theoretical branch Q is given by

$$Q_{\min}\,(\text{bandpass}) = Q_{\min}\,(\text{low-pass}) \times Q_{\text{bp}} \tag{5-17}$$

where $Q_{\text{bp}} = f_0/\text{BW}_{3\text{ dB}}$. In the cases where the filter required is extremely narrow, a branch Q many times higher than the minimum theoretical Q may be difficult to obtain. Predistorted bandpass filters can then be used so that exact theoretical results can be obtained with reasonable branch Qs.

Predistorted bandpass filters can be obtained from the normalized predistorted low-pass filters given in Chapter 11 by the conventional bandpass transformation. The low-pass filters must be of the uniform dissipation type since the lossy-L networks would be transformed to a bandpass filter having losses in the series branches only.

The uniform dissipation networks are tabulated for different values of dissipation factor d. These values relate to the required inductor Q by the relationship

$$Q_L = \frac{Q_{\text{bp}}}{d} \tag{5-39}$$

where $Q_{\text{bp}} = f_0/\text{BW}_{3\text{ dB}}$.

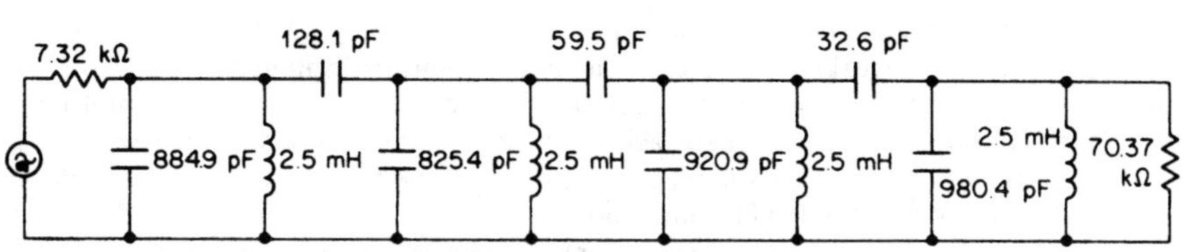

FIGURE 5-15 The capacitive coupled resonator filter of Example 5-7.

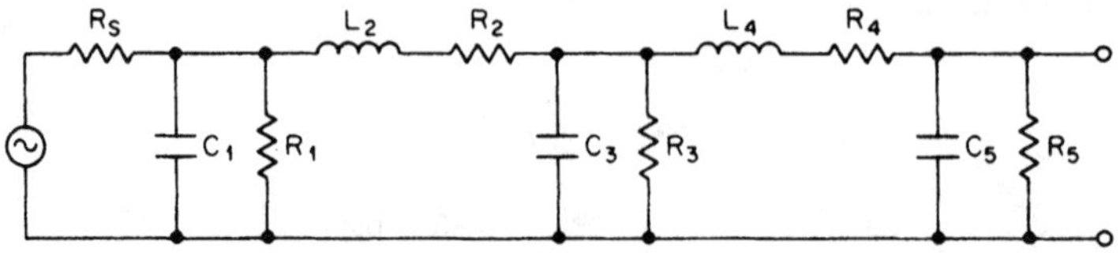

(a)

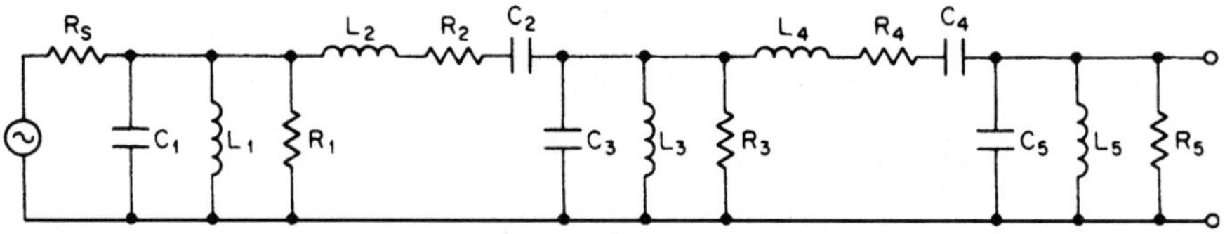

(b)

FIGURE 5-16 The location of losses in uniformly predistorted filters: (*a*) a low-pass filter; and (*b*) a bandpass filter.

The losses of a predistorted low-pass filter having uniform dissipation are evenly distributed and occur as both series losses in the inductors and shunt losses across the capacitors. The equivalent circuit of the filter is shown in Figure 5-16*a*. The inductor losses were previously given by

$$R_L = \frac{\omega L}{Q} \tag{3-2}$$

and the capacitor losses were defined by

$$R_c = \frac{Q}{\omega C} \tag{3-3}$$

When the circuit is transformed to a bandpass filter, the losses are still required to be distributed in series with the series branches and in parallel with the shunt branches, as shown in Figure 5-16*b*. In reality, the capacitor losses are minimal and the inductor losses occur in series with the inductive elements in both the series and shunt branches. Therefore, as a narrowband approximation, the losses may be distributed between the capacitors and inductors in an arbitrary manner. The only restriction is that the combination of inductor and capacitor losses in each branch results in a total branch Q equal to the value computed by Equation (5-39). The combined Q of a lossy inductor and a lossy capacitor in a resonant circuit is given by

$$Q_T = \frac{Q_L Q_C}{Q_L + Q_C} \tag{5-40}$$

where Q_T is the total branch Q, Q_L is the inductor Q, and Q_C is the Q of the capacitor.

The predistorted networks tabulated in Chapter 11 require an infinite termination on one side. In practice, if the resistance used to approximate the infinite termination is large compared with the source termination, satisfactory results will be obtained. If the dual configuration is used, which ideally requires a zero impedance source, the source impedance should be much less than the load termination.

It is usually difficult to obtain inductor Qs precisely equal to the values computed from Equation (5-39). A Q accuracy within 5 or 10 percent at f_0 is usually sufficient. If greater accuracy is required, an inductor Q higher than the calculated value is used. The Q is then degraded to the exact required value by adding resistors.

Example 5-8 Design of a Predistorted *LC* Bandpass Filter

Required:

A bandpass filter
Center frequency of 10 kHz
3 dB at ±250 Hz
60-dB minimum at ±750 Hz
$R_s = 100\ \Omega \qquad R_L = 10\ \text{k}\Omega$ minimum
An available inductor Q of 225

Result:

(*a*) Since the filter is narrow in bandwidth, the requirement is treated in its arithmetically symmetrical form. The bandpass-steepness factor is obtained from

$$A_s = \frac{\text{stopband bandwidth}}{\text{passband bandwidth}} = \frac{1500\ \text{Hz}}{500\ \text{Hz}} = 3 \qquad (2\text{-}19)$$

The curves of Figure 2-43 indicate that a fifth-order ($n = 5$) 0.25-dB Chebyshev filter will meet these requirements. A predistorted design will be used. The corresponding normalized low-pass filters are found in Table 11-33.

(*b*) The specified inductor Q can be used to compute the required d of the low-pass filter as follows:

$$d = \frac{Q_{\text{bp}}}{Q_L} = \frac{20}{225} = 0.0889 \qquad (5\text{-}39)$$

where $Q_{\text{bp}} = f_0/\text{BW}_{3\ \text{dB}}$. The circuit corresponding to $n = 5$ and $d = 0.0919$ will be selected since this d is sufficiently close to the computed value. The schematic is shown in Figure 5-17*a*.

(*c*) Denormalize the low-pass filter using a frequency-scaling factor (FSF) of $2\pi f_c$ or 3140, where $f_c = 500$ Hz, the required bandwidth of the bandpass filter, and an impedance-scaling factor Z of 100.

$$C_1' = \frac{C}{\text{FSF} \times Z} = \frac{1.0397}{3140 \times 100} = 3.309\ \mu\text{F} \qquad (2\text{-}10)$$

$$C_3' = 7.014\ \mu\text{F}$$

$$C_5' = 3.660\ \mu\text{F}$$

and

$$L_2' = \frac{L \times Z}{\text{FSF}} = \frac{1.8181 \times 100}{3140} = 57.87\ \text{mH} \qquad (2\text{-}9)$$

$$L_4' = 55.79\ \text{mH}$$

The denormalized low-pass filter is shown in Figure 5-17*b*, where the termination has been scaled to 100 Ω. The filter has also been turned end for end since the high-impedance termination is required at the output and the 100-Ω source at the input.

(*d*) To transform the circuit into a bandpass filter, resonate each capacitor with an inductor in parallel, and each inductor with a series capacitor using a resonant frequency of $f_0 = 10$ kHz. The parallel inductor is computed from

$$L = \frac{1}{\omega_0^2 C} \tag{5-11}$$

and the series capacitor is calculated by

$$C = \frac{1}{\omega_0^2 L} \tag{5-12}$$

where both formulas are forms of the general relationship for resonance: $\omega_0^2 LC = 1$.The resulting bandpass filter is given in Figure 5-17*c*. The large spread of values can be reduced by applying some of the techniques later discussed in Chapter 8.

Elliptic-Function Bandpass Filters. Elliptic-function low-pass filters were clearly shown to be far superior to the other filter types in terms of achieving a required attenuation within a given frequency ratio. This superiority is mainly the result of the presence of transmission zeros beginning just outside the passband.

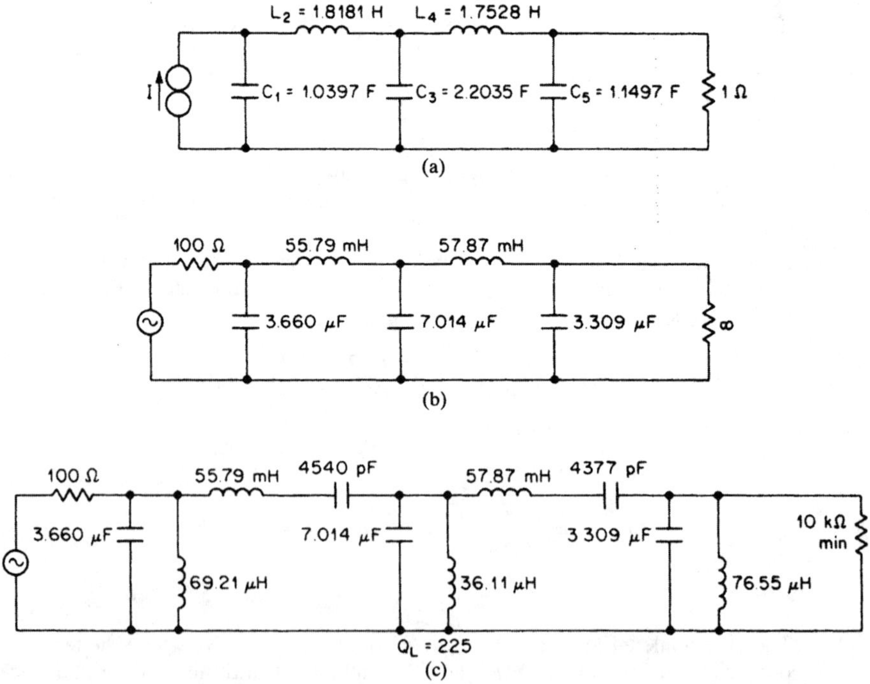

FIGURE 5-17 The predistorted bandpass filter of Example 5-8: (*a*) normalized low-pass filter; (*b*) frequency- and impedance-scaled network; and (*c*) resulting bandpass filter.

Elliptic-function *LC* low-pass filters have been extensively tabulated by Saal and Ulbrich and by Zverev (see the Bibliography). A program called *Filter Solutions* is included on the CD-ROM and allows the design of elliptic-function *LC* filters (up to $n = 10$). These networks can be transformed into bandpass filters in the same manner as the all-pole filter types. The elliptic-function bandpass filters will then exhibit the same superiority over the all-pole types as their low-pass counterparts.

When an elliptic-function low-pass filter is transformed into a bandpass filter, each low-pass transmission zero is converted into a pair of zeros, one above and one below the passband, and are geometrically related to the center frequency. (For the purposes of this discussion, negative zeros will be disregarded.) The low-pass zeros are directly determined by the resonances of the parallel tuned circuits in the series branches. When each series branch containing a parallel tuned circuit is modified by the bandpass transformation, two parallel branch resonances are introduced corresponding to the upper and lower zeros.

A sixth-order elliptic-function low-pass filter structure is shown in Figure 5-18*a*. After frequency- and impedance-scaling the low-pass values, we can make a bandpass transformation by resonating each inductor with a series capacitor, and each capacitor with a parallel inductor, where the resonant frequency is f_0, the filter center frequency. The circuit of Figure 5-18*b* results. The configuration obtained in branches 2 and 4 corresponds to a type III network from Table 5-2.

The type III network realizes two parallel resonances corresponding to a geometrically related pair of transmission zeros above and below the passband. The circuit configuration itself is not very desirable. The elements corresponding to both parallel resonances are not distinctly isolated. Each resonance is determined by the interaction of a number of elements, so tuning is made difficult. Also, for very narrow filters, the values may become unreasonable. Fortunately, an alternate circuit exists that provides a more practical relationship between the coils and capacitors. The two equivalent configurations are shown in Figure 5-19. The alternate configuration utilizes two parallel tuned circuits where each condition of parallel resonance directly corresponds to a transmission zero.

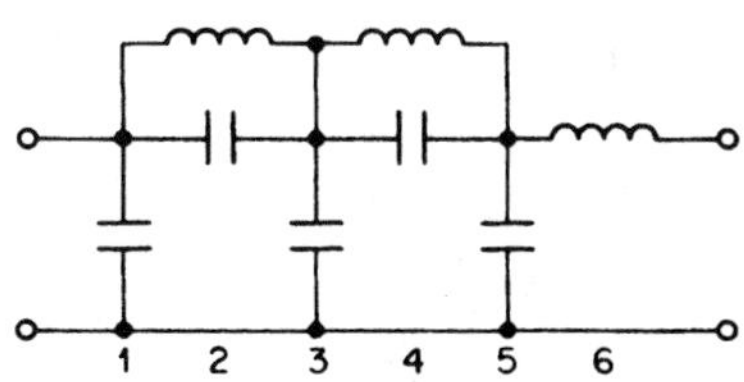

(a)

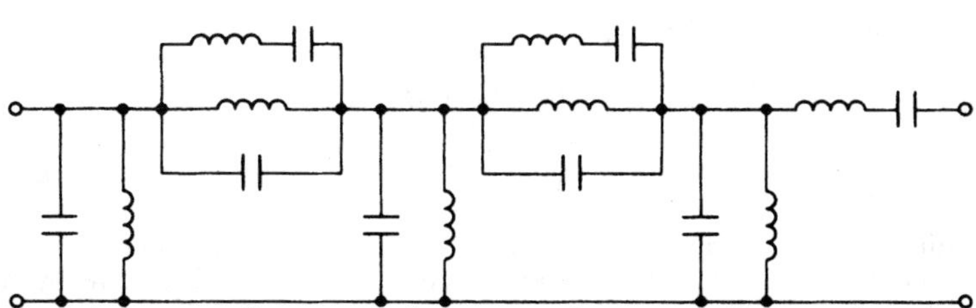

(b)

FIGURE 5-18 The low-pass to bandpass transformation of an elliptic-function filter: (*a*) $n = 6$ low-pass filter; and (*b*) transformed bandpass configuration.

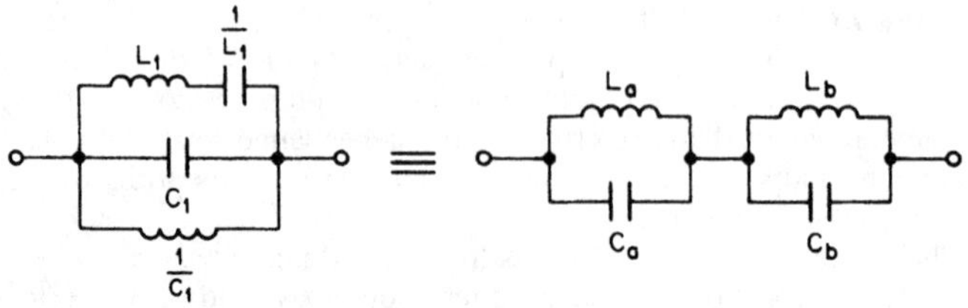

FIGURE 5-19 Equivalent circuit of a type III network.

The type III network of Figure 5-19 is shown with reciprocal element values. These result when we normalize the bandpass filter to a center frequency of 1 rad/s. Since the general equation for resonance $\omega_0^2 LC = 1$ reduces to $LC = 1$ at $\omega_0 = 1$, the resonant elements become reciprocals of each other.

The reason for this normalization is to greatly simplify the equations for the transformation shown in Figure 5-19. Otherwise, the equations relating the two circuits would be significantly more complex. Therefore, elliptic function bandpass filters are first designed, normalized, and then scaled to the required center frequency and impedance.

To obtain the normalized bandpass filter, first multiply all L and C values of the normalized low-pass filter by Q_{bp}, which is equal to f_0/BW, where BW is the passband bandwidth. The network can then be transformed directly into a normalized bandpass filter by resonating each inductor with a series capacitor, and each capacitor with a parallel inductor. The resonant elements are merely reciprocals of each other since $\omega_0 = 1$.

The transformation of Figure 5-19 can now be performed. First calculate

$$\beta = 1 + \frac{1}{2L_1C_1} + \sqrt{\frac{1}{4L_1^2C_1^2} + \frac{1}{L_1C_1}} \tag{5-41}$$

The values are then obtained from

$$L_a = \frac{1}{C_1(\beta + 1)} \tag{5-42}$$

$$L_b = \beta L_a \tag{5-43}$$

$$C_a = \frac{1}{L_b} \tag{5-44}$$

$$C_b = \frac{1}{L_a} \tag{5-45}$$

The resonant frequencies are given by

$$\Omega_{\infty,a} = \sqrt{\beta} \tag{5-46}$$

and

$$\Omega_{\infty,b} = \frac{1}{\Omega_{\infty,a}} \tag{5-47}$$

After the transformation of Figure 5-19 is made wherever applicable, the normalized bandpass filter is scaled to the required center frequency and impedance level by multiplying all inductors by Z/FSF and dividing all capacitors by Z × FSF. The frequency-scaling factor in this case is equal to ω_0 ($\omega_0 = 2\pi f_0$), where f_0 is the desired center frequency of the filter. The resonant frequencies in hertz can be found by multiplying all normalized radian resonant frequencies by f_0.

The design of an elliptic-function bandpass filter is demonstrated by the following example.

Example 5-9 Designing an *LC* Elliptic Function Bandpass Filter

Required:

A bandpass filter

1-dB maximum variation from 15 to 20 kHz

50-dB minimum below 14.06 kHz and above 23 kHz

$R_s = R_L = 10\ \text{k}\Omega$

Result:

(*a*) Convert to a geometrically symmetrical bandpass requirement:

First, calculate the geometric center frequency.

$$f_0 = \sqrt{f_L f_u} = \sqrt{15 \times 20 \times 10^6} = 17.32\ \text{kHz} \tag{2-14}$$

Compute the corresponding geometric frequency for each stopband frequency given, using the relationship

$$f_1 f_2 = f_0^2 \tag{2-18}$$

f_1	f_2	$f_2 - f_1$
14.06 kHz	21.34 kHz	7.28 kHz
13.04 kHz	23.00 kHz	9.96 kHz

The first pair of frequencies has the lesser separation and therefore represents the more severe requirement. Thus, it will be retained. The geometrically symmetrical requirements can be summarized as

$$f_0 = 17.32\ \text{kHz}$$

$$\text{BW}_{1\ \text{dB}} = 5\ \text{kHz}$$

$$\text{BW}_{50\ \text{dB}} = 7.28\ \text{kHz}$$

(*b*) Compute the bandpass steepness factor.

$$A_s = \frac{\text{stopband bandwidth}}{\text{passband bandwidth}} = \frac{7.28\ \text{kHz}}{5\ \text{kHz}} = 1.456 \tag{2-19}$$

(*c*) Open ***Filter Solutions***.

Check the ***Stop Band Freq*** box.

Enter **.18** in the ***Pass Band Ripple(dB)*** box.

Enter **1** in ***Pass Band Freq*** box.

Enter **1.456** in the ***Stop Band Freq*** box.

Check the ***Frequency Scale Rad/Sec*** box.

Enter **1** for ***Source Res*** and ***Load Res***.

(*d*) Click the ***Set Order*** control button to open the second panel.

Enter **50** for the ***Stopband Attenuation (dB)***.

Click the ***Set Minimum Order*** button and then click ***Close***.

6 Order is displayed on the main control panel.

Check the ***Even Order Mod*** box.

(*e*) Click the ***Circuits*** button.

Two schematics are presented. Select **Passive Filter 1**, shown in Figure 5-20*a*.

(*f*) The filter must now be converted to a normalized bandpass filter having a center frequency of $\omega_0 = 1$. The bandpass Q is first computed from

$$Q_{\text{bp}} = \frac{f_0}{\text{BW}} = \frac{17.32\ \text{kHz}}{5\ \text{kHz}} = 3.464$$

Multiply all inductance and capacitance values by Q_{bp}. Then, transform the network into a bandpass filter centered at $\omega_0 = 1$ by resonating each capacitor with a parallel inductor and each inductor with a series capacitor. The resonating elements introduced are simply the reciprocal values, as shown in Figure 5-20*b*.

(*g*) The type III branches will now be transformed in accordance with Figure 5-19.

For the third branch

$$L_1 = 4.451\ \text{H} \quad C_1 = 1.199\ \text{F}$$

First, compute

$$\beta = 1 + \frac{1}{2L_1C_1} + \sqrt{\frac{1}{4L_1^2C_1^2} + \frac{1}{L_1C_1}} = 1.5366 \tag{5-41}$$

then

$$L_a = \frac{1}{C_1(\beta + 1)} = 0.3288\ \text{H} \tag{5-42}$$

$$L_b = \beta L_a = 0.5052\ \text{H} \tag{5-43}$$

$$C_a = \frac{1}{L_b} = 1.9793\ \text{F} \tag{5-44}$$

$$C_b = \frac{1}{L_a} = 3.0414\ \text{F} \tag{5-45}$$

The resonant frequencies are

$$\Omega_{\infty,a} = \sqrt{\beta} = 1.2396 \tag{5-46}$$

$$\Omega_{\infty,b} = \frac{1}{\Omega_{\infty,b}} = 0.8067 \tag{5-47}$$

For the fifth branch:

$$L_1 = 4.43\ \text{H} \quad C_1 = 0.684\ \text{F}$$

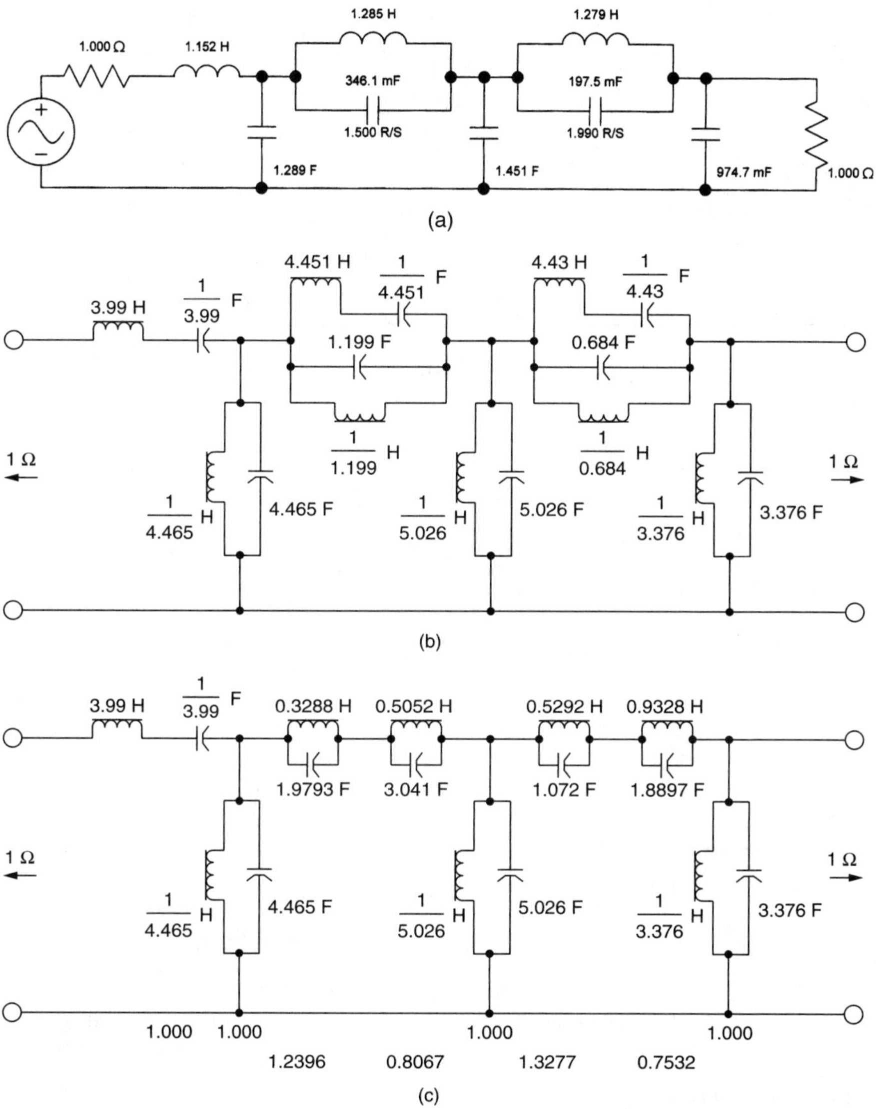

FIGURE 5-20 The elliptic-function bandpass filter: (*a*) normalized low-pass filter; and (*b*) bandpass filter normalized to $\omega_0 = 1$. An elliptic-function bandpass filter; (*c*) transformed type III branches; (*d*) final scaled circuit; and (*e*) frequency response.

then

$$\beta = 1.7627$$

$$L_a = 0.5292 \text{ H}$$

$$L_b = 0.9328 \text{ H}$$

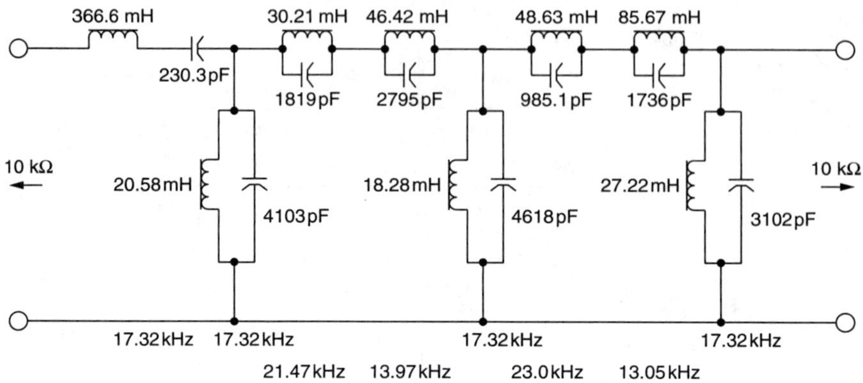

(d)

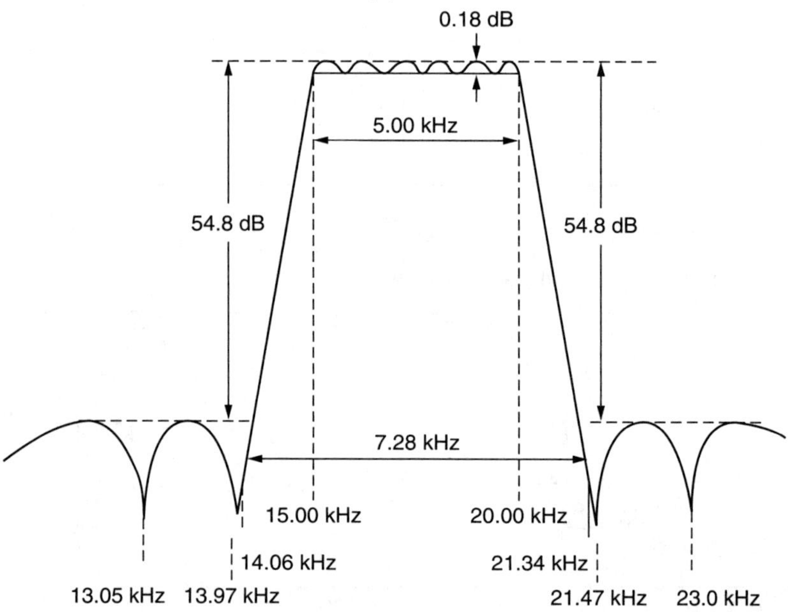

(e)

FIGURE 5-20 (*Continued*)

$$C_a = 1.072 \text{ F}$$

$$C_b = 1.8897 \text{ F}$$

$$\Omega_{\infty,a} = 1.3277$$

$$\Omega_{\infty,a} = 0.7532$$

The transformed filter is shown in Figure 5-20*c*. The resonant frequencies in radians per second are indicated below the schematic.

(*h*) To complete the design, denormalize the filter to a center frequency (f_0) of 17.32 kHz, and an impedance level of 10 kΩ. Multiply all inductors by Z/FSF and divide all capacitors by Z × FSF, where $Z = 10^4$ and FSF $= 2\pi f_0$ or 1.0882×10^5. The final filter is shown in Figure 5-20*d*. The resonant frequencies were obtained by directed multiplication of the normalized resonant frequencies of Figure 5-20*c* by the geometric center frequency f_0 of 17.32 kHz. The frequency response is shown in Figure 5-20*e*.

5.2 ACTIVE BANDPASS FILTERS

When the separation between the upper and lower cutoff frequencies exceeds a ratio of approximately 2, the bandpass filter is considered a wideband type. The specifications are then separated into individual low-pass and high-pass requirements and met by a cascade of active low-pass and high-pass filters.

Narrowband *LC* bandpass filters are usually designed by transforming a low-pass configuration directly into the bandpass circuit. Unfortunately, no such circuit transformation exists for active networks. The general approach involves transforming the low-pass transfer function into a bandpass type. The bandpass poles and zeros are then implemented by a cascade of bandpass filter sections.

Wideband Filters

When *LC* low-pass and high-pass filters were cascaded, care had to be taken to minimize terminal impedance variations so that each filter maintained its individual response in the cascaded form. Active filters can be interconnected with no interaction because of the inherent buffering of the operational amplifiers. The only exception occurs in the case of the elliptic-function VCVS filters of sections 3.2 and 4.2 where the last sections are followed by an *RC* network to provide the real poles. An amplifier must then be introduced for isolation.

Figure 5-21 shows two simple amplifier configurations which can be used after an active elliptic function VCVS filter. The gain of the voltage follower is unity. The noninverting amplifier has a gain equal to $R_2/R_1 + 1$. The resistors R_1 and R_2 can have any convenient values since only their ratio is of significance.

Example 5-10 Designing a Wideband Active Bandpass Filter

Required:

An active bandpass filter

1-dB maximum variation from 3000 to 9000 Hz

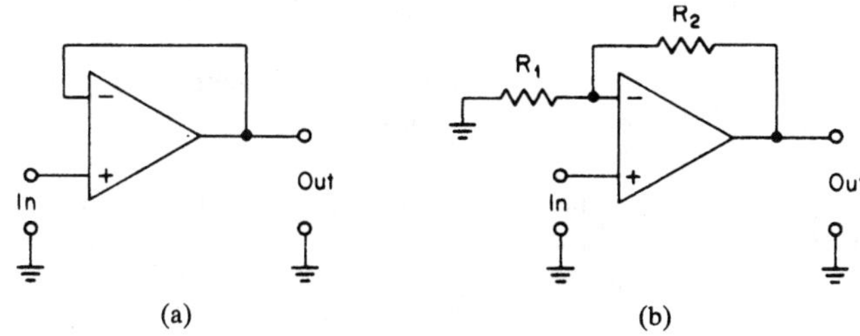

FIGURE 5-21 Isolation amplifiers: (*a*) voltage follower; and (*b*) noninverting amplifier.

35-dB minimum below 1000 Hz and above 18,000 Hz

Gain of +20 dB

Result:

(*a*) Since the ratio of upper cutoff frequency to lower cutoff frequency is well in excess of an octave, the design will be treated as a cascade of low-pass and high-pass filters. The frequency-response requirement can be restated as the following set of individual low-pass and high-pass specifications:

High-pass filter:	***Low-pass filter:***
1-dB maximum at 3000 Hz	1-dB maximum at 9000 Hz
35-dB minimum below 1000 Hz	35-dB minimum above 18,000 Hz

(*b*) To design the high-pass filter, first compute the high-pass steepness factor.

$$A_s = \frac{f_c}{f_s} = \frac{3000 \text{ Hz}}{1000 \text{ Hz}} = 3 \tag{2-13}$$

A normalized low-pass filter must now be chosen that makes the transition from less than 1 dB to more than 35 dB within a frequency ratio of 3:1. An elliptic-function type will be selected. The high-pass filter designed in Example 4-5 illustrates this process and will meet this requirement. The circuit is shown in Figure 5-22*a*.

(*c*) The low-pass filter is now designed. The low-pass steepness factor is computed by

$$A_s = \frac{f_s}{f_c} = \frac{18{,}000 \text{ Hz}}{9{,}000 \text{ Hz}} = 2 \tag{2-11}$$

A low-pass filter must be selected that makes the transition from less than 1 dB to more than 35 dB within a frequency ratio of 2:1. The curves of Figure 2-44 indicate that the attenuation of a normalized 0.5-dB Chebyshev filter of a complexity of $n = 5$ is less than 1 dB at 0.9 rad/s and more than 35 dB at 1.8 rad/s, which satisfies the requirements. The corresponding active filter is found in Table 11-39 and is shown in Figure 5-22*b*.

To denormalize the low-pass circuit, first compute the FSF, which is given by

$$\text{FSF} = \frac{\text{desired reference frequency}}{\text{existing reference frequency}}$$

$$= \frac{2\pi \times 9000 \text{ rad/s}}{0.9 \text{ rad/s}} = 62{,}830 \tag{2-1}$$

The filter is then denormalized by dividing all capacitors by $Z \times \text{FSF}$ and multiplying all resistors by Z, where Z is arbitrarily chosen at 10^4. The denormalized circuit is shown in Figure 5-22*c*.

(*d*) To complete the design, the low-pass and high-pass filters are cascaded. Since the real-pole *RC* network of the elliptic high-pass filter must be buffered and since a gain of +20 dB is required, the noninverting amplifier of Figure 5-21*b* will be used.

The finalized design is shown in Figure 5-22*d*, where the resistors have been rounded off to standard 1-percent values.

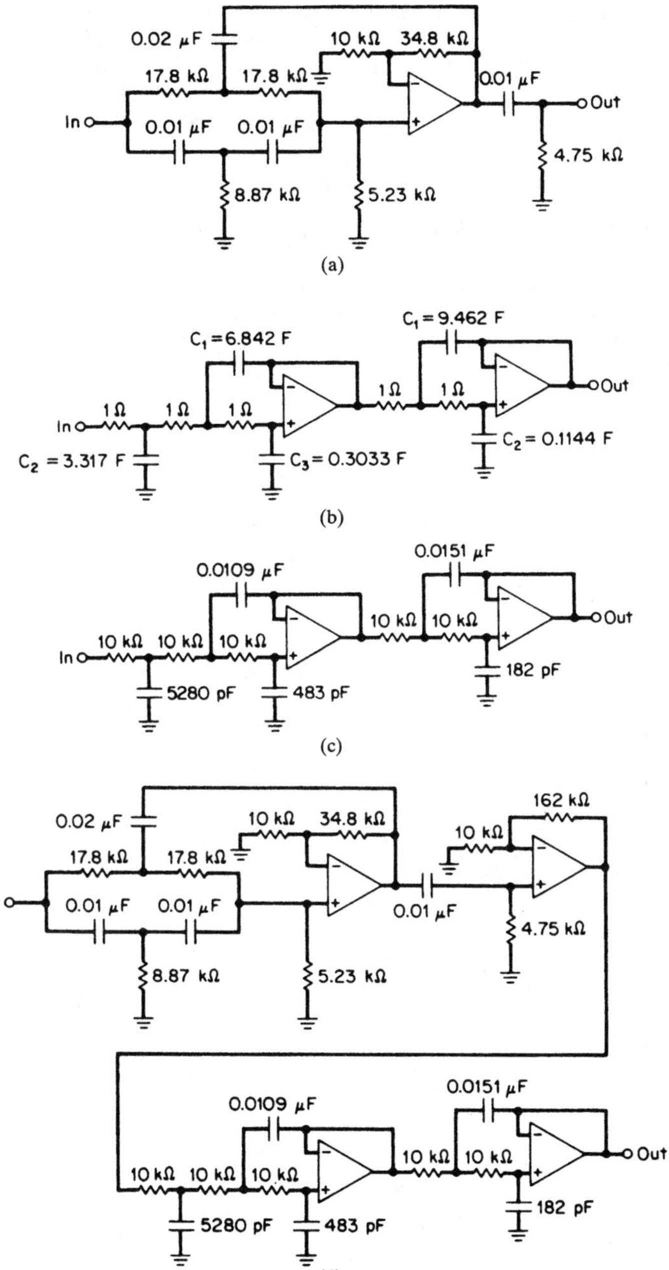

FIGURE 5-22 The wideband bandpass filter of Example 5-10: (*a*) the high-pass filter of Example 4-5; (*b*) normalized low-pass filter; (*c*) denormalized low-pass filter; and (*d*) bandpass filter configuration.

The Bandpass Transformation of Low-Pass Poles and Zeros. Active bandpass filters are designed directly from a bandpass transfer function. To obtain the bandpass poles and zeros from the low-pass transfer function, a low-pass-to-bandpass transformation must be performed. It was shown in Section 5.1 how this transformation can be accomplished by replacing the frequency variable by a new variable, which was given by

$$f_{\text{bp}} = f_0\left(\frac{f}{f_0} - \frac{f_0}{f}\right) \tag{5-1}$$

This substitution maps the low-pass frequency response into a bandpass magnitude characteristic.

Two sets of bandpass poles are obtained from each low-pass complex pole pair. If the low-pass pole is real, a single pair of complex poles results for the bandpass case. Also, each pair of imaginary axis zeros is transformed into two pairs of conjugate zeros. This is shown in Figure 5-23.

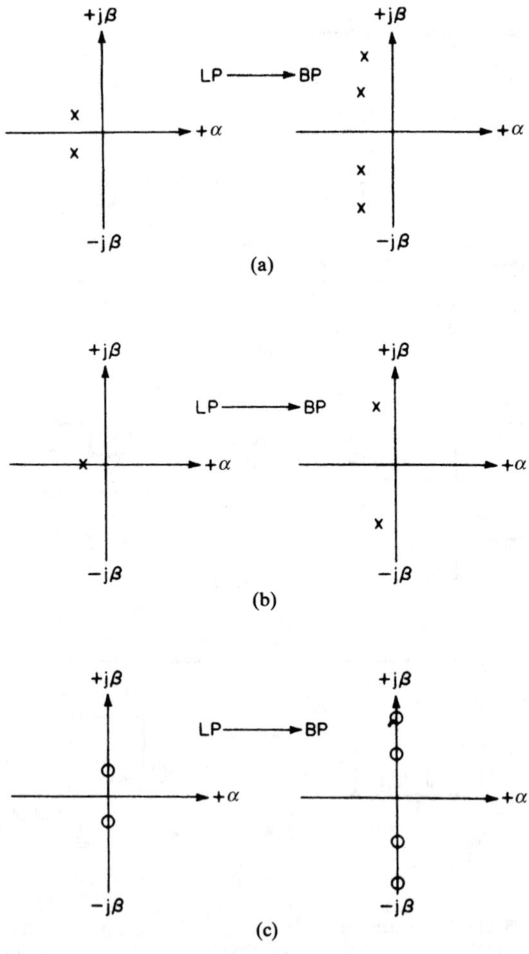

FIGURE 5-23 A low-pass-to-bandpass transformation: (*a*) low-pass complex pole pair; (*b*) low-pass real pole; and (*c*) low-pass pair of imaginary zeros.

Clearly, the total number of poles and zeros is doubled when the bandpass transformation is performed. However, it is conventional to disregard the conjugate bandpass poles and zeros below the real axis. An n-pole low-pass filter is said to result in an nth-order bandpass filter even though the bandpass transfer function is of the order $2n$. An nth-order active bandpass filter will then consist of n bandpass sections.

Each all-pole bandpass section has a second-order transfer function given by

$$T(s) = \frac{Hs}{s^2 + \frac{\omega_r}{Q}s + \omega_r^2} \tag{5-48}$$

where ω_r is equal to $2\pi f_r$, the pole resonant frequency in radians per second, Q is the bandpass section Q, and H is a gain constant.

If transmission zeros are required, the section transfer function will then take the form

$$T(s) = \frac{H(s^2 + \omega_\infty^2)}{s^2 + \frac{\omega_r}{Q}s + \omega_r^2} \tag{5-49}$$

where ω_∞ is equal to $2\pi f_\infty$, the frequency of the transmission zero in radians per second. Active bandpass filters are designed by the following sequence of operations:

1. Convert the bandpass specifications to a geometrically symmetrical requirement as described in Section 2.1.
2. Calculate the bandpass steepness factor A_s using Equation (2-19) and select a normalized filter type from Chapter 2.
3. Look up the corresponding normalized poles (and zeros) from the tables of Chapter 11 or use the *Filter Solutions* program for elliptic function filters and transform these coordinates into bandpass parameters.
4. Select the appropriate bandpass circuit configuration from the types presented in this chapter and cascade the required number of sections.

It is convenient to specify each bandpass filter section in terms of its center frequency and Q. Elliptic-function filters will require zeros. These parameters can be directly transformed from the poles (and zeros) of the normalized low-pass transfer function. A numerical procedure will be described for making this transformation.

First make the preliminary calculation

$$Q_{\text{bp}} = \frac{f_0}{\text{BW}} \tag{2-16}$$

where f_0 is the geometric bandpass center frequency and BW is the passband bandwidth. The bandpass transformation is made in the following manner.

Complex Poles. Complex poles occur in the tables of Chapter 11, having the form

$$-\alpha \pm j\beta$$

where α is the real coordinate and β is the imaginary part.

When using the *Filter Solutions* program for the design of elliptic function filters, the program provides the low-pass parameters Q and ω_0. These two parameters can be converted into α and β by using

$$\alpha = \frac{\omega_0}{2Q}$$

$$\beta = \sqrt{\omega_0^2 - \alpha^2}$$

Given α β, Q_{bp}, and f_0, the following series of calculations results in two sets of values for Q and center frequencies which defines a pair of bandpass filter sections:

$$C = \alpha^2 + \beta^2 \tag{5-50}$$

$$D = \frac{2\alpha}{Q_{bp}} \tag{5-51}$$

$$E = \frac{C}{Q_{bp}^2} + 4 \tag{5-52}$$

$$G = \sqrt{E^2 - 4D^2} \tag{5-53}$$

$$Q = \sqrt{\frac{E + G}{2D^2}} \tag{5-54}$$

$$M = \frac{\alpha Q}{Q_{bp}} \tag{5-55}$$

$$W = M + \sqrt{M^2 - 1} \tag{5-56}$$

$$f_{ra} = \frac{f_0}{W} \tag{5-57}$$

$$f_{rb} = W f_0 \tag{5-58}$$

The two bandpass sections have resonant frequencies of f_{ra} and f_{rb} (in hertz), and identical Qs as given by Equation (5-54).

Real Poles. A normalized low-pass real pole having a real coordinate of magnitude α_0 is transformed into a single bandpass section having a Q defined by

$$Q = \frac{Q_{bp}}{\alpha_0} \tag{5-59}$$

The section is tuned to f_0, the geometric center frequency of the filter.

Imaginary Zeros. Elliptic-function low-pass filters contain transmission zeros of the form $\pm j\omega_\infty$ as well as poles. These zeros must be transformed along with the poles when a bandpass filter is required. The bandpass zeros can be obtained as follows:

$$H = \frac{\omega_\infty^2}{2Q_{bp}^2} + 1 \tag{5-60}$$

$$Z = \sqrt{H + \sqrt{H^2 - 1}} \tag{5-61}$$

$$f_{\infty,a} = \frac{f_0}{Z} \tag{5-62}$$

$$f_{\infty,b} = Z \times f_0 \tag{5-63}$$

A pair of imaginary bandpass zeros are obtained that occur at $f_{\infty,a}$ and $f_{\infty,b}$ (in hertz) from each low-pass zero.

Determining Section Gain. The gain of a single bandpass section at the filter geometric center frequency f_0 is given by

$$A_0 = \frac{A_r}{\sqrt{1 + Q^2\left(\frac{f_0}{f_r} - \frac{f_r}{f_0}\right)^2}} \tag{5-64}$$

where A_r is the section gain at its resonant frequency f_r. The section gain will always be less at f_0 than at f_r since the circuit is peaked to f_r, except for transformed real poles where $f_r = f_0$. Equation (5-64) will then simplify to $A_0 = A_r$. The composite filter gain is determined by the product of the A_0 values of all the sections.

If the section Q is relatively high ($Q > 10$), Equation (5-64) can be simplified to

$$A_0 = \frac{A_r}{\sqrt{1 + \left(\frac{2Q\Delta f}{f_r}\right)^2}} \tag{5-65}$$

where Δf is the frequency separation between f_0 and f_r.

Example 5-11 Computing Bandpass Pole Locations and Section Gains

Required:

Determine the pole locations and section gains for a third-order Butterworth bandpass filter having a geometric center frequency of 1000 Hz, a 3-dB bandwidth of 100 Hz, and a midband gain of +30 dB.

Result:

(*a*) The normalized pole locations for an $n = 3$ Butterworth low-pass filter are obtained from Table 11-1 and are

$-0.500 \pm j0.8660$

-1.000

To obtain the bandpass poles, first compute

$$Q_{bp} = \frac{f_0}{BW_{3\,dB}} = \frac{1000 \text{ Hz}}{100 \text{ Hz}} = 10 \tag{2-16}$$

The low-pass to bandpass pole transformation is performed as follows:

Complex pole:

$$\alpha = 0.500 \quad \beta = 0.8660$$

$$C = \alpha^2 + \beta^2 = 1.000000 \tag{5-50}$$

$$D = \frac{2\alpha}{Q_{bp}} = 0.100000 \tag{5-51}$$

$$E = \frac{C}{Q_{bp}^2} + 4 = 4.010000 \tag{5-52}$$

$$G = \sqrt{E^2 - 4D^2} = 4.005010 \tag{5-53}$$

$$Q = \sqrt{\frac{E + G}{2D^2}} = 20.018754 \tag{5-54}$$

$$M = \frac{\alpha Q}{Q_{bp}} = 1.000938 \tag{5-55}$$

$$W = M + \sqrt{M^2 - 1} = 1.044261 \tag{5-56}$$

$$f_{ra} = \frac{f_0}{W} = 957.6 \text{ Hz} \tag{5-57}$$

$$f_{rb} = Wf_0 = 1044.3 \text{ Hz} \tag{5-58}$$

Real pole:

$$\alpha_0 = 1.0000$$

$$Q = \frac{Q_{bp}}{\alpha_0} = 10 \tag{5-59}$$

$$f_r = f_0 = 1000 \text{ Hz}$$

(*b*) Since a composite midband gain of +30 dB is required, let us distribute the gain uniformly among the three sections. Therefore, $A_0 = 3.162$ for each section corresponding to +10 dB.

The gain at section resonant frequency f_r is obtained from the following form of Equation (5-64):

$$A_r = A_0\sqrt{1 + Q^2\left(\frac{f_0}{f_r} - \frac{f_r}{f_0}\right)^2}$$

The resulting values are

Section 1:

$$f_r = 957.6 \text{ Hz}$$
$$Q = 20.02$$
$$A_r = 6.333$$

Section 2:

$$f_r = 1044.3 \text{ Hz}$$
$$Q = 20.02$$
$$A_r = 6.335$$

Section 3:

$$f_r = 1000.0 \text{ Hz}$$
$$Q = 10.00$$
$$A_r = 3.162$$

The block diagram of the realization is shown in Figure 5-24.

The calculations required for the bandpass pole transformation should be maintained to more than four significant figures after the decimal point to obtain accurate results since differences of close numbers are involved. Equations (5-55) and (5-56) are especially critical, so the value of M should be computed to five or six places after the decimal point.

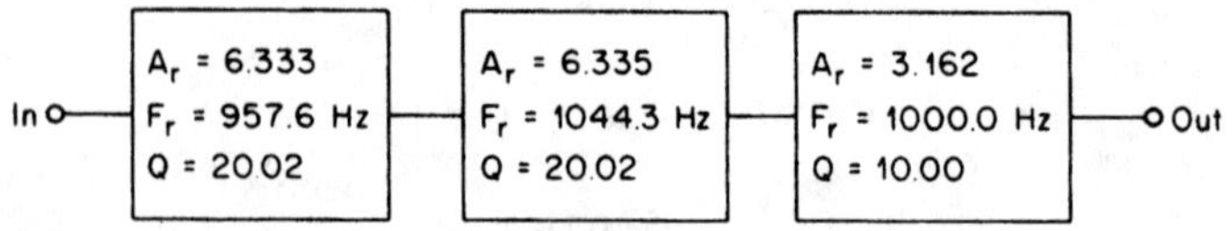

FIGURE 5-24 A block realization of Example 5-11.

Sensitivity in Active Bandpass Circuits. Sensitivity defines the amount of change of a dependent variable that results from the variation of an independent variable. Mathematically, the sensitivity of y with respect to x is expressed as

$$S_x^y = \frac{dy/y}{dx/x} \tag{5-66}$$

Sensitivity is used as a figure of merit to measure the change in a particular filter parameter, such as Q, or the resonant frequency for a given change in a component value.

Deviations of components from their nominal values occur because of the effects of temperature, aging, humidity, and other environmental conditions in addition to errors due to tolerances. These variations cause changes in parameters such as Q and the center frequency from their design values.

As an example, let's assume we are given the parameter $S_{R_1}^Q = -3$ for a particular circuit. This means that for a 1-percent increment of R_1, the circuit Q will change 3 percent in the opposite direction.

In addition to component value sensitivity, the operation of a filter is dependent on the active elements as well. The Q and resonant frequency can be a function of amplifier open-loop gain and phase shift, so the sensitivity to these active parameters is useful in determining an amplifier's suitability for a particular design.

The Q sensitivity of a circuit is a good measure of its stability. With some circuits, the Q can increase to infinity, which implies self-oscillation. Low Q sensitivity of a circuit usually indicates that the configuration will be practical from a stability point of view.

Sometimes the sensitivity is expressed as an equation instead of a numerical value, such as $S_A^Q = 2Q^2$. This expression implies that the sensitivity of Q with respect to amplifier gain A increases with Q^2, so the circuit is not suitable for high Q realizations.

The frequency-sensitivity parameters of a circuit are useful in determining whether the circuit will require resistive trimming and indicate which element should be made variable. It should be mentioned that, in general, only resonant frequency is made adjustable when the bandpass filter is sufficiently narrow. Q variations of 5 or 10 percent are usually tolerable, whereas a comparable frequency error would be disastrous in narrow filters. However, in the case of a state-variable realization, a Q-enhancement effect occurs that's caused by amplifier phase shift. The Q may increase very dramatically, so Q adjustment is usually required in addition to resonant frequency.

All-Pole Bandpass Configurations

Multiple-Feedback Bandpass (MFBP). The circuit of Figure 5-25*a* realizes a bandpass pole pair and is commonly referred to as a multiple-feedback bandpass (MFBP)

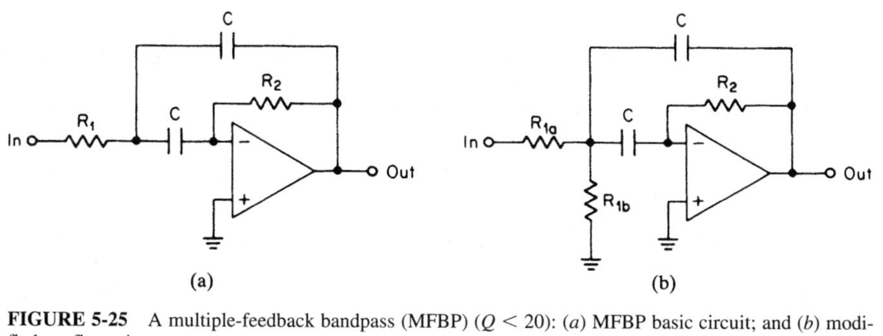

FIGURE 5-25 A multiple-feedback bandpass (MFBP) ($Q < 20$): (*a*) MFBP basic circuit; and (*b*) modified configuration.

configuration. This circuit features a minimum number of components and a low sensitivity to component tolerances. The transfer function is given by

$$T(s) = \frac{sC/R_1}{s^2C^2 + s2C/R_2 + 1/R_1R_2} \tag{5-67}$$

If we equate the coefficients of this transfer function with the general bandpass transfer function of Equation (5-48), we can derive the following expressions for the element values:

$$R_2 = \frac{Q}{\pi f_r C} \tag{5-68}$$

and

$$R_1 = \frac{R_2}{4Q^2} \tag{5-69}$$

where C is arbitrary.

The circuit gain at resonant frequency f_r is given by

$$A_r = 2Q^2 \tag{5-70}$$

The open-loop gain of the operational amplifier at f_r should be well in excess of $2Q^2$ so that the circuit performance is controlled mainly by the passive elements. This requirement places a restriction on realizable Qs to values typically below 20, depending upon the amplifier type and frequency range.

Extremely high gains occur for moderate Q values because of the Q^2 gain proportionality. Thus, there will be a tendency for clipping at the amplifier output with moderate input levels. Also, the circuit gain is fixed by the Q, which limits flexibility.

An alternate and preferred form of the circuit is shown in Figure 5-25*b*. The input resistor R_1 has been split into two resistors, R_{1a} and R_{1b}, to form a voltage divider so that the circuit gain can be controlled. The parallel combination of the two resistors is equal to R_1 in order to retain the resonant frequency. The transfer function of the modified circuit is given by

$$T(s) = -\frac{sR_2C}{s^2R_{1a}R_2C^2 + s2R_{1a}C + (1 + R_{1a}/R_{1b})} \tag{5-71}$$

The values of R_{1a} and R_{1b} are computed from

$$R_{1a} = \frac{R_2}{2A_r} \tag{5-72}$$

and

$$R_{1b} = \frac{R_2/2}{2Q^2 - A_r} \tag{5-73}$$

where A_r is the desired gain at resonant frequency f_r and cannot exceed $2Q^2$. The value of R_2 is still computed from Equation (5-68).

The circuit sensitivities can be determined as follows:

$$S^Q_{R_{1a}} = S^{f_r}_{R_{1a}} = \frac{A_r}{4Q^2} \tag{5-74}$$

$$S^Q_{R_{1b}} = S^{f_r}_{R_{1b}} = \frac{1}{2}(1 + A_r/2Q^2) \tag{5-75}$$

$$S^{f_r}_{R_2} = S^{f_r}_{C} = -\frac{1}{2} \tag{5-76}$$

$$S^Q_{R_2} = \frac{1}{2} \tag{5-77}$$

For $Q^2/A \gg 1$, the resonant frequency can be directly controlled by R_{1b} since $S_{R_{1b}}^{f_r}$ approaches $\frac{1}{2}$. To use this result, let's assume that the capacitors have 2-percent tolerances and the resistors have a tolerance of 1 percent, which could result in a possible 3-percent frequency error. If frequency adjustment is desired, R_{1b} should be made variable over a minimum resistance range of ± 6 percent. This would then permit a frequency adjustment of ± 3 percent, since $S_{R_{1b}}^{f_r}$ is equal to $\frac{1}{2}$. Resistor R_{1b} should be composed of a fixed resistor in series with a single-turn potentiometer to provide good resolution.

Adjustment of Q can be accomplished by making R_2 adjustable. However, this will affect resonant frequency and in any event is not necessary for most filters if 1- or 2-percent tolerance parts are used. The section gain can be varied by making R_{1a} adjustable, but again resonant frequency may be affected.

In conclusion, this circuit is highly recommended for low Q requirements. Although a large spread in resistance values can occur and the Q is limited by amplifier gain, the circuit simplicity, low element sensitivity, and ease of frequency adjustment make it highly desirable.

The following example demonstrates the design of a bandpass filter using the MFBP configuration.

Example 5-12 Design of an Active All-Pole Bandpass Filter Using the MFBP Configuration

Required:

Design an active bandpass filter having the following specifications:

A center frequency of 300 Hz

3 dB at ± 10 Hz

25-dB minimum at ± 40 Hz

Essentially zero overshoot to a 300-Hz carrier pulse step

A gain of ± 12 dB at 300 Hz

Result:

(*a*) Since the bandwidth is narrow, the requirement can be treated on an arithmetically symmetrical basis. The bandpass steepness factor is given by

$$A_s = \frac{\text{stopband bandwidth}}{\text{passband bandwidth}} = \frac{80\text{ Hz}}{20\text{ Hz}} = 4 \tag{2-19}$$

The curves of Figures 2-69 and 2-74 indicate that an $n = 3$ transitional gaussian to 6-dB filter will meet the frequency- and step-response requirements.

(*b*) The pole locations for the corresponding normalized low-pass filter are found in Table 11-50 and are as follows:

$-0.9622 \pm j1.2214$

-0.9776

First compute the bandpass Q:

$$Q_{\text{bp}} = \frac{f_0}{\text{BW}_{3\text{ dB}}} = \frac{300\text{ Hz}}{20\text{ Hz}} = 15 \tag{2-16}$$

The low-pass poles are transformed to the bandpass form in the following manner:

Complex pole:

$$\alpha = 0.9622 \quad \beta = 1.2214$$

$$C = 2.417647 \tag{5-50}$$

$$D = 0.128293 \tag{5-51}$$

$$E = 4.010745 \tag{5-52}$$

$$G = 4.002529 \tag{5-53}$$

$$Q = 15.602243 \tag{5-54}$$

$$M = 1.000832 \tag{5-55}$$

$$W = 1.041630 \tag{5-56}$$

$$f_{ra} = 288.0 \text{ Hz} \tag{5-57}$$

$$f_{rb} = 312.5 \text{ Hz} \tag{5-58}$$

Real pole:

$$\alpha_0 = 0.9776$$

$$Q = 15.34 \tag{5-59}$$

$$f_r = 300.0 \text{ Hz}$$

(*c*) A midband gain of +12 dB is required. Let us allocate a gain of +4 dB to each section corresponding to $A_0 = 1.585$. The value of A_r, the resonant frequency gain for each section, is obtained from Equation (5-64) and is listed in the following table, which summarizes the design parameters of the filters sections:

	f_r	Q	A_r
Section 1	288.0 Hz	15.60	2.567
Section 2	312.5 Hz	15.60	2.567
Section 3	300.0 Hz	15.34	1.585

(*d*) Three MFBP bandpass sections will be connected in tandem. The following element values are computed where C is set equal to 0.1 μF:

Section 1:

$$R_2 = \frac{Q}{\pi f_r C} = \frac{15.6}{\pi \times 288 \times 10^{-7}} = 172.4 \text{ k}\Omega \tag{5-68}$$

$$R_{1a} = \frac{R_2}{2A_r} = \frac{172.4 \times 10^3}{2 \times 2.567} = 33.6 \text{ k}\Omega \tag{5-72}$$

$$R_{1b} = \frac{R_2/2}{2Q^2 - A_r} = \frac{86.2 \times 10^3}{2 \times 15.6^2 - 2.567} = 178 \ \Omega \tag{5-73}$$

Section 2:

$R_2 = 158.9 \text{ k}\Omega$

$R_{1a} = 30.9 \text{ k}\Omega$

$R_{1b} = 164 \ \Omega$

Section 3:

$R_2 = 162.8 \text{ k}\Omega$

$R_{1a} = 51.3 \text{ k}\Omega$

$R_{1b} = 174 \ \Omega$

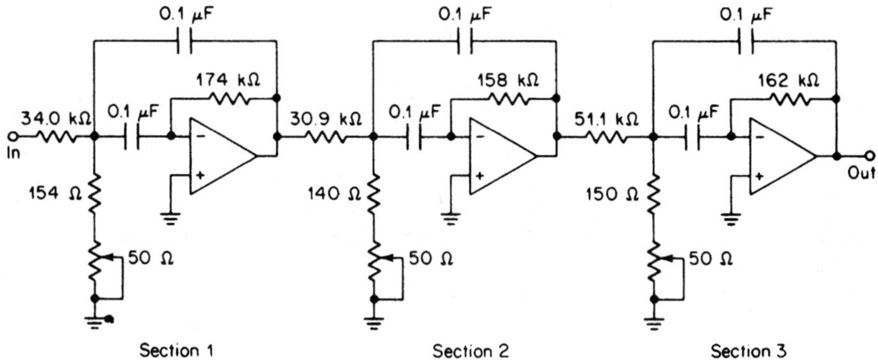

FIGURE 5-26 The MFBP circuit of Example 5-12.

The final circuit is shown in Figure 5-26. Resistor values have been rounded off to standard 1-percent values, and resistor R_{1b} has been made variable in each section for tuning purposes.

Each filter section can be adjusted by applying a sine wave at the section f_r to the filter input. The phase shift of the section being adjusted is monitored by connecting one channel of an oscilloscope to the section input and the other channel to the section output. A Lissajous pattern is thus obtained. Resistor R_{1b} is then adjusted until the ellipse closes to a straight line.

The Dual-Amplifier Bandpass (DABP) Structure. The bandpass circuit of Figure 5-27 was first introduced by Sedra and Espinoza (see Bibliography). Truly remarkable performance in terms of available Q, low sensitivity, and flexibility can be obtained in comparison with alternate schemes involving two amplifiers.

The transfer function is given by

$$T(s) = \frac{s2/R_1C}{s^2 + s1/R_1C + 1/R_2R_3C^2} \tag{5-78}$$

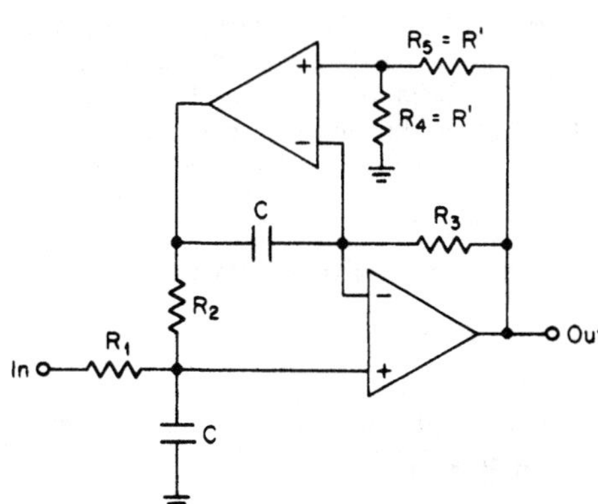

FIGURE 5-27 A dual-amplifier bandpass (DABP) configuration ($Q < 15$).

If we compare this expression with the general bandpass transfer function of Equation (5-48) and let $R_2R_3 = R^2$, the following design equations for the element values can be obtained.
First, compute

$$R = \frac{1}{2\pi f_r C} \tag{5-79}$$

then

$$R_1 = QR \tag{5-80}$$

$$R_2 = R_3 = R \tag{5-81}$$

where C is arbitrary. The value of R' in Figure 5-27 can also be chosen at any convenient value. Circuit gain at f_r is equal to 2.

The following sensitivities can be derived:

$$S_{R_1}^{Q} = 1 \tag{5-82}$$

$$S_{R_2}^{f_r} = S_{R_3}^{f_r} = S_{R_4}^{f_r} = S_{C}^{f_r} = -1/2 \tag{5-83}$$

$$S_{R_5}^{f_r} = 1/2 \tag{5-84}$$

An interesting result of sensitivity studies is that if the bandwidths of both amplifiers are nearly equivalent, extremely small deviations of Q from the design values will occur. This is especially advantageous at higher frequencies where the amplifier poles have to be taken into account. It is then suggested that a dual-type amplifier be used for each filter section since both amplifier halves will be closely matched to each other.

A useful feature of this circuit is that resonant frequency and Q can be independently adjusted. Alignment can be accomplished by first adjusting R_2 for resonance at f_r. Resistor R_1 can then be adjusted for the desired Q without affecting the resonant frequency.

Since each section provides a fixed gain of 2 at f_r, a composite filter may require an additional amplification stage if higher gains are needed. If a gain reduction is desired, resistor R_1 can be split into two resistors to form a voltage divider in the same manner as in Figure 5-25*b*. The resulting values are

$$R_{1a} = \frac{2R_1}{A_r} \tag{5-85}$$

and

$$R_{1b} = \frac{R_{1a}A_r}{2 - A_r} \tag{5-86}$$

where A_r is the desired gain at resonance.

The spread of element values of the MFBP section previously discussed is equal to $4Q^2$. In comparison, this circuit has a ratio of resistances determined by Q, so the spread is much less.

The DABP configuration has been found very useful for designs covering a wide range of Qs and frequencies. Component sensitivity is small, resonant frequency and Q are easily adjustable, and the element spread is low. The following example illustrates the use of this circuit.

Example 5-13 Design of an Active All-Pole Bandpass Filter Using the DABP Configuration

Required:

Design an active bandpass filter to meet the following specifications:

A center frequency of 3000 Hz

3 dB at ±30 Hz

20-dB minimum at ±120 Hz

Result:

(*a*) If we consider the requirement as being arithmetically symmetrical, the bandpass steepness factor becomes

$$A_s = \frac{\text{stopband bandwidth}}{\text{passband bandwidth}} = \frac{240 \text{ Hz}}{60 \text{ Hz}} = 4 \tag{2-19}$$

We can determine from the curve of Figure 2-34 that a second-order Butterworth low-pass filter provides over 20 dB of rejection within a frequency ratio of 4:1. The corresponding poles of the normalized low-pass filter are found in Table 11-1 and are as follows:

$$-0.7071 \pm j0.7071$$

(*b*) To convert these poles to the bandpass form, first compute:

$$Q_{\text{bp}} = \frac{f_0}{\text{BW}_{3\text{ dB}}} = \frac{3000 \text{ Hz}}{60 \text{ Hz}} = 50 \tag{2-16}$$

The bandpass poles transformation is performed in the following manner:

$$\alpha = 0.7071 \quad \beta = 0.7071$$

$$C = 1.000000 \tag{5-50}$$

$$D = 0.028284 \tag{5-51}$$

$$E = 4.000400 \tag{5-52}$$

$$G = 4.000000 \tag{5-53}$$

$$Q = 70.713124 \tag{5-54}$$

$$M = 1.000025 \tag{5-55}$$

$$W = 1.007096 \tag{5-56}$$

$$f_{ra} = 2978.9 \text{ Hz} \tag{5-57}$$

$$f_{rb} = 3021.3 \text{ Hz} \tag{5-58}$$

(*c*) Two DABP sections will be used. The element values are now computed, where C is set equal to 0.01 μF and R' is 10 kΩ.

Section 1:

$$f_r = 2978.9 \text{ Hz}$$

$$Q = 70.7$$

$$R = \frac{1}{2\pi f_r C} = \frac{1}{2\pi \times 2978.9 \times 10^{-8}} = 5343\,\Omega \tag{5-79}$$

$$R_1 = QR = 70.7 \times 5343 = 377.7 \text{ k}\Omega \tag{5-80}$$

$$R_2 = R_3 = R = 5343\ \Omega \tag{5-81}$$

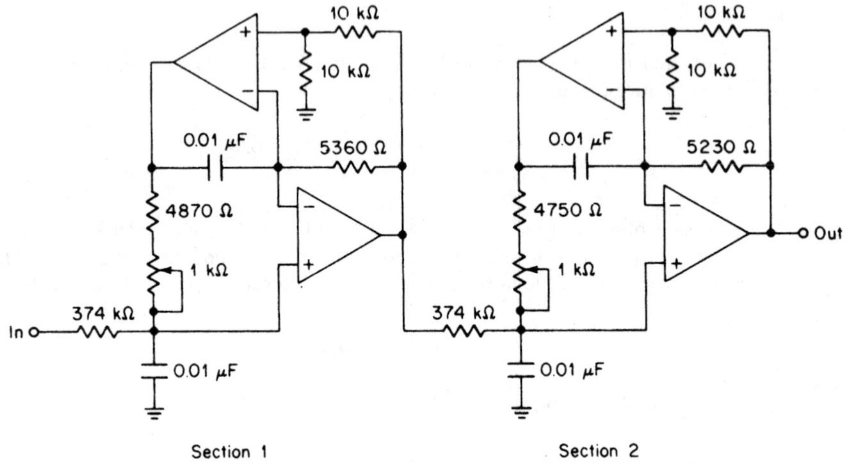

FIGURE 5-28 The DABP filter of Example 5-13.

Section 2:

$$f_r = 3021.3 \text{ Hz}$$

$$Q = 70.7$$

$$R = 5268\ \Omega \tag{5-79}$$

$$R_1 = 372.4\ \text{k}\Omega \tag{5-80}$$

$$R_2 = R_3 = 5268\ \Omega \tag{5-81}$$

The circuit is illustrated in Figure 5-28, where resistors have been rounded off to standard 1-percent values and R_2 is made adjustable for tuning.

Low-Sensitivity Three-Amplifier Configuration. The DABP circuit shown in Figure 5-27 provides excellent performance and is useful for general-purpose bandpass filtering. A modified version utilizing a total of three amplifiers is shown in Figure 5-29. This circuit will exhibit performance superior to the DABP configuration, especially at higher frequencies. Using this structure and off-the-shelf op amps, active bandpass filters having moderate percentage bandwidths can be designed to operate in the frequency range approaching 1–2 MHz.

The section within the dashed line in Figure 5-29 realizes a shunt inductance to ground using a configuration called a *gyrator*.* Both op amps within this block should be closely matched to obtain low op-amp sensitivity at high frequencies, so the use of a dual op amp for this location would be highly recommended. The third op amp serves as a voltage follower or buffer to obtain low output impedance.

*A gyrator is an impedance inverting device which converts an impedance Z into a reciprocal impedance $1/G^2Z$, where G is a constant. Therefore, a capacitor having an impedance $1/SC$ can be converted into an impedance SC/G^2, which corresponds to an inductance of C/G^2. For this circuit $\sigma = 1/R$, so $L = R^2C$.

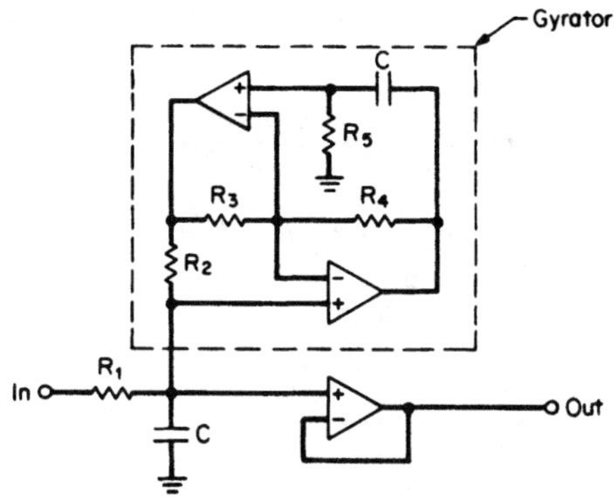

FIGURE 5-29 Low-sensitivity three-amplifier bandpass configuration.

As in the case of the DABP circuit, resonant frequency and Q can be adjusted independently. The design proceeds as follows:

First select a convenient value for C. Then compute

$$R = \frac{1}{2\pi f_r C} \tag{5-87}$$

Let $R_2 = R_3 = R_4 = R_5 = R$

Then

$$R_1 = QR \tag{5-88}$$

Circuit gain at f_r is unity. For additional gain, the voltage follower can be configured as a noninverting amplifier.

For alignment, first adjust R_2 for resonance at f_r. Resistor R_1 can then be adjusted for the desired Q without affecting the resonant frequency.

Example 5-14 Design of an Active All-Pole Bandpass Filter Using the Low-Sensitivity Three-Amplifier Configuration

Required:

Design an active bandpass filter to meet the following specifications:

A center frequency of 30 kHz

3 dB at ±300 Hz

20-dB minimum at ±1200 Hz

Result:

(*a*) Treating the requirement as being arithmetically symmetrical, the bandpass steepness factor is

$$A_s = \frac{\text{stopband bandwidth}}{\text{passband bandwidth}} = \frac{2400\text{ Hz}}{600\text{ Hz}} = 4 \tag{2-19}$$

From Figure 2-34, a second-order Butterworth low-pass filter meets the attenuation. The low-pass poles from Table 11-1 are $-0.7071 \pm j0.7071$.

(*b*) These poles must now be converted to the bandpass form. The procedure is as follows

$$Q_{bp} = \frac{f_0}{BW_{3\,dB}} = \frac{30{,}000\text{ Hz}}{600\text{ Hz}} = 50 \qquad (2\text{-}16)$$

$$\alpha = 0.7071 \quad \beta = 0.7071$$

$$C = 1 \qquad (5\text{-}50)$$

$$D = 0.028284 \qquad (5\text{-}51)$$

$$E = 4.000400 \qquad (5\text{-}52)$$

$$G = 4.000000 \qquad (5\text{-}53)$$

$$Q = 70.713124 \qquad (5\text{-}54)$$

$$M = 1.000025 \qquad (5\text{-}55)$$

$$W = 1.007096 \qquad (5\text{-}56)$$

$$f_{ra} = 29.789\text{ kHz} \qquad (5\text{-}57)$$

$$f_{rb} = 30.213\text{ kHz} \qquad (5\text{-}58)$$

(*c*) Compute the element values for the circuit of Figure 5-29 using $C = 0.01\ \mu$F.

Section 1:

$$f_r = 29.789\text{ kHz}$$

$$Q = 70.71$$

$$R = \frac{1}{2\pi f_r C} = \frac{1}{2\pi \times 29.789 \times 10^{-5}}$$

$$= 534.3\ \Omega \qquad (5\text{-}87)$$

$$R_1 = QR = 70.7 \times 534.3 = 37.8\text{ k}\Omega \qquad (5\text{-}88)$$

Section 2:

$$f_r = 30.213\text{ kHz}$$

$$Q = 70.71$$

$$R = 526.8\ \Omega \qquad (5\text{-}87)$$

$$R_1 = 37.2\text{ k}\Omega \qquad (5\text{-}88)$$

The circuit of this example is illustrated in Figure 5-30 using standard 1-percent resistor values and a potentiometer for frequency adjustment.

The State-Variable (Biquad) All-Pole Circuit. The state-variable or biquad configuration was first introduced in Section 3.2 for use as a low-pass filter section. A bandpass output is available as well. The biquad approach features excellent sensitivity properties and

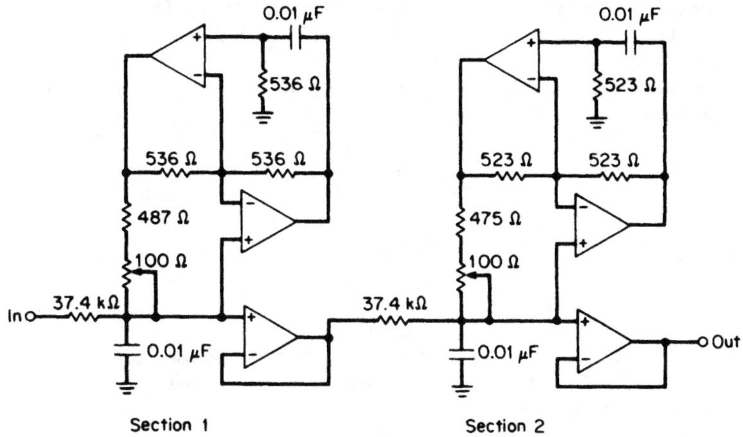

FIGURE 5-30 The bandpass filter of Example 5-14.

the capability to control resonant frequency and Q independently. It is especially suited for constructing precision-active filters in a standard form.

The circuit of Figure 5-31 is the all-pole bandpass form of the general biquadratic configuration. The transfer function is given by

$$T(s) = \frac{s/CR_4}{s^2 + s/CR_1 + 1/R_2R_3C^2} \tag{5-89}$$

If we equate this expression to the general bandpass transfer function of Equation (5-48), the circuit resonant frequency and 3-dB bandwidth can be expressed as

$$f_r = \frac{1}{2\pi C\sqrt{R_2R_3}} \tag{5-90}$$

and

$$\mathrm{BW}_{3\,\mathrm{dB}} = \frac{1}{2\pi R_1C} \tag{5-91}$$

where $\mathrm{BW}_{3\,\mathrm{dB}}$ is equal to f_r/Q.

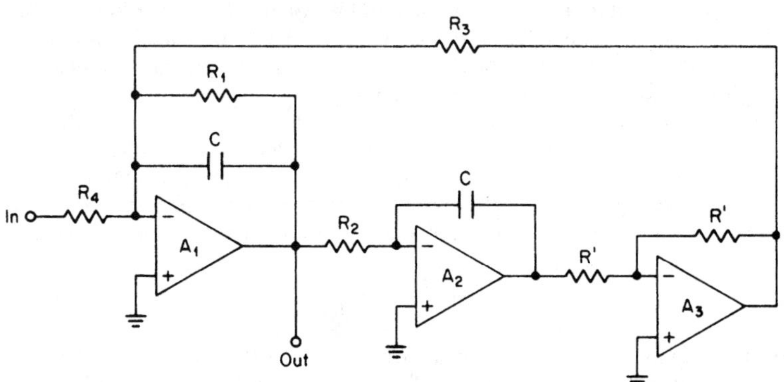

FIGURE 5-31 A biquad all-pole circuit ($Q < 200$).

Equations (5-90) and (5-91) indicate that the resonant frequency and 3-dB bandwidth can be independently controlled. This feature is highly desirable and can lead to many useful applications, such as variable filters.

If we substitute f_r/Q for $BW_{3\text{ dB}}$ and set $R_2 = R_3$, the following design equations can be derived for the section:

$$R_1 = \frac{Q}{2\pi f_r C} \tag{5-92}$$

$$R_2 = R_3 = \frac{R_1}{Q} \tag{5-93}$$

and

$$R_4 = \frac{R_1}{A_r} \tag{5-94}$$

where A_r is the desired gain at resonant frequency f_r. The values of C and R' in Figure 5-31 can be conveniently selected. By making R_3 and R_1 adjustable, the resonant frequency and Q, respectively, can be adjusted.

The sensitivity factors are

$$S_{R_2}^{f_r} = S_{R_3}^{f_r} = S_C^{f_r} = -1/2 \tag{5-95}$$

$$S_{R_1}^{Q} = 1 \tag{5-96}$$

and

$$S_{\mu}^{Q} = \frac{2Q}{\mu} \tag{5-97}$$

where μ is the open-loop gain of amplifiers A_1 and A_2. The section Q is then limited by the finite gain of the operational amplifier.

Another serious limitation occurs because of finite amplifier bandwidth. Thomas (see Bibliography) has shown that, as the resonant frequency increases for a fixed design Q, the actual Q remains constant over a broad band and then begins to increase, eventually becoming infinite (oscillatory). This effect is called Q enhancement.

If we assume that the open-loop transfer function of the amplifier has a single pole, the effective Q can be approximated by

$$Q_{\text{eff}} = \frac{Q_{\text{design}}}{1 - \dfrac{2Q_{\text{design}}}{\mu_0 \omega_c}(2\omega_r - \omega_c)} \tag{5-98}$$

where ω_r is the resonant frequency, ω_c is the 3-dB breakpoint of the open-loop amplifier gain, and μ_0 is the open-loop gain at DC. As ω_r is increased, the denominator approaches zero.

The Q-enhancement effect can be minimized by having a high gain-bandwidth product. If the amplifier requires external frequency compensation, the compensation can be made lighter than the recommended values. The state-variable circuit is well suited for light compensation since the structure contains two integrators which have a stabilizing effect.

A solution suggested by Thomas is to introduce a leading phase component in the feedback loop which compensates for the lagging phase caused by finite amplifier bandwidth. This can be achieved by introducing a capacitor in parallel with resistor R_3 having the value

$$C_p = \frac{4}{\mu_0 \omega_c R_3} \tag{5-99}$$

Probably the most practical solution is to make resistor R_1 variable. The Q may be determined by measuring the 3-dB bandwidth. R_1 is adjusted until the ratio $f_r/BW_{3\text{ dB}}$ is equal to the required Q.

As the Q is enhanced, the section gain is also increased. Empirically it has been found that correcting for the gain enhancement compensates for the Q enhancement as well. R_1 can be adjusted until the measured gain at f_r is equal to the design value of A_r used in Equation (5-94). Although this technique is not as accurate as determining the actual Q from the 3-dB bandwidth, it certainly is much more convenient and will usually be sufficient.

The biquad is a low-sensitivity filter configuration suitable for precision applications. Circuit Qs of up to 200 are realizable over a broad frequency range. The following example demonstrates the use of this structure.

Example 5-15 Design of an Active All-Pole Bandpass Filter Using the Biquad Configuration

Required:

Design an active bandpass filter satisfying the following specifications:

A center frequency of 2500 Hz

3 dB at ±15 Hz

15-dB minimum at ±45 Hz

A gain of +12 dB at 2500 Hz

Result:

(*a*) The bandpass steepness factor is determined from

$$A_s = \frac{\text{stopband bandwidth}}{\text{passband bandwidth}} = \frac{90\text{ Hz}}{30\text{ Hz}} = 3 \tag{2-19}$$

Using Figure 2-42, we find that a second-order 0.1-dB Chebyshev normalized low-pass filter will meet the attenuation requirements. The corresponding poles found in Table 11-23 are as follows:

$$-0.6125 \pm j0.7124$$

(*b*) To transform these low-pass poles to the bandpass form, first compute

$$Q_{\text{bp}} = \frac{f_0}{\text{BW}_{3\text{ dB}}} = \frac{2500\text{ Hz}}{30\text{ Hz}} = 83.33 \tag{2-16}$$

The bandpass poles are determined from the following series of computations. Since the filter is very narrow, an extended number of significant figures will be used in Equations (5-50) through (5-58) to maintain accuracy.

$$\alpha = 0.6125 \quad \beta = 0.7124$$

$$C = 0.882670010 \tag{5-50}$$

$$D = 0.014700588 \tag{5-51}$$

$$E = 4.000127115 \tag{5-52}$$

$$G = 4.000019064 \tag{5-53}$$

$$Q = 136.0502228 \tag{5-54}$$

$$M = 1.000009138 \tag{5-55}$$

$$W = 1.004284182 \tag{5-56}$$

$$f_{ra} = 2489.3\text{ Hz} \tag{5-57}$$

$$f_{rb} = 2510.7\text{ Hz} \tag{5-58}$$

(*c*) Since a midband gain of +12 dB is required, each section will be allocated a midband gain of +6 dB corresponding to $A_0 = 2.000$. The gain A_r at the resonant frequency of each section is determined from Equation (5-65) and is listed in the following table:

	f_r	Q	A_r
Section 1	2489.3 Hz	136	3.069
Section 2	2510.7 Hz	136	3.069

(*d*) Two biquad sections in tandem will be used. C is chosen to be 0.1 μF and R' is 10 kΩ. The element values are computed as follows:

Section 1:

$$R_1 = \frac{Q}{2\pi f_r C} = \frac{136}{2\pi \times 2489.3 \times 10^{-7}} = 86.9\ \text{k}\Omega \tag{5-92}$$

$$R_2 = R_3 = \frac{R_1}{Q} = \frac{86.9 \times 10^3}{136} = 639\ \Omega \tag{5-93}$$

$$R_4 = \frac{R_1}{A_r} = \frac{86.9 \times 10^3}{3.069} = 28.3\ \text{k}\Omega \tag{5-94}$$

Section 2:

$$R_1 = 86.2\ \text{k}\Omega \tag{5-92}$$

$$R_2 = R_3 = 634\ \Omega \tag{5-93}$$

$$R_4 = 28.1\ \text{k}\Omega \tag{5-94}$$

The final circuit is shown in Figure 5-32. Resistors R_3 and R_1 are made variable so that resonant frequency and Q can be adjusted. Standard values of 1-percent resistors have been used.

The Q-Multiplier Approach. Certain active bandpass structures such as the MFBP configuration of Section 5.2 are severely Q-limited because of insufficient amplifier gain or other inadequacies. The technique outlined in this section uses a low-Q-type bandpass circuit within a Q-multiplier structure which increases the circuit Q to the desired value.

A bandpass transfer function having unity gain at resonance can be expressed as

$$T(s) = \frac{\dfrac{\omega_r}{Q}s}{s^2 + \dfrac{\omega_r}{Q}s + \omega_r^2} \tag{5-100}$$

If the corresponding circuit is combined with a summing amplifier in the manner shown in Figure 5-33*a*, where β is an attenuation factor, the following overall transfer function can be derived:

$$T(s) = \frac{\dfrac{\omega_r}{Q}s}{s^2 + \dfrac{\omega_r}{\dfrac{Q}{\dfrac{Q}{1-\beta}}}s + \omega_r^2} \tag{5-101}$$

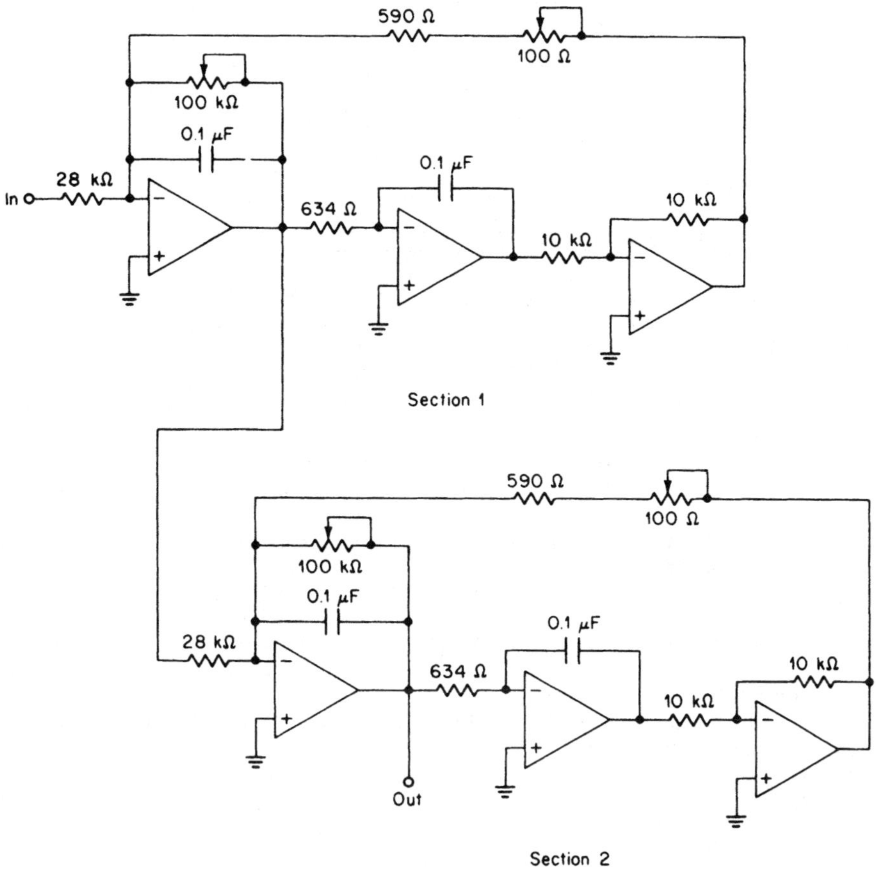

FIGURE 5-32 The biquad circuit of Example 5-15.

The middle term of the denominator has been modified so that the circuit Q is given by $Q/(1 - \beta)$, where $0 < \beta < 1$. By selecting a β sufficiently close to unity, the Q can be increased by the factor $1/(1 - \beta)$. The circuit gain is also increased by the same factor.

If we use the MFBP section for the bandpass circuit, the Q-multiplier configuration will take the form of Figure 5-33*b*. Since the MFBP circuit is inverting, an inverting amplifier can also be used for summing.

The value of β can be found from

$$\beta = 1 - \frac{Q_r}{Q_{\text{eff}}} \tag{5-102}$$

where Q_{eff} is the effective overall Q, and Q_r is the design Q of the bandpass section. The component values are determined by the following equations:

$$R_3 = \frac{R}{\beta} \tag{5-103}$$

$$R_4 = R \tag{5-104}$$

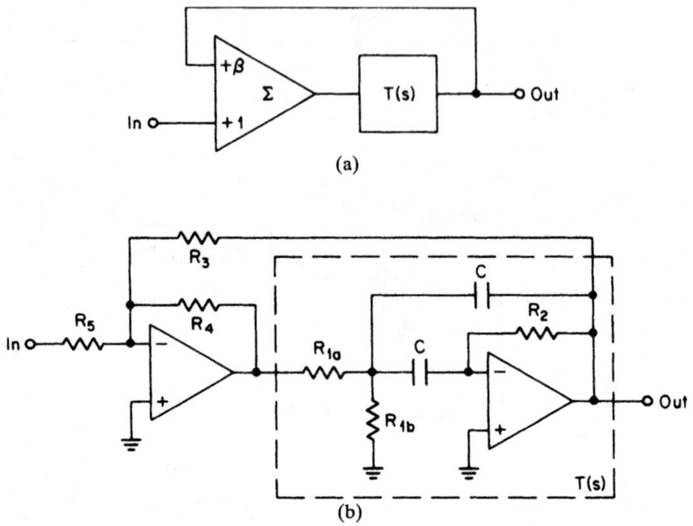

FIGURE 5-33 *Q*-multiplier circuit: (*a*) block diagram; and (*b*) a realization using an MFBP section.

and
$$R_5 = \frac{R}{(1 - \beta)A_r} \tag{5-105}$$

where R can be conveniently chosen and A_r is the desired gain at resonance.

Design equations for the MFBP section were derived in Section 5.2 and are repeated here corresponding to unity gain.

$$R_2 = \frac{Q_r}{\pi f_r C} \tag{5-68}$$

$$R_{1a} = \frac{R_2}{2} \tag{5-72}$$

and
$$R_{1b} = \frac{R_{1a}}{2Q_r^2 - 1} \tag{5-73}$$

The value of C can be freely chosen.

The configuration of Figure 5-33*b* is not restricted to the MFBP section. The state-variable all-pole bandpass circuit may be used instead. The only requirements are that the filter section be of an inverting type and that the gain be unity at resonance. This last requirement is especially critical because of the positive feedback nature of the circuit. Small gain errors could result in large overall Q variations when β is close to 1. It may then be desirable to adjust section gain precisely to unity.

Example 5-16 Design of an Active Bandpass Filter Section Using the *Q*-Multiplier Configuration

Required:

Design a single bandpass filter section having the following characteristics:

A center frequency of 3600 Hz

3-dB bandwidth of 60 Hz

A gain of 3

Result:

(*a*) The bandpass Q is given by

$$Q_r = \frac{f_0}{BW_{3\text{ dB}}} = \frac{3600\text{ Hz}}{60\text{ Hz}} = 60 \tag{2-16}$$

A Q-multiplier implementation using the MFBP section will be employed.

(*b*) Let us use a Q_r of 10 for the MFBP circuit. The following component values are computed where C is set equal to 0.01 μF.

$$R_2 = \frac{Q_r}{\pi f_r C} = \frac{10}{\pi 3600 \times 10^{-8}} = 88.4\text{ k}\Omega \tag{5-68}$$

$$R_{1a} = \frac{R_2}{2} = \frac{88.4 \times 10^3}{2} = 44.2\text{ k}\Omega \tag{5-72}$$

$$R_{1b} = \frac{R_{1a}}{2Q_r^2 - 1} = \frac{44.2 \times 10^3}{2 \times 10^2 - 1} = 222\ \Omega \tag{5-73}$$

The remaining values are given by the following design equations where R is chosen at 10 kΩ and gain A_r is equal to 3:

$$\beta = 1 - \frac{Q_r}{Q_{\text{eff}}} = 1 - \frac{10}{60} = 0.8333 \tag{5-102}$$

$$R_3 = \frac{R}{\beta} = \frac{10^4}{0.8333} = 12.0\text{ k}\Omega \tag{5-103}$$

$$R_4 = R = 10\text{ k}\Omega \tag{5-104}$$

$$R_5 = \frac{R}{(1-\beta)A_r} = \frac{10^4}{(1 - 0.8333)3} = 20\text{ k}\Omega \tag{5-105}$$

The resulting circuit is shown in Figure 5-34 using standard resistor values. R_{1b} has been made adjustable for tuning.

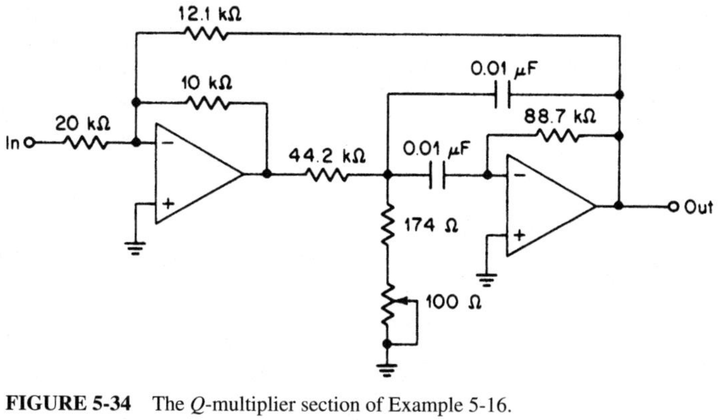

FIGURE 5-34 The Q-multiplier section of Example 5-16.

Elliptic-Function Bandpass Filters. An active elliptic-function bandpass filter is designed by first transforming the low-pass poles and zeros to the bandpass form using the formulas of Section 5.2. The bandpass poles and zeros are then implemented using active structures.

Normalized low-pass poles and zeros for elliptic-function low-pass filters can be obtained in terms Q, ω_0, ω_∞, and α_0 using the *Filter Solutions* program.

The general form of a bandpass transfer function containing zeros was given in Section 5.2 as

$$T(s) = \frac{H(s^2 + \omega_\infty^2)}{s^2 + \frac{\omega_r}{Q}s + \omega_r^2} \tag{5-49}$$

Elliptic-function bandpass filters are composed of cascaded first-order bandpass sections. When n is odd, $n - 1$ zero-producing sections are required, along with a single all-pole section. When n is even, $n - 2$ zero-producing sections are used, along with two all-pole networks.

This section discusses the VCVS and biquad configurations, which have a transfer function in the form of Equation (5-49) and their use in the design of active elliptic-function bandpass filters.

VCVS Network. Section 3.2 discussed the design of active elliptic-function low-pass filters using an *RC* section containing a voltage-controlled voltage source (VCVS). The circuit is repeated in Figure 5-35*a*. This structure is not restricted to the design of low-pass filters extensively. Transmission zeros can be obtained at frequencies either above or below the pole locations as required by the bandpass transfer function.

First, calculate

$$a = \frac{1}{Q} \tag{5-106}$$

$$b = \left(\frac{f_\infty}{f_r}\right)^2 \tag{5-107}$$

$$c = 2\pi f_r \tag{5-108}$$

where Q, f_∞, and f_r are the bandpass parameters corresponding to the general form bandpass transfer function given in Equation 5-49.

The element values are computed as follows:

Select C.

Then

$$C_1 = C \tag{5-109}$$

$$C_3 = C_4 = \frac{C_1}{2} \tag{5-110}$$

and

$$C_2 \geq \frac{C_1(b - 1)}{4} \tag{5-111}$$

$$R_3 = \frac{1}{cC_1\sqrt{b}} \tag{5-112}$$

$$R_1 = R_2 = 2R_3 \tag{5-113}$$

$$R_4 = \frac{4\sqrt{b}}{cC_1(1 - b) + 4cC_2} \tag{5-114}$$

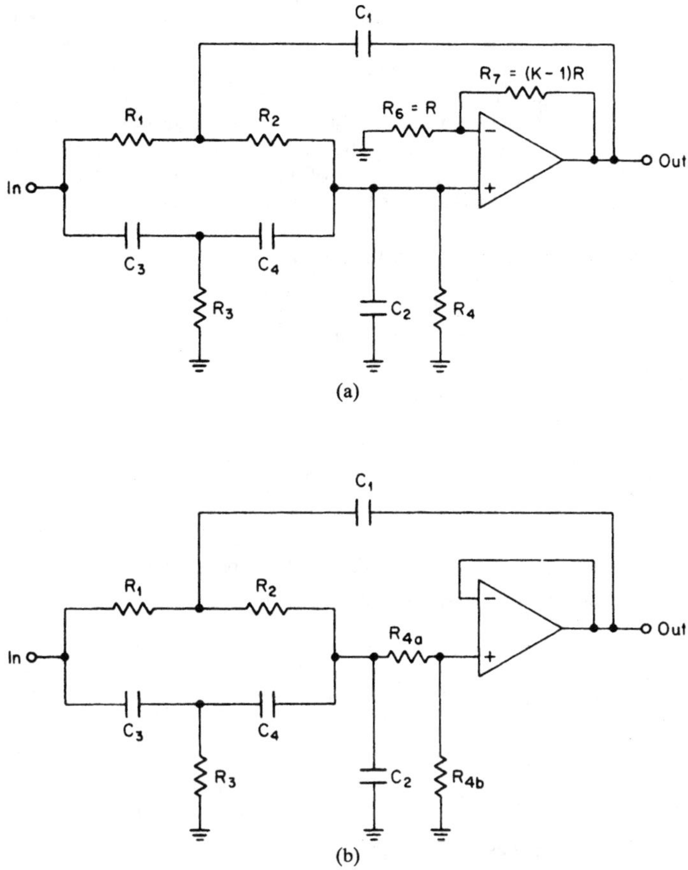

FIGURE 5-35 The VCVS elliptic-function bandpass section: (*a*) circuit for $K > 1$; and (*b*) circuit for $K < 1$.

$$K = 2 + \frac{2C_2}{C_1} - \frac{a}{2\sqrt{b}} + \frac{2}{C_1\sqrt{b}}\left(\frac{1}{cR_4} - aC_2\right) \tag{5-115}$$

$$R_6 = R \tag{5-116}$$

$$R_7 = (K - 1)R \tag{5-117}$$

where R can be arbitrarily chosen.

In the event that K is less than 1, the circuit of Figure 5-35*b* is used. Resistor R_4 is split into two resistors, R_{4a} and R_{4b}, which are given by

$$R_{4a} = (1 - K)\, R_4 \tag{5-118}$$

and

$$R_{4b} = KR_4 \tag{5-119}$$

The section Q can be controlled independently of resonant frequency by making R_6 or R_7 adjustable when $K > 1$. The resonant frequency, however, is not easily adjusted. Experience has shown that with 1-percent resistors and capacitors, section Qs of up to 10 can be realized with little degradation to the overall filter response due to component tolerances.

The actual circuit Q cannot be measured directly since the section's 3-dB bandwidth is determined not only by the design Q but by the transmission zero as well. Nevertheless, the VCVS configuration uses a minimum number of amplifiers and is widely used by low-Q elliptic-function realizations. The design technique is demonstrated in the following example.

Example 5-17 Design of an Active Elliptic-Function Bandpass Filter Using the VCVS Configuration

Required:

An active bandpass filter

A center frequency of 500 Hz

1-dB maximum at ±100 Hz (400 Hz, 600 Hz)

35-dB minimum at ±363 Hz (137 Hz, 863 Hz)

Result:

(*a*) Convert to geometrically symmetrical bandpass requirements:

First, calculate the geometric center frequency

$$f_0 = \sqrt{f_L f_u} = \sqrt{400 \times 600} = 490.0 \text{ Hz} \tag{2-14}$$

Since the stopband requirement is arithmetically symmetrical, compute stopband bandwidth using Equation (5-18).

$$\text{BW}_{35 \text{ dB}} = f_2 - \frac{f_0^2}{f_2} = 863 - \frac{490^2}{863} = 584.8 \text{ Hz}$$

The bandpass steepness factor is given by

$$A_s = \frac{\text{stopband bandwidth}}{\text{passband bandwidth}} = \frac{584.8 \text{ Hz}}{200 \text{ Hz}} = 2.924 \tag{2-19}$$

(*b*) Open ***Filter Solutions***.

Check the ***Stop Band Freq*** box.

Enter **.18** in the ***Pass Band Ripple(dB)*** box.

Enter **1** in the ***Pass Band Freq*** box.

Enter **2.924** in the ***Stop Band Freq*** box.

Check the ***Frequency Scale Rad/Sec*** box.

(*c*) Click the ***Set Order*** control button to open the second panel.

Enter **35** for the ***Stopband Attenuation (dB)***.

Click the ***Set Minimum Order*** button and then click ***Close***.

3 Order is displayed on the main control panel.

(*d*) Click the ***Transfer Function*** button.

Check the ***Casc*** box.

The following is displayed:

Continuous Transfer Function

Wn = 3.351

$$\frac{.1134 \quad (S^2 + 11.23)}{(S + .883) \quad (S^2 + .7697{*}S + 1.441)}$$

Wo = 1.2
Q = 1.56

3rd Order Low Pass Elliptic

Pass Band Frequency = 1.000 Rad/Sec
Pass Band Ripple = 180.0 mdB
Stop Band Ratio = 2.924
Stop Band Frequency = 2.924 Rad/Sec
Stop Band Attenuation = 37.51 dB

(*e*) The normalized low-pass design parameters are summarized as follows:

$$\text{Section } Q = 1.56$$

$$\text{Section } \omega_0 = 1.2$$

$$\text{Section } \omega_\infty = 3.351$$

$$\alpha_0 = 0.883 \text{ (from the denominator)}$$

The pole coordinates in rectangular form are

$$\alpha = \frac{\omega_0}{2Q} = 0.3646$$

$$\beta = \sqrt{\omega_0^2 - \alpha^2} = 1.1367$$

(*f*) To determine the bandpass parameters, first compute

$$Q_{\text{bp}} = \frac{f_0}{\text{BW}_{1\text{ dB}}} = \frac{490 \text{ Hz}}{200 \text{ Hz}} = 2.45 \tag{2-16}$$

The poles and zeros are transformed as follows:

Complex pole:

$$\alpha = 0.3846 \quad \beta = 1.1367$$

$$C = 1.440004 \tag{5-50}$$

$$D = 0.313959 \tag{5-51}$$

$$E = 4.239901 \tag{5-52}$$

$$G = 4.193146 \tag{5-53}$$

$$Q = 6.540396 \tag{5-54}$$

$$M = 1.026709 \tag{5-55}$$

$$W = 1.259369 \tag{5-56}$$

$$f_{ra} = 389 \text{ Hz} \tag{5-57}$$

$$f_{rb} = 617 \text{ Hz} \tag{5-58}$$

Real pole:

$$\alpha_0 = 0.883$$

$$Q = 2.7746 \tag{5-59}$$

$$f_r = 490 \text{ Hz}$$

Zero:

$$\omega_\infty = 3.351$$

$$H = 1.935377 \tag{5-60}$$

$$Z = 1.895359 \tag{5-61}$$

$$f_{\infty,a} = 258.5 \text{ Hz} \tag{5-62}$$

$$f_{\infty,b} = 928.7 \text{ Hz} \tag{5-63}$$

The bandpass parameters are summarized in the following table, where the zeros are arbitrarily assigned to the first two sections:

Section	f_r	Q	f_∞
1	389 Hz	6.54	258.5 Hz
2	617 Hz	6.54	928.7 Hz
3	490 Hz	2.77	

(*g*) Sections 1 and 2 are realized using the VCVS configuration of Figure 5-35. The element values are computed as follows, where R' and R are both 10 kΩ.

Section 1:

$$f_r = 389 \text{ Hz}$$

$$Q = 6.54$$

$$f_\infty = 259 \text{ Hz}$$

$$a = 0.15291 \quad (5\text{-}106)$$

$$b = 0.4433 \quad (5\text{-}107)$$

$$c = 2444 \quad (5\text{-}108)$$

Let $C = 0.02\ \mu\text{F}$

then $C_1 = 0.02\ \mu\text{F}$ (5-109)

$$C_3 = C_4 = 0.01\ \mu\text{F} \quad (5\text{-}110)$$

$$C_2 \geq -0.0027835\ \mu\text{F} \quad (5\text{-}111)$$

Let $C_2 = 0$

$$R_3 = 30.725\ \text{k}\Omega \quad (5\text{-}112)$$

$$R_1 = R_2 = 61.450\ \text{k}\Omega \quad (5\text{-}113)$$

$$R_4 = 97.866\ \text{k}\Omega \quad (5\text{-}114)$$

$$K = 2.5131 \quad (5\text{-}115)$$

Section 2:

$$f_r = 617 \text{ Hz}$$

$$Q = 6.54$$

$$f_\infty = 929 \text{ Hz}$$

$$a = 0.15291 \quad (5\text{-}106)$$

$$b = 2.2671 \quad (5\text{-}107)$$

$$c = 3877 \quad (5\text{-}108)$$

Let $C = 0.02\ \mu\text{F}$

then $C_1 = 0.02\ \mu\text{F}$ (5-109)

$$C_3 = C_4 = 0.01\ \mu\text{F} \quad (5\text{-}110)$$

$$C_2 \geq -0.006335\ \mu\text{F} \quad (5\text{-}111)$$

Let $C_2 = 0.01\ \mu\text{F}$

$$R_3 = 8566\ \Omega \quad (5\text{-}112)$$

$$R_1 = R_2 = 17.132\ \text{k}\Omega \quad (5\text{-}113)$$

$$R_4 = 105.98\ \text{k}\Omega \quad (5\text{-}114)$$

$$K = 3.0093 \quad (5\text{-}115)$$

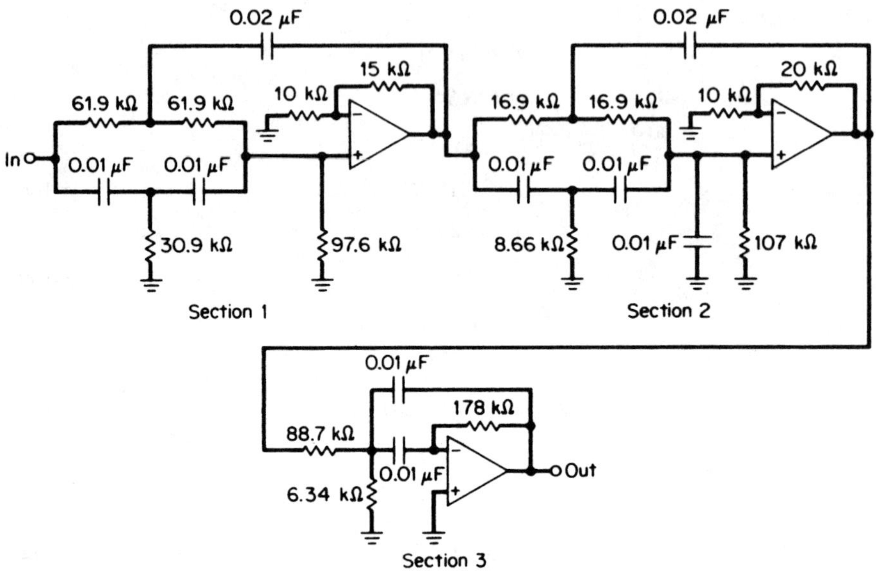

FIGURE 5-36 The circuit of the elliptic-function bandpass filter in Example 5-17.

(*h*) Section 3 is required to be of the all-pole type, so the MFBP configuration of Figure 5-25*b* will be used, where *C* is chosen as 0.01 μF and the section gain A_r is set to unity:

Section 3:

$$f_r = 490 \text{ Hz}$$

$$Q = 2.77$$

$$R_2 = 179.9 \text{ k}\Omega \tag{5-68}$$

$$R_{1a} = 89.97 \text{ k}\Omega \tag{5-72}$$

$$R_{1b} = 6.27 \text{ k}\Omega \tag{5-73}$$

The complete circuit is shown in Figure 5-36 using standard 1-percent resistor values.

State-Variable (Biquad) Circuit. The all-pole bandpass form of the state-variable or biquad section was discussed in Section 5.2. With the addition of an operational amplifier, the circuit can be used to realize transmission zeros as well as poles. The configuration is shown in Figure 5-37. This circuit is identical to the elliptic-function low-pass and high-pass filter configurations of Sections 3.2 and 4.2. By connecting R_5 either to node 1 or to node 2, the zero can be located above or below the resonant frequency.

On the basis of sensitivity and flexibility, the biquad configuration has been found to be the optimum method of constructing precision active elliptic-function bandpass filters. Section *Q*s of up to 200 can be obtained, whereas the VCVS section is limited to *Q*s below 10. Resonant frequency f_r, Q, and notch frequency f_∞ can be independently monitored and adjusted.

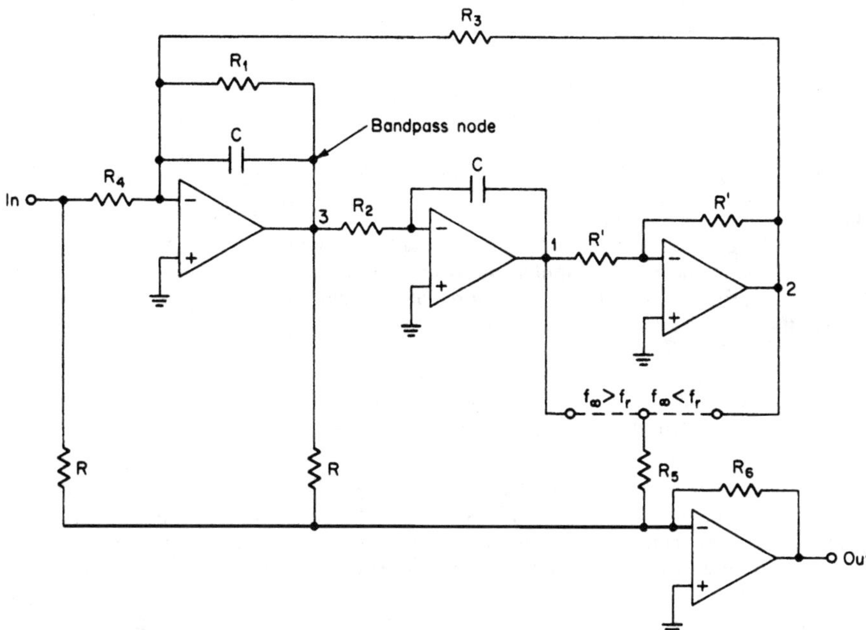

FIGURE 5-37 A biquad elliptic-function bandpass configuration.

For the case where $f_\infty < f_r$, the transfer function is given by

$$T(s) = -\frac{R_6}{R} \frac{s^2 + \frac{1}{R_2 R_3 C^2}\left(1 - \frac{R_3 R}{R_4 R_5}\right)}{s^2 + \frac{1}{R_1 C}s + \frac{1}{R_2 R_3 C^2}} \tag{5-120}$$

and when $f_\infty > f_r$, the corresponding transfer function is

$$T(s) = -\frac{R_6}{R} \frac{s^2 + \frac{1}{R_2 R_3 C^2}\left(1 + \frac{R_3 R}{R_4 R_5}\right)}{s^2 + \frac{1}{R_1 C}s + \frac{1}{R_2 R_3 C^2}} \tag{5-121}$$

If we equate the transfer-function coefficients to those of the general bandpass transfer function (with zeros) of Equation (5-49), the following series of design equations can be derived:

$$R_1 = R_4 = \frac{Q}{2\pi f_r C} \tag{5-122}$$

$$R_2 = R_3 = \frac{R_1}{Q} \tag{5-123}$$

$$R_5 = \frac{f_r^2 R}{Q|f_r^2 - f_\infty^2|} \tag{5-124}$$

for $f_\infty > f_r$:

$$R_6 = \frac{f_r^2 R}{f_\infty^2} \tag{5-125}$$

and when $f_\infty < f_r$:

$$R_6 = R \tag{5-126}$$

where C and R can be conveniently selected. The value of R_6 is based on unity section gain. The gain can be raised or lowered by proportionally changing R_6.

The section can be tuned by implementing the following steps in the indicated sequence. Both resonant frequency and Q are monitored at the bandpass output occurring at node 3, whereas the notch frequency f_∞ is observed at the section output.

1. *Resonance frequency f_r:* If R_3 is made variable, the section resonant frequency can be adjusted. Resonance is monitored at node 3 (see Figure 5-37) and can be determined by the 180° phase shift method.
2. *Q adjustment:* The section Q is controlled by R_1 and can be directly measured at node 3. The configuration is subject to the Q-enhancement effect discussed in Section 5.2 under "All-Pole Bandpass Configurations," so a Q adjustment is normally required. The Q can be monitored in terms of the 3-dB bandwidth at node 3, or R_1 can be adjusted until unity gain occurs between the section input and node 3 with f_r applied.
3. *Notch frequency f_∞:* Adjustment of the notch frequency (transmission zero) usually is not required if the circuit is previously tuned to f_r, since f_∞ will usually then fall in. If an adjustment is desired, the notch frequency can be controlled by making R_5 variable.

The biquad approach is a highly stable and flexible implementation for precision active elliptic-function filters. The independent adjustment capability for resonant frequency, Q, and the notch frequency preclude its use when Qs in excess of 10 are required. Stable Qs of up to 200 are obtainable.

Example 5-18 Design of an Active Elliptic-Function Bandpass Filter Using the Biquad Configuration

Required:

An active bandpass filter

A center frequency 500 Hz

0.2-dB maximum at ±50 Hz (450 Hz, 550 Hz)

30-dB minimum at ±130 Hz (370 Hz, 630 Hz)

Result:

(*a*) Convert to the geometrically symmetrical requirement

$$f_0 = \sqrt{f_L f_u} = \sqrt{450 \times 550} = 497.5 \text{ Hz} \tag{2-14}$$

$$\text{BW}_{30\text{ dB}} = f_2 - \frac{f_0^2}{f_2} = 630 - \frac{497.5^2}{630} = 237.1 \text{ Hz} \tag{5-18}$$

$$A_s = \frac{\text{stopband bandwidth}}{\text{passband bandwidth}} = \frac{237.1 \text{ Hz}}{100 \text{ Hz}} = 2.371 \tag{2-19}$$

(*b*) Open ***Filter Solutions***.
Check the ***Stop Band Freq*** box.
Enter **.18** in the ***Pass Band Ripple(dB)*** box.
Enter **1** in the ***Pass Band Freq*** box.
Enter **2.371** in the ***Stop Band Freq*** box.
Check the ***Frequency Scale Rad/Sec*** box.

(*c*) Click the ***Set Order*** control button to open the second panel.
Enter **30** for the ***Stopband Attenuation (dB)***.
Click the ***Set Minimum Order*** button and then click ***Close***.
3 Order is displayed on the main control panel.

(*d*) Click the ***Transfer Function*** button.
Check the ***Casc*** box.

The following is displayed:

Continuous Transfer Function

$$W_n = 2.705$$

$$\frac{.1784 \quad (S^2 + 7.32)}{(S^2 + .7321{*}S + 1.435) \quad (S + .9105)}$$

$$W_o = 1.198$$
$$Q = 1.636$$

3rd Order Low Pass Elliptic

Pass Band Frequency = 1.000 Rad/Sec
Pass Band Ripple = 180.0 mdB

Stop Band Ratio = 2.371
Stop Band Frequency = 2.371 Rad/Sec
Stop Band Attenuation = 31.59 dB

(*e*) The normalized low-pass design parameters are summarized as follows:

$$\text{Section } Q = 1.636$$

$$\text{Section } \omega_0 = 1.198$$

$$\text{Section } \omega_\infty = 2.705$$

$$\alpha_0 = 0.9105 \text{ (from the denominator)}$$

The pole coordinates in rectangular form are

$$\alpha = \frac{\omega_0}{2Q} = 0.3661$$

$$\beta = \sqrt{\omega_0^2 - \alpha^2} = 1.1408$$

(*f*) The bandpass pole-zero transformation is now performed. First compute

$$Q_{\text{bp}} = \frac{f_0}{\text{BW}_{0.2\text{ dB}}} = \frac{497.5\text{ Hz}}{100\text{ Hz}} = 4.975 \tag{2-16}$$

The transformation proceeds as follows:

Complex pole:

$$\alpha = 0.3661 \quad \beta = 1.1408$$

$$C = 1.435454 \tag{5-50}$$

$$D = 0.147176 \tag{5-51}$$

$$E = 4.05800 \tag{5-52}$$

$$G = 4.047307 \tag{5-53}$$

$$Q = 13.678328 \tag{5-54}$$

$$M = 1.006560 \tag{5-55}$$

$$W = 1.121290 \tag{5-56}$$

$$f_{ra} = 443.6\text{ Hz} \tag{5-57}$$

$$f_{rb} = 557.8\text{ Hz} \tag{5-58}$$

Real pole:

$$\alpha_0 = 0.9105$$

$$Q = 5.464 \tag{5-59}$$

$$f_r = 497.5\text{ Hz}$$

Zero:

$$\omega_\infty = 2.705$$

$$H = 1.147815 \tag{5-60}$$

$$Z = 1.308154 \tag{5-61}$$

$$f_{\infty,a} = 380.31 \text{ Hz} \tag{5-62}$$

$$f_{\infty,b} = 650.81 \text{ Hz} \tag{5-63}$$

The computed bandpass parameters are summarized in the following table. The zeros are assigned to the first two sections.

Section	f_r	Q	f_∞
1	443.6 Hz	13.7	380.3 Hz
2	557.8 Hz	13.7	650.8 Hz
3	497.5 Hz	5.46	

(*g*) Sections 1 and 2 will be realized in the form of the biquad configuration of Figure 5-37 where R' and R are both 10 kΩ and $C = 0.047$ μF.

Section 1:

$$f_r = 443.6 \text{ Hz}$$

$$Q = 13.7$$

$$f_\infty = 380.3 \text{ Hz}$$

$$R_1 = R_4 = \frac{Q}{2\pi f_r C} = 104.58 \text{ k}\Omega \tag{5-122}$$

$$R_2 = R_3 = \frac{R_1}{Q} = 7.63 \text{ k}\Omega \tag{5-123}$$

$$R_5 = \frac{f_r^2 R}{Q|f_r^2 - f_\infty^2|} = 2.76 \text{ k}\Omega \tag{5-124}$$

$$R_6 = R = 10 \text{ k}\Omega \tag{5-126}$$

Section 2:

$$f_r = 557.8 \text{ Hz}$$

$$Q = 13.7$$

$$f_\infty = 650.8 \text{ Hz}$$

$$R_1 = R_4 = 83.17 \text{ k}\Omega \tag{5-122}$$

$$R_2 = R_3 = 6.07 \text{ k}\Omega \tag{5-123}$$

$$R_5 = 2.02 \text{ k}\Omega \tag{5-124}$$

$$R_6 = 7.35 \text{ k}\Omega \tag{5-125}$$

(*h*) The MFBP configuration of Figure 5-25*b* will be used for the all-pole circuit of section 3. The value of C is 0.047 μF, and A_r is unity.

Section 3:

$$f_r = 497.5 \text{ Hz}$$

$$Q = 5.46$$

$$R_2 = 74.3 \text{ k}\Omega \tag{5-68}$$

$$R_{1a} = 37.1 \text{ k}\Omega \tag{5-72}$$

$$R_{1b} = 634 \ \Omega \tag{5-73}$$

The resulting filter is shown in Figure 5-38, where standard 1-percent resistors are used. The resonant frequency and Q of each section have been made adjustable.

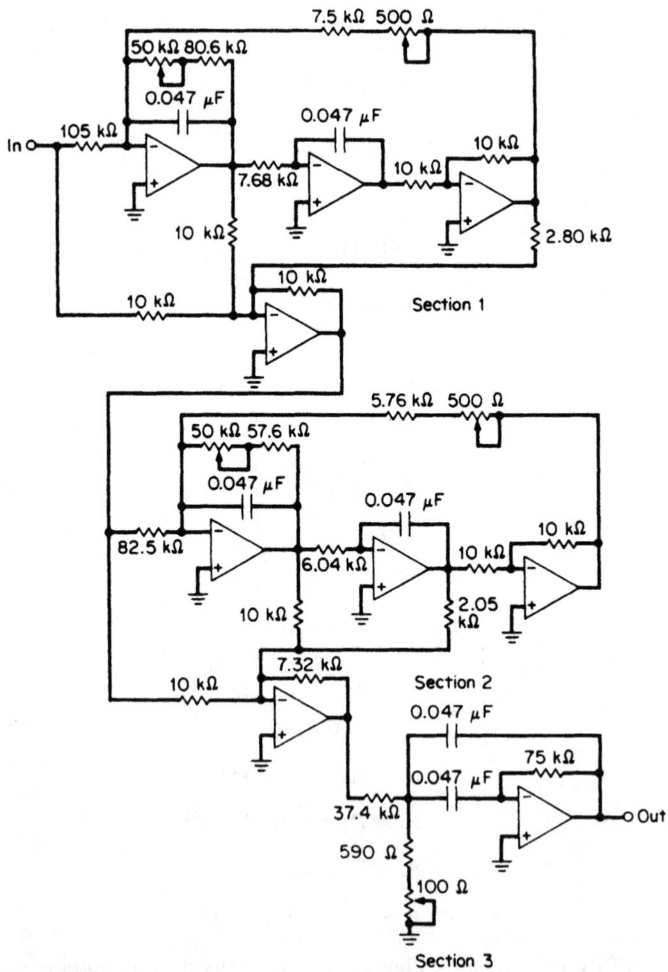

FIGURE 5-38 The biquad elliptic-function bandpass filter of Example 5-18.

BIBLIOGRAPHY

Huelsman, L. P. *Theory and Design of Active RC Circuits*. New York: McGraw-Hill, 1968.

Sedra, A. S., and J. L. Espinoza. "Sensitivity and Frequency Limitations of Biquadratic Active Filters." *IEEE Transactions on Circuits and Systems* CAS-22, no. 2 (February, 1975).

Thomas, L. C. "The Biquad: Part I—Some Practical Design Considerations." *IEEE Transactions on Circuit Theory* CT–18 (May, 1971).

Tow, J. "A Step-by-Step Active Filter Design." *IEEE Spectrum* 6 (December, 1969).

Williams, A. B. *Active Filter Design*. Dedham, Massachusetts: Artech House, 1975.

——————. "Q-Multiplier Techniques Increases Filter Selectivity." *EDN* (October 5, 1975): 74–76.

Zverev, A. I., *Handbook of Filter Synthesis*, John Wiley and Sons, New York, 1967.

CHAPTER 6
BAND-REJECT FILTERS

6.1 LC *BAND-REJECT FILTERS*

Normalization of a band-reject requirement and the definitions of the response shape parameters were discussed in Section 2.1. Like bandpass filters, band-reject networks can also be derived from a normalized low-pass filter by a suitable transformation.

In Section 5.1, we discussed the design of wideband bandpass filters by cascading a low-pass filter and a high-pass filter. In a similar manner, wideband band-reject filters can also be obtained by combining low-pass and high-pass filters. Both the input and output terminals are paralleled, and each filter must have a high input and output impedance in the band of the other filter to prevent interaction. Therefore, the order n must be odd and the first and last branches should consist of series elements. These restrictions make the design of band-reject filters by combining low-pass and high-pass filters undesirable. The impedance interaction between filters is a serious problem unless the separation between cutoffs is many octaves, so the design of band-reject filters is best approached by transformation techniques.

The Band-Reject Circuit Transformation

Bandpass filters were obtained by first designing a low-pass filter with a cutoff frequency equivalent to the required bandwidth and then resonating each element to the desired center frequency. The response of the low-pass filter at DC then corresponds to the response of the bandpass filter at the center frequency.

Band-reject filters are designed by initially transforming the normalized low-pass filter into a high-pass network with a cutoff frequency equal to the required bandwidth, and at the desired impedance level. Every high-pass element is then resonated to the center frequency in the same manner as bandpass filters.

This corresponds to replacing the frequency variable in the high-pass transfer function by a new variable, which is given by

$$f_{br} = f_0\left(\frac{f}{f_0} - \frac{f_0}{f}\right) \tag{6-1}$$

As a result, the response of the high-pass filter at DC is transformed to the band-reject network at the center frequency. The bandwidth response of the band-reject filter is identical to the frequency response of the high-pass filter. The high-pass to band-reject transformation is shown in Figure 6-1. Negative frequencies, of course, are strictly of theoretical interest, so only the response shape corresponding to positive frequencies is applicable. As in the case of bandpass filters, the response curve exhibits geometric symmetry.

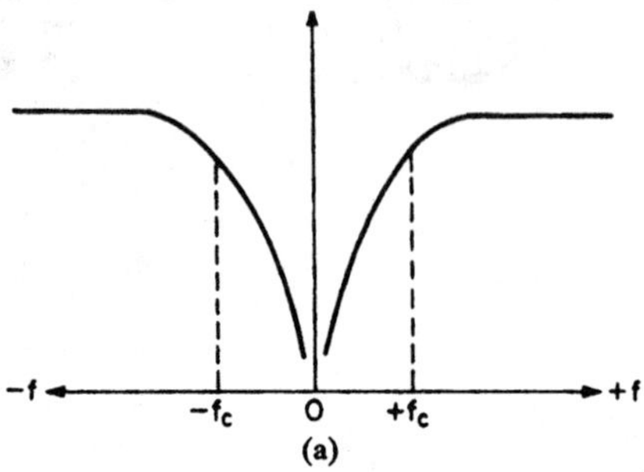

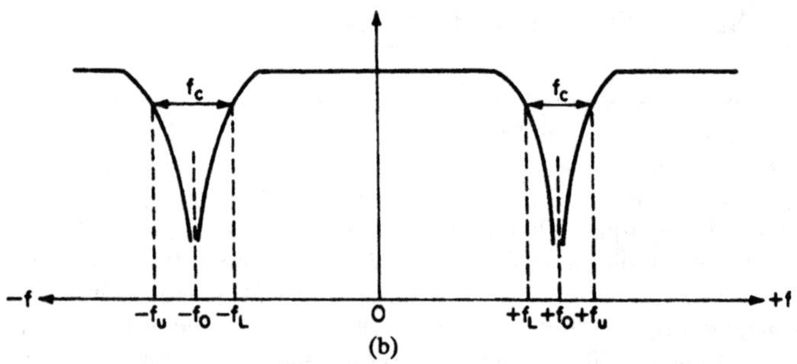

FIGURE 6-1 The band-reject transformation: (*a*) high-pass filter response; and (*b*) transformed band-reject filter response.

The design procedure can be summarized as follows:

1. Normalize the band-reject filter specification and select a normalized low-pass filter that provides the required attenuation within the computed steepness factor.
2. Transform the normalized low-pass filter to a normalized high-pass filter. Then scale the high-pass filter to a cutoff frequency equal to the desired bandwidth and to the preferred impedance level.
3. Resonate each element to the center frequency by introducing a capacitor in series with each inductor and an inductor in parallel with each capacitor to complete the design. The transformed circuit branches are summarized in Table 6-1.

All-Pole Band-Reject Filters. Band-reject filters can be derived from any all-pole or elliptic-function *LC* low-pass network. Although not as efficient as elliptic-function filters, the all-pole approach results in a simpler band-reject structure where all sections are tuned to the center frequency.

The following example demonstrates the design of an all-pole band-reject filter.

Example 6-1 Design of an All-Pole *LC* Band-Reject Filter

Required:

Band-reject filter
A center frequency of 10 kHz
3 dB at ±250 Hz (9.75 kHz, 10.25 kHz)

TABLE 6-1 The High-Pass to Band-Reject Transformation

	High-Pass Branch	Band-Reject Configuration	Circuit Values
Type I	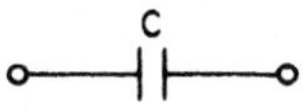	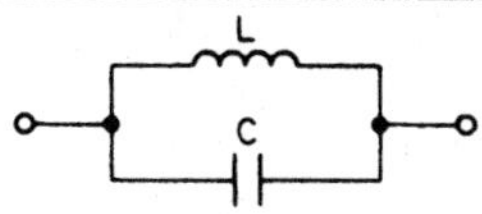	$L=\frac{1}{\omega_0^2 C}$ (6-2)
Type II		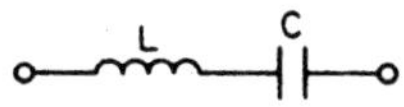	$C=\frac{1}{\omega_0^2 L}$ (6-3)
Type III	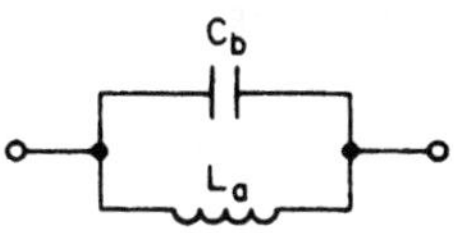	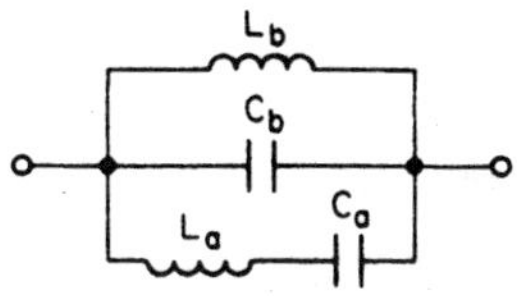	$C_a=\frac{1}{\omega_0^2 L_a}$ (6-4) $L_b=\frac{1}{\omega_0^2 C_b}$ (6-5)
Type IV	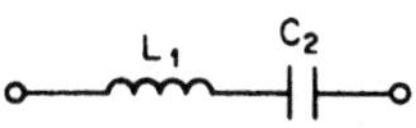	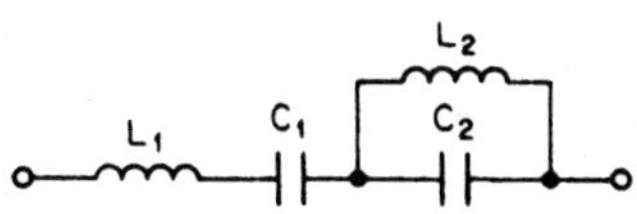	$C_1=\frac{1}{\omega_0^2 L_1}$ (6-6) $L_2=\frac{1}{\omega_0^2 C_2}$ (6-7)

30-dB minimum at ±100 Hz (9.9 kHz, 10.1 kHz)

A source and load impedance of 600 Ω

Result:

(*a*) Convert to a geometrically symmetrical requirement. Since the bandwidth is relatively narrow, the specified arithmetically symmetrical frequencies will determine the following design parameters:

$$f_0 = 10 \text{ kHz}$$
$$\text{BW}_{3 \text{ dB}} = 500 \text{ Hz}$$
$$\text{BW}_{30 \text{ dB}} = 200 \text{ Hz}$$

(*b*) Compute the band-reject steepness factor.

$$A_s = \frac{\text{passband bandwidth}}{\text{stopband bandwidth}} = \frac{500 \text{ Hz}}{200 \text{ Hz}} = 2.5 \tag{2-20}$$

The response curves of Figure 2-45 indicate that an $n = 3$ Chebyshev normalized low-pass filter having a 1-dB ripple provides over 30 dB of attenuation within a frequency ratio of 2.5:1. The corresponding circuit is found in Table 11-31 and is shown in Figure 6-2*a*.

(*c*) To transform the normalized low-pass circuit into a normalized high-pass filter, replace inductors with capacitors and vice versa using reciprocal element values. The transformed structure is shown in Figure 6-2*b*.

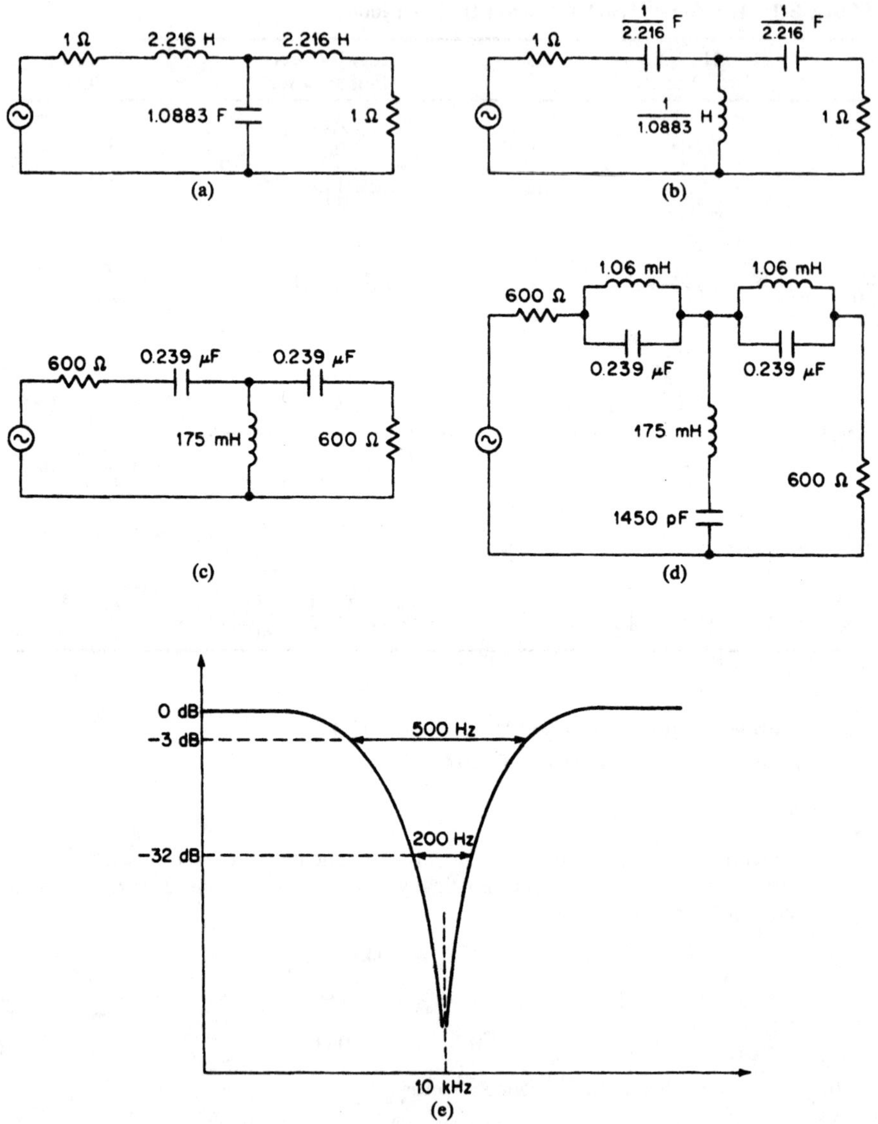

FIGURE 6-2 The band-reject filter of Example 6-1: (*a*) normalized low-pass filter; (*b*) transformed normalized high-pass filter; (*c*) frequency- and impedance-scaled high-pass filter; (*d*) transformed band-reject filter; and (*e*) frequency response.

(*d*) The normalized high-pass filter is scaled to a cutoff frequency of 500 Hz corresponding to the desired bandwidth and to an impedance level of 600 Ω. The capacitors are divided by $Z \times \text{FSF}$ and the inductors are multiplied by Z/FSF, where Z is 600 and the FSF (frequency-scaling factor) is given by $2\pi f_c$, where f_c is 500 Hz. The scaled high-pass filter is illustrated in Figure 6-2*c*.

(*e*) To make the high-pass to band-reject transformation, resonate each capacitor with a parallel inductor and each inductor with a series capacitor. The resonating inductors for the series branches are both given by

$$L = \frac{1}{\omega_0^2 C} = \frac{1}{(2\pi 10 \times 10^3)^2 \times 0.239 \times 10^{-6}} = 1.06 \text{ mH} \tag{6-2}$$

The tuning capacitor for the shunt inductor is determined from

$$C = \frac{1}{\omega_0^2 L} = \frac{1}{(2\pi 10 \times 10^3)^2 \times 0.175} = 1450 \text{ pF} \tag{6-3}$$

The final filter is shown in Figure 6-2*d*, where all peaks are tuned to the center frequency of 10 kHz. The theoretical frequency response is illustrated in Figure 6-2*e*.

When a low-pass filter undergoes a high-pass transformation, followed by a band-reject transformation, the minimum Q requirement is increased by a factor equal to the Q of the band-reject filter. This can be expressed as

$$Q_{\text{min}} \text{ (band-reject)} = Q_{\text{min}} \text{ (low-pass)} \times Q_{\text{br}} \tag{6-8}$$

where values for Q_{min} (low-pass) are given in Figure 3-8 and $Q_{\text{br}} = f_0/\text{BW}_{3\text{ dB}}$. The branch Q should be several times larger than Q_{min} (band-reject) to obtain near-theoretical results.

The equivalent circuit of a band-reject filter at the center frequency can be determined by replacing each parallel tuned circuit by a resistor of $\omega_0 L Q_L$ and each series tuned circuit by a resistor of $\omega_0 L/Q_L$. These resistors correspond to the branch impedances at resonance, where ω_0 is $2\pi f_0$, L is the branch inductance, and Q_L is the branch Q, which is normally determined only by the inductor losses.

It is then apparent that at the center frequency, the circuit can be replaced by a resistive voltage divider. The amount of attenuation that can be obtained is then directly controlled by the branch Qs. Let's determine the attenuation of the circuit of Example 6-1 for a finite value of inductor Q.

Example 6-2 Estimate Maximum Band-Reject Rejection as a Function of Q

Required:

Estimate the amount of rejection obtainable at the center frequency of 10 kHz for the band-reject filter of Example 6-1. An inductor Q of 100 is available and the capacitors are assumed to be lossless. Also determine if the Q is sufficient to retain the theoretical passband characteristics.

Result:

(*a*) Compute the equivalent resistances at resonance for all tuned circuits.

Parallel tuned circuits:

$$R = \omega_0 L Q_L = 2\pi \times 10^4 \times 1.06 \times 10^{-3} \times 100 = 6660\ \Omega \tag{5-21}$$

Series tuned circuits:

$$R = \frac{\omega_0 L}{Q_L} = \frac{2\pi \times 10^4 \times 0.175}{100} = 110\ \Omega \tag{5-30}$$

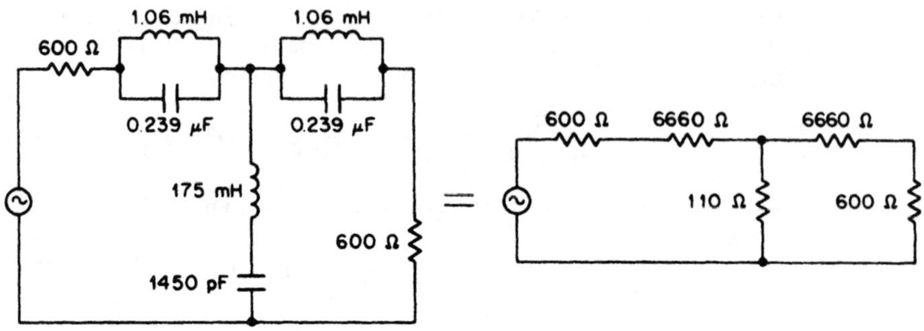

FIGURE 6-3 The equivalent circuit at the center frequency for the filter of Figure 6-2.

(*b*) The equivalent circuit at 10 kHz is shown in Figure 6-3. Using conventional circuit analysis methods such as mesh equations or approximation techniques, the overall loss is found to be 58 dB. Since the flat loss due to the 600-Ω terminations is 6 dB, the relative attenuation at 10 kHz will be 52 dB.

(*c*) The curves of Figure 3-8 indicate that an $n = 3$ Chebyshev filter with a 1-dB ripple has a minimum theoretical Q requirement of 4.5. The minimum Q of the band-reject filter is given by

$$Q_{\min}(\text{band-reject}) = Q_{\min}(\text{low-pass}) \times Q_{\text{br}} = 4.5 \times \frac{10{,}000}{500} = 90 \qquad (6\text{-}8)$$

Therefore, the available Q of 100 is barely adequate, and some passband rounding will occur in addition to the reduced stopband attenuation. The resulting effect on frequency response is shown in Figure 6-4.

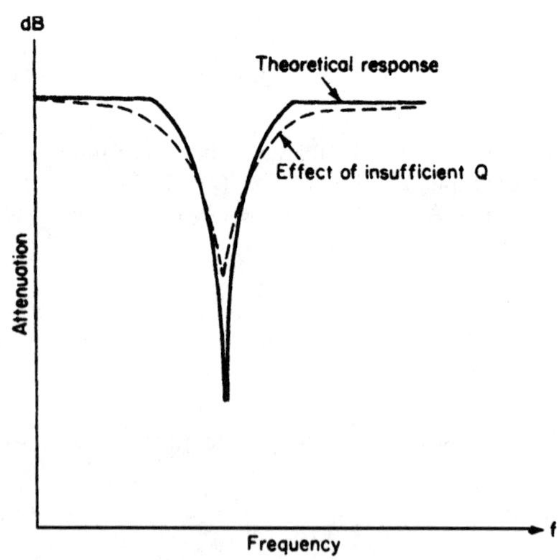

FIGURE 6-4 The effects of insufficient Q upon a band-reject filter.

Elliptic-Function Band-Reject Filters. The superior properties of the elliptic-function family of filters can also be applied to band-reject requirements. Extremely steep characteristics in the transition region between passband and stopband can be achieved much more efficiently than with all-pole filters.

Saal and Ulbrich, as well as Zverev (see Bibliography), have extensively tabulated the *LC* values for normalized elliptic-function low-pass networks. Using the *Filter Solutions* program or the ELI 1.0 program, low-pass filters can be directly designed using the filter requirements as the program input rather than engaging normalized tables. These circuits can then be transformed to high-pass filters, and subsequently to a band-reject filter in the same manner as the all-pole filters.

Since each normalized low-pass filter can be realized in dual forms, the resulting band-reject filters can also take on two different configurations, as illustrated in Figure 6-5.

Branch 2 of the standard band-reject filter circuit corresponds to the type III network shown in Table 6-1. This branch provides a pair of geometrically related zeros, one above and one below the center frequency. These zeros result from two conditions of parallel resonance. However, the circuit configuration itself is not very desirable. The elements corresponding to the individual parallel resonances are not distinctly isolated since each

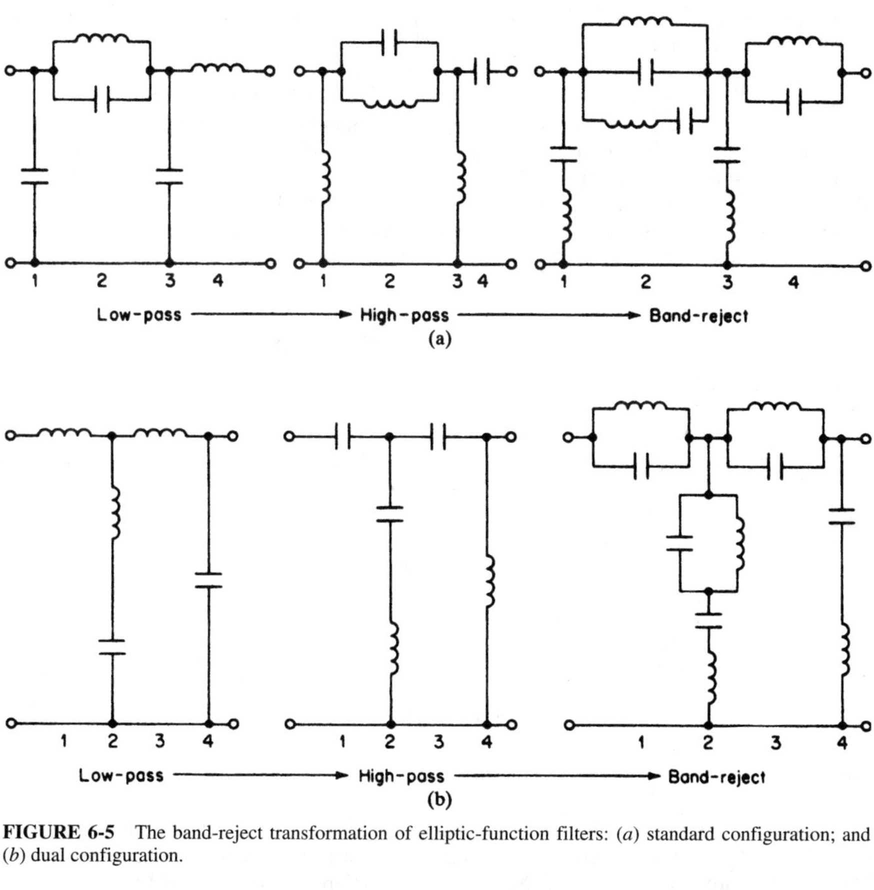

FIGURE 6-5 The band-reject transformation of elliptic-function filters: (*a*) standard configuration; and (*b*) dual configuration.

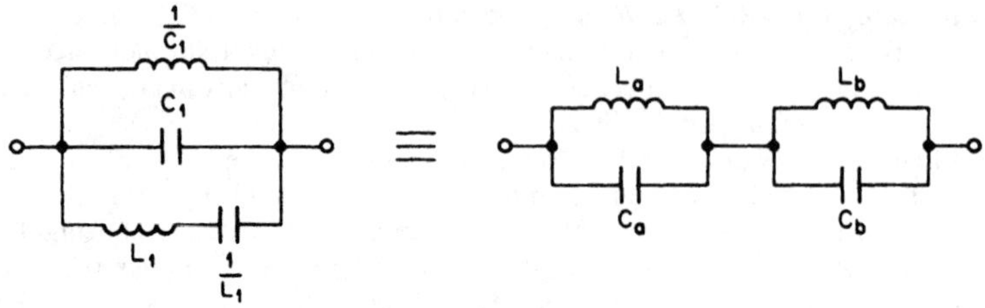

FIGURE 6-6 The equivalent circuit of a type III network.

resonance is determined by the interaction of a number of elements. This makes tuning somewhat difficult. For very narrow filters, the element values also become somewhat unreasonable.

An identical situation occurred during the bandpass transformation of an elliptic-function low-pass filter discussed in Section 5.1. An equivalent configuration was presented as an alternate and is repeated in Figure 6-6.

The type III network of Figure 6-6 has reciprocal element values which occur when the band-reject filter has been normalized to a 1-rad/s center frequency since the equation of resonance, $\omega_0^2 LC = 1$, then reduces to $LC = 1$. The reason for this normalization is to greatly simplify the transformation equations.

To normalize the band-reject filter circuit, first transform the normalized low-pass filter to a normalized high-pass configuration in the conventional manner by replacing inductors with capacitors and vice versa using reciprocal element values. The high-pass elements are then multiplied by the factor Q_{br}, which is equal to f_0/BW, where f_0 is the geometric center frequency of the band-reject filter and BW is the bandwidth. The normalized band-reject filter can be directly obtained by resonating each inductor with a series capacitor and each capacitor with a parallel inductor using reciprocal values.

To make the transformation of Figure 6-6, first compute

$$\beta = 1 + \frac{1}{2L_1C_1} + \sqrt{\frac{1}{4L_1^2C_1^2} + \frac{1}{L_1C_1}} \tag{6-9}$$

The values are then found from

$$L_a = \frac{1}{C_1(\beta + 1)} \tag{6-10}$$

$$L_b = \beta L_a \tag{6-11}$$

$$C_a = \frac{1}{L_b} \tag{6-12}$$

$$C_b = \frac{1}{L_a} \tag{6-13}$$

The resonant frequencies for each tuned circuit are given by

$$\Omega_{\infty,a} = \sqrt{\beta} \tag{6-14}$$

and

$$\Omega_{\infty,b} = \frac{1}{\Omega_{\infty,a}} \tag{6-15}$$

After the normalized band-reject filter has undergone the transformation of Figure 6-6 wherever applicable, the circuit can be scaled to the desired impedance level and frequency. The inductors are multiplied by Z/FSF, and capacitors are divided by $Z \times \text{FSF}$. The value

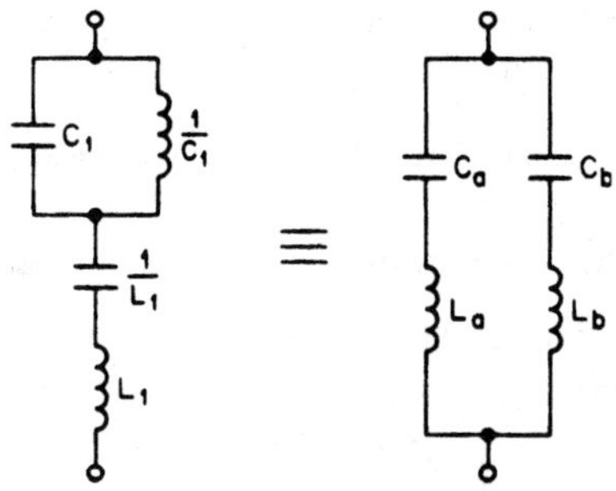

FIGURE 6-7 The equivalent circuit of the type IV network.

of Z is the desired impedance level, and the frequency-scaling factor (FSF) in this case is equal to ω_0 ($\omega_0 = 2\pi f_0$). The resulting resonant frequencies in hertz are determined by multiplying the normalized radian resonant frequencies by f_0.

Branch 2 of the band-reject filter derived from the dual low-pass structure of Figure 6-5*b* corresponds to the type IV network of Table 6-1. This configuration realizes a pair of finite zeros resulting from two conditions of series resonance. However, as in the case of the type III network, the individual resonances are determined by the interaction of all the elements, which makes tuning difficult and can result in unreasonable values for narrow filters. An alternate configuration is shown in Figure 6-7 consisting of two series resonant circuits in parallel.

To simplify the transformation equations, the type IV network requires reciprocal values, so the band-reject filter must be normalized to a 1-rad/s center frequency. This is accomplished as previously described, and the filter is subsequently denormalized after the transformations have been made.

The transformation is accomplished as follows:

First, compute

$$\beta = 1 + \frac{1}{2L_1C_1} + \sqrt{\frac{1}{4L_1^2C_1^2} + \frac{1}{L_1C_1}} \tag{6-16}$$

then

$$L_a = \frac{(\beta + 1)L_1}{\beta} \tag{6-17}$$

$$C_a = \frac{1}{(\beta + 1)L_1} \tag{6-18}$$

$$L_b = \frac{1}{C_a} \tag{6-19}$$

$$C_b = \frac{1}{L_a} \tag{6-20}$$

$$\Omega_{\infty,a} = \sqrt{\beta} \tag{6-21}$$

$$\Omega_{\infty,b} = \frac{1}{\Omega_{\infty,a}} \tag{6-22}$$

The standard configuration of the elliptic-function filter is usually preferred over the dual circuit so that the transformed low-pass zeros can be realized using the structure of Figure 6-6. Parallel tuned circuits are generally more desirable than series tuned circuits since they can be transformed to alternate L/C ratios to optimize Q and reduce capacitor values (see Section 8.2 on tapped inductors).

Example 6-3 Design of an LC Elliptic Function Band-Reject Filter

Required:

Design a band-reject filter to satisfy the following requirements:

1-dB maximum at 2200 and 2800 Hz

50-dB minimum at 2300 and 2700 Hz

A source and load impedance of 600 Ω

Result:

(*a*) Convert to a geometrically symmetrical requirement. First, calculate the geometric center frequency.

$$f_0 = \sqrt{f_L f_u} = \sqrt{2200 \times 2800} = 2482 \text{ Hz} \tag{2-14}$$

Compute the corresponding geometric frequency for each stopband frequency given using Equation (2-18).

$$f_1 f_2 = f_0^2 \tag{2-18}$$

f_1	f_2	$f_2 - f_1$
2300 Hz	2678 Hz	378 Hz
2282 Hz	2700 Hz	418 Hz

The second pair of frequencies is retained since they represent the steeper requirement. The complete geometrically symmetrical specification can be stated as

$$f_0 = 2482 \text{ Hz}$$

$$\text{BW}_{1\text{ dB}} = 600 \text{ Hz}$$

$$\text{BW}_{50\text{ dB}} = 418 \text{ Hz}$$

(*b*) Compute the band-reject steepness factor.

$$A_s = \frac{\text{passband bandwidth}}{\text{stopband bandwidth}} = \frac{600 \text{ Hz}}{418 \text{ Hz}} = 1.435 \tag{2-20}$$

A normalized low-pass filter must be chosen that makes the transition from less than 1 dB to more than 50 dB within a frequency ratio of 1.435. An elliptic-function filter will be used.

(*c*) Open ***Filter Solutions***.

Check the ***Stop Band Freq*** box.

Enter **.18** in the ***Pass Band Ripple (dB)*** box.

Enter **1** in the ***Pass Band Freq*** box.

Enter **1.435** in the ***Stop Band Freq*** box.

The ***Frequency Scale Rad/Sec*** box should be checked.

Enter ***1*** for ***Source Res*** and ***Load Res***.

(*d*) Click the ***Set Order*** control button to open the second panel.

Enter **50** for the ***Stopband Attenuation (dB)***.

Click the ***Set Minimum Order*** button and then click ***Close***.

6 Order is displayed on the main control panel.

Check the ***Even Order Mod*** box.

(*e*) Click the ***Circuits*** button.

Two schematics are presented by *Filter Solutions*. Use Passive Filter 1, which is shown in Figure 6-8*a*.

(*f*) The normalized low-pass filter is now transformed into a normalized high-pass structure by replacing all inductors with capacitors, and vice versa, using reciprocal values. The resulting filter is given in Figure 6-8*b*.

(*g*) To obtain a normalized band-reject filter so that the transformation of Figure 6-6 can be performed, first multiply all the high-pass elements by Q_{br}, which is given by

$$Q_{br} = \frac{f_0}{BW_{1\,dB}} = \frac{2482\text{ Hz}}{600\text{ Hz}} = 4.137$$

The modified high-pass filter is shown in Figure 6-8*c*.

(*h*) Each high-pass inductor is resonated with a series capacitor, and each capacitor is resonated with a parallel inductor to obtain the normalized band-reject filter. Since the center frequency is 1 rad/s, the resonant elements are simply the reciprocal of each other, as illustrated in Figure 6-8*d*.

(*i*) The type III networks of the second and fourth branches are now transformed to the equivalent circuit of Figure 6-6 as follows:

The type III network of third branch:

$$L_1 = 11.428\text{ H} \quad C_1 = 3.270\text{ F}$$

$$\beta = 1 + \frac{1}{2L_1C_1} + \sqrt{\frac{1}{4L_1^2C_1^2} + \frac{1}{L_1C_1}} = 1.1775 \tag{6-9}$$

$$L_a = \frac{1}{C_1(\beta + 1)} = 0.1404\text{ H} \tag{6-10}$$

$$L_b = \beta L_a = 0.1654\text{ H} \tag{6-11}$$

$$C_a = \frac{1}{L_b} = 6.047\text{ F} \tag{6-12}$$

$$C_b = \frac{1}{L_a} = 7.1205\text{ F} \tag{6-13}$$

$$\Omega_{\infty,a} = \sqrt{\beta} = 1.0851 \tag{6-14}$$

$$\Omega_{\infty,b} = \frac{1}{\Omega_{\infty,a}} = 0.92155 \tag{6-15}$$

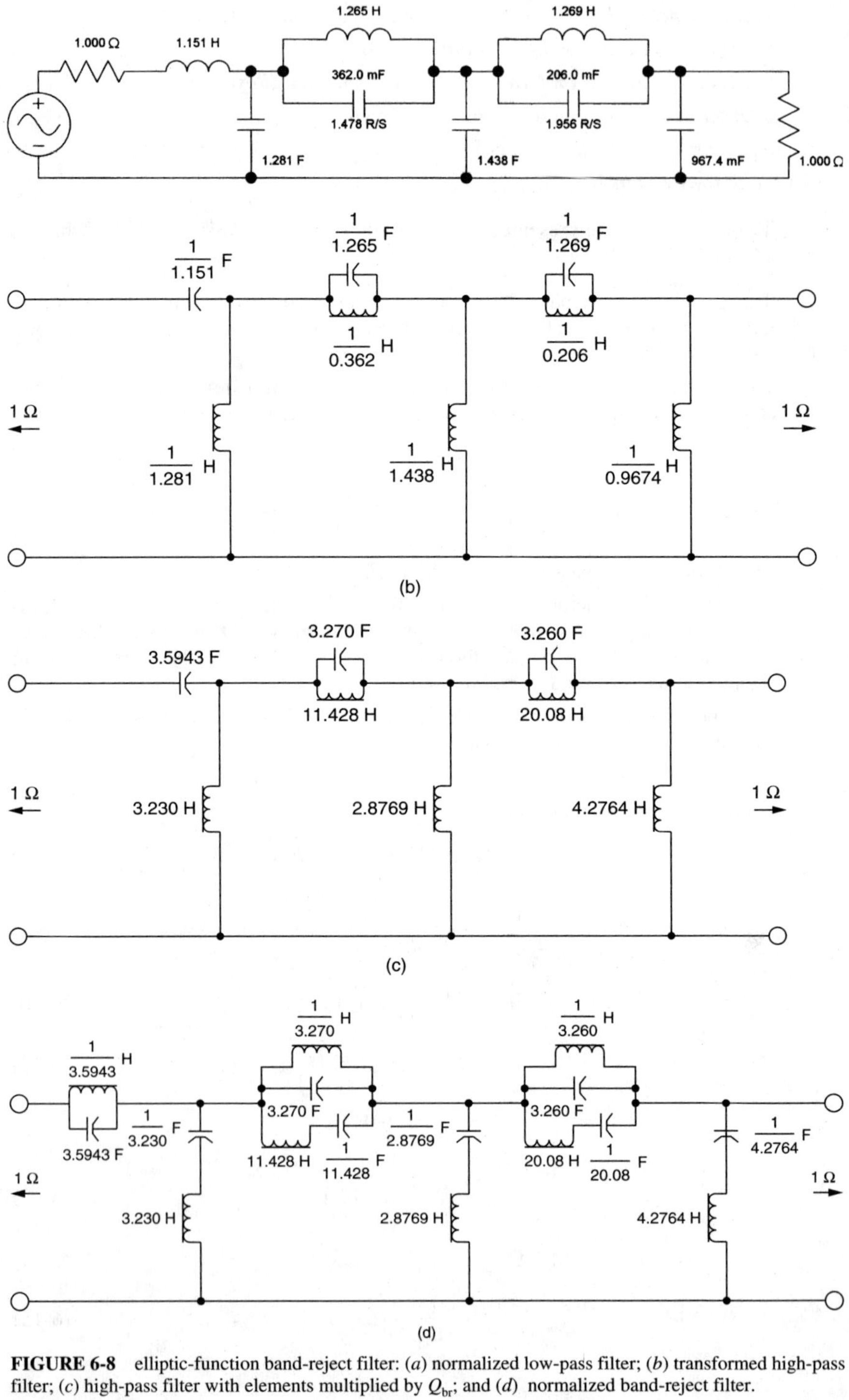

FIGURE 6-8 elliptic-function band-reject filter: (*a*) normalized low-pass filter; (*b*) transformed high-pass filter; (*c*) high-pass filter with elements multiplied by Q_{br}; and (*d*) normalized band-reject filter.

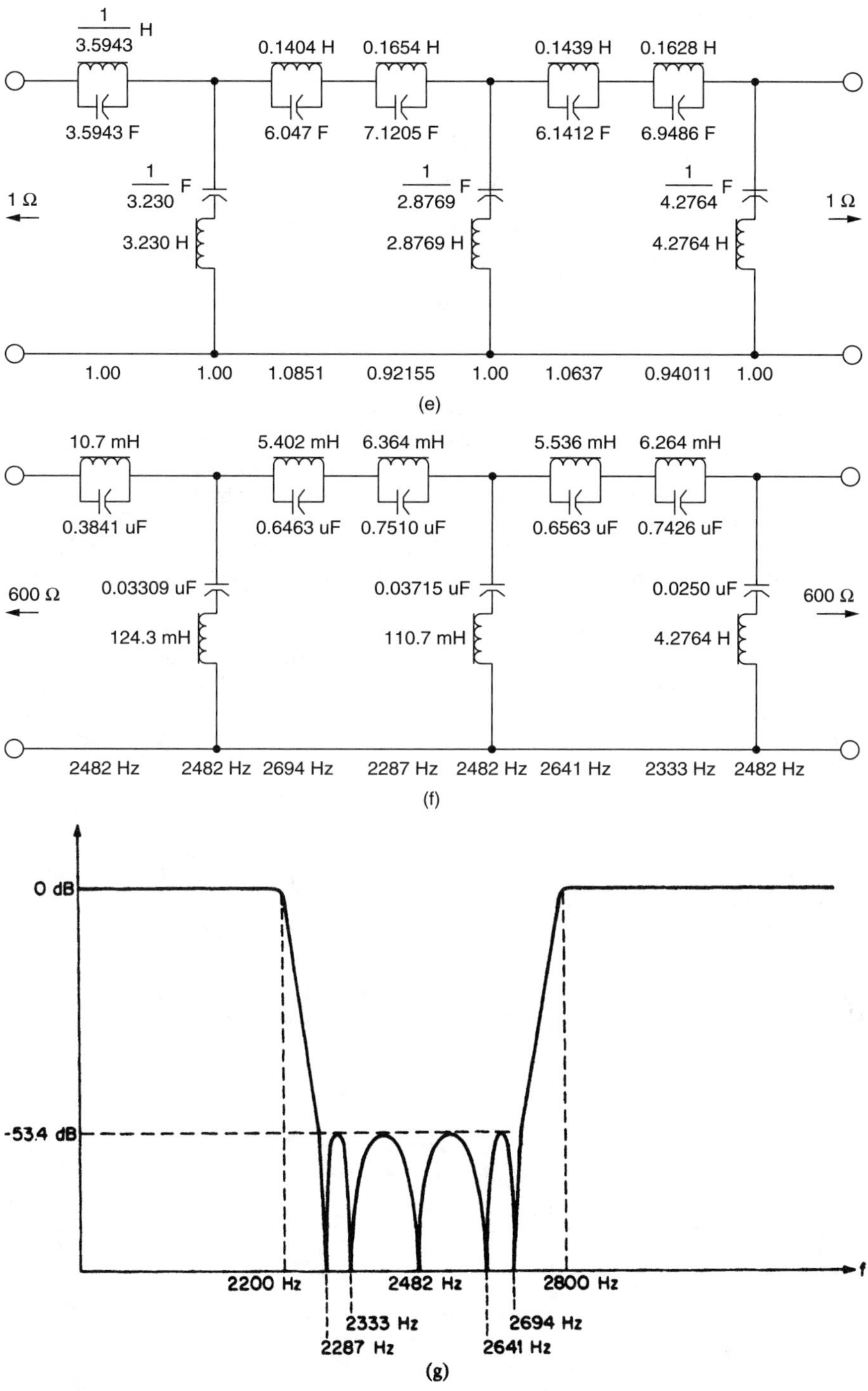

FIGURE 6-8 (*Continued*) elliptic-function band-reject filter: (*e*) transformed type III network; (*f*) frequency- and impedance-scaled circuit; and (*g*) frequency response.

The type III network of fourth branch:

$$L_1 = 20.08 \text{ H} \quad C_1 = 3.260 \text{ F}$$

$$\beta = 1.13147$$

$$L_a = 0.1439 \text{ H}$$

$$L_b = 0.1628 \text{ H}$$

$$C_a = 6.1412 \text{ F}$$

$$C_b = 6.9486 \text{ F}$$

$$\Omega_{\infty,a} = 1.0637$$

$$\omega_{\infty,b} = 0.94011$$

The resulting normalized band-reject filter is shown in Figure 6-8*e*.

(*j*) The final filter can now be obtained by frequency- and impedance-scaling the normalized band-reject filter to a center frequency of 2482 Hz and 600 Ω. The inductors are multiplied by Z/FSF, and the capacitors are divided by Z × FSF, where Z is 600 and the FSF is $2\pi f_0$, where f_0 is 2482 Hz. The circuit is given in Figure 6-8*f*, where the resonant frequencies of each section were obtained by multiplying the normalized frequencies by f_0. The frequency response is illustrated in Figure 6-8*g*.

Null Networks. A null network can be loosely defined as a circuit intended to reject a single frequency or a very narrow band of frequencies, and is frequently referred to as a trap. Notch depth rather than rate of roll-off is the prime consideration, and the circuit is restricted to a single section.

Parallel Resonant Trap. The RC high-pass circuit of Figure 6-9a has a 3-dB cutoff given by

$$f_c = \frac{1}{2\pi RC} \tag{6-23}$$

A band-reject transformation will result in the circuit of Figure 6-9*b*. The value of *L* is computed from

$$L = \frac{1}{\omega_0^2 C} \tag{6-24}$$

where $\omega_0 = 2\pi f_0$. The center frequency is f_0 and the 3-dB bandwidth is f_c.

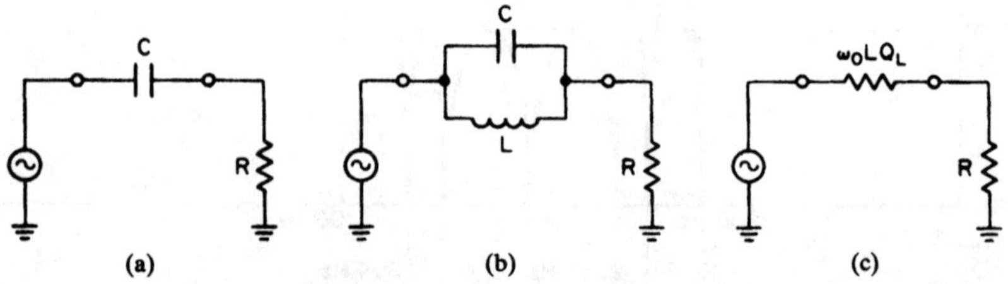

FIGURE 6-9 A parallel resonant trap: (*a*) *RC* high-pass filter; (*b*) results of a band-reject transformation; and (*c*) equivalent circuit at f_0.

The frequency response of a first-order band-reject filter can be expressed as

$$A_{\mathrm{dB}} = 10 \log\left[1 + \left(\frac{\mathrm{BW}_{3\ \mathrm{dB}}}{\mathrm{BW}_{x\ \mathrm{dB}}}\right)^2\right] \tag{6-25}$$

where $\mathrm{BW}_{3\ \mathrm{dB}}$ is the 3-dB bandwidth corresponding to f_c in Equation (6-23), and where $\mathrm{BW}_{x\ \mathrm{dB}}$ is the bandwidth of interest. The response can also be determined from the normalized Butterworth attenuation curves of Figure 2-34 corresponding to $n = 1$, where $\mathrm{BW}_{3\ \mathrm{dB}}/\mathrm{BW}_{x\ \mathrm{dB}}$ is the normalized bandwidth.

The impedance of a parallel tuned circuit at resonance is equal to $\omega_0 L Q_L$, where Q_L is the inductor Q and the capacitor is assumed to be lossless. We can then represent the band-reject filter at f_0 by the equivalent circuit of Figure 6-9*c*. After some algebraic manipulation involving Equations (6-23) and (6-24) and the circuit of Figure 6-9*c*, we can derive the following expression for the attenuation at resonance of the $n = 1$ band-reject filter of Figure 6-9:

$$A_{\mathrm{dB}} = 20\log\left(\frac{Q_L}{Q_{\mathrm{br}}} + 1\right) \tag{6-26}$$

where $Q_{\mathrm{br}} = f_0/\mathrm{BW}_{3\ \mathrm{dB}}$. Equation (6-26) is plotted in Figure 6-10. When Q_{br} is high, the required inductor Q may become prohibitively large in order to attain sufficient attenuation at f_0.

The effect of insufficient inductor Q will not only reduce relative attenuation, but will also cause some rounding of the response near the cutoff frequencies. Therefore, the ratio Q_L/Q_{br} should be as high as possible.

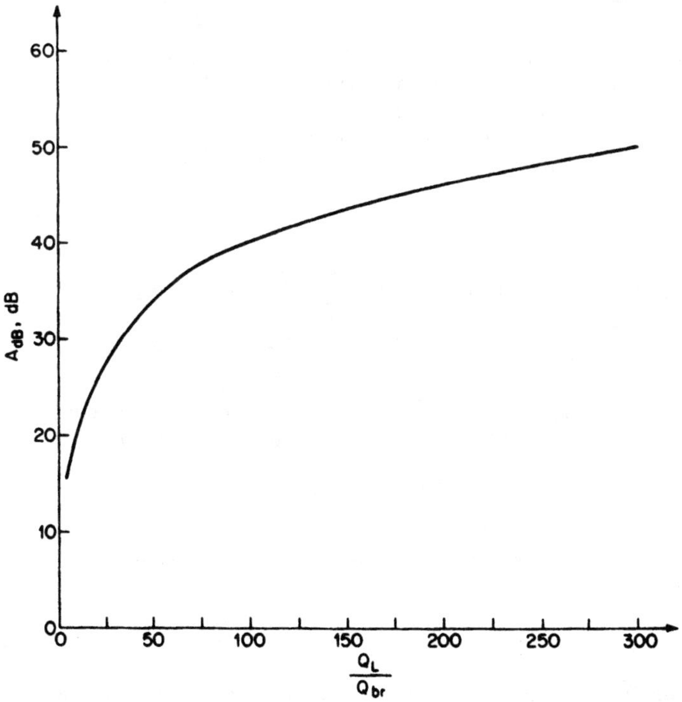

FIGURE 6-10 Attenuation vs. Q_L/Q_{br}.

Example 6-4 Designing a Parallel Resonant Trap

Required:

Design a parallel resonant circuit which has a 3-dB bandwidth of 500 Hz and a center frequency of 7500 Hz. The source resistance is zero and the load is 1 kΩ. Also determine the minimum inductor Q for a relative attenuation of at least 30 dB at 7500 Hz.

Result:

(*a*) Compute the value of the capacitor from

$$C = \frac{1}{2\pi f_c R} = \frac{1}{2\pi 500 \times 1000} = 0.3183\ \mu\text{F} \tag{6-23}$$

The inductance is given by

$$L = \frac{1}{\omega_0^2 C} = \frac{1}{(2\pi 7500)^2 \times 3.183 \times 10^{-7}} - 1.415\ \text{mH} \tag{6-24}$$

The resulting circuit is shown in Figure 6-11.

(*b*) The required ratio of Q_L/Q_{br} for 30-dB attenuation at f_0 can be determined from Figure 6-10 or Equation (6-26), and is approximately 30. Therefore, the inductor Q should exceed 30 Q_{br} or 450, where $Q_{br} = f_0/\text{BW}_{3\ \text{dB}}$.

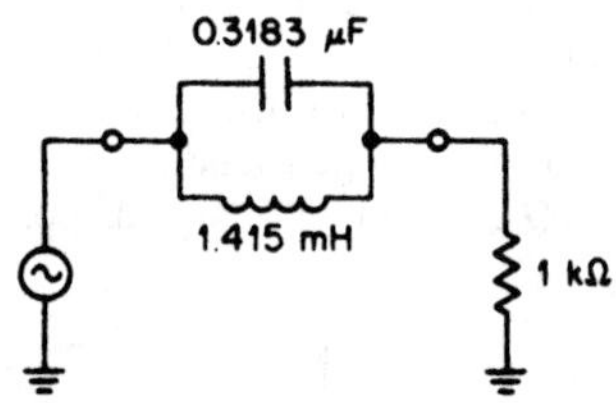

FIGURE 6-11 The parallel resonant trap of Example 6-4.

Frequently it is desirable to operate the band-reject network between equal source and load terminations instead of a voltage source, as in Figure 6-9. If a source and load resistor are specified where both are equal to R, Equation (6-23) is modified to

$$f_c = \frac{1}{4\pi RC} \tag{6-27}$$

When the source and load are unequal, the cutoff frequency is given by

$$f_c = \frac{1}{2\pi(R_s + R_L)C} \tag{6-28}$$

Series Resonant Trap. An $n = 1$ band-reject filter can also be derived from the *RL* high-pass filter of Figure 6-12*a*. The 3-dB cutoff is determined from

$$f_c = \frac{R}{2\pi L} \tag{6-29}$$

The band-reject filter of Figure 6-12*b* is obtained by resonating the coil with a series capacitor where

$$C = \frac{1}{\omega_0^2 L} \tag{6-24}$$

The center frequency is f_0 and the 3-dB bandwidth is equal to f_c. The series losses of an inductor can be represented by a resistor of $\omega_0 L/Q_L$. The equivalent circuit of the band-reject

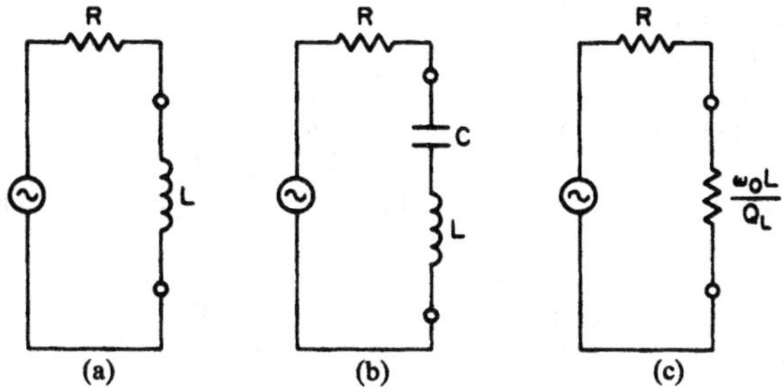

FIGURE 6-12 A series resonant trap: (*a*) *RL* high-pass filter; (*b*) result of band-reject transformation; and (*c*) equivalent circuit at f_0.

network at resonance is given by the circuit of Figure 6-12*c* and the attenuation computed from Equation (6-26) or Figure 6-10.

Example 6-5 Designing a Series Resonant Trap

Required:

Design a series resonant circuit having a 3-dB bandwidth of 500 Hz and a center frequency of 7500 Hz, as in the previous example. The source impedance is 1 kΩ and the load is assumed infinite.

Result:

Compute the element values from the following relationships:

$$L = \frac{R}{2\pi f_c} = \frac{1000}{2\pi 500} = 0.318 \text{ H} \tag{6-29}$$

and

$$C = \frac{1}{\omega_0^2 L} = \frac{1}{(2\pi 7500)^2 0.318} = 1420 \text{ pF} \tag{6-24}$$

The circuit is given in Figure 6-13.

When a series resonant trap is to be terminated with a load resistance equal to the source, the high-pass 3-dB cutoff and resulting 3-dB bandwidth of the band-reject filter are given by

$$f_c = \frac{R}{4\pi L} \tag{6-30}$$

For the more general case where source and load are unequal, the cutoff frequency is determined from

$$f_c = \frac{R_{eq}}{2\pi L} \tag{6-31}$$

where R_{eq} is the equivalent value of the source and load resistors in parallel.

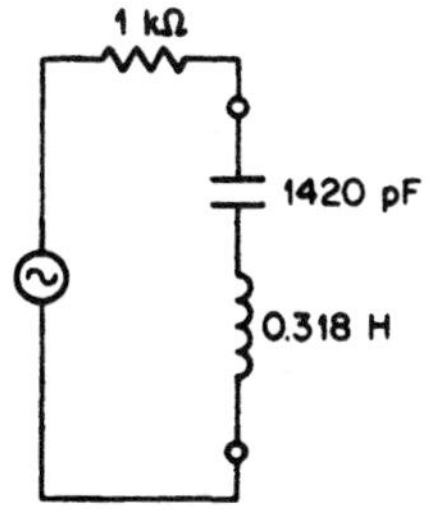

FIGURE 6-13 The series resonant trap of Example 6-5.

The Bridged-T Configuration. The resonant traps previously discussed suffer severe degradation of notch depth unless an inductor Q is many magnitudes greater than Q_{br}. The bridged-T band-reject structure can easily provide rejection of 60 dB or more with practical values of inductor Q. The configuration is shown in Figure 6-14*a*.

To understand the operation of the circuit, let us first consider the equivalent circuit of a center-tapped inductor having a coefficient of magnetic coupling equal to unity, which is shown in Figure 6-14*b*. The inductance between terminals *A* and *C* corresponds to *L* of Figure 6-14*a*. The inductance between *A* and *B* or *B* and *C* is equal to *L*/4 since, as the reader may recall, the impedance across one-half of a center-tapped autotransformer is one-fourth the overall impedance. This occurs because the impedance is proportional to the turns ratio squared.

The impedance of a parallel tuned circuit at resonance was previously determined to be equivalent to a resistor of $\omega_0 L Q_L$. Since the circuit of Figure 6-14*a* is center-tapped, the equivalent three-terminal network is shown in Figure 6-14*c*. The impedance between *A* and *C* is still $\omega_0 L Q_L$. A negative resistor must then exist in the middle shunt branch so that the impedance across one half of the tuned circuit is one-fourth the overall impedance, or $\omega_0 L Q_L/4$. Of course, negative resistors or inductors are physically impossible as individual passive two-terminal elements, but they can be embedded within an equivalent circuit.

If we combine the equivalent circuit of Figure 6-14*c* with the bridged-T network of Figure 6-14*a*, we obtain the circuit of Figure 6-14*d*. The positive and negative resistors in the center branch will cancel, resulting in infinite rejection of center frequency. The degree of rejection actually obtained is dependent upon a variety of factors such as center-tap accuracy, the coefficient of coupling, and the magnitude of Q_L. When the bridged-T configuration is implemented after modifying a parallel trap design of Figure 6-9*b* by adding a center tap and a resistor of $\omega_0 L Q_L/4$, a dramatic improvement in notch depth will usually occur.

A center-tapped inductor is not always available or practical. An alternate form of a bridged-T is given in Figure 6-15. The parallel resonant trap design of Figure 6-9 is modified

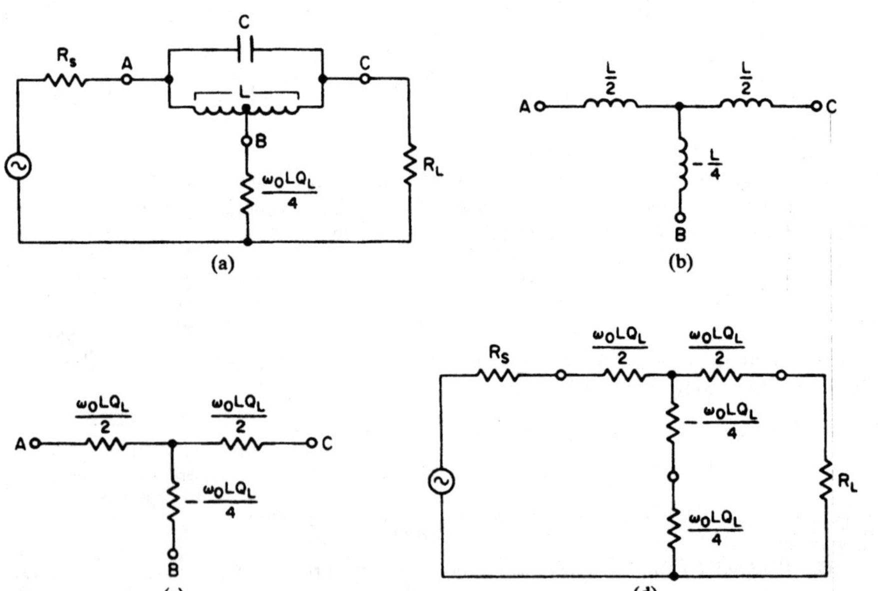

FIGURE 6-14 A bridged-T null network: (*a*) circuit configuration; (*b*) equivalent circuit of center-tapped inductor; (*c*) tuned circuit equivalent at resonance; and (*d*) bridged-T equivalent circuit at resonance.

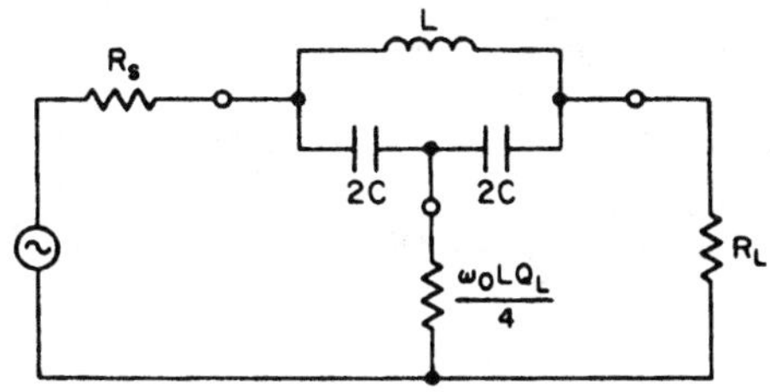

FIGURE 6-15 An alternate form of bridged-T.

by splitting the capacitor into two capacitors of twice the value, and a resistor of $\omega_0 L Q_L/4$ is introduced. The two capacitors should be closely matched.

In conclusion, the bridged-T structure is an economical and effective means of increasing the available notch rejection of a parallel resonant trap without increasing the inductor Q. However, as a final general comment, a single null section can provide high rejection only at a single frequency or relatively narrow band of frequencies for a given 3-dB bandwidth, since $n = 1$. The stability of the circuit then becomes a significant factor. A higher-order band-reject filter design can have a wider stopband and yet maintain the same 3-dB bandwidth.

6.2 ACTIVE BAND-REJECT FILTERS

This section considers the design of active band-reject filters for both wideband and narrowband applications. Active null networks are covered, and the popular twin-T circuit is discussed in detail.

Wideband Active Band-Reject Filters

Wideband filters can be designed by first separating the specification into individual low-pass and high-pass requirements. Low-pass and high-pass filters are then independently designed and combined by paralleling the inputs and summing both outputs to form the band-reject filter.

A wideband approach is valid when the separation between cutoffs is an octave or more for all-pole filters so that minimum interaction occurs in the stopband when the outputs are summed (see Section 2.1 and Figure 2-13). Elliptic-function networks will require less separation since their characteristics are steeper.

An inverting amplifier is used for summing and can also provide gain. Filters can be combined using the configuration of Figure 6-16*a*, where R is arbitrary and A is the desired gain. The individual filters should have a low output impedance to avoid loading by the summing resistors.

The VCVS elliptic-function low-pass and high-pass filters of Sections 3.2 and 4.2 each require an RC termination on the last stage to provide the real pole. These elements can be combined with the summing resistors, resulting in the circuit of Figure 6-16*b*. R_a and C_a correspond to the denormalized values of R_5 for the low-pass filter of Figure 3-20. The denormalized high-pass filter real-pole values are R_b and C_b. If only one filter is of the VCVS type, the summing network of the filter having the low output impedance can be replaced by a single resistor having a value of R.

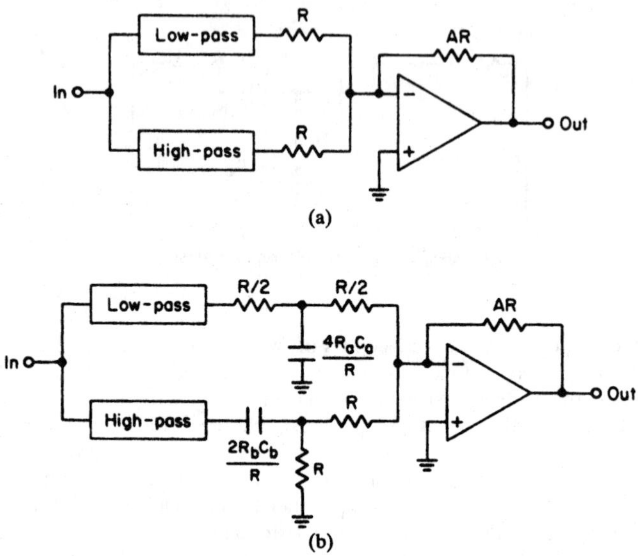

FIGURE 6-16 Wideband band-reject filters: (*a*) the combining of filters with low output impedance; and (*b*) combined filters requiring RC real poles.

When one or both filters are of the elliptic-function type, the ultimate attenuation obtainable is determined by the filter having the lesser value of A_{min} since the stopband output is the summation of the contributions of both filters.

Example 6-6 Design of a Wideband Band-Reject Filter

Required:

Design an active band-reject filter having 3-dB points at 100 and 400 Hz, and greater than 35 dB of attenuation between 175 and 225 Hz.

Result:

(*a*) Since the ratio of upper cutoff to lower cutoff is well in excess of an octave, a wideband approach can be used. First, separate the specification into individual low-pass and high-pass requirements.

Low-pass:	**High-pass:**
3 dB at 100 Hz	3 dB at 400 Hz
35-dB minimum at 175 Hz	35-dB minimum at 225 Hz

(*b*) The low-pass and high-pass filters can now be independently designed as follows:

Low-pass filter:

Compute the steepness factor.

$$A_s = \frac{f_s}{f_c} = \frac{175 \text{ Hz}}{100 \text{ Hz}} = 1.75 \qquad (2\text{-}11)$$

An $n = 5$ Chebyshev filter having a 0.5-dB ripple is chosen using Figure 2-44. The normalized active low-pass filter values are given in Table 11-39, and the circuit is shown in Figure 6-17*a*.

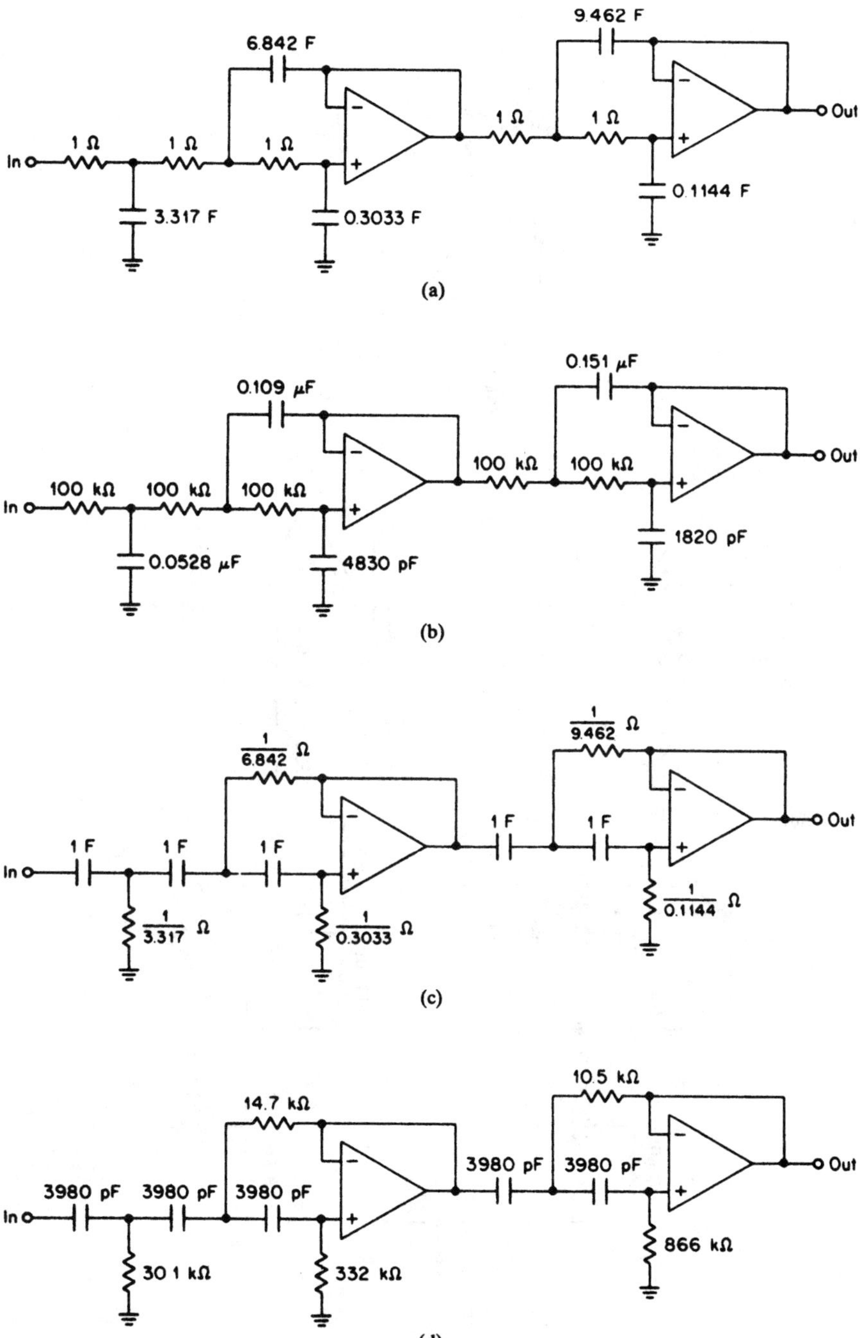

FIGURE 6-17 The wideband band-reject filter of Example 6-6: (*a*) normalized low-pass filter; (*b*) denormalized low-pass filter; (*c*) transformed normalized high-pass filter; and (*d*) denormalized high-pass filter.

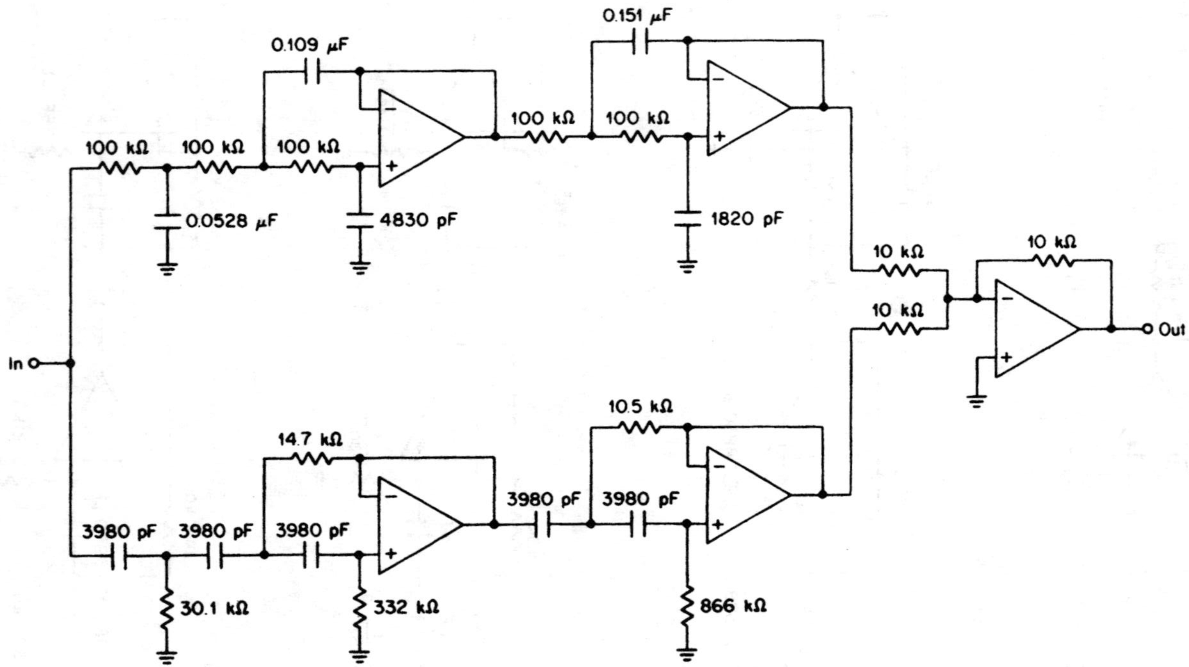

(e)

FIGURE 6-17 (*Continued*) The wideband band-reject filter of Example 6-6: (*e*) combining filters to obtain a band-reject response.

To denormalize the filter, multiply all resistors by Z and divide all capacitors by $Z \times$ FSF, where Z is conveniently selected at 10^5 and the FSF is $2\pi f_c$, where f_c is 100 Hz. The denormalized low-pass filter is given in Figure 6-17*b*.

High-pass filter:

Compute the steepness factor.

$$A_s = \frac{f_c}{f_s} = \frac{400\ \text{Hz}}{225\ \text{Hz}} = 1.78 \qquad (2\text{-}13)$$

An $n = 5$ Chebyshev filter with a 0.5-dB ripple will also satisfy the high-pass requirement. A high-pass transformation can be performed on the normalized low-pass filter of Figure 6-17*a* to obtain the circuit of Figure 6-17*c*. All resistors have been replaced with capacitors and vice versa using reciprocal element values.

The normalized high-pass filter is then frequency- and impedance-scaled by multiplying all resistors by Z and dividing all capacitors by $Z \times$ FSF, where Z is chosen at 10^5 and FSF is $2\pi f_c$, using an f_c of 400 Hz. The denormalized high-pass filter is shown in Figure 6-17*d* using standard 1-percent resistor values.

(*c*) The individual low-pass and high-pass filters can now be combined using the configuration of Figure 6-16*a*. Since no gain is required, A is set equal to unity. The value of R is conveniently selected at 10 kΩ, resulting in the circuit of Figure 6-17*e*.

Band-Reject Transformation of Low-Pass Poles. The wideband approach to the design of band-reject filters using combined low-pass and high-pass networks is applicable to bandwidths of typically an octave or more. If the separation between cutoffs is insufficient, interaction in the stopband will occur, resulting in inadequate stopband rejection (see Figure 2-13).

A more general approach involves normalizing the band-reject requirement and selecting a normalized low-pass filter type that meets these specifications. The corresponding normalized low-pass poles are then directly transformed to the band-reject form and realized using active sections.

A band-reject transfer function can be derived from a low-pass transfer function by substituting the frequency variable f by a new variable given by

$$f_{\text{br}} = \frac{1}{f_0\left(\dfrac{f}{f_0} - \dfrac{f_0}{f}\right)} \qquad (6\text{-}32)$$

This transformation combines the low-pass to high-pass and subsequent band-reject transformation discussed in Section 6.1 so that a band-reject filter can be obtained directly from the low-pass transfer function.

The band-reject transformation results in two pairs of complex poles and a pair of second-order imaginary zeros from each low-pass complex pole pair. A single low-pass real pole is transformed into a complex pole pair and a pair of first-order imaginary zeros. These relationships are illustrated in Figure 6-18. The zeros occur at center frequency and result from the transformed low-pass zeros at infinity.

The band-reject pole-zero pattern of Figure 6-18*a* corresponds to two band-reject sections where each section provides a zero at center frequency and also provides one of the pole pairs. The pattern of Figure 6-18*b* is realized by a single band-reject section where the zero also occurs at the center frequency.

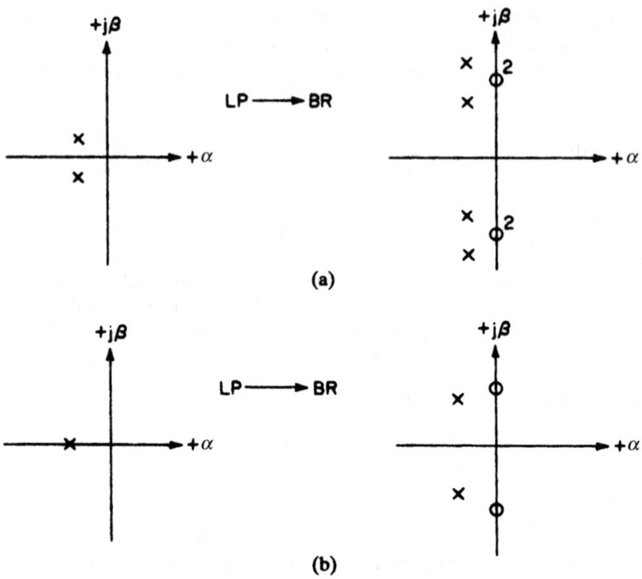

FIGURE 6-18 The band-reject transformation of low-pass poles: (*a*) low-pass complex pole pair; and (*b*) low-pass real pole.

To make the low-pass to band-reject transformation, first compute

$$Q_{br} = \frac{f_0}{BW} \tag{6-33}$$

where f_0 is the geometric center frequency and BW is the passband bandwidth. The transformation then proceeds as follows in the next section for complex poles and real poles.

Complex Poles. The tables of Chapter 11 contain tabulated poles corresponding to the all-pole low-pass filter families discussed in Chapter 2. Complex poles are given in the form $-\alpha \pm j\beta$, where α is the real coordinate and β is the imaginary part. Given α, β, Q_{br}, and f_0, the following computations [see Equations (6-34) through (6-44)] result in two sets of values for Q and frequency which defines two band-reject filter sections. Each section also has a zero at f_0.

$$C = \alpha^2 + \beta^2 \tag{6-34}$$

$$D = \frac{\alpha}{Q_{br} C} \tag{6-35}$$

$$E = \frac{\beta}{Q_{br} C} \tag{6-36}$$

$$F = E^2 - D^2 + 4 \tag{6-37}$$

$$G = \sqrt{\frac{F}{2} + \sqrt{\frac{F^2}{4} + D^2 E^2}} \tag{6-38}$$

$$H = \frac{DE}{G} \tag{6-39}$$

$$K = \frac{1}{2}\sqrt{(D + H)^2 + (E + G)^2} \tag{6-40}$$

$$Q = \frac{K}{D + H} \tag{6-41}$$

$$f_{ra} = \frac{f_0}{K} \tag{6-42}$$

$$f_{rb} = Kf_0 \tag{6-43}$$

$$f_\infty = f_0 \tag{6-44}$$

The two band-reject sections have resonant frequencies of f_{ra} and f_{rb} (in hertz) and identical Qs given by Equation (6-41). In addition, each section has a zero at f_0, the filter geometric center frequency.

Real Poles. A normalized low-pass real pole having a real coordinate of α_0 is transformed into a single band-reject section having a Q given by

$$Q = Q_{\text{br}}\alpha_0 \tag{6-45}$$

This section resonant frequency is equal to f_0. The section must also have a transmission zero at f_0.

Example 6-7 Calculating Pole and Zero Locations for a Band-Reject Filter

Required:

Determine the pole and zero locations for a band-reject filter having the following specifications:

A center frequency of 3600 Hz

3 dB at ±150 Hz

40-dB minimum at ±30 Hz

Result:

(*a*) Since the filter is narrow, the requirement can be treated directly in its arithmetically symmetrical form:

$$f_0 = 3600 \text{ Hz}$$

$$\text{BW}_{3\text{ dB}} = 300 \text{ Hz}$$

$$\text{BW}_{40\text{ dB}} = 60 \text{ Hz}$$

The band-reject steepness factor is given by

$$A_s = \frac{\text{passband bandwidth}}{\text{stopband bandwidth}} = \frac{300 \text{ Hz}}{60 \text{ Hz}} = 5 \tag{2-20}$$

(*b*) An $n = 3$ Chebyshev normalized low-pass filter having a 0.1-dB ripple is selected using Figure 2-42. The corresponding pole locations are found in Table 11-23 and are

$-0.3500 \pm j0.8695$

-0.6999

First, make the preliminary computation using

$$Q_{\text{br}} = \frac{f_0}{\text{BW}_{3\text{ dB}}} = \frac{3600\text{ Hz}}{300\text{ Hz}} = 12 \tag{6-33}$$

The low-pass to band-reject pole transformation is performed as follows:

Complex-pole transformation:

$$\alpha = 0.3500 \quad \beta = 0.8695$$

$$C = \alpha^2 + \beta^2 = 0.878530 \tag{6-34}$$

$$D = \frac{\alpha}{Q_{\text{br}}C} = 0.033199 \tag{6-35}$$

$$E = \frac{\beta}{Q_{\text{br}}C} = 0.082477 \tag{6-36}$$

$$F = E^2 - D^2 + 4 = 4.005700 \tag{6-37}$$

$$G = \sqrt{\frac{F}{2} + \sqrt{\frac{F^2}{4} + D^2E^2}} = 2.001425 \tag{6-38}$$

$$H = \frac{DE}{G} = 0.001368 \tag{6-39}$$

$$K = \frac{1}{2}\sqrt{(D + H)^2 + (E + G)^2} = 1.042094 \tag{6-40}$$

$$Q = \frac{K}{D + H} = 30.15 \tag{6-41}$$

$$f_{ra} = \frac{f_0}{K} = 3455\text{ Hz} \tag{6-42}$$

$$f_{rb} = Kf_0 = 3752\text{ Hz} \tag{6-43}$$

$$f_\infty = f_0 = 3600\text{ Hz} \tag{6-44}$$

Real-pole transformation:

$$\alpha_0 = 0.6999$$

$$Q = Q_{\text{br}}\alpha_0 = 8.40 \tag{6-45}$$

$$f_r = f_\infty = f_0 = 3600\text{ Hz}$$

The block diagram is shown in Figure 6-19.

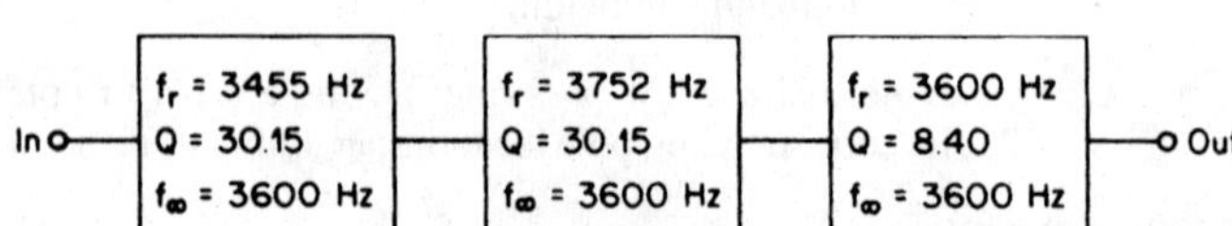

FIGURE 6-19 The block diagram of Example 6-7.

Narrowband Active Band-Reject Filters. Narrowband active band-reject filters are designed by first transforming a set of normalized low-pass poles to the band-reject form. The band-reject poles are computed in terms of resonant frequency f_r, Q, and f_∞ using the results of section 6.2 and are then realized with active band-reject sections.

The VCVS Band-Reject Section. Complex low-pass poles result in a set of band-reject parameters where f_r and f_∞ do not occur at the same frequency. Band-reject sections are then required that permit independent selection of f_r and f_∞ in their design procedure. Both the VCVS and biquad circuits covered in Section 5.2 under "Elliptic-Function Bandpass Filters" have this degree of freedom.

The VCVS realization is shown in Figure 6-20. The design equations were given in Section 5.2 under. "Elliptic-Function Bandpass Filters" and are repeated here for convenience, where f_r, Q, and f_∞ are obtained by the band-reject transformation procedure of Section 6.2. The values are computed as follows:

First, calculate

$$a = \frac{1}{Q} \tag{6-46}$$

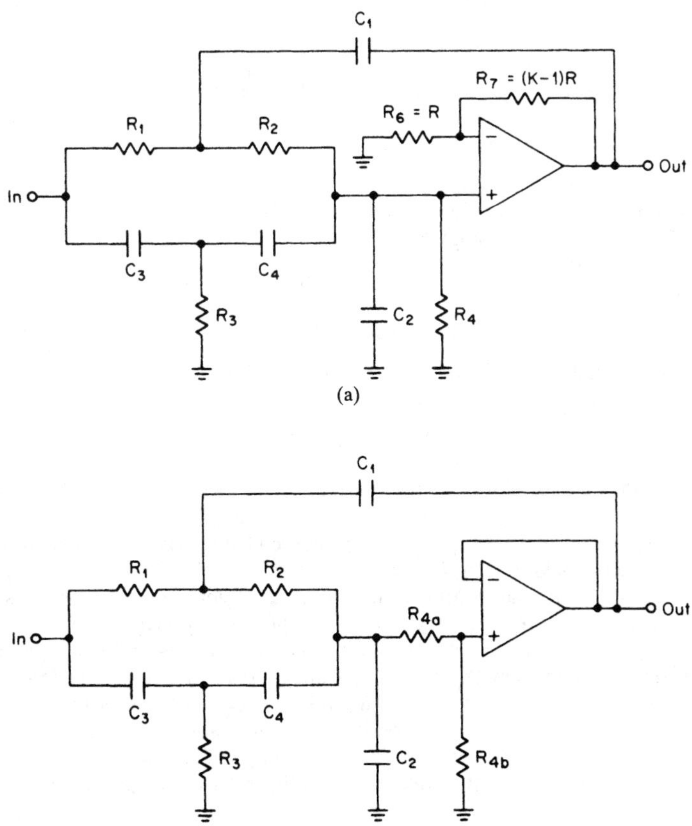

FIGURE 6-20 A VCVS realization for band-reject filters: (*a*) circuit for $K > 1$; and (*b*) circuit for $K < 1$.

$$b = \left(\frac{f_\infty}{f_r}\right)^2 \tag{6-47}$$

$$c = 2\pi f_r \tag{6-48}$$

Select C, then

$$C_1 = C \tag{6-49}$$

$$C_3 = C_4 = \frac{C_1}{2} \tag{6-50}$$

$$C_2 \geq \frac{C_1(b-1)}{4} \tag{6-51}$$

$$R_3 = \frac{1}{cC_1\sqrt{b}} \tag{6-52}$$

$$R_1 = R_2 = 2R_3 \tag{6-53}$$

$$R_4 = \frac{4\sqrt{b}}{cC_1(1-b) + 4cC_2} \tag{6-54}$$

$$K = 2 + \frac{2C_2}{C_1} - \frac{a}{2\sqrt{b}} + \frac{2}{C_1\sqrt{b}}\left(\frac{1}{cR_4} - aC_2\right) \tag{6-55}$$

$$R_6 = R \qquad \text{and} \qquad R_7 = (K-1)R \tag{6-56}$$

where R can be arbitrarily chosen.

The circuit of Figure 6-20*a* is used when $K > 1$. In the cases where $K < 1$, the configuration of Figure 6-20*b* is utilized, where

$$R_{4a} = (1-K)R_4 \tag{6-57}$$

and

$$R_{4b} = KR_4 \tag{6-58}$$

The section gain at DC is given by

$$\text{Section gain} = \frac{bKC_1}{4C_2 + C_1} \tag{6-59}$$

The gain of the composite filter in the passband is the product of the DC gains of all the sections.

The VCVS structure has a number of undesirable characteristics. Although the circuit Q can be adjusted by making R_6 or R_7 variable when $K > 1$, the Q cannot be independently measured since the 3-dB bandwidth at the output is affected by the transmission zero. Resonant frequency f_r or the notch frequency f_∞ cannot be easily adjusted since these parameters are determined by the interaction of a number of elements. Also, the section gain is fixed by the design parameters. Another disadvantage of the circuit is that a large spread in capacitor values* may occur so that standard values cannot be easily used. Nevertheless, the VCVS realization makes effective use of a minimum number of operational amplifiers in comparison with other implementations and is widely used. However, because of its lack of adjustment capability, its application is generally restricted to Qs below 10 and with 1-percent component tolerances.

*The elliptic-function configuration of the VCVS uniform capacitor structure given in Section 3.2 can be used at the expense of additional sensitivity.

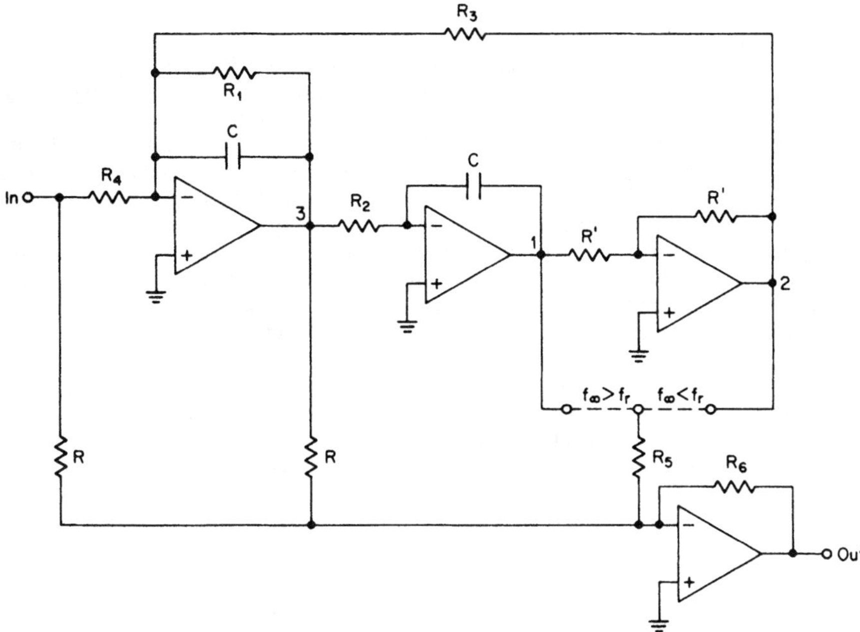

FIGURE 6-21 biquad band-reject realization.

The State-Variable Band-Reject Section. The biquad or state-variable elliptic-function bandpass filter section discussed in Section 5.2 is highly suitable for implementing band-reject transfer functions. The circuit is given in Figure 6-21. By connecting resistor R_5 to either node 1 or node 2, the notch frequency f_∞ will be located above or below the pole resonant frequency f_r.

Section Qs of up to 200 can be obtained. The design parameters f_r, Q, and f_∞, as well as the section gain, can be independently chosen, monitored, and adjusted. From the point of view of low sensitivity and maximum flexibility, the biquad approach is the most desirable method of realization.

The design equations were stated in Section 5.2 under "Elliptic-Function Bandpass Filters" and are repeated here for convenience, where f_r, Q, and f_∞ are given and the values of C, R, and R' can be arbitrarily chosen.

$$R_1 = R_4 = \frac{Q}{2\pi f_r C} \tag{6-60}$$

$$R_2 = R_3 = \frac{R_1}{Q} \tag{6-61}$$

$$R_5 = \frac{f_r^2 R}{Q|f_r^2 - f_\infty^2|} \tag{6-62}$$

for $f_\infty > f_r$:

$$R_6 = \frac{f_r^2 R}{f_\infty^2} \tag{6-63}$$

and when $f_\infty < f_r$:

$$R_6 = R \tag{6-64}$$

The value of R_6 is based on unity section gain at DC. The gain can be raised or lowered by proportionally increasing or decreasing R_6.

Resonance is adjusted by monitoring the phase shift between the section input and node 3 using a Lissajous pattern and adjusting R_3 for 180° phase shift with an input frequency of f_r.

The Q is controlled by R_1 and can be measured at node 3 in terms of section 3-dB bandwidth, or R_1 can be adjusted until unity gain occurs between the input and node 3 with f_r applied. Because of the Q-enhancement effect discussed in Section 5.2 under "All-Pole Bandpass Configuration," a Q adjustment is usually necessary.

The notch frequency is then determined by monitoring the section output for a null. Adjustment is normally not required since the tuning of f_r will usually bring in f_∞ with acceptable accuracy. If an adjustment is desired, R_5 can be made variable.

Sections for Transformed Real Poles. When a real pole undergoes a band-reject transformation, the result is a single pole pair and a single set of imaginary zeros. Complex poles resulted in two sets of pole pairs and two sets of zeros. The resonant frequency f_r of the transformed real pole is exactly equal to the notch frequency f_∞, thus the design flexibility of the VCVS and biquad structures is not required.

A general second-order bandpass transfer function can be expressed as

$$T(s) = \frac{\dfrac{\omega_r}{Q}s}{s^2 + \dfrac{\omega_r}{Q}s + \omega_r^2} \tag{6-65}$$

where the gain is unity at ω_r. If we realize the circuit of Figure 6-22 where $T(s)$ corresponds to the above transfer function, the composite transfer function at the output is given by

$$T(s) = \frac{s^2 + \omega_r^2}{s^2 + \dfrac{\omega_r}{Q}s + \omega_r^2} \tag{6-66}$$

This corresponds to a band-reject transfer function having a transmission zero at f_r (that is, $f_\infty = f_r$). The occurrence of this zero can also be explained intuitively from the structure of Figure 6-22. Since $T(s)$ is unity at f_r, both input signals to the summing amplifier will then cancel, resulting in no output signal.

These results indicate that band-reject sections for transformed real poles can be obtained by combining any of the all-pole bandpass circuits of section 5.2 in the configuration of Figure 6-22. The basic design parameters are the required f_r and Q of the band-reject section, which are directly used in the design equations for the bandpass circuits.

By combining these bandpass sections with summing amplifiers, the three band-reject structures of Figure 6-23 can be derived. The design equations for the bandpass sections were given in Section 5.2 and are repeated here where C, R, and R' can be arbitrarily chosen.

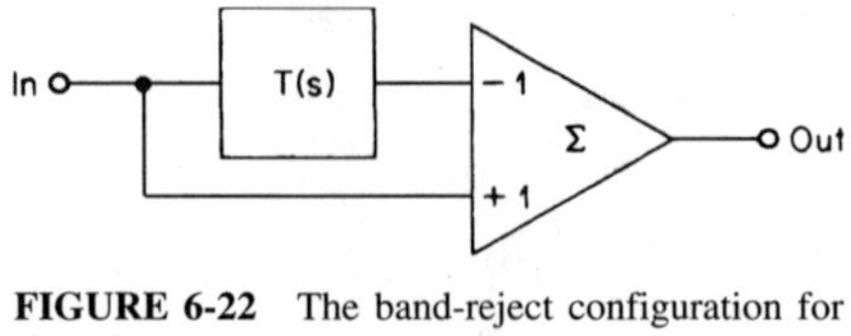

FIGURE 6-22 The band-reject configuration for $f_r = f_\infty$.

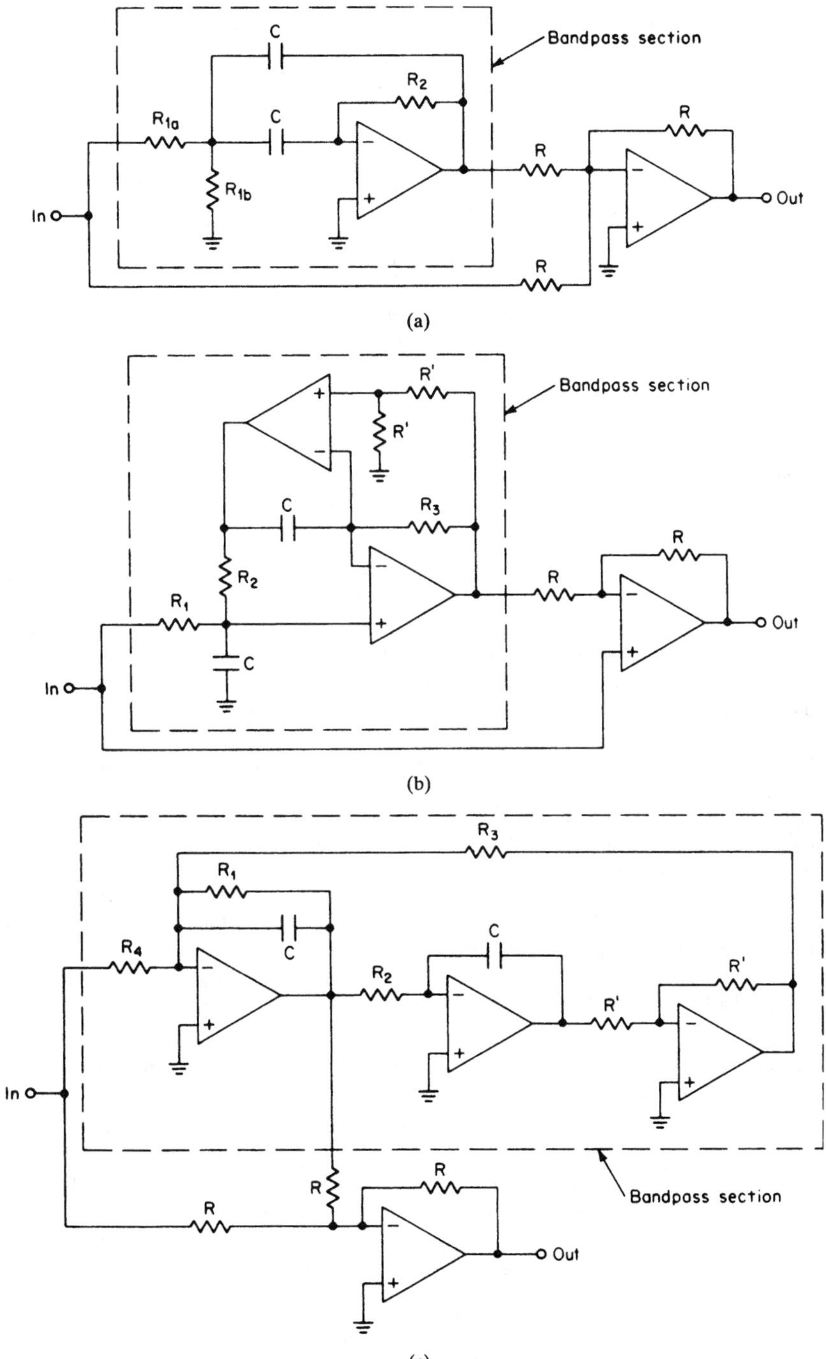

FIGURE 6-23 The band-reject circuits for $f_r = f_\infty$: (*a*) MFBP band-reject section ($Q < 20$); (*b*) DABP band-reject section ($Q < 150$); and (*c*) biquad band-reject section ($Q < 200$).

The MFBP band-reject section ($f_r = f_\infty$) is given by

$$R_2 = \frac{Q}{\pi f_r C} \tag{6-67}$$

$$R_{1a} = \frac{R_2}{2} \tag{6-68}$$

$$R_{1b} = \frac{R_{1a}}{2Q^2 - 1} \tag{6-69}$$

The DABP band-reject section ($f_r = f_\infty$) is given by

$$R_1 = \frac{Q}{2\pi f_r C} \tag{6-70}$$

$$R_2 = R_3 = \frac{R_1}{Q} \tag{6-71}$$

The biquad band-reject section ($f_r = f_\infty$) is given by

$$R_1 = R_4 = \frac{Q}{2\pi f_r C} \tag{6-72}$$

$$R_2 = R_3 = \frac{R_1}{Q} \tag{6-73}$$

These equations correspond to unity bandpass gain for the MFBP and biquad circuits so that cancellation at f_r will occur when the section input and bandpass output signals are equally combined by the summing amplifiers. Since the DABP section has a gain of 2 and has a noninverting output, the circuit of Figure 6-23*b* has been modified accordingly so that cancellation occurs.

Tuning can be accomplished by making R_{1b}, R_2, and R_3 variable in the MFBP, DABP, and biquad circuits, respectively. In addition, the biquad circuit will usually require R_1 to be made adjustable to compensate for the Q-enhancement effect (see Section 5.2 under "All-Pole Bandpass Configurations"). The circuit can be tuned by adjusting the indicated elements for either a null at f_r measured at the circuit output or for 0° or 180° phase shift at f_r observed between the input and the output of the bandpass section. If the bandpass section gain is not sufficiently close to unity for the MFBP and biquad case, and 2 for the DABP circuit, the null depth may be inadequate.

Example 6-8 Design of an Active Band-Reject Filter

Required:

Design an active band-reject filter from the band-reject parameters determined in Example 6-7 having a gain of +6 dB.

Result:

(*a*) The band-reject transformation in Example 6-7 resulted in the following set of requirements for a three-section filter:

Section	f_r	Q	f_∞
1	3455 Hz	30.15	3600 Hz
2	3752 Hz	30.15	3600 Hz
3	3600 Hz	8.40	3600 Hz

(*b*) Two biquad circuits in tandem will be used for sections 1 and 2 followed by a DABP band-reject circuit for section 3. The value of C is chosen at 0.01 μF, and R, as well as R', at 10 kΩ. Since the DABP section has a gain of 2 at DC, which satisfies the 6-dB gain requirement, both biquad sections should then have unity gain. The element values are determined as follows:

Section 1 (biquad of Figure 6-21):

$$f_r = 3455 \text{ Hz} \quad Q = 30.15 \quad f_\infty = 3600 \text{ Hz}$$

$$R_1 = R_4 = \frac{Q}{2\pi f_r C} = \frac{30.15}{2\pi \times 3455 \times 10^{-8}} = 138.9 \text{ k}\Omega \tag{6-60}$$

$$R_2 = R_3 = \frac{R_1}{Q} = \frac{138.9 \times 10^3}{30.15} = 4610\,\Omega \tag{6-61}$$

$$R_5 = \frac{f_r^2 R}{Q\left|f_r^2 - f_\infty^2\right|} = \frac{3455^2 \times 10^4}{30.15/\left|3455^2 - 3600^2\right|} = 3870\,\Omega \tag{6-62}$$

$$R_6 = \frac{f_r^2 R}{f_\infty^2} = \frac{3455^2 \times 10^4}{3600^2} = 9210\,\Omega \tag{6-63}$$

Section 2 (biquad of Figure 6-21):

$$f_r = 3752 \text{ Hz} \quad Q = 30.15 \quad f_\infty = 3600 \text{ Hz}$$

$$R_1 = R_4 = 127.9 \text{ k}\Omega \tag{6-60}$$

$$R_2 = R_3 = 4240\ \Omega \tag{6-61}$$

$$R_5 = 4180\ \Omega \tag{6-62}$$

$$R_6 = 10 \text{ k}\Omega \tag{6-64}$$

Section 3 (DABP of Figure 6-23):

$$f_r = f_\infty = 3600 \text{ Hz} \quad Q = 8.40$$

$$R_1 = \frac{Q}{2\pi f_r C} = \frac{8.40}{2\pi \times 3600 \times 10^{-8}} = 37.1 \text{ k}\Omega \tag{6-70}$$

$$R_2 = R_3 = \frac{R_1}{Q} = \frac{37.1 \times 10^3}{8.40} = 4420\ \Omega \tag{6-71}$$

The final circuit is shown in Figure 6-24 with standard 1-percent resistor values. The required resistors have been made variable so that the resonant frequencies can be adjusted for all sections and, in addition, the Q is variable for the biquad circuits.

Active Null Networks. Active null networks are single sections used to provide attenuation at a single frequency or over a narrow band of frequencies. The most popular sections are of the twin-T form, so this circuit will be discussed in detail along with some other structures.

The Twin-T. The twin-T was first discovered by H. W. Augustadt in 1934. Although this circuit is passive by nature, it is also used in many active configurations to obtain a variety of different characteristics.

The circuit of Figure 6-25*a* is an RC bridge structure where balance or an output null occurs at 1 rad/s when all arms have an equal impedance (0.5 − *j*0.5 Ω). The circuit is

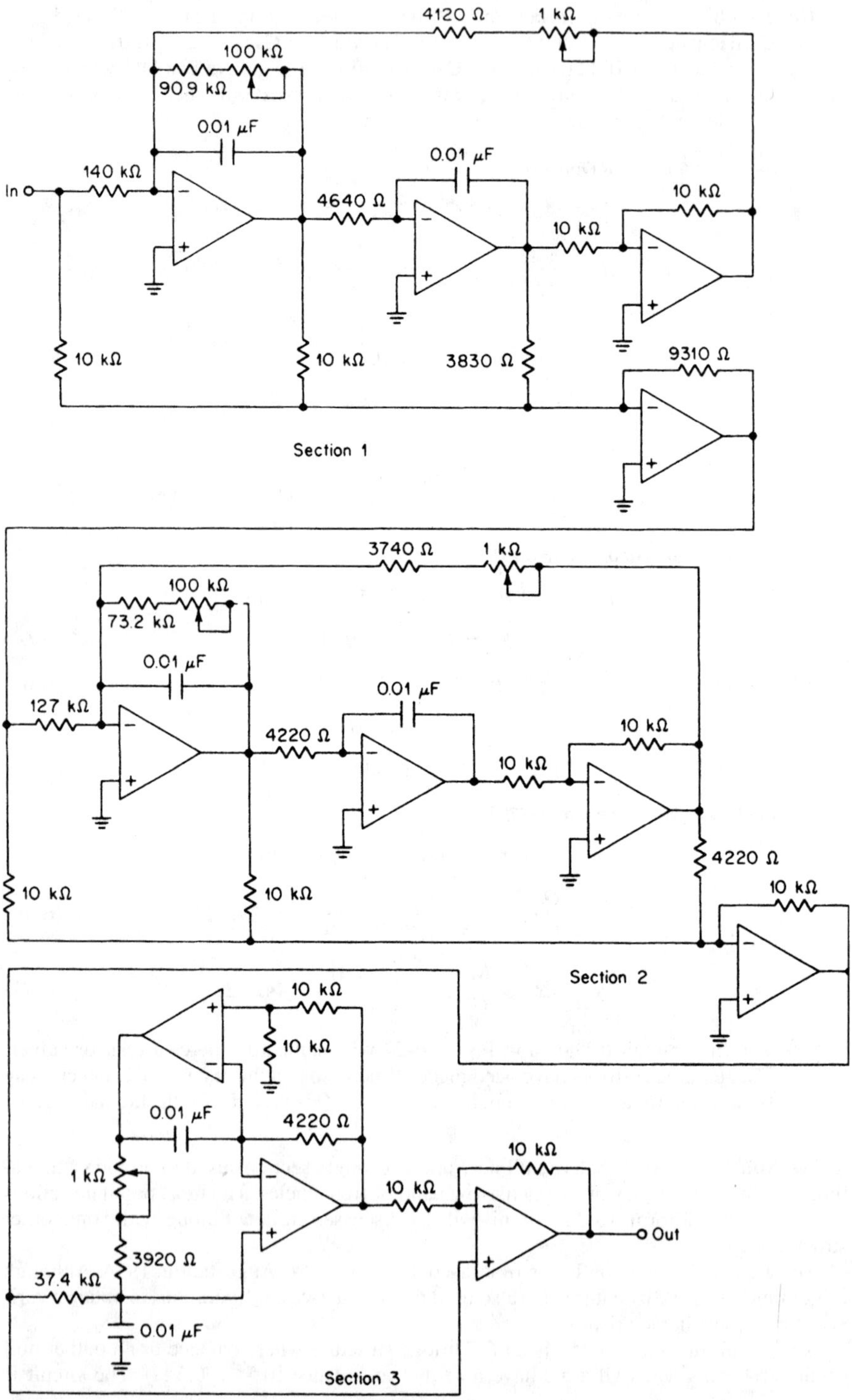

FIGURE 6-24 The band-reject filter of Example 6-8.

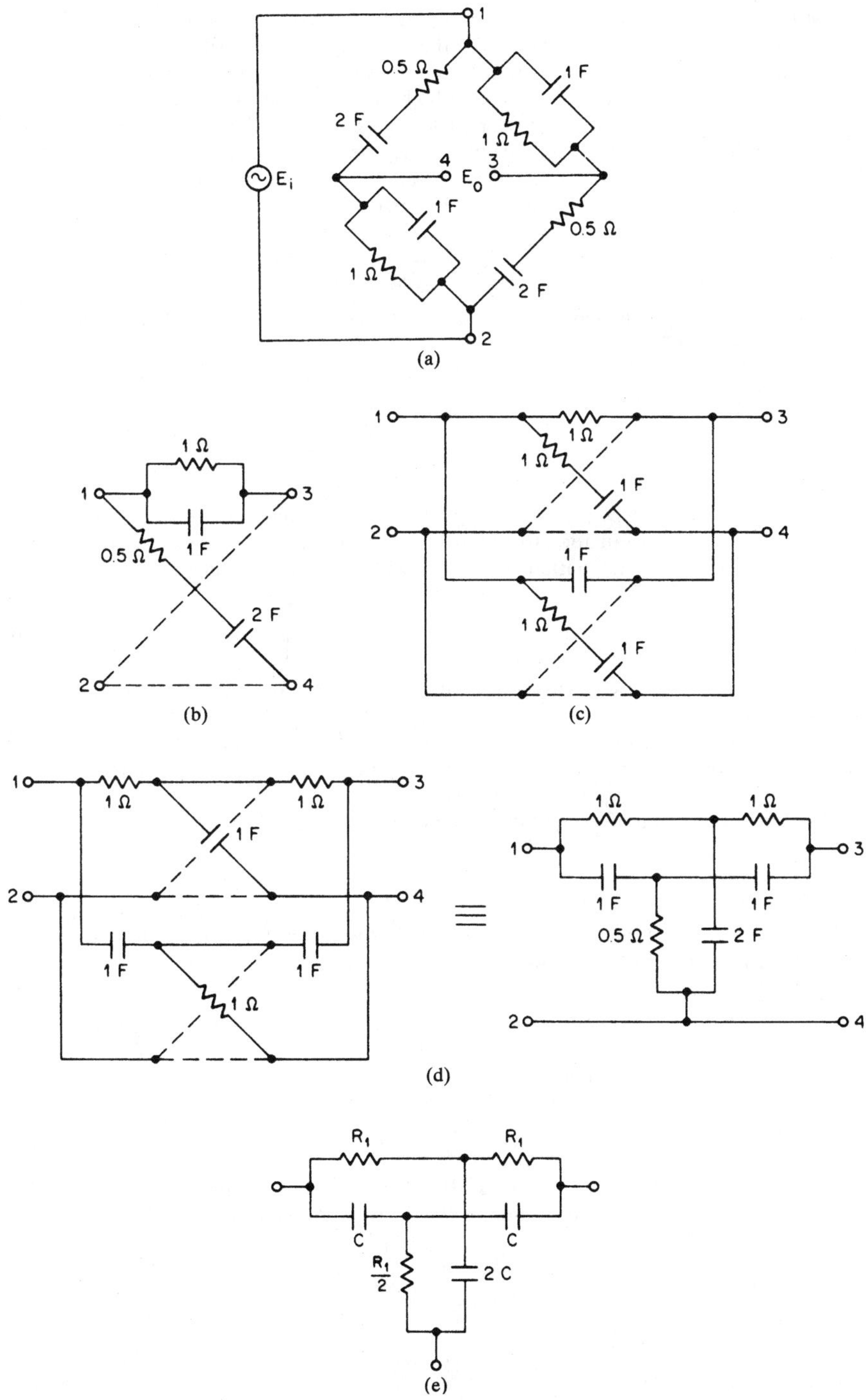

FIGURE 6-25 A derivation of the twin-T: (*a*) *RC* bridge; (*b*) lattice circuit; (*c*) parallel lattice; (*d*) twin-T equivalent; and (*e*) general form of twin-T.

redrawn in the form of a symmetrical lattice in Figure 6-25*b* (refer to Guillemin and Stewart in Bibliography for detailed discussions of the lattice). The lattice of Figure 6-25*b* can be redrawn again in the form of two parallel lattices, as shown in Figure 6-25*c*.

If identical series elements are present in both the series and shunt branches of a lattice, the element may be extracted and symmetrically placed outside the lattice structure. A 1-Ω resistor satisfies the requirement for the upper lattice, and a 1-F capacitor for the lower lattice. Removal of these components to outside the lattice results in the twin-T of Figure 6-25*d*.

The general form of a twin-T is shown in Figure 6-25*e*. The value of R_1 is computed from

$$R_1 = \frac{1}{2\pi f_0 C} \tag{6-74}$$

where C is arbitrary. This denormalizes the circuit of Figure 6-25*d* so that the null now occurs at f_0 instead of at 1 rad/s.

When a twin-T is driven from a voltage source and terminated in an infinite load,* the transfer function is given by

$$T(s) = \frac{s^2 + \omega_0^2}{s^2 + 4\omega_0 s + \omega_0^2} \tag{6-75}$$

If we compare this expression with the general transfer function of a second-order pole-zero section as given by Equation (6-66), we can determine that a twin-T provides a notch at f_0 with a Q of $\frac{1}{4}$. The attenuation at any bandwidth can be computed by

$$A_{\text{dB}} = 10\log\left[1 + \left(\frac{4f_0}{\text{BW}_{x\,\text{dB}}}\right)^2\right] \tag{6-76}$$

The frequency response is shown in Figure 6-26, where the requirement for geometric symmetry applies.

Twin-T with Positive Feedback The twin-T has gained widespread usage as a general-purpose null network. However, a major shortcoming is a fixed Q of $\frac{1}{4}$. This limitation can be overcome by introducing positive feedback.

The transfer function of the circuit of Figure 6-27*a* can be derived as

$$T(s) = \frac{\beta}{1 + K(\beta - 1)} \tag{6-77}$$

If β is replaced by Equation (6-75), the transfer function of a twin-T, the resulting circuit transfer function expression becomes

$$T(s) = \frac{s^2 + \omega_0^2}{s^2 + 4\omega_0(1 - K)s + \omega_0^2} \tag{6-78}$$

The corresponding Q is then

$$Q = \frac{1}{4(1 - K)} \tag{6-79}$$

By selecting a positive K of < 1 and sufficiently close to unity, the circuit Q can be dramatically increased. The required value of K can be determined by

$$K = 1 - \frac{1}{4Q} \tag{6-80}$$

---

*Since the source and load are always finite, the value of R_1 should be in the vicinity of $\sqrt{R_s R_L}$, provided that the ratio R_L/R_s is in excess of 10.

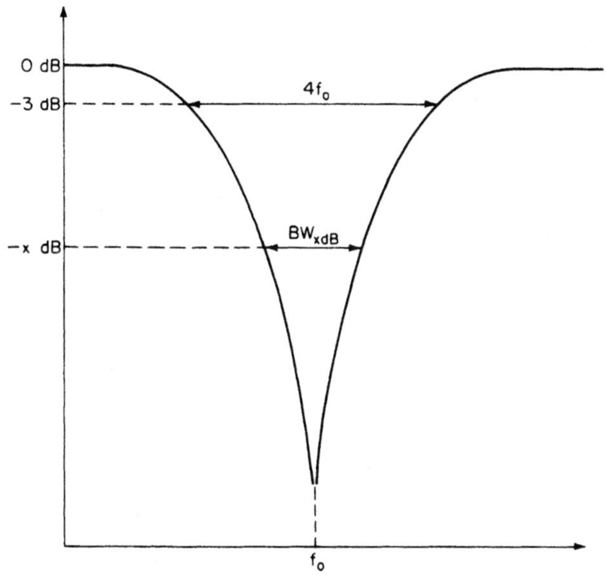

FIGURE 6-26 The frequency response of a twin-T.

The block diagram of Figure 6-27*a* can be implemented using the circuit of Figure 6-27*b*, where R is arbitrary. By choosing C and R so that $R_1 \gg (1 - K)R$, the circuit may be simplified to the configuration of Figure 6-27*c*, which uses only one amplifier.

The attenuation at any bandwidth is given by

$$A_{\text{dB}} = 10\log\left[1 + \left(\frac{f_0}{Q \times \text{BW}_{x\,\text{dB}}}\right)^2\right] \tag{6-81}$$

Equation (6-81) is the general expression for the attenuation of a single band-reject section where the resonant frequency and notch frequency are identical (that is, $f_r = f_\infty$). The attenuation formula can be expressed in terms of the 3-dB bandwidth as follows:

$$A_{\text{dB}} = 10\log\left[1 + \left(\frac{\text{BW}_{3\,\text{dB}}}{\text{BW}_{x\,\text{dB}}}\right)^2\right] \tag{6-82}$$

The attenuation characteristics can also be determined from the frequency-response curve of a normalized $n = 1$ Butterworth low-pass filter (see Figure 2-34) by using the ration $\text{BW}_{3\,\text{dB}}/\text{BW}_{x\,\text{dB}}$ for the normalized frequency.

The twin-T in its basic form or in the positive-feedback configuration is widely used for single-section band-reject sections. However, it suffers from the fact that tuning cannot be easily accomplished. Tight component tolerances may then be required to ensure sufficient accuracy of tuning and adequate notch depth. About a 40- to 60-dB rejection at the notch could be expected using 1-percent components.

Example 6-9 Designing a Twin-T Band-Reject Filter Using Positive Feedback

Required:

Design a single null network having a center frequency of 1000 Hz and a 3-dB bandwidth of 100 Hz. Also determine the attenuation at the 30-Hz bandwidth.

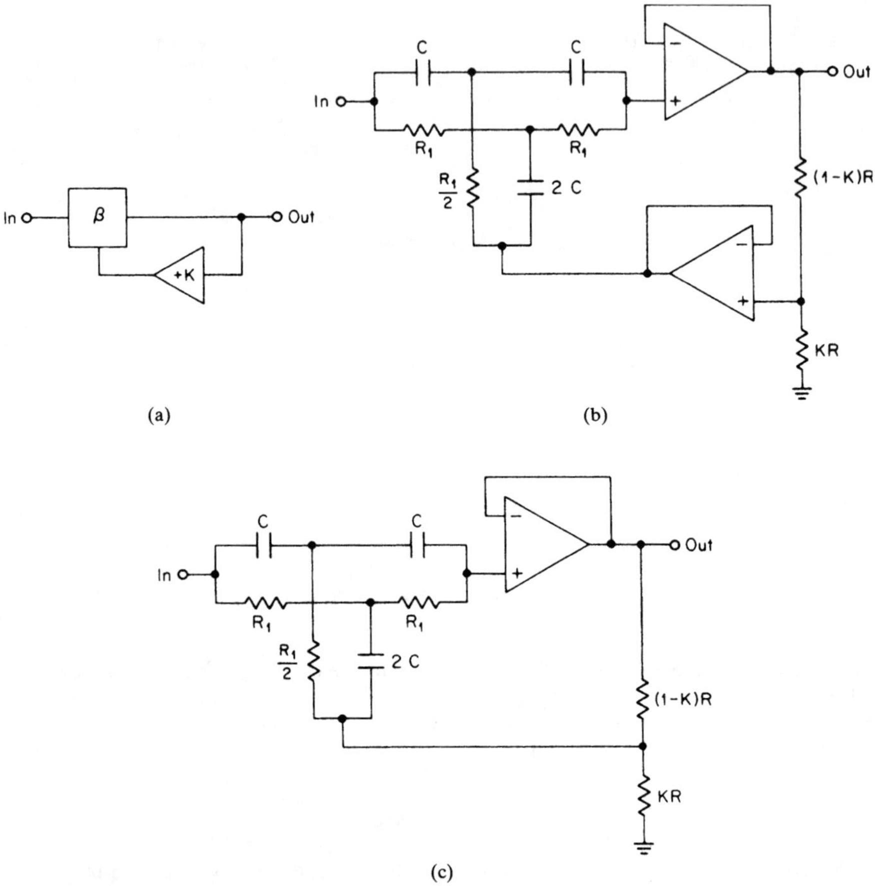

FIGURE 6-27 Twin-T with positive feedback: (*a*) block diagram; (*b*) circuit realization; and (*c*) simplified configuration $R_1 \gg (1 - K)R$.

Result:

(*a*) A twin-T structure with positive feedback will be used. To design the twin-T, first choose a capacitance C of 0.01 μF. The value of R_1 is given by

$$R_1 = \frac{1}{2\pi f_0 C} = \frac{1}{2\pi \times 10^3 \times 10^{-8}} = 15.9 \text{ k}\Omega \qquad (6\text{-}74)$$

(*b*) The required value of K for the feedback network is calculated from

$$K = 1 - \frac{1}{4Q} = 1 - \frac{1}{4 \times 10} = 0.975 \qquad (6\text{-}80)$$

where $Q = f_0/\text{BW}_{3\text{ dB}}$.

(*c*) The single amplifier circuit of Figure 6-27*c* will be used. If R is chosen at 1 kΩ, the circuit requirement for $R_1 \gg (1 - K)R$ is satisfied. The resulting section is shown in Figure 6-28.

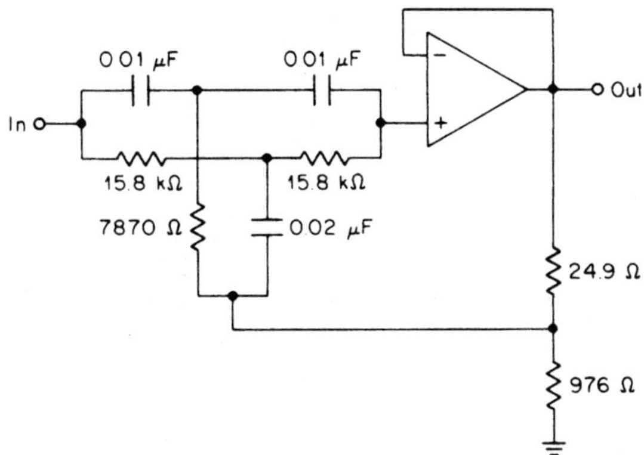

FIGURE 6-28 The twin-T network of Example 6-9.

(*d*) To determine the attenuation at a bandwidth of 30 Hz, calculate

$$A_{dB} = 10\log\left[1 + \left(\frac{BW_{3\,dB}}{BW_{x\,dB}}\right)^2\right] = 10\log\left[1 + \left(\frac{100\text{ Hz}}{30\text{ Hz}}\right)^2\right] = 10.8\text{ dB} \quad (6\text{-}82)$$

Bandpass Structure Null Networks. Section 6.2 under "Narrowband Active Band-Reject Filters" showed how a first-order bandpass section can be combined with a summing amplifier to obtain a band-reject circuit for transformed real poles, where $f_r = f_\infty$. Three types of sections were illustrated in Figure 6-23, corresponding to different Q ranges of operation. These same sections can be used as null networks. They offer more flexibility than the twin-T since the null frequency can be adjusted to compensate for component tolerances. In addition, the DABP and biquad circuits permit Q adjustment as well.

The design formulas were given by Equations (6-67) through (6-73). The values of f_r and Q in the equations correspond to the section center frequency and Q, respectively.

Frequently, a bandpass and band-reject output are simultaneously required. A typical application might involve the separation of signals for comparison of in-band and out-of-band spectral energy. The band-reject sections of Figure 6-23 can each provide a bandpass output from the bandpass section along with the null output signal. An additional feature of this technique is that the bandpass and band-reject outputs will track.

BIBLIOGRAPHY

Guillemin, E. A. *Communication Networks,* Vol 2. New York: John Wiley and Sons, 1935.

Saal, R., and E. Ulbrich. "On the Design of Filters by Synthesis." *IRE Transactions on Circuit Theory* (December, 1958).

Steward, J. L. *Circuit Theory and Design.* New York: John Wiley and Sons, 1956.

Tow, J. "A Step-by-Step Active Filter Design." *IEEE Spectrum* 6 (December, 1969): 64–68.

Williams, A. B. *Active Filter Design.* Dedham, Massachusetts: Artech House, 1975.

Zverev, A. I. *Handbook of Filer Synthesis.* New York: John Wiley and Sons, 1967.

CHAPTER 7
NETWORKS FOR THE TIME DOMAIN

7.1 ALL-PASS TRANSFER FUNCTIONS

Up until now, the networks we've discussed were used to obtain a desired amplitude versus frequency characteristic. No less important is the all-pass family of filters. This class of networks exhibits a flat frequency response but introduces a prescribed phase shift versus frequency. All-pass filters are frequently called *delay equalizers.*

If a network is to be of an all-pass type, the absolute magnitudes of the numerator and denominator of the transfer function must be related by a fixed constant at all frequencies. This condition will be satisfied if the zeros are the images of the poles. Since poles are restricted to the left-half quadrants of the complex frequency plane to maintain stability, the zeros must occur in the right-half plane as the mirror image of the poles about the $j\omega$ axis. Figure 7-1 illustrates the all-pass pole-zero representations in the complex frequency plane for first-order and second-order all-pass transfer functions.

First-Order All-Pass Transfer Functions

The real pole-zero pair of Figure 7-1*a* has a separation of $2\alpha_0$ between the pole and zero and corresponds to the following first-order all-pass transfer function:

$$T(s) = \frac{s - \alpha_0}{s + \alpha_0} \tag{7-1}$$

To determine the absolute magnitude of $T(s)$, compute

$$|T(s)| = \frac{|s - \alpha_0|}{|s + \alpha_0|} = \frac{\sqrt{\alpha_0^2 + \omega^2}}{\sqrt{\alpha_0^2 + \omega^2}} = 1 \tag{7-2}$$

where $s = j\omega$. For any value of frequency, the numerator and denominator of Equation (7-2) are equal, so the transfer function is clearly all-pass and has an absolute magnitude of unity at all frequencies.

The phase shift is given by

$$\beta(\omega) = -2\tan^{-1}\frac{\omega}{\alpha_0} \tag{7-3}$$

where $\beta(\omega)$ is in radians.

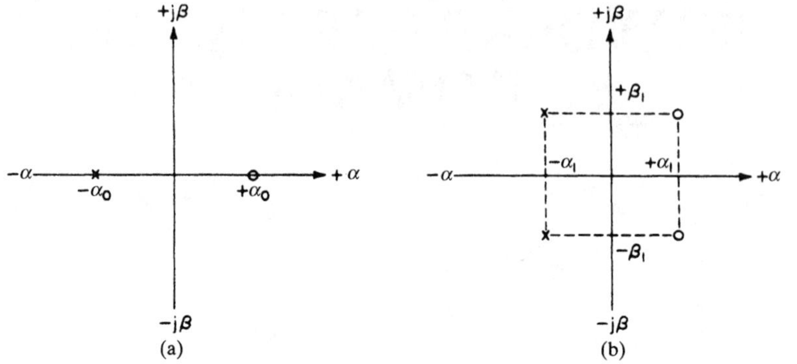

FIGURE 7-1 All-pass pole-zero patterns: (*a*) first-order all-pass transfer function; and (*b*) second-order all-pass transfer function.

The phase shift versus the ratio ω/α_0, as defined by Equation (7-3), is plotted in Figure 7-2. The phase angle is 0° at DC and $-90°$ at $\omega = \alpha_0$. The phase shift asymptotically approaches $-180°$ with increasing frequency.

The group delay was defined in Section 2.2, under "Effect of Nonuniform Time Delay," as the derivative of the phase shift which results in

$$T_{gd} = -\frac{d\beta(\omega)}{d\omega} = \frac{2\alpha_0}{\alpha_0^2 + \omega^2} \tag{7-4}$$

If Equation (7-4) is plotted with respect to ω for different values of α_0, a family of curves is obtained, as shown in Figure 7-3. First-order all-pass sections exhibit maximum delay at DC and decreasing delay with increasing frequency. For small values of α_0, the delay becomes large at low frequencies and decreases quite rapidly above this range. The delay at DC is found by setting ω equal to zero in Equation (7-4), which results in

$$T_{gd}(\text{DC}) = \frac{2}{a_0} \tag{7-5}$$

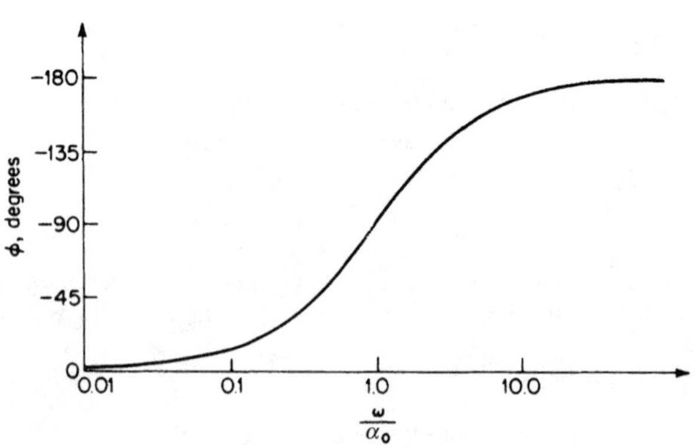

FIGURE 7-2 The phase shift of a first-order all-pass section.

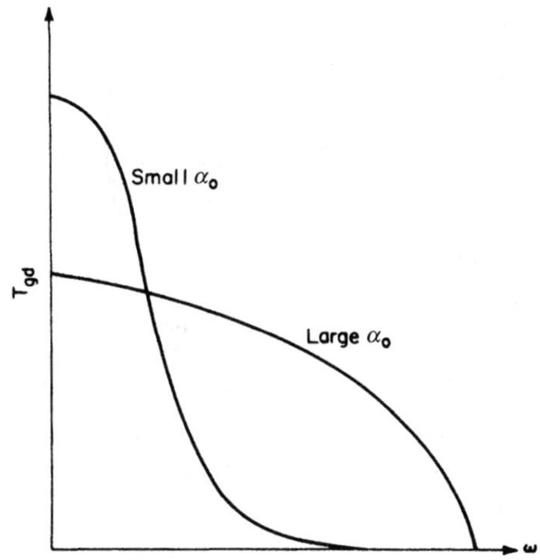

FIGURE 7-3 Group delay of first-order all-pass transfer functions.

Second-Order All-Pass Transfer Functions

The second-order all-pass transfer function represented by the pole-zero pattern of Figure 7-1*b* is given by

$$T(s) = \frac{s^2 - \dfrac{\omega_r}{Q} s + \omega_r^2}{s^2 + \dfrac{\omega_r}{Q} s + \omega_r^2} \tag{7-6}$$

where ω_r and Q are the pole resonant frequency (in radians per second) and the pole Q. These terms may also be computed from the real and imaginary pole-zero coordinates of Figure 7-1*b* by

$$\omega_r = \sqrt{\alpha_1^2 + \beta_1^2} \tag{7-7}$$

and

$$Q = \frac{\omega_r}{2\alpha_1} \tag{7-8}$$

The absolute magnitude of $T(s)$ is found to be

$$|T(s)| = \frac{\sqrt{\left(\omega_r^2 - \omega^2\right)^2 + \dfrac{\omega^2 \omega_r^2}{Q^2}}}{\sqrt{\left(\omega_r^2 - \omega^2\right)^2 + \dfrac{\omega^2 \omega_r^2}{Q^2}}} = 1 \tag{7-9}$$

which is all-pass.

The phase shift in radians is

$$\beta(\omega) = -2\tan^{-1}\left(\frac{\dfrac{\omega\omega_r}{Q}}{\omega_r^2 - \omega^2}\right) \tag{7-10}$$

and the group delay is given by

$$T_{gd} = \frac{2Q\omega_r(\omega^2 + \omega_r^2)}{Q^2(\omega^2 - \omega_r^2)^2 + \omega^2\omega_r^2} \tag{7-11}$$

The phase and delay parameters of first-order transfer functions are relatively simple to manipulate since they are a function of a single design parameter α_0. A second-order type, however, has two design parameters, Q and ω_r.

The phase shift of a second-order transfer function is $-180°$ at $\omega = \omega_r$. At DC, the phase shift is zero and at frequencies well above ω_r the phase asymptotically approaches $-360°$.

The group delay reaches a peak which occurs very close to ω_r. As the Q is made larger, the peak delay increases, the delay response becomes sharper, and the delay at DC decreases, as shown in Figure 7-4.

The frequency of maximum delay is slightly below ω_r and is expressed in radians per second by

$$\omega(T_{gd,\max}) = \omega_r\sqrt{\sqrt{4 - \frac{1}{Q^2}} - 1} \tag{7-12}$$

For all practical purposes, the maximum delay occurs at ω_r for Qs in excess of 2. By setting $\omega = \omega_r$ in Equation (7-11), the delay at ω_r is given by

$$T_{gd,\max} = \frac{4Q}{\omega_r} = \frac{2Q}{\pi f_r} \tag{7-13}$$

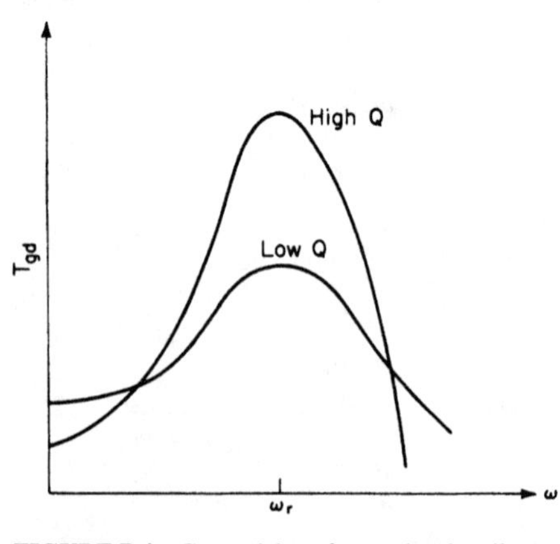

FIGURE 7-4 Group delay of second-order all-pass transfer functions.

If we set $\omega = 0$, the delay at DC is found from

$$T_{gd}(\text{DC}) = \frac{2}{Q\omega_r} = \frac{1}{Q\pi f_r} \tag{7-14}$$

7.2 DELAY EQUALIZER SECTIONS

Passive or active networks that realize first- or second-order all-pass transfer functions are called *delay equalizers*, since they are normally used to provide a required delay characteristic without disturbing the amplitude response. All-pass networks can be realized in a variety of configurations both passive and active. Equalizers with adjustable characteristics can also be designed, and are discussed in Section 7.6.

LC All-Pass Structures

First-Order Constant-Resistance Circuit. The lattice of Figure 7-5*a* realizes a first-order all-pass transfer function. The network is also a constant-resistance type, which means that the input impedance has a constant value of R over the entire frequency range. Constant-resistance networks can be cascaded with no interaction so that composite delay curves can be built up by accumulating the individual delay contributions. The lattice has an equivalent unbalanced form shown in Figure 7-5*b*. The design formulas are given by

$$L = \frac{2R}{\alpha_0} \tag{7-15}$$

$$C = \frac{2}{\alpha_0 R} \tag{7-16}$$

where R is the desired impedance level and α_0 is the real pole-zero coordinate. The phase shift and delay properties were defined by Equations (7-3) through (7-5).

The circuit of Figure 7-5*b* requires a center-tapped inductor having a coefficient of magnetic coupling K equal to unity.

Second-Order Constant-Resistance Sections. A second-order all-pass lattice with constant-resistance properties is shown in Figure 7-6*a*. The circuit may be transformed into the unbalanced bridged-T form of Figure 7-6*b*. The elements are given by

$$L_a = \frac{2R}{\omega_r Q} \tag{7-17}$$

$$C_a = \frac{Q}{\omega_r R} \tag{7-18}$$

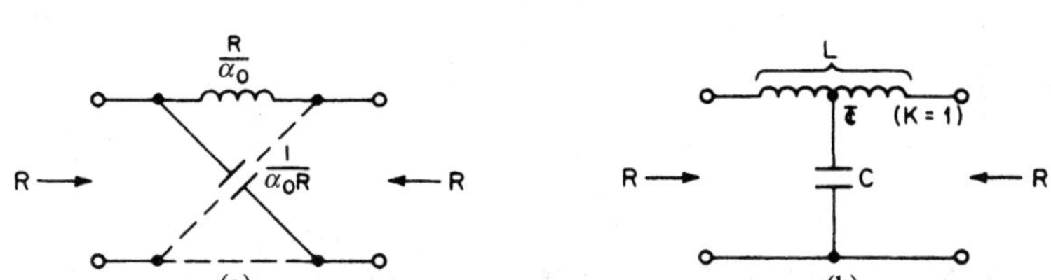

FIGURE 7-5 First-order LC equalizer section: (*a*) lattice form; and (*b*) unbalanced form.

$$L_b = \frac{QR}{2\omega_r} \tag{7-19}$$

$$C_b = \frac{2Q}{\omega_r(Q^2 - 1)R} \tag{7-20}$$

For tuning and test purposes, the section can be split into parallel and series resonant branches by opening the shunt branch and shorting the bridging or series branch, as shown in Figure 7-6*c*. Both circuits will resonate at ω_r.

The T-to-pi transformation was first introduced in Section 4.1. This transformation may be applied to the T of capacitors that are embedded in the section of Figure 7-6*b* to reduce capacitor values if desired. The resulting circuit is given in Figure 7-6*d*. Capacitors C_1 and C_2 are computed as follows:

$$C_1 = \frac{C_a^2}{2C_a + C_b} \tag{7-21}$$

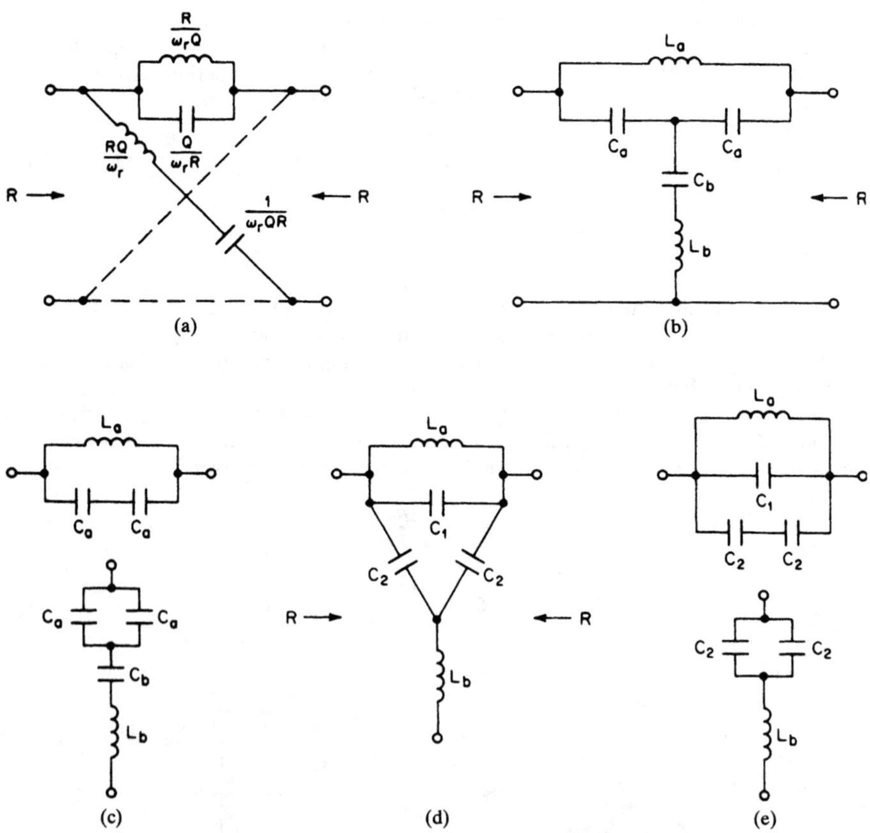

FIGURE 7-6 Second-order section $Q > 1$: (*a*) lattice form; (*b*) unbalanced form; (*c*) circuit for measuring branch resonances; (*d*) circuit modified by a T-to-pi transformation; and (*e*) resonant branches of modified circuit.

$$C_2 = \frac{C_a C_b}{2C_a + C_b} \tag{7-22}$$

The branch resonances are obtained by opening the shunt branch and then shorting the bridging branch, which results in the parallel and series resonant circuits of Figure 7-6*e*. Both resonances occur at ω_r.

Close examination of Equation (7-20) indicates that C_b will be negative if the Q is less than 1. (If $Q = 1$, C_b can be replaced by a short.) This restricts the circuits of Figure 7-6 to those cases where the Q is in excess of unity. Fortunately, this is true in most instances.

In those cases where the Q is below 1, the configurations of Figure 7-7 are used. The circuit of Figure 7-7*a* uses a single inductor with a controlled coefficient of coupling given by

$$K_3 = \frac{1 - Q^2}{1 + Q^2} \tag{7-23}$$

The element values are given by

$$L_{3a} = \frac{(Q^2 + 1)R}{2Q\omega_r} \tag{7-24}$$

$$C_3 = \frac{Q}{2\omega_r R} \tag{7-25}$$

$$C_4 = \frac{2}{Q\omega_r R} \tag{7-26}$$

It is not always convenient to control the coefficient of coupling of a coil to obtain the specific value required by Equation (7-23). A more practical approach uses the circuit of Figure 7-7*b*. The inductor L_{3b} is center-tapped and requires a unity coefficient of coupling (typical values of 0.95 or greater can usually be obtained and are acceptable). The values of L_{3b} and L_4 are computed from

$$L_{3b} = 2(1 + K_3)L_{3a} \tag{7-27}$$

$$L_4 = \frac{(1 - K_3)L_{3a}}{2} \tag{7-28}$$

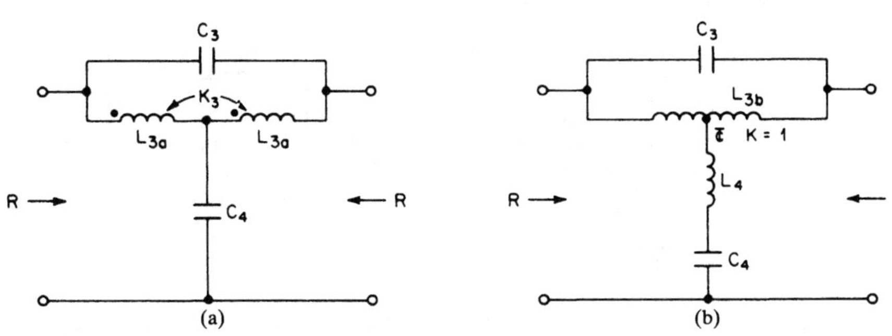

FIGURE 7-7 Second-order section $Q < 1$: (*a*) circuit with a controlled coefficient of coupling; and (*b*) circuit with unity coefficient of coupling.

The sections of Figure 7-7 may be tuned to ω_r in the same manner as in the equalizers in Figure 7-6. A parallel resonant circuit is obtained by opening C_4, and a series resonant circuit will result by placing a short across C_3.

The second-order section of Figures 7-6 and 7-7 may not always be all-pass. If the inductors have insufficient Q, a notch will occur at the resonances and will have a notch depth that can be approximated by

$$A_{\text{dB}} = 20 \log \frac{Q_L + 4Q}{Q_L - 4Q} \tag{7-29}$$

where Q_L is the inductor Q. If the notch is unacceptable, adequate coil Q must be provided or amplitude-equalization techniques are used, as discussed in section 8.4.

Minimum Inductor All-Pass Sections. The bridged-T circuit of Figure 7-8 realizes a second-order all-pass transfer function with a single inductor. The section is not a constant-resistance type, and operates between a zero impedance source and an infinite load. If the ratio of load to source is well in excess of 10, satisfactory results will be obtained. The elements are computed by

$$C = \frac{Q}{4\omega_r R} \tag{7-30}$$

and

$$L = \frac{1}{\omega_r^2 C} \tag{7-31}$$

The value of R can be chosen as the geometric mean of the source and load impedance $\left(\sqrt{R_S R_L}\right)$. The LC circuit is parallel resonant at ω_r.

The reader is reminded that the design parameters Q and ω_r are determined from the delay parameters defined in Section 7.1, which covers all-pass transfer functions.

A notch will occur at resonance due to a finite inductor Q and can be calculated from

$$A_{\text{dB}} = 20 \log \frac{4R + \omega_r L Q_L}{4R - \omega_r L Q_L} \tag{7-32}$$

If R is set equal to $\omega_r L Q_L/4$, the notch attenuation becomes infinite and the circuit is then identical to the bridged-T null network of section 6.1.

Two sets of all pass poles and zeros corresponding to a fourth-order transfer function can also be obtained by using a minimum-inductance-type structure. The circuit configuration is shown in Figure 7-9.

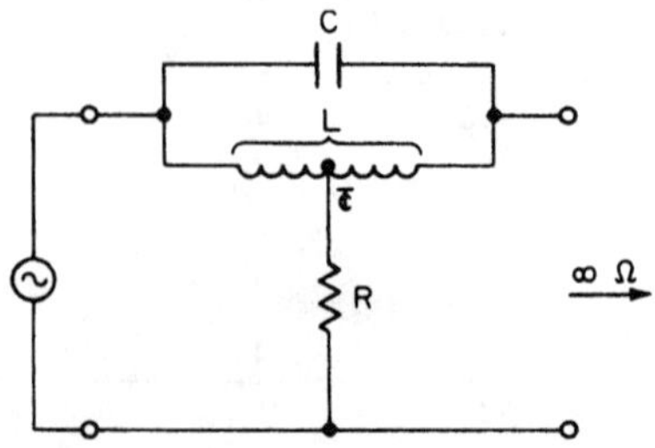

FIGURE 7-8 Minimum inductor type, second-order section.

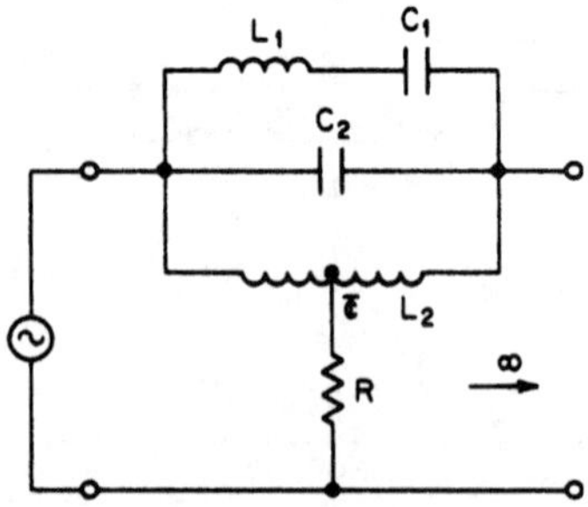

FIGURE 7-9 A fourth-order minimum inductance, all-pass structure.

Upon being given two sets of equalizer parameters Q_1, ω_{r1} and Q_2, ω_{r2}, as defined in Section 7.1, the following equations are used to determine the element values:

First, compute

$$A = \omega_{r1}^2 \omega_{r2}^2 \tag{7-33}$$

$$B = A\left(\frac{1}{\omega_{r2}Q_2} + \frac{1}{\omega_{r1}Q_1}\right) \tag{7-34}$$

$$C = \frac{Q_1 Q_2}{A(Q_2\omega_{r1} + Q_1\omega_{r2})} \tag{7-35}$$

$$D = \frac{Q_1 Q_2\left(\omega_{r1}^2 + \omega_{r2}^2\right) + \omega_{r1}\omega_{r2}}{ABQ_1Q_2} - C - \frac{1}{AB^2C} \tag{7-36}$$

$$E = \frac{1}{ABCD} \tag{7-37}$$

The element values are then given by

$$L_1 = \frac{4ER}{A} \tag{7-38}$$

$$C_1 = \frac{AD}{4R} \tag{7-39}$$

$$L_2 = \frac{4BR}{A} \tag{7-40}$$

$$C_2 = \frac{AC}{4R} \tag{7-41}$$

The value of R is generally chosen as the geometric mean of the source and load terminations, as with the second-order minimum-inductance section. The series and parallel branch resonant frequencies are found from

$$\omega_{L1C1} = \frac{1}{\sqrt{ED}} \tag{7-42}$$

and

$$\omega_{L2C2} = \frac{1}{\sqrt{BC}} \tag{7-43}$$

Active All-Pass Structures. First- and second-order all-pass transfer functions can be obtained by using an active approach. The general form of the active all-pass section is represented by the block diagram of Figure 7-10, where $T(s)$ is a first- or second-order transfer function having a gain of unity.

First-Order Sections. The transfer function of the circuit of Figure 7-10 is given by

$$\frac{E_{\text{out}}}{E_{\text{in}}} = 2T(s) - 1 \tag{7-44}$$

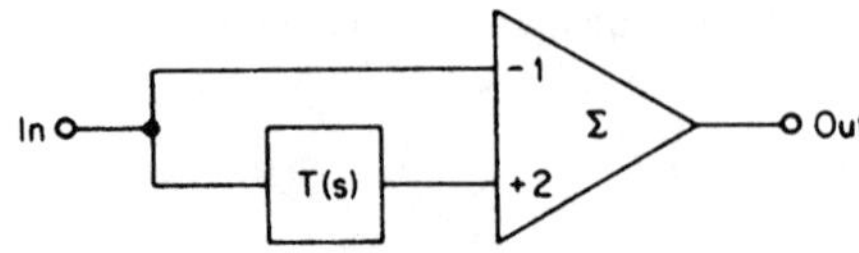

FIGURE 7-10 The general form of an active all-pass section.

If $T(s)$ is a first-order RC high-pass network having the transfer function $sCR/(sCR + 1)$, the composite transfer function becomes

$$\frac{E_{out}}{E_{in}} = \frac{s - 1/RC}{s + 1/RC} \tag{7-45}$$

This expression corresponds to the first-order all-pass transfer function of Equation (7-1), where

$$\alpha_0 = \frac{1}{RC} \tag{7-46}$$

The circuit can be directly implemented by the configuration of Figure 7-11*a*, where R' is arbitrary. The phase shift is then given by

$$\beta(\omega) = -2\tan^{-1}\omega RC \tag{7-47}$$

and the delay is found from

$$T_{gd} = \frac{2RC}{(\omega RC)^2 + 1} \tag{7-48}$$

At DC, the delay is a maximum and is computed from

$$T_{gd}(\text{DC}) = 2RC \tag{7-49}$$

The corresponding phase shift is shown in Figure 7-2. A phase shift of $-90°$ occurs at $\omega = 1/RC$ and approaches $-180°$ and $0°$ at DC and infinity, respectively. By making the element R variable, an all-pass network can be obtained having a phase shift adjustable between 0 and $-180°$.

A sign inversion of the phase will occur if the circuit of Figure 7-11*b* is used. The circuit will remain all-pass and first-order, and the group delay is still defined by Equations (7-48) and (7-49).

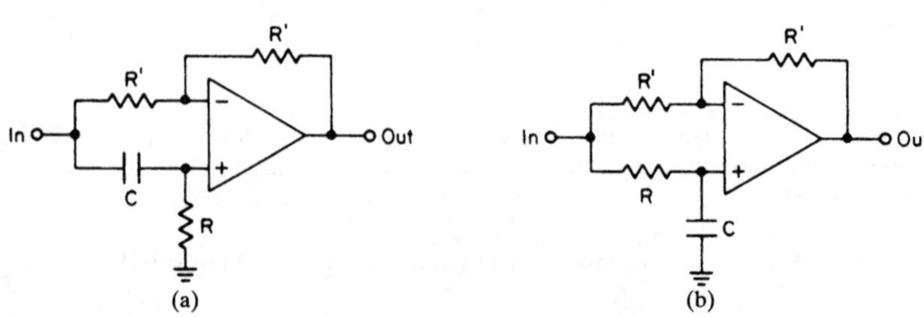

FIGURE 7-11 First-order all-pass sections: (*a*) circuit with lagging phase shift; and (*b*) circuit with leading phase shift.

Second-Order Section. If $T(s)$ in Figure 7-10 is a second-order bandpass network having the general bandpass transfer function

$$T(s) = \frac{\frac{\omega_r}{Q} s}{s^2 + \frac{\omega_r}{Q} s + \omega_r^2} \tag{7-50}$$

the composite transfer function then becomes

$$\frac{E_{\text{out}}}{E_{\text{in}}} = 2T(s) - 1 = -\frac{s^2 - \frac{\omega_r}{Q} s + \omega_r^2}{s^2 + \frac{\omega_r}{Q} s + \omega_r^2} \tag{7-51}$$

which corresponds to the second-order all-pass expression given by Equation (7-6) (except for a simple sign inversion). Therefore, a second-order all-pass equalizer can be obtained by implementing the structure of Figure 7-10 using a single active band-pass section for $T(s)$.

Section 5-2 discussed the MFBP, DABP, and biquad all-pole bandpass sections. Each circuit can be combined with a summing amplifier to generate a delay equalizer.

The MFBP equalizer is shown in Figure 7-12*a*. The element values are given by

$$R_2 = \frac{2Q}{\omega_r C} = \frac{Q}{\pi f_r C} \tag{7-52}$$

$$R_{1a} = \frac{R_2}{2} \tag{7-53}$$

$$R_{1b} = \frac{R_{1a}}{2Q^2 - 1} \tag{7-54}$$

The values of C and R can be arbitrarily chosen, and A in Figure 7-12*a* corresponds to the desired gain.

The maximum delay which occurs at f_r was given by Equation (7-13). This expression can be combined with Equations (7-52) and (7-54), so the element values can alternately by expressed in terms of $T_{gd,\,\max}$ as follows for $Q > 2$:

$$R_2 = \frac{T_{gd,\max}}{2C} \tag{7-55}$$

$$R_{1b} = \frac{R_2}{\left(\pi f_r T_{gd,\max}\right)^2 - 2} \tag{7-56}$$

where R_{1a} remains $R_2/2$.

The MFBP section can be tuned by making R_{1b} variable. R_{1b} can then be adjusted until 180° of phase shift occurs between the input and output of the bandpass section at f_r. In order for the response to be all-pass, the bandpass section gain must be exactly unity at resonance. Otherwise, an amplitude ripple will occur in the frequency-response characteristic in the vicinity of f_r.

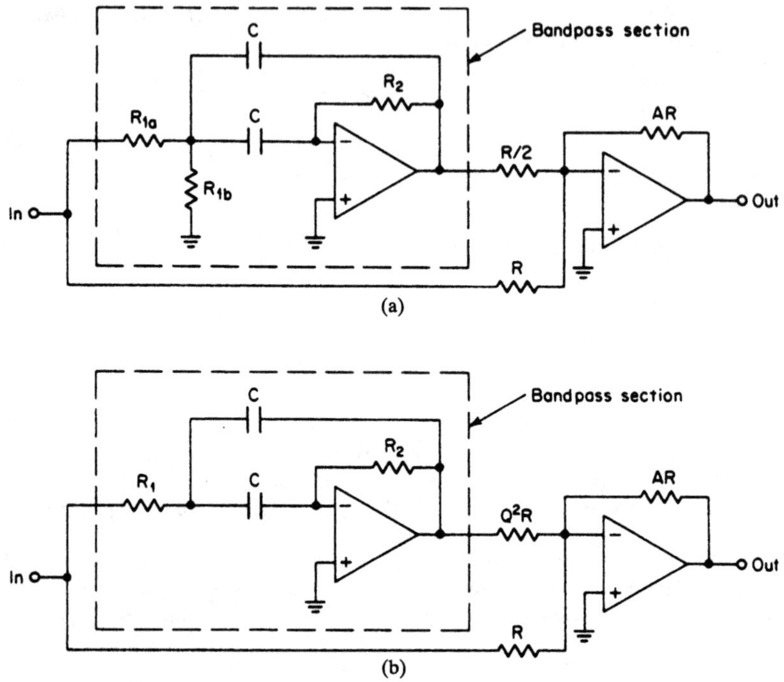

FIGURE 7-12 The MFBP delay equalizer $Q < 20$: (*a*) circuit for $0.707 < Q < 20$; and (*b*) circuit for $Q < 0.707$.

The section Q is limited to values below 20, or is expressed in terms of delay, as in the following:

$$T_{gd,\max} < \frac{40}{\pi f_r} \tag{7-57}$$

Experience has indicated that required Qs are usually well under 20, so this circuit will suffice in most cases. However, if the Q is below 0.707, the value of R_{1b}, as given by Equation (7-54), is negative, so the circuit of Figure 7-12*b* is used. The value of R_1 is given by

$$R_1 = \frac{R_2}{4Q^2} \tag{7-58}$$

In the event that higher Qs are required, the DABP section can be applied to the block diagram of Figure 7-10. Since the DABP circuit has a gain of 2 and is noninverting, the implementation shown in Figure 7-13 is used. The element values are given by

$$R_1 = \frac{Q}{\omega_r C} = \frac{Q}{2\pi f_r C} \tag{7-59}$$

and

$$R_2 = R_3 = \frac{R_1}{Q} \tag{7-60}$$

where C, R, and R' can be conveniently chosen. Resistor R_2 may be made variable if tuning is desired. The Q, and therefore the delay, can also be trimmed by making R_1 adjustable.

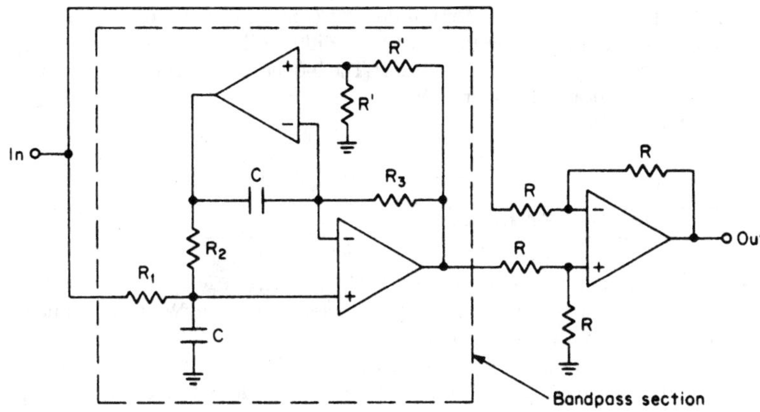

FIGURE 7-13 The DABP delay equalizer $Q < 150$.

The biquad structure can be configured in the form of a delay equalizer. The circuit is shown in Figure 7-14, and the element values are computed from

$$R_1 = R_4 = \frac{Q}{\omega_r C} = \frac{Q}{2\pi f_r C} \tag{7-61}$$

and

$$R_2 = R_3 = \frac{R_1}{Q} \tag{7-62}$$

where C, R, and R' are arbitrary.

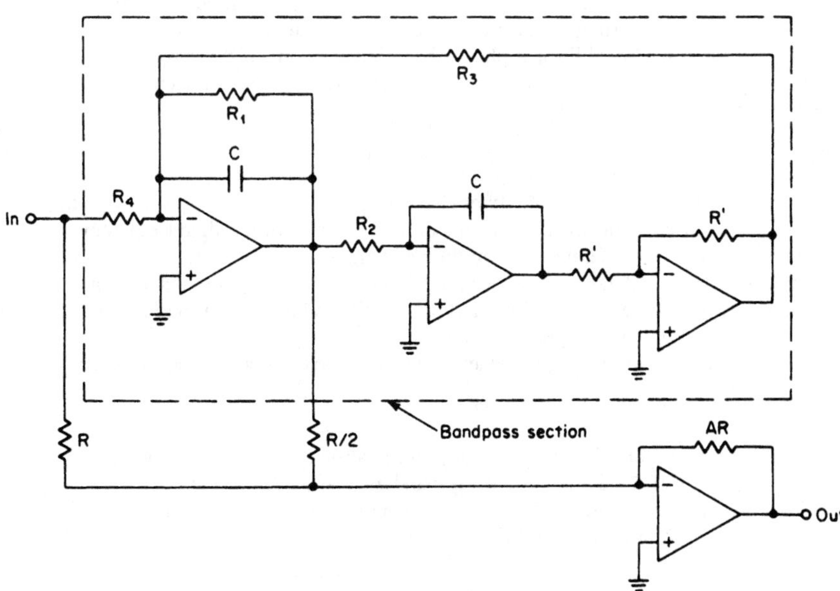

FIGURE 7-14 The biquad delay equalizer $Q < 200$.

Resistor R_3 can be made variable for tuning. The Q is adjusted for the nominal value by making R_1 variable and monitoring the 3-dB bandwidth at the output of the bandpass section, or by adjusting for unity bandpass gain at f_r. The biquad is subject to the Q-enhancement effect discussed in Section 5.2, under "All-Pole Bandpass Configurations," so a Q adjustment is usually required.

7.3 DESIGN OF DELAY LINES

The classical approach to the design of delay lines involves a cascade of identical *LC* sections (except for the end sections) and uses image-parameter theory (see Wallis in Bibliography). This technique is an approximation at best.

Modern network theory permits us to predict the delay of networks accurately and to obtain a required delay in a much more efficient manner than with the classical approach. The Bessel, linear phase with equiripple error and transitional filters all feature a constant delay. The curves in Chapter 2 indicate that for $n > 3$, the flat delay region is extended well into the stopband. If a delay line is desired, a low-pass filter implementation is not a very desirable approach from a delay-bandwidth perspective. A significant portion of the constant delay region would be attenuated.

All the low-pass transfer functions covered can be implemented by using an all-pass realization to overcome the bandwidth limitations. This results in a precise and efficient means of designing delay lines.

The Low-Pass to All-Pass Transformation

A low-pass transfer function can be transformed to an all-pass transfer function simply by introducing zeros in the right-half plane of the $j\omega$ axis corresponding to each pole. If the real and complex poles tabulated in Chapter 11 are realized using the first- and second-order all-pass structures of Section 7.2, complementary zeros will also occur. When a low-pass to all-pass transformation is made, the low-pass delay is increased by a factor of exactly 2 because of the additional phase-shift contributions of the zeros.

An all-pass delay-bandwidth factor can be derived from the delay curves of Chapter 2, which is given by

$$TU = \omega_u T_{gd}(\text{DC}) \tag{7-63}$$

The value of T_{gd} (DC) is the delay at DC, which is twice the delay shown in the curves because of the all-pass transformation, and ω_u is the upper limit radian frequency where the delay deviates a specified amount from the DC value.

Table 7-1 lists the delay at DC, ω_u, and the delay-bandwidth product TU for an all-pass realization of the Bessel maximally flat delay family. Values are provided for both 1- and 10-percent deviations of delay at ω_u.

To choose a transfer-function type and determine the complexity required, first compute

$$TU_{\text{req}} = 2\pi f_{gd} T_{gd} \tag{7-64}$$

where f_{gd} is the maximum desired frequency of operation and T_{gd} is the nominal delay needed. A network is then selected that has a delay-bandwidth factor TU which exceeds TU_{req}.

Compute the delay-scaling factor (DSF), which is the ratio of the normalized delay at DC to the required nominal delay. For example:

$$\text{DSF} = \frac{T_{gd}(\text{DC})}{T_{gd}} \tag{7-65}$$

TABLE 7-1 All-Pass Bessel Delay Characteristics

		1-Percent Deviation		10-Percent Deviation	
N	T_{gd} (DC)	ω_u	TU	ω_u	TU
2	2.72	0.412	1.121	0.801	2.179
3	3.50	0.691	2.419	1.109	3.882
4	4.26	0.906	3.860	1.333	5.679
5	4.84	1.120	5.421	1.554	7.521
6	5.40	1.304	7.042	1.737	9.380
7	5.90	1.478	8.720	1.912	11.280
8	6.34	1.647	10.440	2.079	13.180
9	6.78	1.794	12.160	2.227	15.100

The corresponding poles of the filter selected are denormalized by the DSF and can then be realized by the all-pass circuits of Section 7.2.

A real pole α_0 is denormalized by the formula

$$\alpha_0' = \alpha_0 \times \text{DSF} \tag{7-66}$$

Complex poles tabulated in the form $\alpha + j\beta$ are denormalized and transformed into the all-pass section design parameters ω_r and Q by the relationships

$$\omega_r = \text{DSF}\sqrt{\alpha^2 + \beta^2} \tag{7-67}$$

$$Q = \frac{\omega_r}{2\alpha\text{DSF}} \tag{7-68}$$

The parameters α_0', ω_r, and Q are then directly used in the design equations for the circuits of Section 7.2.

Sometimes the required delay-bandwidth factor TU_{req}, as computed by Equation (7-64), is in excess of the TU factors available from the standard filter families tabulated. The total delay required can then be subdivided into N smaller increments, and realized by N delay lines in cascade, since the delays will add algebraically.

LC *Delay Lines.* *LC* delay lines are designed by first selecting a normalized filter type and then denormalizing the corresponding real and complex poles, all in accordance with Section 7.3, under "The Low-Pass to All-Pass Transformation."

The resulting poles and associated zeros are then realized using the *LC* all-pass circuit of Section 7.2. This procedure is best illustrated by the following design example.

Example 7-1 Design of a 1mS LC Delay Line

Required:

Design a passive delay line to provide 1 ms of delay constant within 10 percent from DC to 3200 Hz. The source and load impedances are both 10 kΩ.

Result:

(*a*) Compute the required delay-bandwidth factor.

$$TU_{\text{req}} = 2\pi f_{gd} T_{gd} = 2\pi 3200 \times 0.001 = 20.1 \tag{7-64}$$

A linear phase design with an equiripple error of 0.5° will be chosen. The delay characteristics for the corresponding low-pass filters are shown in Figure 2-64. The delay at DC of a normalized all-pass network for $n = 9$ is equal to 7.5 s, which is twice the value obtained from the curves. Since the delay remains relatively flat to 3 rad/s, the delay-bandwidth factor is given by

$$TU = \omega_u T_{gd}(\text{DC}) = 3 \times 7.5 = 22.5 \tag{7-63}$$

Since TU is in excess of TU_{req}, the $n = 9$ design will be satisfactory.

(*b*) The low-pass poles are found in Table 11-45 and are as follows:

$$-0.5688 \pm j0.7595$$

$$-0.5545 \pm j1.5089$$

$$-0.5179 \pm j2.2329$$

$$-0.4080 \pm j2.9028$$

$$-0.5728$$

Four second-order all-pass sections and a single first-order section will be required. The delay-scaling factor is given by

$$\text{DSF} = \frac{T_{gd}(\text{DC})}{T_{gd}} = \frac{7.5}{10^{-3}} = 7500 \tag{7-65}$$

The denormalized design parameters ω_r and Q for the second-order sections are computed by Equations (7-67) and (7-68), respectively, and are tabulated as follows:

Section	α	β	ω_r	Q
1	0.5688	0.7595	7117	0.8341
2	0.5545	1.5089	12057	1.450
3	0.5179	2.2329	17191	2.213
4	0.4080	2.9028	21985	3.592

The design parameter α_0' for section 5 corresponding to the real pole is found from

$$\alpha_0' = \alpha_0 \times \text{DSF} = 4296 \tag{7-66}$$

where α_0 is 0.5728.

(*c*) The element values can now be computed as follows:

Section 1:

Since the Q is less than unity, the circuit of Figure 7-7*b* will be used. The element values are found from

$$K_3 = \frac{1 - Q^2}{1 + Q^2} = \frac{1 - 0.8341^2}{1 + 0.8341^2} = 0.1794 \tag{7-23}$$

$$L_{3a} = \frac{(Q^2 + 1)R}{2Q\omega_r} = \frac{(0.8341^2 + 1)10^4}{2 \times 0.8341 \times 7117} = 1.428\,\text{H} \tag{7-24}$$

$$C_3 = \frac{Q}{2\omega_r R} = \frac{0.8341}{2 \times 7117 \times 10^4} = 5860\,\text{pF} \tag{7-25}$$

$$C_4 = \frac{2}{Q\omega_r R} = \frac{2}{0.8341 \times 7117 \times 10^4} = 0.0337\,\mu\text{F} \tag{7-26}$$

$$L_{3b} = 2(1 + K_3)L_{3a} = 3.368\,\text{H} \tag{7-27}$$

$$L_4 = \frac{(1 - K_3)L_{3a}}{2} = 0.586\,\text{H} \tag{7-28}$$

Sections 2 through 4:
Since the Qs are in excess of unity, the circuit of Figure 7-6*b* will be used. The values for section 2 are found from

$$L_a = \frac{2R}{\omega_r Q} = \frac{2 \times 10^4}{12{,}057 \times 1.450} = 1.144\,\text{H} \tag{7-17}$$

$$C_a = \frac{Q}{\omega_r R} = \frac{1.450}{12{,}057 \times 10^4} = 0.012\,\mu\text{F} \tag{7-18}$$

$$L_b = \frac{QR}{2\omega_r} = \frac{1.450 \times 10^4}{2 \times 12{,}057} = 0.601\,\text{H} \tag{7-19}$$

$$C_b = \frac{2Q}{\omega_r(Q^2 - 1)R} = \frac{2 \times 1.450}{12{,}057(1.45^2 - 1)10^4} = 0.0218\,\mu\text{F} \tag{7-20}$$

In the same manner, the remaining element values can be computed, which results in

Section 3:

$$L_a = 0.526\text{ H}$$
$$C_a = 0.0129\ \mu\text{F}$$
$$L_b = 0.644\text{ H}$$
$$C_b = 6606\text{ pF}$$

Section 4:

$$L_a = 0.253\text{ H}$$
$$C_a = 0.0163\ \mu\text{F}$$
$$L_b = 0.817\text{ H}$$
$$C_b = 2745\text{ pF}$$

Section 5:
The remaining first-order all-pass section is realized using the circuit of Figure 7-5*b*. The element values are given by

$$L = \frac{2R}{\alpha_0'} = \frac{2 \times 10^4}{4296} = 4.655\text{ H} \tag{7-15}$$

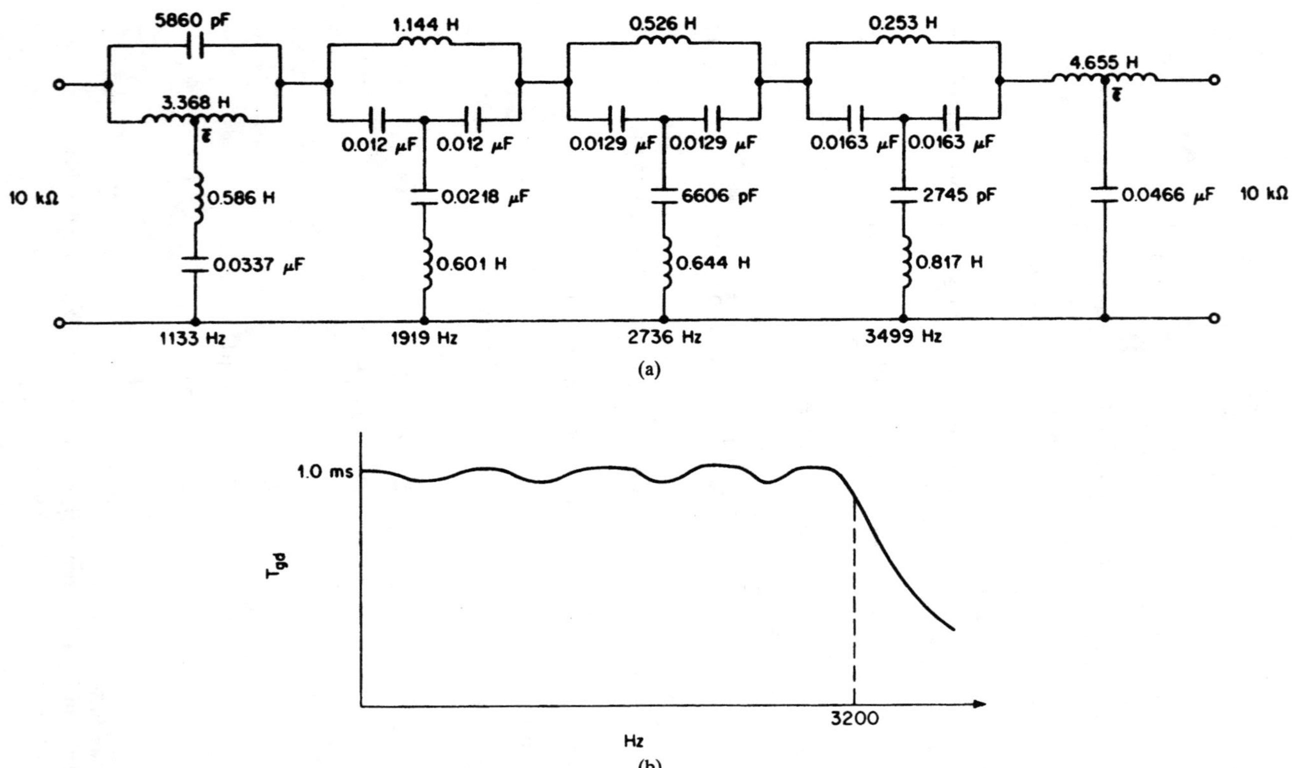

FIGURE 7-15 The 1-ms delay line of Example 7-1: (*a*) delay-line circuit; and (*b*) frequency response.

$$C = \frac{2}{\alpha_0' R} = \frac{2}{4296 \times 10^4} = 0.0466\ \mu\text{F} \tag{7-16}$$

(*d*) The resulting delay line is illustrated in Figure 7-15*a*. The resonant frequencies shown are in hertz, and correspond to $\omega_r/2\pi$ for each section. The center-tapped inductors require a unity coefficient of coupling. The delay characteristics as a function of frequency are also shown in Figure 7-15*b*.

The delay line of Example 7-1 requires a total of nine inductors. If the classical design approach (see Wallis in Bibliography), which is based on image-parameter theory, were used, the resulting delay line would use about twice as many coils. Although the inductors would all be uniform in value (except for the end sections), this feature is certainly not justified by the added cost and complexity.

Active Delay Lines. An active delay line is designed by initially choosing a normalized filter and then denormalizing the associated poles in the same manner as in the case of *LC* delay lines. The resulting all-pass design parameters are implemented using the first- and second-order active structures of Section 7.2.

Active delay lines do not suffer from the *Q* limitations of *LC* delay lines and are especially suited for low-frequency applications where inductor values may become impractical. The following example illustrates the design of an active delay line.

Example 7-2 Design of a 100 μS Active Delay Line

Required:

Design an active delay line having a delay of 100 μs constant within 3 percent to 3 kHz. A gain of 10 is also required.

Result:

(*a*) Compute the required delay-bandwidth factor.

$$TU_{\text{req}} = 2\pi f_{gd} T_{gd} = 2\pi 3000 \times 10^{-4} = 1.885 \tag{7-64}$$

A Bessel-type all-pass network will be chosen. Table 7-1 indicates that for a delay deviation of 1 percent, a complexity of $n = 3$ has a delay-bandwidth factor of 2.419, which is in excess of the required value.

(*b*) The Bessel low-pass poles are given in Table 12-41 and the corresponding values for $n = 3$ are

$$-1.0509 \pm j1.0025$$

$$-1.3270$$

Two sections are required consisting of a first-order and second-order type. The delay-scaling factor is computed to be

$$\text{DSF} = \frac{T_{gd}(\text{DC})}{T_{gd}} = \frac{3.5}{10^{-4}} = 3.5 \times 10^4 \tag{7-65}$$

where T_{gd}(DC) is obtained from Table 7-1 and T_{gd} is 100 μs, the design value.
The second-order section design parameters are

$$\omega_r = \text{DSF}\sqrt{\alpha^2 + \beta^2} = 3.5 \times 10^4\sqrt{1.0509^2 + 1.0025^2} = 50{,}833 \qquad (7\text{-}67)$$

and
$$Q = \frac{\omega_r}{2\alpha\text{DSF}} = \frac{50{,}833}{2 \times 1.509 \times 3.5 \times 10^4} = 6.691 \qquad (7\text{-}68)$$

The first-order section design parameter is given by

$$a_0' = \alpha_0 \times \text{DSF} = 1.327 \times 3.5 \times 10^4 = 46{,}450 \qquad (7\text{-}66)$$

(*c*) The element values are computed as follows:

The second-order section:

The MFBP equalizer section of Figure 7-12*b* will be used corresponding to $Q < 0.707$, where $R = 10$ kΩ, $C = 0.01$ μF, and $A = 10$. The element values are found from

$$R_2 = \frac{2Q}{\omega_r C} = \frac{2 \times 0.691}{50{,}833 \times 10^{-8}} = 2719\ \Omega \qquad (7\text{-}52)$$

$$R_1 = \frac{R_2}{4Q^2} = \frac{2719}{4 \times 0.691^2} = 1424\ \Omega \qquad (7\text{-}58)$$

The first-order section:

The first-order section of Figure 7-11*a* will be used, where R' is chosen at 10 kΩ, C at 0.01 μF, and α_0' is 46,450. The value of R is given by

$$R = \frac{1}{\alpha_0' C} = \frac{1}{46{,}450 \times 10^{-8}} = 2153\ \Omega \qquad (7\text{-}46)$$

(*d*) The resulting 100-μs active delay line is shown in Figure 7-16 using standard 1-percent resistor values.

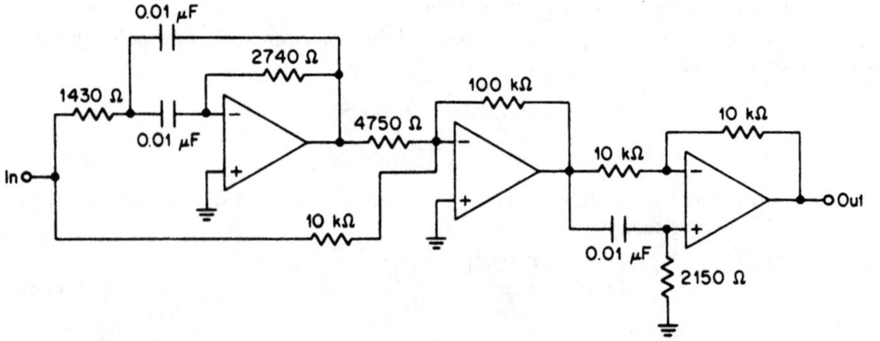

FIGURE 7-16 The 100-μs delay line of Example 7-2.

7.4 DELAY EQUALIZATION OF FILTERS

The primary emphasis in previous chapters has been the attenuation characteristics of filters. However, if the signal consists of a modulated waveform, the delay characteristics are also of importance. To minimize distortion of the input signal, a constant delay over the frequency range of interest is desirable. Typically, this would be 6 dB or so. The greater the attenuation, the less significant the impact of delay variation (delay distortion) since the spectral contributions of the attenuated signals are reduced.

The Bessel, linear phase with equiripple error, and transitional filter families all exhibit a flat delay. However, the amplitude response is less selective than that of other families. Frequently, the only solution to an attenuation requirement is a Butterworth, Chebyshev, or elliptic-function filter type. To also maintain the delay constant, delay equalizers would be required.

It is important to recognize that there are trade-offs between steep attenuation requirements and flatness of delay. For example, the higher the ripple of a Chebyshev filter, the steeper the rate of attenuation, but also the larger the delay deviation from flatness, especially around the corner frequency. Delay distortion also grows larger with increasing order n and steepness of elliptic function filters, as well as ripple. Steep elliptic function filters and high-order Chebyshev filters (see Section 2.4) are especially difficult to equalize since their delay characteristics near cutoff exhibit sharp delay peaks (horn-like in appearance).

Delay equalizer networks are frequently at least as complex as the filter being equalized. The number of sections required is dependent on the initial delay curve, the portion of the curve to be equalized, and the degree of equalization necessary. A very crude approximation to the number of equalizer sections required is given by

$$n = 2\Delta_{\text{BW}}\Delta_T + 1 \tag{7-69}$$

where Δ_{BW} is the bandwidth of interest in hertz, and Δ_T is the delay distortion over Δ_{BW} in seconds.

The approach to delay equalization discussed in this section is graphical rather than analytical. A closed-form solution to the delay equalization of filters is not available. However, computer programs can be obtained that achieve a least-squares approximation to the required delay specifications, and are preferred to trial-and-error techniques. (See Note 1.)

Simply stated, delay equalization of a filter involves designing a network that has a delay shape which complements the delay curve of the filter being equalized. The composite delay will then exhibit the required flatness. Although the absolute delay increases as well, this result is usually of little significance, since it is the delay variation over the band of interest that disperses the spectral components of the signal. Typical delay curves of a bandpass filter, the delay equalizer network, and the composite characteristics are shown in Figure 7-17.

To equalize the delay of a low-pass filter graphically, the highest frequency of interest and corresponding delay should be scaled to 1 rad/s so that the lower portion of the curve falls within the frequency region between DC and $\omega = 1$. This is accomplished by multiplying the delay axis by $2\pi f_h$, where f_h is the highest frequency to be equalized. The frequency axis is also divided by f_h and interpreted in radians per second so that f_h is transformed to 1 rad/s and all other frequencies are normalized to this point.

Note 1: The full version of *Filter Solutions* available from Nuhertz Technologies® (www.nuhrtz.com), uses a proprietary approach to automatically perform equalization of low-pass and bandpass filters. The final results can then be manually "tweaked" by adjusting the pole-zero locations while observing the changes in group delay in real time.

The normalized low-pass filter delay curves shown in section 2 for the various filter families may also be used directly. In either case, the required equalizer delay characteristic is

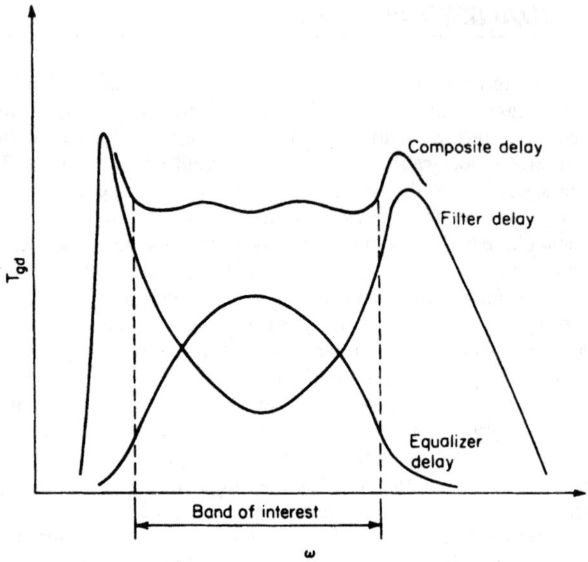

FIGURE 7-17 The delay equalization of a bandpass filter.

obtained by subtracting the delay curve from a constant equal to the maximum delay that occurs over the band. The resulting curve is then approximated by adding the delay contributions of equalizer sections. A sufficient number of sections is used to obtain the required composite delay flatness.

When a suitable match to the required curve is found, the equalizer parameters may be directly used to design the equalizer, and the circuit is then denormalized to the required frequency range and impedance level. Alternatively, the equalizer parameters can first be denormalized and the equalizer designed directly.

Bandpass filters are equalized in a manner similar to low-pass filters. The delay curve is first normalized by multiplying the delay axis by $2\pi f_0$, where f_0 is the filter center frequency. The frequency axis is divided by f_0 and interpreted in radians per second so that the center frequency is 1 rad/s and all other frequencies are normalized to the center frequency. A complementary curve is found, and appropriate equalizer sections are used until a suitable fit occurs. The equalizer is then denormalized.

First-Order Equalizers. First-order all-pass transfer functions were first introduced in Section 7.1. The delay of a first-order all-pass section is characterized by a maximum delay at low frequencies, and decreasing delay with increasing frequency. As the value of α_0 is reduced, the delay tends to peak at DC and will roll off more rapidly with increasing frequencies.

The delay of a first-order all-pass section was given in Section 7.1 by

$$T_{gd} = \frac{2\alpha_0}{\alpha_0^2 + \omega^2} \tag{7-4}$$

Working directly with Equation (7-4) is somewhat tedious, so a table of delay values for α_0 ranging between 0.05 and 2.00 at frequencies from $\omega = 0$ to $\omega = 1$ is provided in Table 7-2. This table can be directly used to determine the approximate α_0 necessary to equalize the normalized filter delay. A more exact value of α_0 can then be determined from Equation (7-4) if desired.

TABLE 7-2 The First-Order Equalizer Delay in Seconds

	ω, rad/s										
α_0	0	0.1	0.2	0.3	0.4	0.5	0.6	0.7	0.8	0.9	1.0
0.05	40.00	8.00	2.35	1.08	0.62	0.40	0.28	0.20	0.16	0.12	0.10
0.10	20.00	10.00	4.00	2.00	1.18	0.77	0.54	0.40	0.31	0.24	0.20
0.15	13.33	9.23	4.80	2.67	1.64	1.10	0.78	0.59	0.45	0.36	0.29
0.20	10.00	8.00	5.00	3.08	2.00	1.38	1.00	0.75	0.59	0.47	0.38
0.25	8.00	6.90	4.88	3.28	2.25	1.60	1.18	0.91	0.71	0.57	0.47
0.30	6.67	6.00	4.62	3.33	2.40	1.76	1.33	1.03	0.82	0.67	0.55
0.35	5.71	5.28	4.31	3.29	2.48	1.88	1.45	1.14	0.92	0.75	0.62
0.40	5.00	4.71	4.00	3.20	2.50	1.95	1.54	1.23	1.00	0.82	0.69
0.45	4.44	4.24	3.71	3.08	2.48	1.99	1.60	1.30	1.07	0.89	0.75
0.50	4.00	3.85	3.45	2.94	2.44	2.00	1.64	1.35	1.12	0.94	0.80
0.55	3.64	3.52	3.21	2.80	2.38	1.99	1.66	1.39	1.17	0.99	0.84
0.60	3.33	3.24	3.00	2.67	2.31	1.97	1.67	1.41	1.20	1.03	0.88
0.65	3.08	3.01	2.81	2.54	2.23	1.93	1.66	1.42	1.22	1.05	0.91
0.70	2.86	2.80	2.64	2.41	2.15	1.89	1.65	1.43	1.24	1.08	0.94
0.75	2.67	2.62	2.49	2.30	2.08	1.85	1.63	1.43	1.25	1.09	0.96
0.80	2.50	2.46	2.35	2.19	2.00	1.80	1.60	1.42	1.25	1.10	0.98
0.85	2.35	2.32	2.23	2.09	1.93	1.75	1.57	1.40	1.25	1.11	0.99
0.90	2.22	2.20	2.12	2.00	1.86	1.70	1.54	1.38	1.24	1.11	0.99
0.95	2.11	2.08	2.02	1.91	1.79	1.65	1.50	1.36	1.23	1.11	1.00
1.00	2.00	1.98	1.92	1.83	1.72	1.60	1.47	1.34	1.22	1.10	1.00
1.25	1.60	1.59	1.56	1.51	1.45	1.38	1.30	1.22	1.14	1.05	0.98
1.50	1.33	1.33	1.31	1.28	1.24	1.20	1.15	1.09	1.04	0.98	0.92
1.75	1.14	1.14	1.13	1.11	1.09	1.06	1.02	0.99	0.95	0.90	0.86
2.00	1.00	1.00	0.99	0.98	0.97	0.94	0.92	0.89	0.86	0.83	0.80

Use of Table 7-2 is best illustrated by an example, as follows.

Example 7-3 Design of an *LC* and Active Delay Equalizer for a Low-Pass Filter

Required:

Design a delay equalizer for an $n = 5$ Butterworth low-pass filter having a 3-dB cutoff of 1600 Hz. The delay variation should not exceed 75 μs from DC to 1600 Hz.

Result:

(*a*) The Butterworth normalized delay curves of Figure 2-35 can be used directly since the region between DC and 1 rad/s corresponds to the frequency range of interest. The curve for $n = 5$ indicates that the peak delay occurs near 0.9 rad/s and is approximately 1.9 s greater than the value at DC. This corresponds to a denormalized variation of $1.9/2\pi f_h$, or 190 μs, where f_h is 1600 Hz, so an equalizer is required.

(*b*) Examination of Table 7-2 indicates that a first-order equalizer with an α_0 of 0.7 has a delay at DC that is approximately 1.8 s greater than the delay at 0.9 rad/s, so a reasonable fit to the required shape should occur.

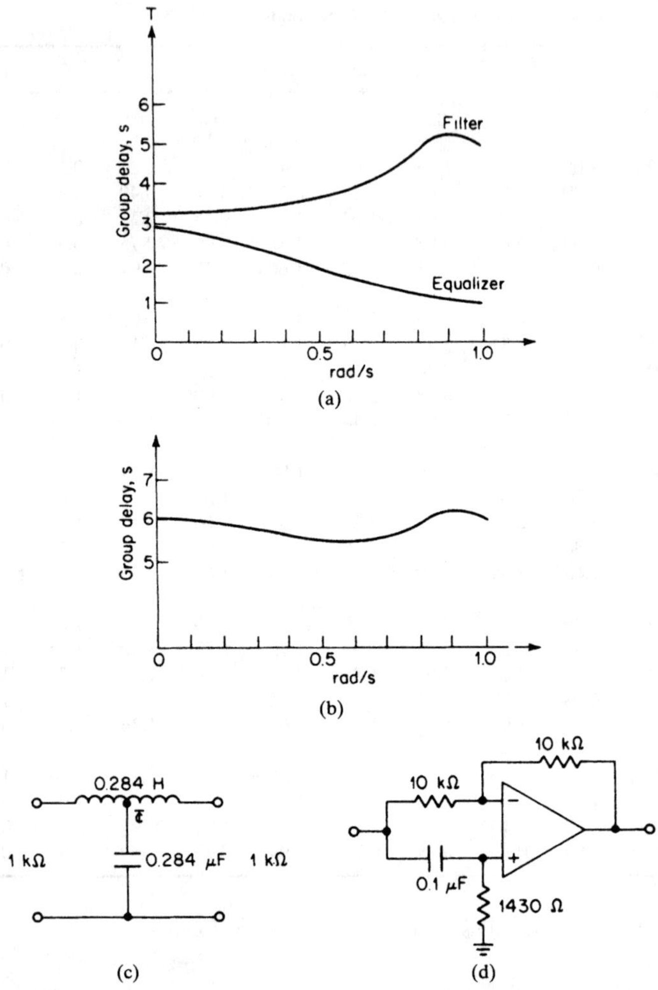

FIGURE 7-18 The delay equalization of Example 7-3: (*a*) filter and equalizer delay curves; (*b*) composite delay curve; (*c*) *LC* equalizer; and (*d*) active equalizer.

The delay of the normalized filter and the first-order equalizer for $\alpha_0 = 0.7$ is shown in Figure 7.18*a*. The combined delay is given in Figure 7-18*b*. The peak-to-peak delay variation is about 0.7 s, which corresponds to a denormalized delay variation of $0.7/2\pi f_h$ or 70 μs.

(*c*) The first-order equalizer parameter $\alpha_0 = 0.7$ is denormalized by the factor $2\pi f_h$, resulting in $\alpha_0' = 7037$. The corresponding passive equalizer is designed as follows, where the impedance level R is chosen to be 1 kΩ:

$$L = \frac{2R}{\alpha_0'} = \frac{2 \times 10^3}{7037} = 0.284\,\text{H} \tag{7-15}$$

$$C = \frac{2}{\alpha_0' R} = \frac{2}{7037 \times 10^3} = 0.284\ \mu\text{F} \tag{7-16}$$

The first-order *LC* equalizer section is shown in Figure 7-18*c*.

(*d*) An active first-order equalizer section can also be designed using the circuit of Figure 7-11*a*. If we select a C of 0.1 μF, where $R' = 10\ k\Omega$, the value of R is given by

$$R = \frac{1}{\alpha_0' C} = \frac{1}{7037 \times 10^{-7}} = 1421\ \Omega \tag{7-46}$$

The active equalizer circuit is illustrated in Figure 7-18*d*.

Highly selective low-pass filters such as the elliptic-function type have a corresponding delay characteristic that increases very dramatically near cutoff. First-order all-pass sections cannot then provide a complementary delay shape, so they are limited to applications involving low-pass filters of moderate selectivity.

Second-Order Equalizers. First-order equalizers have a maximum delay at DC and a single design parameter α_0 which limits their use. Second-order sections have two design parameters, ω_r and Q. The delay shape is bandpass in nature and can be made broad or sharp by changing the Q. The peak delay frequency is determined by the design parameter ω_r. As a result of this flexibility, second-order sections can be used to equalize virtually any type of delay curve. The only limitation is in the number of sections the designer is willing to use, and the effort required to achieve a given degree of equalization.

The group delay of a second-order all-pass section was given by

$$T_{gd} = \frac{2Q\omega_r\left(\omega^2 + \omega_r^2\right)}{Q^2\left(\omega^2 - \omega_r^2\right)^2 + \omega^2\omega_r^2} \tag{7-11}$$

If we normalize this expression by setting ω_r equal to 1, we obtain

$$T_{gd} = \frac{2Q(\omega^2 + 1)}{Q^2(\omega^2 - 1)^2 + \omega^2} \tag{7-70}$$

To determine the delay at DC, we can set equal to zero, which results in

$$T_{gd}(\text{DC}) = \frac{2}{Q} \tag{7-71}$$

For Qs below 0.577, the maximum delay occurs at DC. As the Q is increased, the frequency where maximum delay occurs approaches 1 rad/s and is given by

$$\omega\left(T_{gd,\max}\right) = \sqrt{\sqrt{4 - \frac{1}{Q^2}} - 1} \tag{7-72}$$

For Qs of 2 or more, the maximum delay can be assumed to occur at 1 rad/s and may be determined from

$$T_{gd,\max} = 4Q \tag{7-73}$$

Equations (7-70) through (7-72) are evaluated in Table 7-6 for Qs ranging from 0.25 to 10.

To use Table 7-6 directly, first normalize the curve to be equalized so that the minimum delay occurs at 1 rad/s. Then, select an equalizer from the table that provides the best fit for a complementary curve.

A composite curve is then plotted. If the delay ripple is excessive, additional equalizer sections are required to fill in the delay gaps. The data of Table 7-6 can again be used by scaling the region to be equalized to a 1-rad/s center and selecting a complementary equalizer shape from the table. The equalizer parameters can then be shifted to the region of interest by scaling.

The procedure described is an oversimplification of the design process. The equalizer responses will interact with each other, so each delay region to be filled in cannot be treated independently. Every time a section is added, the previous sections may require an adjustment of their design parameters.

Delay equalization generally requires considerably more skill than the actual design of filters. Standard pole-zero patterns are defined for the different filter families, whereas the design of equalizers involves the approximation problem where a pole-zero pattern must be determined for a suitable fit to a curve. The following example illustrates the use of second-order equalizer sections to equalize delay.

Example 7-4 Design of a Delay Equalizer for a Band-Pass Filter

Required:

A bandpass filter with the delay measurements of Table 7-3 must be equalized to within 700 μs.

The corresponding delay curve is plotted in Figure 7-19*a*.

Result:

(*a*) Since the minimum delay occurs at 1000 Hz, normalize the curve by dividing the frequency axis by 1000 Hz and multiplying the delay axis by 2π 1000. The results are shown in Table 7-4 and plotted in Figure 7-19*b*.

(*b*) An equalizer is required that has a nominal delay peak of 10 s at 1 rad/s relative to the delay at 0.5 and 1.5 rad/s. Examination of Table 7-6 indicates that the delay corresponding to a Q of 2.75 will meet this requirement.

If we add this delay, point by point, to the normalized delay of Table 7-4, the values of Table 7-5 will be obtained.

TABLE 7-3 Specified Delay

Frequency, Hz	Delay, μs
500	1600
600	960
700	640
800	320
900	50
1000	0
1100	160
1200	480
1300	800
1400	1120
1500	1500

TABLE 7-4 Normalized Delay

Frequency, rad/s	Delay, s
0.5	10.1
0.6	6.03
0.7	4.02
0.8	2.01
0.9	0.31
1.0	0
1.1	1.01
1.2	3.02
1.3	5.03
1.4	7.04
1.5	9.42

TABLE 7-5 Equalized Delay

Frequency, rad/s	Delay, s
0.5	11.6
0.6	8.19
0.7	7.36
0.8	7.58
0.9	9.50
1.0	11.0
1.1	8.89
1.2	7.64
1.3	7.83
1.4	8.86
1.5	10.7

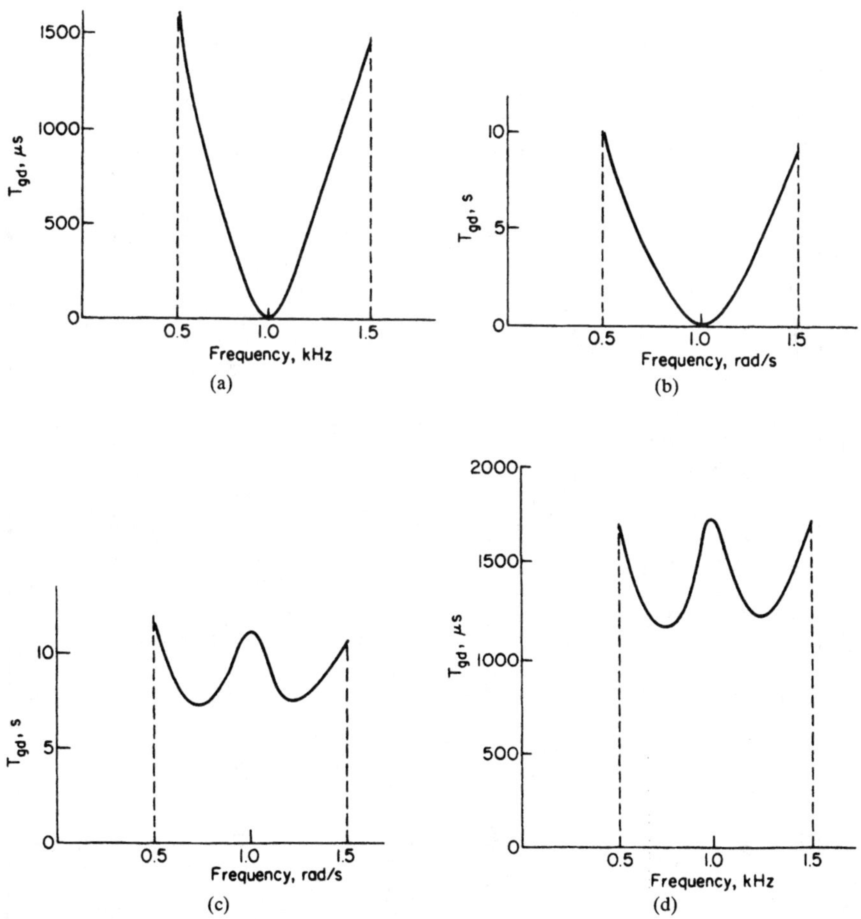

FIGURE 7-19 The delay equalization of Example 7-4: (*a*) unequalized delay; (*b*) normalized delay; (*c*) normalized equalized delay; and (*d*) denormalized equalized delay.

TABLE 7-6 The Delay of Normalized Second-Order Section ($\omega_r = 1$ rad/s)

Q	T_{gd}(DC)	$\omega(T_{gd,\max})$	$T_{gd,\max}$	0.1	0.2	0.3	0.4	0.5	0.6	0.7	0.8	0.9	1.0	1.1	1.2	1.3	1.4	1.5	1.6	1.7	1.8	1.9	2.0	3.0	4.0	5.0
0.25	8.00	DC	8.00	7.09	5.33	3.85	2.84	2.19	1.76	1.42	1.27	1.11	1.00	0.91	0.84	0.78	0.73	0.69	0.66	0.62	0.60	0.57	0.55	0.38	0.28	0.21
0.50	4.00	DC	4.00	3.96	3.85	3.67	3.45	3.20	2.94	2.69	2.44	2.21	2.00	1.81	1.64	1.49	1.35	1.23	1.12	1.03	0.94	0.87	0.80	0.40	0.24	0.15
0.75	2.67	0.700	3.51	2.70	2.79	2.94	3.12	3.31	3.46	3.51	3.45	3.27	3.00	2.69	2.36	2.06	1.79	1.56	1.36	1.19	1.05	0.93	0.83	0.33	0.18	0.11
1.00	2.00	0.856	4.31	2.04	2.16	2.37	2.68	3.08	3.53	3.97	4.26	4.28	4.00	3.52	2.99	2.48	2.05	1.71	1.43	1.20	1.03	0.88	0.77	0.27	0.14	0.09
1.25	1.60	0.913	5.23	1.64	1.76	1.97	2.30	2.77	3.40	4.16	4.87	5.22	5.00	4.32	3.50	2.76	2.18	1.73	1.40	1.15	0.96	0.81	0.69	0.23	0.11	0.07
1.50	1.33	0.941	6.18	1.37	1.47	1.67	1.99	2.47	3.18	4.16	5.28	6.09	6.00	5.06	3.90	2.92	2.20	1.69	1.33	1.07	0.88	0.73	0.62	0.20	0.10	0.06
1.75	1.14	0.957	7.15	1.17	1.27	1.45	1.75	2.22	2.95	4.05	5.54	6.88	7.00	5.75	4.20	2.99	2.17	1.62	1.24	0.98	0.80	0.66	0.55	0.17	0.08	0.05
2.00	1.00	0.968	8.13	1.03	1.12	1.28	1.56	2.00	2.72	3.89	5.66	7.59	8.00	6.38	4.41	2.99	1.10	1.53	1.16	0.91	0.73	0.60	0.50	0.15	0.07	0.04
2.25	0.89	0.975	9.12	0.91	0.99	1.15	1.40	1.82	2.52	3.71	5.69	8.20	9.00	6.94	4.54	2.95	2.01	1.44	1.08	0.83	0.66	0.54	0.45	0.14	0.06	0.04
2.50	0.80	0.980	10.1	0.82	0.90	1.04	1.27	1.66	2.33	3.52	5.66	8.74	10.00	7.44	4.60	2.88	1.92	1.35	1.00	0.77	0.61	0.50	0.41	0.12	0.06	0.04
2.75	0.73	0.983	11.1	0.75	0.82	0.94	1.16	1.53	2.16	3.34	5.57	9.19	11.0	7.88	4.62	2.80	1.82	1.27	0.93	0.72	0.57	0.46	0.38	0.11	0.05	0.03
3.00	0.67	0.986	12.1	0.69	0.75	0.87	1.07	1.41	2.02	3.16	5.45	9.57	12.0	8.25	4.60	2.70	1.73	1.20	0.87	0.67	0.53	0.43	0.35	0.10	0.05	0.03
3.25	0.61	0.988	13.1	0.63	0.69	0.80	0.99	1.31	1.89	2.99	5.31	9.88	13.0	8.57	4.55	2.60	1.65	1.13	0.82	0.62	0.49	0.40	0.33	0.09	0.05	0.03
3.50	0.57	0.990	14.1	0.59	0.64	0.75	0.92	1.23	1.77	2.84	5.15	10.1	14.0	8.84	4.48	2.50	1.56	1.06	0.77	0.58	0.46	0.37	0.31	0.09	0.04	0.03
3.75	0.53	0.991	15.1	0.55	0.60	0.70	0.86	1.15	1.67	2.69	5.00	10.3	15.0	9.06	4.40	2.41	1.49	1.01	0.73	0.55	0.43	0.35	0.29	0.08	0.04	0.02
4.00	0.50	0.992	16.1	0.51	0.56	0.65	0.81	1.08	1.57	2.56	4.84	10.4	16.0	9.23	4.30	2.31	1.42	0.95	0.69	0.52	0.41	0.33	0.27	0.08	0.04	0.02
4.25	0.47	0.993	17.1	0.48	0.53	0.62	0.76	1.02	1.49	2.44	4.68	10.5	17.0	9.36	4.20	2.22	1.35	0.91	0.65	0.49	0.38	0.31	0.26	0.07	0.04	0.02
4.50	0.44	0.994	18.1	0.46	0.50	0.58	0.72	0.97	1.41	2.33	4.52	10.6	18.0	9.46	4.10	2.14	1.29	0.86	0.62	0.47	0.36	0.29	0.24	0.07	0.03	0.02
4.75	0.42	0.994	19.1	0.43	0.47	0.55	0.69	0.92	1.35	2.22	4.37	10.6	19.0	9.52	3.99	2.06	1.24	0.82	0.59	0.44	0.35	0.29	0.23	0.07	0.03	0.02
5.00	0.40	0.995	20.1	0.41	0.45	0.52	0.65	0.87	1.28	2.13	4.23	10.6	20.0	9.56	3.89	1.98	1.18	0.79	0.56	0.42	0.33	0.27	0.22	0.06	0.03	0.02
6.00	0.33	0.997	24.0	0.34	0.38	0.44	0.54	0.73	1.08	1.82	3.71	10.3	24.0	9.48	3.48	1.71	1.01	0.67	0.47	0.36	0.28	0.22	0.18	0.05	0.03	0.02
7.00	0.29	0.997	28.0	0.29	0.32	0.38	0.47	0.63	0.93	1.58	3.28	9.83	28.0	9.18	3.13	1.51	0.88	0.58	0.41	0.31	0.24	0.19	0.16	0.04	0.02	0.01
8.00	0.25	0.998	32.0	0.26	0.28	0.33	0.41	0.55	0.82	1.39	2.94	9.28	32.0	8.77	2.82	1.34	0.78	0.51	0.36	0.27	0.21	0.17	0.14	0.04	0.02	0.01
9.00	0.22	0.999	36.0	0.23	0.25	0.29	0.36	0.49	0.73	1.24	2.65	8.73	36.0	8.32	2.57	1.20	0.70	0.45	0.32	0.24	0.19	0.15	0.12	0.03	0.02	0.01
10.00	0.20	0.999	40.0	0.21	0.23	0.26	0.33	0.44	0.66	1.13	2.41	8.19	40.0	7.87	2.35	1.09	0.69	0.41	0.30	0.22	0.17	0.13	0.11	0.03	0.02	0.01

The corresponding curve is plotted in Figure 7-19*c*. This curve can be denormalized by dividing the delay by 2π 1000 and multiplying the frequency axis by 1000, resulting in the final curve of Figure 7-19*d*. The differential delay variation over the band is about 675 μs.

The equalizer of Example 7-4 provides over a 2:1 reduction in the differential delay. Further equalization can be obtained with two addition equalizers to fill in the concave regions around 750 and 1250 Hz.

7.5 WIDEBAND 90° PHASE-SHIFT NETWORKS

Wideband 90° phase-shift networks have a single input and two output ports. Both outputs maintain a constant phase difference of 90° within a prescribed error over a wide range of frequencies. The overall transfer function is all-pass. These networks are widely used in the design of single-sideband systems and in other applications requiring 90° phase splitting.

Bedrosian (see the Bibliography) solved the approximation problem for this family of networks on a computer. The general structure is shown in Figure 7-20*a* and consists of *N* and *P* networks. Each network provides real-axis pole-zero pairs and is all-pass. The transfer function is of the form

$$T(s) = \frac{(s - \alpha_1)(s - \alpha_2)\cdots(s - \alpha_{n/2})}{(s + \alpha_1)(s + \alpha_2)\cdots(s + \alpha_{n/2})} \tag{7-74}$$

where $n/2$ is the order of the numerator and denominator polynomials. The total complexity of both networks is then n.

Real-axis all-pass transfer functions can be realized using a cascade of passive or active first-order sections. Both versions are shown in Figure 7-20*b* and *c*.

The transfer functions tabulated in Table 7-7 approximate a 90° phase difference in an equiripple manner. This approximation occurs within the bandwidth limits ω_L and ω_u, as shown in Figure 7-21. These frequencies are normalized so that $\sqrt{\omega_L\omega_u} = 1$. For a specified bandwidth ratio ω_u/ω_L, the individual band limits can be found from

$$\omega_L = \sqrt{\frac{\omega_L}{\omega_u}} \tag{7-75}$$

and

$$\omega_u = \sqrt{\frac{\omega_u}{\omega_L}} \tag{7-76}$$

As the total complexity n is made larger, the phase error decreases for a mixed bandwidth ratio or, for a fixed phase error, the bandwidth ratio will increase.

To use Table 7-7, first determine the required bandwidth ratio from the frequencies given. A network is then selected that has a bandwidth ratio ω_u/ω_L that exceeds the requirements, and a phase error $\pm\Delta\phi$ that is acceptable.

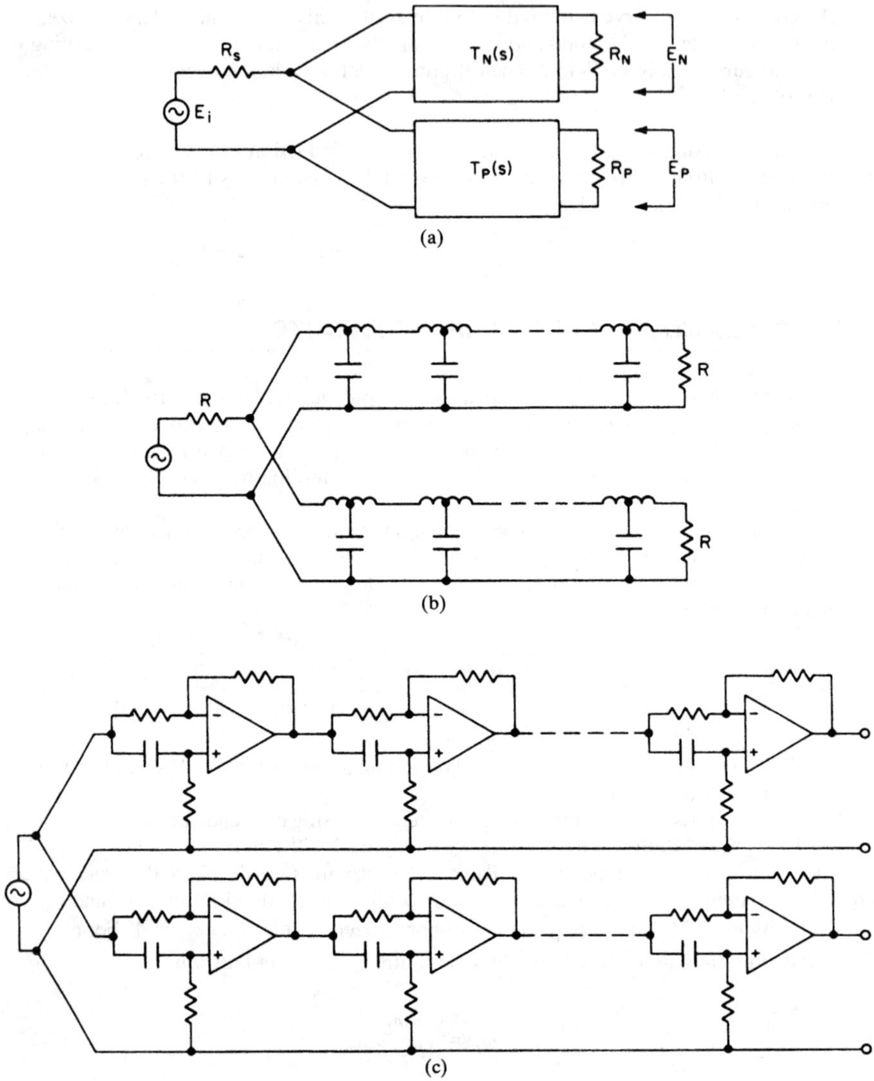

FIGURE 7-20 Wideband 90° phase-shift networks: (*a*) the general structure; (*b*) a passive realization; and (*c*) an active realization.

A frequency-scaling factor (FSF) is determined from

$$\text{FSF} = 2\pi f_0 \tag{7-77}$$

where f_0 is the geometric mean of the specified band limits or $\sqrt{f_L f_u}$. The tabulated α's are then multiplied by the FSF for denormalization. The resulting pole-zero pairs can be realized by a cascade of active or passive first-order sections for each network.

TABLE 7-7 Pole-Zero Locations for 90° Phase-Shift Networks*

n	$\Delta\phi$	α_N	α_P
		$\omega_u/\omega_L = 1146$	
6	6.84°	43.3862	8.3350
		2.0264	0.4935
		0.1200	0.0231
8	2.12°	59.7833	14.4159
		4.8947	1.6986
		0.5887	0.2043
		0.0694	0.0167
10	0.66°	75.8845	20.4679
		8.3350	3.5631
		1.5279	0.6545
		0.2807	0.1200
		0.0489	0.0132
		$\omega_u/\omega_L = 573.0$	
6	4.99°	34.3132	7.0607
		1.9111	0.5233
		0.1416	0.0291
8	1.39°	47.0857	11.8249
		4.3052	1.6253
		0.6153	0.2323
		0.0846	0.0212
10	0.39°	59.6517	16.5238
		7.0607	3.2112
		1.4749	0.6780
		0.3114	0.1416
		0.0605	0.0168
		$\omega_u/\omega_L = 286.5$	
4	13.9°	16.8937	2.4258
		0.4122	0.0592
6	3.43°	27.1337	5.9933
		1.8043	0.5542
		0.1669	0.0369
8	0.84°	37.0697	9.7136
		3.7944	1.5566
		0.6424	0.2636
		0.1030	0.0270
10	0.21°	46.8657	13.3518
		5.9933	2.8993
		1.4247	0.7019
		0.3449	0.1669
		0.0749	0.0213

(Continued)

TABLE 7-7 Pole-Zero Locations for 90° Phase-Shift Networks (*Continued*)

n	$\Delta\phi$	α_N	α_P
		$\omega_u/\omega_L = 143.2$	
4	10.2°	13.5875	2.2308
		0.4483	0.0736
8	0.46°	29.3327	8.0126
		3.3531	1.4921
		0.6702	0.2982
		0.1248	0.0341
10	0.10°	37.0091	10.8375
		5.1050	2.6233
		1.3772	0.7261
		0.3812	0.1959
		0.0923	0.0270
		$\omega_u/\omega_L = 81.85$	
4	7.58°	11.4648	2.0883
		0.4789	0.0918
6	1.38°	18.0294	4.5017
		1.6316	0.6129
		0.2221	0.0555
8	0.25°	24.4451	6.8929
		3.0427	1.4432
		0.6929	0.3287
		0.1451	0.0409
10	0.046°	30.7953	9.2085
		4.5017	2.4248
		1.3409	0.7458
		0.4124	0.2221
		0.1086	0.0325
		$\omega_u/\omega_L = 57.30$	
4	6.06°	10.3270	2.0044
		0.4989	0.0968
6	0.99°	16.1516	4.1648
		1.5873	0.6300
		0.2401	0.0619
8	0.16°	21.8562	6.2817
		2.8648	1.4136

TABLE 7-7 Pole-Zero Locations for 90° Phase-Shift Networks (*Continued*)

n	$\Delta\phi$	α_N	α_P
		0.7074	0.3491
		0.1592	0.0458
10	0.026°	27.5087	8.3296
		4.1648	2.3092
		1.3189	0.7582
		0.4331	0.2401
		0.1201	0.0364
		$\omega_u/\omega_L = 28.65$	
4	3.57°	8.5203	1.6157
		0.5387	0.1177
6	0.44°	13.1967	3.6059
		1.5077	0.6633
		0.2773	0.0758
8	0.056°	17.7957	5.2924
		2.5614	1.3599
		0.7354	0.3904
		0.1890	0.0562
10	0.0069°	22.3618	6.9242
		3.6059	2.1085
		1.2786	0.7821
		0.4743	0.2773
		0.1444	0.0447
		$\omega_u/\omega_L = 11.47$	
4	1.31°	5.9339	1.5027
		0.5055	0.1280
6	0.10°	10.4285	3.0425
		1.4180	0.7052
		0.3287	0.0959
8	0.0075°	14.0087	4.3286
		2.2432	1.2985
		0.7701	0.4458
		0.2310	0.0714

*Numerical values for this table is obtained from S. D. Bedrosian, "Normalized Design of 90° Phase-Difference Networks," *IRE Transactions on Circuit Theory,* June 1960.

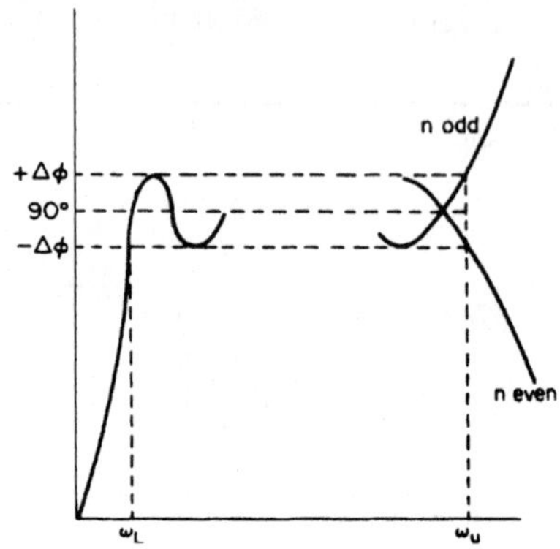

FIGURE 7-21 Wideband 90° phase-shift approximation.

The following example illustrates the design of a 90° phase-shift network.

Example 7-5 Design of an Active Wideband 90° Phase Splitter

Required:
Design a network having dual outputs which maintain a phase difference of 90° within ±0.2° over the frequency range of 300–3000 Hz. The circuit should be all-pass and active.

Result:

(*a*) Since a 10:1 bandwidth ratio is required (3000 Hz/300 Hz), the design corresponding to $n = 6$ and $\omega_u/\omega_L = 11.47$ is chosen. The phase-shift error will be ±0.1°.

(*b*) The normalized real pole-zero coordinates for both networks are given as follows:

P Network	*N* Network
$\alpha_1 = 10.4285$	$\alpha_4 = 3.0425$
$\alpha_2 = 1.4180$	$\alpha_5 = 0.7052$
$\alpha_3 = 0.3287$	$\alpha_6 = 0.0959$

The frequency-scaling factor is

$$\text{FSF} = 2\pi f_0 = 2\pi \times 948.7 = 5961 \tag{7-77}$$

where f_0 is $\sqrt{300 \times 3000}$. The pole-zero coordinates are multiplied by the FSF, resulting in the following set of denormalized values for α:

P Network	*N* Network
$\alpha_1' = 62164$	$\alpha_4' = 18136$
$\alpha_2' = 8453$	$\alpha_5' = 4204$
$\alpha_3' = 1959$	$\alpha_6' = 571.7$

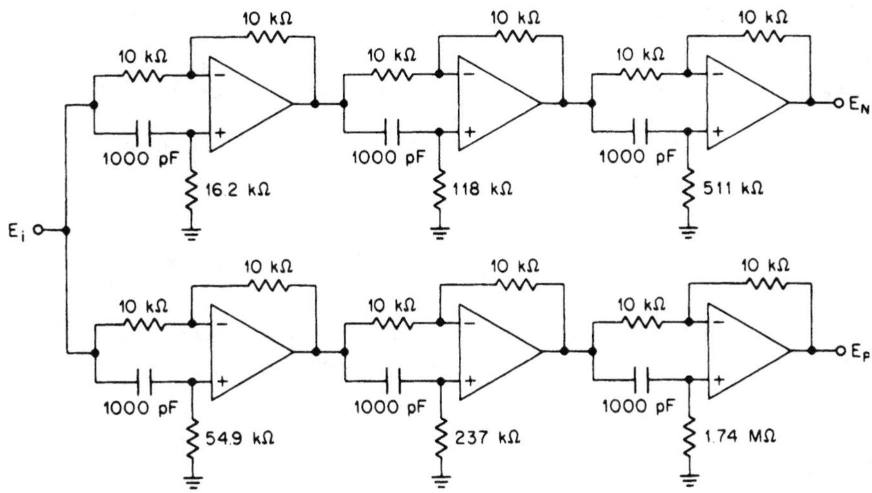

FIGURE 7-22 The wideband 90° phase-shift network of Example 7-5.

(*c*) The *P* and *N* networks can now be realized using the active first-order all-pass circuit of Section 7.2 and Figure 7-11*a*.

If we let $R' = 10\ \text{k}\Omega$ and $C = 1000$ pF, the value of *R* is given by

$$R = \frac{1}{\alpha_0 C} \tag{7-46}$$

Using the denormalized α's for the *P* and *N* networks, the following values are obtained:

Section	*P* Network	*N* Network
1	$R = 16.09\ \text{k}\Omega$	$R = 55.14\ \text{k}\Omega$
2	$R = 118.3\ \text{k}\Omega$	$R = 237.9\ \text{k}\Omega$
3	$R = 510.5\ \text{k}\Omega$	$R = 1.749\ \text{M}\Omega$

The final circuit is shown in Figure 7-22 using standard 1-percent resistor values.

7.6 ADJUSTABLE DELAY AND AMPLITUDE EQUALIZERS

Delay equalizers were discussed in Section 7.2 and applied to the delay equalization of filters in Section 7.4. Frequently, a transmission channel must be equalized to reduce the delay and amplitude variation. This process is called *line conditioning*. Since the initial parameters of lines vary and the line characteristics may change from time to time, the equalizer will consist of multiple sections where each stage is required to be adjustable.

LC Delay Equalizers

The circuit of Figure 7-23*a* illustrates a simplified adjustable *LC* delay equalizer section. The emitter and collector load resistors R_e and R_c are equal, so Q_1 serves as a phase splitter. Transistor Q_2 is an emitter follower output stage.

The equivalent circuit is shown in Figure 7-23*b*. The transfer function can be determined by superposition as

$$T(s) = \frac{s^2 - \dfrac{1}{RC}s + \dfrac{1}{LC}}{s^2 + \dfrac{1}{RC}s + \dfrac{1}{LC}} \tag{7-78}$$

This expression is of the same form as the general second-order all-pass transfer function of Equation (7-6). By equating coefficients, we obtain

$$\omega_r = \frac{1}{\sqrt{LC}} \tag{7-79}$$

and

$$Q = \omega_r RC \tag{7-80}$$

Equation (7-80) can be substituted in Equation (7-13) for the maximum delay of a second-order section, resulting in

$$T_{gd,\,\max} = 4RC \tag{7-81}$$

By making *R* variable, the delay can be directly controlled while retaining the all-pass properties. The peak delay will occur at or near the *LC* resonant frequency.

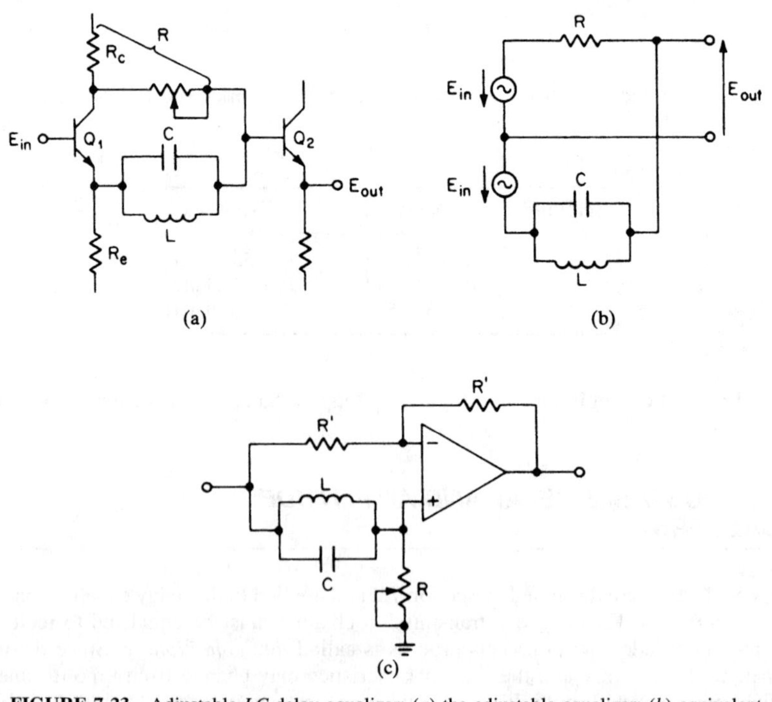

FIGURE 7-23 Adjustable *LC* delay equalizer: (*a*) the adjustable equalizer; (*b*) equivalent circuit; and (*c*) operational-amplifier realization.

The all-pass transfer function of Equation (7-78) can also be implemented using an operational amplifier. This configuration is shown in Figure 7-23*c*, where R' is arbitrary. Design Equations (7-79) through (7-81) still apply.

Example 7-6 Design of an Adjustable *LC* Delay Equalizer using the Two-Transistor Circuit

Required:

Design an adjustable *LC* delay equalizer using the two-transistor circuit of Figure 7-23*a*. The delay should be variable from 0.5 to 2.5 ms with a center frequency of 1700 Hz.

Result:

Using a capacitor C of 0.05 μF, the range of resistance R is given by

$$R_{\text{min}} = \frac{T_{gd,\text{max}}}{4C} = \frac{0.5 \times 10^{-3}}{4.05 \times 10^{-6}} = 2500\,\Omega \qquad (7\text{-}81)$$

$$R_{\text{max}} = \frac{2.5 \times 10^{-3}}{4 \times 0.05 \times 10^{-6}} = 12.5\,\text{k}\Omega$$

The inductor is computed by the general formula for resonance $\omega^2 LC = 1$, resulting in an inductance of 175 mH. The circuit is shown in Figure 7-24*a*. The emitter resistor R_e is

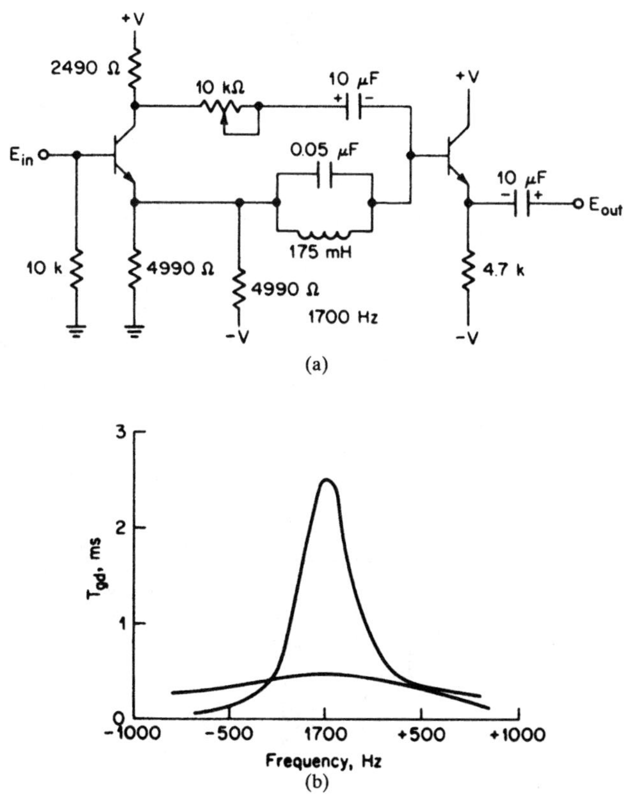

FIGURE 7-24 The adjustable delay equalizer of Example 7-6: (*a*) equalizer circuit; and (*b*) the delay adjustment range.

composed of two resistors for proper biasing of phase splitter Q_1, and electrolytic capacitors are used for DC blocking. The delay extremes are shown in the curves of Figure 7-24*b*.

LC ***Delay and Amplitude Equalizers.*** Frequently, the magnitude response of a transmission channel must be equalized along with the delay. An equalizer circuit featuring both adjustable amplitude and delay is shown in Figure 7-25*a*. Transistor Q_1 serves as a phase splitter where the signal applied to emitter follower Q_2 is K times the input signal. The equivalent circuit is illustrated in Figure 7-25*b*. The transfer function can be determined by superposition as

$$T(s) = \frac{s^2 - \dfrac{K}{RC}s + \dfrac{1}{LC}}{s^2 + \dfrac{K}{RC}s + \dfrac{1}{LC}} \tag{7-82}$$

If K is set equal to unity, the expression is then equivalent to Equation (7-78) corresponding to a second-order all-pass transfer function. As K increases or decreases from unity, a boost or null occurs at midfrequency with an asymptotic return to unity gain at DC and infinity.

The amount of amplitude equalization at midfrequency in decibels is given by

$$A_{\text{dB}} = 20 \log K \tag{7-83}$$

The maximum delay occurs at the LC resonant frequency and can be derived as

$$T_{gd,\max} = \frac{2RC}{K} + 2RC \tag{7-84}$$

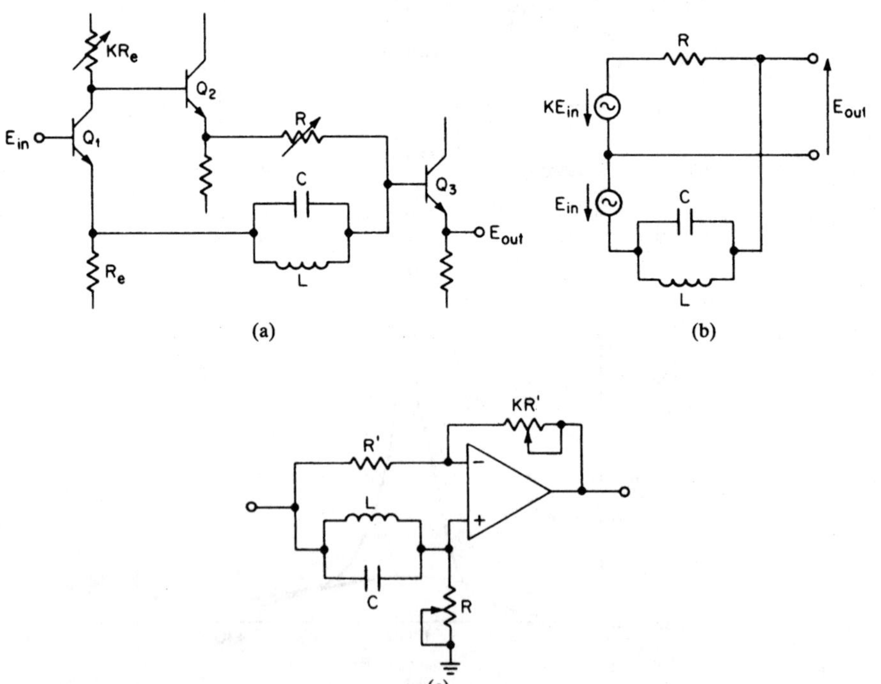

FIGURE 7-25 An adjustable *LC* delay and amplitude equalizer: (*a*) adjustable delay and amplitude equalizer; (*b*) equivalent circuit; and (*c*) an operational-amplifier realization.

If K is unity, Equation (7-84) reduces to 4 RC, which is equivalent to Equation (7-81) for the all-pass circuit of Figure 7-23.

An operational-amplifier implementation is also shown in Figure 7-25*c*. The value of R' is arbitrary, and Equations (7-83) and (7-84) are still applicable.

The following conclusions may be reached based on the evaluation of Equations (7-82) through (7-84):

1. The maximum delay is equal to 4 RC for $K = 1$, so R is a delay magnitude control.
2. The maximum delay will be minimally affected by a nonunity K, as is evident from Equation (7-84).
3. The amount of amplitude equalization at the LC resonant frequency is independent of the delay setting and is strictly a function of K. However, the selectivity of the amplitude response is a function of the delay setting and becomes more selective with increased delay.

The curves of Figure 7-26 show some typical delay and amplitude characteristics. The interaction between delay and amplitude is not restricted to LC equalizers and will occur whenever the same resonant element, either passive or active, is used to provide both the

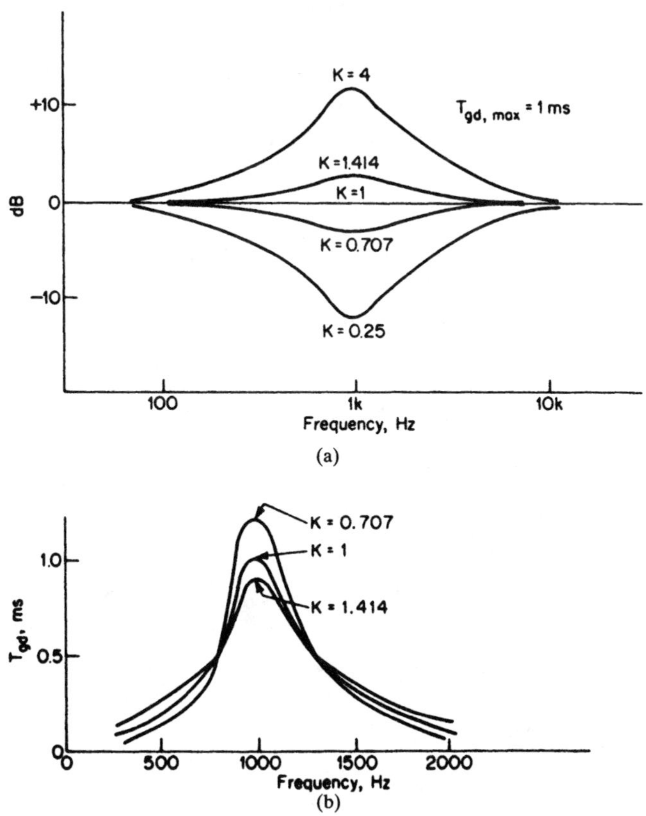

FIGURE 7-26 A typical delay and amplitude response for an LC delay and amplitude equalizer: (*a*) amplitude characteristics for a fixed delay; and (*b*) the delay variation for ±3 dB of amplitude equalization.

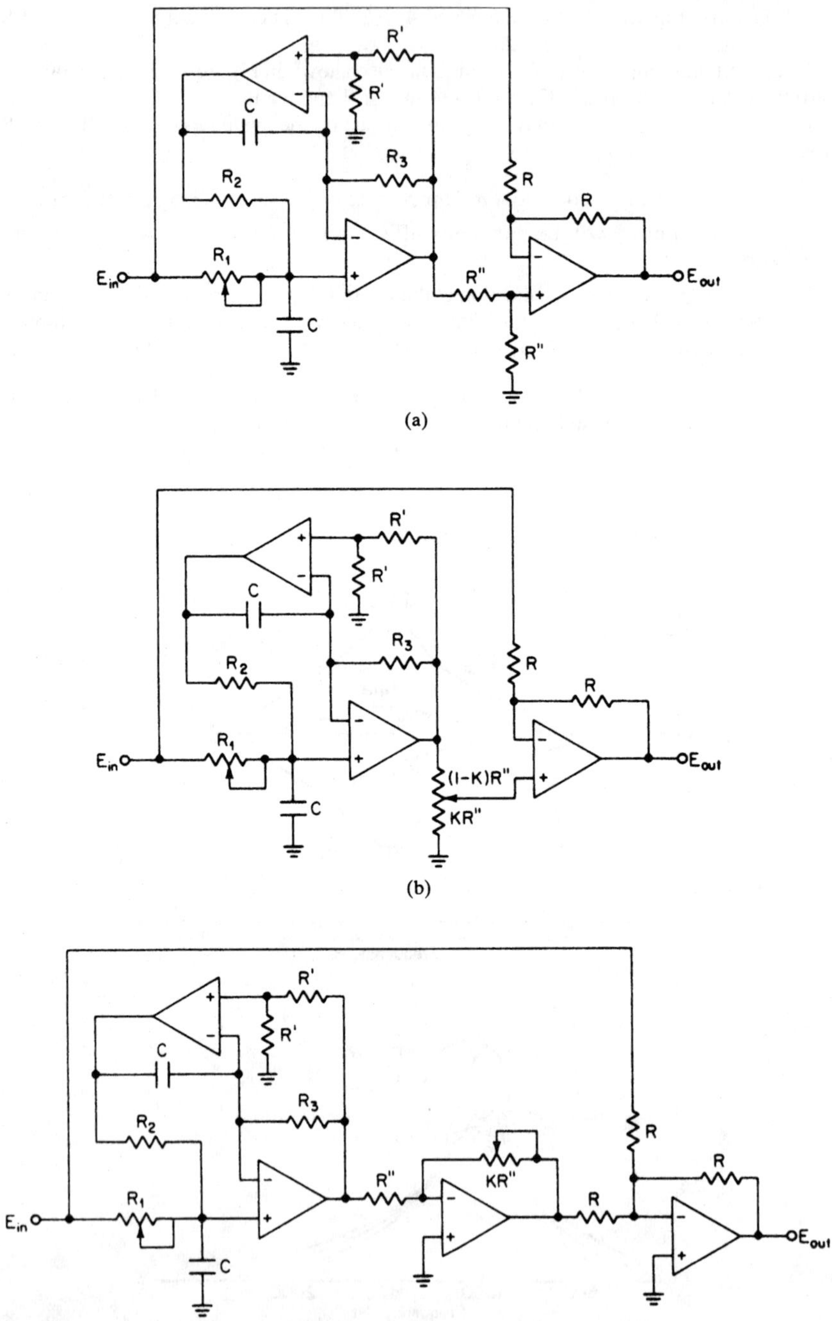

FIGURE 7-27 A DABP delay and amplitude equalizer: (*a*) adjustable delay equalizer; (*b*) adjustable delay and amplitude equalizer; and (*c*) adjustable delay and amplitude equalizer with extended amplitude range.

amplitude and delay equalization. However, for small amounts of amplitude correction, such as ±3 dB, the effect on the delay is minimal.

Active Delay and Amplitude Equalizers. The dual-amplifier bandpass (DABP) delay equalizer structure of Section 7.2, under "Active All-Pass Structures," and Figure 7-13 has a fixed gain and remains all-pass regardless of the design Q. If resistor R_1 is made variable, the Q, and therefore the delay, can be directly adjusted with no effect on resonant frequency or the all-pass behavior. The adjustable delay equalizer is shown in Figure 7-27*a*. The design equations are

$$T_{gd,\,\max} = 4R_1C \tag{7-85}$$

and

$$R_2 = R_3 = \frac{1}{\omega_r C} \tag{7-86}$$

where C, R, R', and R'', can be conveniently chosen. Resistor R_2 can be made variable for frequency trimming.

If amplitude equalization capability is also desired, a potentiometer can be introduced, resulting in the circuit of Figure 7-27*b*. The amplitude equation at ω_r is given by

$$A_{\text{dB}} = 20 \log (4K - 1) \tag{7-87}$$

where a K variation of 0.25 to 1 covers an amplitude equalization range of $-\infty$ to +9.5 dB.

To extend the equalization range above +9.5 dB, an additional amplifier can be introduced, as illustrated in Figure 7-27*c*. The amplitude equalization at ω_r is then obtained from

$$A_{\text{dB}} = 20 \log (2K - 1) \tag{7-88}$$

where a K variation of 0.5 to ∞ results in an infinite range of equalization capability. In reality, a ±15-dB maximum range has been found to be more than adequate for most equalization requirements.

Example 7-7 Design of an Active Adjustable Delay and Amplitude Equalizer

Required:

Design an adjustable active delay and amplitude equalizer that has a delay adjustment range of 0.5–3 ms, an amplitude range of ±12 dB, and a center frequency of 1000 Hz.

Result:

The circuit of Figure 7-27*c* will provide the required delay and amplitude adjustment capability.

If we choose $C = 0.01\ \mu\text{F}$ and $R = R' = R'' = 10\ \text{k}\Omega$, the element values are computed as follows:

$$R_{1,\min} = \frac{T_{gd,\max}}{4C} = \frac{0.5 \times 10^{-3}}{4 \times 10^{-8}} = 12.5\,\text{k}\Omega \tag{7-85}$$

$$R_{1,\max} = \frac{3 \times 10^{-3}}{4 \times 10^{-8}} = 75\,\text{k}\Omega$$

$$R_2 = R_3 = \frac{1}{\omega_r C} = \frac{1}{2\pi \times 1000 \times 10^{-8}} = 15.9\,\text{k}\Omega \tag{7-86}$$

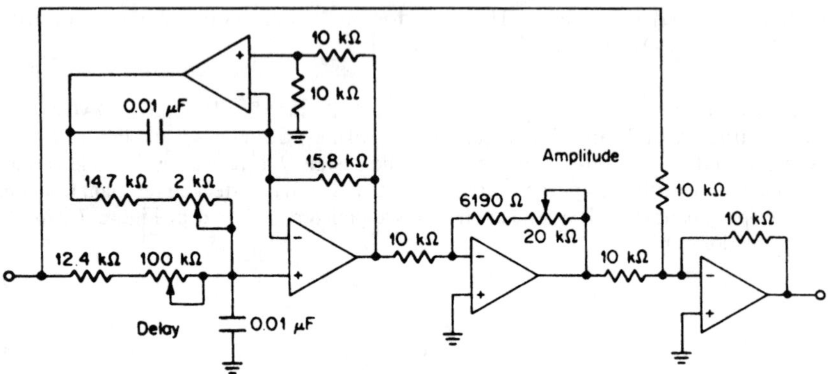

FIGURE 7-28 The adjustable delay and amplitude equalizer of Example 7-7.

The extreme values of K for ± 12 dB of amplitude equalization are found from

$$K = \frac{1}{2}\left[\log^{-1}\left(\frac{A_{\text{dB}}}{20}\right) + 1\right] \tag{7-88}$$

The range of K is then 0.626 to 2.49. The equalizer section is shown in Figure 7-28. Resistor R_2 has also been made adjustable for frequency trimming.

An active delay equalizer having adjustable delay was implemented by combining a second-order bandpass section with a summing amplifier. The bandpass section was required to have a fixed gain and a resonant frequency which were both independent of the Q setting. If amplitude equalization alone is needed, the bandpass section can operate with a fixed design Q. The low-complexity MFBP delay equalizer section of Figure 7-12*a* can then be used as an adjustable amplitude equalizer by making one of the summing resistors variable.

This circuit is shown in Figure 7-29*a*. The design equations are given by

$$R_2 = \frac{2Q}{\omega_r C} \tag{7-89}$$

$$R_{1a} = \frac{R_2}{2} \tag{7-90}$$

$$R_{1b} = \frac{R_{1a}}{2Q^2 - 1} \tag{7-91}$$

The amount of amplitude equalization at ω_r is computed from

$$A_{\text{dB}} = 20 \log\left(\frac{1}{K} - 1\right) \tag{7-92}$$

where K will range from 0 to 1 for an infinite range of amplitude equalization.

If the Q is below 0.707, the value of R_{1b} becomes negative, so the circuit of Figure 7-29*b* is used. R_2 is given by Equation (7-89), and R_1 is found from

$$R_1 = \frac{R_2}{4Q^2} \tag{7-93}$$

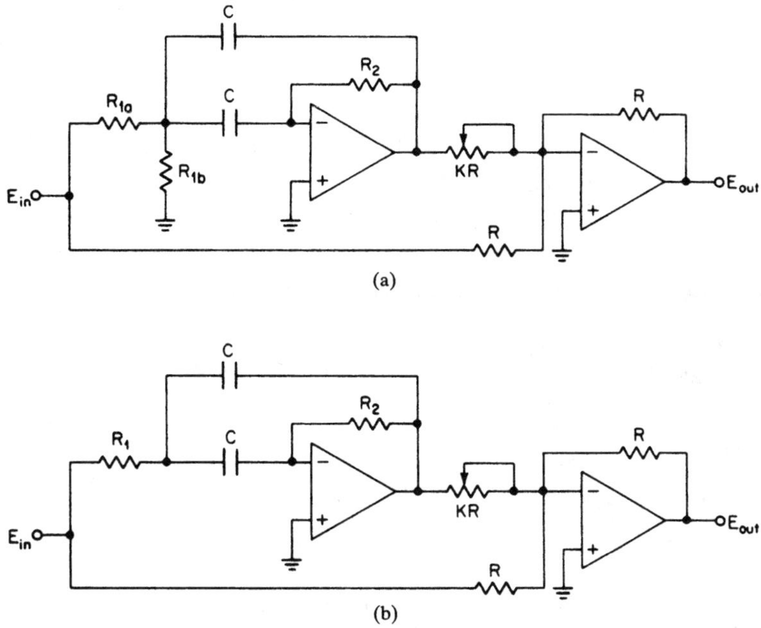

FIGURE 7-29 MFBP amplitude equalizer: (*a*) amplitude equalizer $0.707 < Q < 20$; and (*b*) amplitude equalizer $0 < Q < 20$.

The attenuation or boost at resonance is computed from

$$A_{\text{dB}} = 20 \log\left(\frac{2Q^2}{K} - 1\right) \tag{7-94}$$

The magnitude of Q determines the selectivity of the response in the region of resonance and is limited to values typically below 20 because of amplifier limitations.

If higher Qs are required, or if a circuit featuring independently adjustable Q and amplitude equalization is desired, the DABP circuits of Figure 7-27 may be used, where R_1 becomes the Q adjustment and is given by

$$R_1 = QR_2 \tag{7-95}$$

To compute the required Q of an amplitude equalizer, first define f_b, which is the frequency corresponding to one-half the pad loss (in decibels). The Q is then given by

$$Q = \frac{f_b b^2 \sqrt{K_r}}{f_r(b^2 - 1)} \tag{7-96}$$

where

$$K_r = \log^{-1}\left(\frac{A_{\text{dB}}}{20}\right) = 10^{A_{\text{dB}}/20} \tag{7-97}$$

and
$$b = \frac{f_b}{f_r} \tag{7-98}$$

or
$$b = \frac{f_r}{f_b} \tag{7-99}$$

whichever b is greater than unity.

Example 7-8 Design of an Active Fixed Amplitude Equalizer

Required:

Design a fixed active amplitude equalizer that provides a +12-dB boost at 3200 Hz and has a boost of +6 dB at 2500 Hz.

Result:

(*a*) First, compute

$$K_r = 10^{A_{\text{dB}/20}} = 10^{12/20} = 3.98 \tag{7-97}$$

and
$$b = \frac{f_r}{f_b} = \frac{3200\text{ Hz}}{2500\text{ Hz}} = 1.28 \tag{7-99}$$

The Q is then found from

$$Q = \frac{f_b b^2 \sqrt{K_r}}{f_r(b^2 - 1)} = \frac{2500 \times 1.28^2 \sqrt{3.98}}{3200(1.28^2 - 1)} = 4.00 \tag{7-96}$$

(*b*) The MFBP amplitude equalizer circuit of Figure 7-29*a* will be used. Using a C of 0.0047 μF and an R of 10 kΩ, the element values are given by

$$R_2 = \frac{2Q}{\omega_r C} = \frac{2 \times 4}{2\pi 3200 \times 4.7 \times 10^{-9}} = 84.6\text{ k}\Omega \tag{7-89}$$

$$R_{1a} = \frac{R_2}{2} = 42.3\text{ k}\Omega \tag{7-90}$$

$$R_{1b} = \frac{R_{1a}}{2Q^2 - 1} = \frac{42.3 \times 10^3}{2 \times 4^2 - 1} = 1365\ \Omega \tag{7-91}$$

$$K = \frac{1}{1 + 10^{A_{\text{dB}/20}}} = \frac{1}{1 + 10^{12/20}} = 0.200 \tag{7-92}$$

The equalizer circuit and corresponding frequency response are shown in Figure 7-30.

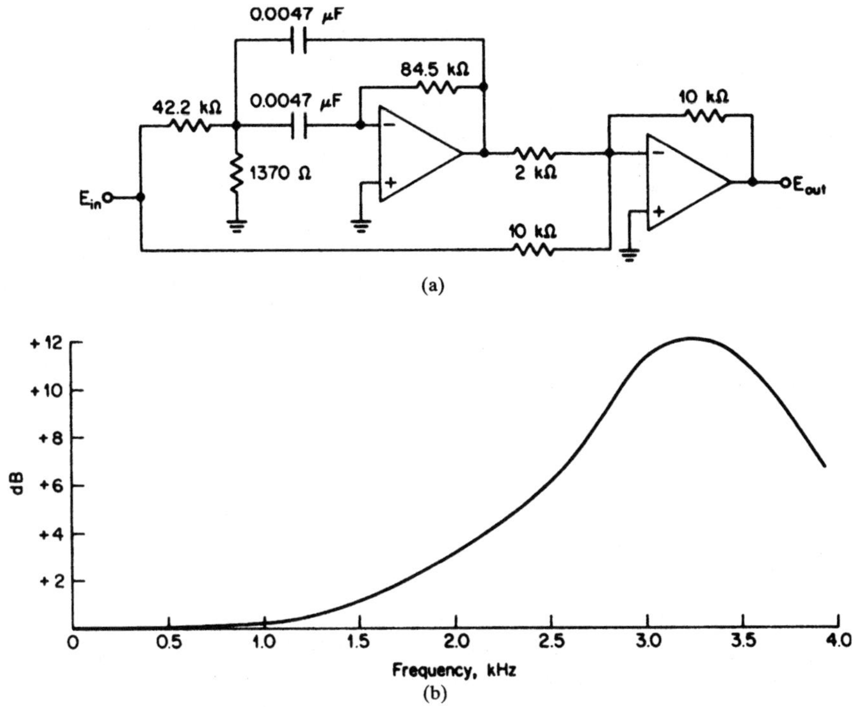

FIGURE 7-30 The amplitude equalizer of Example 7-8: (a) amplitude equalizer circuit; and (*b*) frequency response.

BIBLIOGRAPHY

Bedrosian, S. D. "Normalized Design of 90° Phase-Difference Networks." *IRE Transactions on Circuit Theory* CT-7 (June, 1960).

Geffe, P. R. *Simplified Modern Filter Design*. New York: John F. Rider, 1963.

Lindquist, C. S. *Active Network Design*. Long Beach, California: Steward and Sons, 1977.

Wallis, C. M. "Design of Low-Frequency Constant Time Delay Lines." *AIEE Proceedings* 71 (1952).

Williams, A. B. "An Active Equalizer with Adjustable Amplitude and Delay." *IEEE Transactions on Circuit Theory* CT-16 (November, 1969).

CHAPTER 8
REFINEMENTS IN *LC* FILTER DESIGN AND THE USE OF RESISTIVE NETWORKS

8.1 INTRODUCTION

The straightforward application of the design techniques outlined for *LC* filters will not always result in practical element values or desirable circuit configurations. Extreme cases of impedance or bandwidth can produce designs which may be extremely difficult or even impossible to realize. This chapter is concerned mainly with circuit transformations so that impractical designs can be transformed into alternate configurations having the identical response and using more practical elements. Also, the use of resistive networks to supplement *LC* filters or function independently is covered.

8.2 TAPPED INDUCTORS

An extremely useful tool for eliminating impractical element values is the transformer. As the reader may recall from introductory AC circuit analysis, a transformer having a turns ratio N will transform an impedance by a factor of N^2. A parallel element can be shifted between the primary and secondary at will, provided that its impedance is modified by N^2.

Figure 8-1 illustrates how a tapped inductor is used to reduce the value of a resonating capacitor. The tuned circuit of Figure 8-1*a* is first modified by introducing an impedance step-up transformer, as shown in Figure 8-1*b*, so that capacitor C can be moved to the secondary and reduced by a factor of N^2. This can be carried a step further, resulting in the circuit of Figure 8-1*c*. The transformer has been absorbed as a continuation of the inductor, resulting in an autotransformer. The ratio of the overall inductance to the tap inductance becomes N^2.

As an example, let's modify the tuned circuit of Example 5-4, shown in Figure 8-2*a*. To reduce the capacitor from 0.354 μF to 0.027 μF, the overall inductance is increased by the impedance ratio 0.354 μF/0.027 μF, resulting in the circuit of Figure 8-2*b*. The resonant frequency remains unchanged since the overall *LC* product is still the same.

As a further example, let's consider *LC* elliptic-function low-pass filters. The parallel resonant circuits may also contain high-capacity values which can be reduced by this method. Figure 8-3 shows a section of a low-pass filter. To reduce the resonating capacitor to 0.1 μF, the overall inductance is increased by the factor 1.055 μF/0.1 μF and a tap is provided at the original inductance value.

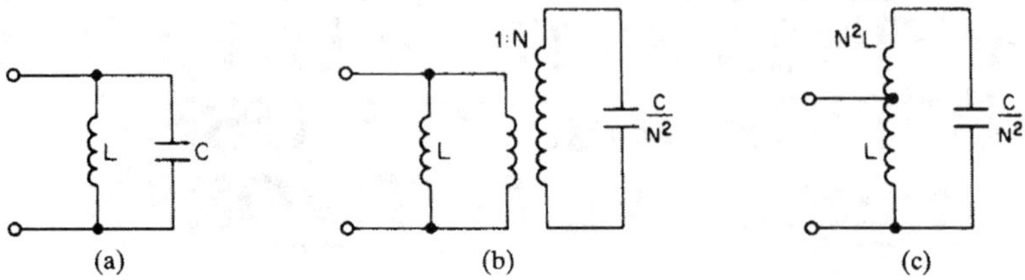

FIGURE 8-1 The tapped inductor: (*a*) basic tuned circuit; (*b*) the introduction of a transformer; and (*c*) absorbed transformer.

The tapped coil is useful not only for reducing resonating capacitors, but also for transforming entire sections of a filter including terminations. The usefulness of the tapped inductor is limited only by the ingenuity and resourcefulness of the designer. Figure 8-4 illustrates some applications of this technique using designs from previous examples. In the case of Figure 8-4*a*, where a tapped coil enables operation from unequal terminations, the same result could have been achieved using Bartlett's bisection theorem or other methods (see Section 3.1). However, the transformer approach results in maximum power transfer (minimum insertion loss). The circuits of Figure 8-4*b* and *c* demonstrate how element values can be manipulated by taps. The tapped inductance values shown are all measured from the grounded end of the shunt inductors. Series branches can be manipulated up or down in impedance level by multiplying the shunt inductance taps on both sides of the branch by the desired impedance-scaling factor.

Transformers or autotransformers are by no means ideal. Imperfect coupling within the magnetic structure will result in a leakage inductance which can cause spurious responses at higher frequencies, as shown in Figure 8-5. These effects can be minimized by using near-unity

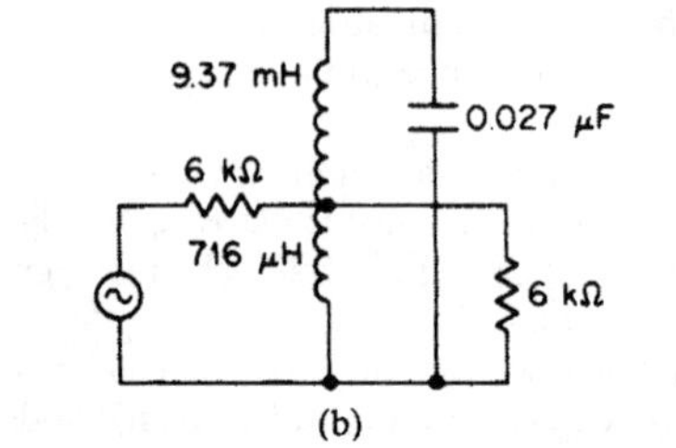

FIGURE 8-2 Reducing the resonant capacitor value: (*a*) tuned circuit; and (*b*) modified circuit.

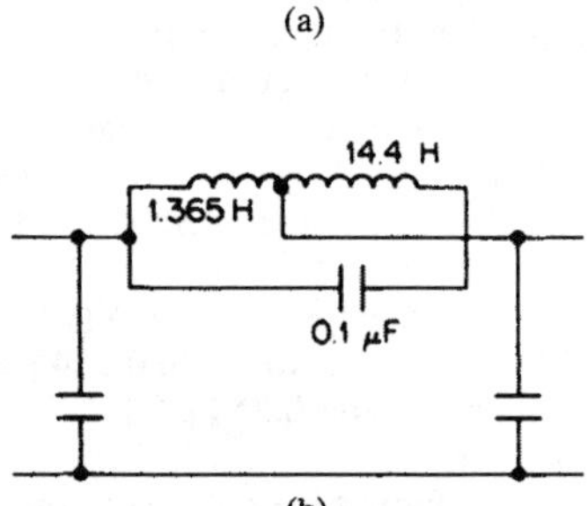

FIGURE 8-3 The application of a tapped inductor in elliptic-function low-pass filters: (*a*) a filter section; and (*b*) tapped inductor.

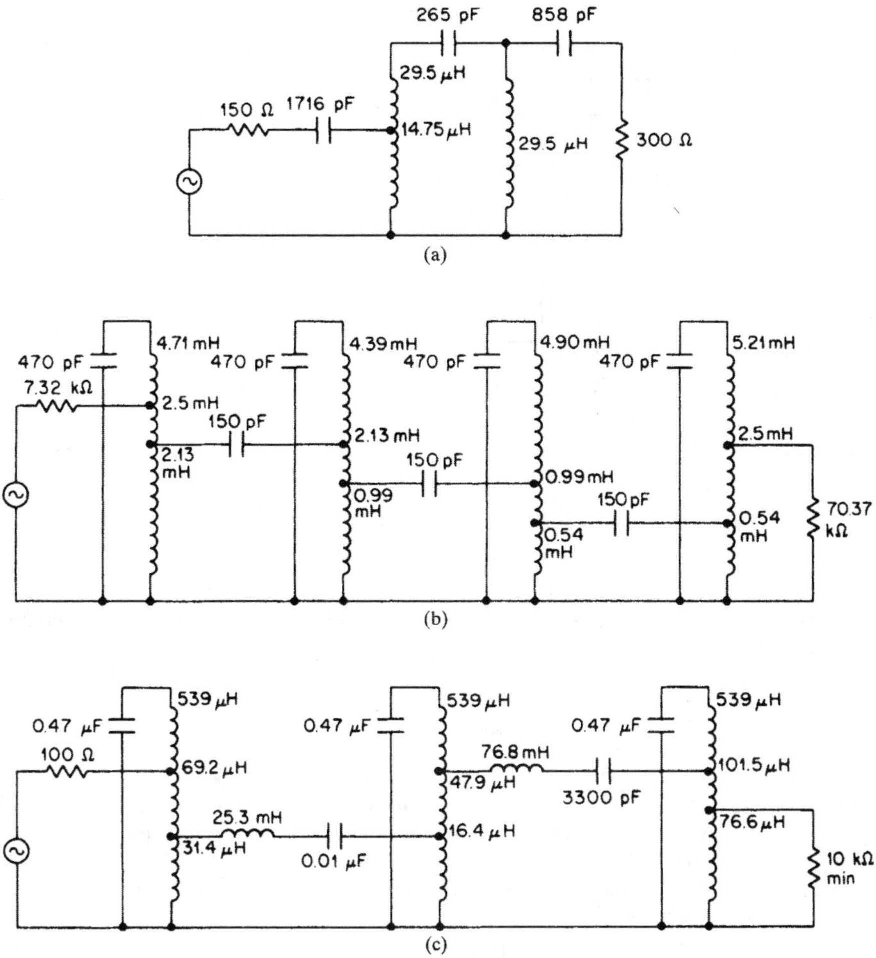

FIGURE 8-4 Applications of tapped inductors: (*a*) the high-pass filter of Example 4-1 modified for unequal terminations; (*b*) the filter of Example 5-7 modified for standard capacitor values; and (*c*) the filter of Example 5-8 modified for standard capacitor values.

turns ratios. Another solution is to leave a portion of the original capacity at the tap for high-frequency bypassing. This method is shown in Figure 8-6.

8.3 CIRCUIT TRANSFORMATIONS

Circuit transformations fall into two categories: equivalent circuits or narrowband approximations. The impedance of a circuit branch can be expressed as a ratio of two polynomials in *s*, similar to a transfer function. If two branches are equivalent, their impedance expressions are identical. A narrowband approximation to a particular filter branch is valid only over a small frequency range. Outside of this region, the impedances depart considerably.

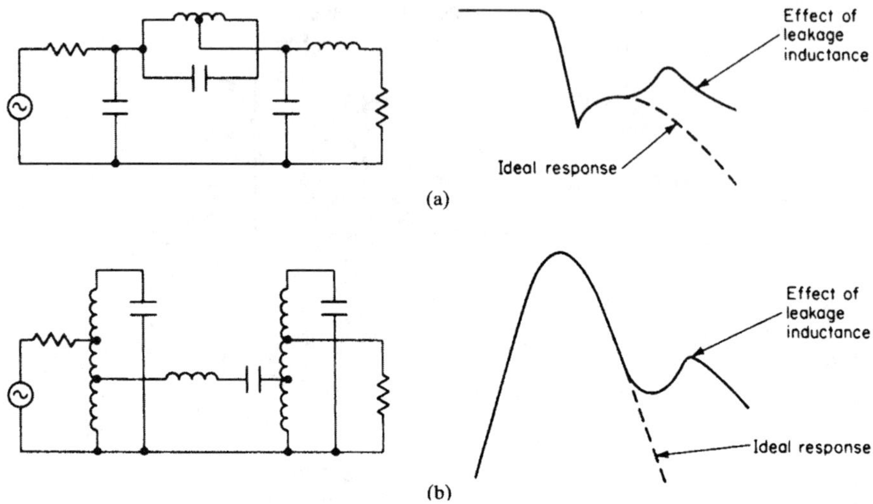

FIGURE 8-5 Spurious responses from leakage inductance: (*a*) in a low-pass filter; and (*b*) in a bandpass filter.

Thus, the filter response is affected. As a result, narrowband approximations are essentially limited to small percentage bandwidth bandpass filters.

Norton's Capacitance Transformer

Let's consider the circuit of Figure 8-7*a*, consisting of impedance Z interconnected between impedances Z_1 and Z_2. If it is desired to raise impedance Z_2 by a factor of N^2 without disturbing an overall transfer function (except for possibly a constant multiplier), a transformer can be introduced, as shown in Figure 8-7*b*.

Determinant manipulation can provide us with an alternate approach. The nodal determinant of a two-port network is given by

$$\begin{vmatrix} Y_{11} & -Y_{12} \\ -Y_{21} & Y_{22} \end{vmatrix}$$

where Y_{11} and Y_{22} are the input and output nodal admittance, respectively, and Y_{12} and Y_{21} are the transfer admittances, which are normally equal to each other.

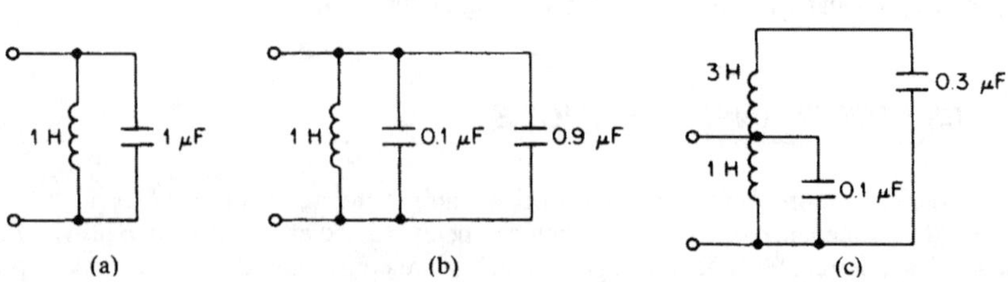

FIGURE 8-6 Preventing spurious response from leakage inductance: (*a*) initial circuit; (*b*) split capacity; and (*c*) transformed circuit.

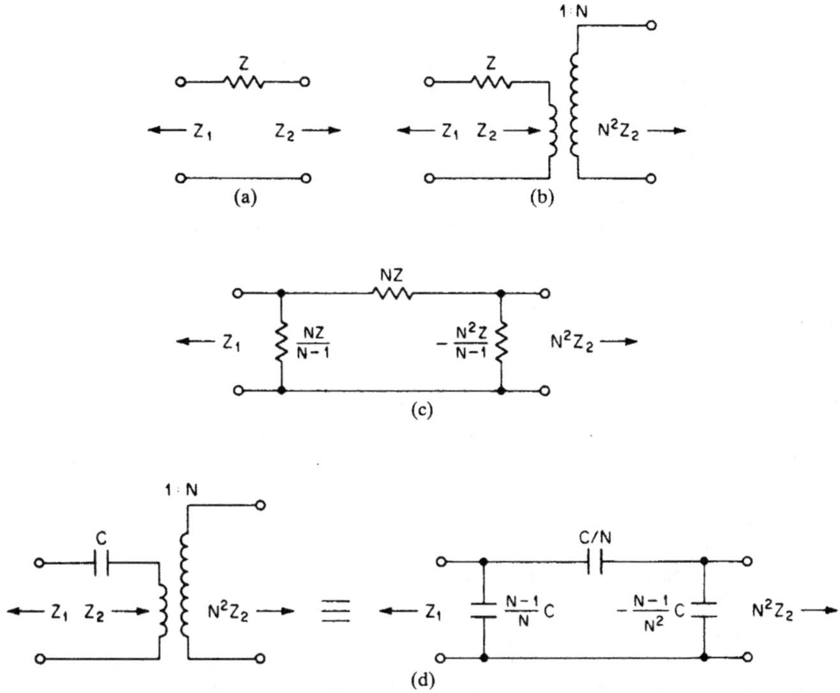

FIGURE 8-7 Norton's capacitance transformer: (*a*) a general two-port network; (*b*) transformer step-up of output impedance; (*c*) the Norton impedance transformation; and (*d*) the Norton capacitance transformation.

If we consider the two-port network of Figure 8-7*a*, the nodal determinant becomes

$$\begin{vmatrix} \frac{1}{Z_1} + \frac{1}{Z} & -\frac{1}{Z} \\ -\frac{1}{Z} & \frac{1}{Z_2} + \frac{1}{Z} \end{vmatrix}$$

To raise the impedance of the output or Y_{22} node by N^2, the second row and second column are multiplied by $1/N$, resulting in

$$\begin{vmatrix} \frac{1}{Z_1} + \frac{1}{Z} & -\frac{1}{NZ} \\ -\frac{1}{NZ} & \frac{1}{N^2 Z_2} + \frac{1}{N^2 Z} \end{vmatrix}$$

This determinant corresponds to the circuit of Figure 8-7*c*. The Y_{11} total nodal admittance is unchanged, and the Y_{22} total nodal admittance has been reduced by N^2, or the impedance has been increased by N^2. This result was originated by Norton and is called *Norton's transformation*.

If the element Z is a capacitor C, this transformation can be applied to obtain the equivalent circuit of Figure 8-7*d*. This transformation is important since it can be used to modify the impedance on one side of a capacitor by a factor of N^2 without a transformer. However, the output shunt capacitor introduced is negative. A positive capacitor must then be present external to the network so that the negative capacitance can be absorbed.

If an N^2 of less than unity is used, the impedance at the output node will be reduced. The shunt capacitor at the input node will then become negative and must be absorbed by an external positive capacitor across the input.

The following Example illustrates the use of the capacitance transformer.

Example 8-1 Using the Capacitance Transformation to Lower Inductor Values

Required:

Using the capacitance transformation, modify the bandpass filter circuit of Figure 5-3*c* so that the 1.91-H inductor is reduced to 100 mH. The source and load impedances should remain 600 Ω.

Result:

(*a*) The circuit to be transformed is shown in Figure 8-8*a*. To facilitate the capacitance transformation, the 0.01329 μF series capacitor is split into two equal capacitors of twice the value and redrawn in Figure 8-8*b*.

To reduce the 1.91-H inductor to 100 mH, first lower the impedance of the network to the right of the dashed line in Figure 8-8*b* by a factor of 100 mH/1.91 H, or 0.05236. Using the capacitance transformation of Figure 8-7*d*, where $N^2 = 0.05236$, the circuit of Figure 8-8*c* is obtained where the input negative capacitor has been absorbed.

(*b*) To complete the transformation, the output node must be transformed back up in impedance to restore the 600-Ω termination. Again using the capacitance transformation with an N^2 of 600 Ω/31.42 Ω or 19.1, the final circuit of Figure 8-8*d* is obtained. Because of the symmetrical nature of the circuit of Figure 8-8*b*, both capacitor transformations are also symmetrical.

(*c*) Each parallel resonant circuit is tuned by opening the inductors of the adjacent series resonant circuits, and each series resonant circuit is resonated by shorting the inductors of the adjacent parallel tuned circuits, as shown in Figure 8-8*e*.

Narrowband Approximations

A narrowband approximation to a circuit branch consists of an alternate network which is theoretically equivalent only at a single frequency. Nevertheless, good results can be obtained with bandpass filters having small percentage bandwidths typically of up to 20 percent.

The series and parallel *RL* and *RC* circuits of Table 8-1 are narrowband approximations which are equivalent at ω_0. This frequency is generally set equal to the bandpass center frequency in Equations (8-1) through (8-8). These equations were derived simply by determining the expressions for the network impedances and then equating the real parts and the imaginary parts to solve for the resistive and reactive components, respectively.

Narrowband approximations can be used to manipulate the source and load terminations of bandpass filters. If a parallel *RC* network is converted to a series *RC* circuit, it is apparent from Equation (8-8) that the resistor value decreases. When we apply this approximation to a bandpass filter having a parallel resonant circuit as the terminating branch, the source or load resistor can be made smaller. To control the degree of reduction so that a desired termination can be obtained, the shunt capacitor is first subdivided into two capacitors where only one capacitor is associated with the termination.

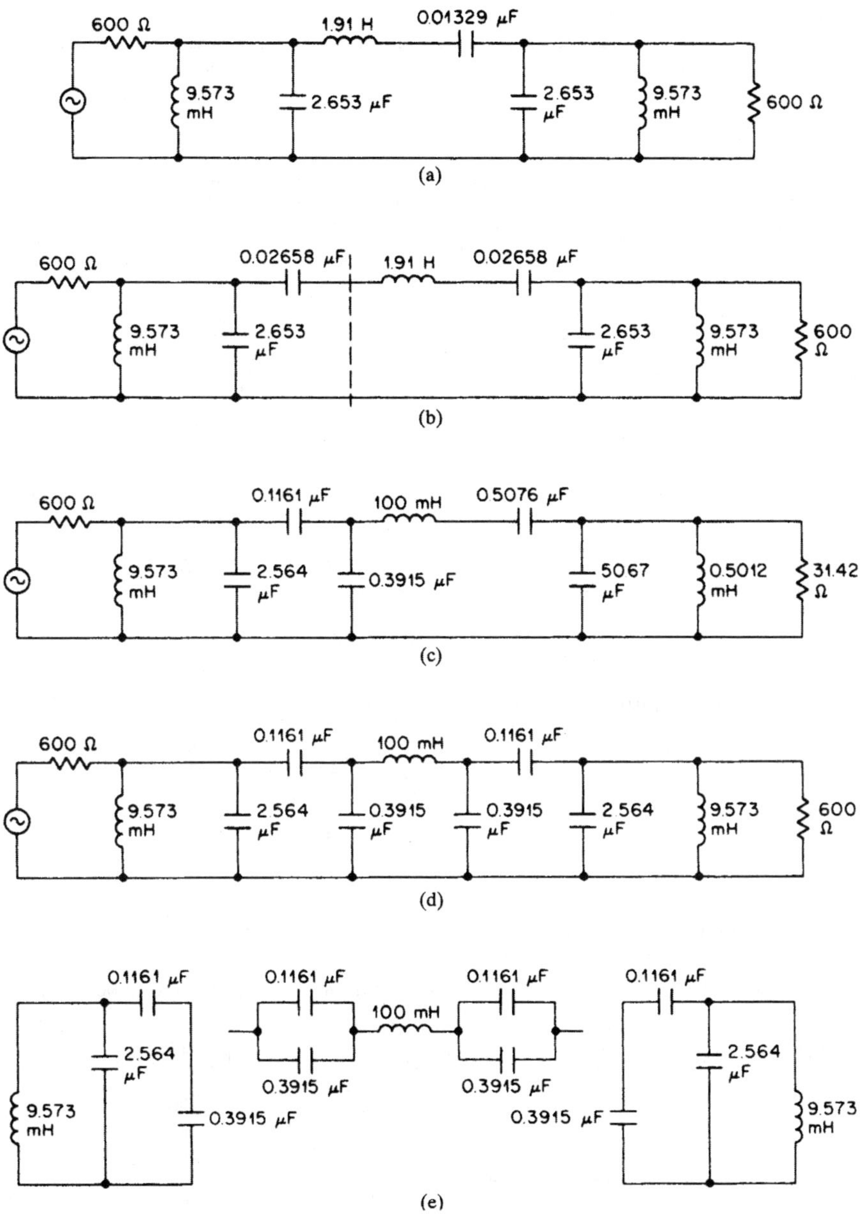

FIGURE 8-8 Capacitance transformation applied to the filter of Example 5-2: (*a*) the bandpass filter of Example 5-2; (*b*) split series capacitors; (*c*) the reduction of the 1.91-H inductor using capacitance transformation; (*d*) the restoration of the 600-Ω output impedance using capacitance transformation; and (*e*) equivalent circuits for tuning.

TABLE 8-1 Narrowband Approximations

Circuit	Design Equations	
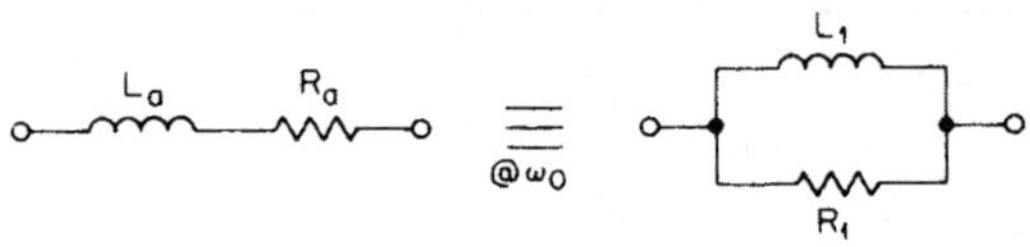	$L_1 = L_a + \frac{R_a^2}{\omega_0^2 L_a}$	(8-1)
	$R_1 = R_a + \frac{\omega_0^2 L_a^2}{R_a}$	(8-2)
	$L_a = \frac{L_1 R_1^2}{R_1^2 + \omega_0^2 L_1^2}$	(8-3)
	$R_a = \frac{\omega_0^2 L_1^2 R_1}{R_1^2 + \omega_0^2 L_1^2}$	(8-4)
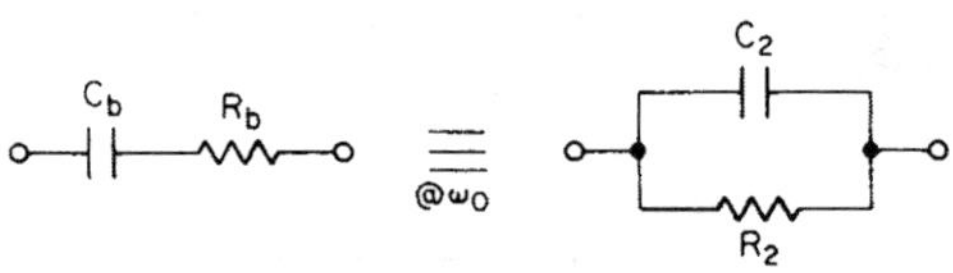	$C_2 = \frac{C_b}{1 + \omega_0^2 C_b^2 R_b^2}$	(8-5)
	$R_2 = R_b + \frac{1}{\omega_0^2 C_b^2 R_b}$	(8-6)
	$C_b = C_2 + \frac{1}{\omega_0^2 R_2^2 C_2}$	(8-7)
	$R_b = \frac{R_2}{1 + \omega_0^2 C_2^2 R_2^2}$	(8-8)

These results are illustrated in Figure 8-9. The element values are given by

$$C_2 = \frac{1}{\omega_0 \sqrt{R_1 R_2 - R_2^2}} \tag{8-9}$$

and

$$C_1 = C_T - \frac{1}{\omega_0}\sqrt{\frac{R_1 - R_2}{R_1^2 R_2}} \tag{8-10}$$

where the restrictions $R_2 < R_1$ and $(R_1 - R_2)/(R_1^2 R_2) < \omega_0^2 C_T^2$ apply.

Example 8-2 Using a Narrowband Transformation to Lower Source Impedance

Required:
Modify the 100 kHz bandpass filter of Example 5.7 for a source impedance of 600 Ω.

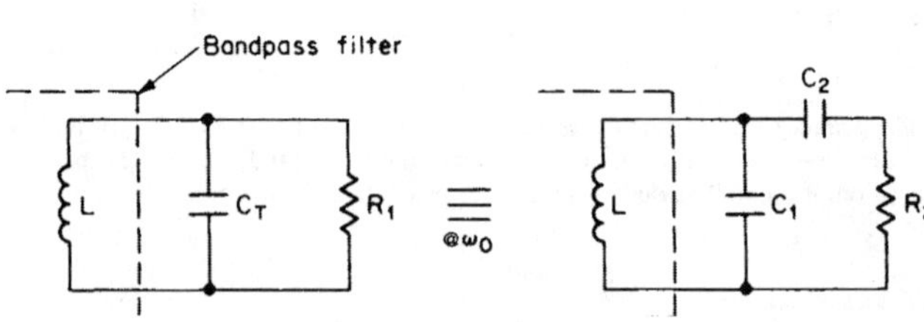

FIGURE 8-9 A narrowband transformation of terminations.

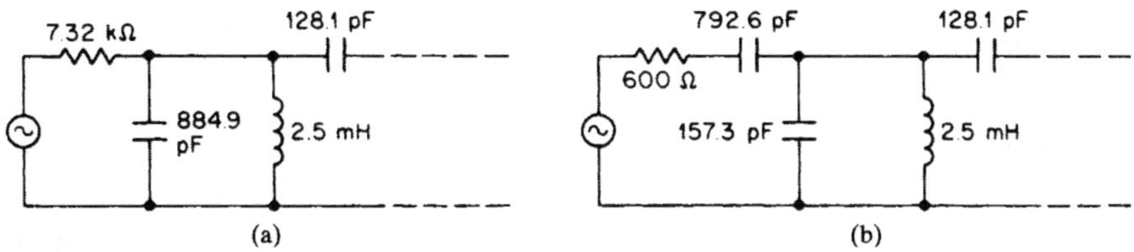

FIGURE 8-10 The narrowband source transformation of Example 8-2: (*a*) source input to the filter of Example 5-7; and (*b*) transformed source.

Result:

The filter is shown in Figure 8-10*a*. If we use the narrowband source transformation of Figure 8-9, the values are given by

$$C_2 = \frac{1}{\omega_0\sqrt{R_1R_2 - R_2^2}} = \frac{1}{2\pi \times 10^5\sqrt{7.32 \times 6 \times 10^5 - 600^2}}$$

$$= 792.6 \text{ pF} \tag{8-9}$$

$$C_1 = C_T - \frac{1}{\omega_0}\sqrt{\frac{R_1 - R_2}{R_1^2R_2}} = 884.9 \times 10^{-12}$$

$$-\frac{1}{2\pi \times 10^5}\sqrt{\frac{7.32 \times 10^3 - 600}{7320^2 \times 600}} = 157.3 \text{ pF} \tag{8-10}$$

The resulting filter is illustrated in Figure 8-10*b*.

8.4 DESIGNING WITH PARASITIC CAPACITANCE

As a first approximation, inductors and capacitors are considered pure lumped reactive elements. Most physical capacitors are nearly perfect reactances. Inductors, on the other hand, have impurities which can be detrimental in many cases. In addition to the highly critical resistive losses, distributed capacity across the coil will occur because of interturn capacitance of the coil winding and other stray capacities involving the core. The equivalent circuit of an inductor is shown in Figure 8-11.

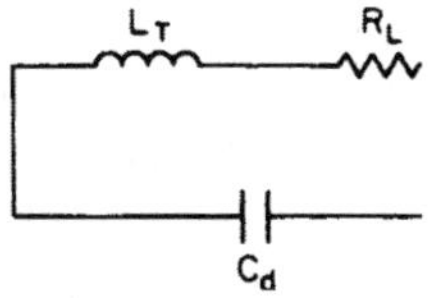

FIGURE 8-11 Equivalent circuit of an inductor.

The result of this distributed capacitance is to create the effect of a parallel resonant circuit instead of an inductor. If the coil is to be located in shunt with an external capacitance, the external capacitor value can be decreased accordingly, thus absorbing the distributed capacitance.

The distributed capacity across the inductor in a series resonant circuit causes parallel resonances resulting in nulls in the frequency response. If the self-resonant frequency is too low, the null may even occur in the passband, thus severely distorting the expected response.

To determine the effective inductance of a practical inductor, the coil is resonated to the frequency of interest with an external capacitor, and the effective inductance is

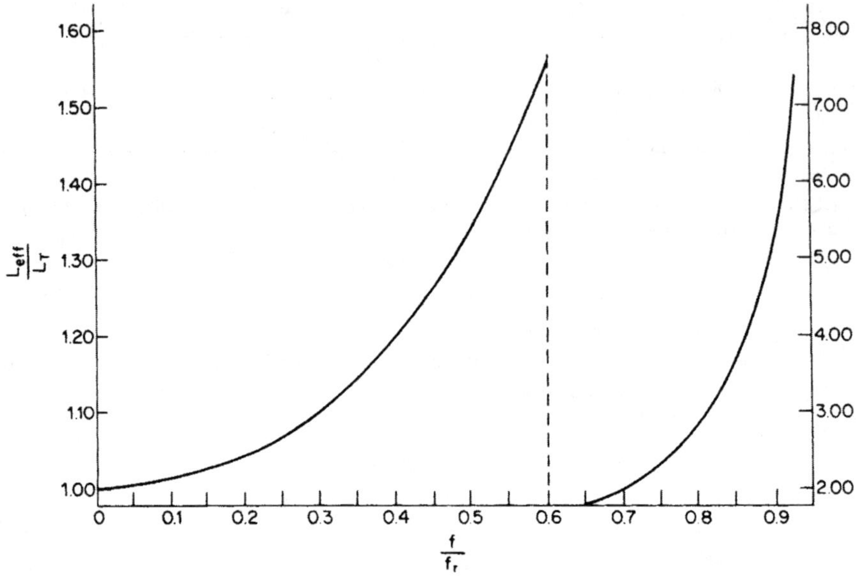

FIGURE 8-12 The effective inductance with frequency.

calculated using the standard formula for resonance. The effective inductance can also be found from

$$L_{\text{eff}} = \frac{L_T}{1 - \left(\dfrac{f}{f_r}\right)^2} \tag{8-11}$$

where L_T is the true (low-frequency) inductance, f is the frequency of interest, and f_r is the inductor's self-resonant frequency. As f approaches f_r, the value of L_{eff} will increase quite dramatically and will become infinite at self-resonance. Equation (8-11) is plotted in Figure 8-12.

To compensate for the effect of distributed capacity in a series resonant circuit, the true inductance L_T can be appropriately decreased so that the effective inductance given by Equation (8-11) is the required value. However, the Q of a practical series resonant circuit is given by

$$Q_{\text{eff}} = Q_L\left[1 - \left(\frac{f}{f_r}\right)^2\right] \tag{8-12}$$

where Q_L is the Q of the inductor as determined by the series losses (that is, $Q_L = \omega L_T/R_L$). The effective Q is therefore reduced by the distributed capacity.

Distributed capacity is determined by the mechanical parameters of the core and winding, and as a result is subject to change due to mechanical stresses, and so on. Therefore, for maximum stability, the distributed capacity should be kept as small as possible. Techniques for minimizing inductor capacity are discussed in Chapter 9.

Another form of parasitic capacity is stray capacitance between the circuit nodes and ground. These strays may be especially harmful at high frequencies and with high-impedance nodes. In the case of low-pass filters where the circuit nodes already have shunt

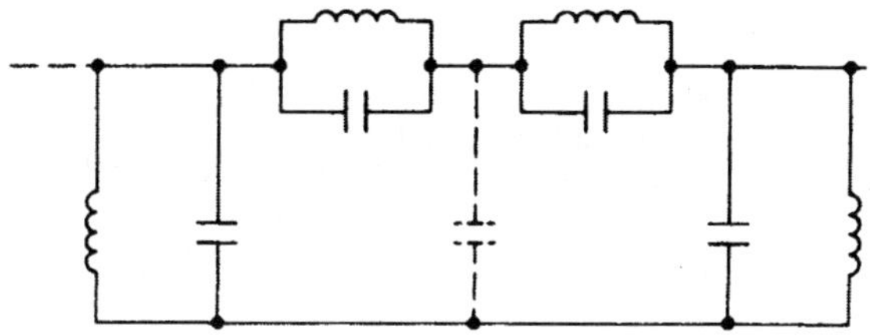

FIGURE 8-13 An elliptic-function bandpass filter.

capacitors to ground, these strays can usually be neglected, especially when the impedance levels are low.

A portion of an elliptic-function bandpass filter is shown in Figure 8-13. The stray capacity at nodes not connected to ground by a design capacitor may cause problems since these nodes have high impedances.

Geffe (see Bibliography) has derived a transformation to introduce a design capacitor from the junction of the parallel tuned circuits to the ground. The stray capacity can then be absorbed. The design of elliptic-function bandpass filters was discussed in Section 5.1. This transformation is performed upon the filter while it is normalized to a 1-rad/s center frequency and a 1-Ω impedance level. A section of the normalized network is shown in Figure 8-14*a*. The transformation proceeds as follows:

Choose an arbitrary value of $m < 1$, then

$$n = 1 - \frac{L_b}{L_c} \frac{1 - m}{m^2} \tag{8-13}$$

$$C_0 = \frac{1 - n}{n^2 L_c} - \frac{1 - m}{mn^2 L_b} \tag{8-14}$$

$$C_1 = \frac{1}{L_a} - \frac{1 - n}{nL_c} \tag{8-15}$$

$$C_2 = \frac{1}{nL_C} \tag{8-16}$$

$$C_3 = \frac{1}{mn^2 L_b} \tag{8-17}$$

$$C_4 = \frac{1 - m}{m^2 n^2 L_b} + \frac{1}{n^2 L_d} \tag{8-18}$$

$$L_1 = \frac{L_a}{1 - \frac{L_a}{L_b} \frac{1 - n}{n}} \tag{8-19}$$

$$L_2 = nL_b \tag{8-20}$$

$$L_3 = mn^2 L_c \tag{8-21}$$

$$L_4 = \frac{m^2 n^2 L_c}{1 + \frac{L_c}{L_d} \frac{m^2}{1 - m}} \tag{8-22}$$

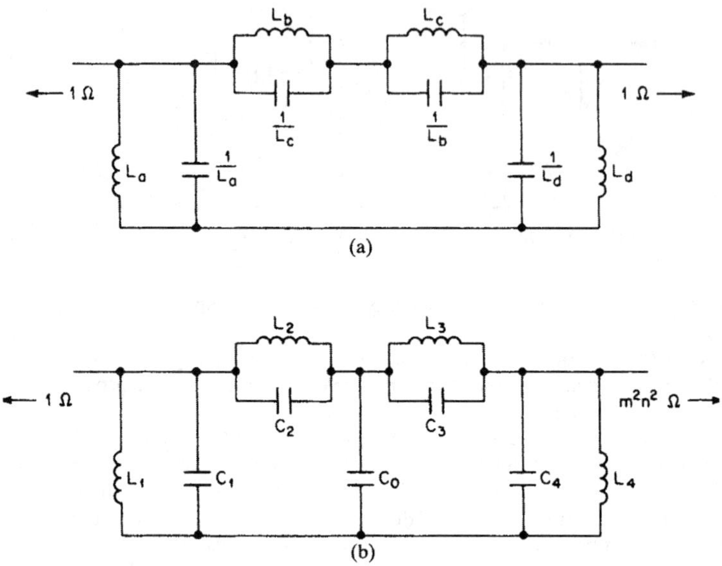

FIGURE 8-14 A transformation to absorb stray capacitance: (*a*) normalized filter section; and (*b*) transformed circuit.

The resulting network is given in Figure 8-14*b*. The output node has been transformed to an impedance level of m^2n^2 Ω. Therefore, all the circuitry to the right of this node, up to and including the termination, must be impedance-scaled by this same factor. The filter is subsequently denormalized by scaling to the desired center frequency and impedance level.

8.5 AMPLITUDE EQUALIZATION FOR INADEQUATE Q

Insufficient element Q will cause a sagging or rounding of the frequency response in the region of cutoff. Some typical cases are shown in Figure 8-15, where the solid curve represents the theoretical response. Finite Q will also result in less rejection in the vicinity of any stopband zeros and increased filter insertion loss.

Amplitude-equalization techniques can be applied to compensate for the sagging response near cutoff. A passive amplitude equalizer will not actually "boost" the corner response,

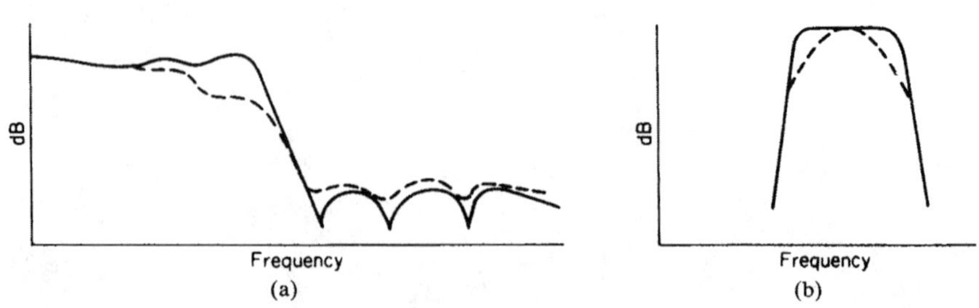

FIGURE 8-15 The effects of insufficient Q: (*a*) low-pass response; and (*b*) bandpass response.

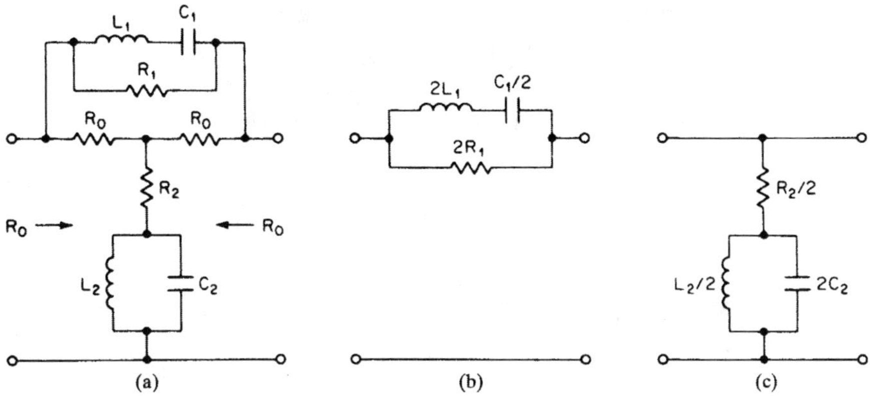

FIGURE 8-16 Bandpass-type amplitude equalizers: (*a*) constant-impedance type; (*b*) series nonconstant-impedance type; and (*c*) shunt nonconstant-impedance type.

since a gain cannot be achieved as with active equalizer circuits. However, the equalizer will introduce attenuation except in the region of interest, therefore resulting in a boost in terms of the relative response.

Amplitude equalizers used for low Q compensation are of the bandpass type. They have either constant-impedance or nonconstant-impedance characteristics. The constant-impedance types can be cascaded with each other and the filter with no interaction. The nonconstant-impedance equalizer sections are less complex but will result in some interaction when cascaded with other networks. However, for a boost of 1 or 2 dB, these effects are usually minimal and can be neglected.

Both types of equalizers are shown in Figure 8-16. The nonconstant-impedance type can be used in either the series or shunt form. In general, the shunt form is preferred since the resonating capacitor may be reduced by tapping the inductor.

To design a bandpass equalizer, the following characteristics must be determined from the curve to be equalized:

A_{dB} = total amount of equalization required in decibels
f_r = frequency corresponding to A_{dB}
f_b = frequency corresponding to $A_{\mathrm{dB}}/2$

These parameters are illustrated in Figure 8-17, where the corner response and corresponding equalizer are shown for both upper and lower cutoff frequencies.

To design the equalizer, first compute K from

$$A_{\mathrm{dB}} = 20 \log K \qquad (8\text{-}23)$$

Then calculate b where

$$b = \frac{f_r}{f_b} \qquad (8\text{-}24)$$

or

$$b = \frac{f_b}{f_r} \qquad (8\text{-}25)$$

selecting whichever b is greater than unity.

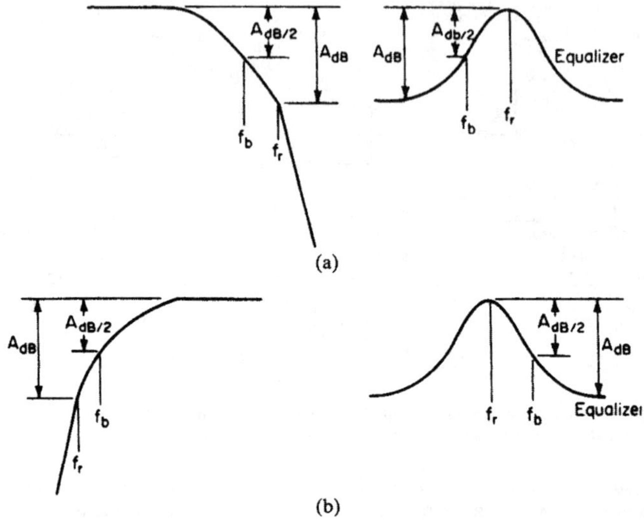

FIGURE 8-17 Bandpass equalization of the corner response: (*a*) equalization of the upper cutoff; and (*b*) equalization of the lower cutoff.

The element values corresponding to the sections of Figure 8-16 are found as follows:

$$L_1 = \frac{R_0(K - 1)}{2\pi f_b(b^2 - 1)\sqrt{K}} \tag{8-26}$$

$$C_1 = \frac{1}{(2\pi f_r)^2 L_1} \tag{8-27}$$

$$L_2 = \frac{R_0(b^2 - 1)\sqrt{K}}{2\pi f_b b^2(K - 1)} \tag{8-28}$$

$$C_2 = \frac{1}{(2\pi f_r)^2 L_2} \tag{8-29}$$

$$R_1 = R_0(K - 1) \tag{8-30}$$

$$R_2 = \frac{R_0}{K - 1} \tag{8-31}$$

where R_0 is the terminating impedance of the filter.

To equalize a low-pass or high-pass filter, a single equalizer is required at the cutoff. For bandpass or band-reject filters, a pair of equalizer sections is needed for the upper and lower cutoff frequencies.

The following example illustrates the design of an equalizer to compensate for low Q.

Example 8-3 Using an Amplitude Equalizer to Compensate for Low Q

Required:

A low-pass filter should have a theoretical roll-off of 0.1 dB at 2975 Hz but has instead the following response in the vicinity of cutoff due to insufficient Q:

2850 Hz	−0.5 dB
2975 Hz	−1.0 dB

Design a shunt nonconstant-impedance equalizer to restore the sagging response. The filter impedance level is 1000 Ω.

Result:

First, make the following preliminary computations:

$$K = 10^{A_{\text{dB}}/20} = 10^{1/20} = 1.122 \tag{8-23}$$

$$b = \frac{f_r}{f_b} = \frac{2975\text{ Hz}}{2850\text{ Hz}} = 1.0439 \tag{8-24}$$

then

$$L_2 = \frac{R_0(b^2 - 1)\sqrt{K}}{2\pi f_b b^2(K - 1)} = \frac{10^3(1.0439^2 - 1)\sqrt{1.122}}{2\pi 2850 \times 1.0439^2(1.122 - 1)} = 40.0\text{ mH} \tag{8-28}$$

$$C_2 = \frac{1}{(2\pi f_r)^2 L_2} = \frac{1}{(2\pi 2975)^2 \times 0.04} = 0.0715\ \mu\text{F} \tag{8-29}$$

$$R_2 = \frac{R_0}{K - 1} = \frac{1000}{1.122 - 1} = 8197\ \Omega \tag{8-31}$$

The resulting equalizer is shown in Figure 8-18 using the circuit of Figure 8-16*c*.

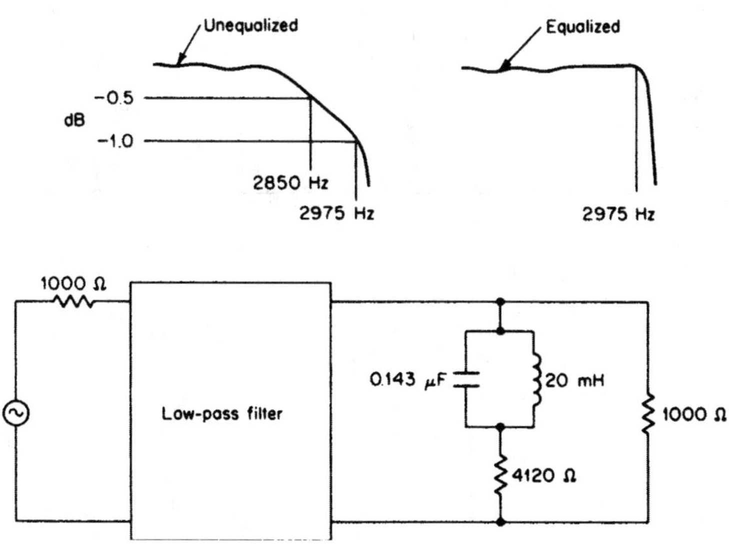

FIGURE 8-18 The equalizer of Example 8-3.

8.6 COIL-SAVING ELLIPTIC-FUNCTION BANDPASS FILTERS

If an even-order elliptic-function low-pass filter (as shown in Figure 8-19*a*) is transformed into a bandpass filter using the methods of Section 5.1, the bandpass circuit of Figure 8-19*b* is obtained.

A method has been developed to transform the low-pass filter into the configuration of Figure 8-19*c*. The transfer function is unchanged except for a constant multiplier, and 1/2 $(n-2)$ coils are saved in comparison with the conventional transformation. These structures are called minimum-inductance or zigzag bandpass filters. However, this transformation requires a very large number of calculations (see Saal and Ulbrich in Bibliography) and is therefore considered impractical without a computer.

Geffe (see Bibliography) has presented a series of formulas so that this transformation can be performed on an $n = 4$ low-pass network. The normalized low-pass element values for $n = 4$ can be determined with the *Filter Solutions* program or found in either Zverev's *Handbook of Filter Synthesis* or Saal's "Der Entwurf von Filtern mit Hilfe des Kataloges

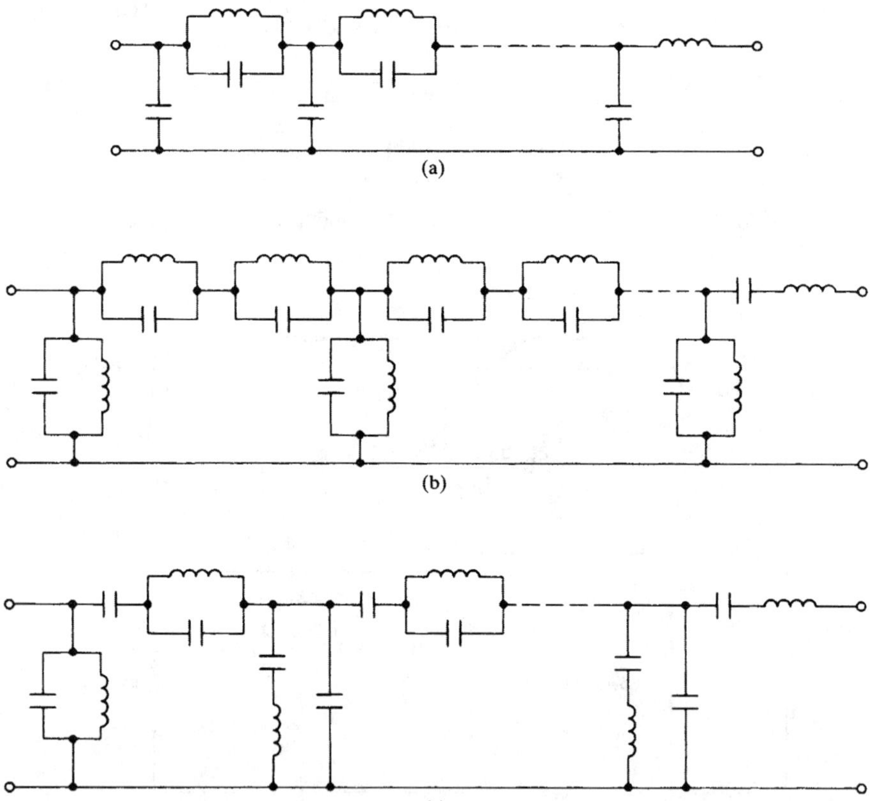

FIGURE 8-19 Coil-saving bandpass transformation: (*a*) an even order elliptic-function low-pass filter; (*b*) conventional bandpass transformation; and (*c*) minimum-inductance bandpass transformation.

Normierter Tiefpasse" (see Bibliography). Although these calculations are laborious, the EXCEL spreadsheet of formulas on the CD-ROM performs these computations.

The low-pass filter and corresponding bandpass network are shown in Figure 8-20. The following preliminary computations are required.

$$Q_{bp} = \frac{f_0}{BW} \tag{8-32}$$

$$a = \frac{\omega_\infty}{2Q_{bp}} \tag{8-33}$$

$$x = a + \sqrt{a^2 + 1} \tag{8-34}$$

$$t_1 = 1 + \frac{C_3}{C_2} \tag{8-35}$$

$$T = \frac{1 + t_1 x^2}{t_1 + x^2} \tag{8-36}$$

$$k = \frac{Q_{bp} T}{t_1} \tag{8-37}$$

$$t_2 = \frac{x^2}{x^2 + t_1} \tag{8-38}$$

$$t_3 = \frac{t_1 t_2}{T} \tag{8-39}$$

$$\alpha = 1 - \frac{1}{x^2} \tag{8-40}$$

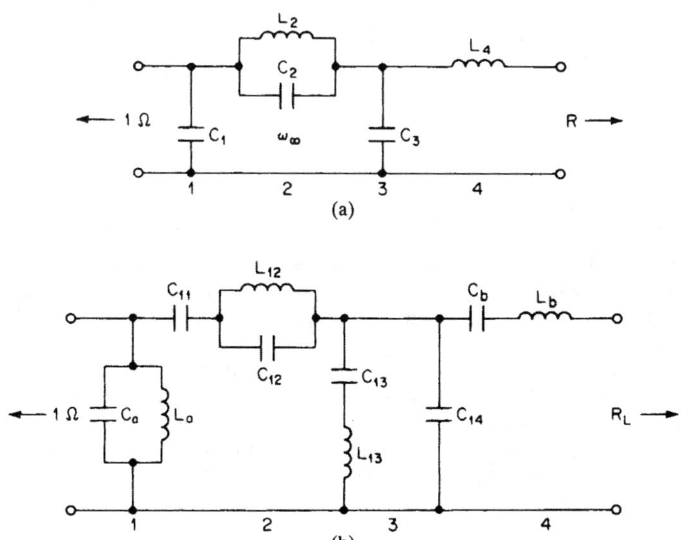

FIGURE 8-20 The minimum-inductance transformation for $n = 4$: (*a*) an $n = 4$ low-pass filter; and (*b*) the transformed bandpass filter.

$$\beta = x^2 - 1 \tag{8-41}$$

$$A = \frac{C_3 k\alpha}{T} \tag{8-42}$$

$$B = \frac{t_2 t_3}{C_3 k\beta} \tag{8-43}$$

The bandpass element values can now be computed as follows:

$$R_L = t_3^2 R \tag{8-44}$$

$$C_{11} = \frac{C_3 k\beta}{t_1 t_2} \tag{8-45}$$

$$C_{12} = \frac{C_3 k\alpha}{T - 1} \tag{8-46}$$

$$L_{12} = \frac{1}{x^2 C_{12}} \tag{8-47}$$

$$C_{13} = \frac{C_{11}(T - 1)}{t_2} \tag{8-48}$$

$$L_{13} = \frac{x^2}{C_{13}} \tag{8-49}$$

$$C_{14} = \frac{C_3 k\alpha}{t_2} \tag{8-50}$$

$$L_a = \frac{1}{Q_{\text{bp}}\left(C_1 + \dfrac{C_3}{t_1}\right)} \tag{8-51}$$

$$C_a = \frac{1}{L_a} - A \tag{8-52}$$

$$L_b = t_3^2 Q_{\text{bp}} L_4 \tag{8-53}$$

$$C_b = \frac{1}{L_b - B} \tag{8-54}$$

The bandpass filter of Figure 8-20*b* must be denormalized to the required impedance level and center frequency f_0. Since the source and load impedance levels are unequal, either the tapped inductor or the capacitance transformation can be used to obtain equal terminations if required.

The transmission zero above the passband is provided by the parallel resonance of $L_{12}C_{12}$ in branch 2, and the lower zero corresponds to the series resonance of $L_{13}C_{13}$ in branch 3. The circuits of branches 2 and 3 each have conditions of both series and parallel resonance and can be transformed from one form to the other. The following equations relate the type 1 and 2 networks shown in Figure 8-21.

For a type 1 network:

$$L_1 = L_a\left(1 + \frac{C_a}{C_b}\right)^2 \tag{8-55}$$

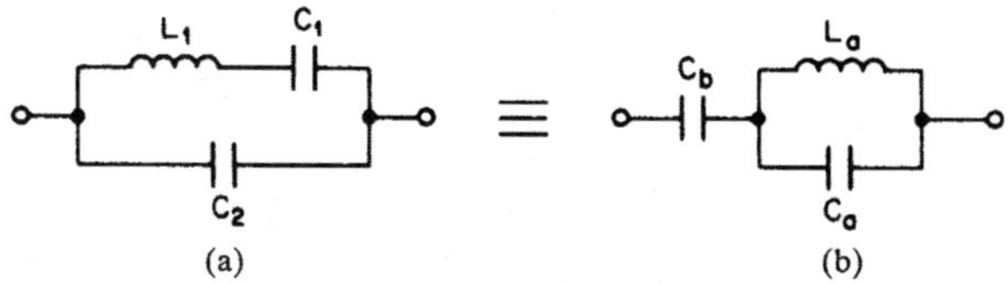

FIGURE 8-21 Equivalent branches: (*a*) type 1 network; and (*b*) type 2 network.

$$C_1 = C_b \frac{1}{1 + \dfrac{C_a}{C_b}} \tag{8-56}$$

$$C_2 = C_a \frac{1}{1 + \dfrac{C_a}{C_b}} \tag{8-57}$$

$$f_{\text{series}} = \frac{1}{2\pi\sqrt{L_1 C_1}} \tag{8-58}$$

$$f_{\text{par}} = \frac{1}{2\pi\sqrt{\dfrac{L_1 C_1 C_2}{C_1 + C_2}}} \tag{8-59}$$

For a type 2 network:

$$L_a = L_1 \frac{1}{\left(1 + \dfrac{C_2}{C_1}\right)^2} \tag{8-60}$$

$$C_a = C_2\left(1 + \frac{C_2}{C_1}\right) \tag{8-61}$$

$$C_b = C_1 + C_2 \tag{8-62}$$

$$f_{\text{series}} = \frac{1}{2\pi\sqrt{L_a(C_a + C_b)}} \tag{8-63}$$

$$f_{\text{par}} = \frac{1}{2\pi\sqrt{L_a C_a}} \tag{8-64}$$

In general, the bandpass series arms are of the type 2 form, and the shunt branches are of the type 1 form, as in Figure 8-19c. The tuning usually consists of adjusting the parallel resonances of the series branches and the series resonances of the shunt branches—in other words, the transmission zeros.

8.7 FILTER TUNING METHODS

LC filters are typically assembled using elements with 1- or 2-percent tolerances. For many applications, the deviation in the desired response caused by component variations may be unacceptable, so the adjustment of elements will be required. It has been found that wherever

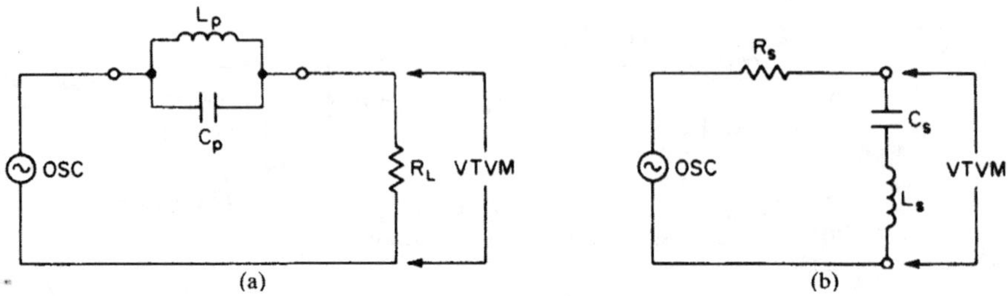

FIGURE 8-22 Test circuits for adjusting resonant frequencies: (*a*) adjusting parallel resonance; and (*b*) adjusting series resonance.

resonances occur, the *LC* product of the resonant circuit is significantly more critical than the *L*/*C* ratio. As a result, filter adjustment normally involves adjusting each tuned circuit for resonance at the specified frequency.

Adjustment techniques are based on the impedance extremes that occur at resonance. In the circuit of Figure 8-22*a*, an output null will occur at parallel resonance because of voltage-divider action. Series *LC* circuits are tuned using the circuit of Figure 8-22*b*, where an output null will also occur at resonance.

The adjustment method in both cases involves setting the oscillator for the required frequency and adjusting the variable element, usually the inductor, for an output null. Resistors R_L and R_s are chosen so that an approximately 20- to 30-dB drop occurs between the oscillator and the output at resonance. These values can be estimated from

$$R_L \approx \frac{2\pi f_r L_p Q_L}{20} \tag{8-65}$$

and

$$R_s \approx \frac{40\pi f_r L_s}{Q_L} \tag{8-66}$$

where Q_L is the inductor Q. A feature of this technique is that no tuning errors result from stray capacity across the VTVM. (Note that the term "VTVM" is used to represent a high input-impedance voltage (dB) measuring meter and not necessarily a *vacuum tube* meter.) Care should be taken that the oscillator does not have excessive distortion, since a sharp null may then be difficult to obtain. Also, excessive levels should be avoided, as detuning can occur from inductor saturation effects. The oscillator and VTVM can be replaced by the tracking generator output and the signal input of a network analyzer. A swept measurement can be made to determine resonant frequency.

When inductor *Q*s are below 10, sharp nulls cannot be obtained. A more desirable tuning method is to adjust for the condition of zero phase shift at resonance, which will be more distinct than the null. The circuits of Figure 8-22 can still be used in conjunction with an oscilloscope having both vertical and horizontal inputs. One channel monitors the oscillator and the other channel is connected to the output instead of using the VTVM. A Lissajous pattern is obtained, and the tuned circuit is then adjusted for a closed ellipse. Some network analyzers can display phase shift and can be used as well.

Certain construction practices must be used so that the assembled filter can be tuned. There must be provision for access to each tuned circuit on an individual basis. This is usually accomplished by leaving all the grounds disconnected until after tuning so that each branch can be individually inserted into the tuning configurations of Figure 8-22 with all the other branches present.

8.8 MEASUREMENT METHODS

This section discusses some major filter parameters and describes techniques for their measurement. Also, some misconceptions associated with these characteristics are clarified. All measurements should be made using rated operating levels so that the results are meaningful. After fabrication, filters should be subjected to insertion-loss and frequency-response measurements as a minimum production test.

Insertion Loss and Frequency Response

The frequency response of filters is always considered as relative to the attenuation occurring at a particular reference frequency. The actual attenuation at this reference is called *insertion loss*.

The classical definition of insertion loss is the decrease in power delivered to the load when a filter is inserted between the source and the load. Using Figure 8-23, the insertion loss is given by

$$\mathrm{IL}_{\mathrm{dB}} = 10 \log \frac{P_{L1}}{P_{L2}} \tag{8-67}$$

where P_{L1} is the power delivered to the load with both switches in position 1 (filter bypassed) and P_{L2} is the output power with both switches in position 2. Equation (8-67) can also be expressed in terms of a voltage ratio as

$$\mathrm{IL}_{\mathrm{dB}} = 10 \log \frac{E_{L1}^2/R_L}{E_{L2}^2/R_L} = 20 \log \frac{E_{L1}}{E_{L2}} \tag{8-68}$$

so a decibel meter can be used at the output to measure insertion loss directly in terms of output voltage.

The classical definition of insertion loss may be somewhat inapplicable when the source and load terminations are unequal. In reality, if the filter were not used, the source and load would probably be connected through an impedance-matching transformer instead of a direct connection. Therefore, the comparison of Figure 8-23 would be invalid.

An alternate definition is *transducer loss*, which is defined as the decrease in power delivered to the load when an ideal impedance-matching transformer is replaced by the filter. The test circuit of Figure 8-23 can still be used if a correction factor is added to Equation (8-68). The resulting expression then becomes

$$\mathrm{IL}_{\mathrm{dB}} = 20 \log \frac{E_{L1}}{E_{L2}} + 20 \log \frac{R_s + R_L}{2\sqrt{R_s R_L}} \tag{8-69}$$

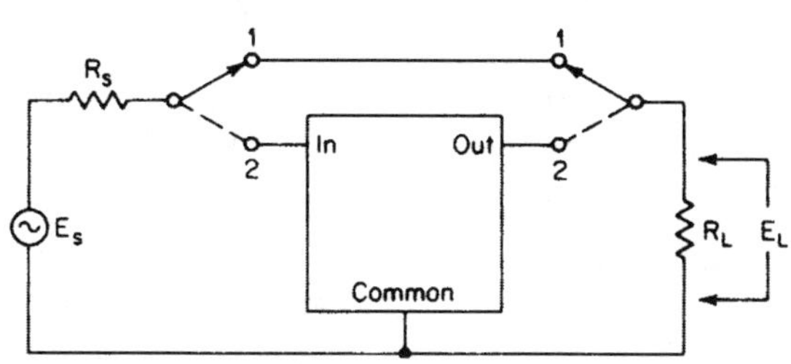

FIGURE 8-23 The test circuit for insertion loss.

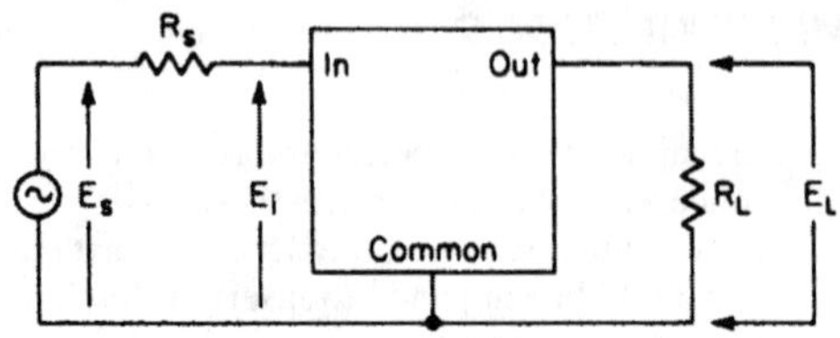

FIGURE 8-24 The test circuit for frequency response.

Frequency response or relative attenuation is measured using the test circuit of Figure 8-24. The input source E_s is set to the reference frequency and the level is arbitrarily set for a 0-dB reference at the output. As the input frequency is changed, the variation in output level is the relative attenuation.

It must be understood that the variation of the ratio E_L/E_s is the frequency response. The ratio E_L/E_i is of no significance since it reflects the frequency response of the filter when driven by a voltage source. As a source frequency is varied, the voltage E_s must be kept constant. Any attempt to keep E_i constant will distort the response shape since voltage-divider action between R_s and the filter input impedance must normally occur to satisfy the transfer function.

The oscillator source itself, E_s, may contain some internal impedance. Nevertheless, the value of R_s should correspond to the design source impedance since the internal impedance of E_s is allowed for by maintaining the terminal voltage of E_s constant.

A network analyzer is commonly used in the industry for measurement of frequency response and can generate test results in various electronic forms which can then be saved and manipulated.

Input Impedance of Filter Networks. The input or output impedance of filters must frequently be determined to ensure compatibility with external circuitry. The input impedance of a filter is the impedance measured at the input terminals with the output appropriately terminated. Conversely, the output impedance can be measured by terminating the input.

Let us first consider the test circuit of Figure 8-25*a*. A common fallacy is to adjust the value of *R* until $|E_2|$ is equal to 1/2 $|E_1|$—that is, a 6-dB drop.

The input impedance Z_{11} is then said to be equal to *R*. However, this will be true only if Z_{11} is purely resistive. As an example, if Z_{11} is purely reactive and its absolute magnitude is equal to *R*, the value of E_2 will be 0.707 dB or 3 dB below E_1 and not 6 dB.

Using the circuit of Figure 8-25*a*, an alternate approach will result in greater accuracy. If *R* is adjusted until a 20-dB drop occurs between $|E_2|$ and $|E_1|$, then $|Z_1|$ is determined by *R*/10. The accuracy will be within 10 percent. For even more accurate results, the 40-dB method can be used where *R* is adjusted for a 40-dB drop. The magnitude of Z_{11} is then given by *R*/100.

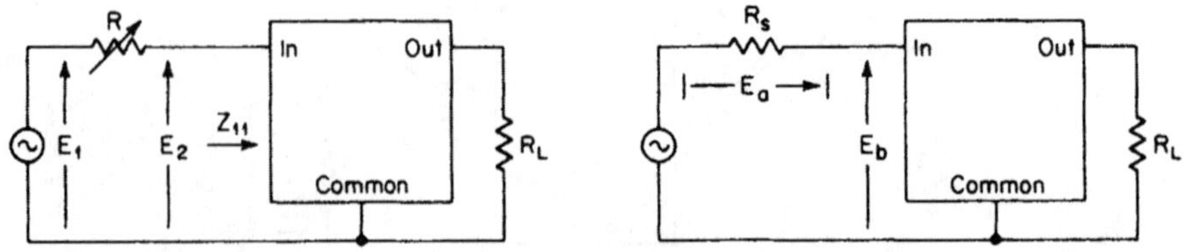

FIGURE 8-25 The measurement of input impedance: (*a*) an indirect method; and (*b*) a direct method.

If a more precise measurement is required, a floating meter can be used in the configuration of Figure 8-25*b*. The input impedance is then directly given by

$$|Z_{11}| = \frac{|E_b|}{|E_a|} R_s \tag{8-70}$$

Most network analyzers can also measure impedance.

Return Loss. Return loss is a figure of merit which indicates how closely a measured impedance matches a standard impedance, both in magnitude and in phase angle. Return loss is expressed as

$$A_p = 20 \log \left| \frac{Z_s + Z_x}{Z_s - Z_x} \right| \tag{8-71}$$

where Z_s is the standard impedance and Z_x is the measured impedance. For a perfect match, the return loss would be infinite.

Return loss can be directly measured using the bridge arrangement of Figure 8-26. The return loss is given by

$$A_p = 20 \log \left| \frac{V_{01}}{V_{02}} \right| \tag{8-72}$$

where V_{01} is the output voltage with the switch closed and V_{02} is the output voltage with the switch open. The return loss can then be read directly using a decibel meter or by using a network analyzer. The value of R is arbitrary, but both resistors must be closely matched to each other.

The family of curves in Figure 8-27 represents the return loss using a standard impedance of 600 Ω with the phase angle of impedance as a parameter. Clearly, the return loss is very sensitive to the phase angle. If the impedance were 600 Ω at an angle of only 10°, the return loss would be 21 dB. If there was no phase shift, an impedance error of as much as 100 Ω would correspond to 21 dB of return loss.

Time-Domain Characteristics

Step Response. The step response of a filter network is a useful criterion since low transient distortion is a necessary requirement for good transmission of modulated signals. To determine the step response of a low-pass filter, an input DC step is applied. For bandpass filters, a carrier step is used where the carrier frequency is equal to the filter center frequency f_0. Since it is difficult to view a single transient on an oscilloscope unless it's of the storage type, a square-wave generator is used instead of the DC step and a tone-burst generator is substituted for the carrier step. However, a repetition rate must be chosen that is

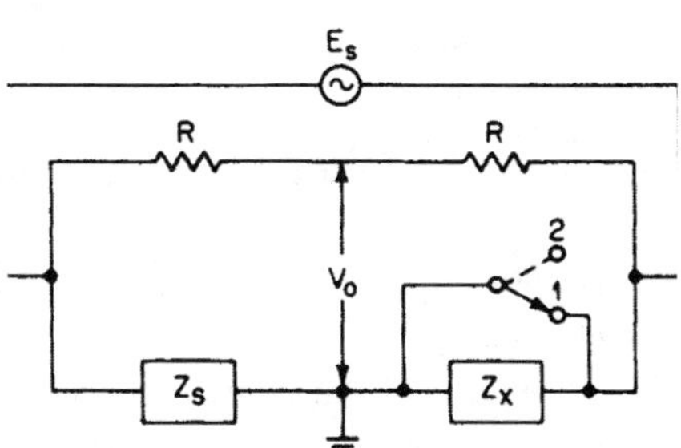

FIGURE 8-26 The measurement of return loss.

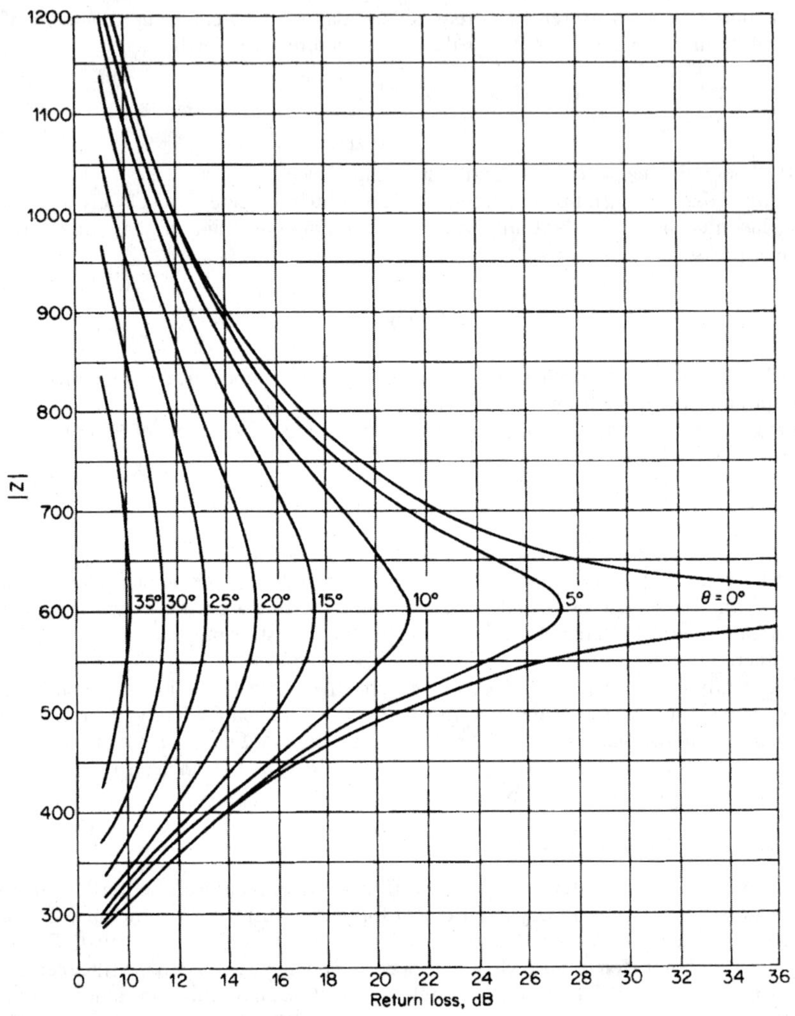

FIGURE 8-27 Return loss versus $|Z|$.

slow enough so that the transient behavior has stabilized prior to the next pulse in order to obtain meaningful results.

The test configuration is shown in Figure 8-28*a*. The output waveforms are depicted in Figure 8-28*b* and *c* for a DC step and tone burst, respectively. The following definitions are applicable:

Percent overshoot P_T: The difference between the peak response and the final steady-state value expressed as a percentage.

Rise time T_r: The interval between 10 and 90 percent of the final value.

Settling time T_s: The time required for the response to settle within a specified percent of its final value.

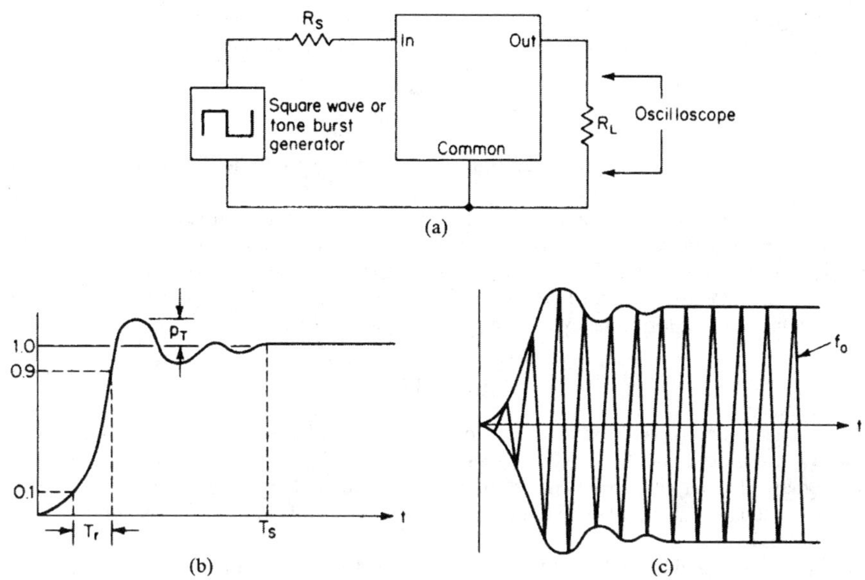

FIGURE 8-28 Step response of networks: (*a*) the test circuit; (*b*) the step response to a DC step; and (*c*) the step response to a tone burst.

The waveform definitions shown in Figure 8-28*b* also apply to Figure 8-28*c* if we consider the envelope of the carrier waveform instead of the instantaneous values.

Group Delay. The phase shift of a filter can be measured by using the Lissajous pattern method. By connecting the vertical channel of an oscilloscope to the input source, and the horizontal channel to the load, an ellipse is obtained, as shown in Figure 8-29. The phase angle in degrees is given by

$$\phi = \sin^{-1}\frac{Y_{\text{int}}}{Y_{\text{max}}} \tag{8-73}$$

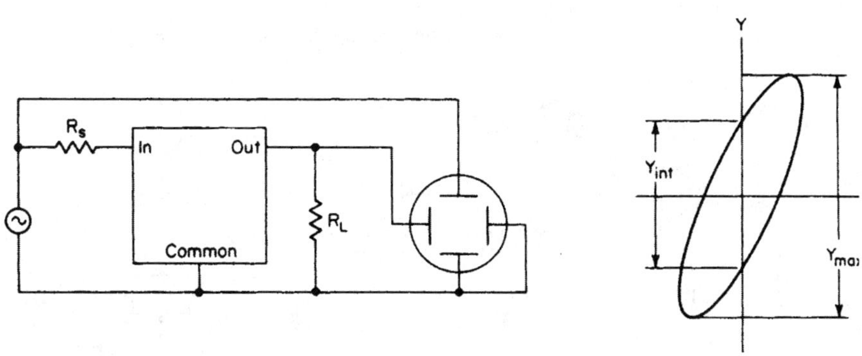

FIGURE 8-29 The measurement of phase shift.

Since group delay is the derivative with respect to frequency of the phase shift, we can measure the phase shift at two closely spaced frequencies and approximate the delay as follows:

$$T_{gd} \approx \frac{\Delta\phi}{360\Delta f} \tag{8-74}$$

where $\Delta\phi$ is $\phi_2 - \phi_1$ in degrees, Δf is $f_2 - f_1$ in hertz, and T_{gd} is the group delay at the midfrequency—in other words, $(f_1 + f_2)/2$.

A less accurate method involves determining the 180° phase-shift points. As the frequency is varied throughout the passband of a high-order filter, the phase shift will go through many integer multiples of 180° where the Lissajous pattern adopts a straight line at either 45° or 225°. If we record the separation between adjacent 180° points, the nominal group delay at the midfrequency can be approximated by

$$T_{gd} \approx \frac{1}{2\Delta f} \tag{8-75}$$

The classical approach for the measurement of group delay is shown in Figure 8-30. A sine-wave source, typically 25 Hz, is applied to an amplitude modulator along with a carrier signal. The output consists of an amplitude-modulated signal comprising the carrier and two sidebands at ±25 Hz on either side of the carrier. The signal is then applied to the network under test.

The output signal from the network is of the same form as the input, but the 25 Hz envelope has been shifted in time by an amount equal to the group delay at the carrier frequency. The output envelope is recovered by an AM detector and applied to a phase detector along with a reference 25 Hz signal.

The phase detector output is a DC signal proportional to the phase shift between the 25 Hz reference and the demodulated 25-Hz carrier envelope. As the carrier is varied in frequency, the DC signal will vary in accordance with the change in group delay (differential delay distortion) and the delay can be displayed on a DC meter having the proper calibration. If an adjustable phase-shift network is interposed between the AM detector and phase detector, the meter indication can be adjusted to establish a reference level at a desired reference frequency.

The theoretical justification for this scheme is based on the fact that the delay of the envelope is determined by the slope of a line segment interconnecting the phase shift of the two sidebands at ±25 Hz about the carrier—that is, $(\phi_2 - \phi_1)/(\omega_2 - \omega_1)$. This definition is sometimes called the *envelope delay*, for obvious reasons. As the separation between sidebands is decreased, the envelope approaches the theoretical group delay at the carrier frequency since group delay is defined as the derivative of the phase shift. For most measurements, a modulation rate of 25 Hz is adequate.

Some network analyzers have a feature where group delay can be measured.

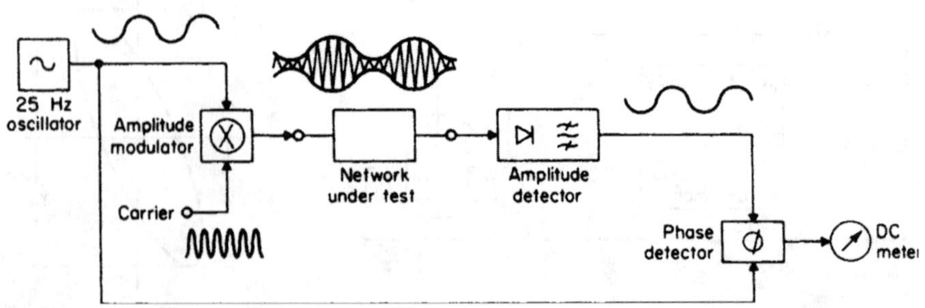

FIGURE 8-30 Direct measurement of group delay.

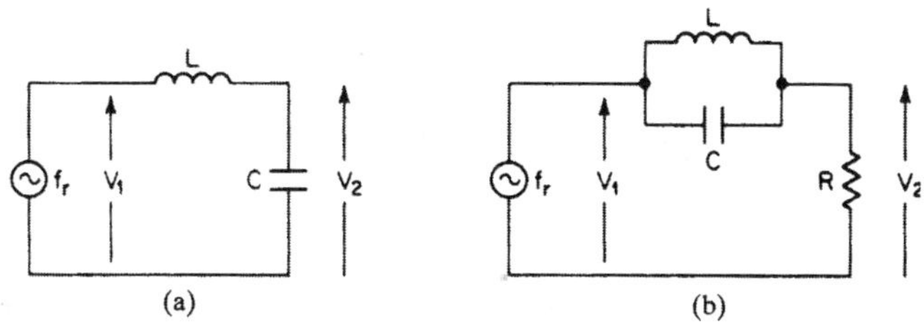

FIGURE 8-31 Measuring coil Q: (*a*) the Q meter method; and (*b*) the parallel resonant circuit method.

Measuring the Q of Inductors. A device called *Q meter* is frequently used to measure the Q of inductors at a specified frequency. The principle of the Q meter is based on the fact that in a series resonant circuit, the voltage across each reactive element is Q times the voltage applied to the resonant circuit. The Q can then be directly determined by the ratio of two voltages.

A test circuit is shown in Figure 8-31*a*. Care should be taken that the applied voltage does not result in an excessive voltage developed across the inductor during the measurement. Also, the resonating capacitor should have a much higher Q than the inductor, and the meter used for the voltage measurements should be a high-impedance type to avoid loading errors.

An alternate approach involves measuring the impedance of a parallel resonant circuit consisting of the inductor and the required resonating capacitor for the frequency of interest. Using the voltage divider of Figure 8-31*b*, the Q at resonance is found from

$$Q = \frac{R}{2\pi f_r L}\left(\frac{V_1}{V_2} - 1\right) \tag{8-76}$$

A null will occur in V_2 at resonance.

For meaningful Q measurements, it's important that the measurement frequency corresponds to the frequency of interest of the filter since coil Q can decrease quite dramatically outside a particular range. In low-pass and high-pass filters, the Q should be measured at the cutoff, and for bandpass and band-reject filters, the center frequency is the frequency of interest.

8.9 DESIGNING FOR UNEQUAL IMPEDANCES

Impedance Matching

Exponentially Tapered Impedance Scaling. Bartlett's Bisection Theorem, discussed in Section 3.1, allows us to modify a design for unequal impedances if the initial schematic (including values) is totally symmetrical around a center line (mirror image). However, that is not always the case. In addition, the tables in section 11 for all-pole filters having unequal impedances are limited to finite ratios of termination impedances.

The method shown in Figure 8-32, although not theoretically precise, allows us to gradually taper the impedance of each element of the ladder from source to load. The more the number of branches of the ladder, the more gradual becomes the tapering, and the less the change of the original transfer function. Experience has indicated that this works quite well with at least six branches of the ladder, (excluding source and load resistors). Note that the load can be higher or lower than the source.

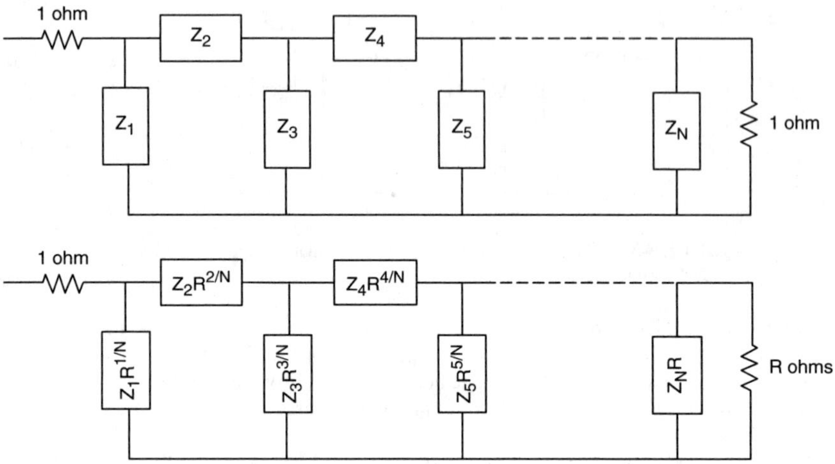

FIGURE 8-32 A tapered network for unequal impedances.

Minimum Loss Resistive Pad for Impedance Matching. The circuit of Figure 8-33 matches resistive source impedance R_s to resistive load impedance R_L with minimum loss. It can be useful when two different filter designs having different impedances are cascaded, or to match a filter to a termination impedance it is not designed for. The design equations are

$$R_1 = R_s\sqrt{1 - \frac{R_L}{R_s}} \tag{8-77}$$

$$R_2 = \frac{R_L}{\sqrt{1 - \frac{R_L}{R_s}}} \tag{8-78}$$

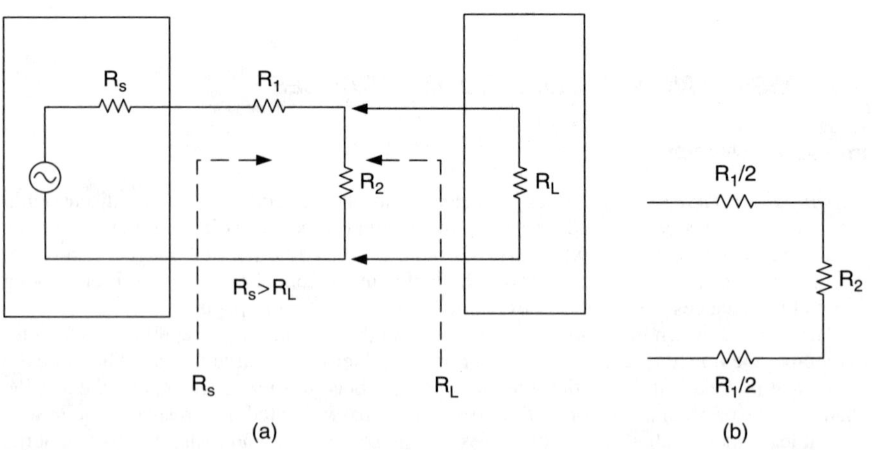

FIGURE 8-33 Minimum loss pads: (*a*) single ended; and (*b*) the balanced version.

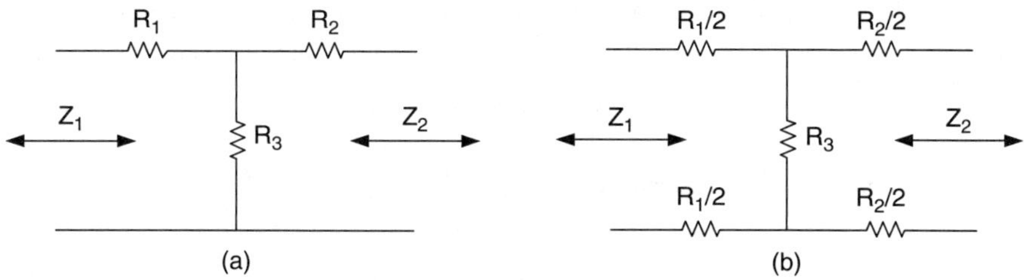

FIGURE 8-34 Unsymmetrical *T* attenuators $Z_2 > Z_1$: (*a*) single ended; and (*b*) the balanced version.

Note that R_s must be higher than R_L. To form a balanced circuit, split R_1 into two $R_1/2$ resistors as shown. The voltage loss in dB is given by

$$\text{Voltage Loss dB} = 20 \text{ Log}_{10}\left(\frac{R_1(R_2 + R_L)}{R_2R_L} + 1\right) \tag{8-79}$$

In terms of power, the loss in dB is

$$\text{Power Loss dB} = \text{Voltage Loss dB} - 10 \text{ Log}_{10}\frac{R_s}{R_L} \tag{8-80}$$

Design of Unsymmetrical Resistive T and π Attenuators for Impedance Matching. The circuit of Figure 8-33 matches two unequal impedances with minimum loss. To obtain a fixed amount of attenuation between two *unequal* resistive impedances, a *T* or *π* attenuator can be used. The values are computed as follows for the unbalanced and balanced circuits of Figures 8-34 and 8-35: *Note that* Z_1 *must be greater than* Z_2.

For an unsymmetrical T attenuator:
First compute

$$K_{\min} = \frac{2Z_1}{Z_2} - 1 + 2\sqrt{\frac{Z_1}{Z_2}\left(\frac{Z_1}{Z_2} - 1\right)} \tag{8-81}$$

For a required voltage loss in dB

$$\text{dB}_{\text{min-voltage loss}} = 10 \text{ Log}_{10}K_{\min} + 10 \text{ Log}_{10}\frac{Z_1}{Z_2} \tag{8-82}$$

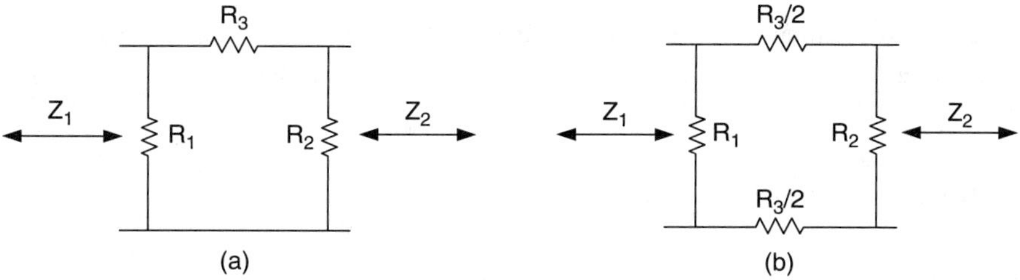

FIGURE 8-35 Unsymmetrical *π* attenuators $Z_2 > Z_1$: (*a*) single ended; and (*b*) the balanced version.

This is the minimum voltage loss in dB that results in positive resistor values.
Select a desired voltage loss ($dB_{voltage\ loss}$) > Minimum Voltage Loss ($dB_{min\text{-}voltage\ loss}$)

$$K = 10^{dB\ voltage\ loss/10 + Log(Z_2/Z_1)} \tag{8-83}$$

The resistor values are obtained from

$$R_1 = \frac{Z_1(K+1) - 2\sqrt{KZ_1Z_2}}{K-1} \tag{8-84}$$

$$R_2 = \frac{Z_2(K+1) - 2\sqrt{KZ_1Z_2}}{K-1} \tag{8-85}$$

$$R_3 = \frac{2\sqrt{KZ_1Z_2}}{K-1} \tag{8-86}$$

For a required *power loss in dB*, first compute:

$$dB_{min\text{-}voltage\ loss} = 10\ Log_{10}K_{min} \tag{8-87}$$

Select a desired power loss ($dB_{power\ loss}$) > Minimum power Loss ($dB_{min\text{-}power\ loss}$)

$$K = 10^{dB/10} \tag{8-88}$$

Resistors R_1, R_2, and R_3 are computed using Equations (8-84), (8-85), and (8-86), respectively. The resulting voltage loss is

$$dB_{voltage\ loss} = dB_{power\ loss} + 10\ Log_{10}\frac{Z_1}{Z_2} \tag{8-89}$$

For an unsymmetrical π attenuator:
First, compute

$$K_{min} = \frac{2Z_1}{Z_2} - 1 + 2\sqrt{\frac{Z_1}{Z_2}\left(\frac{Z_1}{Z_2} - 1\right)} \tag{8-81}$$

For a required *voltage loss in dB*

$$dB_{min\text{-}voltage\ loss} = 10\ Log_{10}K_{min} + 10\ Log_{10}\frac{Z_1}{Z_2} \tag{8-82}$$

This is the minimum voltage loss in dB that results in positive resistor values.
Select a desired voltage loss ($dB_{voltage\ loss}$) > Minimum Voltage Loss ($dB_{min\text{-}voltage\ loss}$)

$$K = 10^{dB\ voltage\ loss/10 + Log(Z_2/Z_1)} \tag{8-83}$$

The resistor values are computed by

$$R_1 = \frac{Z_1(K-1)\sqrt{Z_2}}{(K+1)\sqrt{Z_2} - 2\sqrt{KZ_1}} \tag{8-90}$$

$$R_2 = \frac{Z_2(K-1)\sqrt{Z_1}}{(K+1)\sqrt{Z_1} - 2\sqrt{KZ_2}} \tag{8-91}$$

$$R_3 = \frac{K-1}{2}\sqrt{\frac{Z_1 Z_2}{K}} \tag{8-92}$$

For a required *power loss in dB*, first compute

$$dB_{\text{min-power loss}} = 10\ \text{Log}_{10} K_{\min} \tag{8-87}$$

Select a desired power loss ($dB_{\text{power loss}}$) > Minimum power Loss ($dB_{\text{min-power loss}}$)

$$K = 10^{dB/10} \tag{8-88}$$

Resistors R_1, R_2, and R_3 are computed using Equations (8-90), (8-91), and (8-92), respectively.

The resulting voltage loss is

$$dB_{\text{voltage loss}} = dB_{\text{power loss}} + 10\ \text{Log}_{10}\frac{Z_1}{Z_2} \tag{8-89}$$

8-10 SYMMETRICAL ATTENUATORS

Symmetrical T and π Attenuators. When source- and load-resistive impedances are equal, a symmetrical *T* or *π* attenuator can be used to symmetrically (bidirectionally) introduce fixed loss where needed. Figures 8-36 and 8-37 illustrate symmetrical *T* and *π* attenuators, respectively.

Section 8.8 discusses return loss, which is a figure of merit that indicates how closely a measured impedance matches a standard impedance, both in magnitude and in phase angle. Return loss is expressed as

$$A_\rho = 20\log\left|\frac{Z_s + Z_x}{Z_s - Z_x}\right| \tag{8-71}$$

where Z_s is the standard impedance and Z_x is the measured impedance. For a perfect match, the return loss would be infinite. If the standard impedance is resistive Z_s, and the network is preceded with a symmetrical attenuator of *X* dB at an impedance level of Z_s, *a minimum return loss of* 2X *dB is guaranteed* even if the network has impedance extremes of zero or infinity. The attenuator will smooth any impedance gyrations.

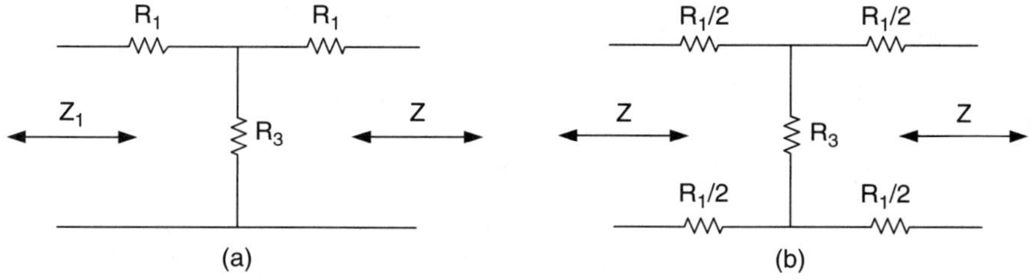

FIGURE 8-36 Symmetrical *T* attenuators: (*a*) single ended; and (*b*) the balanced version.

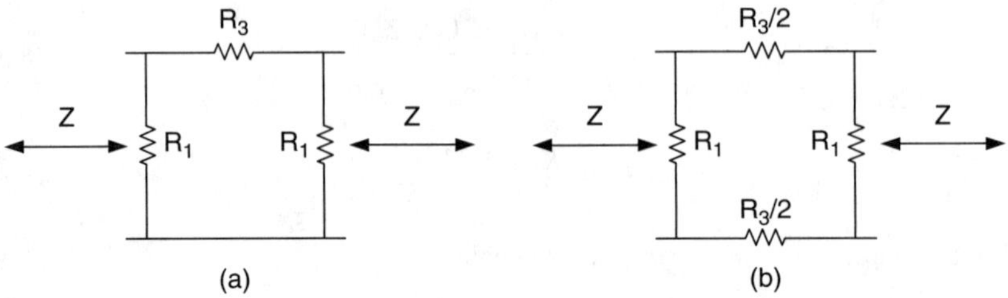

FIGURE 8-37 Symmetrical π attenuators: (*a*) single ended; and (*b*) the balanced version.

For a given loss in dB (the power loss in dB is equal to the voltage loss in dB since the impedances are equal on both sides), first compute

$$K = 10^{\mathrm{dB}/20} \tag{8-93}$$

For a symmetrical T attenuator:

$$R_1 = Z\frac{K-1}{K+1} \tag{8-94}$$

$$R_3 = \frac{2\,ZK}{K^2-1} \tag{8-95}$$

For a symmetrical π attenuator:

$$R_1 = Z\frac{K+1}{K-1} \tag{8-96}$$

$$R_3 = Z\frac{K^2-1}{2K} \tag{8-97}$$

Bridged T Attenuator. Figure 8-38 shows a symmetrical attenuator in a bridged-T form. One advantage of this configuration is that only two resistor values have to be changed to vary attenuation. Two of the resistors always remain at the source and load impedances R_0.

The resistors R_1 and R_2 are calculated from

$$R_1 = \frac{R_0}{K-1} \tag{8-98}$$

and

$$R_2 = R_0(K-1) \tag{8-99}$$

where

$$K = 10^{\mathrm{dB}/20} \tag{8-93}$$

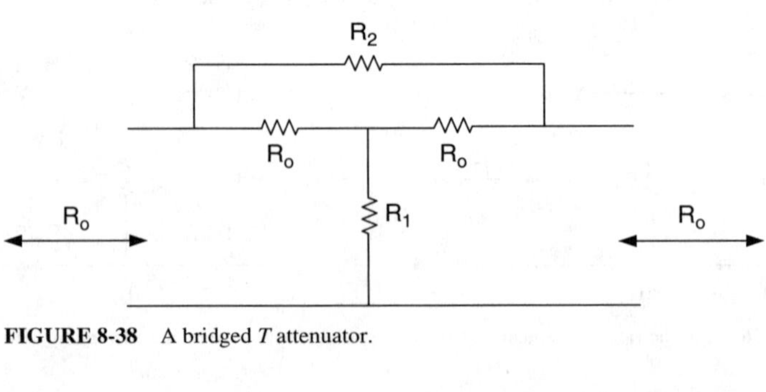

FIGURE 8-38 A bridged *T* attenuator.

8-11 POWER SPLITTERS

Resistive Power Splitters

Resistive power splitters are essentially resistive voltage dividers that distribute a signal in multiple directions while maintaining the same impedance at all ports. They provide no isolation between ports. In other words, even under ideal perfectly terminated conditions a signal arriving at any one port appears at all other ports.

Figure 8-39 illustrates an "N" way splitter, where

$$N = \text{total number of ports} - 1 \tag{8-100}$$

All resistors are equal to R, which is calculated by

$$R = R_0 \frac{N - 1}{N + 1} \tag{8-101}$$

where R_0 is the impedance at all ports.

Resistive power splitters by their nature are very inefficient. The loss in dB is

$$\text{Power Loss dB} = 10 \text{ Log}_{10} \frac{1}{N^2} \tag{8-102}$$

So for a two-way splitter (three ports, $N = 2$) the power loss is 6 dB, a four-way splitter has 12-dB loss, and so forth. Since all impedances are matched, the voltage loss in dB is exactly equal to the power loss computed by Equation 8-102. This loss in dB is also the isolation between ports.

A Magic-T Splitter. The circuit of Figure 8-40 is commonly called a magic-T splitter or two-way splitter/combiner. To some extent, it has the functionality of the resistive power splitter, but it differs in two major ways. First, it has *a power loss of 3 dB rather than 6 dB* for the equivalent function of the resistive power splitter. Second, it can have theoretical *infinite isolation between ports A and B*. Thus, a signal entering port A will be prevented from arriving at port B, and vice versa. However, any signals entering either port A or port B will arrive at port S with 3 dB of loss, and any signal applied to port S will arrive at ports A and B with 3 dB of loss.

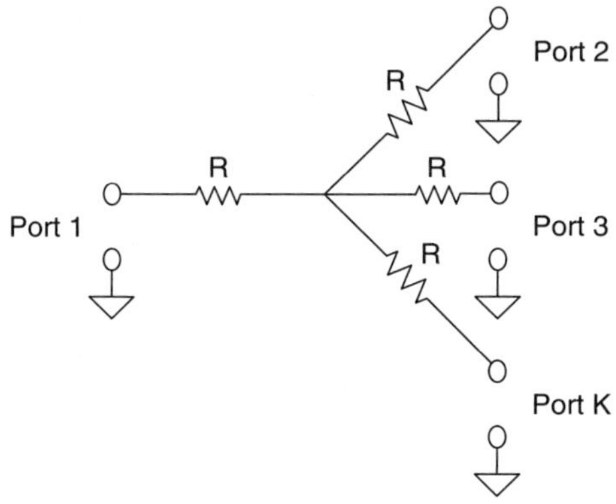

FIGURE 8-39 A resistive power splitter.

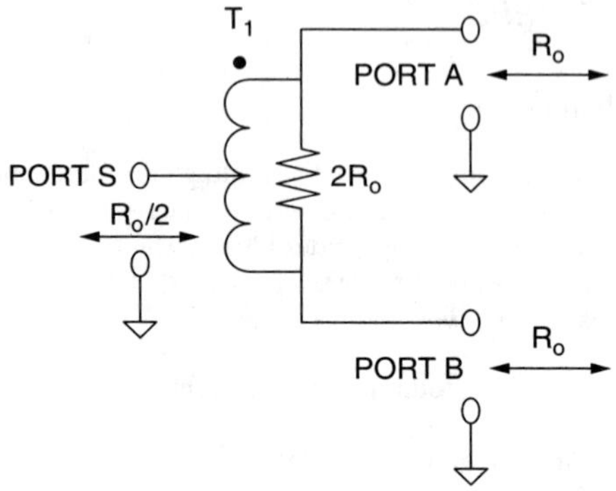

FIGURE 8-40 The magic-T splitter/combiner.

Observe that port *S* has an impedance of $R_0/2$. A 2:1 impedance ratio transformer can step this up to R_0.

The output signal at port *B* will not change when port *A* is terminated with an impedance other than $R_{0,}$ even on a short or open, (Ports *A* and *B* are interchangeable). An impedance mismatch at port *S* will cause a reflection of the signal applied to port *A* onto port *B*. The amount of this reflection is the return loss of the impedance mismatch at port *S*, plus 6 dB.

$$\text{Reflection Loss dB} = 20\,\text{Log}_{10}\left|\frac{R_0 + R_x}{R_0 - R_x}\right| + 6\text{ dB} \tag{8-103}$$

where R_x is the value of the termination of port *S*.

Operation of the magic-T splitter is not very intuitive from the schematic. Let's examine Figure 8-41, where the circuit has been redrawn showing a signal applied to port *S*.

In the circuit of Figure 8-41, V_s results in equal currents, I_1 and I_2, in opposite directions through T_1. The resulting voltages, E_1 and E_2, are equal in magnitude, but opposite in polarity, so they cancel. From the symmetry of the circuit, the signal at port *S* appears equally at

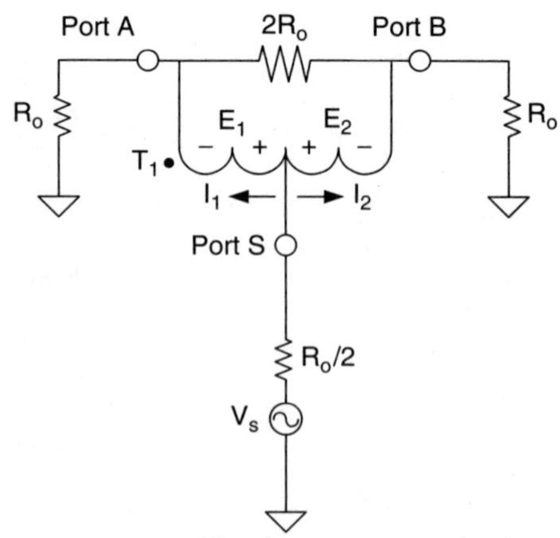

FIGURE 8-41 A magic-T with a signal at the *S* port.

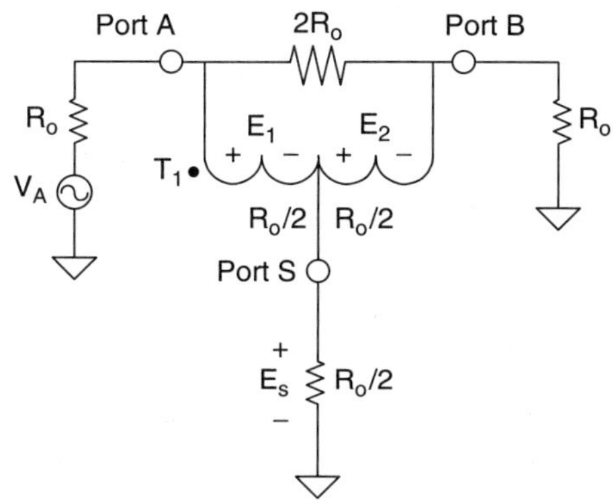

FIGURE 8-42 The magic-T isolation between ports *A* and *B*.

ports *A* and *B*. The impedance seen at port *S* is $R_0/2$ since the terminations of both ports *A* and *B* are reflected to port *S* and are in parallel.

The isolation mechanism between ports *A* and *B* is illustrated in Figure 8-42. A signal V_A is applied to port *A*. The impedance measured at the input of T_1 between the port *A* and port *S* terminals is $R_0/2$ since T_1 is terminated with $2R_0$ and a four-to-one impedance step-down occurs between the port *A* and port *S* terminals (center tapped T_1). As a result, the applied voltage divides evenly so E_1 and E_S are equal.

Because of transformer action, E_1 results in E_2, which is equal in magnitude. As measured from the input of port *B*, E_2 and E_S are equal and in series but *opposite* in polarity so they cancel. Therefore, under these conditions, V_A will not appear at port *B* so total isolation will occur. V_A *will* appear at port S attenuated 3 dB in power.

The final circuit is shown in Figure 8-43. An additional transformer (autotransformer) having a turns ratio of 1.414:1 has been added to step up $R_0/2$ to R_0. The impedance ratio is 2:1.

Splitter-combiners can be cascaded in a tree-like fashion to create additional ports. The circuit of Figure 8-44 illustrates how a four-way splitter-combiner can be made from three two-way splitter combiners. All ports must have equal impedances, however. The insertion loss for Figure 8-44 is 6 dB.

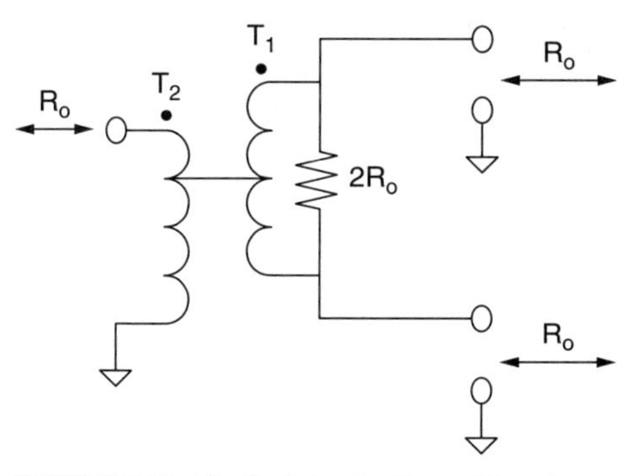

FIGURE 8-43 The final circuit with equal impedances at all ports.

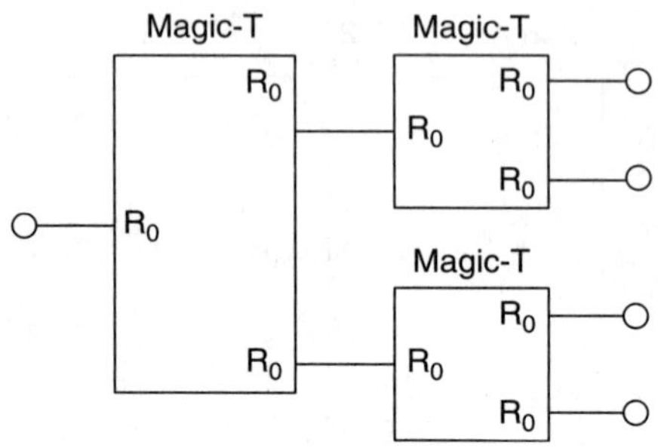

FIGURE 8-44 A four-way splitter-combiner (with a loss of 6 dB).

Care should be taken when cascading magic-T circuits that transformers T_1 and T_2 have sufficient bandwidth since the roll-off will be cumulative. As a good rule-of-thumb the inductance of T_1 and T_2 should be a minimum of $R_0/(\pi F_L)$, where F_L is the lowest frequency of interest. In addition, the center-tap accuracy should be in the vicinity of 1 percent.

BIBLIOGRAPHY

Geffe, P. R. *Simplified Modern Filter Design.* New York: John F. Rider, 1963.

Saal, R. "Der Entwurf von Filtern mit Hilfe des Kataloges Normierter Tiefpasse." *Telefunken GMBH* (1963).

Saal, R., and E. Ulbrich. "On the Design of Filters by Synthesis." *IRE Transactions on Circuit Theory* CT-5 (December, 1958).

Zverev, A. I. *Handbook of Filter Synthesis.* New York: John Wiley and Sons, 1967.

CHAPTER 9
DESIGN AND SELECTION OF INDUCTORS FOR *LC* FILTERS

9.1 BASIC PRINCIPLES OF MAGNETIC-CIRCUIT DESIGN

Units of Measurement

Magnetic permeability is represented by the symbol μ and is defined by

$$\mu = \frac{B}{H} \tag{9-1}$$

B is the magnetic flux density in lines per square centimeter and is measured in gauss, while H is the magnetizing force in oersteds that produced the flux. Permeability is dimensionless and can be considered a figure of merit of a particular magnetic material since it represents the ease of producing a magnetic flux for a given input. The permeability of air, or that of a vacuum, is 1.

Magnetizing force is caused by current flowing through turns of wire. Thus, H can be determined from ampere-turns by

$$H = \frac{4\pi NI}{10\ \text{mL}} \tag{9-2}$$

where N is the number of turns, I is the current in amperes, and mL is the mean length of the magnetic path in centimeters.

The inductance of a coil is directly proportional to the number of flux linkages per unit current. The total flux is found from

$$\phi = BA = \mu HA = \frac{4\pi NI\mu A}{10\ \text{mL}} \tag{9-3}$$

where A is the cross-sectional area in square centimeters.

The inductance proportionality may then be expressed as

$$L \propto \frac{4\pi NI\mu A}{10\ \text{mL}} \frac{N}{I} \tag{9-4}$$

or directly in henrys by

$$L = \frac{4\pi N^2 \mu A}{\text{mL}} 10^{-9} \tag{9-5}$$

A number of things should be apparent from Equation (9-5). First of all, the inductance of a coil is directly proportional to the permeability of the core material. If an iron core is inserted into an air-core inductor, the inductance will increase in direct proportion to the iron core's permeability. The inductance is also proportional to N^2.

All the previous design equations make the assumption that the magnetic path is uniform and closed with negligible leakage flux in the surrounding air such as would occur with a single-layer toroidal coil structure. However, this assumption is really never completely valid, so some deviations from the theory can be expected.

The induced voltage of an inductor can be related to the flux density by

$$E_{\mathrm{rms}} = 4.44BNfA \times 10^{-8} \tag{9-6}$$

where B is the maximum flux density in Gauss, N is the number of turns, f is the frequency in hertz, and A is the cross-sectional area of the core in square centimeters. This important equation is derived from Faraday's law.

As a point of information, the units of flux density B in gauss and the magnetizing force H in oersteds are CGS units. Mainly outside of the United States, SI units are gaining in usage. Flux density is then expressed in mT, which is a milli-Tesla and is equivalent to 10 gauss. The magnetizing force is expressed in A/m (ampere/meter) and is equivalent to $4\pi/10^3$ oersteds.

Saturation and DC Polarization. A plot of B vs. H is shown in Figure 9-1. Let's start at point A and increase the magnetizing force to point B. A decrease in magnetizing force will pass through point C, and then D and E as it becomes negative. An increasing magnetizing force, again in the positive direction, will travel to B through point F. The enclosed area formed by the curve is called a *hysteresis loop* and results from the energy required to reverse the magnetic molecules of the core. The magnitude of H between points D and A is called *coercive force* and is the amount of H necessary to reduce the residual magnetism in the core to zero.

Permeability was defined as the ratio B/H and can be obtained from the slope of the BH curve. Since we normally deal with low-level AC signals, the region of interest is restricted to a relatively narrow range. We can then assume that the permeability is determined by the derivative of the curve at the origin. The derivative of a B/H curve is sometimes called *incremental permeability.*

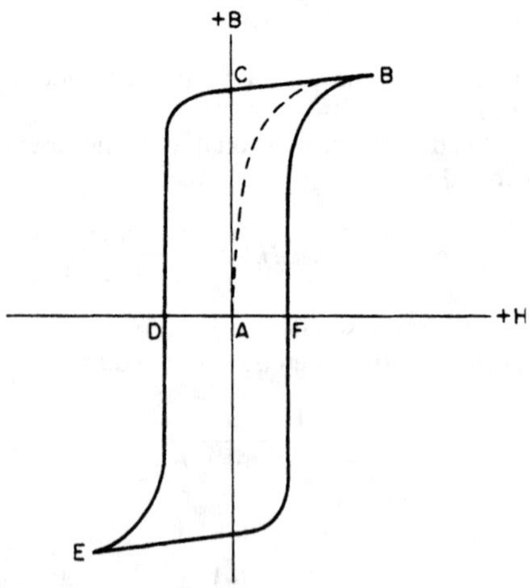

FIGURE 9-1 The hysteresis loop.

If a DC bias is introduced, the quiescent point will move from the origin to a point farther out on the curve. Since the curve tends to flatten out with higher values of H, the incremental permeability will decrease, which reduces the inductance. This effect is known as *saturation* and can also occur without a DC bias for large AC signals. Severe waveform distortion usually accompanies saturation. The B/H curve for an air core is a straight line at a 45° angle through the origin. The permeability is unity, and no saturation can occur.

Inductor Losses. The Q of a coil can be found from

$$Q = \frac{\omega L}{R_{dc} + R_{ac} + R_d} \tag{9-7}$$

where R_{dc} is the copper loss, R_{ac} is the core loss, and R_d is the dielectric loss. Copper loss consists strictly of the DC winding resistance and is determined by the wire size and total length of wire required. The core loss is composed mostly of losses due to eddy currents and hysteresis. Eddy currents are induced in the core material by changing magnetic fields. These circulating currents produce losses that are proportional to the square of the inducing frequency.

When a core is subjected to an AC or pulsating DC magnetic field, the B vs. H characteristics can be represented by the curve of Figure 9-1. The enclosed area was called a hysteresis loop and resulted from the energy required to reverse the magnetic domains in the core material. These core losses increase in direct proportion to frequency since each cycle traverses the hysteresis loop.

The dielectric losses are important at higher frequencies and are determined by the power factor of the distributed capacity. Keeping the distributed capacity small as well as using wire insulation with good dielectric properties will minimize dielectric losses.

Above approximately 50 kHz, the current will tend to travel on the surface of a conductor rather than through the cross section. This phenomenon is called *skin effect.* To reduce this effect, litz wire is commonly used. This wire consists of many braided strands of insulated conductors so that a larger surface area is available in comparison with a single solid conductor of the equivalent cross section. Above 1 or 2 MHz, solid wire can again be used.

A figure of merit of the efficiency of a coil at low frequencies is the ratio of ohms per henry (Ω/H), where the ohms correspond to R_{dc}—that is, the copper losses. For a given coil structure and permeability, the ratio Ω/H is a constant independent of the total number of turns, provided that the winding cross-sectional area is kept constant.

Effect of an Air Gap. If an ideal toroidal core has a narrow air gap introduced, the flux will decrease and the permeability will be reduced. The resulting effective permeability can be found from

$$\mu_e = \frac{\mu_i}{1 + \mu_i\left(\dfrac{g}{\mathrm{mL}}\right)} \tag{9-8}$$

where μ_i is the initial permeability of the core and g/mL is the ratio of gap to length of the magnetic path. Equation (9-8) applies to closed magnetic structures of any shape if the initial permeability is high and the gap ratio small.

The effect of an air gap is to reduce the permeability and make the coil's characteristics less dependent upon the initial permeability of the core material. A gap will prevent saturation with large AC signals or DC bias and allow tighter control of inductance. However, lower permeability due to the gap requires more turns and, thus, associated copper losses, so a suitable compromise is required.

The Design of Coil Windings. Inductors are normally wound using insulated copper wire. The general method used to express wire size is the American Wire Gauge (AWG) system. As the wire size numerically decreases, the diameter increases, where the ratio of the diameter of one size to the next larger size is 1.1229. The ratio of cross-sectional areas of adjacent wire sizes corresponds to the square of the diameter, or 1.261. Therefore, for an available cross-sectional winding area, reducing the wire by one size permits 1.261 times as many turns. Two wire sizes correspond to a factor of $(1.261)^2$, or 1.59, and three wire sizes permit twice as many turns.

Physical and electrical properties of a range of wire sizes are given in the wire chart of Table 9-1. This data is based on using a standard heavy film for the insulation. In the past, enamel insulations were used which required acid or abrasives for stripping the insulation from the wire ends to make electrical connections. Solder-able insulations have been available for the last 30 years so that the wire ends can be easily tinned. In the event that litz wire is required, a cross reference between litz wire sizes and their solid equivalent is given in Table 9-2. The convention for specifying litz wire is the number of strands/wire size. For example, using the chart, a litz equivalent to No. 31 solid is 20 strands of No. 44, or 20/44. In general, a large number of strands is desirable for a more effective surface area.

The temperature coefficient of resistance for copper is 0.393-percent per degree Celsius. The DC resistance of a winding at a particular temperature is given by

$$R_{t1} = R_t[1 + 0.00393(t_1 - t)] \tag{9-9}$$

where t is the initial temperature and t_1 is the final temperature, both in degrees Celsius. The maximum permitted temperature of most wire insulations is about 130°C.

To compute the number of turns for a required inductance, the inductance factor for the coil structure must be used. This factor is generally called A_L and is the nominal inductance per 1000 turns. Since inductance is proportional to the number of turns squared, the required number of turns N for an inductance L is given by

For A_L = mH/1,000 turns
$$N = 10^3\sqrt{\frac{L}{A_L}} \tag{9-10}$$

Other commonly used expressions in some data sheets are

For A_L = μH/100 turns
$$N = 10^2\sqrt{\frac{L}{A_L}} \tag{9-11}$$

And for A_L = nH/turn
$$N = \sqrt{\frac{L}{A_L}} \tag{9-12}$$

In all cases, L and A_L must be in identical units.

After the required number of turns is computed, a wire size must be chosen. For each winding structure, an associated chart can be tabulated which indicates the maximum number of turns for each wire size. A wire size can then be chosen which results in the maximum utilization of the available winding cross-sectional area.

Coil winding methods are very diverse since the winding techniques depend upon the actual coil structure, the operating frequency range, and so on. Coil winding techniques are discussed in the remainder of this chapter on an individual basis for each coil structure type.

TABLE 9-1 Wire Chart—Round Heavy Film Insulated Solid Copper*

	Diameter over Bare, in			Insulation Additions		Diameter over Insulation			Weight		Resistance at 20°C (68°F)			Turns		
AWG	Minimum	Nominal	Maximum	Minimum	Maximum	Minimum	Maximum	Pounds per 1000 Feet	Feet per Pound	Pounds per Cubic Inch	Ohms per 1000 Feet	Ohms per Pounds	Ohms per Cubic Inch	Per Linear Inch	Per Square Inch	AWG
4	0.2023	0.2043	0.2053	0.0037	0.0045	0.2060	0.2098	127.20	7.86	0.244	0.2485	0.001954	0.0004768	4.80	24.0	4
5	0.1801	0.1819	0.1828	0.0036	0.0044	0.1837	0.1872	100.84	9.92	0.243	0.3134	0.003108	0.0007552	5.38	28.9	5
6	0.1604	0.1620	0.1628	0.0035	0.0043	0.1639	0.1671	80.00	12.50	0.242	0.3952	0.004940	0.001195	6.03	36.4	6
7	0.1429	0.1443	0.1450	0.0034	0.0041	0.1463	0.1491	63.51	15.75	0.241	0.4981	0.007843	0.001890	6.75	45.6	7
8	0.1272	0.1285	0.1292	0.0033	0.0040	0.1305	0.1332	50.39	19.85	0.240	0.6281	0.01246	0.002791	7.57	57.3	8
9	0.1133	0.1144	0.1150	0.0032	0.0039	0.1165	0.1189	39.98	25.0	0.239	0.7925	0.01982	0.004737	8.48	71.9	9
10	0.1009	0.1019	0.1024	0.0031	0.0037	0.1040	0.1061	31.74	31.5	0.238	0.9988	0.03147	0.007490	9.50	90.3	10
11	0.0898	0.0907	0.0912	0.0030	0.0036	0.0928	0.0948	25.16	39.8	0.237	1.26	0.0501	0.0119	10.6	112	11
12	0.0800	0.0808	0.0812	0.0029	0.0035	0.0829	0.0847	20.03	49.9	0.236	1.59	0.0794	0.0187	11.9	142	12
13	0.0713	0.0720	0.0724	0.0028	0.0033	0.0741	0.0757	15.89	62.9	0.235	2.00	0.126	0.0296	13.3	177	13
14	0.0635	0.0641	0.0644	0.0032	0.0038	0.0667	0.0682	12.60	82.9	0.230	2.52	0.200	0.0460	14.8	219	14
15	0.0565	0.0571	0.0574	0.0030	0.0035	0.0595	0.0609	10.04	99.6	0.229	3.18	0.317	0.0726	16.6	276	15
16	0.0503	0.0508	0.0511	0.0029	0.0034	0.0532	0.0545	7.95	126	0.228	4.02	0.506	0.115	18.5	342	16
17	0.0448	0.0453	0.0455	0.0028	0.0033	0.0476	0.0488	6.33	158	0.226	5.05	0.798	0.180	20.7	428	17
18	0.0399	0.0403	0.0405	0.0026	0.0032	0.425	0.0437	5.03	199	0.224	6.39	1.27	0.284	23.1	534	18
19	0.0355	0.0359	0.0361	0.0025	0.0030	0.0380	0.0391	3.99	251	0.223	8.05	2.02	0.450	25.9	671	19
20	0.0317	0.0320	0.0322	0.0023	0.0029	0.0340	0.0351	3.18	314	0.221	10.1	3.18	0.703	28.9	835	20
21	0.0282	0.0285	0.0286	0.0022	0.0028	0.0302	0.0314	2.53	395	0.219	12.8	5.06	1.11	32.3	1,043	21
22	0.0250	0.0253	0.0254	0.0021	0.0027	0.0271	0.0281	2.00	500	0.217	16.2	8.10	1.76	36.1	1,303	22
23	0.0224	0.0226	0.0227	0.0020	0.0026	0.0244	0.0253	1.60	625	0.215	20.3	12.7	2.73	40.2	1,616	23

(Continued)

TABLE 9-1 Wire Chart—Round Heavy Film Insulated Solid Copper* (*Continued*)

	Diameter over Bare, in			Insulation Additions		Diameter over Insulation			Weight		Resistance at 20°C (68°F)			Turns		
AWG	Minimum	Nominal	Maximum	Minimum	Maximum	Minimum	Maximum	Pounds per 1000 Feet	Feet per Pound	Pounds per Cubic Inch	Ohms per 1000 Feet	Ohms per Pounds	Ohms per Cubic Inch	Per Linear Inch	Per Square Inch	AWG
24	0.0199	0.0201	0.0202	0.0019	0.0025	0.0218	0.0227	1.26	794	0.211	25.7	20.4	4.30	44.8	2,007	24
25	0.0177	0.0179	0.0180	0.0018	0.0023	0.0195	0.0203	1.00	1,000	0.210	32.4	32.4	6.80	50.1	2,510	25
26	0.0157	0.0159	0.0160	0.0017	0.0022	0.0174	0.0182	0.794	1,259	0.208	41.0	51.6	10.7	56.0	3,136	26
27	0.0141	0.0142	0.0143	0.0016	0.0021	0.0157	0.0164	0.634	1,577	0.205	51.4	81.1	16.6	62.3	3,831	27
28	0.0125	0.0126	0.0127	0.0016	0.0020	0.0141	0.0147	0.502	1,992	0.202	65.3	130	26.3	69.4	4,816	28
29	0.0112	0.0113	0.0114	0.0015	0.0019	0.0127	0.0133	0.405	2,469	0.200	81.2	200	40.0	76.9	5,914	29
30	0.0099	0.0100	0.0101	0.0014	0.0018	0.0113	0.0119	0.318	3,145	0.197	104	327	64.4	86.2	7,430	30
31	0.0088	0.0089	0.0090	0.0013	0.0018	0.0101	0.0108	0.253	4,000	0.193	131	520	100	96	9,200	31
32	0.0079	0.0080	0.0081	0.0012	0.0017	0.0091	0.0098	0.205	4,900	0.191	162	790	151	106	11,200	32
33	0.0070	0.0071	0.0072	0.0011	0.0016	0.0081	0.0088	0.162	6,200	0.189	206	1,270	240	118	13,900	33
34	0.0062	0.0063	0.0064	0.0010	0.0014	0.0072	0.0078	0.127	7,900	0.189	261	2,060	388	133	17,700	34
35	0.0055	0.0056	0.0057	0.0009	0.0013	0.0064	0.0070	0.101	9,900	0.187	331	3,280	613	149	22,200	35
36	0.0049	0.0050	0.0051	0.0008	0.0012	0.0057	0.0063	0.0805	12,400	0.186	415	5,150	959	167	27,900	36
37	0.0044	0.0045	0.0046	0.0008	0.0011	0.0052	0.0057	0.0655	15,300	0.184	512	7,800	1,438	183	33,500	37
38	0.0039	0.0040	0.0041	0.0007	0.0010	0.0046	0.0051	0.0518	19,300	0.183	648	12,500	2,289	206	42,400	38
39	0.0034	0.0035	0.0036	0.0006	0.0009	0.0040	0.0045	0.0397	25,200	0.183	847	21,300	3,904	235	52,200	39
40	0.0030	0.0031	0.0032	0.0006	0.0008	0.0036	0.0040	0.0312	32,100	0.183	1,080	34,600	6,335	263	69,200	40
41	0.0027	0.0028	0.0029	0.0005	0.0007	0.0032	0.0036	0.0254	39,400	0.183	1,320	52,000	9,510	294	86,400	41
42	0.0024	0.0025	0.0026	0.0004	0.0006	0.0028	0.0032	0.0203	49,300	0.182	1,660	81,800	14,883	328	107,600	42

*courtesy Belden Corp.

TABLE 9-2 Stranded Wire Equivalent Chart

Solid Equivalent	Size per Strand										
	34	35	36	37	38	39	40	41	42	43	44
15	80	100									
16	64	80	100								
17	50	64	80	100							
18	40	50	64	80	100						
19	32	40	50	64	80	100					
20	25	32	40	50	64	80	100				
21	20	25	32	40	50	64	80	100			
22	16	20	25	32	40	50	64	80	100		
23	12	16	20	25	32	40	50	64	80	100	
24	10	12	16	20	25	32	40	50	64	80	100
25	8	10	12	16	20	25	32	40	50	64	80
26	6	8	10	12	16	20	25	32	40	50	64
27	5	6	8	10	12	16	20	25	32	40	50
28	4	5	6	8	10	12	16	20	25	32	40
29		4	5	6	8	10	12	16	20	25	32
30			4	5	6	8	10	12	16	20	25
31				4	5	6	8	10	12	16	20
32					4	5	6	8	10	12	16
33						4	5	6	8	10	12
34							4	5	6	8	10
35								4	5	6	8
36									4	5	6
37										4	5
38											4

9.2 MPP TOROIDAL COILS

MPP Toroidal cores are manufactured by pulverizing a magnetic alloy consisting of approximately 2-percent molybdenum, 81-percent nickel, and 17-percent iron into a fine powder, insulating the powder with a ceramic binder to form a uniformly distributed air gap, and then compressing it into a toroidal core at extremely high pressures. Finally, the cores are coated with an insulating finish.

Molypermalloy powder cores (MPP cores) result in extremely stable inductive components for use below a few hundred kilohertz. Core losses are low over a wide range of available permeabilities. Inductance remains stable with large changes in flux density, frequency, temperature, and DC magnetization due to high resistivity, low hysteresis, and low eddy-current losses.

There are mainly two dominant manufacturers of MPP toroids: Magnetics Incorporated and Arnold Engineering.

Characteristics of Cores

MPP cores are categorized according to size, permeability, and temperature stability. Generally, the largest core size that physical and economical considerations permit should be chosen. Larger cores offer higher *Q*s, since flux density is lower due to the larger

TABLE 9-3 Toroidal Core Dimensions

OD, in	ID, in	HT, in	Cross Section, cm^2	Path Length, cm	Window Area, circular mils	Wound OD, in	Coil HT, in
0.310	0.156	0.125	0.0615	1.787	18,200	$^{11}/_{32}$	$^{3}/_{16}$
0.500	0.300	0.187	0.114	3.12	75,600	$^{19}/_{32}$	$^{9}/_{32}$
0.650	0.400	0.250	0.192	4.11	140,600	$^{25}/_{32}$	$^{3}/_{8}$
0.800	0.500	0.250	0.226	5.09	225,600	1	$^{3}/_{8}$
0.900	0.550	0.300	0.331	5.67	277,700	$1^{3}/_{32}$	$^{1}/_{2}$
1.060	0.580	0.440	0.654	6.35	308,000	$1^{1}/_{4}$	$^{5}/_{8}$
1.350	0.920	0.350	0.454	8.95	788,500	$1^{5}/_{8}$	$^{5}/_{8}$
1.570	0.950	0.570	1.072	9.84	842,700	$1^{7}/_{8}$	$^{7}/_{8}$
2.000	1.250	0.530	1.250	12.73	1,484,000	$2^{3}/_{8}$	$1^{1}/_{8}$

NOTE: Core dimensions are before finish.

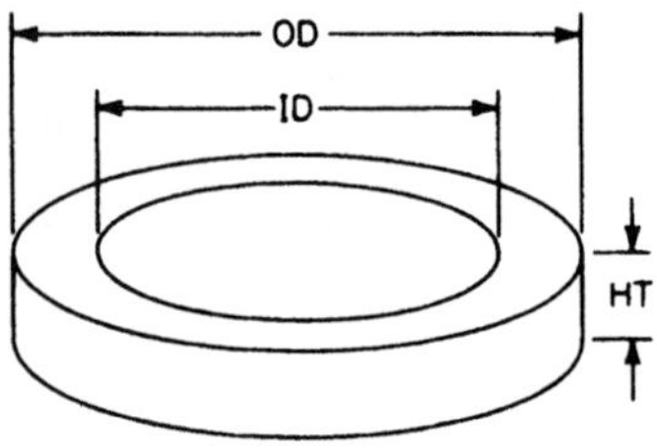

cross-sectional area, resulting in lower core losses. The larger window area also reduces the copper losses.

Cores range in size from an OD of 0.140 to 5.218 in. Table 9-3 contains the physical data for some selected core sizes, as well as the approximate overall dimensions for the wound coil.

Available core permeabilities range from 14 to 550. The lower permeabilities are more suitable for use at the higher frequencies since the core losses are lower. Table 9-4 lists the A_L (inductance per 1000 turns) and ohms per henry for cores with a permeability of 125. For other permeabilities, the A_L is directly proportional to μ and the ohms per henry is inversely proportional to μ. The ohms per henry corresponds to the DC resistance factor when the core

TABLE 9-4 Electrical Properties for $\mu = 125$

OD, in	A_L, mH	Ω/H
0.310	52	900
0.500	56	480
0.650	72	160
0.800	68	220
0.900	90	150
1.060	157	110
1.350	79	80
1.570	168	45
2.000	152	30

TABLE 9-5 Wire Capacities of Standard Toroidal Cores

	Core OD, in								
Wire Size	0.310	0.500	0.650	0.800	0.900	1.060	1.350	1.570	2.000
25		75	148	189	257	284	750	930	1,735
26		95	186	238	323	357	946	1,172	2,180
27		119	235	300	406	450	1,190	1,470	2,750
28		150	295	377	513	567	1,500	1,860	3,470
29		190	375	475	646	714	1,890	2,350	4,365
30	59	238	472	605	814	925	2,390	2,960	5,550
31	74	300	595	765	1,025	1,180	3,000	3,720	7,090
32	94	376	750	985	1,290	1,510	3,780	4,700	9,000
33	118	475	945	1,250	1,625	1,970	4,763	5,920	11,450
34	150	600	1,190	1,580	2,050	2,520	6,000	7,440	14,550
35	188	753	1,500	2,000	2,585	3,170	7,560	9,400	18,500
36	237	950	1,890	2,520	3,245	4,000	9,510	11,840	23,500
37	300	1,220	2,380	3,170	4,100	5,050	12,000	14,880	30,000
38	378	1,550	3,000	4,000	5,175	6,300	15,150	18,800	38,000
39	476	1,970	3,780	5,050	6,510	8,000	19,050	23,680	48,500
40	600	2,500	4,750	6,300	8,200	10,100	24,000	30,000	61,300
41	755								
42	950								
43	1,200								
44	1,510								
45	1,900								

window is approximately 50-percent utilized. Full window utilization is not possible, however, because a hole must be provided for a shuttle in the coil winding machine which applies the turns. The corresponding wire chart for each size core is given in Table 9-5.

DC Bias and AC Flux Density. Under conditions of DC bias current, MPP cores may exhibit a reduction in permeability because of the effects of saturation. Figure 9-2 illustrates

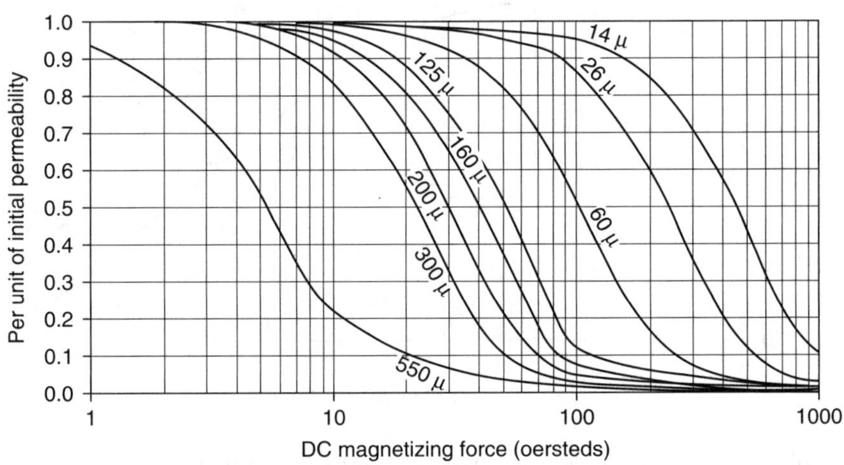

FIGURE 9-2 Permeability versus DC bias. *(Courtesy Magnetics Inc.)*

this effect, which is more pronounced for higher permeabilities. To use these curves, the magnetizing force for the design is computed using Equation (9-2). If the reduction in permeability and resulting inductance decrease is no more than about 30 percent, the turns can be increased to compensate for the effect of the bias. If the decrease in permeability is more than 30 percent, the permeability may further decrease faster than N^2 if turns are added. A larger core would then be required.

The core permeability will also change as a function of the AC flux density. This effect is shown in Figure 9-3*a*. The flux density can be computed using Equation (9-6). As the AC flux density is increased, the permeability will rise initially and then fall beyond approximately 2000 G. Flux density can be reduced by going to a larger core size. Frequency will also impact permeability for higher permeability cores. This effect is illustrated in Figure 9-3*b*.

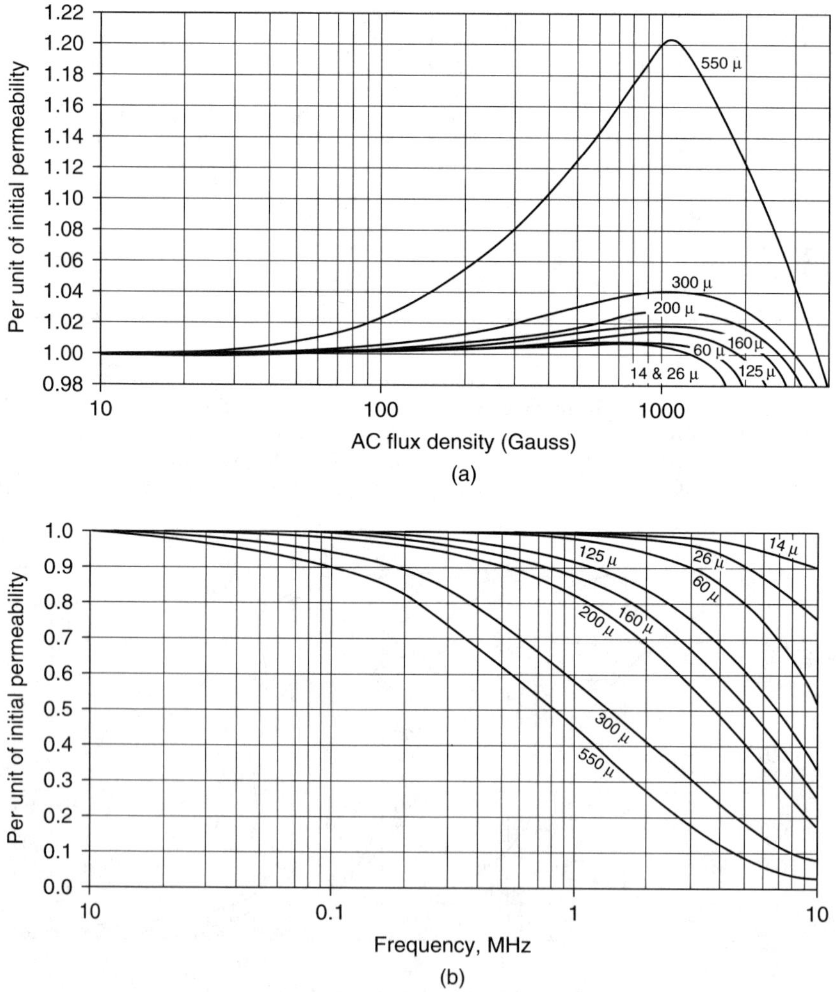

FIGURE 9-3 Permeability versus AC flux density and frequency. *(Courtesy Magnetics Inc.)*

TABLE 9-6 Stabilized Cores

Stabilization	Temperature Range, °C	Inductance Stability, 1%
M4	−65 to +125	±0.25
W4	−55 to +85	±0.25
D4	0 to +55	±0.1

The core losses can be assumed to be relatively constant for flux densities below 200 G. For higher excitations, the Q may be adversely affected, however.

Temperature Stability. MPP cores are available in three categories of temperature stabilities: standard, stabilized, and linear (where linear is a subset of stabilized). Stabilizing techniques are based on the addition of a small amount of special compensating alloys having Curie points within the temperature range of operation. (The Curie point is the temperature where the material becomes nonmagnetic.) As each Curie point is passed, the particles act as distributed air gaps which can be used to compensate for permeability changes of the basic alloy so as to maintain the inductance at almost a constant value.

Standard cores, also called "A" stabilization, have no guaranteed limits for permeability variations with temperature. Typically, the temperature coefficient of permeability is positive and ranges from 25 to 100 ppm/°C, but these limits are not guaranteed. Stabilized cores, on the other hand, have guaranteed limits over wide temperature ranges. These cores are available in three degrees of stabilization: M4, W4, and D4 from Magnetics Incorporated. The limits are as given in Table 9-6.

Because of the nature of the compensation technique, the slope of inductance change with temperature can be either positive or negative within the temperature range of operation, but will not exceed the guaranteed limits.

Polystyrene capacitors maintain a precise temperature coefficient of −120 ppm/°C from −55 to +85°C. Cores are available which provide a matching positive temperature coefficient so that a constant LC product can be maintained over the temperature range of operation. These cores are said to have linear temperature characteristics. Two types of linear temperature characteristics can be obtained. The $L6$ degree of stabilization has a positive temperature coefficient ranging between +25 and +90 ppm/°C, from −55° to +25°C, and between +65 and +150 ppm/°C over the temperature range +25 to +85°C.

In general, the temperature stability of MPP cores is affected by factors such as winding stresses and moisture. To minimize these factors, suitable precautions can be taken. Before adjustment, the coils should be temperature-cycled from −55 to +100°C at least once. After cooling to room temperature, the coils are adjusted and should be kept dry until encapsulated. Encapsulation compounds should be chosen carefully to minimize mechanical stresses. A common technique involves dipping the coils in a silicon rubber compound for a cushioning and sealing effect prior to encapsulation.

Typically, stabilization performance is not guaranteed above excitations of around 100 G.

Winding Methods for Q Optimization. At low frequencies, the dominant losses are caused by the DC resistance of the winding. The major consideration then is to utilize the maximum possible winding area. Distributed capacity is of little consequence unless the inductance is extremely high, causing self-resonance to occur near the operating frequency range.

The most efficient method of packing the most turns on a toroidal core is to rotate the core continuously in the same direction in the winding machine until maximum capacity is obtained. This technique is called the *360° method.*

At medium frequencies, special winding techniques are required to minimize distributed capacity. By winding half the turns over a 180° sector of the core in a back-and-forth manner and then applying the remaining turns over the second half in the same fashion, the capacity will be reduced. This technique is called the *two-section method.* If after winding half the turns the coil is removed from the machine, turned over, reinserted, and completed, the two starts can be tied together, and the resulting coil using the two finishes as terminals has even less capacity. This modified two-section winding structure is commonly referred to as *two-section reversed.*

If the core is divided into four 90° quadrants and each sector is completed in a back-and-forth winding fashion using one-fourth the total turns, a four-section coil will be obtained. A four-section winding structure has lower distributed capacity than the two-section method. However, whereas the two-section and 360° methods correspond to the wire chart of Table 9-5, the wire must be reduced one size for a four-section coil, resulting in more copper losses.

At frequencies near 50 kHz or higher, the distributed capacity becomes a serious limiting factor of the obtainable Q from both self-resonance and dielectric losses of the winding insulation. The optimum winding method for minimizing distributed capacity is *back-to-back progressive.* Half the total number of turns are applied over a 180° sector of the core by gradually filling up a 30° sector at a time. The core is then removed from the machine, turned over, reinserted in the machine, and completed in the same manner. The two starts are then joined and the two finish leads are used for the coil terminals. A barrier is frequently used to separate the two finish leads for further capacity reduction. Litz wire can be combined with the back-to-back progressive winding method to reduce skin-effect losses. As with a four-section winding, the wire must be reduced one size from the wire chart of Table 9-5. If the core is not removed from the winding machine after completion of a 180° sector but instead is continued for approximately 360°, a *straight-progressive* type of winding is obtained. Although slightly inferior to the back-to-back progressive, the reduction in winding time will frequently warrant its use.

When a coil includes a tap, a coefficient of magnetic coupling near unity is desirable to avoid leakage inductance. The 360° winding method will have a typical coefficient of coupling of about 0.99 for permeabilities of 125 and higher. A two-section winding has a coefficient of coupling near 0.8. The four-section and progressive winding methods result in coupling coefficients of approximately 0.3, which is usually unacceptable. A compromise between the 360° and the progressive method to improve coupling for tapped inductors involves applying the turns up to the tap in a straight-progressive fashion over the total core. The remaining turns are then distributed completely over the initial winding, also using the straight-progressive method.

In general, to obtain good coupling, the portion of the winding up to the tap should be in close proximity to the remainder of the winding. However, this results in higher distributed capacity, so a compromise may be desirable. Higher core permeability will increase the coupling but can sometimes result in excessive core loss. The higher the ratio of overall to tap inductance, the lower the corresponding coefficient of coupling. Inductance ratios of 10 or more should be avoided if possible.

Designing MPP Toroids from Q Curves. The Q curves given in Figure 9-4 are based on empirical data using the 360° winding method. The range of distributed capacity is typically between 10 and 25 pF for cores under 0.500 in OD, 25 to 50 pF for cores between 0.500 and 1.500 in OD, and 50 to 80 pF for cores over 1.500 in OD.

Curves are presented for permeabilities of 60 and 125 and for a range of core sizes. For a given size and permeability, the Q curves converge on the low-frequency portion of the

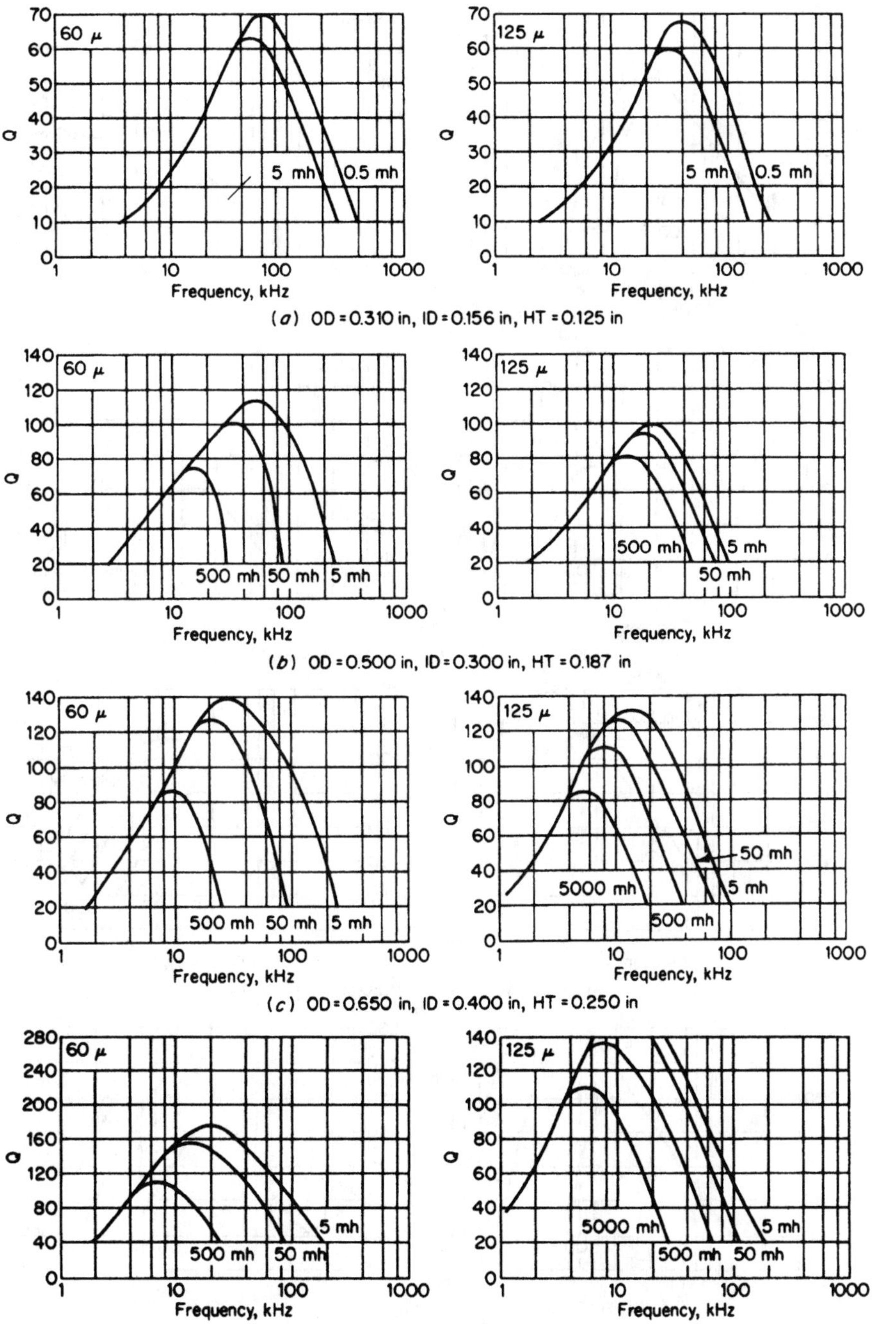

FIGURE 9-4 *Q* curves of MPP toroidal cores. *(Courtesy Magnetics Inc.)*

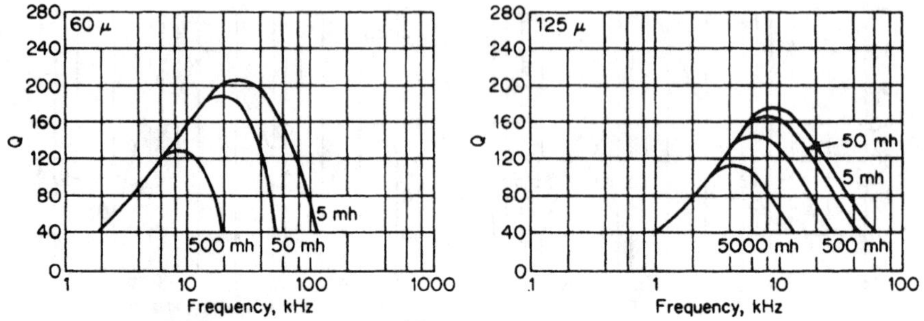

(*e*) OD = 0.900 in, ID = 0.550 in, HT = 0.300 in

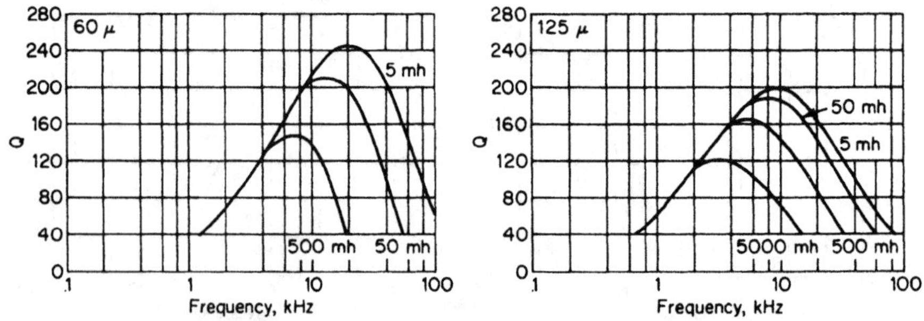

(*f*) OD = 1.060 in, ID = 0.580 in, HT = 0.440 in

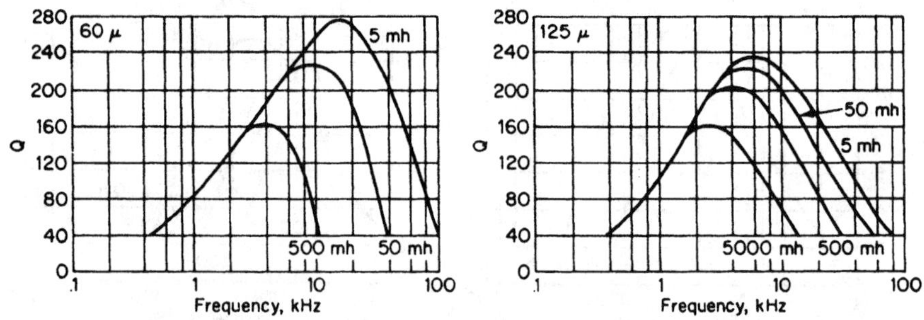

(*g*) OD = 1.350 in, ID = 0.920 in, HT = 0.350 in

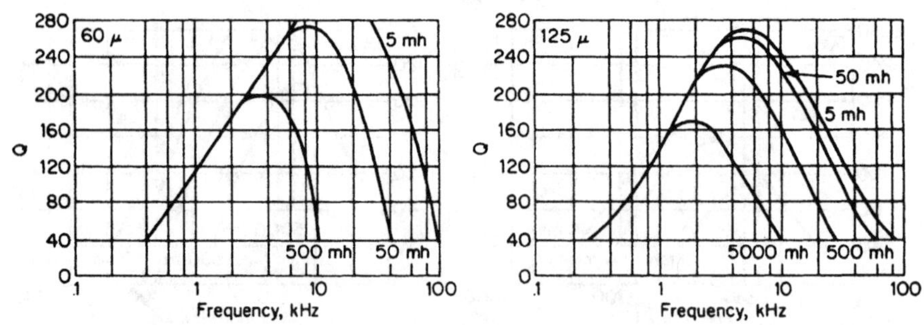

(*h*) OD = 1.570 in, ID = 0.950 in, HT = 0.570 in

FIGURE 9-4 (*Continued*)

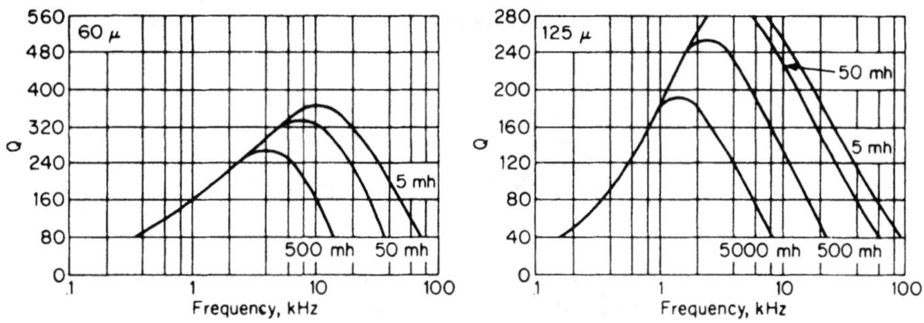

(*i*) OD = 2.000 in, ID = 1.250 in, HT = 0.530 in

FIGURE 9-4 (*Continued*)

curve, where the losses are determined almost exclusively by the DC resistance of the winding. The Q in this region can be approximated by

$$Q = \frac{2\pi f}{\Omega/\text{H}} \tag{9-13}$$

where f is the frequency of interest and Ω/H is the rated ohms per henry of the core (as given by Table 9-4) and modified for permeability if other than 125.

As the frequency is increased, the curves start to diverge and reach a *maximum* at a frequency where the *copper and core losses are equal*. Beyond this region, the core losses begin to dominate along with increased dielectric losses as self-resonance is approached, causing the Q to roll off dramatically. It is always preferable to operate on the rising portion of Q curves since the losses can be tightly controlled and the effective inductance remains relatively constant with frequency.

Example 9-1 Design of a Toroidal Inductor

Required:

Design a toroidal inductor having an inductance of 1.5 H and a minimum Q of 55 at 1 kHz. The coil must pass a DC current of up to 10 mA and operate with AC signals as high as 10 V_{rms} with negligible effect.

Result:

(*a*) Using the Q curves of Figure 9-4, a 1.060-in-diameter core having a μ of 125 will have a Q of approximately 60 at 1 kHz. The required number of turns is found from

$$N = 10^3\sqrt{\frac{L}{A_L}} = 10^3\sqrt{\frac{1.5}{0.157}} = 3090 \tag{9-10}$$

where the value of A_L was given in Table 9-4. The corresponding wire size is determined from Table 9-5 as No. 35 AWG.

(*b*) To estimate the effect of the DC current, compute the magnetizing force from

$$H = \frac{4\pi NI}{10\ \text{mL}} = \frac{4\pi \times 3090 \times 0.01}{10 \times 6.35} = 6.1\ \text{Oe} \tag{9-2}$$

where mL, the magnetic path length, is obtained from Table 9-3. According to Figure 9-2, the permeability will remain essentially constant.

(*c*) An excitation level of 10 V_{rms} at 1000 Hz results in a flux density of

$$B = \frac{E_{rms}}{4.44NfA \times 10^{-8}}$$

$$= \frac{10}{4.44 \times 3090 \times 1000 \times 0.654 \times 10^{-8}}$$

$$= 111 \text{ G} \qquad (9\text{-}6)$$

where *A*, the core's cross-sectional area, is also found in Table 9-3. Figure 9-3 indicates that at the calculated flux density the permeability change will be of little consequence.

9.3 FERRITE POT CORES

Ferrites are ceramic structures created by combining iron oxide with oxides or carbonates of other metals such as manganese, nickel, or magnesium. The mixtures are pressed, fired in a kiln at very high temperatures, and machined into the required shapes.

The major advantage of ferrites over MPP cores is their high resistivity so that core losses are extremely low even at higher frequencies where eddy-current losses become critical. Additional properties such as high permeability and good stability with time and temperature often make ferrites the best core-material choice for frequencies from 10 kHz to well in the megahertz region.

The Pot Core Structure

A typical pot core assembly is shown in Figure 9-5. A winding supported on a bobbin is mounted in a set of symmetrical ferrite pot core halves. The assembly is held rigid by a metal clamp. An air gap is introduced in the center post of each half since only the outside surfaces of the pot core halves mate with each other. By introducing an adjustment slug containing a ferrite sleeve, the effect of the gap can be partially neutralized as the slug is inserted into the gap region.

A ferrite pot core has a number of distinct advantages over other approaches. Since the wound coil is contained within the ferrite core, the structure is self-shielding because stray magnetic fields are prevented from entering or leaving the structure.

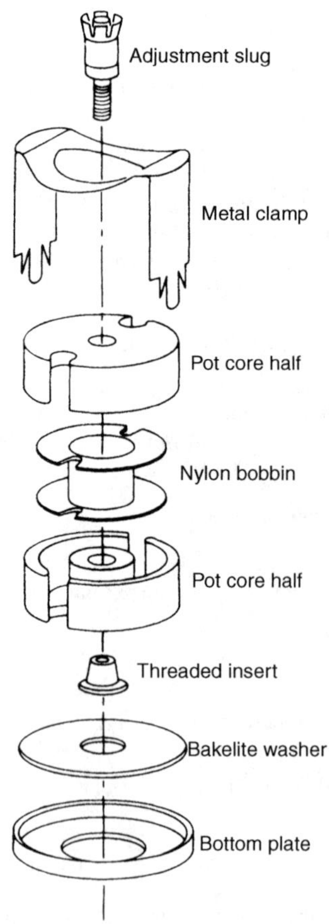

FIGURE 9-5 A typical ferrite pot core assembly.

TABLE 9-7 Standard Pot Core Sizes

Manufacturer Core Size Designation	Diameter (max) in	Diameter (max) mm	Height (max) in	Height (max) mm	Cross Section cm^2	Path Length cm
905	0.366	9.3	0.212	5.4	0.101	1.25
1107	0.445	11.3	0.264	6.7	0.167	1.55
1408	0.559	14.2	0.334	8.5	0.251	1.98
1811	0.727	18.5	0.426	10.8	0.433	2.58
2213	0.858	21.8	0.536	13.6	0.635	3.15
2616	1.024	26.0	0.638	16.2	0.948	3.76
3019	1.201	30.5	0.748	19.0	1.38	4.52
3622	1.418	36.0	0.864	22.0	2.02	5.32
4229	1.697	43.1	1.162	29.5	2.66	6.81

Very high *Q*s and good temperature stability can be obtained by selecting the appropriate materials and by controlling the effective permeability μ_e through the air gap. Fine adjustment of the effective permeability is accomplished using the adjustment slug.

Compared with other structures, ferrite pot cores are more economical. Bobbins can be rapidly wound using multiple-bobbin winding machines in contrast to toroids, which must be individually wound. Assembly and mounting is easily accomplished using a variety of hardware available from the ferrite manufacturer. Printed circuit-type brackets and bobbins facilitate the use of pot cores on printed circuit boards.

Pot cores have been standardized into nine international sizes ranging from 9 by 5 mm to 42 by 29 mm, where these dimensions represent the diameter and height, respectively, of a pot core pair. These sizes are summarized in Table 9-7. Two of the major suppliers of ferrite pot cores are Magnetics Inc. and Ferroxcube Corporation.

Electrical Properties of Ferrite Pot Cores. Ferrite materials are available having a wide range of electrical properties. Table 9-8 lists some representative materials from Magnetics Inc. and Ferroxcube Corporation and their properties.

Let's first consider initial permeability as a function of frequency. A material must be chosen that provides uniform permeability over the frequency range of interest. It is evident from Figure 9-6 that for that particular material shown, operation must be restricted to below 1.5 MHz. *Even though a relatively large variation may occur even over this restricted range, a gapped core will show much less variation.* As a rule-of-thumb, lower permeability materials have wider frequency ranges of operation.

Another factor of importance is the stability of permeability versus temperature. Figure 9-7 illustrates the behavior of permeability versus temperature for Magnetics Inc. type R, F, and P materials. Although the temperature coefficient appears to have a high positive temperature coefficient, for a gapped core it is much less—depending on gap size, of course, the larger the gap the lower the temperature coefficient.

The temperature coefficient of permeability is an important factor in *LC*-tuned circuits and filters. Upon the proper selection of size, material, and gap, a positive linear temperature coefficient can sometimes be obtained to precisely match the negative temperature coefficient of polystyrene or polypropylene capacitors to maintain a constant *LC* product. In other cases, the objective would be to obtain as low a combined temperature coefficient as possible.

TABLE 9-8 Ferrite Material Properties

Magnetics Inc.						
Material	A	D	G	R	P	F
Initial permeability μ_i	750 ±20%	2000 ±20%	2300 ±20%	2300 ±25%	2500 ±25%	3000 ±20%
Maximum usable frequency (50% roll-off in μ_I)	<9 MHz	<4 MHz	<4 MHz	<1.5 MHz	<1.2 MHz	<1.3 MHz
Saturation flux density at 15 Oe in gauss (B_m)	4600	3800	4600	5000	5000	4900

Ferroxcube Corporation			
Material	3B7	3D3	3H3
Initial permeability μ_i	2300	750	2000
Maximum usable frequency (50% roll-off in μ_I)	<0.1 MHz	<2 MHz	<0.2 MHz
Saturation flux density at 15 Oe in gauss (B_m)	4400	3800	3600

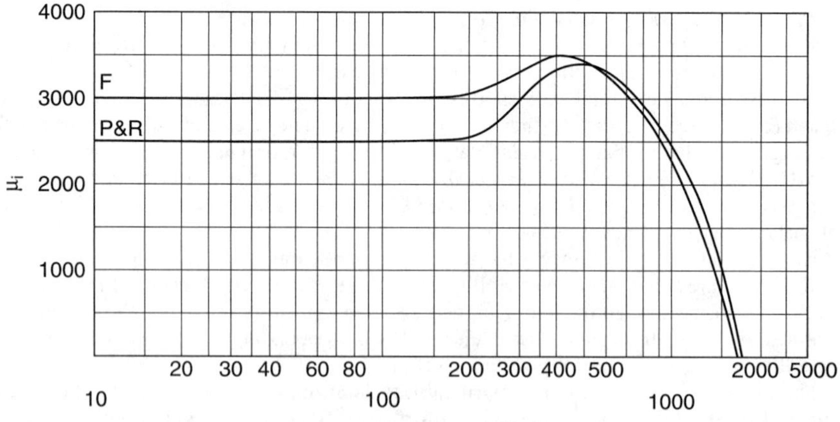

FIGURE 9-6 Initial permeability μ_i versus frequency for F, P, and R materials. (*Courtesy Magnetics Inc.*)

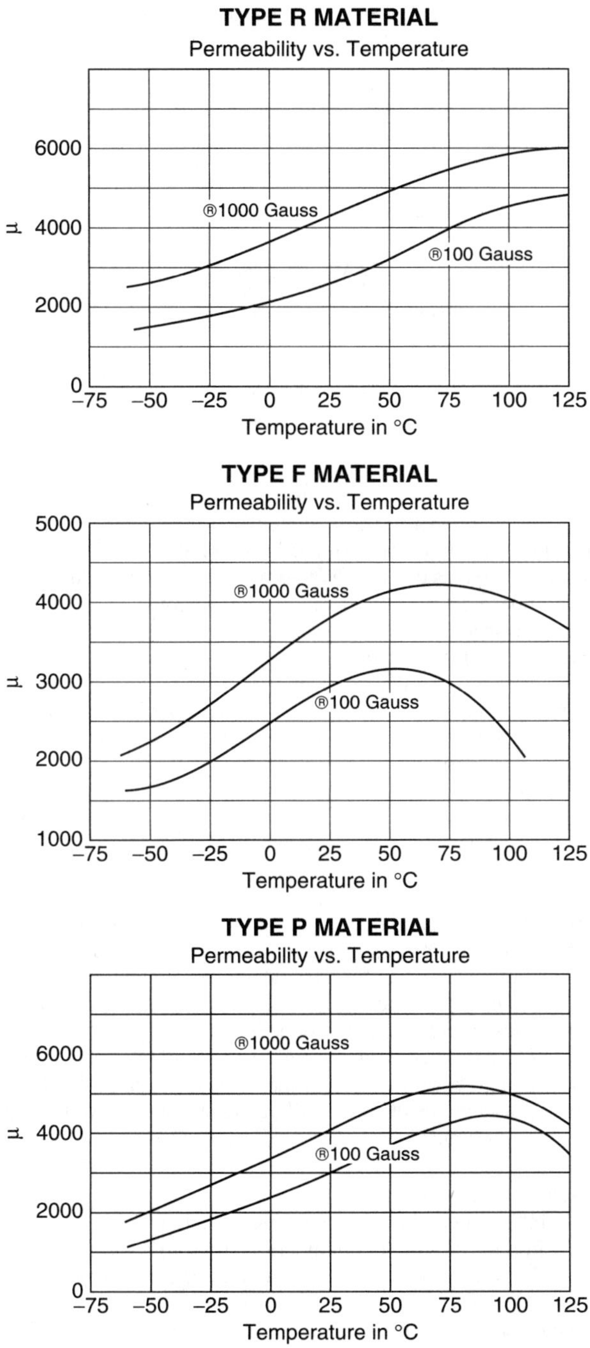

FIGURE 9-7 Initial permeability μ_i versus temperature for R, F, and P Materials. *(Courtesy Magnetics Inc.)*

TABLE 9-9 The Electrical Properties of Standard Core Sizes for $A_L = 250$ mH

Size Designation	μ_e	Ω/H
905	230	700
1107	184	570
1408	156	380
1811	119	260
2213	98	190
2616	78	170
3019	64	140
3622	52	130
4229	51	80

For a given core, a wide range of A_L factors are available. This parameter is determined by the effective permeability of the core, which in turn is controlled by the initial permeability of the ferrite material, the core dimensions, and the size of the air gap introduced in the center post. Table 9-9 lists the effective permeability and the ohms-per-henry rating for all the standard core sizes, where the A_L is maintained constant at 250 mH. For other values, the effective permeability is proportional to the rated A_L and the ohms per henry is inversely proportional to this same factor.

Pot cores are also available that have no gap. However, since the resulting electrical characteristics are essentially equivalent to the initial properties of the ferrite material, these cores should be restricted to applications where high permeability is the major requirement, such as for wideband transformers.

Most ferrite materials saturate with AC flux densities in the region of 3000–4000 G. AC excitation levels should be kept below a few hundred gauss to avoid nonlinear saturation effects and to minimize core losses. Superimposing a DC bias will also cause saturation when the ampere-turns exceed a critical value dependent upon the core size, material, and gap.

For a specific pot core size and rated A_L, a variety of different adjusters is available featuring ranges of adjustment from a few percent to over 25 percent. For most applications, an adjustment range of about 10 or 15 percent will be satisfactory since this should cover both the capacitor and core tolerances. The larger-range slugs should be avoided for precision requirements since the adjustment may be too coarse.

To compensate for the effect of an adjuster, the nominal A_L rating of the core should be increased by a percentage equal to one-half the adjustment range for the actual turns computation. This is because the resulting A_L will be higher by one-half the range with the slug set for the midrange position.

Winding of Bobbins. Bobbins are normally wound by guiding the wire feed back and forth as the bobbin is rotated. In the event that the inductor's self-resonant frequency becomes a problem, a multisection bobbin can be used to reduce distributed capacitance. One section at a time is filled to capacity. Most bobbin sizes are available with up to three sections.

A wire capacity chart for standard single-section bobbins is shown in Table 9-10. Multisection bobbins have slightly less winding area than single-section bobbins, so the wire gauge may require reduction by one size in some cases.

TABLE 9-10 Wire Capacities of Standard Bobbins

Wire Size	905	1107	1408	1811	2213	2616	3019	3622	4229
20							63	100	160
21						50	79	125	200
22						63	100	160	250
23					50	79	125	200	315
24					63	100	160	250	400
25				50	79	125	200	315	500
26		17		63	100	160	250	400	630
27		21	50	79	125	200	315	500	794
28		27	63	100	160	250	400	630	1,000
29		34	79	125	200	315	500	794	1,260
30		43	100	160	250	400	630	1,000	1,588
31	50	54	125	200	315	500	794	1,260	2,000
32	63	68	160	250	400	630	1,000	1,588	2,520
33	79	86	200	315	500	794	1,260	2,000	3,176
34	100	108	250	400	630	1,000	1,588	2,520	4,000
35	125	136	315	500	794	1,260	2,000	3,176	5,042
36	160	171	400	630	1,000	1,588	2,520	4,000	6,353
37	200	216	500	794	1,260	2,000	3,176	5,042	8,000
38	250	272	630	1,000	1,588	2,520	4,000	6,353	10,086
39	315	343	794	1,260	2,000	3,176	5,042	8,000	
40	400	432	1,000	1,588	2,520	4,000	6,353	10,086	
41	500	544	1,260	2,000	3,176	5,042	8,000		
42	630	686	1,588	2,520	4,000	6,353	10,086		
43	794	864	2,000	3,176	5,042	8,000			
44	1,000	1,089	2,520	4,000	6,353	10,086			
45	1,260	1,372	3,176	5,042	8,000				

Because of the air gap, the permeability is not uniform throughout the interior of the pot core. If the bobbin is only partially filled, the A_L will be reduced as a function of the relative winding height. The decrease in A_L for a 2616 size core can be approximated from the graph of Figure 9-8. This effect becomes more pronounced as the gap is made larger.

To maximize Q over the frequency range of 50 kHz to 1 or 2 MHz, litz wire should be used. The litz wire equivalents to solid wire gauges can be determined from Table 9-2.

Various methods are used by designers to determine the optimum core size, material, and gap for a particular application. Depending upon the requirements, some parameters are more significant than others. Generally, Q and the temperature coefficient are of prime importance. A standard set of Q curves can be obtained from Magnetics Inc. and is also provided on the CD-ROM. These Q curves facilitate the rapid selection of the optimum core for a given Q requirement. In general, larger cores provide a higher Q, a lower temperature coefficient, and better immunity from saturation.

For a given core size, the lowest A_L which provides sufficient Q should be chosen to minimize temperature coefficient and saturation effects. Operation should be restricted to the rising portion of the curves to avoid the effects of core and dielectric losses as well as those of self-resonance.

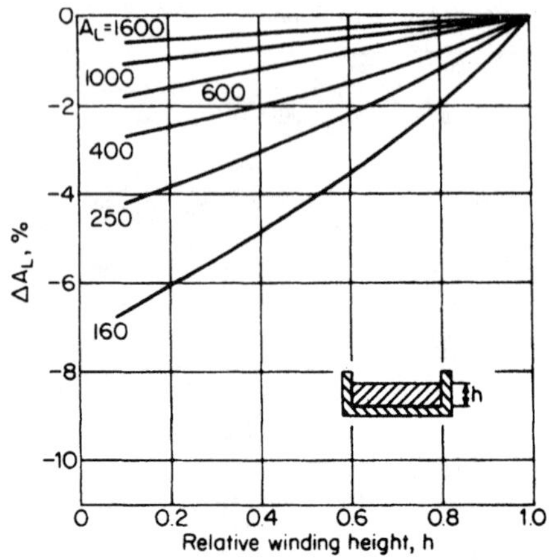

FIGURE 9-8 Change in A_L versus the relative winding height for a 2616-size pot core.

RM Cores. Ferrite cores come in many other shapes besides pot cores. A popular family is the RM cores where RM is an abbreviation for *rectangular module*. These are square or rectangular in shape so they have a more efficient form factor as far as PC board utilization than round pot cores. Multiple RM cores can be physically packed more tightly together than the round pot cores. The magnetic properties are similar although RM cores offer less of a closed magnetic path than the round pot cores so more flux leakage can occur between cores. Figure 9-9 illustrates a cutaway view of an RM core.

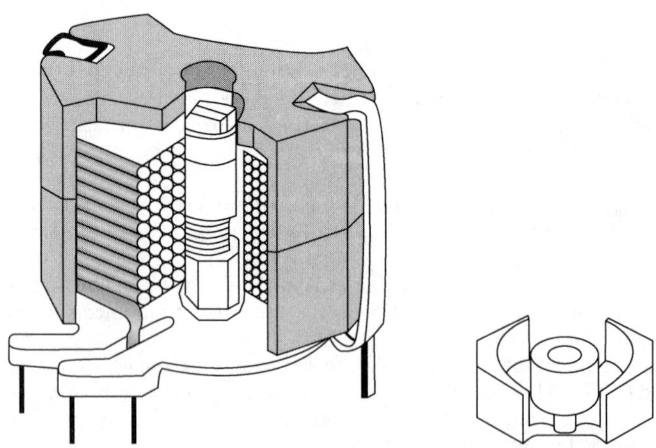

FIGURE 9-9 RM cores: (*a*) a cutaway view; and (*b*) a drawing of the core half. *(Courtesy Magnetics Inc.)*

9.4 HIGH-FREQUENCY COIL DESIGN

Powdered-Iron Toroids

Above 1 or 2 MHz, the core losses of most ferrite materials become prohibitive. Toroidal cores composed of compressed iron powder known as carbonyl iron then become desirable for use up to the VHF range.

These cores consist of finely divided iron particles which are insulated and then compressed at very high pressures into a toroidal form in a manner similar to MPP cores. The high resistance in conjunction with the very small particles results in good high-frequency performance. Permeability can thus be controlled to very tight tolerances, with low temperature coefficients, while the saturation levels of iron are quite high compared to other materials such as ferrite.

Material Selection. A variety of iron powders within the carbonyl iron family are available which are suitable for use over different frequency bands. The more commonly used materials are listed in Table 9-11, along with the nominal permeability, temperature coefficient, and maximum frequency of operation above which the core losses can become excessive. A major supplier of these cores is Micrometals Inc. The Micrometals nomenclature for the various materials is provided as well. In general, a material should be selected that has the highest permeability over the frequency range of operation.

Maximizing Q. For optimum Q, the distributed and stray capacities should be minimized. By maximizing permeability, the number of turns can be kept small. In addition, the turns should be applied as a single-layer winding using the largest wire size that the core will accommodate. However, there is usually a point of diminishing return with increasing wire gauge, so unwieldy wire sizes should be avoided. The turns should be spread evenly around the core except that the start and finish should be separated by a small unwound sector. Litz wire can be used up to about 2 MHz. To approximate the

TABLE 9-11 Iron-Powder Core Materials. *(Courtesy Micrometals Inc.)*

Micrometals Designation	Type of Iron Powder	Frequency Range of Operation	Material Permeability μ	Temperature Stability ppm/°C
−1	Carbonyl C	150 KHz to 3 MHz	20	280
−2	Carbonyl E	250 KHz to 10 MHz	10	95
−3	Carbonyl HP	20 KHz to 1 MHz	35	370
−4	Carbonyl J	3 MHz to 40 MHz	9	280
−6	Carbonyl SF	3 MHz to 40 MHz	8.5	35
−7	Carbonyl TH	1 MHz to 25 MHz	9	30
−8	Carbonyl GQ4	20 KHz to 1 MHz	35	255
−10	Carbonyl W	15 MHz to 100 MHz	6	150
−15	Carbonyl GS6	150 KHz to 3 MHz	25	190
−17	Carbonyl	20 MHz to 200 MHz	4	50
−42	Hydrogen Reduced	300 KHz to 80 MHz	40	550
−0	Phenolic	50 MHz to 350 MHz	1	0

required wire size, the inside circumference of the core can be computed and then divided by the number of turns to yield an approximate wire diameter. A gauge can then be chosen using Table 9-1. Generally, it should be about two sizes smaller than computed. The maximum Q occurs when all of the core losses are equal to the copper loss. At this frequency, the peak of the Q curve occurs. As a rule-of-thumb, the maximum Q frequency is inversely proportional to core size, core permeability, and the square root of core loss.

Winding Methods. The capacitance of a coil winding is known as *Distributed Capacity* and should be kept as low as possible in order to maximize Q. This is mainly because the wire insulation that forms this capacitance is quite lossy, and as a result circuit Q is reduced. The capacitance between adjacent turns is dependant on wire size, spacing, and the location of the turns on the core. The self-capacitance of a toroidal core is least when the coil is wound over a single layer.

The resulting inductance of a single-layer powdered iron toroid can not be precisely predicted due to the effect of leakage inductance. The further apart the turns, the lower the resulting inductance due to the changing flux linkage between turns. For best results in terms of Q, the turns should be evenly spread around the core. In practice, toroidal coils are sometimes tuned by adjusting the separation between turns. Figure 9-10 illustrates the variation of inductance for a 0.5-in OD toroid of carbonyl material having 10 turns of 20 AWG wire as the turns are compressed in a single layer.

To minimize dielectric losses, Teflon-coated wire is occasionally used since this material has an extremely small power factor in comparison with the more commonly used polyurethane insulation. Also, wrapping the core with Teflon tape prior to applying the winding helps to keep stray capacity low. If potting compounds are used, they should be carefully chosen since inductor Qs can be easily degraded upon impregnation by lossy materials.

Table 9-12 lists some physical properties of selected standard-sized powdered-iron toroidal cores. For convenience, the A_L is specified in terms of μH/100 turns. The required number of turns can then be computed from

$$N = 10^2\sqrt{\frac{L}{A_L}} \tag{9-11}$$

where both L and A_L are in microhenrys. Some selected Q curves are illustrated in Figure 9-11.

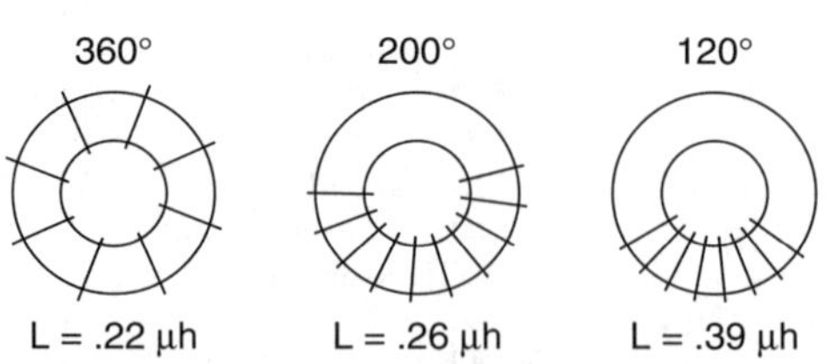

FIGURE 9-10 The effect of an uneven turn distribution around the core. *(Courtesy Micrometals Inc.)*

TABLE 9-12 Powdered-Iron Toroidal Cores

OD, in	ID, in	HT, in	Cross Section, cm	Path Length, cm	Material	A_L, μH/100 turns
0.120	0.062	0.035	0.0065	0.73	HP	28
					C	28
					E	11
					TH	10
					SF	9
					W	8
0.197	0.092	0.082	0.0278	1.57	HP	76
					C	76
					E	30
					TH	26
					SF	24
					W	23
0.309	0.156	0.125	0.0615	1.86	HP	105
					C	105
					E	42
					TH	36
					SF	33
					W	31
0.500	0.300	0.187	0.114	3.19	HP	118
					C	118
					E	47
					TH	40
					SF	38
					W	35
0.800	0.500	0.250	0.226	5.19	HP	146
					C	146
					E	58
					TH	50
					SF	47
					W	42
1.060	0.580	0.440	0.654	6.54	HP	327
					C	327
					E	130
					TH	111
					SF	104
					W	92
1.570	0.965	0.570	1.072	10.11	HP	345
					C	345
					E	138
					TH	117
					SF	110
					W	98

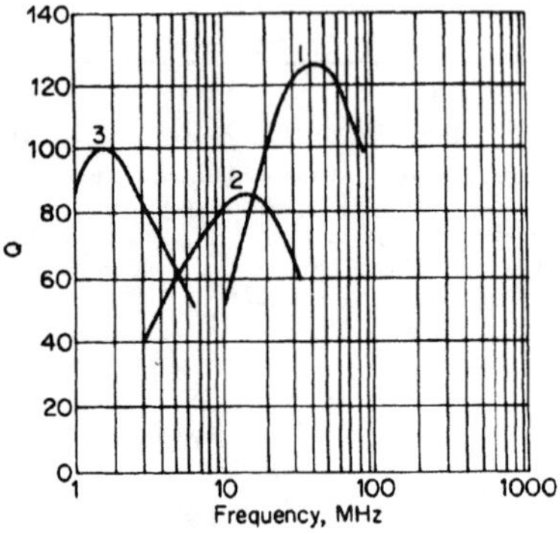

Curve	Material	Winding	L
1	W	11T#30	0.1 μH
2	TH	25T#36	0.6 μH
3	C	50T#40	7 μH

OD = 0.120 in, ID = 0.062 in, HT = 0.035 in

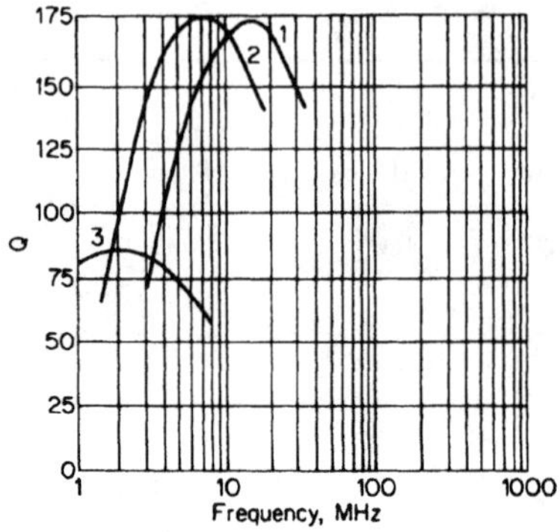

Curve	Material	Winding	L
1	SF	17T#26	1 μH
2	E	29T#28	5 μH
3	C	50T#32	26 μH

OD = 0.309 in, ID = 0.156 in, HT = 0.125 in

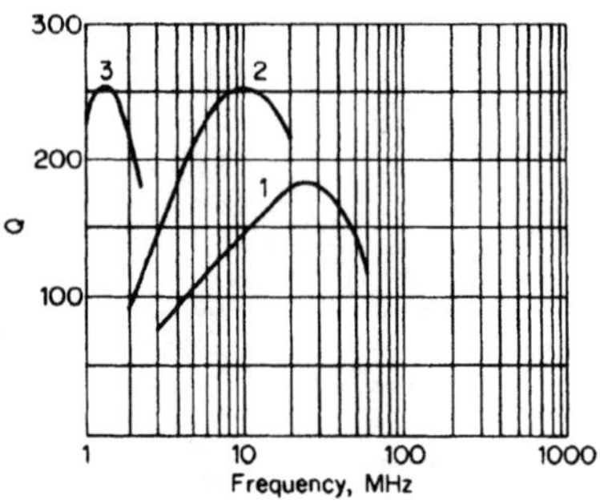

Curve	Material	Winding	L
1	W	10T#20	0.35 μH
2	SF	19T#20	1.4 μH
3	E	125T#15/44	73 μH

OD = 0.500 in, ID = 0.300 in, HT = 0.187 in

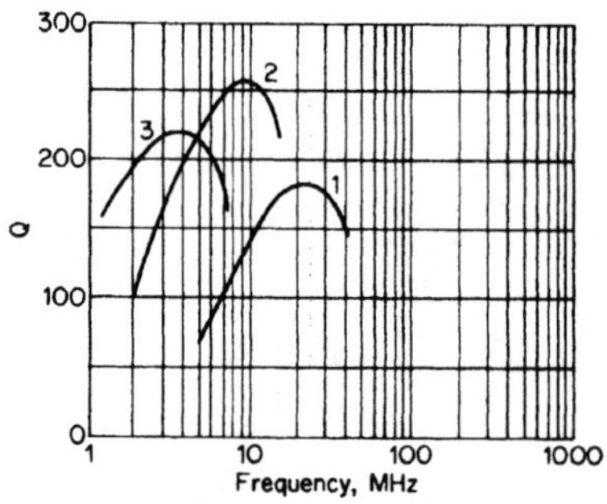

Curve	Material	Winding	L
1	W	10T#20	0.4 μH
2	SF	25T#22	3 μH
3	E	45T#26	12 μH

OD = 0.800 in, ID = 0.500 in, HT = 0.250 in

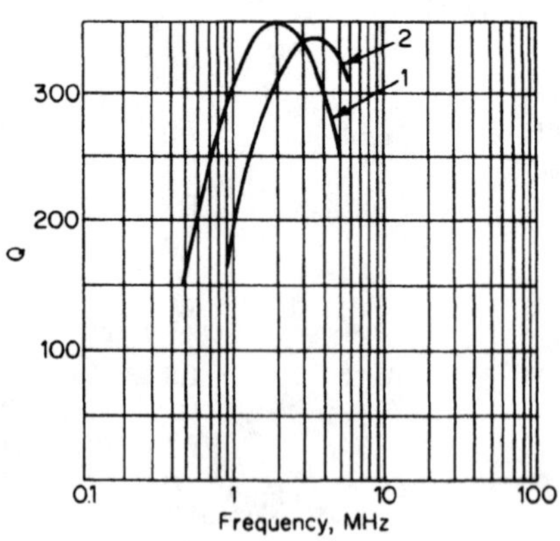

Curve	Material	Winding	L
1	E	40T#20	21 μH
2	E	20T#16	5 μH

OD = 1.06 in, ID = 0.58 in, HT = 0.44 in

FIGURE 9-11 *Q* curves of powdered-iron toroidal cores.

Air-Core Inductors. Coils containing no magnetic materials are said to have an air core. Inductors of this form are useful above 10 MHz. Air has a μ of 1, will not saturate, and has no core losses, so the Q is strictly dependent upon the winding.

For a single-layer solenoid wound on a nonmagnetic material such as ceramic or phenolic, the inductance in henrys can be approximated within a reasonable accuracy by

$$L = N^2 \frac{r^2}{9r + 10l} \times 10^{-6} \tag{9-14}$$

where N is the number of turns, r is the radius (diameter/2), and l is the coil length, with r and l both in inches.

Air-core solenoid inductors have high leakage inductance, so they are easily affected by nearby metallic surfaces. A toroidal shape will have less leakage, since the magnetic field will be better contained. The expression for the inductance of an air-core toroid with a single-layer winding was given by Equation (9-5) with $\mu = 1$.

$$L = \frac{4\pi N^2 \mu A}{\text{mL}} 10^{-9} \tag{9-5}$$

Surface Mount RF Inductors. Surface mount RF inductors (sometimes referred to as "chip" inductors) are wound on a solenoid-like core and encapsulated. The two ends of the coil winding are terminated in metal end caps which are then soldered to the board during assembly. The core itself can consist of phenolic, ceramic, powdered iron, or ferrite depending on the inductance values and core size. Figure 9-12 shows a cross-sectional view of a wound surface mount RF inductor.

Surface mount RF inductors are available in standard EIA sizes such as 0402, 0603, 0805, 1206, and 1210. These numbers describe the approximate body length and width as a multiple of 10 mil inches. For example, 1206 size is 120 mil inches (0.120″) long and 60 mil inches (0.060″) wide. The height varies with each design. Some manufacturers also offer non-standard sizes.

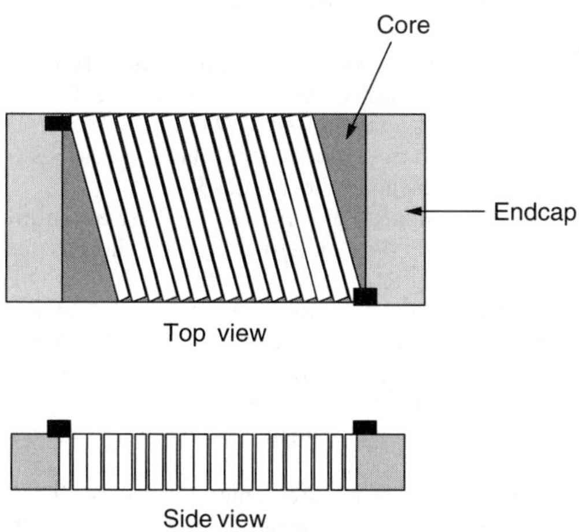

FIGURE 9-12 Cross-sectional view of a wound surface mount RF inductor.

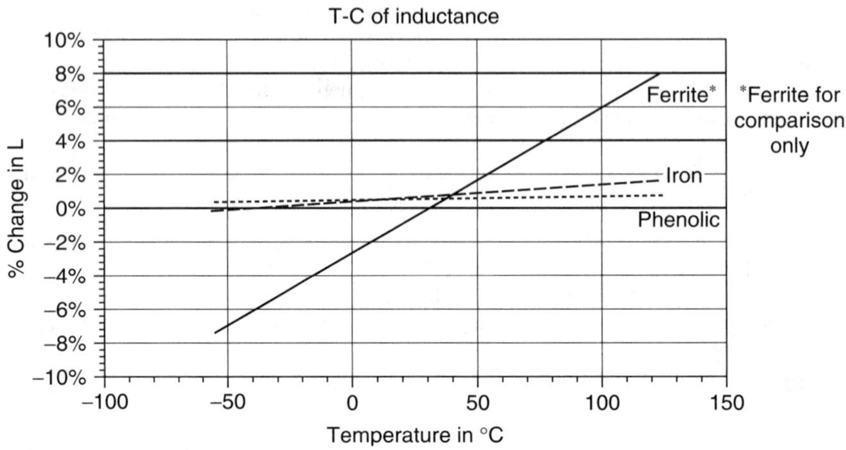

FIGURE 9-13 Comparison of materials for the Gowanda SM3 series of RF inductors. *(Courtesy of Gowanda Electronics Inc.)*

Each core material has unique properties. Phenolic or ceramic is essentially an air core having a permeability of 1. It never saturates with DC current or large AC signals, and has a zero temperature coefficient of inductance, but because of a μ of only 1 it is best suited for low-inductance high-frequency requirements. Powdered iron, on the other hand, offers higher permeabilities and a very low temperature coefficient. Last comes ferrite which has the highest permeability but also the highest temperature coefficient. It also sometimes has non-linear properties with higher levels of excitation. Typically, the higher the permeability, the more the tendency to saturate, and the higher the temperature coefficient. Figure 9-13 compares the temperature stability for the Gowanda Electronics Inc. SM3 series of temperature stable surface mount RF inductors consisting of either phenolic or iron core material to that of ferrite.

The major factors in selecting an RF inductor are *Q* (quality factor), self-resonant frequency (SRF), and inductance tolerance. The *Q* will dramatically affect performance of a filter, so an inductor having the highest possible *Q* over the frequency range of operation should be chosen. Figure 9-14 shows some representative *Q* curves for ceramic and powdered iron RF inductors from Gowanda Electronics Inc.

RF inductors using the construction of Figure 9-12 are essentially solenoids where, in theory, lines of magnetic flux extend to infinity, whereas torroidal construction results in closed (contained) magnetic paths. As a result, this type of construction results in more magnetic coupling to other inductors than torroidal devices. One solution would be to place these inductors at 90° relative to each other to minimize coupling. This is not always possible or desired, so shielded RF inductors are available. They can reduce magnetic fields by 10 to 20 dB or so, and hence reduce coupling. An internal shield usually consisting of the same core material surrounds the basic core containing the winding.

Self-resonant frequency (SRF) is the frequency where the distributed capacitance of the inductor forms a resonant circuit with the actual inductance. As the frequency of operation approaches this self-resonant frequency, *Q* will dramatically be decreased since the distributed capacitance results from very lossy dielectric material. Therefore, operation near the SRF should be avoided as much as possible. As a rule-of-thumb, the SRF should be at least ten times the highest frequency of operation, if possible.

Another effect of the SRF is to increase the effective inductance so it appears higher than the true inductance. For example, if a 1 μH inductor has 10 pF of distributed capacitance, the actual resonating capacitance *needed* for a particular frequency would be 10 pF *lower* due to the distributed capacitance. Therefore, the effective inductance appears higher than 1 μH. The effective inductance is given by

$$L_{\text{eff}} = \frac{L_T}{1 - \left(\dfrac{f}{f_r}\right)^2} \tag{9-15}$$

where L_T is the true (low-frequency) inductance, f is the frequency of interest, and f_r is the inductor's self-resonant frequency. As f approaches f_r, the value of L_{eff} will increase quite dramatically and will become infinite at self-resonance. This is shown in Figure 9-15.

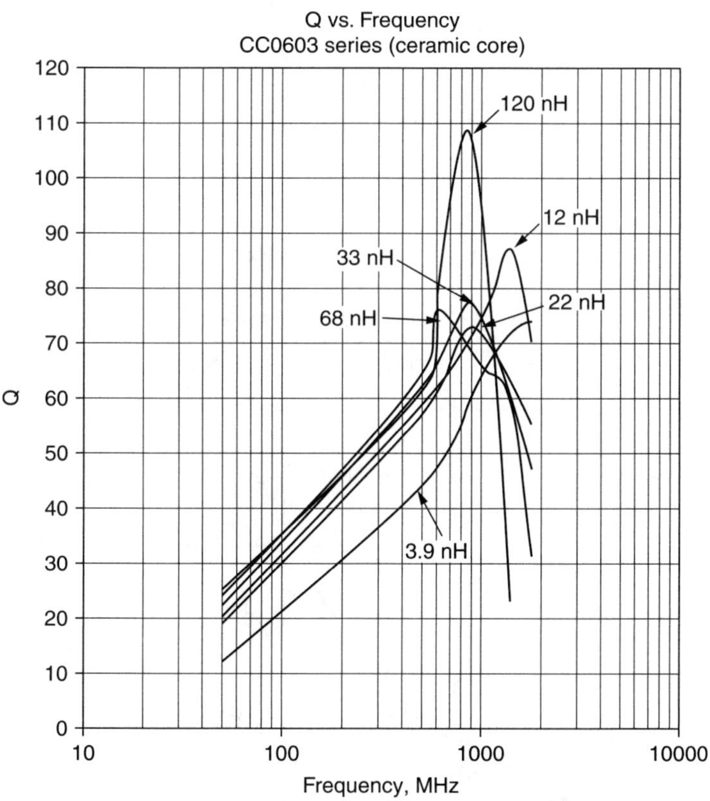

FIGURE 9-14 Some *Q* curves for ceramic and powdered-iron RF inductors. *(Courtesy Gowanda Electronics Inc.)*

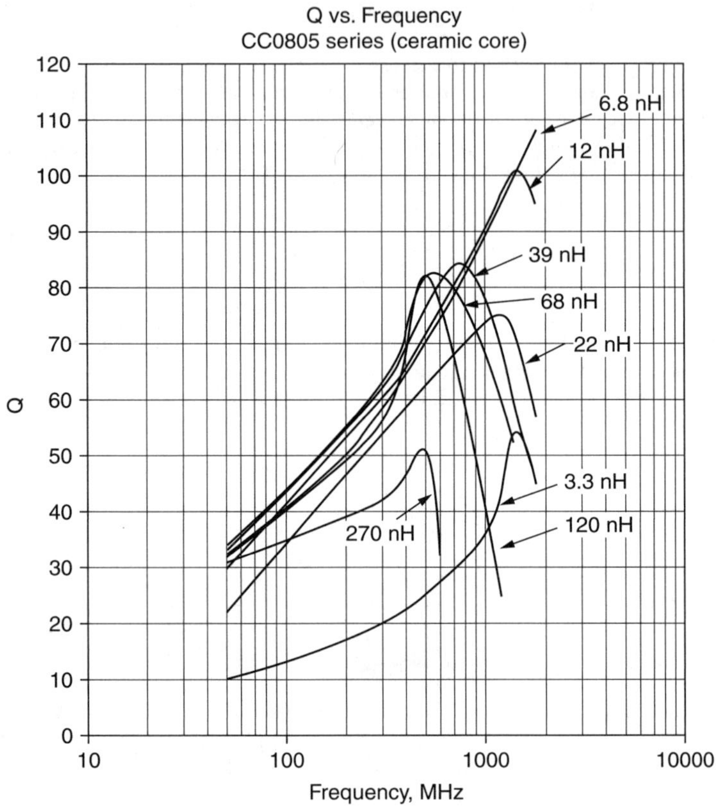

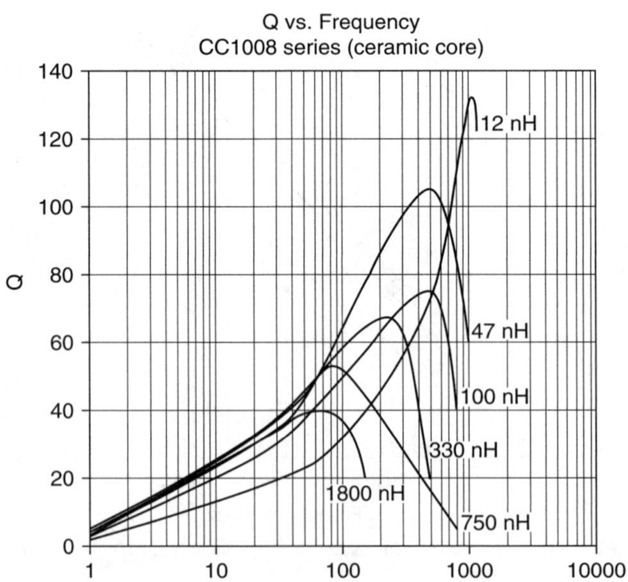

FIGURE 9-14 (*Continued*)

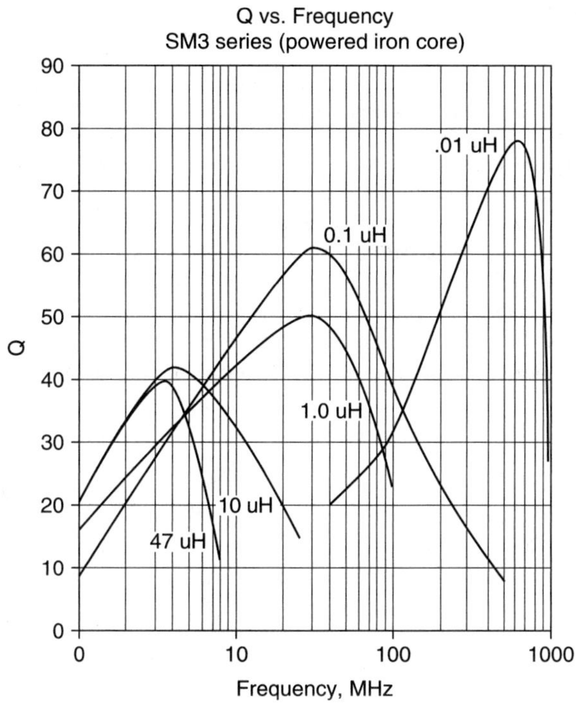

FIGURE 9-14 (*Continued*)

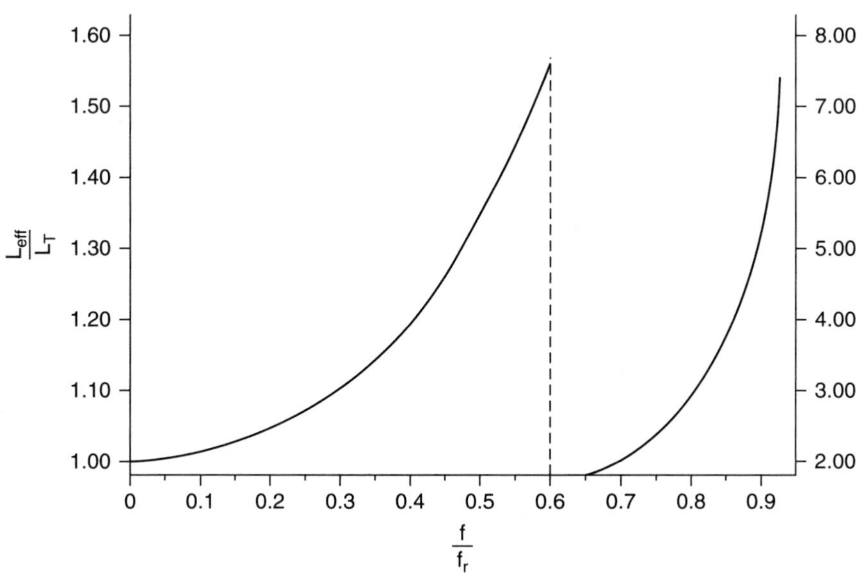

FIGURE 9-15 Effective inductance with frequency.

BIBLIOGRAPHY

Arnold Engineering Co. "Magnetic Powder Cores." *Marengo* III (February, 2003).

Ferroxcube Corp. "Linear Ferrite Magnetic Design Manual—Bulletin 550." Saugerties, New York: Ferroxcube Corp.

———. "Linear Ferrite Materials and Components." Saugerties, New York: Ferroxcube Corp.

Gowanda Electronics Corp., www.gowanda.com.

Magnetics Inc. *Ferrite Cores*. Butler, Pennsylvania: Magnetics Inc.

———. *Powder Cores*. Catalog 2004/2005, Butler, Pennsylvania: Magnetics Inc.

———. *Molypermalloy Powder Cores*. Catalog MPP-303S. Butler, Pennsylvania: Magnetics Inc.

Micrometals. *Q Curves for Iron Powder Toroidal Cores*. Anaheim, California: Micrometals.

———, Jim Cox, ed. *Iron Powder Cores for High Q Inductors*. Anaheim, California: Micrometals.

Welsby, V. G. *The Theory and Design of Inductance Coils*. London: Macdonald and Sons, 1950.

CHAPTER 10
COMPONENT SELECTION FOR *LC* AND ACTIVE FILTERS

10.1 CAPACITOR SELECTION

An extensive selection of capacitor types is available for the designer to choose from. They differ in terms of construction and electrical characteristics. The abundance of different capacitors often results in a dilemma in choosing the appropriate type for a specific application. Some of the factors to consider are stability, size, losses, voltage rating, tolerances, cost, and construction.

The initial step in the selection process is to determine the capacitor dielectric. These substances include air, glass, ceramic, mica, plastic films, aluminum, and tantalum. The type of mechanical construction must also be chosen. Since miniaturization is usually a prime consideration, the smallest possible capacitors will require thin dielectrics and efficient packaging without degrading performance.

Properties of Dielectrics

Capacitors in their most fundamental form consist of a pair of metallic plates or electrodes separated by an insulating substance called the dielectric. The capacitance in farads is given by

$$C = \frac{kA}{D} 8.85 \times 10^{-12} \tag{10-1}$$

where k is the dielectric constant, A is the plate area in square meters, and D is the separation between plates in meters. The dielectric constant of air is 1. Since capacity is proportional to the dielectric constant, the choice of dielectric highly influences the physical size of the capacitor.

A practical capacitor can be represented by the equivalent circuit of Figure 10-1*a*, where L_s is the series inductance, R_s is the series resistance, and R_p is the parallel resistance. L_s, R_s, and R_p are all parasitic elements. If we assume L_s negligible and R_p infinite, the equivalent impedance can be found by the vector addition of R_s and the capacitive reactance X_c, as shown in Figure 10-1*b*. R_s is often referred to as *ESR* or *Equivalent Series Resistance*.

An important characteristic of a capacitor is the dissipation factor. This parameter is the reciprocal of capacitor Q and is given by

$$d = \frac{1}{Q} = 2\pi f C R_s = \cot \theta \tag{10-2}$$

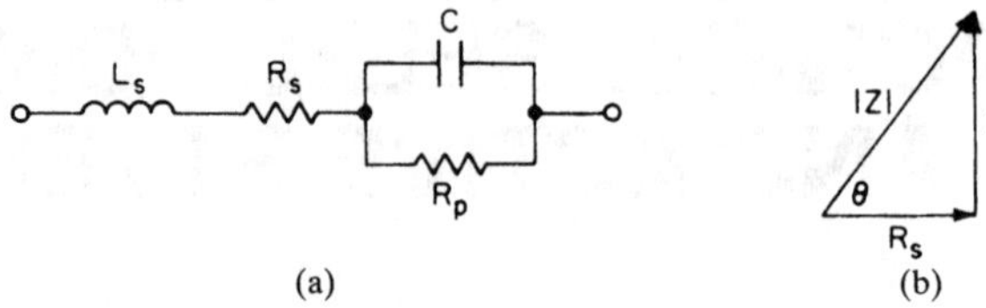

FIGURE 10-1 The equivalent representation of a capacitor: (*a*) an equivalent circuit; and (*b*) a simplified vector diagram.

where f is the frequency of operation. The dissipation factor is an important figure of merit and should be as low as possible. This is especially true for selective *LC* filters where branch *Q*s should be high. Capacitor *Q* must be sufficiently higher than inductor *Q* so that the effective *Q* is not degraded.

The power factor is another figure of merit similar to the dissipation factor. Referring to Figure 10-1, the power factor is defined as

$$\mathrm{PF} = \frac{R_s}{|Z|} = \cos\theta \qquad (10\text{-}3)$$

For capacitor *Q* in excess of 10, we can state $d = \mathrm{PF} = 1/Q$. The dissipation factor or power factor is sometimes expressed as a percentage.

The shunt resistive element R_p in Figure 10-1 is often referred to as insulation resistance and results from dielectric leakage currents. A commonly used figure of merit is the R_pC time constant, normally given in M$\Omega \times \mu$F. It is frequently convenient to combine all the losses in terms of a single resistor in parallel with *C*, which is given by

$$R = \frac{1}{2\pi fCd} \qquad (10\text{-}4)$$

The temperature coefficient (TC) is the rate of change of capacity with temperature and is usually given in parts per million per degree Celsius (ppm/°C). This parameter is extremely important when filter stability is critical. TC should either be minimized or of a specific nominal value so that cancellation will occur with an inductor or resistor's TC of the opposite sign. A dielectric material will also have an operating temperature range beyond which the material can undergo permanent molecular changes.

Another important parameter is *retrace*, which is defined as the capacity deviation from the initial value after temperature cycling. To maintain long-term stability in critical filters that are subjected to temperature variation, retrace should be small.

Table 10-1 summarizes some of the properties of the more commonly used dielectric materials when applied to capacitors.

Capacitor Construction

Surface Mount Capacitors. Surface-mount technology provides circuit densities previously unattainable using older through-hole methods. This technology allows the placement of subminiature passive and active components on both sides of a circuit board. As a result, densities can be nearly quadrupled in practice since the components are much smaller than their through-hole counterparts and the available board area for their placement is essentially doubled. Figure 10-2 illustrates the construction of a typical surface mount ceramic type capacitor. End terminations allow soldering to pads on the PCB. Most forms of surface mount construction do not result in hermetically sealed capacitors. Surface mount components are sometimes called *chip* components such as chip capacitors, chip resistors, chip inductors, and so on.

TABLE 10-1 Properties of Capacitor Dielectrics

Capacitor Type	Dielectric Constant	TC, ppm/°C	Dissipation Factor, %	Insulation Resistance, MΩ-μF	Temperature Range, °C
Aluminum	8	+2500	10	100	−40 to +85
Ceramic NPO/COG	65	±30	0.02	5×10^3	−55 to +125
Ceramic X7R	2000	±15%[1]	1	5×10^3	−55 to +125
Glass	5	±140	0.001	10^6	−55 to +125
Mica	6	±50	0.001	2.5×10^4	−55 to +150
Paper	3	±800	1.0	5×10^3	−55 to +125
Polycarbonate	3	±50	0.2	5×10^5	−55 to +125
PEN (Polyethylene Napthalate)	3	+200	0.40	10^5	−55 to +125
PET (Polyethylene Terephtalate/Polyester/ Mylar)	3.2	+400	0.50	10^5	−55 to +100
PPS (Polyphenylene Sulphide)	3	+80	0.10	5×10^5	−55 to +150
Polypropylene	2.2	−250	0.05	10^5	−55 to +105
Polystyrene	2.5	−120	0.01	3.5×10^7	−55 to +85
Polysulfone	3.1	+80	0.3	10^5	−55 to +150
Porcelain	5	+120	0.1	5×10^5	−55 to +125
Tantalum	28	+800	4.0	20	−55 to +85
Teflon	2.1	−200	0.04	2.5×10^5	−70 to +250

[1]Over entire temperature range

Surface mount capacitors are categorized in terms of size by length and width, as follows in Table 10-2.

The first two significant figures for standard values are given in Table 10-3. The bold values are more readily available. Frequently, a design may require nonstandard values. It is nearly always far more economical to parallel a few off-the-shelf capacitors of standard values when possible than to order a custom-manufactured component unless the quantities involved are substantial.

Axial and Radial Lead Construction. Another method of packaging capacitors is axial lead construction (illustrated in Figure 10-3).

For film type capacitors such as Mylar, polypropylene, or polystyrene, the electrodes are either a conductive foil such as aluminum or a layer of metallization placed directly on the dielectric film. The electrode and dielectric combination is tightly rolled onto a core. The alternate layers of foil or metallization are slightly offset so that leads can be attached.

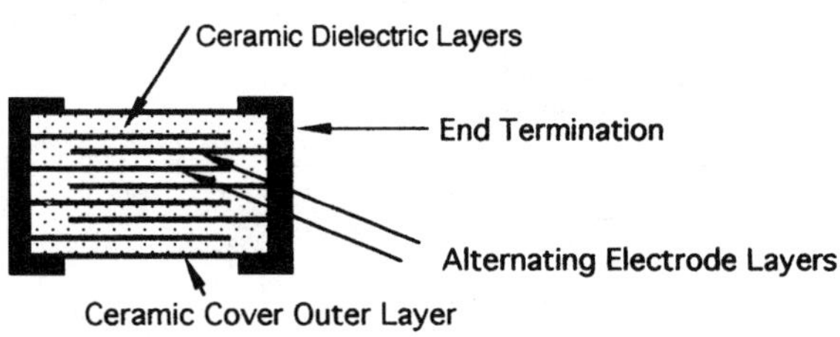

FIGURE 10-2 The typical construction of surface-mount ceramic capacitors.

TABLE 10-2 Preferred Sizes

EIA Size Code	Metric Size Code	Length L	Width W
0201	0603	0.02" (0.6 mm)	0.01" (0.3 mm)
0402	1005	0.04" (1.0 mm)	0.02" (0.5 mm)
0603	1608	0.06" (1.6 mm)	0.03" (0.8 mm)
0805	2012	0.08" (2.0 mm)	0.05" (1.2 mm)
1206	3216	0.12" (3.2 mm)	0.06" (1.6 mm)
1210	3225	0.12" (3.2 mm)	0.10" (2.5 mm)
1812	4532	0.18" (4.5 mm)	0.12" (3.2 mm)
2225	5764	0.22" (5.7 mm)	0.25" (6.4 mm)

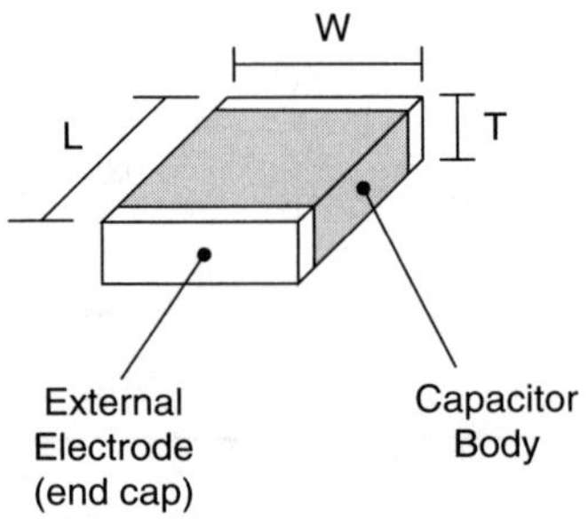

TABLE 10-3 Standard Capacitor Values

10	**18**	**33**	**56**
11	20	36	62
12	**22**	**39**	**68**
13	24	43	75
15	**27**	**47**	**82**
16	30	51	91

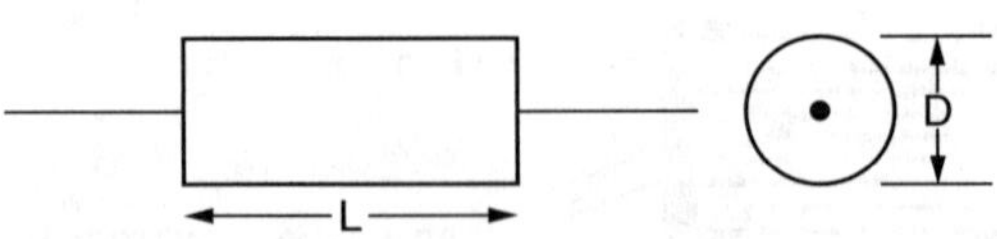

FIGURE 10-3 Axial lead construction.

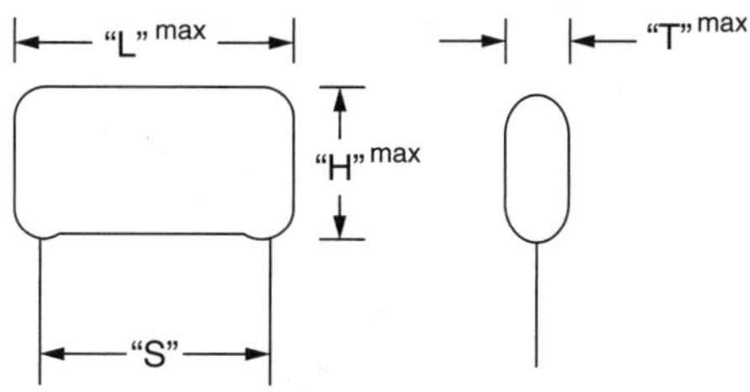

FIGURE 10-4 Radial lead construction.

The assembly is heat-shrunk to form a tight package. It can then be sealed by coating it with epoxy or by encapsulating it in a molded package. A more economical form of sealing is to wrap the capacitor in a plastic film and fill the ends with epoxy.

The most economical form of construction involves inserting the wound capacitor in a polystyrene sleeve and heat shrinking the entire assembly to obtain a rigid package. This is called *wrap-and-fill*. However, the capacitor is not truly sealed and can be affected by humidity and cleaning solvents entering through the porous end seals. Increased protection against humidity or chemical effects is obtained by sealing the ends with epoxy. If a true hermetic seal is required for extremely harsh environments, the capacitor can be encased in a metal can having glass-to-metal end seals.

An alternate to axial leads (where the leads are in the same plane as the capacitor axis) is *radial leads*. An example of radial leads is shown in Figure 10-4. The leads are *radial* to the body of the capacitor. The critical dimensions are lead spacing "*S*," height "*H*," length "*L*," and thickness "*T*." Axial and radial leaded components are known as "through-hole" since they mount through holes in a printed circuit board rather than mount on the surface like surface mount components.

Capacitors such as ceramic, mica, and porcelain contain a more rigid dielectric substance and cannot be rolled like the film capacitors. They are constructed in stacks or layers. Leads are attached to the end electrodes and the entire assembly is molded or dipped in epoxy. More commonly, they are packaged using surface mount construction.

Electrolytic Capacitor Construction. Capacitors having the highest possible capacitance per unit volume are the electrolytics made of either aluminum or tantalum. Basically, these capacitors consist of two electrodes immersed in a liquid electrolyte (see Figure 10-5). One or both electrodes are coated with an extremely thin oxide layer of aluminum or tantalum, forming a film having a high dielectric constant and good electrical characteristics. The electrolyte liquid makes contact between the film and the electrodes. The entire unit is housed in a leakproof metal can. The dielectric film is "formed" by applying a DC voltage

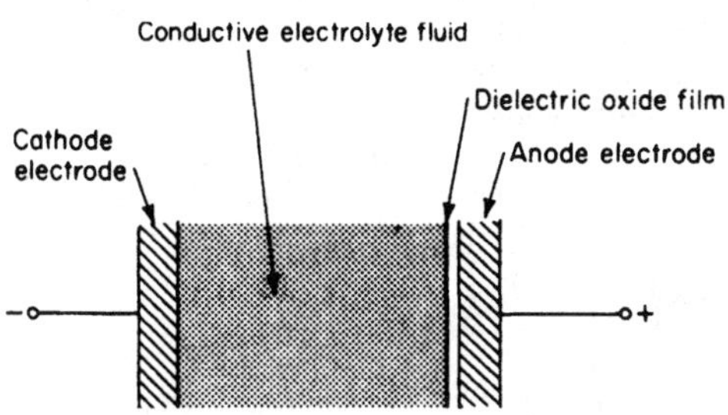

FIGURE 10-5 An electrolytic capacitor.

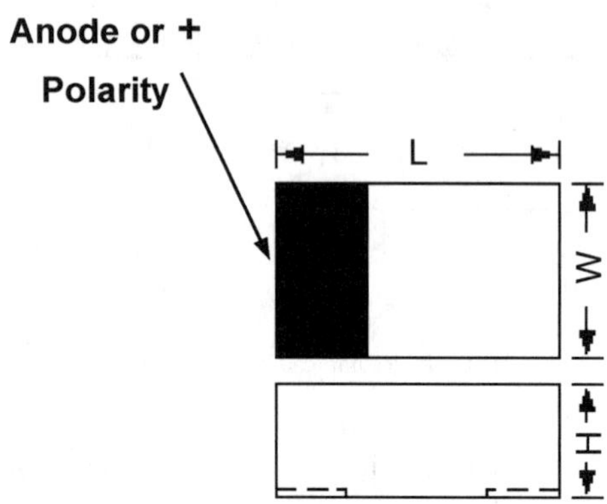

FIGURE 10-6 A tantalum capacitor in a surface mount package.

between cathode and anode, resulting in a *permanent polarization* of the electrodes. If both plates are "formed," a nonpolar unit will result having half the capacitance of the equivalent polarized type.

A solid-anode tantalum capacitor is composed of a sintered anode pellet on which is a tantalum oxide layer. The pellet is coated with a solid electrolyte of manganese dioxide which also becomes the cathode. This construction is superior electrically to the other forms of electrolytic construction. Figure 10-6 shows the surface mount version of a solid tantalum capacitor. Note that the polarity is always indicated on the body of the capacitor.

Selecting Capacitors for Filter Applications. The performance of a capacitor in its application is highly dependant on the choice of dielectric material. Table 10-1 provides some of these properties. The decision is governed by a variety of factors such as the value of the capacitor, operating temperature range and stability, frequency range, desired accuracy (tolerance), dissipation factor (Q), voltage rating, and so on. The most popular capacitor families are film and ceramic for both active and passive (LC) filters.

Film Capacitors. Film capacitors in surface mount configurations use mainly two construction methods. The most common method of manufacture involves stacking sheets of film dielectric which are metallized on one side and are called stacked-film chips. The other form of construction involves wound chips rather than stacked, and are called MELF chips. Also the through-hole axial and radial type construction methods mentioned previously can be used.

Polyester (Mylar/PET) capacitors are the smallest and most economical of the film types. They should be considered first for general-purpose filters operating below a few hundred kilohertz and at temperatures up to 125°C. The value range is from 1000 pF up to 10 μF. Values below 1000 pF are not recommended unless the voltage rating is intentionally increased to a few hundred volts, which forces a thicker and therefore more robust film. PEN (polyethylene napthalate) film capacitors are similar to PET but are available in higher voltage ratings and tighter tolerances. PPS (polyphenylene sulphide) capacitors have the best qualities compared to the PET and PEN types. They can support higher temperatures during both operation and soldering. They can be provided with tighter tolerances and are generally more expensive than the PET and PEN types.

Polystyrene capacitors probably have the best electric properties of all film capacitors. The temperature coefficient is precisely controlled, almost perfectly linear, and has a nominal value of −120 ppm/°C. Because of the predictable temperature characteristics, these

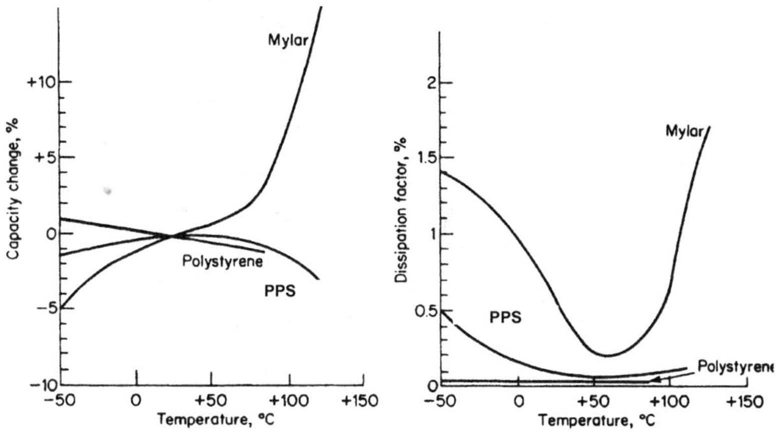

FIGURE 10-7 The capacitance and dissipation factor vs. ambient temperature of film capacitors.

capacitors are highly suited for *LC* resonant circuits, where the inductors have corresponding positive temperature coefficients. Capacity retrace is typically 0.1 percent. Losses are extremely small, resulting in a dissipation factor of approximately 0.01 percent. The maximum temperature, however, is limited to 85°C. Polypropylene capacitors are comparable in performance to polystyrene, although they have a slightly higher dissipation factor and temperature coefficient. Their maximum temperature rating, however, is 105°C. They are more economical than polystyrene and priced to be nearly competitive with Mylar.

Figure 10-7 compares the capacitance and dissipation factor versus temperature for Mylar (PET), PPS, and polystyrene capacitors. Clearly, polystyrene is superior, although PPS exhibits slightly less capacity variation with temperature over a limited temperature range. The dissipation factor is also a function of frequency, so at higher frequencies the dielectric losses will increase *d*, thus reducing circuit *Q*s.

Film capacitors are available with standard tolerances of 1, $2\frac{1}{2}$, 5, 10, and 25 percent. For most applications, the $2\frac{1}{2}$ or 5 percent tolerances will be adequate. This is especially true when the resonant frequency is adjustable by tuning the inductor, or when using a potentiometer in active filters. Precision capacitors tend to become more expensive with decreasing tolerance.

If the differential voltage across a capacitor becomes excessive, the dielectric will break down and may become permanently damaged. Most film capacitors are available with DC voltage ratings ranging from 33 to 600 V or more. Since most filters process signals of a few volts, voltage rating is normally not a critical requirement. Since capacitor volume increases with voltage rating, unnecessarily high ratings should be avoided. However, polystyrene capacitors below 0.01 μF with a low voltage rating will have a tendency to change value from the printed circuit-board soldering process because of distortion of the dielectric film from heat conducted through the leads. Capacitors rated at 100 V or more will generally be immune, since the dielectric film will be sufficiently thick.

In addition to resistive losses, the equivalent circuit of Figure 10-1*a* contains a parasitic inductance L_s. As the frequency of operation is increased, the series resonance of L_s and C is approached, resulting in a dramatic drop in impedance, as shown in Figure 10-8. Above self-resonance, the reactance becomes inductive. The self-resonant frequency is highly dependent upon construction and value, but in general, film capacitors are limited to operation below a few megahertz.

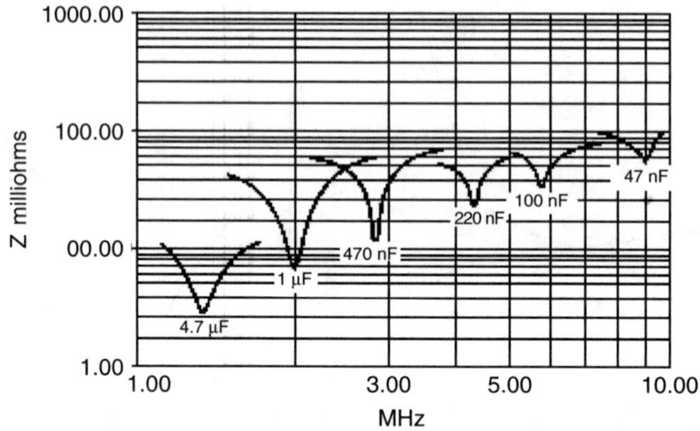

FIGURE 10-8 The self-resonant effect in film capacitors.

Film capacitors do not have piezoelectric properties. This means that any mechanical shock such as vibration will not result in the generation of small voltages.

Ceramic Capacitors. Ceramic capacitors are formed in stacks or layers using the construction illustrated in Figure 10-2. The ceramic plates are formed by first casting a film slurry of barium titanate and binders, which is then coated with a metallic ink to form electrodes. This results in a high dielectric constant. Another type uses electrically stable paraelectric material which has lower dielectric constants but highly superior properties. (Barium titanates are not a major part of the composition.) The plates are stacked into layers, fired in ovens, and separated into individual capacitors.

The basic dielectric material may have a dielectric constant k as high as 3000 at room temperature, which can be become as high as 10,000 at 125°C. The temperature at which the maximum k occurs is called the *Curie point.* By introducing additives, the Curie point can be lowered, the temperature variation reduced, and negative temperature coefficients obtained.

Ceramic capacitors are manufactured in a variety of shapes including monolithic chips which can be directly bonded to a metallized substrate for surface mounting (see Table 10-2 for standard case sizes). They are also available in the traditional disc form.

Ceramic capacitors are available in specific temperature coefficients. Some typical values are shown in Figure 10-9, where the P100 has a positive temperature coefficient of 100 ppm/°C and the negative types range from NPO (an approximately zero temperature coefficient) through N4700 (4700 ppm/°C). These capacitors become very useful for temperature compensation of passive and active filter networks.

There are essentially two classes of ceramic capacitors. The first and most popular class, EIA *Class 1* (and highly recommended for filters), is the temperature compensation type, and more specifically the NPO category (also known as COG), which is extremely stable with temperature.

The second class, *EIA Class 2*, has much higher dielectric constants to achieve higher volumetric efficiencies for high capacitance values. They are known as the X7R, X5R, Z5U, and Y5V types, which have much higher temperature coefficients, higher dissipation factors, and a piezoelectric characteristic. This means that any shock or vibration will result in a piezoelectric effect where small voltages will be generated. This could wreak havoc where small signals are present, such as in preamplifiers, and so on. In addition, low levels of third harmonic distortion can occur due to the capacitors having a high voltage coefficient (change of capacitance with instantaneous voltage). Therefore, this class of ceramic

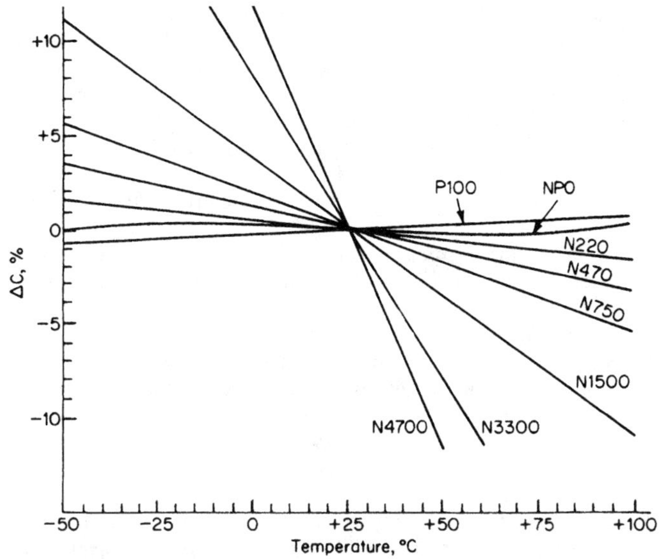

FIGURE 10-9 Temperature characteristics of ceramic capacitors.

capacitors should be limited to brute-force filtering applications such as those for power supplies or IC decoupling and should not be used for precision passive or active filters.

Table 10-4 compares the electrical properties of ceramic capacitors.

Capacitance values range from 0.5 pF to values as high as 2.2 μF. The DC working voltages may be has high as a few kilovolts, or as little as 3 V for the larger capacitance values. Tolerances for the NPO (COG) types can be as close as ±1 percent.

Ceramic capacitors exhibit minimum parasitic inductance and little variation in the dissipation factor with frequency. These properties make them particularly suited for high frequency use even into the GHz range. The surface mount package is best suited for high frequencies since packaging parasitic effects are minimized.

Mica Capacitors. Mica capacitors, although more costly than the film or ceramic types, have virtually unequaled electrical properties. (Mica itself is a rock that is extremely stable

TABLE 10-4 Properties of Ceramic Capacitors

Parameter	NPO or COG	X7R	X5R	Z5U	Y5V
Operating Temperature Range	−55 to +125°C	−55 to +125°C	−55 to +85°C	10 to +85°C	−30 to +85°C
Temperature Coeff.	0 ±50 ppm/°C	±15%*	±15%*	+22%, –56%*	+22%, −56%*
Dissipation Factor	<0.1%	<2.5%	<3%	<3%	<5%
Insulation Resistance	>1000 MΩ-μF	>1000 MΩ-μF	>1000 MΩ-μF	>100 MΩ-μF	>100 MΩ-μF

*Over the entire temperature range

and inert.) Typical temperature coefficients are ±25 ppm/°C. Retrace is better than 0.1 percent and the dissipation factor is typically 0.01 percent ($Q = 10{,}000$). Mica capacitors can operate to 150°C. Values above 10,000 pF become prohibitively expensive and should be avoided.

Mica capacitors are available in both surface mount and dipped construction. The dipped type package is similar to that shown in Figure 10-4.

Dipped mica capacitors are also known as silvered mica. Mica plates are silvered on both sides to form electrodes, stacked, and thermally bonded together. This results in a mechanically stable assembly with characteristics indicative of the mica itself. A coating of epoxy resin is then applied by dipping or other means.

The stacked construction of mica capacitors results in excellent performance well into the gigahertz region. The dissipation factor remains low at high frequencies, and parasitic inductance is small.

Electrolytic Capacitors. Aluminum electrolytic capacitors are intended for low-frequency bypassing or non-precision timing and are generally unsuitable for active or passive filters. They have unsymmetrical and broad tolerances such as +80 percent/−20 percent and require a DC polarization. (If the polarity is reversed, electrolytic capacitors can violently explode.) They also have poor stability and a shelf-life limitation. Large parasitic inductances and series resistances preclude usage at high frequencies.

Tantalum capacitors, on the other hand, can be used in low-frequency passive or active filters which require very large capacity values in a small volume. They have fairly high temperature coefficients (approximately +800 ppm/°C) and high dissipation factors (approximately 4 percent). However, these limitations may not be serious when applied to low-selectivity filters. The high-frequency characteristics are superior to those of aluminum electrolytics. Tolerances of 10 or 20 percent are standard.

Tantalum capacitors are formed in a similar manner to aluminum electrolytics and are polarized. However, they have no shelf-life restrictions and can even operate indefinitely without DC polarization. A momentary reverse polarity usually will not damage the capacitor, which is not true in the case of the aluminum electrolytics.

A polarization voltage may not always be present. Nonpolar tantalums can be obtained but are somewhat more expensive and not always available off the shelf. If two identical tantalum capacitors are series-connected back to back, as shown in Figure 10-10, a nonpolar type will result. The total value will be $C/2$, corresponding to capacitors in series.

Tantalum capacitors are available in three forms: dry foil, wet, and solid. The dry form consists of foil anodes and cathodes which are stacked or rolled using a paper spacer and impregnated with electrolyte. The wet forms are constructed using a porous slug of tantalum for the anode electrode and the silver-plated case as the cathode. The unit is filled with sulfuric acid for the electrolyte. Solid tantalums consist of a slab of compressed tantalum powder for the anode with a lead attached. A layer of tantalum pentoxide is formed on the surface for the dielectric, which is then connected to a lead to form the cathode, and the entire assembly is encapsulated in epoxy.

Tantalum capacitors are not recommended where any degree of accuracy is required. Filter designs should be scaled up in impedance so that film or ceramic capacitors can be used.

Trimmer Capacitors. In *LC* filters, particularly those for RF use, it is sometimes found to be more convenient to resonate a tuned circuit by adjusting capacity rather than inductance. A smaller trimmer is then placed in parallel with a fixed resonating capacitor. These trimmers usually consist of air, ceramic, mica, or glass as the dielectric.

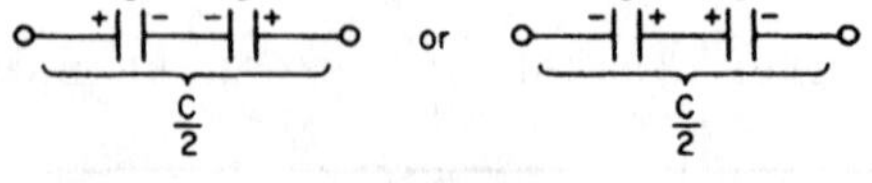

FIGURE 10-10 Nonpolar tantalum capacitors.

Air capacitors consist of two sets of plates, one called the *rotor*, which is mounted on a shaft, and the other the *stator*, which is fixed. As the rotor is revolved, the plates intermesh or overlap without making contact, resulting in increasing capacity. For high-capacity values, the plate size and number must increase dramatically, since the dielectric constant of air is only 1. This becomes a serious limitation when the available room is restricted. Standard air trimmers usually range from maximum values of 15 or 20 pF for small packages and up to 150 pF for larger physical size trimmers.

Ceramic trimmers are smaller than air for comparable values due to increased dielectric constant. They are usually composed of a single pair of ceramic disks joined at the center in a manner which permits rotation of one of the disks. A silvered region covers part of each disk, forming the plates of the capacitor. As the disk is rotated, the silvered areas begin to overlap, resulting in increasing capacity. Ceramic trimmers are available with maximum capacity values up to 50 pF. Good performance is obtained well into the VHF frequency range this way. Surface mount packages are also available.

Piston trimmers are composed of a glass or quartz tube with an outside conductive coating corresponding to one electrode. The other plate or electrode is a piston which by rotation is inserted deeper into the outside tube, resulting in increased capacitance. Multiturn construction results in excellent resolution. Piston trimmers have the best electrical properties of all trimmer types, but are also the most costly. They are suitable for use even at microwave frequencies.

When using a trimmer capacitor, a fixed capacitor is normally placed in parallel, having a larger value in order to obtain a finer resolution of adjustment.

10.2 RESISTORS

A fundamental component of active filters are resistors. Sensitivity studies show that resistors are usually at least as important as capacitors, so their proper selection is crucial to the success of a particular design.

Resistors are formed by connecting leads across a resistive element. The resistance in ohms is determined by

$$R = \frac{\rho L}{A} \tag{10-5}$$

where ρ is the resistivity of the element in ohm-centimeters, L is the length of the element in centimeters, and A is the cross-sectional area in square centimeters. By using materials of particular resistivities and special geometries, resistors can be manufactured that have the desired properties. Resistors fall into one of two general categories: fixed or variable.

Fixed Resistors

Fixed resistors are normally classified as carbon composition, carbon film, cermet film, metal film, wirewound, thick film, and thin film according to the resistive element. Of all these types, thick and thin film resistors in chip form are preferred for use in filters and are most readily available. Table 10-5 summarizes some typical properties of fixed resistors.

Thick and Thin Film. Thick film resistors are fabricated by depositing formulated pastes onto a ceramic substrate using a silk screen method. It is essentially an *additive* process where layers of material, including terminations, are added to a substrate. Thin film on the other hand is a *subtractive* process. A layer of metallization is sputtered onto a substrate, and material is removed by etching a pattern into the previously applied metallization layer.

TABLE 10-5 Typical Properties of Fixed Resistors

Type	Range, Ω	Standard Tolerances, %	Wattage Rating	Temperature Coefficient, ppm/°C
Carbon composition	1–100 M	5, 10	$^1/_8$, $^1/_4$, $^1/_2$, 1, 2	±1000
Carbon film	1–10 M	2, 5	$^1/_8$, $^1/_4$, $^1/_2$, 1, 2	±200
Cermet film	10–22 M	0.5, 1	$^1/_4$, $^1/_2$	±100
Metal film	0.1–1 M	0.1, 0.25, 0.5, 1	$^1/_8$, $^1/_4$, $^1/_2$, 1, 2	±25
Wirewound	1–100 K	5, 10, 20	3, 5, 10, 20	±50
Thick film	1–20 M	0.5, 1, 2, 5	$^1/_8$, $^1/_4$, $^1/_2$, 1	±100
Thin film	10–100 K	0.05, 1, 2, 5	$^1/_8$, $^1/_4$, $^1/_2$, 1, 2	±25

Thin film processes are photolithographic, which can result in very precise tolerance resistors with better temperature properties than thick film. However thick film resistors can handle more power and are more economical.

Thick and thin film resistors for surface mount applications come in the standard sizes and power ratings as indicated in Table 10-6 and Figure 10-11.

Thick and thin film resistors can also be obtained in other packing formats besides surface mount (such as axial leaded packages), but for medium- or low-power applications, surface mount is recommended since the least parasitic effects will occur and the packaging is the most economical.

Standard values for precision film resistors are given in Table 10-7. The value is represented by four digits, where the first three digits are significant and are selected from the following table, and the fourth digit (not shown) is the number of zeros. The bold values in this table specifically correspond to standard 1-percent values, and all listed numbers are standard for 0.1-, 0.25-, and 0.5-percent tolerances.

Metal Film. Metal film resistors are manufactured by depositing nichrome alloys on a rod substrate. Exceptional characteristics can be obtained. Normal tolerances are ±1 percent, but tolerances of ±0.1, ±0.25, and ±0.5 percent, are available. Temperature coefficients can be as low as ±15 ppm/°C. Retrace and aging result in changes usually not exceeding 0.25 percent. Typical construction is shown in Figure 10-12.

TABLE 10-6 Sizes and Wattage Ratings of Surface Mount Film Resistors

EIA Size Code	Metric Size Code	Length L	Width W	Power
0201	0603	0.02" (0.6 mm)	0.01" (0.3 mm)	1/20 watt
0402	1005	0.04" (1.0 mm)	0.02" (0.5 mm)	1/16 watt
0603	1608	0.06" (1.6 mm)	0.03" (0.8 mm)	1/10 watt
0805	2012	0.08" (2.0 mm)	0.05" (1.2 mm)	1/8 watt
1206	3216	0.12" (3.2 mm)	0.06" (1.6 mm)	1/4 watt
1210	3225	0.12" (3.2 mm)	0.10" (2.5 mm)	1/4 watt
1812	4532	0.18" (4.5 mm)	0.12" (3.2 mm)	1/2 watt
2010	5025	0.20" (5.0 mm)	0.10" (2.5 mm)	1/2 watt
2225	5764	0.22" (5.7 mm)	0.25" (6.4 mm)	1 watt

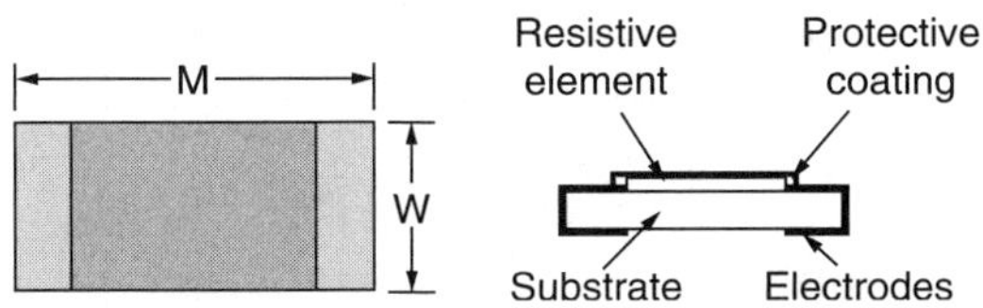

FIGURE 10-11 Surface mount film resistors.

Metal film resistors have many highly desirable features. In addition to low temperature coefficients and good long-term stability, they exhibit the lowest noise attainable in resistors. Parasitic effects are minimal and have no significant effect below 10 MHz. Metal film resistors are rugged in design, have excellent immunity to environmental stress, and have high reliability. Although tolerances to 0.1 percent are available, 1 percent values are used almost exclusively because of their lower cost.

Carbon Composition. The carbon composition type of a resistor in axial lead format is shown in Figure 10-13. It consists of a solid cylinder, composed mainly of carbon with wire leads, which is molded under high pressure and temperature in an insulated jacket. By changing the proportion of carbon powder and filler, different resistance values can be obtained. Standard wattage ratings are 1/8, 1/4, 1/2, 1, and 2 W. Resistance values are maintained to about ± 10 percent by the manufacturing process. Values within a ± 5-percent tolerance band are obtained using automatic sorting equipment.

A series of standard values has been adopted by RETMA based on $10^{1/12}$ and $10^{1/24}$ and is widely used for 10- and 5-percent resistors. These preferred values are rounded to two significant figures which are given in Table 10-8. All 10-percent values are also available in 5-percent tolerances, but not conversely.

The value and tolerance of carbon composition resistors are indicated by a series of color-coded bands beginning on one end of the resistor body. The first and second bands determine the two significant figures, the third band the multiplier, and the fourth band the tolerance. Sometimes a fifth band is present to establish a reliability rating, as shown in Table 10-9.

TABLE 10-7 Standard Values for Film Resistors

10.0	**12.1**	**14.7**	**17.8**	**21.5**	**26.1**	**31.6**	**38.3**	**46.4**	**56.2**	**68.1**	**82.5**
10.1	12.3	14.9	18.0	21.8	26.4	32.0	38.8	47.0	56.9	69.0	83.5
10.2	**12.4**	**15.0**	**18.2**	**22.1**	**26.7**	**32.4**	**39.2**	**47.5**	**57.6**	**69.8**	**84.5**
10.4	12.6	15.2	18.4	22.3	27.1	32.8	39.7	48.1	58.3	70.6	85.6
10.5	**12.7**	**15.4**	**18.7**	**22.6**	**27.4**	**33.2**	**40.2**	**48.7**	**59.0**	**71.5**	**86.6**
10.6	12.9	15.6	18.9	22.9	27.7	33.6	40.7	49.3	59.7	72.3	87.6
10.7	**13.0**	**15.8**	**19.1**	**23.2**	**28.0**	**34.0**	**41.2**	**49.9**	**60.4**	**73.2**	**88.7**
10.9	13.2	16.0	19.3	23.4	28.4	34.4	41.7	50.5	61.2	74.1	89.8
11.0	**13.3**	**16.2**	**19.6**	**23.7**	**28.7**	**34.8**	**42.2**	**51.1**	**61.9**	**75.0**	**90.9**
11.1	13.5	16.4	19.8	24.0	29.1	35.2	42.7	51.7	62.6	75.9	92.0
11.3	**13.7**	**16.5**	**20.0**	**24.3**	**29.4**	**35.7**	**43.2**	**52.3**	**63.4**	**76.8**	**93.1**
11.4	13.8	16.7	20.3	24.6	29.8	36.1	43.7	53.0	64.2	77.7	94.2
11.5	**14.0**	**16.9**	**20.5**	**24.9**	**30.1**	**36.5**	**44.2**	**53.6**	**64.9**	**78.7**	**95.3**
11.7	14.2	17.2	20.8	25.2	30.5	37.0	44.8	54.2	65.7	79.6	96.5
11.8	**14.3**	**17.4**	**21.0**	**25.5**	**30.9**	**37.4**	**45.3**	**54.9**	**66.5**	**80.6**	**97.6**
12.0	14.5	17.6	21.3	25.8	31.2	37.9	45.9	55.6	67.3	81.6	98.8

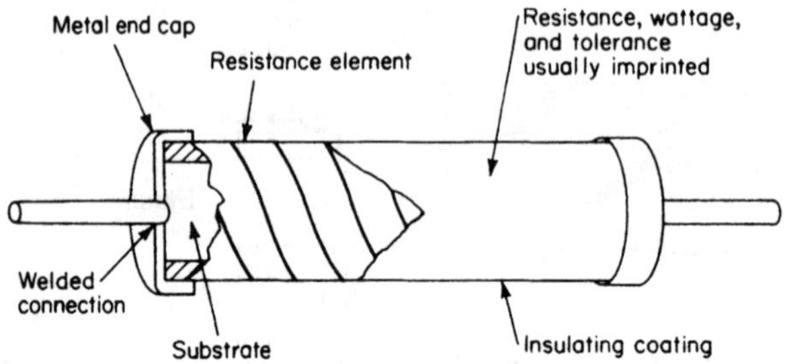

FIGURE 10-12 The composition of film resistors.

The power rating of resistors corresponds to the maximum power that can be continually dissipated with no permanent damage. This rating is normally applicable up to ambient temperatures of 70°C, above which derating is required. In general, a safety margin corresponding to a factor of 2 is desirable to obtain a high level of reliability.

Carbon composition resistors have a high temperature coefficient, typically 1,000 ppm/°C. In addition, permanent resistance changes of a few percent will occur due to poor retrace and aging. Generally, the lower resistance values exhibit better stability. Carbon composition resistors exhibit few parasitic effects and then only above 10 MHz. Carbon composition resistors are also the most economical of the various resistor types. They can be used in general-purpose active filters, where stability and accuracy is not a critical requirement. They are not recommended for use in precision filters since thick and thin film resistors are far superior. Carbon composition resistors are also available in surface mount packages.

Carbon Film. Film resistors are manufactured by depositing a thin layer of a resistance element on a nonconductive substrate. This form of construction was illustrated in Figure 10-12. The resistive element is usually spiral in form to increase the net resistance.

Carbon film resistors use this form of construction. They are manufactured by heating carbon-bearing gases so that a deposit of carbon film forms on the ceramic substrate. Temperature coefficients are in the region of 200 ppm/°C, which is a significant improvement over carbon composition. Resistance changes from aging and temperature cycling are typically specified at 1 percent or less. Normal tolerances are ±5 percent, although ±1 percent tolerance units can be obtained. The values are in accordance with Table 10-8. Carbon film resistors are especially suited for general-purpose applications requiring better performance than can be obtained using the carbon composition type at competitive pricing.

Cermet Film. Cermet film resistors are manufactured by screening a layer of combined metal and ceramic or glass particles on a ceramic core and firing it at high temperatures. They can provide higher resistance values for a particular size than most other types,

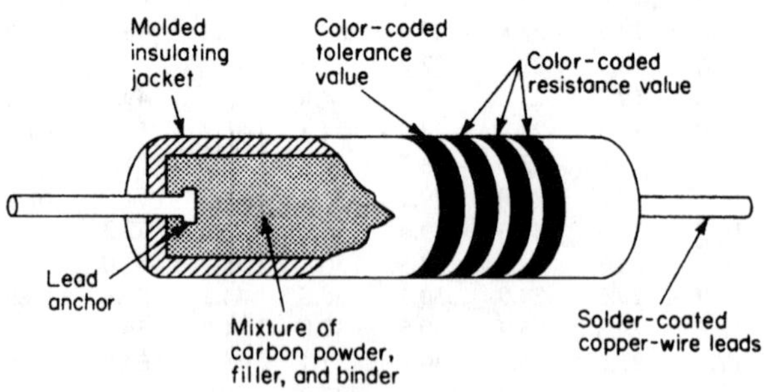

FIGURE 10-13 Carbon composition resistor construction.

TABLE 10-8 Preferred 5- and 10-Percent Resistor Values

±10%	±5%	±10%	±5%
10	10	33	33
	11		36
12	12	39	39
	13		43
15	15	47	47
	16		51
18	18	56	56
	20		62
22	22	68	68
	24		75
27	27	82	82
	30		91

up to a few hundred megohms. However, they are somewhat inferior electrically to the metal film type. Their temperature coefficients are higher (typically 200 ppm/°C). Retrace and long-term stability are typically 0.5 percent. Tolerances of 1 percent are standard.

Precision resistors are also available in wirewound form. They consist essentially of resistance wire, such as nichrome, wound on an insulated core. They are costlier than the metal film type and comparable in performance except for higher parasitics. Wirewound resistors are best suited for applications with higher power requirements.

TABLE 10-9 The Standard Color Code for Carbon Composition Resistors

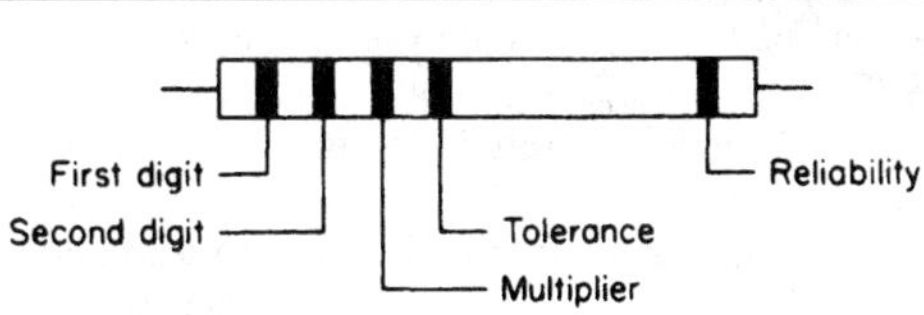

Color	Digit	Multiplier	Tolerance, %	Reliability Level (Failures per 1000 h), %
Black	0	1		
Brown	1	10		1
Red	2	100		0.1
Orange	3	1000		0.01
Yellow	4	10,000		0.001
Green	5	100,000		
Blue	6	1,000,000		
Violet	7	10,000,000		
Gray	8			
White	9			
Gold		0.1	5	
Silver			10	

TABLE 10-10 The Typical Properties of Potentiometers.

Type	Range, Ω	Standard Tolerances, %	Wattage Rating	Temperature Coefficient, ppm/°C
Carbon composition	100–10 M	10, 20	0.5, 1, 2,	±1000
Cermet	100–1 M	5, 10, 20	0.5, 1	±100
Wirewound	10–100 M	5, 10	0.5, 1, 5, 10	±100

Variable Resistors. Variable resistors are commonly referred to as potentiometers or trimmers. They are classified by the type of resistance element, such as carbon, cermet, or wirewound, and also by whether they are single- or multiple-turn. The electrical properties of the three basic element types are given in Table 10-10.

Potentiometers are always three-terminal devices, as depicted in Figure 10-14*a*. (CW indicates the direction of travel of the wiper for clockwise rotation.) Since most applications require a two-terminal variable resistor or rheostat rather than a voltage divider, the wiper is normally externally joined to one of the end terminals, as shown in Figure 10-14*b*.

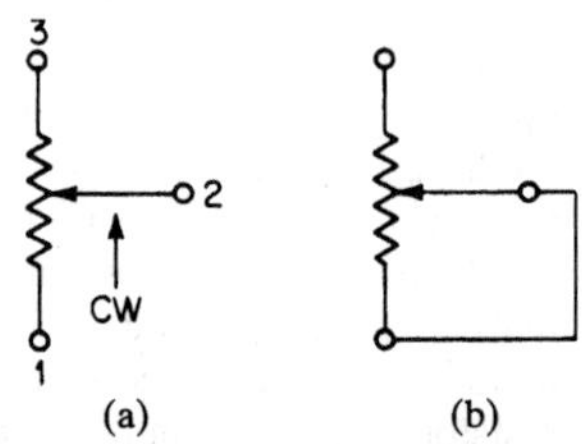

FIGURE 10-14 Potentiometer connections: (*a*) a potentiometer; and (*b*) a rheostat configuration.

Construction. The basic types of construction are the single-turn and multiturn forms. Single-turn potentiometers are rotary devices having a centrally located adjustment hub which contains the movable contact for the wiper. The moveable contact rests upon the circular resistive element. The entire assembly may be exposed or enclosed in a plastic case containing PC pins or surface mount contacts for the external connections. The adjustable hub is usually slotted for screwdriver access. Some types have a toothed thumbwheel for manual adjustment.

Single-turn trimpots require 270° of rotation to fully traverse the entire resistance element. Increasing the number of turns will improve the operator's ability to make very fine adjustments. Multiturn trimmers contain a threaded shaft. A threaded collar, which functions as a wiper, travels along this shaft, making contact with the resistive element. This general construction is shown in Figure 10-15.

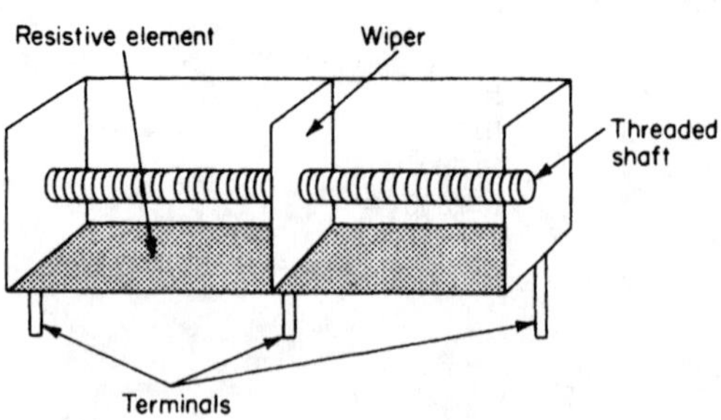

FIGURE 10-15 Multiturn potentiometer construction.

Trimmer controls, especially the single-turn variety, are subject to movement of the wiper adjustment due to vibration. To assure stability, it is desirable to prevent movement by placing a small amount of a rigid sealer such as Glyptal on the adjuster after circuit alignment.

Types of Resistance Elements. Carbon composition potentiometers are formed by molding a carbon mixture on a nonconductive disk or base containing previously embedded leads. The wiper mechanism is then attached to complete the assembly.

Carbon composition potentiometers are available in both the single- and multiple-turn configurations. They are the most economical of all types and also have the poorest characteristics. TC is about 1000 ppm/°C. Retrace and long-term stability are poor, unfortunately. Therefore, carbon composition potentiometers find limited usage in filter circuits.

Cermet film potentiometers are the most commonly used type for filter networks. They are moderate in cost, have a wide range of available values, and have good temperature characteristics and stability. Parasitic effects are minimal. They are manufactured by depositing cermet film on a nonconductive disk or base in thicknesses varying from 0.0005 to 0.005 in.

Wirewound potentiometers are formed by winding resistance wire (usually nichrome) on an insulated base. They have comparable temperature characteristics to cermet and are higher in cost. Their major attribute is high power capability. However, for most filter requirements this feature is of little importance.

Wirewound potentiometers have quite different adjustment characteristics than the other types. As the other type wipers are rotated, the resistance varies linearly with degrees of rotation, providing nearly infinite resolution. (Nonlinear tapers are also available, such as logarithmic.) The resistance of wirewound potentiometers, however, changes in discrete steps as the wiper moves from turn to turn. The resolution (or settability) therefore is not as good as the cermet or carbon composition types.

Ratings. Since potentiometers are almost always used as variable elements, the overall tolerances are not critical and are generally 10 or 20 percent. For the same reason, many different standard values are not required. Standard values are given by one significant figure—either 1, 2, or 5—and range from 10 Ω to 10 MΩ.

Resistor Johnson (Thermal) Noise. At any temperature above absolute zero (−273°C or 0°K) electrons in any substance are in constant motion. Since the direction of motion is random, there is no steady current that can be detected because the randomness results in decorrelation of any short-term current flow. However a continuous series of random noise pulses occur which result in a noise signal known as Johnson noise or thermal noise. The magnitude of noise in a resistor is related to the magnitude of resistance by the following relationship:

$$V_n^2 = 4\,K_b \text{TRB in } V^2/\text{Hz} \tag{10-6}$$

where V_n = noise voltage in volts

K_b = Boltzmanns' constant of 1.38×10^{-23} J/°K

T = temperature in °K

R = resistance in ohms

B = bandwidth in Hz

We can greatly simplify this expression if we assume room temperature as follows:

$$V_{\text{noise}} = 4\sqrt{R} \tag{10-7}$$

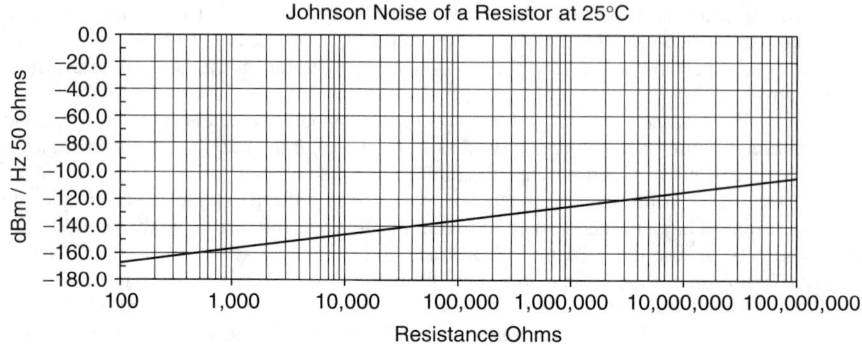

FIGURE 10-16 Johnson noise versus resistance.

where R is in KΩ and V_{noise} is in nanovolts per square-root Hz. Therefore, a 1 MΩ resistor would have a Johnson (thermal) noise of 126 nV/√Hz. Figure 10-16 illustrates the Johnson noise generated by a resistor at 25°C in dBm/Hz with a 50-Ω resistive termination.

Although the noise voltage and resulting power level may be extremely low, if this resistor was part of an active filter having a high gain, the noise could become significant. For low noise requirements, it is best to use as low a resistor value as practical. Noise will increase with temperature and resistance as a square-root proportionality.

The wider the bandwidth, the more the total noise power, so even though the magnitude of the dBm/Hz appears insignificant, the *total* power for a given bandwidth could be much higher. If we convert V_{noise} into watts using V_{noise}^2/R_{term} where R_{term} is the termination resistance for the noise, and then multiply the result by the total bandwidth in Hz, we get the *total* noise power over that bandwidth which could become unacceptable for wide bandwidths and applications with low noise requirements.

10.3 OPERATIONAL AMPLIFIERS

The versatility and low cost of integrated-circuit (IC) operational amplifiers have made them one of the most popular building blocks in the industry. The op amp is capable of performing many mathematical processes upon signals. For active filters, op amps are specifically used to provide gain and isolation.

IC op amps have evolved from the invention of the Fairchild μA 709 in the mid sixties to the many different types available today having a variety of special features. This section reviews some of the essential characteristics, discusses some important considerations, and provides a survey of some popular IC amplifier types. An extensive formal analysis is covered by many standard texts and will not be repeated here. The reader is encouraged to visit the many web sites of op amp manufacturers for detailed technical data, selection guides, and application assistance, if needed.

A Review of Basic Operational-Amplifier Theory

A simplified equivalent circuit of an operational amplifier is shown in Figure 10-17. The output voltage e_0 is the difference of the input voltages at the two input terminals amplified by amplifier gain A. A positive changing signal applied to the positive (+) input terminals

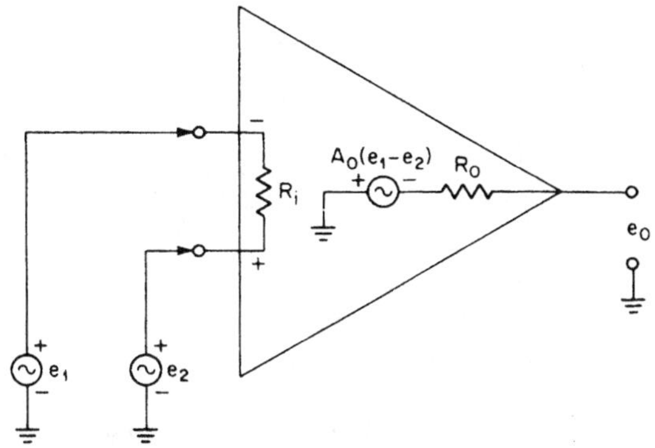

FIGURE 10-17 The equivalent circuit of an operational amplifier.

results in a positive change at the output, whereas a positive changing signal applied to the negative input terminal (−) results in a negative change at the output. Hence, the positive input terminal is called the *noninverting* input and the negative input terminal is referred to as the *inverting* input. If we consider the amplifier ideal, the input impedance R_i is infinite, the output impedance R_0 is zero, and the voltage gain A_0 is infinite.

Negative feedback applied from the output to the inverting input results in zero differential input voltage if A_0 is infinite. This is called the *virtual ground effect*. This property permits a wide variety of different amplifier configurations and simplifies circuit analysis.

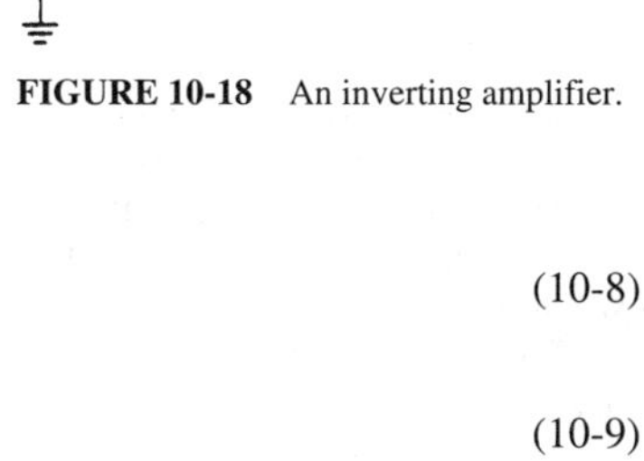

FIGURE 10-18 An inverting amplifier.

Let us first consider the basic inverting amplifier circuit of Figure 10-18. If we consider the amplifier ideal, the differential input voltage becomes zero because of the negative feedback path through R_2. Therefore, the inverting input terminal is at ground potential. The currents through resistors R_1 and R_2 are

$$I_1 = \frac{E_{in}}{R_1} \tag{10-8}$$

and

$$I_2 = -\frac{E_{out}}{R_2} \tag{10-9}$$

If the amplifier input impedance is infinite, no current can flow into the inverting terminal, so $I_1 = I_2$. If we equate Equations (10-8) and (10-9) and solve for the overall transfer function, we obtain

$$\frac{E_{out}}{E_{in}} = -\frac{R_2}{R_1} \tag{10-10}$$

The circuit amplification is determined directly from the ratio of two resistors and is *independent* of the amplifier itself. Also, the input impedance is R_1 and the output impedance is zero.

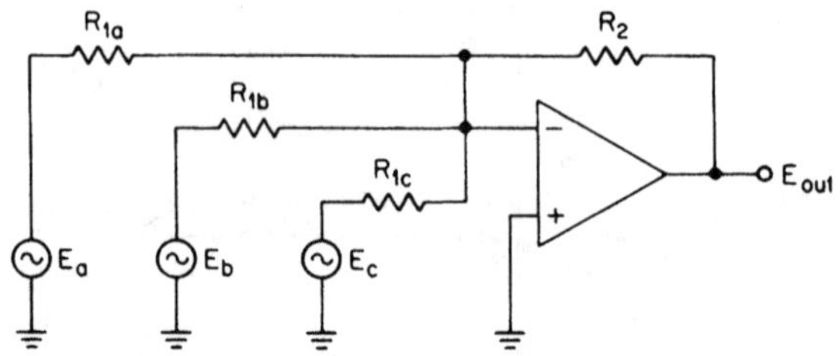

FIGURE 10-19 A summing amplifier.

Multiple inputs can be summed at the inverting input terminal as a direct result of the virtual ground effect. The triple-input summing amplifier of Figure 10-19 has the following output based on superposition:

$$E_{out} = -\frac{R_2}{R_{1a}}E_a - \frac{R_2}{R_{1b}}E_b - \frac{R_2}{R_{1c}}E_c \qquad (10\text{-}11)$$

A noninverting amplifier can be configured using the circuit of Figure 10-20. Since the differential voltage between the amplifier input terminal is zero, the voltage across R_1 is E_{in}. Since R_1 and R_2 form a voltage divider, we can state

$$\frac{E_{out}}{E_{in}} = \frac{R_1 + R_2}{R_1} \qquad (10\text{-}12)$$

or the more popular form

$$\frac{E_{out}}{E_{in}} = 1 + \frac{R_2}{R_1} \qquad (10\text{-}13)$$

where the input impedance is infinite and the output impedance is zero.

If we set R_1 to infinity and R_2 to zero, the gain becomes unity, which corresponds to the voltage follower configuration of Figure 10-21.

The applications of operational amplifiers are by no means restricted to summing and amplification. If R_2 in Figure 10-18, for example, were replaced by a capacitor, the circuit would serve as an integrator. Alternatively, a capacitor for R_1 would result in a differentiator. Nonlinear functions can be performed by introducing nonlinear elements into the feedback paths.

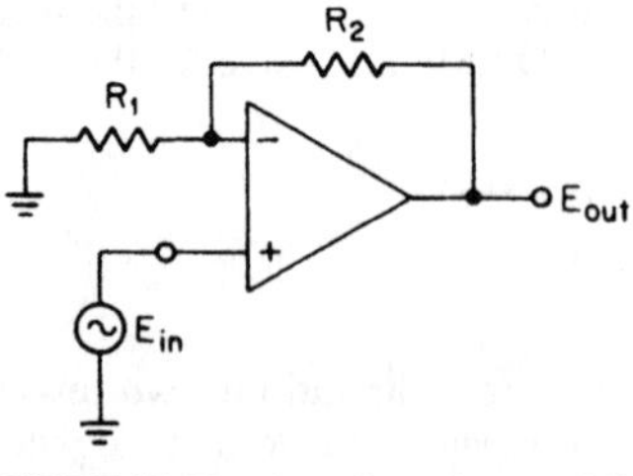

FIGURE 10-20 A noninverting amplifier.

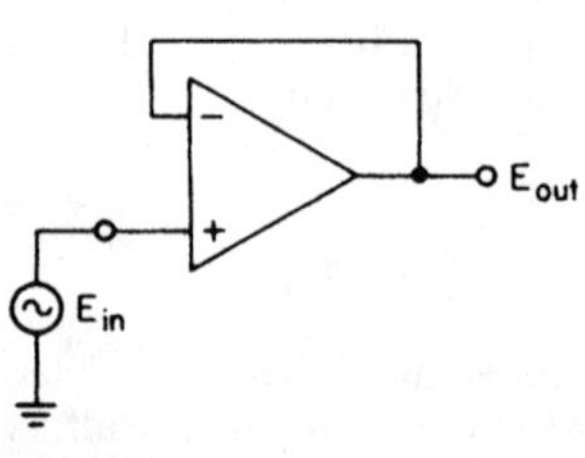

FIGURE 10-21 A voltage follower.

An Analysis of Non-Ideal Amplifiers

The fundamental equation for the closed-loop gain of the noninverting amplifier of Figure 10-20 is given by

$$A_c = \frac{A_0}{1 + A_0\beta} \tag{10-14}$$

where A_0 is the amplifier's open-loop gain and β is the feedback factor. This expression should be familiar to those who have studied feedback systems or servo theory. The closed-loop gain of the inverting amplifier structure of Figure 10-18 is expressed as

$$A_c = \frac{A_0(\beta - 1)}{1 + A_0\beta} \tag{10-15}$$

In both cases, the feedback factor, which corresponds to the portion of the output that is fed back to the input, is determined by

$$\beta = \frac{R_1}{R_1 + R_2} \tag{10-16}$$

Let us examine the term $1 + A_0\beta$ corresponding to the denominator of the closed-loop gain expressions. The open-loop gain of practical amplifiers is neither infinite nor real (zero phase shift). The magnitude and phase of A_0 will be a function of frequency. If at some frequency $A_0\beta$ were equal to -1, the denominator of Equations (10-14) and (10-15) would vanish. The closed-loop gain would then become infinite, which implies an oscillatory condition.

To prevent oscillations, the amplifier open-loop gain must be band-limited so that the product $A_0\beta$ is less than 1 *below* the frequency where the amplifier phase shift reaches 180°. This is achieved by introducing a gain roll-off beginning at low-frequencies and continuing at a 6-dB-per-octave rate. This technique of ensuring stability is called *frequency compensation*. It is evident from the closed-loop gain equations that for high closed-loop gains, β is diminished so that less frequency compensation will be required. Conversely, the voltage follower will need the most compensation. An amplifier that will be stable with unity gain is referred to as *unity-gain stable*.

The Effects of Finite Amplifier Gain. The most critical factor in most op amp applications is the open-loop gain. In order to maintain stability, the open-loop gain must be band-limited. This is usually accomplished by introducing a real pole at a relatively low frequency so that the gain rolls off at 6 dB per octave.

A typical open-loop gain plot is shown in Figure 10-22*a*. The gain has two breakpoints. The low-frequency breakpoint is caused by a real pole resulting from the frequency compensation. The output phase lag increases to 45° at the breakpoint and asymptotically approaches 90° as the frequency is increased. The amplitude response rolls off at a rate of 6 dB per octave.

Another amplifier breakpoint occurs near 100 kHz. Above this second pole, the gain rolls off at 12 dB per octave, an additional 45° of phase shift occurs, and the asymptotic phase limit becomes 180°. The corresponding phase curve is shown in Figure 10-22*b*.

Most operational amplifiers have a built-in frequency compensation network. These values correspond to the worst case for guaranteed stability, which is the voltage follower configuration. The penalty paid for this convenience is that, in the case of high closed-loop gain ($\beta \ll 1$), the open-loop gain is lower than it really could be with less compensation and yet retain stability. Thus, some unnecessary closed-loop gain degradation will occur.

Amplifiers are frequently specified in terms of their unity gain-bandwidth product which is, the frequency at which the open-loop gain is unity.

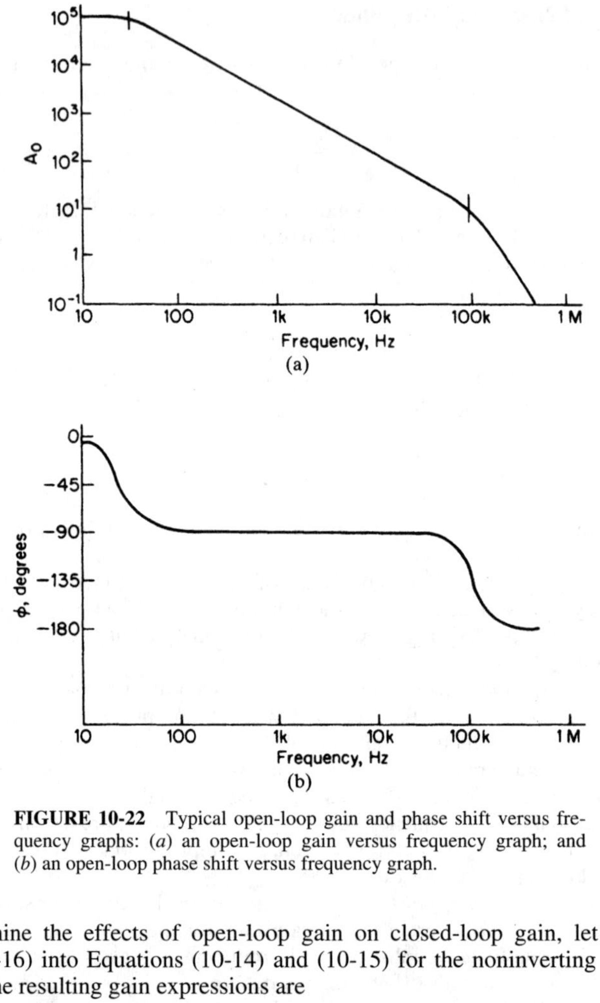

FIGURE 10-22 Typical open-loop gain and phase shift versus frequency graphs: (*a*) an open-loop gain versus frequency graph; and (*b*) an open-loop phase shift versus frequency graph.

To determine the effects of open-loop gain on closed-loop gain, let us substitute Equation (10-16) into Equations (10-14) and (10-15) for the noninverting and inverting amplifiers. The resulting gain expressions are

a noninverting amplifier:
$$A_c = \frac{1 + \dfrac{R_2}{R_1}}{\dfrac{1}{A_0}\left(1 + \dfrac{R_2}{R_1}\right) + 1} \tag{10-17}$$

and an inverting amplifier:
$$A_c = -\frac{\dfrac{R_2}{R_1}}{\dfrac{1}{A_0}\left(1 + \dfrac{R_2}{R_1}\right) + 1} \tag{10-18}$$

If A_0 were infinite, both denominators would reduce to unity. Equations (10-17) and (10-18) would then be equal to Equations (10-13) and (10-10), the fundamental expressions for the gain of a noninverting and inverting amplifier.

To reduce the degrading effect of open-loop gain upon closed-loop gain, A_0 should be much higher than the desired A_c. Since the open-loop phase shift is usually 90° over most of the band of interest, the error term in the denominator of Equations (10-17) and (10-18) is in quadrature with unity. This relationship minimizes the effect of open-loop gain upon closed-loop gain when the feedback network is purely resistive. An open-loop to closed-loop gain ratio of 10:1 will result in an error of only 0.5 percent, which is more than adequate for most requirements.

Finite open-loop gain also affects circuit input and output impedance. The input impedance of the noninverting amplifier can be derived as

$$R_{in} = (1 + A_0\beta)\,R_i \tag{10-19}$$

and the output impedance is given by

$$R_{out} = \frac{R_0}{A_0\beta} \tag{10-20}$$

where R_i and R_0 are the amplifiers' input and output impedance, respectively. Usually the closed-loop input and output impedance will have a negligible effect on circuit operation with moderate values of $A_0\beta$.

Practical Amplifier Considerations

DC Offsets. An inverting amplifier is shown in Figure 10-23 with the addition of two bias currents I_a and I_b, and an input offset voltage V_{dc}. Bipolar transistors in the input stage always draw some bias current in order to operate. A CMOS type input has a much higher input impedance and draws significantly less bias current but has a higher offset voltage than the bipolar structure.

The input offset voltage results in an offset voltage at the output equal to V_{dc} times the closed-loop gain. The polarity of V_{dc} is random and is typically less than 10 mV.

The two bias currents are nearly equal except for a small difference or offset current I_0. In the circuit of Figure 10-23, I_a produces an additional error voltage at the input given by I_aR_{eq} where

$$R_{eq} = \frac{R_1R_2}{R_1 + R_2} \tag{10-21}$$

The noninverting input bias current I_b has no effect. To minimize the effect of I_a, high values of R_{eq} should be avoided.

A more commonly used approach involves introducing a resistor having the value R_{eq} between the noninverting input and ground, as shown in Figure 10-24. This has no effect on the overall gain. However, a DC offset voltage of I_bR_{eq} is introduced at the noninverting input. Because the amplifier is a differential device, the net error voltage due to the offset currents is $(I_a - I_b)R_{eq}$ or I_0R_{eq}. Since I_a and I_b may each be 80 nA, for example, and the offset current I_0 is in the range of 20 nA, a 4:1 reduction is obtained.

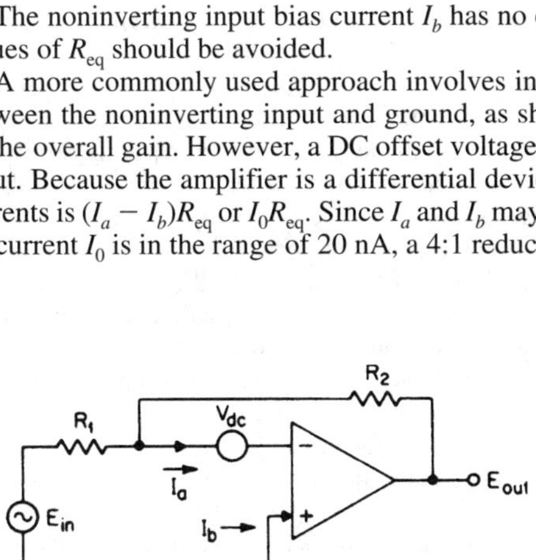

FIGURE 10-23 An inverting amplifier with offsets.

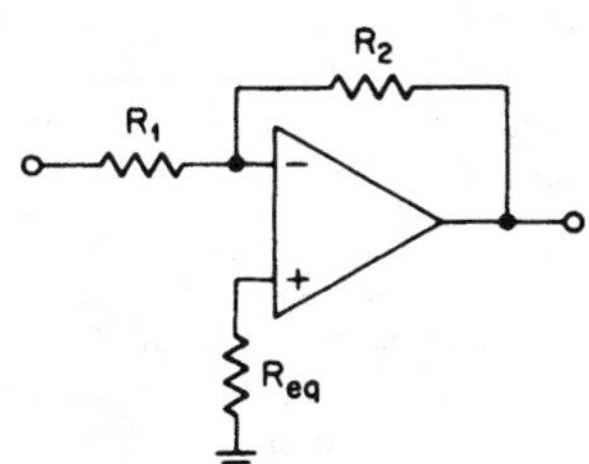

FIGURE 10-24 The minimization of DC offsets due to bias currents.

In the case of the noninverting amplifier configuration of Figure 10-20, the ratio R_2/R_1 is determined by the gain Equation (10-13). However, the actual values of R_1 and R_2 are nearly arbitrary and can be chosen so that their parallel combination (R_{eq}) is approximately equal to the DC loading on the noninverting input—in other words, the parallel combination of all resistance connected between the noninverting input and AC ground.

In general, for moderate closed-loop gains or AC-coupled circuits, the effects of DC offsets are of little consequence. For critical applications such as precision active low-pass filters for the recovery of low-level DC components, the methods discussed can be implemented. Some amplifiers will provide an input terminal for the nulling of output DC offsets.

Bandwidth and Gain. Bandwidth is normally specified as the frequency where the open-loop gain is one—the *unity gain bandwidth.* Since the gain is one, this number is also the *gain-bandwidth product.* As a rule of thumb, the bandwidth should be at least ten times the highest frequency of interest. The open-loop gain should be at least 20 dB more than the closed loop gain desired in order to minimize the limiting effects of the amplifier itself. *However, excessive gain or bandwidth can result in more sensitivity of the amplifier to PC board layout and instability with reactive loads. In many applications, the more general-purpose moderate gain and limited bandwidth amplifiers are best suited and most robust.*

It is important to recognize that the gain (and other parameters) shown on manufacturers' data sheets are for small signal conditions. Larger signal swings degrade performance.

Slew-Rate Limiting. When an operational amplifier is used to provide a high-level sine wave at high frequency, the output will tend to approach a triangular waveform. This effect is called *slew-rate limiting*, and the slope of the triangular waveform is referred to as the *slew rate.* Typical values range from 1 to 100 V/μs, depending upon the amplifier type. If the peak-to-peak output voltage is small, the effects of slew rate will be minimized. Bandwidth can also be extended by using the minimum frequency compensation required for the given closed-loop gain when the compensation is external.

Noise. There are two equivalent noise sources at the input of an op amp: noise voltage and noise current. Noise voltage and current are typically in nV/√Hz and pA/√Hz, respectively. The noise current results in a noise voltage by multiplication with the Thévenin (equivalent) impedance at the op-amp input. Both of these noise sources have to be reflected to the output of the op amp by the transfer function of the entire circuit, including the effect of the feedback elements as a function of frequency, to determine if the noise floor is acceptable. There are classes of op amps that have low noise features. Also, keeping the resistor values as low as possible will minimize their Johnson (thermal) noise contribution (see Figure 10-16).

Power-Supply Considerations. Most IC op amps specify dual supply voltages, typically ranging from ±5V to ±18 V. The actual voltage magnitude is not critical, provided that the output swing from the amplifier is a few volts less than the supply voltages to avoid clipping. Maximum supply-voltage ratings should not be exceeded, of course. CMOS type amplifiers can have rail-to-rail output swings and provide single supply operation. In reality, op amp power supply voltages do not have to be bipolar or symmetrical. A ±15 V supply voltage to an op amp is the same as a single 30 V supply with the op amps' negative supply terminal grounded and +30 V applied to the positive terminal. The inputs have to be biased within the common mode range of the amplifier. However, since most signals fluctuate around a ground reference, dual symmetrical-around-ground supplies are convenient since no biasing is required. Op amps that use voltages as low as ±1.5 V are also called single-supply devices since they can operate off battery power.

To ensure stability, both the positive and negative power-supply voltages should be adequately bypassed to ground for high frequencies. Bypass capacitors can be 0.1 μF ceramic or 10 μF tantalum in most cases. Regulated supplies are desirable but not required. Most amplifiers have a typical supply-voltage sensitivity of 30 μV/V so a 1-V power-supply variation will result in only a 30 μV change reflected at the amplifier input.

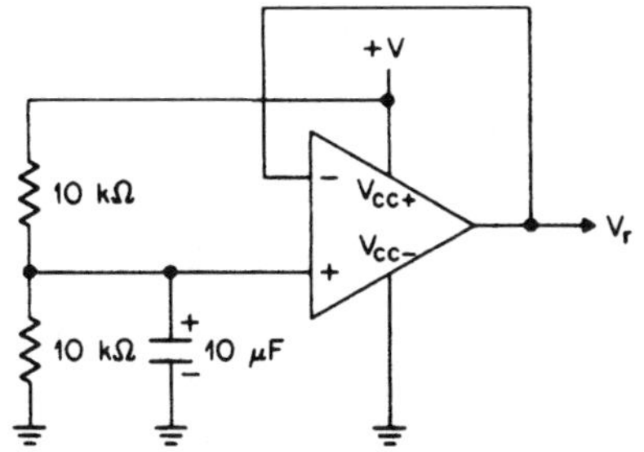

FIGURE 10-25 The generation of a reference voltage.

In many cases, only a single positive supply voltage is available. Dual-voltage–type op amps can still be used by generating a reference voltage V_r, which replaces the circuit's ground connections. The amplifier's negative power terminal is returned to ground and the positive power terminal is connected to the positive supply voltage.

The reference voltage should be midway between the positive supply voltage and the ground and should come from a low-impedance source. A convenient means of generating V_r directly from the positive supply is shown in Figure 10-25 using the voltage follower configuration.

Input signals referenced to ground must be AC-coupled and then superimposed upon V_r. The signals are decoupled at the output to restore the ground reference. A low-pass filter design using this method will not pass low-frequency components near DC.

This technique is illustrated in Figure 10-26 as applied to some previous design examples. The circuit of Figure 10-25 is used to generate V_r and all amplifiers are powered using single-ended power supplies.

Operational Amplifier Selection

Bipolar Input. There are a variety of families of operational amplifiers. Some are differentiated by the type of input stage used within the monolithic integrated circuit. Bipolar inputs are the most commonly used type, where the entire device, including the input stage, consists of bipolar transistors. Input bias and offset currents are a few hundred nanoamperes, and offset voltages are typically under 10 mV. The open-loop input impedance is a few hundred K-ohms.

CMOS Op Amps. CMOS op amps need very little supply current, have high input impedances, and very low bias currents. Offset voltages are somewhat higher than bipolar amplifiers. CMOS amplifiers can operate rail-to-rail and are suitable for single supply operation and low voltage battery powered applications since they draw little power. CMOS amplifiers are typically more noisy than the bipolar type.

BiFET Op Amp. The term *BiFET* is an abbreviation for bipolar-field-effect transistor. It combines two technologies where the front end or input stages use FETs and the remainder of the circuit is bipolar. The result is wider bandwidths than bipolar devices, lower input offset currents, higher input impedances, and more output drive capability. The input offset voltages are typically higher than bipolar op amps.

Op Amp Packaging. Op amps are single, dual, or quad. Dual unit packages will save space and cost. Another useful feature is that both amplifiers in the same package are virtually identical electrically—in other words, matched—because they are on the same substrate. They will typically track very closely with temperature. Quad op amps offer four op amps in the same package. This is very useful for high-density applications, but the designer

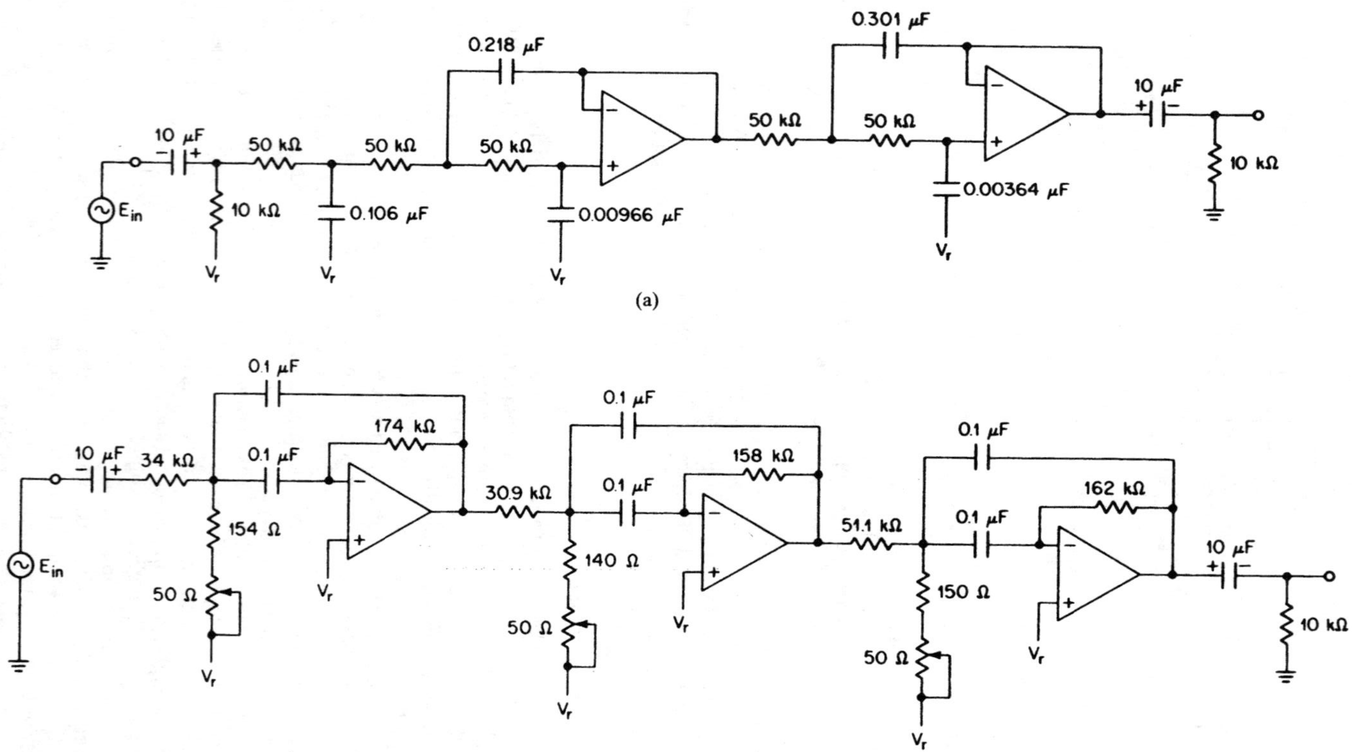

FIGURE 10-26 Designing filters for single supply operation: (*a*) Modified low-pass filter of Example 3-8; and (*b*) Modified bandpass filter of Example 5-12.

should be aware of limited flexibility when designing the PC board since all signal traces have to be routed to a single area to connect to all four op amps, which may result in crosstalk.

Virtually all IC operational amplifiers are available in surface-mount packages. Most are still available in the more traditional dual-in-line package. In most cases, it is the same die or chip that is also used in a through-hole part. Figure 10-27 shows the most popular package, called *small outline* (SO). These packages consist of a circuit attached to a frame and encapsulated with a plastic compound. This compound can withstand the high temperatures resulting from soldering without deforming, and provides a hermetic seal.

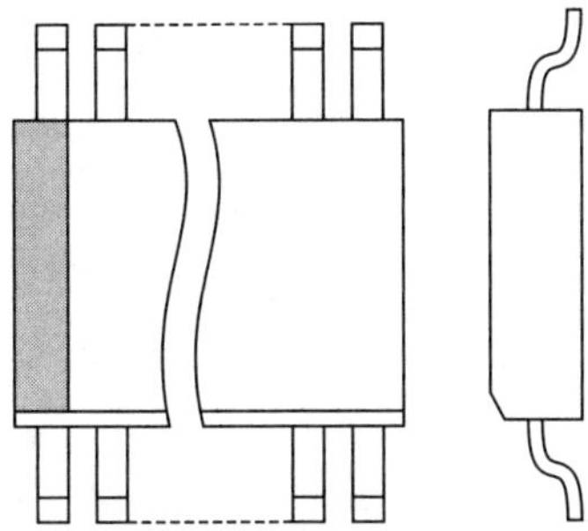

FIGURE 10-27 The plastic small outline (SO) package.

A Survey of Popular Amplifier Types. Integrated Circuit (IC) operational amplifiers range from economical general-purpose devices to the more costly high-performance units A selection guide is provided in Table 10-11 listing some of the numerous op amps that are available from multiple manufacturers and a few parameters. They are sorted by descending bandwidth and whether they are single, dual, or quad. Although manufacturers are listed for each device, many devices are available from multiple manufacturers.

TABLE 10-11 An Operational Amplifier Selection Guide

		Single Op Amps							
		Bias Current nA max	Offset Current nA max	Offset Voltage mV max	Unity Gain BW MHz Typ	Slew rate V/uS Typ	Supply Voltage	Manuf. Code	Features Code
Single	OPA843	37 μA	1 μA	1.5	800	600	±4 to ±6	TI	WB, LN, LH
	MAX4450	20 μA	4 μA	26	210	485	+4.5 to +11	MAX	WB, SS, LH, RR
	AD8065	6 pA	10 pA	1.5	145	180	±2.5 to ±12	AD	JF, WB, LN, SS, RR
	AD8033	11 pA	—	2	80	80	±2.5 to ±13	AD	JF, WB, LN, SS, RR
	LT1363	2 μA	350 nA	1.5	70	1000	±2.5 to ±18	LT	WB, LN, OA
	AD8610	10 pA	10 pA	100 μV	25	50	±5 to ±13	AD	JF, WB, LN, OA
	OPA725	200	50	3	20	30	±2 to ±6	TI	JF, WB, LN, LH
	OPA727	100 pA	100 pA	150 μV	20	30	±2 to ±6	TI	WB, LN, SS, RR
	LM318	250	50	10	15	70	±5 to ±20	TI, NA	GP,WB
	NE5534	1500	500	4	10	13	±3 to ±20	TI	GP, WB, LN, LH
	OPA134	100 pA	50 pA	2	8	20	±2.5 to ±18	TI	JF, LN, LH

(*Continued*)

TABLE 10-11 An Operational Amplifier Selection Guide (*Continued*)

		Single Op Amps							
		Bias Current nA max	Offset Current nA max	Offset Voltage mV max	Unity Gain BW MHz Typ	Slew rate V/uS Typ	Supply Voltage	Manuf. Code	Features Code
	MAX4237	500 pA	1 pA	50 μV	7.5	1.3	+2.4 to +5.5	MAX	SS, RR
	LF351	200 pA	100 pA	10	4	16	±5 to ±18	ST	GP, JF, OA
	TL071	200 pA	100 pA	6	3	13	±3.5 to ±18	TI	GP, JF, LH
	TL080	400 pA	100 pA	15	3	13	±3.5 to ±18	TI	GP, JF, LH, UN
	TL081	400 pA	100 pA	15	3	13	±3.5 to ±18	TI	GP, JF, LH
	LF356	200 pA	50 pA	10	2.5	15	±5 to ±18	NA	GP, JF
	MAX4236	500 pA	1 pA	50 μV	1.7	0.3	+2.4 to +5.5	MAX	SS, RR
	TL061	200 pA	200 pA	15	1	3.5	±3.5 to ±18	TI	GP, JF

		Dual Op Amps							
		Bias Current nA max	Offset Current nA max	Offset Voltage mV max	Unity Gain BW MHz Typ	Slew rate V/uS Typ	Supply Voltage	Manuf. Code	Features Code
Dual	MAX4451	20 μA	4 μA	26	210	485	+4.5 to +11	MAX	WB, SS, LH, RR
	AD8066	6 pA	10 pA	1.5	145	180	±2.5 to ±12	AD	JF, WB, LN, SS, RR
	LT1813	4 μA	400nA	1.5	100	750	±2 to ±6	LT	WB, LN, SS
	AD8034	11 pA		2	80	80	±2.5 to ±13	AD	JF, WB, LN, SS, LH, RR
	OPA2725	200	50	3	20	30	±2 to ±6	TI	JF, WB, LN, SS, RR
	OPA2727	100 pA	100 pA	150 μV	20	30	±2 to ±6	TI	JF, WB, LN, SS, LH, RR
	NE5532	800	500	4	10	9	±3 to ±20	TI	GP, WB, LN
	OPA2134	100 pA	50 pA	2	8	20	±2.5 to ±18	TI	JF, LN, LH
	LF353	200 pA	100 pA	10	4	13	±5 to ±18	TI	GP, JF

	TL072	200 pA	100 pA	6	3	13	±3.5 to ±18	TI	GP, JF, LH
	TL082	400 pA	100 pA	15	3	13	±3.5 to ±18	TI	GP, JF, LH
	MC4558	500	200	6	2.8	1.6	±3 to ±18	ST	GP
	MC1458	700	300	10	1.1	0.8	±3 to ±18	TI	GP
	TL062	200 pA	200 pA	15	1	3.5	±3.5 to ±18	TI	GP, JF
	LM358	250	50	6	1	0.6	±1.5 to ±18	TI	GP, SS

Quad Op Amps

		Bias Current nA max	Offset Current nA max	Offset Voltage mV max	Unity Gain BW MHz Typ	Slew rate V/uS Typ	Supply Voltage	Manuf. Code	Features Code
Quad	OPA4820	22 μA	800	1.5	600	184	±2 to ±6.3	TI	WB, LN, SS
	LT1814	4 μA	400 nA	1.5	100	750	±2 to ±6	LT	WB, SS
	OPA4727	100 pA	100 pA	150 μV	20	30	±2 to ±6	TI	JF, WB, LN, SS, LH, RR
	OPA4134	100 pA	50 pA	2	8	20	±2.5 to ±18	TI	JF, LN, LH
	LF347	200 pA	100 pA	10	4	13	±5 to ±18	TI	GP, JF, LH
	TL074	200 pA	100 pA	6	3	13	±3.5 to ±18	TI	GP, JF, LN, LH
	TL084	400 pA	100 pA	15	3	13	±3.5 to ±18	TI	GP, JF, LH
	TL054	200 pA	1	4	2.7	16	±3.5 to ±18	TI	GP, JF
	TL064	200 pA	200 pA	15	1	3.5	±3.5 to ±18	TI	GP, JF
	LM324	250	50	6	1	0.6	±1.5 to ±16	TI	GP, SS
	LM348	200	50	6	1	0.5	±3 to ±18	TI	GP

Manufacturers' Codes

Code	Manufacturer
TI	Texas Instruments
NA	National Semiconductor
ST	ST Microelectronics
LT	Linear Technology
AD	Analog Devices
MAX	Maxim

(Continued)

Feature Codes

Code	Feature
GP	General Purpose
JF	Junction FET Input
WB	Wide Band
LN	Low Noise
SS	Single Supply
LH	Low Harmonic Distortion
RR	Rail-to-Rail Output
OA	Output Offset Adjust
UN	Uncompensated

The parameters of this table are intended as a general guideline and correspond to commercial-grade op amps at a 25°C ambient temperature. This table is not meant to be all-inclusive and some op amps are better suited than others for specific applications. The manufacturers' data sheets provide more detailed and specific information. Each manufacturer has an applications engineering department which can provide guidance to select the best op amp for a given set of requirements. The designer should keep in mind that excessive gain or bandwidth can result in more sensitivity of the amplifier to PC board layout and instability, so in many applications the more general purpose moderate gain and limited bandwidth amplifiers are the best suited and most robust.

10.4 GENERAL MANUFACTURING CONSIDERATIONS

Specific mechanical considerations determine the robustness of the connections to the circuit board for surface-mount parts. In most cases, the coefficient of expansion of the components differs from that of the board itself. For example, the expansion coefficient of the ceramic dielectric of a chip capacitor is different than that of an epoxy glass printed circuit board. Since the part is held in place only by placing it on top of pads and applying solder, the absence of a further mechanical connection could cause the part to break free with time.

Pad preparation is dependent on which of two soldering processes is used, *reflow*, or *wave*. In the reflow process, the pad is coated by a dried solder paste. The chip is placed on these pads and the board is gradually heated until this solder melts and reflows, joining the part to the pad. With the wave solder method, the parts are first held in place with glue and then the board is covered by a wave of solder. Each method imposes unique requirements on the pads' design to ensure reliability and longevity of the connections.

Although use of surface-mount technology results in high circuit densities, there is a penalty to be paid. Through-hole parts allow the use of the component leads to convey signals from the surface of the board to the inner layers, as well as to the other outside surface of the board. Since the leads of surface-mount components lie flat on the board surface attached to the pads, extra feed-throughs (referred to as *vias*) are required on the board to interconnect layers, which sometimes results in extra layers as well. The end product is a costlier PC board.

RoHS. RoHS is an abbreviation for *Restriction of Hazardous Substances*, which is European Union (EU) directive 2002/95/EG (effective after July 1, 2006). This directive is also known as the *Lead-Free* directive in the electronics industry even though it also restricts usage of other hazardous substances such as Mercury, and so on. The major impact on component selection is to insure that selected parts meet this directive, which should be stated on the data sheet. It also impacts manufacturing processes by dictating lead-free solder requirements.

BIBLIOGRAPHY

Electronic Design Magazine, Gene Heftman, ed., and Analog Devices. "Basics of Op Amps Design." *Electronic Design Magazine* (October 27, 2003).

Lindquist, C. S. *Active Network Design.* Long Beach, California: Steward and Sons, 1977.

Maxim Inc. "Active Filters for Video." *Dallas Semiconductor/Maxim* (July 4, 2003).

National Semiconductor Corp., M. Sauerwald. "Designing Active High Speed Filters." *OA-26 Application Note AN015012* (February, 2005).

Stout, D. F., and M. Kaufman. *Handbook of Operational Amplifier Circuit Design.* New York: McGraw-Hill, 1976.

Texas Instruments. *Amplifier and Data Converter Selection Guide.* Texas Instruments, 3Q, 2005.

CHAPTER 11
NORMALIZED FILTER DESIGN TABLES

TABLE 11-1 Butterworth Pole Locations

Order n	Real Part $-\alpha$	Imaginary Part $\pm j\beta$
2	0.7071	0.7071
3	0.5000	0.8660
	1.0000	
4	0.9239	0.3827
	0.3827	0.9239
5	0.8090	0.5878
	0.3090	0.9511
	1.0000	
6	0.9659	0.2588
	0.7071	0.7071
	0.2588	0.9659
7	0.9010	0.4339
	0.6235	0.7818
	0.2225	0.9749
	1.0000	
8	0.9808	0.1951
	0.8315	0.5556
	0.5556	0.8315
	0.1951	0.9808
9	0.9397	0.3420
	0.7660	0.6428
	0.5000	0.8660
	0.1737	0.9848
	1.0000	
10	0.9877	0.1564
	0.8910	0.4540
	0.7071	0.7071
	0.4540	0.8910
	0.1564	0.9877

TABLE 11-2 Butterworth *LC* Element Values*

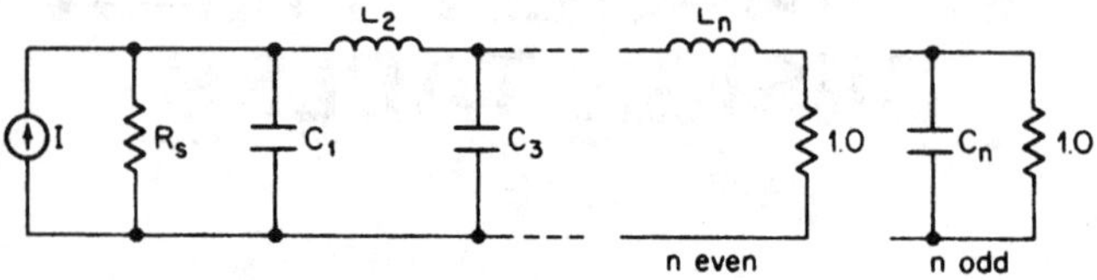

n	R_s	C_1	L_2	C_3	L_4
2	1.0000	1.4142	1.4142		
	1.1111	1.0353	1.8352		
	1.2500	0.8485	2.1213		
	1.4286	0.6971	2.4387		
	1.6667	0.5657	2.8284		
	2.0000	0.4483	3.3461		
	2.5000	0.3419	4.0951		
	3.3333	0.2447	5.3126		
	5.0000	0.1557	7.7067		
	10.0000	0.0743	14.8138		
	Inf.	1.4142	0.7071		
3	1.0000	1.0000	2.0000	1.0000	
	0.9000	0.8082	1.6332	1.5994	
	0.8000	0.8442	1.3840	1.9259	
	0.7000	0.9152	1.1652	2.2774	
	0.6000	1.0225	0.9650	2.7024	
	0.5000	1.1811	0.7789	3.2612	
	0.4000	1.4254	0.6042	4.0642	
	0.3000	1.8380	0.4396	5.3634	
	0.2000	2.6687	0.2842	7.9102	
	0.1000	5.1672	0.1377	15.4554	
	Inf.	1.5000	1.3333	0.5000	
4	1.0000	0.7654	1.8478	1.8478	0.7654
	1.1111	0.4657	1.5924	1.7439	1.4690
	1.2500	0.3882	1.6946	1.5110	1.8109
	1.4286	0.3251	1.8618	1.2913	2.1752
	1.6667	0.2690	2.1029	1.0824	2.6131
	2.0000	0.2175	2.4524	0.8826	3.1868
	2.5000	0.1692	2.9858	0.6911	4.0094
	3.3333	0.1237	3.8826	0.5072	5.3381
	5.0000	0.0804	5.6835	0.3307	7.9397
	10.0000	0.0392	11.0942	0.1616	15.6421
	Inf.	1.5307	1.5772	1.0824	0.3827
n	$1/R_s$	L_1	C_2	L_3	C_4

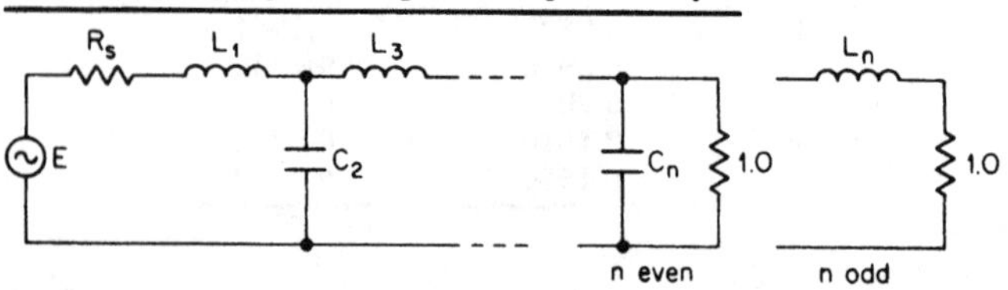

* Reprinted from A. I. Zverev, *Handbook of Filter Synthesis,* John Wiley and Sons, New York, 1967.

TABLE 11-2 Butterworth *LC* Element Values (*Continued*)

n	R_s	C_1	L_2	C_3	L_4	C_5	L_6	C_7
5	1.0000	0.6180	1.6180	2.0000	1.6180	0.6180		
	0.9000	0.4416	1.0265	1.9095	1.7562	1.3887		
	0.8000	0.4698	0.8660	2.0605	1.5443	1.7380		
	0.7000	0.5173	0.7313	2.2849	1.3326	2.1083		
	0.6000	0.5860	0.6094	2.5998	1.1255	2.5524		
	0.5000	0.6857	0.4955	3.0510	0.9237	3.1331		
	0.4000	0.8378	0.3877	3.7357	0.7274	3.9648		
	0.3000	1.0937	0.2848	4.8835	0.5367	5.3073		
	0.2000	1.6077	0.1861	7.1849	0.3518	7.9345		
	0.1000	3.1522	0.0912	14.0945	0.1727	15.7103		
	Inf.	1.5451	1.6944	1.3820	0.8944	0.3090		
6	1.0000	0.5176	1.4142	1.9319	1.9319	1.4142	0.5176	
	1.1111	0.2890	1.0403	1.3217	2.0539	1.7443	1.3347	
	1.2500	0.2445	1.1163	1.1257	2.2389	1.5498	1.6881	
	1.4286	0.2072	1.2363	0.9567	2.4991	1.3464	2.0618	
	1.6667	0.1732	1.4071	0.8011	2.8580	1.1431	2.5092	
	2.0000	0.1412	1.6531	0.6542	3.3687	0.9423	3.0938	
	2.5000	0.1108	2.0275	0.5139	4.1408	0.7450	3.9305	
	3.3333	0.0816	2.6559	0.3788	5.4325	0.5517	5.2804	
	5.0000	0.0535	3.9170	0.2484	8.0201	0.3628	7.9216	
	10.0000	0.0263	7.7053	0.1222	15.7855	0.1788	15.7375	
	Inf.	1.5529	1.7593	1.5529	1.2016	0.7579	0.2588	
7	1.0000	0.4450	1.2470	1.8019	2.0000	1.8019	1.2470	0.4450
	0.9000	0.2985	0.7111	1.4043	1.4891	2.1249	1.7268	1.2961
	0.8000	0.3215	0.6057	1.5174	1.2777	2.3338	1.5461	1.6520
	0.7000	0.3571	0.5154	1.6883	1.0910	2.6177	1.3498	2.0277
	0.6000	0.4075	0.4322	1.9284	0.9170	3.0050	1.1503	2.4771
	0.5000	0.4799	0.3536	2.2726	0.7512	3.5532	0.9513	3.0640
	0.4000	0.5899	0.2782	2.7950	0.5917	4.3799	0.7542	3.9037
	0.3000	0.7745	0.2055	3.6706	0.4373	5.7612	0.5600	5.2583
	0.2000	1.1448	0.1350	5.4267	0.2874	8.5263	0.3692	7.9079
	0.1000	2.2571	0.0665	10.7004	0.1417	16.8222	0.1823	15.7480
	Inf.	1.5576	1.7988	1.6588	1.3972	1.0550	0.6560	0.2225
n	$1/R_s$	L_1	C_2	L_3	C_4	L_5	C_6	L_7

TABLE 11-2 Butterworth *LC* Element Values (*Continued*)

n	R_s	C_1	L_2	C_3	L_4	C_5	L_6	C_7	L_8	C_9	L_{10}
8	1.0000	0.3902	1.1111	1.6629	1.9616	1.9616	1.6629	1.1111	0.3902		
	1.1111	0.2075	0.7575	0.9925	1.6362	1.5900	2.1612	1.7092	1.2671		
	1.2500	0.1774	0.8199	0.8499	1.7779	1.3721	2.3874	1.5393	1.6246		
	1.4286	0.1513	0.9138	0.7257	1.9852	1.1760	2.6879	1.3490	2.0017		
	1.6667	0.1272	1.0455	0.6102	2.2740	0.9912	3.0945	1.1530	2.4524		
	2.0000	0.1042	1.2341	0.5003	2.6863	0.8139	3.6678	0.9558	3.0408		
	2.5000	0.0822	1.5201	0.3945	3.3106	0.6424	4.5308	0.7594	3.8825		
	3.3333	0.0608	1.9995	0.2919	4.3563	0.4757	5.9714	0.5650	5.2400		
	5.0000	0.0400	2.9608	0.1921	6.4523	0.3133	8.8538	0.3732	7.8952		
	10.0000	0.0198	5.8479	0.0949	12.7455	0.1547	17.4999	0.1846	15.7510		
	Inf.	1.5607	1.8246	1.7287	1.5283	1.2588	0.9371	0.5776	0.1951		
9	1.0000	0.3473	1.0000	1.5321	1.8794	2.0000	1.8794	1.5321	1.0000	0.3473	
	0.9000	0.2242	0.5388	1.0835	1.1859	1.7905	1.6538	2.1796	1.6930	1.2447	
	0.8000	0.2434	0.4623	1.1777	1.0200	1.9542	1.4336	2.4189	1.5318	1.6033	
	0.7000	0.2719	0.3954	1.3162	0.8734	2.1885	1.2323	2.7314	1.3464	1.9812	
	0.6000	0.3117	0.3330	1.5092	0.7361	2.5124	1.0410	3.1516	1.1533	2.4328	
	0.5000	0.3685	0.2735	1.7846	0.6046	2.9734	0.8565	3.7426	0.9579	3.0223	
	0.4000	0.4545	0.2159	2.2019	0.4775	3.6706	0.6771	4.6310	0.7624	3.8654	
	0.3000	0.5987	0.1600	2.9006	0.3539	4.8373	0.5022	6.1128	0.5680	5.2249	
	0.2000	0.8878	0.1054	4.3014	0.2333	7.1750	0.3312	9.0766	0.3757	7.8838	
	0.1000	1.7558	0.0521	8.5074	0.1153	14.1930	0.1638	17.9654	0.1862	15.7504	
	Inf.	1.5628	1.8424	1.7772	1.6202	1.4037	1.1408	0.8414	0.5155	0.1736	
10	1.0000	0.3129	0.9080	1.4142	1.7820	1.9754	1.9754	1.7820	1.4142	0.9080	0.3129
	1.1111	0.1614	0.5924	0.7853	1.3202	1.3230	1.8968	1.6956	2.1883	1.6785	1.2267
	1.2500	0.1388	0.6452	0.6762	1.4400	1.1420	2.0779	1.4754	2.4377	1.5245	1.5861
	1.4286	0.1190	0.7222	0.5797	1.6130	0.9802	2.3324	1.2712	2.7592	1.3431	1.9646
	1.6667	0.1004	0.8292	0.4891	1.8528	0.8275	2.6825	1.0758	3.1895	1.1526	2.4169
	2.0000	0.0825	0.9818	0.4021	2.1943	0.6808	3.1795	0.8864	3.7934	0.9588	3.0072
	2.5000	0.0652	1.2127	0.3179	2.7108	0.5384	3.9302	0.7018	4.7002	0.7641	3.8512
	3.3333	0.0484	1.5992	0.2358	3.5754	0.3995	5.1858	0.5211	6.2118	0.5700	5.2122
	5.0000	0.0319	2.3740	0.1556	5.3082	0.2636	7.7010	0.3440	9.2343	0.3775	7.8738
	10.0000	0.0158	4.7005	0.0770	10.5104	0.1305	15.2505	0.1704	18.2981	0.1872	15.7481
	Inf.	1.5643	1.8552	1.8121	1.6869	1.5100	1.2921	1.0406	0.7626	0.4654	0.1564
n	$1/R_s$	L_1	C_2	L_3	C_4	L_5	C_6	L_7	C_8	L_9	C_{10}

TABLE 11-3 Butterworth Uniform Dissipation Network*

$n = 2$

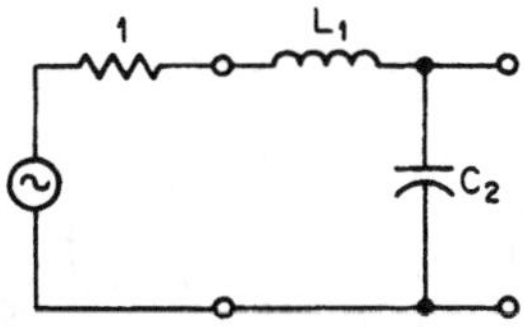

d	L_1	C_2	a_0, dB
0	0.7071	1.414	0
0.05	0.7609	1.410	0.614
0.10	0.8236	1.398	1.22
0.15	0.8974	1.374	1.83
0.20	0.9860	1.340	2.42
0.25	1.094	1.290	2.99
0.30	1.228	1.223	3.53
0.35	1.400	1.138	4.05
0.40	1.628	1.034	4.52
0.45	1.944	0.9083	4.94
0.50	2.414	0.7630	5.30
0.55	3.183	0.5989	5.59
0.60	4.669	0.4188	5.82
0.65	8.756	0.2267	5.96
d	C_2	L_1	a_0, dB

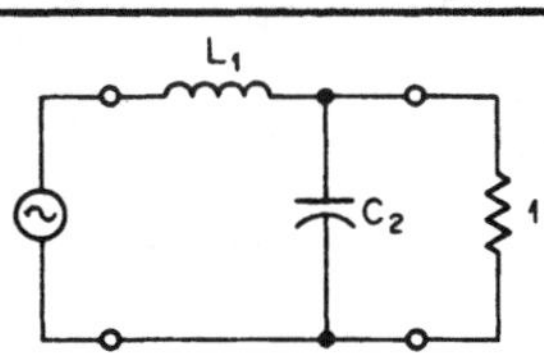

* By permission of P. R. Geffe.

TABLE 11-4 Butterworth Uniform Dissipation Network*

$n = 3$

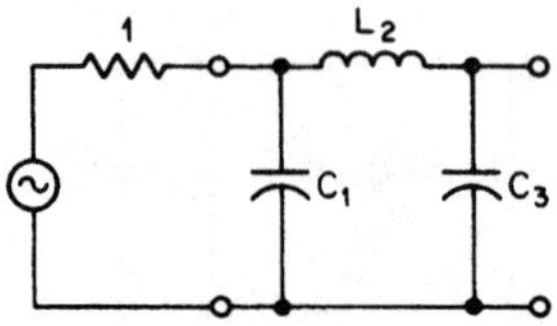

d	C_1	L_2	C_3	α_0, dB
0	0.5000	1.333	1.500	0
0.05	0.5405	1.403	1.457	0.868
0.10	0.5882	1.481	1.402	1.73
0.15	0.6452	1.567	1.334	2.60
0.20	0.7143	1.667	1.250	3.45
0.25	0.8000	1.786	1.149	4.30
0.30	0.9091	1.939	1.026	5.15
0.35	1.053	2.164	0.8743	5.98
0.40	1.250	2.581	0.6798	6.82
0.45	1.538	3.806	0.4126	7.66
d	L_3	C_2	L_1	α_0, dB

* By permission of P. R. Geffe.

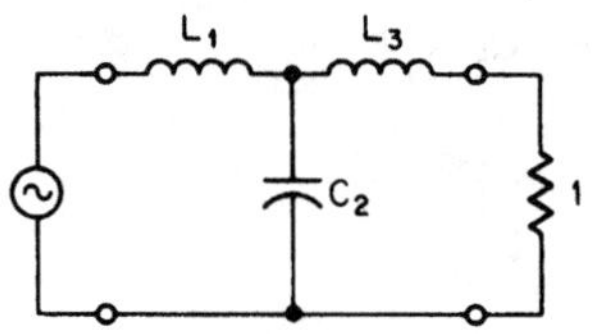

TABLE 11-5 Butterworth Uniform Dissipation Network*

$n = 4$

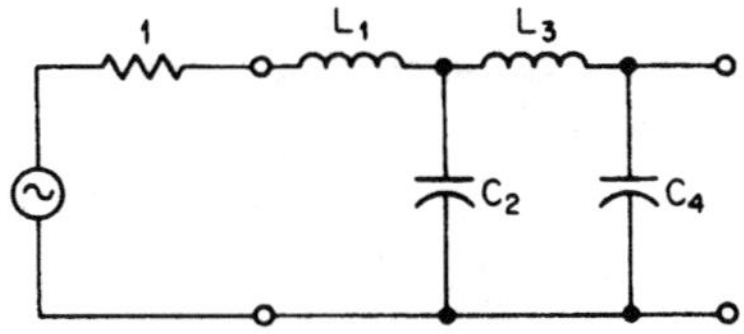

d	L_1	C_2	L_3	C_4	a_0, dB
0	0.3827	1.082	1.577	1.531	0
0.05	0.4144	1.156	1.636	1.454	1.13
0.10	0.4518	1.240	1.701	1.362	2.27
0.15	0.4967	1.339	1.777	1.250	3.39
0.20	0.5515	1.459	1.879	1.113	4.51
0.25	0.6199	1.609	2.039	0.9400	5.63
0.30	0.7077	1.812	2.384	0.7099	6.73
0.35	0.8243	2.124	3.848	0.3651	7.82
d	C_4	L_3	C_2	L_1	a_0, dB

* By permission of P. R. Geffe.

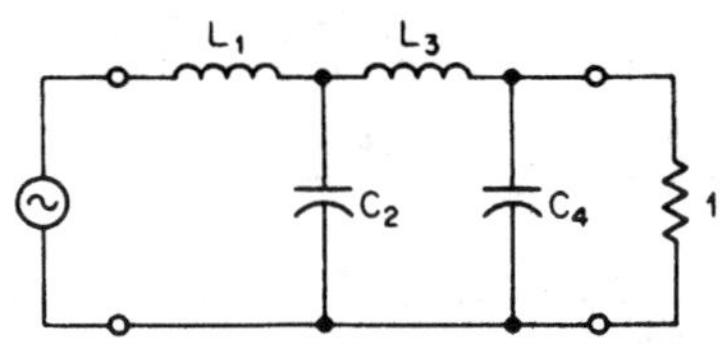

TABLE 11-6 Butterworth Uniform Dissipation Network*

$n = 5$

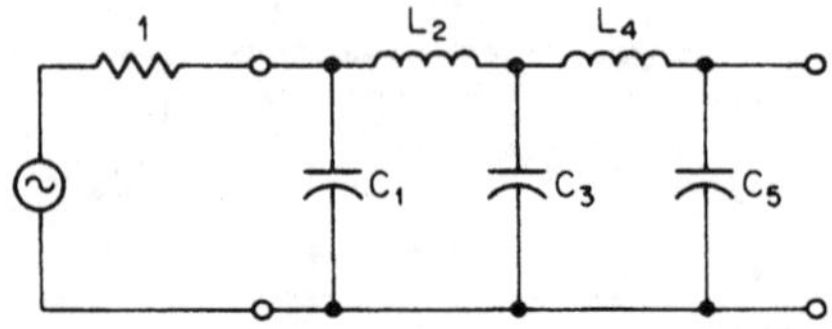

d	C_1	L_2	C_3	L_4	C_5	a_0, dB
0	0.3090	0.8944	1.382	1.694	1.545	0
0.02	0.3189	0.9199	1.412	1.712	1.504	0.562
0.04	0.3294	0.9468	1.443	1.730	1.461	1.12
0.06	0.3406	0.9754	1.476	1.750	1.414	1.69
0.08	0.3526	1.006	1.512	1.771	1.364	2.25
0.10	0.3654	1.038	1.549	1.794	1.309	2.81
0.12	0.3794	1.073	1.589	1.822	1.250	3.37
0.14	0.3943	1.111	1.633	1.854	1.184	3.93
0.16	0.4104	1.151	1.681	1.894	1.113	4.48
0.18	0.4281	1.195	1.734	1.946	1.034	5.04
0.20	0.4472	1.243	1.796	2.018	0.9452	5.59
0.22	0.4681	1.296	1.867	2.124	0.8434	6.15
0.24	0.4911	1.354	1.953	2.300	0.7242	6.70
0.26	0.5165	1.419	2.061	2.631	0.5798	7.25
0.28	0.5446	1.493	2.204	3.453	0.3965	7.79
0.30	0.5760	1.578	2.409	8.084	0.1476	8.34
d	L_5	C_4	L_3	C_2	L_1	a_0, dB

* By permission of P. R. Geffe.

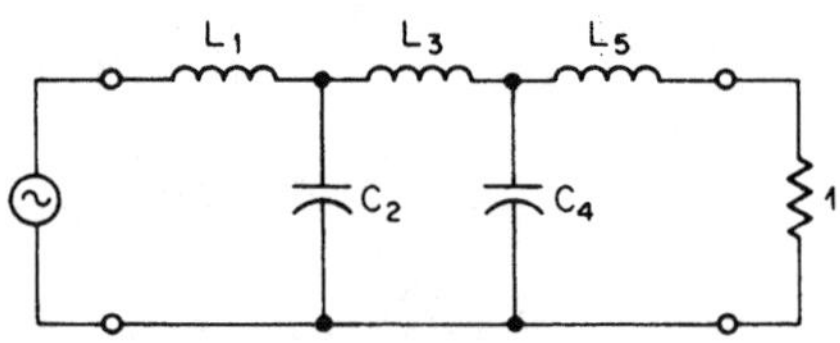

TABLE 11-7 Butterworth Uniform Dissipation Network*

$n = 6$

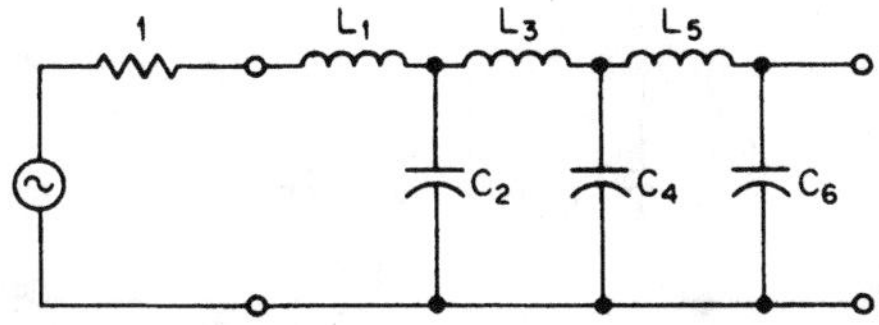

d	L_1	C_2	L_3	C_4	L_5	C_6	a_0, dB
0	0.2588	0.7579	1.202	1.553	1.759	1.533	0
0.02	0.2671	0.7804	1.232	1.581	1.727	1.502	0.671
0.04	0.2760	0.8043	1.264	1.611	1.786	1.446	1.34
0.06	0.2854	0.8297	1.297	1.643	1.802	1.386	2.01
0.08	0.2955	0.8569	1.333	1.679	1.821	1.321	2.68
0.10	0.3064	0.8860	1.372	1.714	1.844	1.250	3.35
0.12	0.3181	0.9172	1.413	1.755	1.874	1.171	4.02
0.14	0.3307	0.9508	1.458	1.802	1.917	1.083	4.69
0.16	0.3443	0.9871	1.508	1.860	1.979	0.9839	5.30
0.18	0.3594	1.027	1.558	1.923	2.080	0.8690	6.00
0.20	0.3754	1.070	1.621	2.008	2.258	0.7313	6.68
0.22	0.3931	1.117	1.690	2.122	2.646	0.5586	7.34
d	C_6	L_5	C_4	L_3	C_2	L_1	a_0, dB

* By permission of P. R. Geffe.

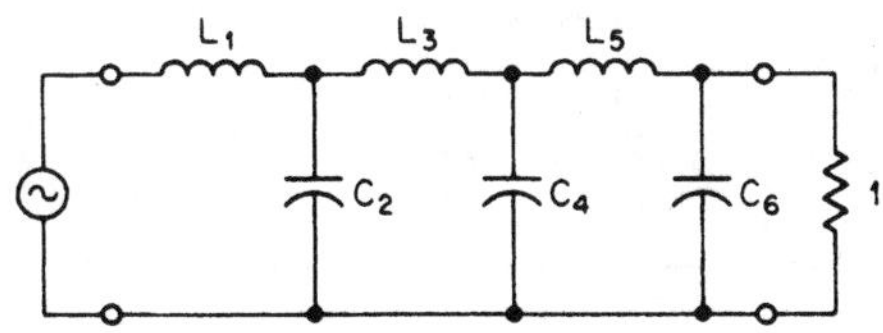

TABLE 11-8 Butterworth Uniform Dissipation Network*

$n = 7$

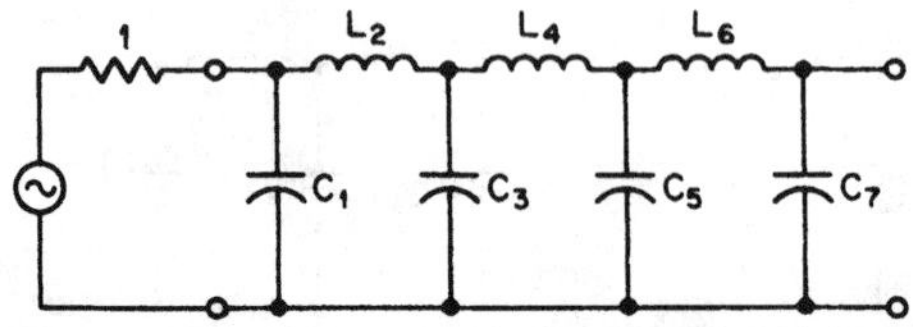

d	C_1	L_2	C_3	L_4	C_5	L_6	C_7	a_0, dB
0	0.2225	0.6560	1.054	1.397	1.659	1.799	1.588	0
0.02	0.2297	0.6759	1.084	1.428	1.684	1.808	1.496	0.781
0.04	0.2373	0.6972	1.114	1.461	1.712	1.818	1.428	1.56
0.06	0.2454	0.7198	1.146	1.496	1.742	1.832	1.354	2.34
0.08	0.2542	0.7440	1.180	1.533	1.775	1.851	1.274	3.12
0.10	0.2636	0.7699	1.217	1.573	1.813	1.878	1.184	3.90
0.12	0.2739	0.7980	1.254	1.614	1.860	1.923	1.085	4.68
0.14	0.2846	0.8281	1.294	1.659	1.910	1.992	0.9701	5.45
0.16	0.2966	0.8608	1.344	1.715	1.979	2.111	0.8350	6.23
0.18	0.3091	0.8960	1.394	1.778	2.073	2.356	0.6679	7.00
0.20	0.3232	0.9243	1.453	1.862	2.233	3.177	0.4220	7.77
d	L_7	C_6	L_5	C_4	L_3	C_2	L_1	a_0, dB

* By permission of P. R. Geffe.

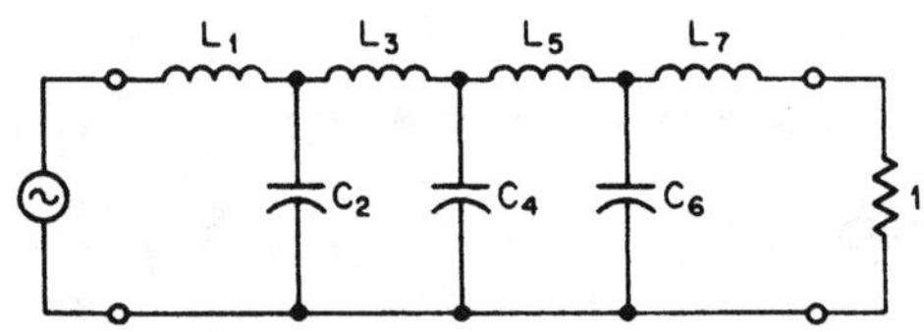

TABLE 11-9 Butterworth Uniform Dissipation Network*

$n=8$

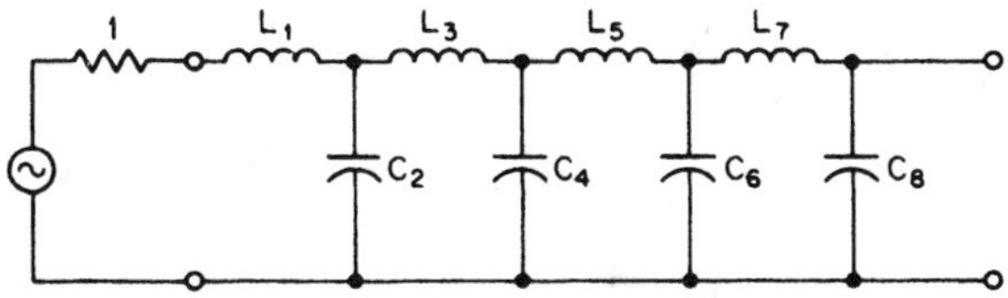

d	L_1	C_2	L_3	C_4	L_5	C_6	L_7	C_8	a_0, dB
0	0.1951	0.5776	0.9371	1.259	1.528	1.729	1.824	1.561	0
0.02	0.2014	0.5954	0.9636	1.290	1.558	1.752	1.830	1.488	0.890
0.04	0.2081	0.6144	0.9918	1.323	1.590	1.777	1.838	1.409	1.78
0.06	0.2152	0.6347	1.022	1.357	1.624	1.806	1.851	1.321	2.67
0.08	0.2229	0.6564	1.054	1.394	1.622	1.839	1.872	1.224	3.56
0.10	0.2312	0.6796	1.088	1.434	1.703	1.880	1.908	1.114	4.45
0.12	0.2400	0.7046	1.124	1.478	1.750	1.932	1.972	0.9856	5.33
0.14	0.2496	0.7316	1.164	1.526	1.804	2.003	2.101	0.8305	6.22
0.16	0.2600	0.7608	1.208	1.579	1.869	2.110	2.414	0.6307	7.10
0.18	0.2713	0.7926	1.255	1.639	1.951	2.294	3.683	0.3439	7.98
d	C_8	L_7	C_6	L_5	C_4	L_3	C_2	L_1	a_0, dB

* By permission of P. R. Geffe.

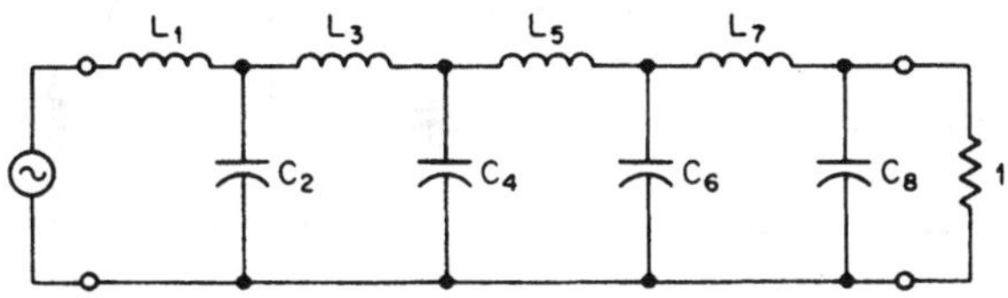

TABLE 11-10 Butterworth Uniform Dissipation Network*

$n = 9$

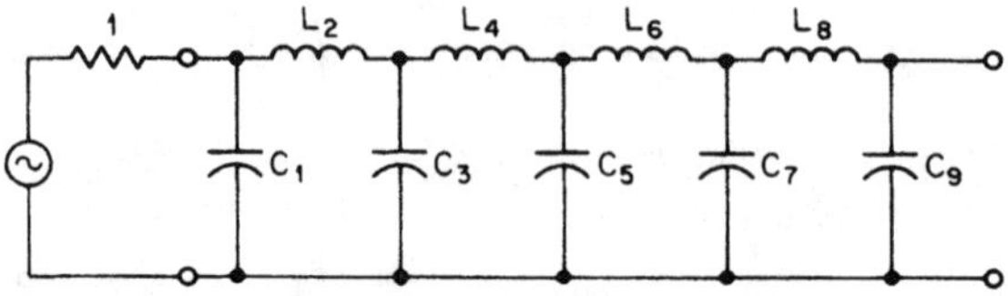

d	C_1	L_2	C_3	L_4	C_5	L_6	C_7	L_8	C_9	a_0, dB
0	0.1736	0.5155	0.8414	1.141	1.404	1.620	1.777	1.842	1.563	0
0.02	0.1793	0.5316	0.8659	1.171	1.435	1.649	1.798	1.845	1.480	1.00
0.04	0.1852	0.5488	0.8921	1.202	1.469	1.680	1.822	1.851	1.388	2.00
0.06	0.1916	0.5671	0.9199	1.236	1.504	1.713	1.850	1.864	1.286	3.00
0.08	0.1984	0.5867	0.9496	1.272	1.543	1.751	1.884	1.891	1.171	4.00
0.10	0.2058	0.6077	0.9814	1.311	1.584	1.794	1.931	1.942	1.036	5.00
0.12	0.2137	0.6303	1.016	1.353	1.630	1.844	1.997	2.054	0.8735	5.99
0.14	0.2223	0.6547	1.053	1.398	1.682	1.907	2.101	2.340	0.6614	6.99
0.16	0.2315	0.6812	1.093	1.448	1.742	1.991	2.293	3.620	0.3486	7.98
d	L_9	C_8	L_7	C_6	L_5	C_4	L_3	C_2	L_1	a_0, dB

* By permission of P. R. Geffe.

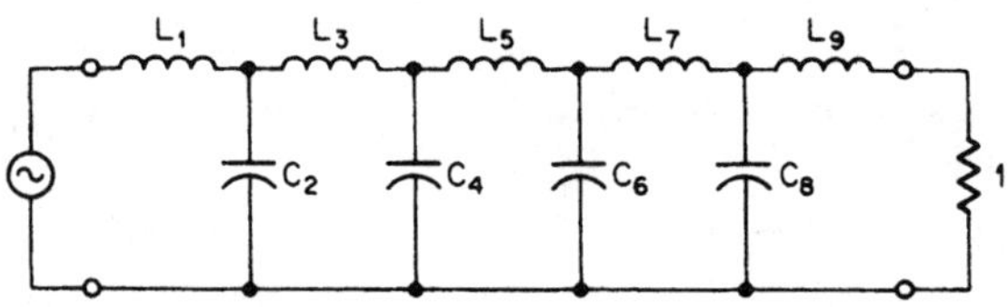

TABLE 11-11 Butterworth Uniform Dissipation Network*

$n = 10$

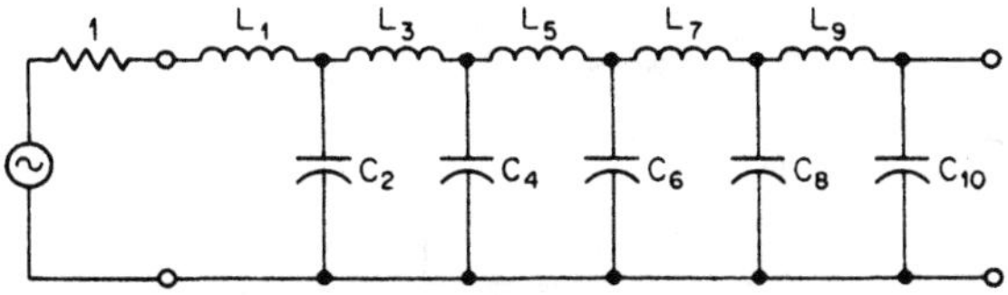

d	L_1	C_2	L_3	C_4	L_5	C_6	L_7	C_8	L_9	C_{10}	a_0, dB
0	0.1564	0.4654	0.7626	1.041	1.292	1.510	1.687	1.812	1.855	1.564	0
0.02	0.1614	0.4800	0.7854	1.069	1.324	1.541	1.714	1.831	1.855	1.471	1.11
0.04	0.1669	0.4956	0.8096	1.099	1.357	1.574	1.744	1.853	1.860	1.367	2.22
0.06	0.1726	0.5123	0.8353	1.132	1.392	1.610	1.777	1.882	1.875	1.249	3.33
0.08	0.1788	0.5301	0.8629	1.166	1.430	1.648	1.814	1.920	1.910	1.114	4.44
0.10	0.1854	0.5493	0.8924	1.203	1.471	1.692	1.860	1.976	1.991	0.9508	5.55
0.12	0.1926	0.5698	0.9242	1.243	1.516	1.741	1.918	2.067	2.201	0.7409	6.65
0.14	0.2003	0.5921	0.9584	1.286	1.566	1.798	1.997	2.239	3.051	0.4349	7.76
d	C_{10}	L_9	C_8	L_7	C_6	L_5	C_4	L_3	C_2	L_1	a_0, dB

* By permission of P. R. Geffe.

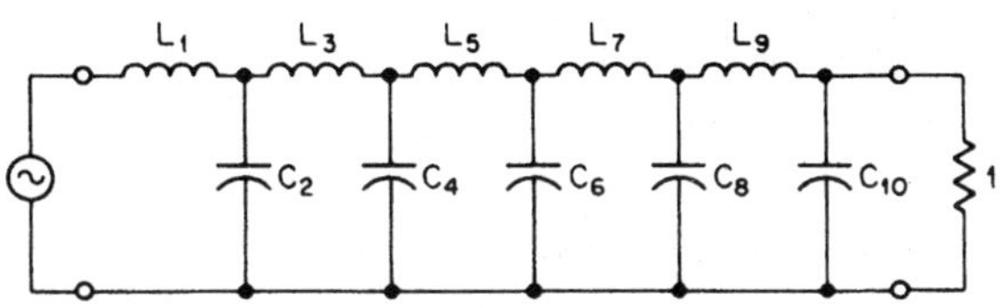

TABLE 11-12 Butterworth Lossy-L Network*

$n = 2$

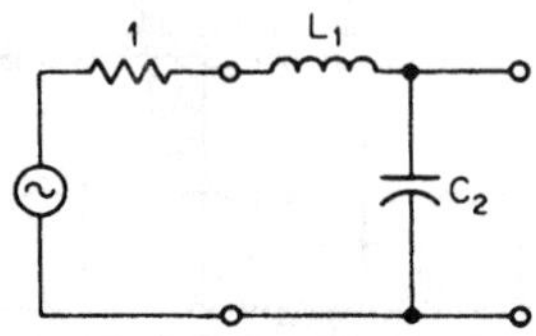

d	L_1	C_2
0	0.7071	1.414
0.05	0.7330	1.364
0.10	0.7609	1.314
0.15	0.7910	1.264
0.20	0.8236	1.214
0.25	0.8589	1.164
0.30	0.8975	1.114
0.35	0.9397	1.064
0.40	0.9860	1.014
0.45	1.037	0.9642
0.50	1.094	0.9142
0.55	1.157	0.8642
0.60	1.228	0.8142
0.65	1.309	0.7642

* By permission of P. R. Geffe.

TABLE 11-13 Butterworth Lossy-L Network*

$n = 3$

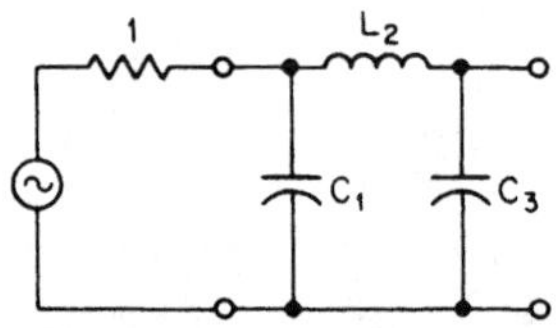

d	C_1	L_2	C_3
0	0.5000	1.333	1.500
0.05	0.5128	1.403	1.390
0.10	0.5263	1.480	1.284
0.15	0.5405	1.565	1.182
0.20	0.5556	1.660	1.084
0.25	0.5714	1.766	0.9911
0.30	0.5882	1.885	0.9018
0.35	0.6061	2.021	0.8164
0.40	0.6250	2.177	0.7350
0.45	0.6452	2.358	0.6573

* By permission of P. R. Geffe.

TABLE 11-14 Butterworth Lossy-L Network*

$n=4$

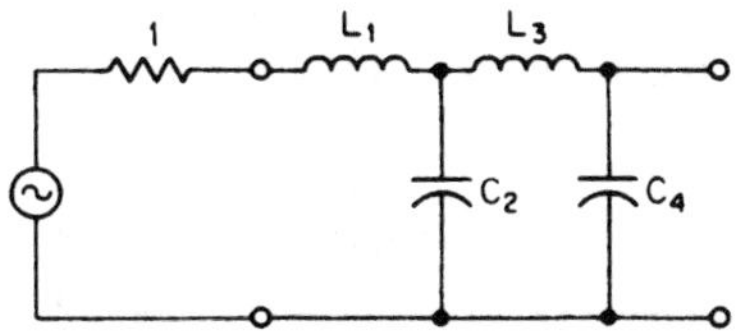

d	L_1	C_2	L_3	C_4
0	0.3827	1.082	1.577	1.531
0.05	0.3979	1.087	1.698	1.362
0.10	0.4144	1.091	1.834	1.205
0.15	0.4323	1.095	1.990	1.061
0.20	0.4518	1.098	2.170	0.9289
0.25	0.4732	1.100	2.380	0.8072
0.30	0.4967	1.102	2.628	0.6955
0.35	0.5227	1.102	2.926	0.5933

* By permission of P. R. Geffe.

TABLE 11-15 Butterworth Lossy-L Network*

$n=5$

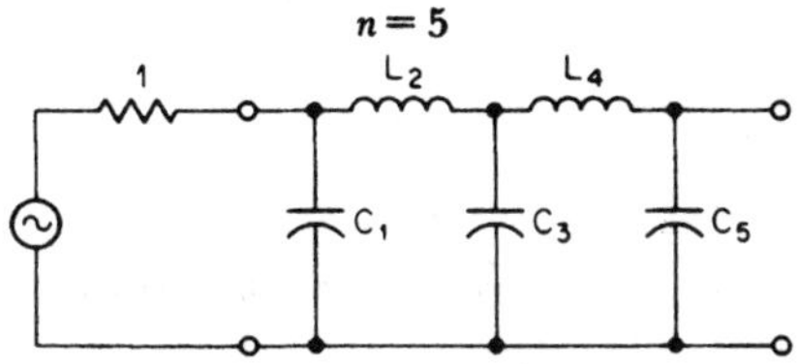

d	C_1	L_2	C_3	L_4	C_5
0	0.3090	0.8944	1.382	1.694	1.545
0.02	0.3129	0.9127	1.369	1.762	1.452
0.04	0.3168	0.9316	1.355	1.834	1.363
0.06	0.3209	0.9514	1.342	1.911	1.278
0.08	0.3251	0.9719	1.327	1.993	1.197
0.10	0.3294	0.9934	1.313	2.080	1.119
0.12	0.3338	1.016	1.298	2.173	1.046
0.14	0.3383	1.039	1.283	2.273	0.9754
0.16	0.3429	1.063	1.268	2.380	0.9086
0.18	0.3477	1.089	1.253	2.494	0.8450
0.20	0.3526	1.116	1.237	2.620	0.7844
0.22	0.3576	1.144	1.221	2.754	0.7269
0.24	0.3628	1.173	1.204	2.901	0.6721
0.26	0.3682	1.204	1.188	3.061	0.6201
0.28	0.3737	1.237	1.171	3.237	0.5076
0.30	0.3794	1.271	1.154	3.431	0.5236

* By permission of P. R. Geffe.

TABLE 11-16 Butterworth Lossy-L Network*

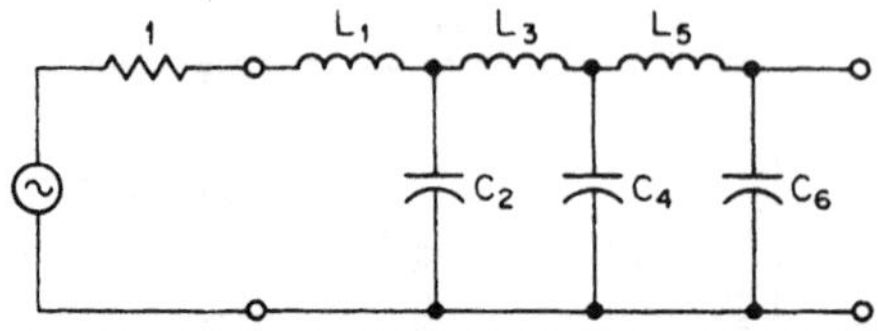

d	L_1	C_2	L_3	C_4	L_5	C_6
0	0.2588	0.7579	1.202	1.553	1.759	1.553
0.02	0.2629	0.7631	1.235	1.519	1.850	1.436
0.04	0.2671	0.7683	1.271	1.485	1.947	1.326
0.06	0.2714	0.7736	1.308	1.451	2.052	1.223
0.08	0.2760	0.7789	1.347	1.417	2.165	1.125
0.10	0.2806	0.7843	1.388	1.383	2.228	1.034
0.12	0.2854	0.7897	1.432	1.349	2.421	0.9487
0.14	0.2904	0.7952	1.478	1.315	2.565	0.8684
0.16	0.2955	0.8007	1.527	1.281	2.723	0.7932
0.18	0.3009	0.8063	1.579	1.248	2.896	0.7227
0.20	0.3064	0.8118	1.634	1.214	3.807	0.6567
0.22	0.3121	0.8174	1.692	1.181	3.298	0.5949

* By permission of P. R. Geffe.

TABLE 11-17 Butterworth Lossy-L Network*

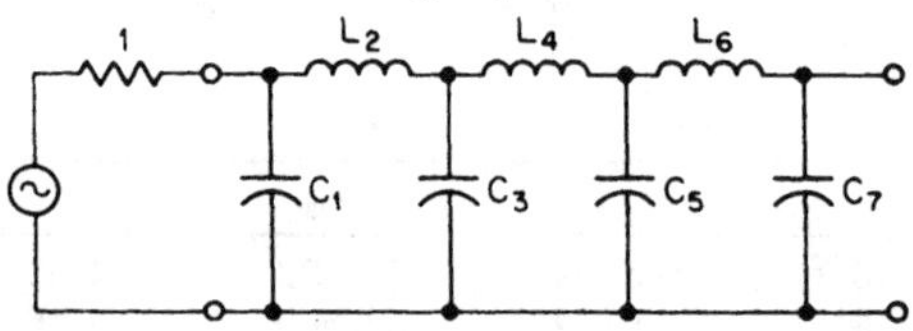

d	C_1	L_2	C_3	L_4	C_5	L_6	C_7
0	0.2225	0.6560	1.054	1.397	1.659	1.799	1.588
0.02	0.2255	0.6688	1.053	1.449	1.602	1.913	1.417
0.04	0.2286	0.6822	1.051	1.504	1.546	2.038	1.288
0.06	0.2318	0.6960	1.048	1.564	1.490	2.173	1.167
0.08	0.2351	0.7104	1.045	1.627	1.436	2.322	1.056
0.10	0.2384	0.7255	1.043	1.694	1.382	2.484	0.9532
0.12	0.2419	0.7412	1.039	1.766	1.330	2.664	0.8581
0.14	0.2454	0.7575	1.036	1.842	1.278	2.862	0.7703
0.16	0.2491	0.7746	1.032	1.924	1.228	3.083	0.6892
0.18	0.2529	0.7924	1.028	2.013	1.178	3.330	0.6144
0.20	0.2568	0.8110	1.024	2.108	1.130	3.609	0.5454

* By permission of P. R. Geffe.

TABLE 11-18 Butterworth Lossy-L Network*

$n=8$

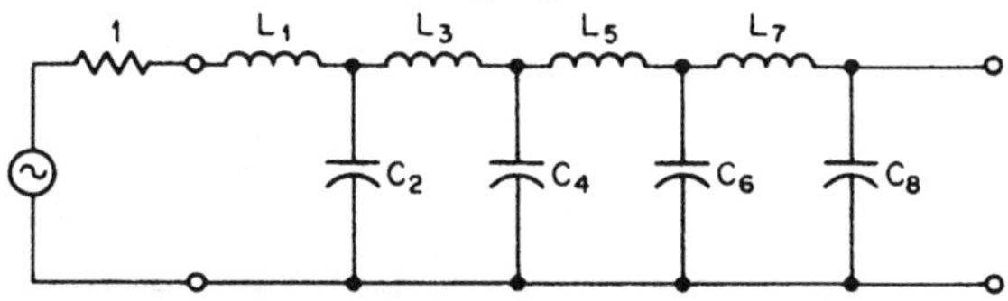

d	L_1	C_2	L_3	C_4	L_5	C_6	L_7	C_8
0	0.1951	0.5776	0.9371	1.259	1.528	1.729	1.824	1.561
0.02	0.1982	0.5829	0.9613	1.243	1.602	1.648	1.963	1.398
0.04	0.2014	0.5884	0.9868	1.227	1.680	1.569	2.116	1.249
0.06	0.2047	0.5939	1.014	1.211	1.764	1.493	2.285	1.113
0.08	0.2081	0.5996	1.042	1.194	1.856	1.419	2.472	0.9894
0.10	0.2116	0.6053	1.071	1.178	1.954	1.347	2.681	0.8768
0.12	0.2152	0.6111	1.102	1.160	2.061	1.278	2.914	0.7743
0.14	0.2190	0.6170	1.134	1.143	2.177	1.211	3.178	0.6810
0.16	0.2229	0.6231	1.169	1.124	2.302	1.147	3.477	0.5962
0.18	0.2270	0.6292	1.206	1.107	2.440	1.084	3.819	0.5191

* By permission of P. R. Geffe.

TABLE 11-19 Butterworth Lossy-L Network*

$n = 9$

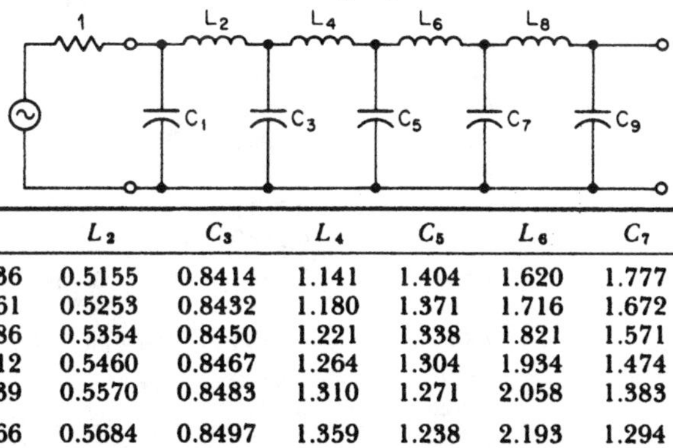

d	C_1	L_2	C_3	L_4	C_5	L_6	C_7	L_8	C_9
0	0.1736	0.5155	0.8414	1.141	1.404	1.620	1.777	1.842	1.563
0.02	0.1761	0.5253	0.8432	1.180	1.371	1.716	1.672	2.006	1.377
0.04	0.1786	0.5354	0.8450	1.221	1.338	1.821	1.571	2.189	1.211
0.06	0.1812	0.5460	0.8467	1.264	1.304	1.934	1.474	2.393	1.061
0.08	0.1839	0.5570	0.8483	1.310	1.271	2.058	1.383	2.623	0.9261
0.10	0.1866	0.5684	0.8497	1.359	1.238	2.193	1.294	2.884	0.8054
0.12	0.1894	0.5802	0.8510	1.412	1.204	2.342	1.211	3.180	0.6971
0.14	0.1924	0.5926	0.8522	1.467	1.171	2.505	1.132	3.521	0.6001
0.16	0.1954	0.6054	0.8533	1.527	1.137	2.686	1.057	3.917	0.5132

* By permission of P. R. Geffe.

TABLE 11-20 Butterworth Lossy-L Network*

$n = 10$

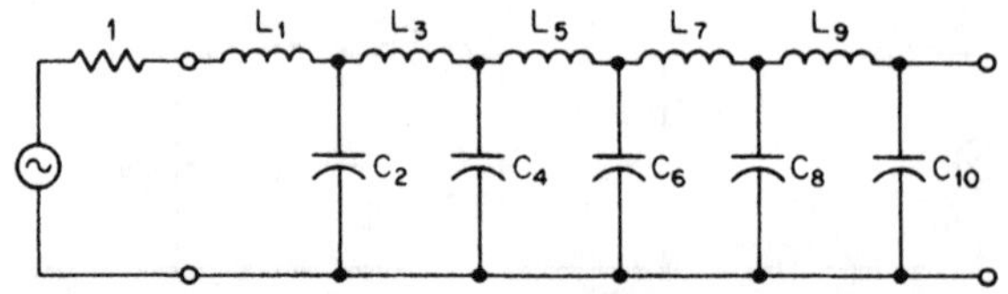

d	L_1	C_2	L_3	C_4	L_5	C_6	L_7	C_8	L_9	C_{10}
0	0.1564	0.4654	0.7626	1.041	1.292	1.510	1.687	1.812	1.855	1.564
0.02	0.1589	0.4704	0.7812	1.034	1.348	1.457	1.807	1.682	2.044	1.357
0.04	0.1614	0.4754	0.8006	1.027	1.408	1.404	1.939	1.560	2.258	1.173
0.06	0.1641	0.4806	0.8209	1.019	1.472	1.353	2.084	1.444	2.501	1.010
0.08	0.1669	0.4859	0.8422	1.011	1.541	1.302	2.245	1.336	2.778	0.8660
0.10	0.1697	0.4913	0.8603	1.003	1.614	1.251	2.423	1.234	3.096	0.7387
0.12	0.1726	0.4969	0.8880	0.9949	1.694	1.201	2.621	1.139	3.466	0.6263
0.14	0.1757	0.5024	0.9127	0.9861	1.780	1.152	2.842	1.050	3.901	0.5270

* By permission of P. R. Geffe.

TABLE 11-21 Butterworth Active Low-Pass Values*

Order n	C_1	C_2	C_3
2	1.414	0.7071	
3	3.546	1.392	0.2024
4	1.082	0.9241	
	2.613	0.3825	
5	1.753	1.354	0.4214
	3.235	0.3090	
6	1.035	0.9660	
	1.414	0.7071	
	3.863	0.2588	
7	1.531	1.336	0.4885
	1.604	0.6235	
	4.493	0.2225	
8	1.020	0.9809	
	1.202	0.8313	
	1.800	0.5557	
	5.125	0.1950	
9	1.455	1.327	0.5170
	1.305	0.7661	
	2.000	0.5000	
	5.758	0.1736	
10	1.012	0.9874	
	1.122	0.8908	
	1.414	0.7071	
	2.202	0.4540	
	6.390	0.1563	

* Reprinted from *Electronics,* McGraw-Hill, Inc., August 18, 1969.

TABLE 11-22 0.01-dB Chebyshev Pole Locations

Order n	Real Part $-\alpha$	Imaginary Part $\pm j\beta$
2	0.6743	0.7075
3	0.4233	0.8663
	0.8467	
4	0.6762	0.3828
	0.2801	0.9241
5	0.5120	0.5879
	0.1956	0.9512
	0.6328	
6	0.5335	0.2588
	0.3906	0.7072
	0.1430	0.9660
7	0.4393	0.4339
	0.3040	0.7819
	0.1085	0.9750
	0.4876	
8	0.4268	0.1951
	0.3618	0.5556
	0.2418	0.8315
	0.08490	0.9808
9	0.3686	0.3420
	0.3005	0.6428
	0.1961	0.8661
	0.06812	0.9848
	0.3923	

TABLE 11-23 0.1-dB Chebyshev Pole Locations

Order n	Real Part $-\alpha$	Imaginary Part $\pm j\beta$
2	0.6104	0.7106
3	0.3490	0.8684
	0.6979	
4	0.2177	0.9254
	0.5257	0.3833
5	0.3842	0.5884
	0.1468	0.9521
	0.4749	
6	0.3916	0.2590
	0.2867	0.7077
	0.1049	0.9667
7	0.3178	0.4341
	0.2200	0.7823
	0.0785	0.9755
	0.3528	
8	0.3058	0.1952
	0.2592	0.5558
	0.1732	0.8319
	0.06082	0.9812
9	0.2622	0.3421
	0.2137	0.6430
	0.1395	0.8663
	0.04845	0.9852
	0.2790	

TABLE 11-24 0.25-dB Chebyshev Pole Locations

Order n	Real Part $-\alpha$	Imaginary Part $\pm j\beta$
2	0.5621	0.7154
3	0.3062	0.8712
	0.6124	
4	0.4501	0.3840
	0.1865	0.9272
5	0.3247	0.5892
	0.1240	0.9533
	0.4013	
6	0.3284	0.2593
	0.2404	0.7083
	0.08799	0.9675
7	0.2652	0.4344
	0.1835	0.7828
	0.06550	0.9761
	0.2944	
8	0.2543	0.1953
	0.2156	0.5561
	0.1441	0.8323
	0.05058	0.9817
9	0.2176	0.3423
	0.1774	0.6433
	0.1158	0.8667
	0.04021	0.9856
	0.2315	

TABLE 11-25 0.5-dB Chebyshev Pole Locations

Order n	Real Part $-\alpha$	Imaginary Part $\pm j\beta$
2	0.5129	0.7225
3	0.2683	0.8753
	0.5366	
4	0.3872	0.3850
	0.1605	0.9297
5	0.2767	0.5902
	0.1057	0.9550
	0.3420	
6	0.2784	0.2596
	0.2037	0.7091
	0.07459	0.9687
7	0.2241	0.4349
	0.1550	0.7836
	0.05534	0.9771
	0.2487	
8	0.2144	0.1955
	0.1817	0.5565
	0.1214	0.8328
	0.04264	0.9824
9	0.1831	0.3425
	0.1493	0.6436
	0.09743	0.8671
	0.03383	0.9861
	0.1949	

TABLE 11-26 1-dB Chebyshev Pole Locations

Order n	Real Part $-\alpha$	Imaginary Part $\pm j\beta$
2	0.4508	0.7351
3	0.2257	0.8822
	0.4513	
4	0.3199	0.3868
	0.1325	0.9339
5	0.2265	0.5918
	0.08652	0.9575
	0.2800	
6	0.2268	0.2601
	0.1660	0.7106
	0.06076	0.9707
7	0.1819	0.4354
	0.1259	0.7846
	0.04494	0.9785
	0.2019	
8	0.1737	0.1956
	0.1473	0.5571
	0.09840	0.8337
	0.03456	0.9836
9	0.1482	0.3427
	0.1208	0.6442
	0.07884	0.8679
	0.02739	0.9869
	0.1577	

TABLE 11-27 0.01-dB Chebyshev *LC* Element Values*

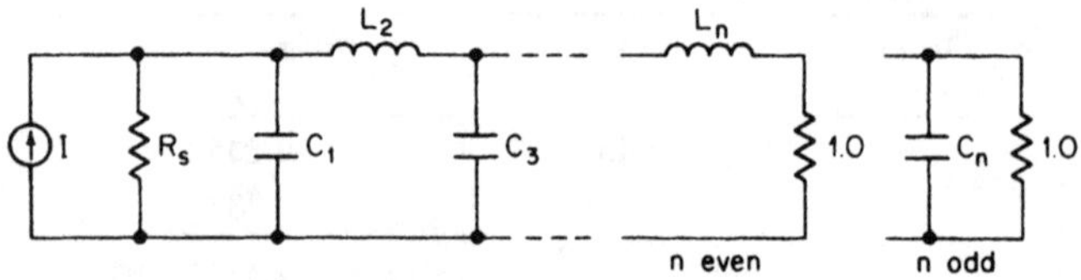

n	R_s	C_1	L_2	C_3	L_4
2	1.1007	1.3472	1.4829		
	1.1111	1.2472	1.5947		
	1.2500	0.9434	1.9974		
	1.4286	0.7591	2.3442		
	1.6667	0.6091	2.7496		
	2.0000	0.4791	3.2772		
	2.5000	0.3634	4.0328		
	3.3333	0.2590	5.2546		
	5.0000	0.1642	7.6498		
	10.0000	0.0781	14.7492		
	Inf.	1.4118	0.7415		
3	1.0000	1.1811	1.8214	1.1811	
	0.9000	1.0917	1.6597	1.4802	
	0.8000	1.0969	1.4431	1.8057	
	0.7000	1.1600	1.2283	2.1653	
	0.6000	1.2737	1.0236	2.5984	
	0.5000	1.4521	0.8294	3.1644	
	0.4000	1.7340	0.6452	3.9742	
	0.3000	2.2164	0.4704	5.2800	
	0.2000	3.1934	0.3047	7.8338	
	0.1000	6.1411	0.1479	15.3899	
	Inf.	1.5012	1.4330	0.5905	
4	1.1000	0.9500	1.9382	1.7608	1.0457
	1.1111	0.8539	1.9460	1.7439	1.1647
	1.2500	0.6182	2.0749	1.5417	1.6170
	1.4286	0.4948	2.2787	1.3336	2.0083
	1.6667	0.3983	2.5709	1.1277	2.4611
	2.0000	0.3156	2.9943	0.9260	3.0448
	2.5000	0.2418	3.6406	0.7293	3.8746
	3.3333	0.1744	4.7274	0.5379	5.2085
	5.0000	0.1121	6.9102	0.3523	7.8126
	10.0000	0.0541	13.4690	0.1729	15.5100
	Inf.	1.5287	1.6939	1.3122	0.5229
n	$1/R_s$	L_1	C_2	L_3	C_4

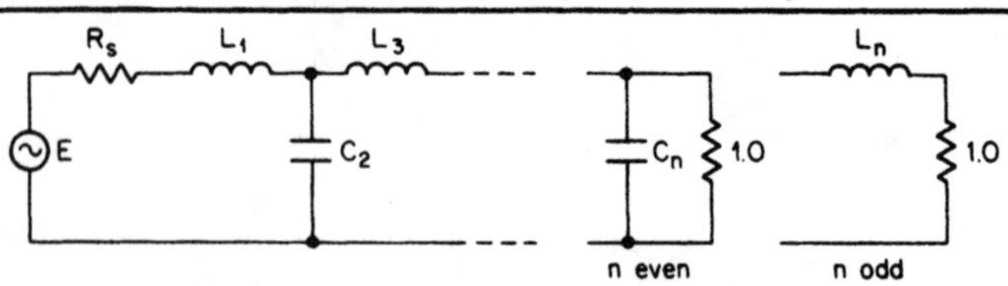

* Reprinted from A. I. Zverev, *Handbook of Filter Synthesis,* John Wiley and Sons, New York, 1967.

TABLE 11-27 0.01-dB Chebyshev *LC* Element Values (*Continued*)

n	R_s	C_1	L_2	C_3	L_4	C_5	L_6	C_7
5	1.0000	0.9766	1.6849	2.0366	1.6849	0.9766		
	0.9000	0.8798	1.4558	2.1738	1.6412	1.2739		
	0.8000	0.8769	1.2350	2.3785	1.4991	1.6066		
	0.7000	0.9263	1.0398	2.6582	1.3228	1.9772		
	0.6000	1.0191	0.8626	3.0408	1.1345	2.4244		
	0.5000	1.1658	0.6985	3.5835	0.9421	3.0092		
	0.4000	1.3983	0.5442	4.4027	0.7491	3.8453		
	0.3000	1.7966	0.3982	5.7721	0.5573	5.1925		
	0.2000	2.6039	0.2592	8.5140	0.3679	7.8257		
	0.1000	5.0406	0.1266	16.7406	0.1819	15.6126		
	Inf.	1.5466	1.7950	1.6449	1.2365	0.4883		
6	1.1007	0.8514	1.7956	1.8411	2.0266	1.6312	0.9372	
	1.1111	0.7597	1.7817	1.7752	2.0941	1.6380	1.0533	
	1.2500	0.5445	1.8637	1.4886	2.4025	1.5067	1.5041	
	1.4286	0.4355	2.0383	1.2655	2.7346	1.3318	1.8987	
	1.6667	0.3509	2.2978	1.0607	3.1671	1.1451	2.3568	
	2.0000	0.2786	2.6781	0.8671	3.7683	0.9536	2.9483	
	2.5000	0.2139	3.2614	0.6816	4.6673	0.7606	3.7899	
	3.3333	0.1547	4.2448	0.5028	6.1631	0.5676	5.1430	
	5.0000	0.0997	6.2227	0.3299	9.1507	0.3760	7.7852	
	10.0000	0.0483	12.1707	0.1623	18.1048	0.1865	15.5950	
	Inf.	1.5510	1.8471	1.7897	1.5976	1.1904	0.4686	
7	1.0000	0.9127	1.5947	2.0021	1.8704	2.0021	1.5947	0.9127
	0.9000	0.8157	1.3619	2.0886	1.7217	2.2017	1.5805	1.2060
	0.8000	0.8111	1.1504	2.2618	1.5252	2.4647	1.4644	1.5380
	0.7000	0.8567	0.9673	2.5158	1.3234	2.8018	1.3066	1.9096
	0.6000	0.9430	0.8025	2.8720	1.1237	3.2496	1.1310	2.3592
	0.5000	1.0799	0.6502	3.3822	0.9276	3.8750	0.9468	2.9478
	0.4000	1.2971	0.5072	4.1563	0.7350	4.8115	0.7584	3.7900
	0.3000	1.6692	0.3716	5.4540	0.5459	6.3703	0.5682	5.1476
	0.2000	2.4235	0.2423	8.0565	0.3604	9.4844	0.3776	7.8019
	0.1000	4.7006	0.1186	15.8718	0.1784	18.8179	0.1879	15.6523
	Inf.	1.5593	1.8671	1.8657	1.7651	1.5633	1.1610	0.4564
n	$1/R_s$	L_1	C_2	L_3	C_4	L_5	C_6	L_7

TABLE 11-27 0.01-dB Chebyshev *LC* Element Values (*Continued*)

n	R_s	C_1	L_2	C_3	L_4	C_5	L_6	C_7	L_8	C_9	L_{10}
8	1.1007	0.8145	1.7275	1.7984	2.0579	1.8695	1.9796	1.5694	0.8966		
	1.1111	0.7248	1.7081	1.7239	2.1019	1.8259	2.0595	1.5827	1.0111		
	1.2500	0.5176	1.7772	1.4315	2.3601	1.5855	2.4101	1.4754	1.4597		
	1.4286	0.4138	1.9422	1.2141	2.6686	1.3723	2.7734	1.3142	1.8544		
	1.6667	0.3336	2.1896	1.0169	3.0808	1.1660	3.2393	1.1369	2.3136		
	2.0000	0.2650	2.5533	0.8313	3.6598	0.9639	3.8820	0.9518	2.9073		
	2.5000	0.2036	3.1118	0.6537	4.5303	0.7653	4.8393	0.7627	3.7524		
	3.3333	0.1474	4.0539	0.4826	5.9828	0.5697	6.4287	0.5718	5.1118		
	5.0000	0.0951	5.9495	0.3170	8.8889	0.3770	9.6002	0.3804	7.7668		
	10.0000	0.0462	11.6509	0.1562	17.6067	0.1870	19.1009	0.1895	15.6158		
	Inf.	1.5588	1.8848	1.8988	1.8556	1.7433	1.5391	1.1412	0.4483		
9	1.0000	0.8854	1.5513	1.9614	1.8616	2.0717	1.8616	1.9614	1.5513	0.8854	
	0.9000	0.7886	1.3192	2.0330	1.6941	2.2249	1.7402	2.1774	1.5478	1.1764	
	0.8000	0.7834	1.1127	2.1959	1.4930	2.4614	1.5603	2.4565	1.4423	1.5076	
	0.7000	0.8273	0.9353	2.4404	1.2924	2.7808	1.3 62	2.8093	1.2927	1.8793	
	0.6000	0.9109	0.7761	2.7852	1.0962	3.2140	1.1688	3.2747	1.1233	2.3295	
	0.5000	1.0436	0.6290	3.2805	0.9045	3.8249	0.9710	3.9223	0.9436	2.9193	
	0.4000	1.2542	0.4910	4.0329	0.7167	4.7444	0.7739	4.8900	0.7582	3.7637	
	0.3000	1.6151	0.3599	5.2951	0.5325	6.2792	0.5780	6.4989	0.5697	5.1254	
	0.2000	2.3468	0.2349	7.8274	0.3518	9.3504	0.3835	9.7114	0.3797	7.7882	
	0.1000	4.5556	0.1150	15.4334	0.1743	18.5641	0.1908	19.3382	0.1895	15.6645	
	Inf.	1.5646	1.8884	1.9242	1.8977	1.8425	1.7261	1.5217	1.1273	0.4427	
10	1.1007	0.7970	1.6930	1.7690	2.0395	1.8827	2.0724	1.8529	1.9472	1.5380	0.8773
	1.1111	0.7083	1.6714	1.6921	2.0763	1.8281	2.1308	1.8167	2.0310	1.5541	0.9910
	1.2500	0.5049	1.7353	1.4005	2.3184	1.5706	2.4371	1.5953	2.3952	1.4574	1.4381
	1.4286	0.4037	1.8958	1.1871	2.6178	1.3552	2.7830	1.3895	2.7685	1.3027	1.8327
	1.6667	0.3255	2.1375	0.9942	3.0205	1.1497	3.2370	1.1863	3.2448	1.1300	2.2923
	2.0000	0.2586	2.4932	0.8128	3.5878	0.9497	3.8698	0.9849	3.9004	0.9484	2.8867
	2.5000	0.1988	3.0398	0.6394	4.4418	0.7538	4.8173	0.7849	4.8757	0.7617	3.7333
	3.3333	0.1440	3.9619	0.4723	5.8678	0.5612	6.3951	0.5863	6.4939	0.5722	5.0955
	5.0000	0.0930	5.8175	0.3103	8.7220	0.3715	9.5486	0.3893	9.7217	0.3814	7.7563
	10.0000	0.0451	11.3993	0.1530	17.2866	0.1844	19.0046	0.1938	19.3905	0.1904	15.6234
	Inf.	1.5625	1.8978	1.9323	1.9288	1.8907	1.8309	1.7128	1.5088	1.1173	0.4386
n	$1/R_s$	L_1	C_2	L_3	C_4	L_5	C_6	L_7	C_8	L_9	C_{10}

TABLE 11-28 0.1-dB Chebyshev *LC* Element Values*

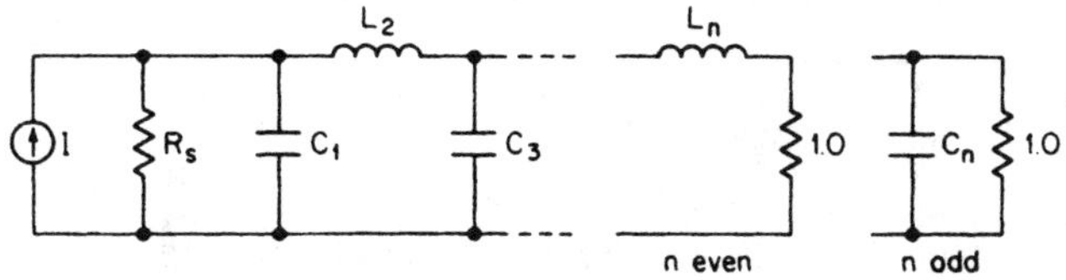

n	R_s	C_1	L_2	C_3	L_4
2	1.3554	1.2087	1.6382		
	1.4286	0.9771	1.9824		
	1.6667	0.7326	2.4885		
	2.0000	0.5597	3.0538		
	2.5000	0.4169	3.8265		
	3.3333	0.2933	5.0502		
	5.0000	0.1841	7.4257		
	10.0000	0.0868	14.4332		
	Inf.	1.3911	0.8191		
3	1.0000	1.4328	1.5937	1.4328	
	0.9000	1.4258	1.4935	1.6219	
	0.8000	1.4511	1.3557	1.8711	
	0.7000	1.5210	1.1927	2.1901	
	0.6000	1.6475	1.0174	2.6026	
	0.5000	1.8530	0.8383	3.1594	
	0.4000	2.1857	0.6603	3.9675	
	0.3000	2.7630	0.4860	5.2788	
	0.2000	3.9418	0.3172	7.8503	
	0.1000	7.5121	0.1549	15.4656	
	Inf.	1.5133	1.5090	0.7164	
4	1.3554	0.9924	2.1476	1.5845	1.3451
	1.4286	0.7789	2.3480	1.4292	1.7001
	1.6667	0.5764	2.7304	1.1851	2.2425
	2.0000	0.4398	3.2269	0.9672	2.8563
	2.5000	0.3288	3.9605	0.7599	3.6976
	3.3333	0.2329	5.1777	0.5602	5.0301
	5.0000	0.1475	7.6072	0.3670	7.6143
	10.0000	0.0704	14.8873	0.1802	15.2297
	Inf.	1.5107	1.7682	1.4550	0.6725
n	$1/R_s$	L_1	C_2	L_3	C_4

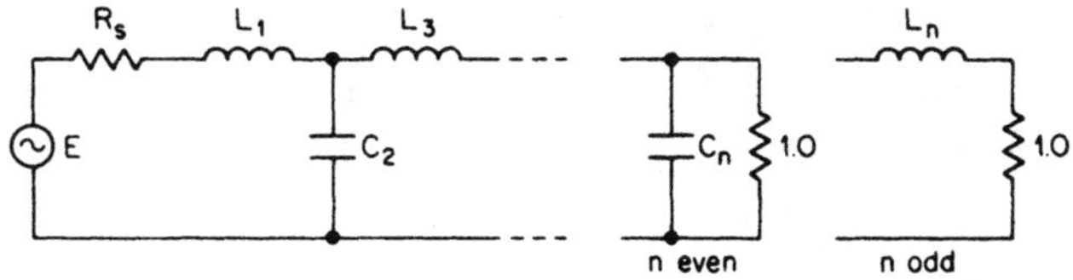

* Reprinted from A. I. Zverev, *Handbook of Filter Synthesis*, John Wiley and Sons, New York, 1967.

TABLE 11-28 0.1-dB Chebyshev *LC* Element Values (*Continued*)

n	R_s	C_1	L_2	C_3	L_4	C_5	L_6	C_7
5	1.0000	1.3013	1.5559	2.2411	1.5559	1.3013		
	0.9000	1.2845	1.4329	2.3794	1.4878	1.4883		
	0.8000	1.2998	1.2824	2.5819	1.3815	1.7384		
	0.7000	1.3580	1.1170	2.8679	1.2437	2.0621		
	0.6000	1.4694	0.9469	3.2688	1.0846	2.4835		
	0.5000	1.6535	0.7777	3.8446	0.9126	3.0548		
	0.4000	1.9538	0.6119	4.7193	0.7333	3.8861		
	0.3000	2.4765	0.4509	6.1861	0.5503	5.2373		
	0.2000	3.5457	0.2950	9.1272	0.3659	7.8890		
	0.1000	6.7870	0.1447	17.9569	0.1820	15.7447		
	Inf.	1.5613	1.8069	1.7659	1.4173	0.6507		
6	1.3554	0.9419	2.0797	1.6581	2.2473	1.5344	1.2767	
	1.4286	0.7347	2.2492	1.4537	2.5437	1.4051	1.6293	
	1.6667	0.5422	2.6003	1.1830	3.0641	1.1850	2.1739	
	2.0000	0.4137	3.0679	0.9575	3.7119	0.9794	2.7936	
	2.5000	0.3095	3.7652	0.7492	4.6512	0.7781	3.6453	
	3.3333	0.2195	4.9266	0.5514	6.1947	0.5795	4.9962	
	5.0000	0.1393	7.2500	0.3613	9.2605	0.3835	7.6184	
	10.0000	0.0666	14.2200	0.1777	18.4267	0.1901	15.3495	
	Inf.	1.5339	1.8838	1.8306	1.7485	1.3937	0.6383	
7	1.0000	1.2615	1.5196	2.2392	1.6804	2.2392	1.5196	1.2615
	0.9000	1.2422	1.3946	2.3613	1.5784	2.3966	1.4593	1.4472
	0.8000	1.2550	1.2449	2.5481	1.4430	2.6242	1.3619	1.6967
	0.7000	1.3100	1.0826	2.8192	1.2833	2.9422	1.2326	2.0207
	0.6000	1.4170	0.9169	3.2052	1.1092	3.3841	1.0807	2.4437
	0.5000	1.5948	0.7529	3.7642	0.9276	4.0150	0.9142	3.0182
	0.4000	1.8853	0.5926	4.6179	0.7423	4.9702	0.7384	3.8552
	0.3000	2.3917	0.4369	6.0535	0.5557	6.5685	0.5569	5.2167
	0.2000	3.4278	0.2862	8.9371	0.3692	9.7697	0.3723	7.8901
	0.1000	6.5695	0.1405	17.6031	0.1838	19.3760	0.1862	15.8127
	Inf.	1.5748	1.8577	1.9210	1.8270	1.7340	1.3786	0.6307
n	$1/R_s$	L_1	C_2	L_3	C_4	L_5	C_6	L_7

TABLE 11-28 0.1-dB Chebyshev *LC* Element Values (*Continued*)

n	R_s	C_1	L_2	C_3	L_4	C_5	L_6	C_7	L_8	C_9	L_{10}
8	1.3554	0.9234	2.0454	1.6453	2.2826	1.6841	2.2300	1.5091	1.2515		
	1.4286	0.7186	2.2054	1.4350	2.5554	1.4974	2.5422	1.3882	1.6029		
	1.6667	0.5298	2.5459	1.1644	3.0567	1.2367	3.0869	1.1769	2.1477		
	2.0000	0.4042	3.0029	0.9415	3.6917	1.0118	3.7619	0.9767	2.7690		
	2.5000	0.3025	3.6859	0.7365	4.6191	0.7990	4.7388	0.7787	3.6240		
	3.3333	0.2147	4.8250	0.5421	6.1483	0.5930	6.3423	0.5820	4.9811		
	5.0000	0.1364	7.1050	0.3554	9.1917	0.3917	9.5260	0.3863	7.6164		
	10.0000	0.0652	13.9469	0.1749	18.3007	0.1942	19.0437	0.1922	15.3880		
	Inf.	1.5422	1.9106	1.9008	1.9252	1.8200	1.7231	1.3683	0.6258		
9	1.0000	1.2446	1.5017	2.2220	1.6829	2.2957	1.6829	2.2220	1.5017	1.2446	
	0.9000	1.2244	1.3765	2.3388	1.5756	2.4400	1.5870	2.3835	1.4444	1.4297	
	0.8000	1.2361	1.2276	2.5201	1.4365	2.6561	1.4572	2.6168	1.3505	1.6788	
	0.7000	1.2898	1.0670	2.7856	1.2751	2.9647	1.3019	2.9422	1.2248	2.0029	
	0.6000	1.3950	0.9035	3.1653	1.1008	3.3992	1.1304	3.3937	1.0761	2.4264	
	0.5000	1.5701	0.7419	3.7166	0.9198	4.0244	0.9494	4.0377	0.9121	3.0020	
	0.4000	1.8566	0.5840	4.5594	0.7359	4.9750	0.7630	5.0118	0.7382	3.8412	
	0.3000	2.3560	0.4307	5.9781	0.5509	6.5700	0.5736	6.6413	0.5579	5.2068	
	0.2000	3.3781	0.2822	8.8291	0.3661	9.7699	0.3827	9.9047	0.3737	7.8891	
	0.1000	6.4777	0.1386	17.3994	0.1823	19.3816	0.1912	19.6976	0.1873	15.8393	
	Inf.	1.5804	1.8727	1.9584	1.9094	1.9229	1.8136	1.7150	1.3611	0.6223	
10	1.3554	0.9146	2.0279	1.6346	2.2777	1.6963	2.2991	1.6805	2.2155	1.4962	1.2397
	1.4286	0.7110	2.1837	1.4231	2.5425	1.5002	2.5915	1.5000	2.5322	1.3789	1.5903
	1.6667	0.5240	2.5194	1.1536	3.0362	1.2349	3.1229	1.2444	3.0839	1.1717	2.1351
	2.0000	0.3998	2.9713	0.9326	3.6647	1.0089	3.7923	1.0214	3.7669	0.9741	2.7572
	2.5000	0.2993	3.6476	0.7295	4.5843	0.7962	4.7673	0.8090	4.7547	0.7779	3.6136
	3.3333	0.2124	4.7758	0.5370	6.1022	0.5907	6.3734	0.6020	6.3758	0.5822	4.9735
	5.0000	0.1350	7.0347	0.3522	9.1248	0.3902	9.5681	0.3987	9.5942	0.3871	7.6148
	10.0000	0.0646	13.8141	0.1734	18.1739	0.1935	19.1282	0.1981	19.2158	0.1929	15.4052
	Inf.	1.5460	1.9201	1.9216	1.9700	1.9102	1.9194	1.8083	1.7090	1.3559	0.6198
n	$1/R_s$	L_1	C_2	L_3	C_4	L_5	C_6	L_7	C_8	L_9	C_{10}

TABLE 11-29 0.25-dB Chebyshev *LC* Element Values

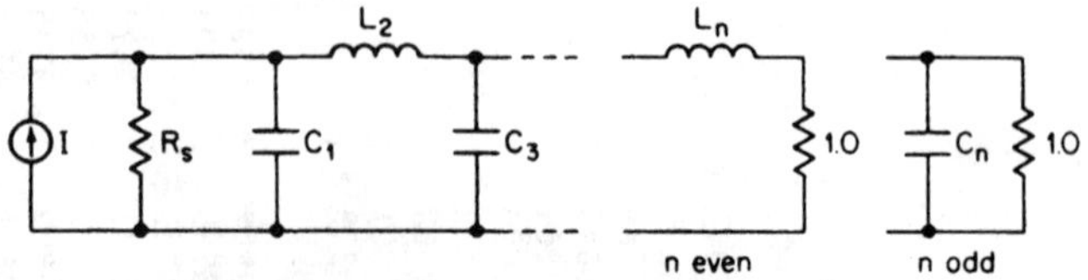

n	R_s	C_1	L_2	C_3	L_4
2	2	0.6552	2.7632		
	3	0.3740	4.3118		
	4	0.2637	5.7389		
	8	0.1215	11.2589		
	Inf.	1.3584	0.8902		
3	1	1.6325	1.4360	1.6325	
	0.5	3.2663	1.0775	1.6325	
	0.333	4.8988	0.9572	1.6325	
	0.25	6.5326	0.8971	1.6325	
	0.125	13.0639	0.8081	1.6325	
	Inf.	1.5348	1.5285	0.8169	
4	2	0.6747	3.6860	1.0247	1.8806
	3	0.4149	6.2744	0.7682	2.1302
	4	0.3020	8.8161	0.6667	2.2533
	8	0.1448	19.0204	0.5334	2.4516
	Inf.	1.4817	1.8213	1.5068	0.7853
n	$1/R_s$	L_1	C_2	L_3	C_4

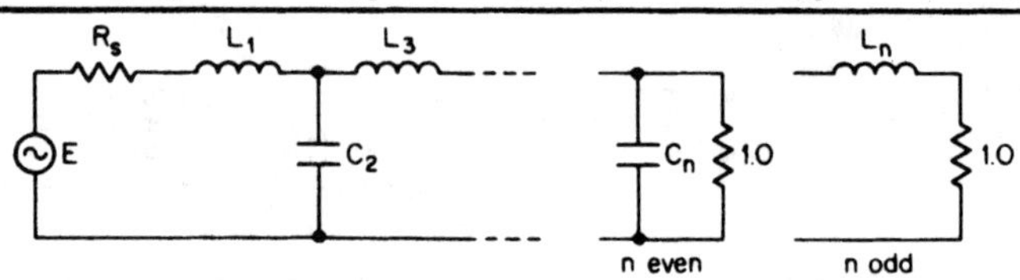

TABLE 11-29 0.25-dB Chebyshev *LC* Element Values (*Continued*)

n	R_s	C_1	L_2	C_3	L_4	C_5	L_6	C_7
5	1	1.5046	1.4436	2.4050	1.4436	1.5046		
	0.5	3.0103	0.7218	3.6080	1.4436	1.5046		
	0.333	4.5149	0.4812	4.8100	1.4436	1.5046		
	0.25	6.0196	0.3615	6.0130	1.4436	1.5046		
	0.125	12.0402	0.1807	10.8230	1.4436	1.5046		
	Inf.	1.5765	1.7822	1.8225	1.4741	0.7523		
6	2	0.6867	3.2074	0.9308	3.8102	1.2163	1.7088	
	3	0.4330	5.0976	0.5392	6.0963	1.0804	1.8393	
	4	0.3173	6.9486	0.3821	8.2530	1.0221	1.8987	
	8	0.1539	14.3100	0.1762	16.7193	0.9393	1.9868	
	Inf.	1.5060	1.9221	1.8191	1.8329	1.4721	0.7610	
7	1	1.5120	1.4169	2.4535	1.5350	2.4535	1.4169	1.5120
	0.5	3.024	0.7085	4.9069	1.1515	2.4535	1.4169	1.5120
	0.333	4.5361	0.4723	7.3596	1.0230	2.4535	1.4169	1.5120
	0.25	6.0471	0.3542	9.8120	0.9593	2.4535	1.4169	1.5120
	0.125	12.0952	0.1776	19.6251	0.8631	2.4535	1.4169	1.5120
	Inf.	1.6009	1.8287	1.9666	1.8234	1.8266	1.4629	0.7555
n	$1/R_s$	L_1	C_2	L_3	C_4	L_5	C_6	L_7

TABLE 11-30 0.5-dB Chebyshev *LC* Element Values*

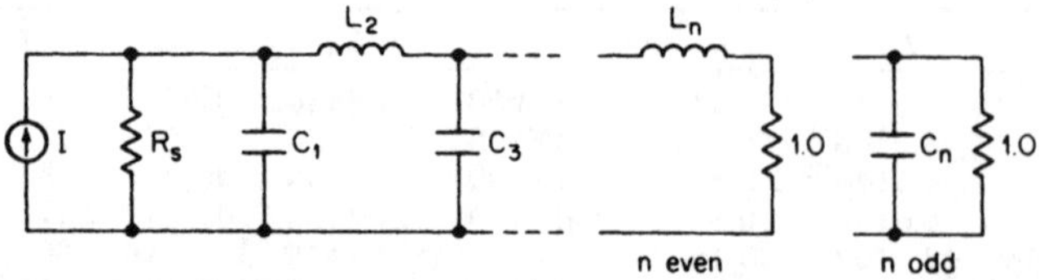

n	R_s	C_1	L_2	C_3	L_4	C_5	L_6
2	1.9841	0.9827	1.9497				
	2.0000	0.9086	2.1030				
	2.5000	0.5635	3.1647				
	3.3333	0.3754	4.4111				
	5.0000	0.2282	6.6995				
	10.0000	0.1052	13.3221				
	Inf.	1.3067	0.9748				
3	1.0000	1.8636	1.2804	1.8636			
	0.9000	1.9175	1.2086	2.0255			
	0.8000	1.9965	1.1203	2.2368			
	0.7000	2.1135	1.0149	2.5172			
	0.6000	2.2889	0.8937	2.8984			
	0.5000	2.5571	0.7592	3.4360			
	0.4000	2.9854	0.6146	4.2416			
	0.3000	3.7292	0.4633	5.5762			
	0.2000	5.2543	0.3087	8.2251			
	0.1000	9.8899	0.1534	16.1177			
	Inf.	1.5720	1.5179	0.9318			
4	1.9841	0.9202	2.5864	1.3036	1.8258		
	2.0000	0.8452	2.7198	1.2383	1.9849		
	2.5000	0.5162	3.7659	0.8693	3.1205		
	3.3333	0.3440	5.1196	0.6208	4.4790		
	5.0000	0.2100	7.7076	0.3996	6.9874		
	10.0000	0.0975	15.3520	0.1940	14.2616		
	Inf.	1.4361	1.8888	1.5211	0.9129		

TABLE 11-30 0.5-dB Chebyshev *LC* Element Values* (*Continued*)

n	R_s	C_1	L_2	C_3	L_4	C_5	L_6
5	1.0000	1.8068	1.3025	2.6914	1.3025	1.8068	
	0.9000	1.8540	1.2220	2.8478	1.2379	1.9701	
	0.8000	1.9257	1.1261	3.0599	1.1569	2.1845	
	0.7000	2.0347	1.0150	3.3525	1.0582	2.4704	
	0.6000	2.2006	0.8901	3.7651	0.9420	2.8609	
	0.5000	2.4571	0.7537	4.3672	0.8098	3.4137	
	0.4000	2.8692	0.6091	5.2960	0.6640	4.2447	
	0.3000	3.5877	0.4590	6.8714	0.5075	5.6245	
	0.2000	5.0639	0.3060	10.0537	0.3430	8.3674	
	0.1000	9.5560	0.1525	19.6465	0.1731	16.5474	
	Inf.	1.6299	1.7400	1.9217	1.5138	0.9034	
6	1.9841	0.9053	2.5774	1.3675	2.7133	1.2991	1.7961
	2.0000	0.8303	2.7042	1.2912	2.8721	1.2372	1.9557
	2.5000	0.5056	3.7219	0.8900	4.1092	0.8808	3.1025
	3.3333	0.3370	5.0554	0.6323	5.6994	0.6348	4.4810
	5.0000	0.2059	7.6145	0.4063	8.7319	0.4121	7.0310
	10.0000	0.0958	15.1862	0.1974	17.6806	0.2017	14.4328
	Inf.	1.4618	1.9799	1.7803	1.9253	1.5077	0.8981
n	$1/R_s$	L_1	C_2	L_3	C_4	L_5	C_6

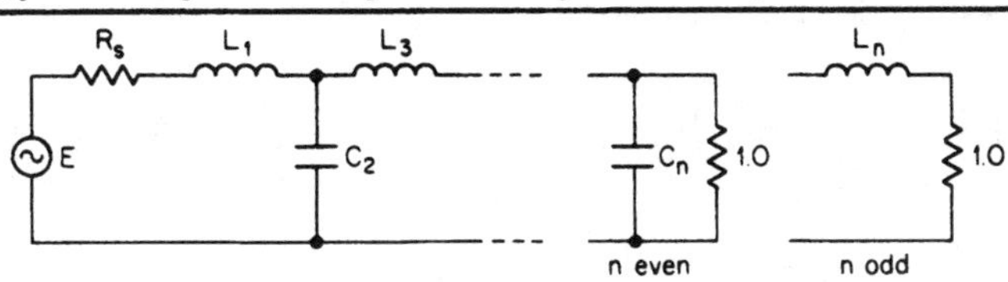

* Reprinted from A. I. Zverev, *Handbook of Filter Synthesis,* John Wiley and Sons, New York, 1967.

TABLE 11-30 0.5-dB Chebyshev *LC* Element Values (*Continued*)

n	R_s	C_1	L_2	C_3	L_4	C_5	L_6	C_7	L_8	C_9	L_{10}
7	1.0000	1.7896	1.2961	2.7177	1.3848	2.7177	1.2961	1.7896			
	0.9000	1.8348	1.2146	2.8691	1.3080	2.8829	1.2335	1.9531			
	0.8000	1.9045	1.1182	3.0761	1.2149	3.1071	1.1546	2.1681			
	0.7000	2.0112	1.0070	3.3638	1.1050	3.4163	1.0582	2.4554			
	0.6000	2.1744	0.8824	3.7717	0.9786	3.8524	0.9441	2.8481			
	0.5000	2.4275	0.7470	4.3695	0.8377	4.4886	0.8137	3.4050			
	0.4000	2.8348	0.6035	5.2947	0.6846	5.4698	0.6690	4.2428			
	0.3000	3.5456	0.4548	6.8674	0.5221	7.1341	0.5129	5.6350			
	0.2000	5.0070	0.3034	10.0491	0.3524	10.4959	0.3478	8.4041			
	0.1000	9.4555	0.1513	19.6486	0.1778	20.6314	0.1761	16.6654			
	Inf.	1.6464	1.7772	2.0306	1.7892	1.9239	1.5034	0.8948			
8	1.9841	0.8998	2.5670	1.3697	2.7585	1.3903	2.7175	1.2938	1.7852		
	2.0000	0.8249	2.6916	1.2919	2.9134	1.3160	2.8800	1.2331	1.9449		
	2.5000	0.5017	3.6988	0.8878	4.1404	0.9184	4.1470	0.8815	3.0953		
	3.3333	0.3344	5.0234	0.6304	5.7323	0.6577	5.7761	0.6370	4.4807		
	5.0000	0.2044	7.5682	0.4052	8.7771	0.4257	8.8833	0.4146	7.0453		
	10.0000	0.0951	15.1014	0.1969	17.7747	0.2081	18.0544	0.2035	14.4924		
	Inf.	1.4710	2.0022	1.8248	2.0440	1.7911	1.9218	1.5003	0.8926		
9	1.0000	1.7822	1.2921	2.7162	1.3922	2.7734	1.3922	2.7162	1.2921	1.7822	
	0.9000	1.8267	1.2103	2.8658	1.3135	2.9353	1.3165	2.8834	1.2302	1.9458	
	0.8000	1.8955	1.1139	3.0709	1.2189	3.1565	1.2246	3.1102	1.1523	2.1611	
	0.7000	2.0013	1.0028	3.3565	1.1075	3.4635	1.1157	3.4232	1.0568	2.4489	
	0.6000	2.1634	0.8786	3.7621	0.9801	3.8985	0.9900	3.8647	0.9436	2.8426	
	0.5000	2.4150	0.7436	4.3573	0.8385	4.5355	0.8493	4.5087	0.8140	3.4010	
	0.4000	2.8203	0.6008	5.2792	0.6850	5.5207	0.6957	5.5023	0.6700	4.2416	
	0.3000	3.5279	0.4528	6.8474	0.5223	7.1951	0.5318	7.1876	0.5142	5.6390	
	0.2000	4.9830	0.3021	10.0212	0.3526	10.5818	0.3600	10.5925	0.3491	8.4189	
	0.1000	9.4131	0.1507	19.5995	0.1779	20.8006	0.1822	20.8588	0.1770	16.7140	
	Inf.	1.6533	1.7890	2.0570	1.8383	2.0481	1.7910	1.9199	1.4981	0.8911	
10	1.9841	0.8972	2.5610	1.3683	2.7631	1.4009	2.7795	1.3927	2.7148	1.2908	1.7801
	2.0000	0.8223	2.6845	1.2901	2.9166	1.3246	2.9390	1.3191	2.8783	1.2306	1.9398
	2.5000	0.4999	3.6868	0.8858	4.1383	0.9216	4.2020	0.9238	4.1540	0.8812	3.0919
	3.3333	0.3332	5.0071	0.6289	5.7274	0.6594	5.8399	0.6631	5.7948	0.6376	4.4804
	5.0000	0.2037	7.5446	0.4042	8.7695	0.4266	8.9727	0.4300	8.9249	0.4154	7.0518
	10.0000	0.0948	15.0578	0.1965	17.7624	0.2086	18.2313	0.2107	18.1644	0.2041	14.5199
	Inf.	1.4753	2.0107	1.8386	2.0733	1.8432	2.0494	1.7904	1.9183	1.4965	0.8900
n	$1/R_s$	L_1	C_2	L_3	C_4	L_5	C_6	L_7	C_8	L_9	C_{10}

TABLE 11-31 1-dB Chebyshev *LC* Element Values

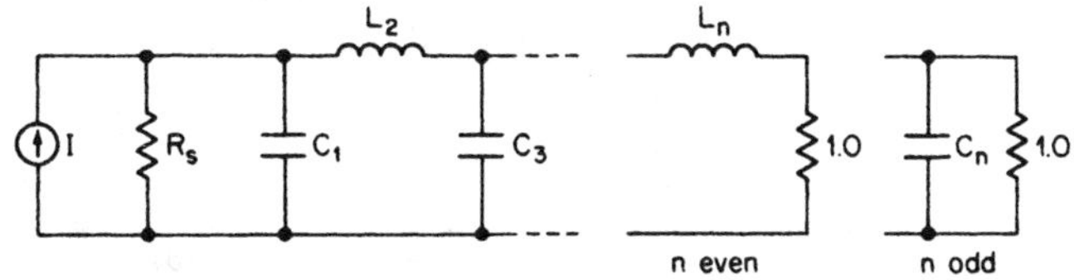

n	R_s	C_1	L_2	C_3	L_4
2	3	0.5723	3.1317		
	4	0.3653	4.6002		
	8	0.1571	9.6582		
	Inf.	1.2128	1.1093		
3	1	2.2160	1.0883	2.2160	
	0.5	4.4309	0.8168	2.2160	
	0.333	6.6469	0.7259	2.2160	
	0.25	8.8619	0.6799	2.2160	
	0.125	17.7248	0.6120	2.2160	
	Inf.	1.6522	1.4595	1.1080	
4	3	0.6529	4.4110	0.8140	2.5346
	4	0.4517	7.0825	0.6118	2.8484
	8	0.2085	17.1639	0.4275	3.2811
	Inf.	1.3499	2.0102	1.4879	1.1057
n	$1/R_s$	L_1	C_2	L_3	C_4

n	R_s	C_1	L_2	C_3	L_4	C_5	L_6	C_7
5	1	2.2072	1.1279	3.1025	1.1279	2.2072		
	0.5	4.4144	0.5645	4.6532	1.1279	2.2072		
	0.333	6.6216	0.3763	6.2050	1.1279	2.2072		
	0.25	8.8288	0.2822	7.7557	1.1279	2.2072		
	0.125	17.6565	0.1406	13.9606	1.1279	2.2072		
	Inf.	1.7213	1.6448	2.0614	1.4928	1.1031		
6	3	0.6785	3.8725	0.7706	4.7107	0.9692	2.4060	
	4	0.4810	5.6441	0.4759	7.3511	0.8494	2.5820	
	8	0.2272	12.3095	0.1975	16.740	0.7256	2.7990	
	Inf.	1.3775	2.0969	1.6896	2.0744	1.4942	1.1022	
7	1	2.2043	1.1311	3.1472	1.1942	3.1472	1.1311	2.2043
	0.5	4.4075	0.5656	6.2934	0.8951	3.1472	1.1311	2.2043
	0.333	6.6118	0.3774	9.4406	0.7955	3.1472	1.1311	2.2043
	0.25	8.8151	0.2828	12.5879	0.7466	3.1472	1.1311	2.2043
	0.125	17.6311	0.1414	25.175	0.6714	3.1472	1.1311	2.2043
	Inf.	1.7414	1.6774	2.1554	1.7028	2.0792	1.4943	1.1016
n	$1/R_s$	L_1	C_2	L_3	C_4	L_5	C_6	L_7

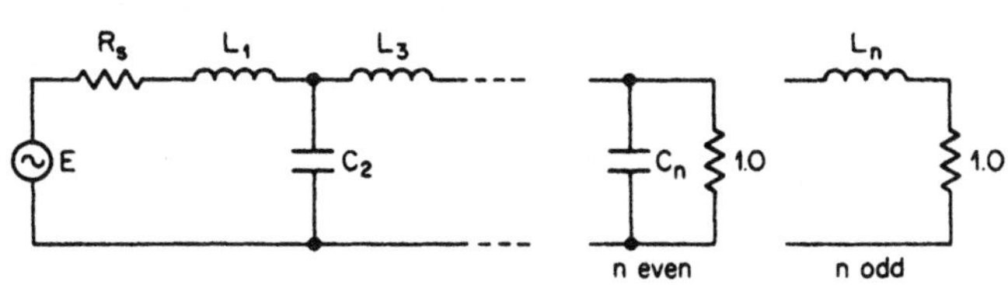

TABLE 11-32 0.1-dB Chebyshev Uniform Dissipation Network

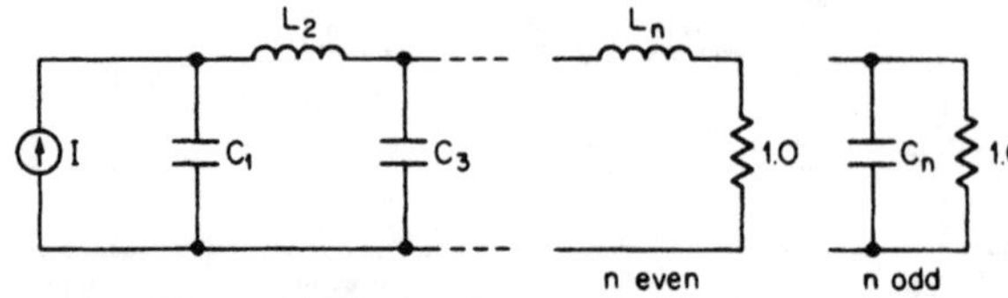

n	d	C_1	L_2	C_3	L_4	C_5	L_6	C_7
2	0.0172	1.3855	0.8433					
	0.0257	1.3816	0.8550					
	0.0515	1.3680	0.8939					
3	0.024	1.4848	1.5390	0.7556				
	0.036	1.4696	1.5543	0.7765				
	0.072	1.4168	1.6015	0.8473				
4	0.0275	1.4375	1.7978	1.5103	0.7266			
	0.0412	1.3975	1.8148	1.5394	0.7570			
	0.0824	1.2556	1.8767	1.6353	0.8637			
5	0.0294	1.4558	1.8064	1.8280	1.4933	0.7194		
	0.0441	1.3945	1.8076	1.8643	1.5352	0.7591		
	0.0881	1.1449	1.8416	2.0209	1.6839	0.9123		
6	0.0305	1.3672	1.8874	1.8612	1.8361	1.4907	0.7224	
	0.0457	1.2645	1.8973	1.8842	1.8907	1.5454	0.7738	
	0.0915	0.6579	2.3639	2.1574	2.1803	1.7574	0.9825	
7	0.0312	1.3628	1.8252	1.9694	1.8797	1.8455	1.4963	0.7316
	0.0468	1.2079	1.8220	2.0207	1.9213	1.9192	1.5646	0.7957
n	d	L_1	C_2	L_3	C_4	L_5	C_6	L_7

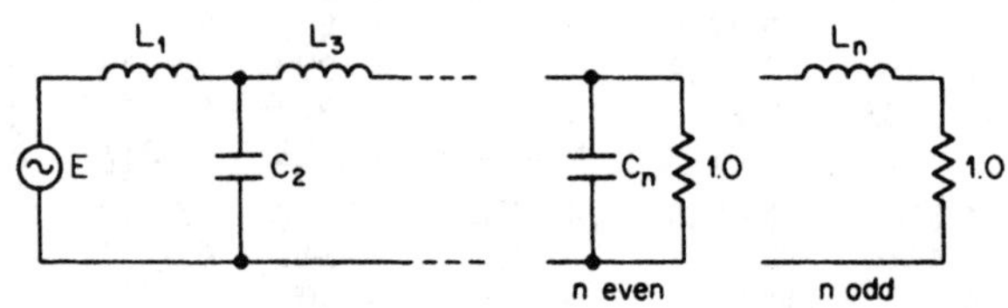

TABLE 11-33 0.25-dB Chebyshev Uniform Dissipation Network

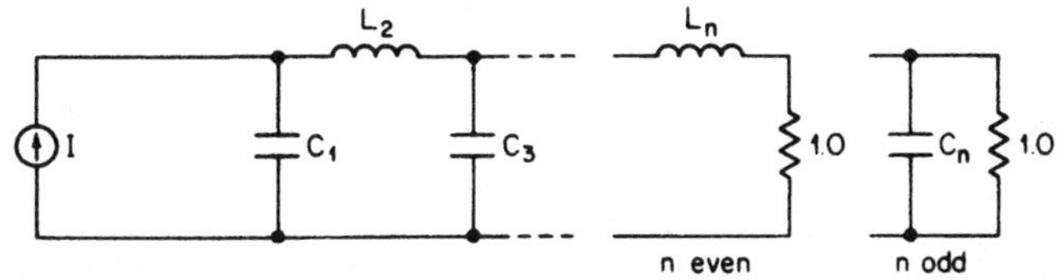

n	d	C_1	L_2	C_3	L_4	C_5	L_6	C_7
2	0.0209	1.3504	0.9157					
	0.0313	1.3376	0.9413					
	0.0626	1.3120	1.0004					
3	0.0266	1.5022	1.5548	0.8733				
	0.0399	1.4834	1.5674	0.9046				
	0.0798	1.4220	1.6062	1.0149				
4	0.0292	1.3894	1.8590	1.5593	0.8651			
	0.0439	1.3370	1.8818	1.5866	0.9107			
	0.0877	1.1444	1.9764	1.6823	1.0839			
5	0.0306	1.4599	1.7670	1.8976	1.5503	0.8503		
	0.0459	1.3881	1.7604	1.9455	1.5917	0.9102		
	0.0919	1.0397	1.8181	2.2035	1.7528	1.1497		
6	0.0314	1.3054	1.9347	1.8339	1.9443	1.5697	0.8883	
	0.0471	1.1696	1.9560	1.8541	2.0218	1.6259	0.9700	
7	0.0319	1.3584	1.7680	2.0376	1.8610	1.9707	1.5820	0.9091
	0.0479	1.1264	1.7722	2.1452	1.9132	2.0814	1.6541	1.0125
n	d	L_1	C_2	L_3	C_4	L_5	C_6	L_7

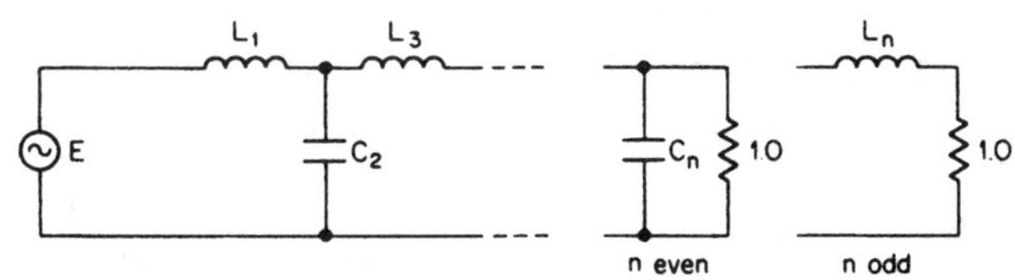

TABLE 11-34 0.5-dB Chebyshev Uniform Dissipation Network

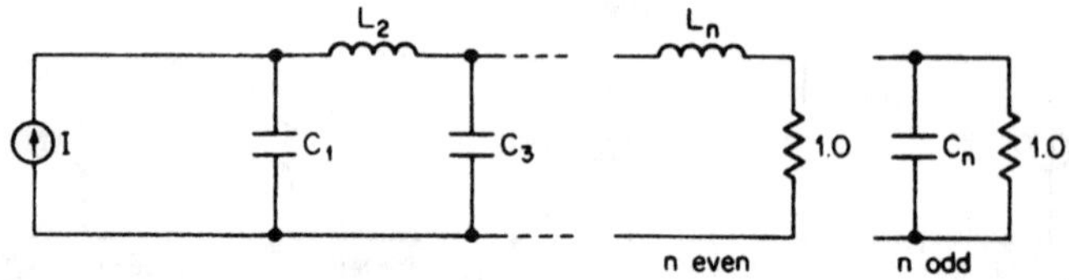

n	d	C_1	L_2	C_3	L_4	C_5	L_6	C_7
2	0.0240	1.2855	1.0228					
	0.0360	1.2730	1.0478					
	0.0720	1.2313	1.1340					
3	0.0286	1.5376	1.5341	1.0122				
	0.0428	1.5189	1.5423	1.0589				
	0.0856	1.4489	1.5621	1.2247				
4	0.0305	1.3205	1.9413	1.5631	1.0275			
	0.0457	1.2549	1.9741	1.5850	1.0964			
	0.0915	0.9991	2.1359	1.6692	1.3707			
5	0.0315	1.5031	1.6980	2.0264	1.5773	1.0529		
	0.0472	1.4162	1.6768	2.0995	1.6133	1.1482		
	0.0944	0.7139	2.0994	2.7297	1.8007	1.5751		
6	0.0320	1.2200	2.0123	1.7707	2.0758	1.5927	1.0858	
	0.0480	1.0389	2.0612	1.7895	2.1976	1.6448	1.2117	
7	0.0324	1.3659	1.6801	2.1488	1.8047	2.1230	1.6090	1.1228
	0.0485	0.9024	1.8171	2.4475	1.8985	2.3126	1.6811	1.2856
n	d	L_1	C_2	L_3	C_4	L_5	C_6	L_7

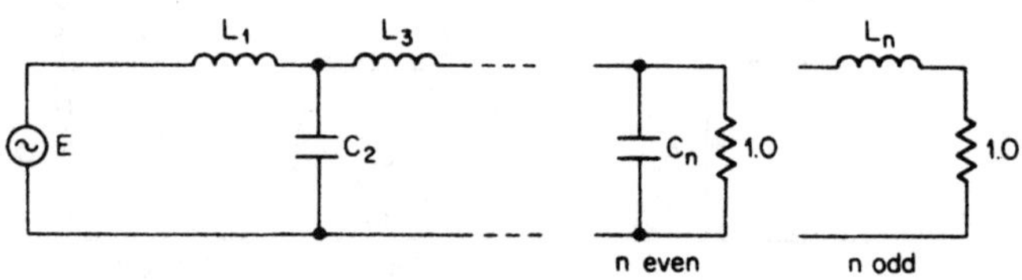

TABLE 11-35 1-dB Chebyshev Uniform Dissipation Network

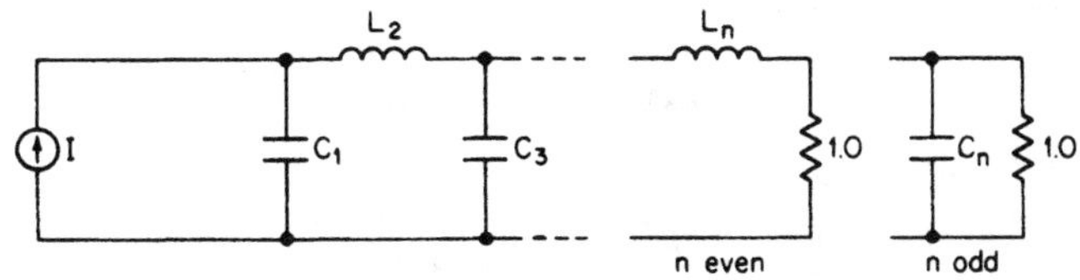

n	d	C_1	L_2	C_3	L_4	C_5	L_6	C_7
2	0.0274	1.1762	1.1811					
	0.0411	1.1020	1.2201					
	0.0821	1.0898	1.3564					
3	0.0304	1.6215	1.4562	1.2328				
	0.0457	1.6029	1.4518	1.3062				
	0.0913	1.5241	1.4387	1.5908				
4	0.0317	1.2015	2.0965	1.5037	1.2847			
	0.0475	1.1141	2.1544	1.5132	1.3984			
	0.0950	0.7434	2.5398	1.5890	1.9049			
5	0.0322	1.5869	1.5610	2.2258	1.5269	1.3419		
	0.0484	1.4680	1.5218	2.3591	1.5507	1.5052		
6	0.0326	1.0736	2.1789	1.6385	2.3150	1.5454	1.4052	
	0.0489	0.7747	2.3611	1.6958	2.5658	1.5894	1.6293	
7	0.0328	1.3610	1.5258	2.4016	1.6865	2.4067	1.5634	1.4749
n	d	L_1	C_2	L_3	C_4	L_5	C_6	L_7

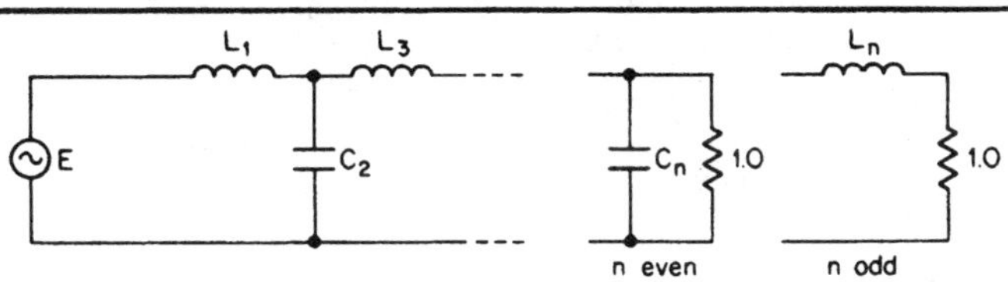

TABLE 11-36 0.01-dB Chebyshev Active Low-Pass Values

Order n	C_1	C_2
2	1.4826	0.7042
4	1.4874	1.1228
	3.5920	0.2985
6	1.8900	1.5249
	2.5820	0.5953
	7.0522	0.1486
8	2.3652	1.9493
	2.7894	0.8196
	4.1754	0.3197
	11.8920	0.08672

TABLE 11-37 0.1-dB Chebyshev Active Low-Pass Values*

Order n	C_1	C_2	C_3
2	1.638	0.6955	
3	6.653	1.825	0.1345
4	1.900	1.241	
	4.592	0.2410	
5	4.446	2.520	0.3804
	6.810	0.1580	
6	2.553	1.776	
	3.487	0.4917	
	9.531	0.1110	
7	5.175	3.322	0.5693
	4.546	0.3331	
	12.73	0.08194	
8	3.270	2.323	
	3.857	0.6890	
	5.773	0.2398	
	16.44	0.06292	
9	6.194	4.161	0.7483
	4.678	0.4655	
	7.170	0.1812	
	20.64	0.04980	
10	4.011	2.877	
	4.447	0.8756	
	5.603	0.3353	
	8.727	0.1419	
	25.32	0.04037	

* Reprinted from *Electronics,* McGraw-Hill, Inc., August 18, 1969.

TABLE 11-38 0.25-dB Chebyshev Active Low-Pass Values*

Order n	C_1	C_2	C_3
2	1.778	0.6789	
3	8.551	2.018	0.1109
4	2.221	1.285	
	5.363	0.2084	
5	5.543	2.898	0.3425
	8.061	0.1341	
6	3.044	1.875	
	4.159	0.4296	
	11.36	0.09323	
7	6.471	3.876	0.5223
	5.448	0.2839	
	15.26	0.06844	
8	3.932	2.474	
	4.638	0.6062	
	6.942	0.2019	
	19.76	0.05234	
9	7.766	4.891	0.6919
	5.637	0.3983	
	8.639	0.1514	
	24.87	0.04131	
10	4.843	3.075	
	5.368	0.7725	
	6.766	0.2830	
	10.53	0.1181	
	30.57	0.03344	

* Reprinted from *Electronics,* McGraw-Hill, Inc., August 18, 1969.

TABLE 11-39 0.5-dB Chebyshev Active Low-Pass Values*

Order n	C_1	C_2	C_3
2	1.950	0.6533	
3	11.23	2.250	0.0895
4	2.582	1.300	
	6.233	0.1802	
5	6.842	3.317	0.3033
	9.462	0.1144	
6	3.592	1.921	
	4.907	0.3743	
	13.40	0.07902	
7	7.973	4.483	0.4700
	6.446	0.2429	
	18.07	0.05778	
8	4.665	2.547	
	5.502	0.5303	
	8.237	0.1714	
	23.45	0.04409	
9	9.563	5.680	0.6260
	6.697	0.3419	
	10.26	0.1279	
	29.54	0.03475	
10	5.760	3.175	
	6.383	0.6773	
	8.048	0.2406	
	12.53	0.09952	
	36.36	0.02810	

* Reprinted from *Electronics,* McGraw-Hill, Inc., August 18, 1969.

TABLE 11-40 1-dB Chebyshev Active Low-Pass Values*

Order n	C_1	C_2	C_3
2	2.218	0.6061	
3	16.18	2.567	0.06428
4	3.125	1.269	
	7.546	0.1489	
5	8.884	3.935	0.2540
	11.55	0.09355	
6	4.410	1.904	
	6.024	0.3117	
	16.46	0.06425	
7	10.29	5.382	0.4012
	7.941	0.1993	
	22.25	0.04684	
8	5.756	2.538	
	6.792	0.4435	
	10.15	0.1395	
	28.94	0.03568	
9	12.33	6.853	0.5382
	8.281	0.2813	
	12.68	0.1038	
	36.51	0.02808	
10	7.125	3.170	
	7.897	0.5630	
	9.952	0.1962	
	15.50	0.08054	
	44.98	0.02269	

* Reprinted from *Electronics*, McGraw-Hill, Inc., August 18, 1969.

TABLE 11-41 Bessel Pole Locations

Order n	Real Part $-\alpha$	Imaginary Part $\pm j\beta$
2	1.1030	0.6368
3	1.0509	1.0025
	1.3270	
4	1.3596	0.4071
	0.9877	1.2476
5	1.3851	0.7201
	0.9606	1.4756
	1.5069	
6	1.5735	0.3213
	1.3836	0.9727
	0.9318	1.6640
7	1.6130	0.5896
	1.3797	1.1923
	0.9104	1.8375
	1.6853	
8	1.7627	0.2737
	0.8955	2.0044
	1.3780	1.3926
	1.6419	0.8253
9	1.8081	0.5126
	1.6532	1.0319
	1.3683	1.5685
	0.8788	2.1509
	1.8575	

TABLE 11-42 Bessel *LC* Element Values*

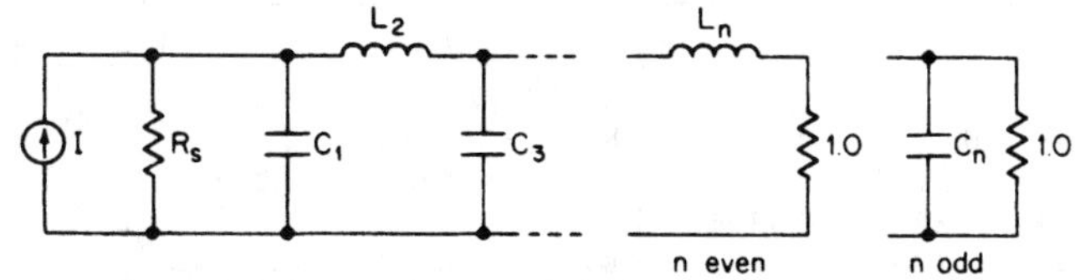

n	R_s	C_1	L_2	C_3	L_4
2	1.0000	0.5755	2.1478		
	1.1111	0.5084	2.3097		
	1.2500	0.4433	2.5096		
	1.4286	0.3801	2.7638		
	1.6667	0.3191	3.0993		
	2.0000	0.2601	3.5649		
	2.5000	0.2032	4.2577		
	3.3333	0.1486	5.4050		
	5.0000	0.0965	7.6876		
	10.0000	0.0469	14.5097		
	Inf.	1.3617	0.4539		
3	1.0000	0.3374	0.9705	2.2034	
	0.9000	0.3708	0.8650	2.3745	
	0.8000	0.4124	0.7609	2.5867	
	0.7000	0.4657	0.6584	2.8575	
	0.6000	0.5365	0.5576	3.2159	
	0.5000	0.6353	0.4587	3.7144	
	0.4000	0.7829	0.3618	4.4573	
	0.3000	1.0283	0.2673	5.6888	
	0.2000	1.5176	0.1752	8.1403	
	0.1000	2.9825	0.0860	15.4697	
	Inf.	1.4631	0.8427	0.2926	
4	1.0000	0.2334	0.6725	1.0815	2.2404
	1.1111	0.2085	0.7423	0.9670	2.4143
	1.2500	0.1839	0.8292	0.8534	2.6304
	1.4286	0.1596	0.9406	0.7410	2.9066
	1.6667	0.1356	1.0886	0.6299	3.2727
	2.0000	0.1120	1.2952	0.5202	3.7824
	2.5000	0.0887	1.6040	0.4120	4.5430
	3.3333	0.0658	2.1174	0.3056	5.8048
	5.0000	0.0434	3.1416	0.2013	8.3185
	10.0000	0.0214	6.2086	0.0993	15.8372
	Inf.	1.5012	0.9781	0.6127	0.2114
n	$1/R_s$	L_1	C_2	L_3	C_4

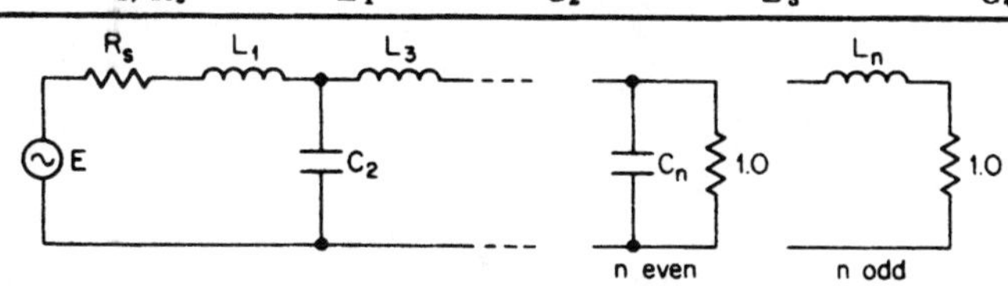

* Reprinted from A. I. Zverev, *Handbook of Filter Synthesis,* John Wiley and Sons, New York, 1967.

TABLE 11-42 Bessel *LC* Element Values (*Continued*)

n	R_s	C_1	L_2	C_3	L_4	C_5	L_6	C_7
5	1.0000	0.1743	0.5072	0.8040	1.1110	2.2582		
	0.9000	0.1926	0.4542	0.8894	0.9945	2.4328		
	0.8000	0.2154	0.4016	0.9959	0.8789	2.6497		
	0.7000	0.2447	0.3494	1.1323	0.7642	2.9272		
	0.6000	0.2836	0.2977	1.3138	0.6506	3.2952		
	0.5000	0.3380	0.2465	1.5672	0.5382	3.8077		
	0.4000	0.4194	0.1958	1.9464	0.4270	4.5731		
	0.3000	0.5548	0.1457	2.5768	0.3174	5.8433		
	0.2000	0.8251	0.0964	3.8352	0.2095	8.3747		
	0.1000	1.6349	0.0478	7.6043	0.1036	15.9487		
	Inf.	1.5125	1.0232	0.7531	0.4729	0.1618		
6	1.0000	0.1365	0.4002	0.6392	0.8538	1.1126	2.2645	
	1.1111	0.1223	0.4429	0.5732	0.9456	0.9964	2.4388	
	1.2500	0.1082	0.4961	0.5076	1.0600	0.8810	2.6554	
	1.4286	0.0943	0.5644	0.4424	1.2069	0.7665	2.9325	
	1.6667	0.0804	0.6553	0.3775	1.4022	0.6530	3.3001	
	2.0000	0.0666	0.7824	0.3131	1.6752	0.5405	3.8122	
	2.5000	0.0530	0.9725	0.2492	2.0837	0.4292	4.5770	
	3.3333	0.0395	1.2890	0.1859	2.7633	0.3193	5.8467	
	5.0000	0.0261	1.9209	0.1232	4.1204	0.2110	8.3775	
	10.0000	0.0130	3.8146	0.0612	8.1860	0.1045	15.9506	
	Inf.	1.5124	1.0329	0.8125	0.6072	0.3785	0.1287	
7	1.0000	0.1106	0.3259	0.5249	0.7020	0.8690	1.1052	2.2659
	0.9000	0.1224	0.2923	0.5815	0.6302	0.9630	0.9899	2.4396
	0.8000	0.1372	0.2589	0.6521	0.5586	1.0803	0.8754	2.6556
	0.7000	0.1562	0.2257	0.7428	0.4873	1.2308	0.7618	2.9319
	0.6000	0.1815	0.1927	0.8634	0.4163	1.4312	0.6491	3.2984
	0.5000	0.2168	0.1599	1.0321	0.3457	1.7111	0.5374	3.8090
	0.4000	0.2698	0.1274	1.2847	0.2755	2.1304	0.4269	4.5718
	0.3000	0.3579	0.0951	1.7051	0.2058	2.8280	0.3177	5.8380
	0.2000	0.5338	0.0630	2.5448	0.1365	4.2214	0.2100	8.3623
	0.1000	1.0612	0.0313	5.0616	0.0679	8.3967	0.1040	15.9166
	Inf.	1.5087	1.0293	0.8345	0.6752	0.5031	0.3113	0.1054
n	$1/R_s$	L_1	C_2	L_3	C_4	L_5	C_6	L_7

TABLE 11-42 Bessel *LC* Element Values (*Continued*)

n	R_s	C_1	L_2	C_3	L_4	C_5	L_6	C_7	L_8	C_9	L_{10}
8	1.0000	0.0919	0.2719	0.4409	0.5936	0.7303	0.8695	1.0956	2.2656		
	1.1111	0.0825	0.3013	0.3958	0.6580	0.6559	0.9639	0.9813	2.4388		
	1.2500	0.0731	0.3380	0.3509	0.7385	0.5817	1.0816	0.8678	2.6541		
	1.4286	0.0637	0.3850	0.3061	0.8418	0.5078	1.2328	0.7552	2.9295		
	1.6667	0.0545	0.4477	0.2616	0.9794	0.4342	1.4340	0.6435	3.2949		
	2.0000	0.0452	0.5354	0.2173	1.1718	0.3608	1.7153	0.5329	3.8041		
	2.5000	0.0360	0.6667	0.1732	1.4599	0.2878	2.1367	0.4233	4.5645		
	3.3333	0.0269	0.8852	0.1294	1.9396	0.2151	2.8380	0.3151	5.8271		
	5.0000	0.0179	1.3218	0.0859	2.8981	0.1429	4.2389	0.2083	8.3441		
	10.0000	0.0089	2.6307	0.0427	5.7710	0.0711	8.4376	0.1032	15.8768		
	Inf.	1.5044	1.0214	0.8392	0.7081	0.5743	0.4253	0.2616	0.0883		
9	1.0000	0.0780	0.2313	0.3770	0.5108	0.6306	0.7407	0.8639	1.0863	2.2649	
	0.9000	0.0864	0.2077	0.4180	0.4588	0.6994	0.6655	0.9578	0.9730	2.4376	
	0.8000	0.0970	0.1841	0.4691	0.4069	0.7854	0.5905	1.0750	0.8604	2.6524	
	0.7000	0.1105	0.1607	0.5348	0.3553	0.8957	0.5157	1.2255	0.7488	2.9271	
	0.6000	0.1286	0.1373	0.6222	0.3038	1.0427	0.4411	1.4258	0.6380	3.2915	
	0.5000	0.1538	0.1141	0.7445	0.2525	1.2483	0.3667	1.7059	0.5283	3.7993	
	0.4000	0.1916	0.0910	0.9278	0.2014	1.5563	0.2926	2.1256	0.4197	4.5578	
	0.3000	0.2545	0.0680	1.2329	0.1506	2.0692	0.2189	2.8241	0.3124	5.8171	
	0.2000	0.3803	0.0452	1.8426	0.1000	3.0941	0.1455	4.2196	0.2065	8.3276	
	0.1000	0.7573	0.0225	3.6704	0.0498	6.1666	0.0725	8.4023	0.1023	15.8408	
	Inf.	1.5006	1.0127	0.8361	0.7220	0.6142	0.4963	0.3654	0.2238	0.0754	
10	1.0000	0.0672	0.1998	0.3270	0.4454	0.5528	0.6493	0.7420	0.8561	1.0781	2.2641
	1.1111	0.0604	0.2216	0.2937	0.4941	0.4967	0.7205	0.6668	0.9492	0.9656	2.4365
	1.2500	0.0536	0.2488	0.2606	0.5548	0.4408	0.8093	0.5918	1.0654	0.8539	2.6508
	1.4286	0.0467	0.2836	0.2275	0.6327	0.3850	0.9233	0.5170	1.2147	0.7430	2.9249
	1.6667	0.0400	0.3301	0.1945	0.7366	0.3294	1.0753	0.4423	1.4134	0.6331	3.2885
	2.0000	0.0332	0.3951	0.1617	0.8818	0.2739	1.2879	0.3678	1.6913	0.5242	3.7953
	2.5000	0.0265	0.4924	0.1290	1.0995	0.2186	1.6064	0.2936	2.1076	0.4164	4.5521
	3.3333	0.0198	0.6546	0.0965	1.4620	0.1635	2.1369	0.2197	2.8007	0.3099	5.8087
	5.0000	0.0132	0.9786	0.0641	2.1864	0.1087	3.1971	0.1461	4.1854	0.2049	8.3137
	10.0000	0.0066	1.9499	0.0319	4.3583	0.0542	6.3759	0.0728	8.3359	0.1015	15.8108
	Inf.	1.4973	1.0045	0.8297	0.7258	0.6355	0.5401	0.4342	0.3182	0.1942	0.0653
n	$1/R_s$	L_1	C_2	L_3	C_4	L_5	C_6	L_7	C_8	L_9	C_{10}

TABLE 11-43 Bessel Active-Low-Pass Values

Order n	C_1	C_2	C_3
2	0.9066	0.6800	
3	1.423	0.9880	0.2538
4	0.7351	0.6746	
	1.012	0.3900	
5	1.010	0.8712	0.3095
	1.041	0.3100	
6	0.6352	0.6100	
	0.7225	0.4835	
	1.073	0.2561	
7	0.8532	0.7792	0.3027
	0.7250	0.4151	
	1.100	0.2164	
8	0.5673	0.5540	
	0.6090	0.4861	
	0.7257	0.3590	
	1.116	0.1857	
9	0.7564	0.7070	0.2851
	0.6048	0.4352	
	0.7307	0.3157	
	1.137	0.1628	
10	0.5172	0.5092	
	0.5412	0.4682	
	0.6000	0.3896	
	0.7326	0.2792	
	1.151	0.1437	

TABLE 11-44 Linear Phase with Equiripple Error of 0.05° Pole Locations

Order n	Real Part $-\alpha$	Imaginary Part $\pm j\beta$
2	1.0087	0.6680
3	0.8541	1.0725
	1.0459	
4	0.9648	0.4748
	0.7448	1.4008
5	0.8915	0.8733
	0.6731	1.7085
	0.9430	
6	0.8904	0.4111
	0.8233	1.2179
	0.6152	1.9810
7	0.8425	0.7791
	0.7708	1.5351
	0.5727	2.2456
	0.8615	
8	0.8195	0.3711
	0.7930	1.1054
	0.7213	1.8134
	0.5341	2.4761
9	0.7853	0.7125
	0.7555	1.4127
	0.6849	2.0854
	0.5060	2.7133
	0.7938	
10	0.7592	0.3413
	0.7467	1.0195
	0.7159	1.6836
	0.6475	2.3198
	0.4777	2.9128

TABLE 11-45 Linear Phase with Equiripple Error of 0.5° Pole Locations

Order n	Real Part $-\alpha$	Imaginary Part $\pm j\beta$
2	0.8590	0.6981
3	0.6969	1.1318
	0.8257	
4	0.7448	0.5133
	0.6037	1.4983
5	0.6775	0.9401
	0.5412	1.8256
	0.7056	
6	0.6519	0.4374
	0.6167	1.2963
	0.4893	2.0982
7	0.6190	0.8338
	0.5816	1.6453
	0.4598	2.3994
	0.6283	
8	0.5791	0.3857
	0.5665	1.1505
	0.5303	1.8914
	0.4184	2.5780
9	0.5688	0.7595
	0.5545	1.5089
	0.5179	2.2329
	0.4080	2.9028
	0.5728	
10	0.5249	0.3487
	0.5193	1.0429
	0.5051	1.7261
	0.4711	2.3850
	0.3708	2.9940

TABLE 11-46 Linear Phase with Equiriple Error of 0.05° *LC* Element Values*

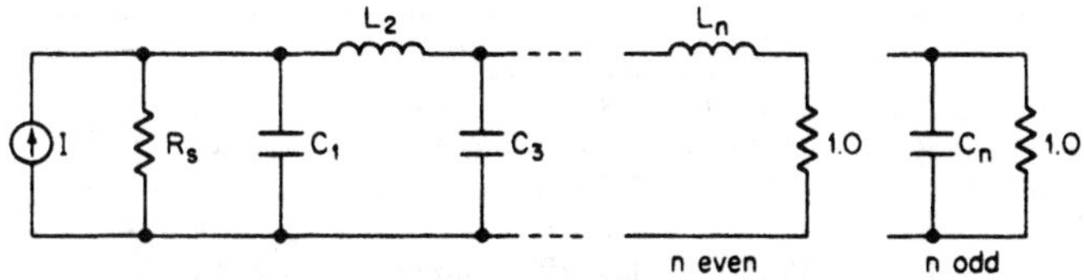

n	R_s	C_1	L_2	C_3	L_4
2	1.0000	0.6480	2.1085		
	1.1111	0.5703	2.2760		
	1.2500	0.4955	2.4817		
	1.4286	0.4235	2.7422		
	1.6667	0.3544	3.0848		
	2.0000	0.2880	3.5589		
	2.5000	0.2244	4.2630		
	3.3333	0.1637	5.4270		
	5.0000	0.1059	7.7400		
	10.0000	0.0513	14.6480		
	Inf.	1.3783	0.4957		
3	1.0000	0.4328	1.0427	2.2542	
	0.9000	0.4745	0.9330	2.4258	
	0.8000	0.5262	0.8238	2.6400	
	0.7000	0.5925	0.7153	2.9146	
	0.6000	0.6805	0.6078	3.2795	
	0.5000	0.8032	0.5015	3.7884	
	0.4000	0.9865	0.3967	4.5487	
	0.3000	1.2910	0.2938	5.8106	
	0.2000	1.8983	0.1931	8.3253	
	0.1000	3.7161	0.0950	15.8472	
	Inf.	1.5018	0.9328	0.3631	
4	1.0000	0.3363	0.7963	1.1428	2.2459
	1.1111	0.2993	0.8810	1.0212	2.4241
	1.2500	0.2631	0.9865	0.9012	2.6445
	1.4286	0.2275	1.1216	0.7826	2.9254
	1.6667	0.1926	1.3009	0.6657	3.2970
	2.0000	0.1584	1.5509	0.5502	3.8138
	2.5000	0.1250	1.9244	0.4364	4.5844
	3.3333	0.0923	2.5448	0.3242	5.8626
	5.0000	0.0606	3.7818	0.2139	8.4091
	10.0000	0.0298	7.4845	0.1058	16.0266
	Inf.	1.5211	1.0444	0.7395	0.2925
n	$1/R_s$	L_1	C_2	L_3	C_4

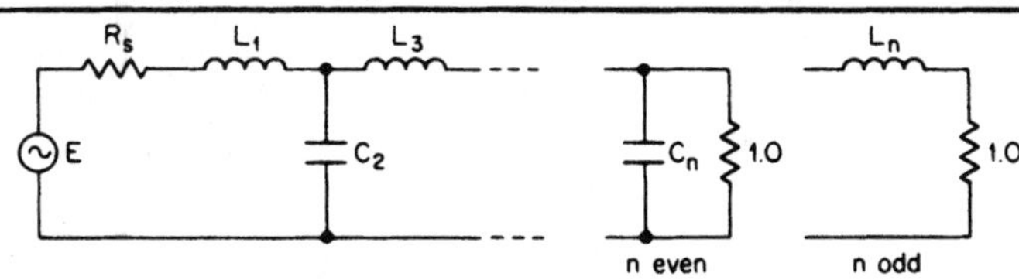

* Reprinted from A. I. Zverev, *Handbook of Filter Synthesis*, John Wiley and Sons, New York, 1967.

TABLE 11-46 Linear Phase with Equiriple Error of 0.05° *LC* Element Values (*Continued*)

n	R_s	C_1	L_2	C_3	L_4	C_5	L_6	C_7
5	1.0000	0.2751	0.6541	0.8892	1.1034	2.2873		
	0.9000	0.3031	0.5868	0.9841	0.9904	2.4589		
	0.8000	0.3380	0.5197	1.1026	0.8774	2.6733		
	0.7000	0.3827	0.4529	1.2548	0.7648	2.9484		
	0.6000	0.4420	0.3865	1.4575	0.6526	3.3144		
	0.5000	0.5248	0.3204	1.7408	0.5410	3.8254		
	0.4000	0.6486	0.2549	2.1651	0.4302	4.5896		
	0.3000	0.8544	0.1899	2.8713	0.3205	5.8595		
	0.2000	1.2649	0.1257	4.2817	0.2120	8.3922		
	0.1000	2.4940	0.0624	8.5082	0.1051	15.9739		
	Inf.	1.5144	1.0407	0.8447	0.6177	0.2456		
6	1.0000	0.2374	0.5662	0.7578	0.8760	1.1163	2.2448	
	1.1111	0.2120	0.6272	0.6799	0.9726	0.9977	2.4214	
	1.2500	0.1870	0.7032	0.6023	1.0931	0.8807	2.6396	
	1.4286	0.1622	0.8008	0.5253	1.2475	0.7652	2.9174	
	1.6667	0.1378	0.9306	0.4487	1.4530	0.6512	3.2849	
	2.0000	0.1138	1.1118	0.3725	1.7401	0.5387	3.7958	
	2.5000	0.0901	1.3830	0.2969	2.1698	0.4277	4.5579	
	3.3333	0.0669	1.8340	0.2217	2.8849	0.3182	5.8220	
	5.0000	0.0441	2.7343	0.1472	4.3129	0.2103	8.3408	
	10.0000	0.0218	5.4312	0.0732	8.5924	0.1041	15.8769	
	Inf.	1.5050	1.0306	0.8554	0.7283	0.5389	0.2147	
7	1.0000	0.2085	0.4999	0.6653	0.7521	0.8749	1.0671	2.2845
	0.9000	0.2302	0.4488	0.7374	0.6768	0.9687	0.9580	2.4538
	0.8000	0.2573	0.3978	0.8274	0.6013	1.0861	0.8489	2.6655
	0.7000	0.2919	0.3470	0.9431	0.5258	1.2369	0.7400	2.9375
	0.6000	0.3380	0.2964	1.0972	0.4503	1.4381	0.6314	3.2996
	0.5000	0.4023	0.2461	1.3127	0.3749	1.7196	0.5235	3.8051
	0.4000	0.4986	0.1960	1.6356	0.2995	2.1416	0.4163	4.5613
	0.3000	0.6585	0.1463	2.1734	0.2242	2.8445	0.3101	5.8180
	0.2000	0.9778	0.0970	3.2480	0.1492	4.2496	0.2052	8.3246
	0.1000	1.9340	0.0482	6.4698	0.0744	8.4623	0.1017	15.8281
	Inf.	1.4988	1.0071	0.8422	0.7421	0.6441	0.4791	0.1911
n	$1/R_s$	L_1	C_2	L_3	C_4	L_5	C_6	L_7

TABLE 11-46 Linear Phase with Equiripple Error of 0.05° *LC* Element Values (*Continued*)

n	R_s	C_1	L_2	C_3	L_4	C_5	L_6	C_7	L_8	C_9	L_{10}
8	1.0000	0.1891	0.4543	0.6031	0.6750	0.7590	0.8427	1.0901	2.2415		
	1.1111	0.1691	0.5035	0.5415	0.7500	0.6813	0.9362	0.9735	2.4176		
	1.2500	0.1494	0.5650	0.4802	0.8435	0.6041	1.0527	0.8588	2.6349		
	1.4286	0.1298	0.6438	0.4191	0.9637	0.5272	1.2019	0.7459	2.9113		
	1.6667	0.1105	0.7487	0.3583	1.1237	0.4508	1.4004	0.6345	3.2767		
	2.0000	0.0914	0.8953	0.2978	1.3475	0.3748	1.6776	0.5247	3.7846		
	2.5000	0.0725	1.1148	0.2376	1.6827	0.2991	2.0927	0.4164	4.5418		
	3.3333	0.0539	1.4801	0.1776	2.2411	0.2237	2.7833	0.3096	5.7978		
	5.0000	0.0356	2.2095	0.1180	3.3568	0.1488	4.1627	0.2046	8.3004		
	10.0000	0.0176	4.3954	0.0588	6.7021	0.0742	8.2969	0.1013	15.7878		
	Inf.	1.4953	1.0018	0.8264	0.7396	0.6688	0.5858	0.4369	0.1743		
9	1.0000	0.1718	0.4146	0.5498	0.6132	0.6774	0.7252	0.8450	1.0447	2.2834	
	0.9000	0.1900	0.3724	0.6097	0.5519	0.7513	0.6529	0.9352	0.9382	2.4512	
	0.8000	0.2125	0.3302	0.6846	0.4905	0.8436	0.5805	1.0481	0.8314	2.6613	
	0.7000	0.2415	0.2882	0.7807	0.4291	0.9624	0.5079	1.1933	0.7247	2.9315	
	0.6000	0.2800	0.2463	0.9088	0.3676	1.1207	0.4352	1.3870	0.6184	3.2914	
	0.5000	0.3337	0.2046	1.0880	0.3062	1.3424	0.3624	1.6581	0.5125	3.7941	
	0.4000	0.4141	0.1631	1.3565	0.2448	1.6749	0.2897	2.0647	0.4075	4.5462	
	0.3000	0.5478	0.1219	1.8038	0.1834	2.2289	0.2170	2.7420	0.3035	5.7960	
	0.2000	0.8148	0.0809	2.6977	0.1222	3.3369	0.1445	4.0960	0.2007	8.2890	
	0.1000	1.6146	0.0403	5.3782	0.0610	6.6602	0.0721	8.1556	0.0995	15.7520	
	Inf.	1.4907	0.9845	0.8116	0.7197	0.6646	0.6089	0.5359	0.4003	0.1598	
10	1.0000	0.1601	0.3867	0.5125	0.5702	0.6243	0.6557	0.7319	0.8178	1.0767	2.2387
	1.1111	0.1433	0.4288	0.4604	0.6336	0.5609	0.7290	0.6567	0.9089	0.9608	2.4151
	1.2500	0.1267	0.4812	0.4084	0.7127	0.4977	0.8205	0.5820	1.0221	0.8471	2.6323
	1.4286	0.1102	0.5486	0.3567	0.8143	0.4348	0.9380	0.5079	1.1672	0.7354	2.9082
	1.6667	0.0939	0.6383	0.3051	0.9498	0.3721	1.0944	0.4342	1.3600	0.6254	3.2727
	2.0000	0.0778	0.7637	0.2537	1.1392	0.3096	1.3131	0.3609	1.6291	0.5170	3.7791
	2.5000	0.0618	0.9515	0.2024	1.4232	0.2473	1.6408	0.2880	2.0320	0.4102	4.5340
	3.3333	0.0460	1.2641	0.1515	1.8961	0.1852	2.1866	0.2154	2.7022	0.3049	5.7860
	5.0000	0.0304	1.8885	0.1007	2.8416	0.1232	3.2775	0.1433	4.0406	0.2014	8.2806
	10.0000	0.0151	3.7600	0.0502	5.6766	0.0615	6.5485	0.0714	8.0520	0.0997	15.7441
	Inf.	1.4905	0.9858	0.8018	0.7123	0.6540	0.6141	0.5669	0.5003	0.3741	0.1494
n	$1/R_s$	L_1	C_2	L_3	C_4	L_5	C_6	L_7	C_8	L_9	C_{10}

TABLE 11-47 Linear Phase with Equiripple Error of 0.5° *LC* Element Values*

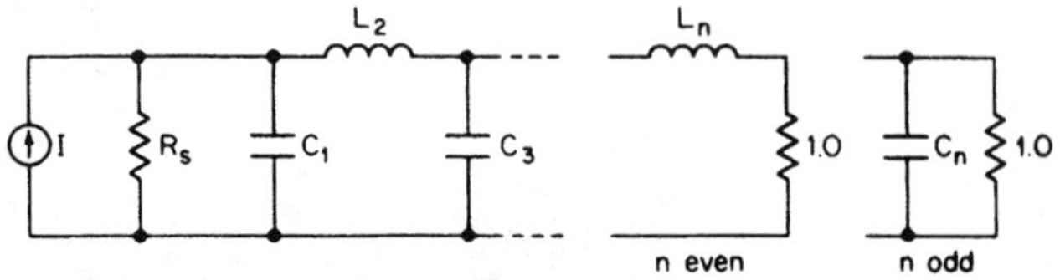

n	R_s	C_1	L_2	C_3	L_4
2	1.0000	0.8245	1.9800		
	1.1111	0.7166	2.1640		
	1.2500	0.6160	2.3850		
	1.4286	0.5216	2.6603		
	1.6667	0.4327	3.0181		
	2.0000	0.3489	3.5088		
	2.5000	0.2700	4.2329		
	3.3333	0.1956	5.4242		
	5.0000	0.1258	7.7842		
	10.0000	0.0606	14.8185		
	Inf.	1.4022	0.5821		
3	1.0000	0.5534	1.0218	2.4250	
	0.9000	0.6059	0.9213	2.5929	
	0.8000	0.6710	0.8197	2.8046	
	0.7000	0.7540	0.7173	3.0787	
	0.6000	0.8639	0.6141	3.4462	
	0.5000	1.0168	0.5105	3.9625	
	0.4000	1.2448	0.4068	4.7385	
	0.3000	1.6231	0.3034	6.0326	
	0.2000	2.3771	0.2008	8.6197	
	0.1000	4.6332	0.0994	16.3738	
	Inf.	1.5495	0.9820	0.4506	
4	1.0000	0.4526	0.7967	1.2669	2.0504
	1.1111	0.3996	0.8889	1.1137	2.2502
	1.2500	0.3486	1.0028	0.9699	2.4866
	1.4286	0.2995	1.1481	0.8333	2.7788
	1.6667	0.2521	1.3405	0.7024	3.1576
	2.0000	0.2062	1.6083	0.5762	3.6769
	2.5000	0.1618	2.0079	0.4542	4.4438
	3.3333	0.1190	2.6708	0.3358	5.7080
	5.0000	0.0777	3.9916	0.2207	8.2168
	10.0000	0.0380	7.9426	0.1088	15.7068
	Inf.	1.4944	1.0715	0.7889	0.3708
n	$1/R_s$	L_1	C_2	L_3	C_4

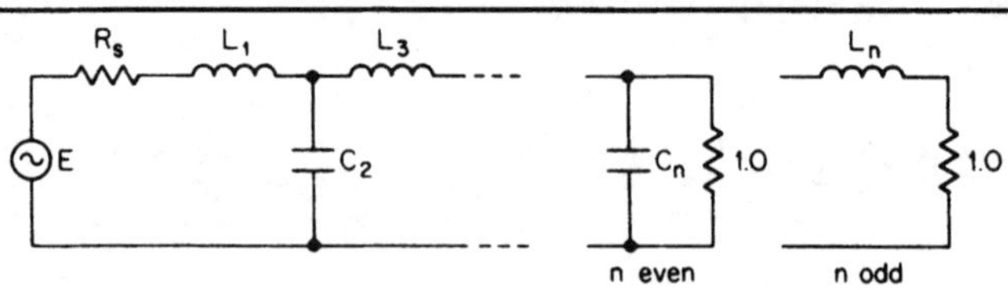

* Reprinted from A. I. Zverev, *Handbook of Filter Synthesis,* John Wiley and Sons, New York, 1967.

TABLE 11-47 Linear Phase with Equiripple Error of 0.5° *LC* Element Values (*Continued*)

n	R_s	C_1	L_2	C_3	L_4	C_5	L_6	C_7
5	1.0000	0.3658	0.6768	0.9513	1.0113	2.4446		
	0.9000	0.4027	0.6099	1.0486	0.9157	2.6062		
	0.8000	0.4485	0.5427	1.1700	0.8182	2.8114		
	0.7000	0.5069	0.4752	1.3260	0.7189	3.0787		
	0.6000	0.5843	0.4074	1.5341	0.6181	3.4387		
	0.5000	0.6921	0.3395	1.8253	0.5160	3.9462		
	0.4000	0.8530	0.2714	2.2623	0.4130	4.7108		
	0.3000	1.1201	0.2033	2.9908	0.3094	5.9881		
	0.2000	1.6524	0.1352	4.4478	0.2057	8.5444		
	0.1000	3.2454	0.0674	8.8185	0.1024	16.2117		
	Inf.	1.5327	1.0180	0.8740	0.6709	0.3182		
6	1.0000	0.3313	0.5984	0.8390	0.7964	1.2734	2.0111	
	1.1111	0.2934	0.6667	0.7446	0.8985	1.1050	2.2282	
	1.2500	0.2571	0.7515	0.6542	1.0223	0.9549	2.4742	
	1.4286	0.2219	0.8600	0.5666	1.1787	0.8164	2.7718	
	1.6667	0.1876	1.0040	0.4812	1.3848	0.6859	3.1529	
	2.0000	0.1541	1.2051	0.3976	1.6709	0.5615	3.6720	
	2.5000	0.1216	1.5058	0.3155	2.0972	0.4420	4.4362	
	3.3333	0.0898	2.0058	0.2347	2.8044	0.3266	5.6935	
	5.0000	0.0589	3.0038	0.1553	4.2137	0.2146	8.1871	
	10.0000	0.0290	5.9928	0.0771	8.4320	0.1058	15.6296	
	Inf.	1.4849	1.0430	0.8427	0.7651	0.5972	0.2844	
7	1.0000	0.2826	0.5332	0.7142	0.6988	0.9219	0.9600	2.4404
	0.9000	0.3118	0.4802	0.7896	0.6322	1.0137	0.8718	2.5953
	0.8000	0.3481	0.4271	0.8836	0.5649	1.1287	0.7809	2.7936
	0.7000	0.3945	0.3739	1.0043	0.4967	1.2768	0.6875	3.0535
	0.6000	0.4560	0.3206	1.1650	0.4277	1.4750	0.5919	3.4051
	0.5000	0.5416	0.2671	1.3899	0.3580	1.7531	0.4947	3.9025
	0.4000	0.6695	0.2136	1.7271	0.2874	2.1714	0.3961	4.6534
	0.3000	0.8819	0.1601	2.2890	0.2163	2.8700	0.2969	5.9091
	0.2000	1.3054	0.1066	3.4127	0.1445	4.2690	0.1974	8.4236
	0.1000	2.5731	0.0532	6.7835	0.0724	8.4691	0.0983	15.9666
	Inf.	1.5079	0.9763	0.8402	0.7248	0.6741	0.5305	0.2532
n	$1/R_s$	L_1	C_2	L_3	C_4	L_5	C_6	L_7

TABLE 11-47 Linear Phase with Equiripple Error of 0.5° *LC* Element Values (*Continued*)

n	R_s	C_1	L_2	C_3	L_4	C_5	L_6	C_7	L_8	C_9	L_{10}
8	1.0000	0.2718	0.4999	0.6800	0.6312	0.8498	0.7447	1.3174	1.9626		
	1.1111	0.2408	0.5567	0.6045	0.7116	0.7452	0.8529	1.1169	2.2146		
	1.2500	0.2114	0.6271	0.5324	0.8086	0.6506	0.9780	0.9551	2.4766		
	1.4286	0.1828	0.7173	0.4622	0.9315	0.5612	1.1331	0.8117	2.7837		
	1.6667	0.1549	0.8373	0.3934	1.0939	0.4753	1.3355	0.6795	3.1715		
	2.0000	0.1276	1.0049	0.3256	1.3201	0.3920	1.6148	0.5550	3.6960		
	2.5000	0.1009	1.2559	0.2589	1.6580	0.3107	2.0297	0.4362	4.4654		
	3.3333	0.0747	1.6734	0.1930	2.2194	0.2311	2.7164	0.3220	5.7294		
	5.0000	0.0492	2.5074	0.1279	3.3400	0.1530	4.0835	0.2114	8.2345		
	10.0000	0.0242	5.0066	0.0636	6.6971	0.0760	8.1733	0.1042	15.7101		
	Inf.	1.4915	1.0265	0.8169	0.7548	0.6709	0.6318	0.4995	0.2387		
9	1.0000	0.2347	0.4493	0.5914	0.5747	0.7027	0.6552	0.8944	0.9255	2.4332	
	0.9000	0.2594	0.4045	0.6547	0.5193	0.7754	0.5943	0.9809	0.8427	2.5822	
	0.8000	0.2900	0.3597	0.7336	0.4635	0.8662	0.5322	1.0895	0.7566	2.7745	
	0.7000	0.3291	0.3148	0.8348	0.4073	0.9829	0.4690	1.2299	0.6673	3.0283	
	0.6000	0.3810	0.2699	0.9695	0.3505	1.1388	0.4046	1.4183	0.5753	3.3734	
	0.5000	0.4533	0.2249	1.1580	0.2932	1.3572	0.3392	1.6834	0.4812	3.8629	
	0.4000	0.5613	0.1799	1.4405	0.2355	1.6854	0.2727	2.0828	0.3855	4.6032	
	0.3000	0.7407	0.1348	1.9111	0.1772	2.2331	0.2054	2.7508	0.2889	5.8424	
	0.2000	1.0986	0.0898	2.8522	0.1185	3.3299	0.1373	4.0895	0.1921	8.3246	
	0.1000	2.1702	0.0448	5.6749	0.0594	6.6230	0.0688	8.1099	0.0956	15.7718	
	Inf.	1.4888	0.9495	0.8044	0.6892	0.6589	0.5952	0.5645	0.4475	0.2141	
10	1.0000	0.2359	0.4369	0.5887	0.5428	0.7034	0.5827	0.8720	0.6869	1.4317	1.8431
	1.1111	0.2081	0.4866	0.5218	0.6141	0.6141	0.6729	0.7394	0.8187	1.1397	2.1907
	1.2500	0.1827	0.5480	0.4601	0.6972	0.5376	0.7708	0.6394	0.9483	0.9616	2.4734
	1.4286	0.1582	0.6267	0.3999	0.8024	0.4651	0.8922	0.5487	1.1042	0.8122	2.7907
	1.6667	0.1343	0.7314	0.3407	0.9416	0.3948	1.0514	0.4631	1.3052	0.6777	3.1847
	2.0000	0.1108	0.8777	0.2823	1.1356	0.3263	1.2719	0.3811	1.5809	0.5525	3.7138
	2.5000	0.0877	1.0969	0.2247	1.4258	0.2591	1.6003	0.3017	1.9888	0.4338	4.4876
	3.3333	0.0651	1.4619	0.1676	1.9085	0.1931	2.1451	0.2242	2.6628	0.3200	5.7573
	5.0000	0.0429	2.1910	0.1112	2.8724	0.1280	3.2311	0.1483	4.0033	0.2100	8.2726
	10.0000	0.0212	4.3764	0.0553	5.7612	0.0636	6.4830	0.0736	8.0118	0.1035	15.7776
	Inf.	1.4973	1.0192	0.8005	0.7312	0.6498	0.6331	0.5775	0.5501	0.4369	0.2091
n	$1/R_s$	L_1	C_2	L_3	C_4	L_5	C_6	L_7	C_8	L_9	C_{10}

TABLE 11-48 Linear Phase with Equiripple Error of 0.05° Active Low-Pass Values

Order n	C_1	C_2
2	0.9914	0.6891
4	1.0365	0.8344
	1.3426	0.2959
6	1.1231	0.9257
	1.2146	0.3810
	1.6255	0.1430
8	1.2203	1.0126
	1.2610	0.4285
	1.3864	0.1894
	1.8723	0.08324
10	1.3172	1.0957
	1.3392	0.4676
	1.3968	0.2139
	1.5444	0.1116
	2.0934	0.05483

TABLE 11-49 Linear Phase with Equiripple Error of 0.5° Active Low-Pass Values

Order n	C_1	C_2
2	1.1641	0.7011
4	1.3426	0.9103
	1.6565	0.2314
6	1.5340	1.0578
	1.6215	0.2993
	2.0437	0.1054
8	1.7268	1.1962
	1.7652	0.3445
	1.8857	0.1374
	2.3901	0.06134
10	1.9051	1.3218
	1.9257	0.3826
	1.9798	0.1562
	2.1227	0.07971
	2.6969	0.04074

TABLE 11-50 Transitional Gaussian to 6-dB Pole Locations

Order n	Real Part $-\alpha$	Imaginary Part $\pm j\beta$
3	0.9622	1.2214
	0.9776	
4	0.7940	0.5029
	0.6304	1.5407
5	0.6190	0.8254
	0.3559	1.5688
	0.6650	
6	0.5433	0.3431
	0.4672	0.9991
	0.2204	1.5067
7	0.4580	0.5932
	0.3649	1.1286
	0.1522	1.4938
	0.4828	
8	0.4222	0.2640
	0.3833	0.7716
	0.2878	1.2066
	0.1122	1.4798
9	0.3700	0.4704
	0.3230	0.9068
	0.2309	1.2634
	0.08604	1.4740
	0.3842	
10	0.3384	0.2101
	0.3164	0.6180
	0.2677	0.9852
	0.1849	1.2745
	0.06706	1.4389

TABLE 11-51 Transitional Gaussian to 12-dB Pole Locations

Order n	Real Part $-\alpha$	Imaginary Part $\pm j\beta$
3	0.9360	1.2168
	0.9630	
4	0.9278	1.6995
	0.9192	0.5560
5	0.8075	0.9973
	0.7153	2.0532
	0.8131	
6	0.7019	0.4322
	0.6667	1.2931
	0.4479	2.1363
7	0.6155	0.7703
	0.5486	1.5154
	0.2905	2.1486
	0.6291	
8	0.5441	0.3358
	0.5175	0.9962
	0.4328	1.6100
	0.1978	2.0703
9	0.4961	0.6192
	0.4568	1.2145
	0.3592	1.7429
	0.1489	2.1003
	0.5065	
10	0.4535	0.2794
	0.4352	0.8289
	0.3886	1.3448
	0.2908	1.7837
	0.1136	2.0599

TABLE 11-52 Transitional Gaussian to 6-dB *LC* Element Values*

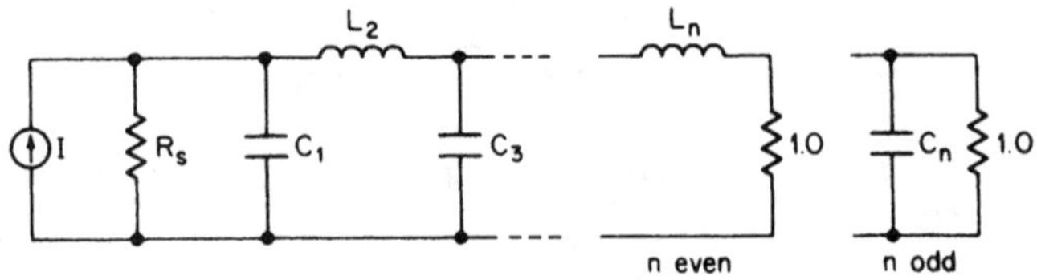

n	R_s	C_1	L_2	C_3	L_4	C_5	L_6	C_7
3	1.0000	0.4042	0.8955	2.3380				
	0.9000	0.4440	0.8038	2.5027				
	0.8000	0.4935	0.7121	2.7088				
	0.7000	0.5568	0.6205	2.9739				
	0.6000	0.6407	0.5292	3.3275				
	0.5000	0.7575	0.4384	3.8223				
	0.4000	0.9319	0.3482	4.5635				
	0.3000	1.2213	0.2590	5.7972				
	0.2000	1.7980	0.1709	8.2605				
	0.1000	3.5236	0.0845	15.6391				
	Inf.	1.4742	0.8328	0.3446				
4	1.0000	0.4198	0.7832	1.1598	2.1427			
	1.1111	0.3720	0.8717	1.0279	2.3286			
	1.2500	0.3256	0.9816	0.9010	2.5539			
	1.4286	0.2804	1.1220	0.7781	2.8367			
	1.6667	0.2365	1.3083	0.6587	3.2069			
	2.0000	0.1938	1.5678	0.5424	3.7179			
	2.5000	0.1524	1.9552	0.4289	4.4761			
	3.3333	0.1122	2.5982	0.3180	5.7296			
	5.0000	0.0733	3.8797	0.2095	8.2221			
	10.0000	0.0359	7.7138	0.1035	15.6717			
	Inf.	1.4871	1.0222	0.7656	0.3510			
5	1.0000	0.4544	0.8457	1.0924	1.0774	2.4138		
	0.9000	0.4991	0.7622	1.2046	0.9769	2.5746		
	0.8000	0.5543	0.6781	1.3452	0.8739	2.7797		
	0.7000	0.6247	0.5936	1.5263	0.7687	3.0475		
	0.6000	0.7179	0.5086	1.7683	0.6615	3.4087		
	0.5000	0.8476	0.4233	2.1077	0.5527	3.9183		
	0.4000	1.0411	0.3379	2.6176	0.4428	4.6863		
	0.3000	1.3621	0.2526	3.4680	0.3321	5.9690		
	0.2000	2.0019	0.1677	5.1693	0.2212	8.5360		
	0.1000	3.9166	0.0833	10.2723	0.1103	16.2354		
	Inf.	1.5392	1.0993	1.0203	0.8269	0.3824		
6	1.0000	0.5041	0.9032	1.2159	1.0433	1.4212	2.0917	
	1.1111	0.4427	1.0079	1.0739	1.1892	1.2274	2.3324	
	1.2500	0.3853	1.1364	0.9415	1.3611	1.0620	2.5935	
	1.4286	0.3306	1.2999	0.8145	1.5753	0.9111	2.9053	
	1.6667	0.2779	1.5162	0.6914	1.8557	0.7692	3.3032	
	2.0000	0.2271	1.8169	0.5713	2.2433	0.6333	3.8456	
	2.5000	0.1780	2.2654	0.4534	2.8200	0.5016	4.6459	
	3.3333	0.1308	3.0091	0.3376	3.7758	0.3730	5.9662	
	5.0000	0.0853	4.4902	0.2235	5.6803	0.2468	8.5904	
	10.0000	0.0416	8.9199	0.1109	11.3810	0.1225	16.4352	
	Inf.	1.5664	1.2166	1.1389	1.1010	0.8844	0.4062	

TABLE 11-52 Transitional Gaussian to 6-dB *LC* Element Values* (*Continued*)

n	R_s	C_1	L_2	C_3	L_4	C_5	L_6	C_7
7	1.0000	0.4918	0.9232	1.2146	1.1224	1.3154	1.1407	2.5039
	0.9000	0.5403	0.8318	1.3393	1.0196	1.4426	1.0434	2.6575
	0.8000	0.6001	0.7399	1.4950	0.9141	1.6040	0.9401	2.8593
	0.7000	0.6760	0.6474	1.6952	0.8061	1.8144	0.8317	3.1285
	0.6000	0.7763	0.5545	1.9626	0.6956	2.0986	0.7190	3.4967
	0.5000	0.9157	0.4613	2.3373	0.5829	2.5004	0.6029	4.0203
	0.4000	1.1236	0.3681	2.9002	0.4684	3.1072	0.4844	4.8129
	0.3000	1.4685	0.2750	3.8389	0.3524	4.1232	0.3644	6.1397
	0.2000	2.1560	0.1823	5.7166	0.2354	6.1604	0.2433	8.7977
	0.1000	4.2137	0.0905	11.3483	0.1178	12.2787	0.1217	16.7743
	Inf.	1.5950	1.2166	1.2240	1.1784	1.1260	0.8975	0.4110
n	$1/R_s$	L_1	C_2	L_3	C_4	L_5	C_6	L_7

* Reprinted from A. I. Zverev, *Handbook of Filter Synthesis,* John Wiley and Sons, New York, 1967.

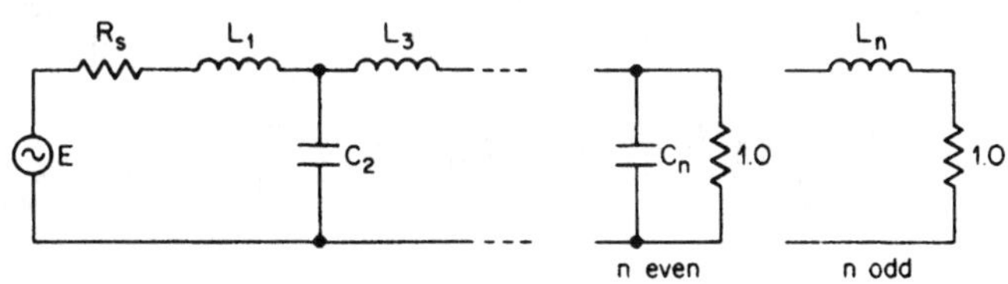

TABLE 11-52 Transitional Gaussian to 6-dB *LC* Element Values (*Continued*)

n	R_s	C_1	L_2	C_3	L_4	C_5	L_6	C_7	L_8	C_9	L_{10}
8	1.0502	0.5031	0.9699	1.2319	1.1324	1.4262	1.0449	1.6000	1.9285		
	1.1111	0.4586	1.0338	1.1286	1.2497	1.2635	1.2099	1.3372	2.2286		
	1.2500	0.3964	1.1670	0.9831	1.4404	1.0842	1.4259	1.1197	2.5453		
	1.4286	0.3392	1.3351	0.8487	1.6698	0.9299	1.6706	0.9502	2.8771		
	1.6667	0.2848	1.5571	0.7195	1.9674	0.7863	1.9808	0.7989	3.2846		
	2.0000	0.2325	1.8656	0.5939	2.3776	0.6487	2.4039	0.6569	3.8326		
	2.5000	0.1822	2.3255	0.4710	2.9870	0.5151	3.0295	0.5204	4.6374		
	3.3333	0.1337	3.0879	0.3504	3.9962	0.3840	4.0640	0.3874	5.9636		
	5.0000	0.0872	4.6062	0.2318	6.0063	0.2547	6.1243	0.2566	8.5995		
	10.0000	0.0425	9.1467	0.1150	12.0217	0.1268	12.2919	0.1276	16.4808		
	Inf.	1.5739	1.2698	1.2325	1.2633	1.2017	1.1404	0.9066	0.4148		
9	1.0000	0.4979	0.9367	1.2371	1.1589	1.3845	1.1670	1.3983	1.1422	2.5277	
	0.9000	0.5475	0.8439	1.3648	1.0517	1.5194	1.0673	1.5233	1.0527	2.6698	
	0.8000	0.6083	0.7505	1.5238	0.9424	1.6894	0.9625	1.6850	0.9540	2.8635	
	0.7000	0.6854	0.6567	1.7278	0.8306	1.9103	0.8527	1.8996	0.8472	3.1279	
	0.6000	0.7870	0.5624	1.9998	0.7165	2.2081	0.7383	2.1929	0.7342	3.4938	
	0.5000	0.9280	0.4679	2.3811	0.6002	2.6288	0.6202	2.6105	0.6166	4.0174	
	0.4000	1.1383	0.3732	2.9537	0.4820	3.2641	0.4993	3.2438	0.4960	4.8118	
	0.3000	1.4873	0.2788	3.9087	0.3625	4.3274	0.3763	4.3062	0.3734	6.1429	
	0.2000	2.1829	0.1848	5.8191	0.2421	6.4590	0.2518	6.4381	0.2496	8.8109	
	0.1000	4.2652	0.0918	11.5490	0.1211	12.8601	0.1262	12.8438	0.1250	16.8186	
	Inf.	1.6014	1.2508	1.2817	1.2644	1.2805	1.2103	1.1456	0.9096	0.4160	
10	1.1372	0.4682	1.0839	1.1516	1.2991	1.3293	1.2748	1.4216	1.1730	1.5040	2.1225
	1.1372	0.4682	1.0839	1.1516	1.2991	1.3293	1.2748	1.4216	1.1730	1.5040	2.1225
	1.2500	0.4087	1.1987	1.0148	1.4855	1.1389	1.5155	1.1705	1.4593	1.1798	2.5537
	1.4286	0.3489	1.3718	0.8744	1.7253	0.9733	1.7813	0.9908	1.7344	0.9878	2.9155
	1.6667	0.2928	1.6000	0.7409	2.0334	0.8219	2.1124	0.8338	2.0664	0.8275	3.3380
	2.0000	0.2389	1.9169	0.6114	2.4574	0.6776	2.5622	0.6868	2.5129	0.6799	3.8995
	2.5000	0.1872	2.3893	0.4848	3.0868	0.5377	3.2264	0.5451	3.1699	0.5387	4.7218
	3.3333	0.1373	3.1723	0.3606	4.1290	0.4007	4.3241	0.4065	4.2549	0.4011	6.0762
	5.0000	0.0895	4.7317	0.2385	6.2048	0.2657	6.5094	0.2698	6.4154	0.2659	8.7681
	10.0000	0.0437	9.3953	0.1183	12.4165	0.1322	13.0503	0.1345	12.8837	0.1323	16.8178
	Inf.	1.6077	1.3178	1.2927	1.3406	1.3070	1.3160	1.2409	1.1733	0.9311	0.4257
n	$1/R_s$	L_1	C_2	L_3	C_4	L_5	C_6	L_7	C_8	L_9	C_{10}

TABLE 11-53 Transitional Gaussian to 12-dB *LC* Element Values*

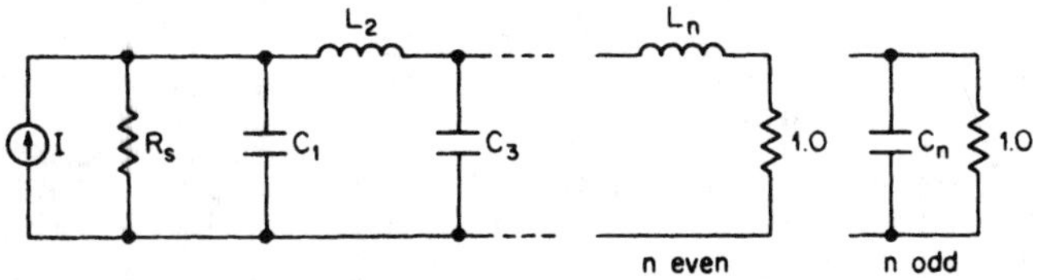

n	R_s	C_1	L_2	C_3	L_4	C_5	L_6	C_7
3	1.0000	0.4152	0.9050	2.3452				
	0.9000	0.4560	0.8126	2.5101				
	0.8000	0.5067	0.7202	2.7166				
	0.7000	0.5715	0.6278	2.9825				
	0.6000	0.6573	0.5356	3.3372				
	0.5000	0.7769	0.4438	3.8336				
	0.4000	0.9554	0.3526	4.5775				
	0.3000	1.2517	0.2623	5.8157				
	0.2000	1.8420	0.1732	8.2884				
	0.1000	3.6083	0.0856	15.6955				
	Inf.	1.4800	0.8440	0.3527				
4	1.0000	0.3097	0.6545	1.0598	2.1518			
	1.1111	0.2757	0.7262	0.9418	2.3289			
	1.2500	0.2423	0.8156	0.8268	2.5459			
	1.4286	0.2096	0.9300	0.7146	2.8203			
	1.6667	0.1775	1.0821	0.6050	3.1814			
	2.0000	0.1461	1.2944	0.4980	3.6812			
	2.5000	0.1153	1.6118	0.3934	4.4241			
	3.3333	0.0853	2.1393	0.2913	5.6532			
	5.0000	0.0560	3.1917	0.1916	8.0979			
	10.0000	0.0276	6.3425	0.0944	15.4048			
	Inf.	1.4585	0.9300	0.6294	0.2707			
5	1.0000	0.2909	0.5837	0.8112	0.9660	2.3745		
	0.9000	0.3207	0.5253	0.8961	0.8707	2.5377		
	0.8000	0.3577	0.4667	1.0019	0.7746	2.7433		
	0.7000	0.4051	0.4081	1.1379	0.6777	3.0092		
	0.6000	0.4680	0.3495	1.3192	0.5804	3.3650		
	0.5000	0.5556	0.2908	1.5727	0.4827	3.8642		
	0.4000	0.6865	0.2322	1.9528	0.3850	4.6138		
	0.3000	0.9038	0.1738	2.5859	0.2875	5.8631		
	0.2000	1.3372	0.1155	3.8515	0.1907	8.3597		
	0.1000	2.6347	0.0575	7.6464	0.0947	15.8420		
	Inf.	1.4953	0.9388	0.7587	0.5724	0.2592		

TABLE 11-53 Transitional Gaussian to 12-dB *LC* Element Values* (*Continued*)

n	R_s	C_1	L_2	C_3	L_4	C_5	L_6	C_7
6	1.0000	0.3164	0.6070	0.7962	0.7880	1.1448	2.1154	
	1.1111	0.2813	0.6750	0.7108	0.8826	1.0087	2.3076	
	1.2500	0.2470	0.7597	0.6273	0.9994	0.8804	2.5365	
	1.4286	0.2135	0.8681	0.5452	1.1481	0.7580	2.8209	
	1.6667	0.1807	1.0123	0.4644	1.3451	0.6402	3.1908	
	2.0000	0.1487	1.2136	0.3847	1.6194	0.5263	3.6993	
	2.5000	0.1174	1.5148	0.3060	2.0292	0.4157	4.4522	
	3.3333	0.0868	2.0154	0.2282	2.7098	0.3080	5.6952	
	5.0000	0.0570	3.0146	0.1513	4.0679	0.2028	8.1654	
	10.0000	0.0280	6.0071	0.0753	8.1355	0.1002	15.5460	
	Inf.	1.4732	0.9894	0.8129	0.7484	0.5979	0.2752	
7	1.0000	0.3207	0.6267	0.8091	0.7753	0.9241	0.9649	2.3829
	0.9000	0.3534	0.5641	0.8946	0.7016	1.0176	0.8750	2.5374
	0.8000	0.3940	0.5015	1.0015	0.6270	1.1350	0.7824	2.7351
	0.7000	0.4458	0.4387	1.1388	0.5513	1.2867	0.6876	2.9937
	0.6000	0.5146	0.3758	1.3218	0.4747	1.4899	0.5910	3.3428
	0.5000	0.6102	0.3128	1.5781	0.3972	1.7755	0.4931	3.8355
	0.4000	0.7531	0.2498	1.9626	0.3189	2.2054	0.3943	4.5779
	0.3000	0.9902	0.1869	2.6034	0.2399	2.9236	0.2951	5.8175
	0.2000	1.4630	0.1243	3.8851	0.1603	4.3624	0.1961	8.2974
	0.1000	2.8780	0.0619	7.7299	0.0803	8.6825	0.0976	15.7326
	Inf.	1.4861	0.9693	0.8643	0.8040	0.7689	0.6157	0.2826
n	$1/R_s$	L_1	C_2	L_3	C_4	L_5	C_6	L_7

* Reprinted from A. I. Zverev, *Handbook of Filter Synthesis,* John Wiley and Sons, New York, 1967.

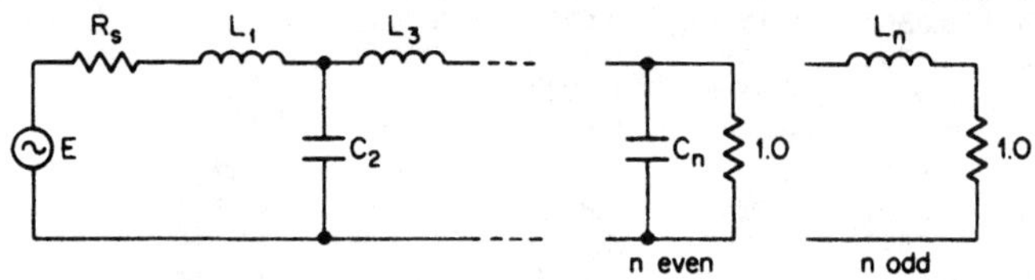

TABLE 11-53 Transitional Gaussian to 12-dB *LC* Element Values (*Continued*)

n	R_s	C_1	L_2	C_3	L_4	C_5	L_6	C_7	L_8	C_9	L_{10}
8	1.0000	0.3449	0.6565	0.8686	0.8028	0.9701	0.8182	1.2503	2.0612		
	1.1111	0.3053	0.7304	0.7729	0.9044	0.8550	0.9339	1.0753	2.2930		
	1.2500	0.2674	0.8221	0.6810	1.0276	0.7489	1.0694	0.9272	2.5445		
	1.4286	0.2308	0.9394	0.5913	1.1837	0.6480	1.2378	0.7929	2.8444		
	1.6667	0.1952	1.0953	0.5034	1.3898	0.5504	1.4579	0.6673	3.2267		
	2.0000	0.1604	1.3128	0.4168	1.6764	0.4553	1.7620	0.5476	3.7473		
	2.5000	0.1265	1.6381	0.3314	2.1045	0.3621	2.2143	0.4323	4.5145		
	3.3333	0.0934	2.1788	0.2471	2.8155	0.2702	2.9640	0.3203	5.7789		
	5.0000	0.0613	3.2575	0.1638	4.2343	0.1794	4.4583	0.2111	8.2899		
	10.0000	0.0301	6.4874	0.0814	8.4843	0.0894	8.9330	0.1044	15.7917		
	Inf.	1.4974	1.0324	0.8943	0.8908	0.8494	0.8098	0.6452	0.2955		
9	1.0000	0.3318	0.6500	0.8467	0.8167	0.9426	0.8239	0.9857	0.9630	2.4140	
	0.9000	0.3657	0.5852	0.9363	0.7389	1.0390	0.7492	1.0803	0.8785	2.5608	
	0.8000	0.4078	0.5201	1.0480	0.6602	1.1599	0.6725	1.2003	0.7894	2.7524	
	0.7000	0.4614	0.4550	1.1914	0.5806	1.3160	0.5936	1.3568	0.6963	3.0070	
	0.6000	0.5324	0.3897	1.3825	0.4999	1.5251	0.5127	1.5681	0.6001	3.3538	
	0.5000	0.6312	0.3243	1.6500	0.4183	1.8192	0.4301	1.8667	0.5015	3.8460	
	0.4000	0.7787	0.2590	2.0512	0.3358	2.2620	0.3460	2.3177	0.4016	4.5897	
	0.3000	1.0234	0.1938	2.7200	0.2526	3.0021	0.2607	3.0729	0.3009	5.8332	
	0.2000	1.5113	0.1288	4.0574	0.1687	4.4850	0.1744	4.5875	0.2000	8.3219	
	0.1000	2.9716	0.0641	8.0692	0.0845	8.9384	0.0875	9.1379	0.0996	15.7849	
	Inf.	1.4917	0.9908	0.9105	0.8770	0.8910	0.8457	0.8022	0.6376	0.2917	
10	1.0139	0.3500	0.6698	0.8817	0.8148	1.0183	0.7949	1.0929	0.7508	1.4303	1.8322
	1.1111	0.3092	0.7364	0.7856	0.9200	0.8864	0.9293	0.9147	0.9187	1.1138	2.2110
	1.2500	0.2701	0.8290	0.6907	1.0477	0.7734	1.0708	0.7890	1.0712	0.9404	2.4914
	1.4286	0.2328	0.9474	0.5992	1.2075	0.6681	1.2428	0.6777	1.2503	0.7978	2.8009
	1.6667	0.1968	1.1046	0.5099	1.4179	0.5671	1.4663	0.5734	1.4791	0.6690	3.1853
	2.0000	0.1616	1.3239	0.4221	1.7103	0.4689	1.7746	0.4734	1.7920	0.5483	3.7034
	2.5000	0.1274	1.6517	0.3355	2.1467	0.3728	2.2329	0.3761	2.2552	0.4326	4.4640
	3.3333	0.0941	2.1966	0.2501	2.8714	0.2781	2.9923	0.2806	3.0215	0.3206	5.7162
	5.0000	0.0617	3.2837	0.1658	4.3171	0.1845	4.5061	0.1863	4.5478	0.2114	8.2022
	10.0000	0.0303	6.5385	0.0824	8.6473	0.0919	9.0397	0.0929	9.1181	0.1046	15.6293
	Inf.	1.4826	1.0350	0.9134	0.9263	0.9061	0.9159	0.8654	0.8190	0.6502	0.2973
n	$1/R_s$	L_1	C_2	L_3	C_4	L_5	C_6	L_7	C_8	L_9	C_{10}

TABLE 11-54 Transitional Gaussian to 6-dB Active Low-Pass Values

Order n	C_1	C_2
4	1.2594	0.8989
	1.5863	0.2275
6	1.8406	1.3158
	2.1404	0.3841
	4.5372	0.09505
8	2.3685	1.7028
	2.6089	0.5164
	3.4746	0.1870
	8.9127	0.05094
10	2.9551	2.1329
	3.1606	0.6564
	3.7355	0.2568
	5.4083	0.1115
	14.9120	0.03232

TABLE 11-55 Transitional Gaussian to 12-dB Active Low-Pass Values

Order n	C_1	C_2
4	1.0778	0.2475
	1.0879	0.7965
6	1.4247	1.0330
	1.5000	0.3150
	2.2326	0.09401
8	1.8379	1.3309
	1.9324	0.4106
	2.3105	0.1557
	5.0556	0.04573
10	2.2051	1.5984
	2.2978	0.4965
	2.5733	0.1983
	3.4388	0.08903
	8.8028	0.02669

TABLE 11-56 Maximally Flat Delay with Chebyshev Stopband

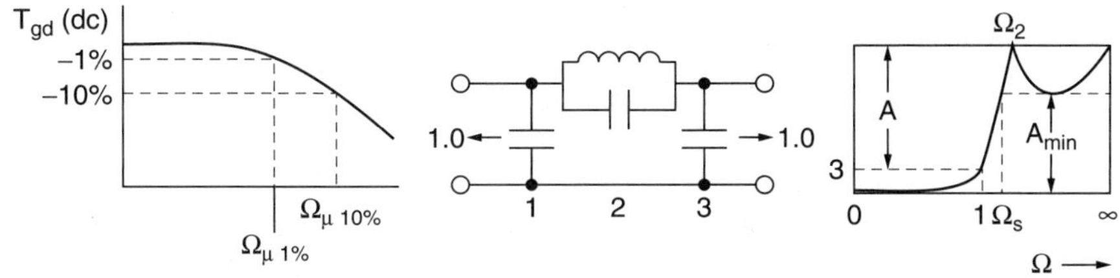

N = 3

A_{min}	Ω_s	T_{gd} (dc)	$\Omega_{\mu\ 1\%}$	$\Omega_{\mu\ 10\%}$	C_1	C_2	L_2	Ω_2	C_3
18	2.152	1.325	.9091	1.461	2.124	.2769	.5242	2.625	.001434
26	2.721	1.493	.8010	1.304	2.144	.1399	.6931	3.211	.1489
34	3.514	1.602	.7522	1.206	2.166	.07387	.8068	4.096	.2320
42	4.627	1.668	.7122	1.161	2.180	.03958	.8769	5.368	.2791
50	6.178	1.706	.6959	1.134	2.189	.02133	.9178	7.147	.3052
58	8.309	1.727	.6886	1.121	2.193	.01151	.9407	9.610	.3197
66	11.23	1.739	.6850	1.112	2.196	.006226	.9537	12.98	.3275
70	13.07	1.743	.6838	1.109	2.198	.004579	.9580	15.10	.3301

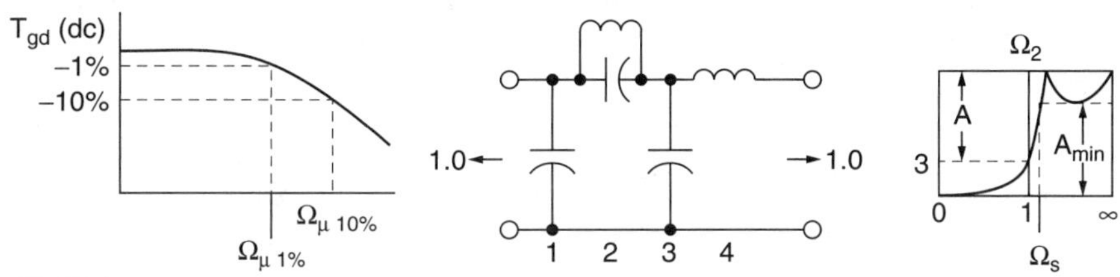

N = 4

A_{min}	Ω_s	T_{gd} (dc)	$\Omega_{\mu\ 1\%}$	$\Omega_{\mu\ 10\%}$	C_1	C_2	L_2	Ω_2	C_3	L_4
18	2.070	1.471	1.303	1.923	2.107	.3324	.5088	2.432	.1690	.1575
26	2.466	1.662	1.164	1.704	2.127	.1915	.6744	2.783	.3419	.1795
34	2.988	1.807	1.068	1.568	.1756	.1586	.5905	3.268	2.439	.4078
42	3.548	1.910	1.007	1.432	.2445	.09775	.6647	3.923	2.472	.4383
50	4.341	1.980	.9720	1.432	.1420	.08106	.5386	4.786	1.033	2.245
58	5.363	2.027	.9509	1.396	.1760	.04895	.5862	5.903	1.048	2.244
66	6.665	2.057	.9382	1.377	.1972	.03014	.6171	7.332	1.059	2.240
70	7.447	2.068	.9297	1.369	.2046	.02374	.6281	8.189	1.063	2.240

TABLE 11-56 Maximally Flat Delay with Chebyshev Stopband (*Continued*)

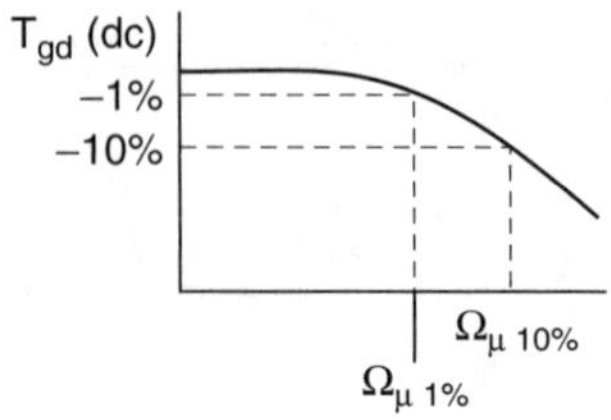

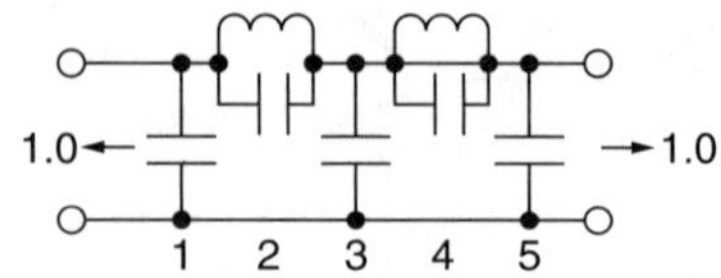

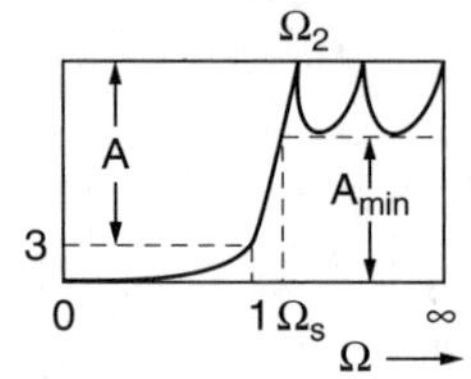

N = 5

A_{min}	Ω_s	T_{gd} (dc)	$\Omega_{\mu\ 1\%}$	$\Omega_{\mu\ 10\%}$	C_1	C_2	L_2
34	2.6802	1.745	1.550	2.142	.02930	.3317	.3665
42	3.0263	1.904	1.408	1.968	.1338	.2216	.4381
50	3.4467	2.035	1.316	1.840	2.1836	.08543	.8830
58	3.9701	2.139	1.304	1.750	2.203	.06060	.9418
66	4.6213	2.218	1.215	1.688	2.218	.04274	.9872
70	5.0044	2.250	1.191	1.665	2.224	.03582	1.006

Ω_2	C_3	C_4	L_4	Ω_4	C_5
2.868	2.420	.07329	.5638	4.919	.1111
3.209	2.435	.05333	.6496	5.373	.1513
3.641	.5958	.08228	.3358	6.016	.07096
4.186	.6503	.05559	.3818	6.864	.1007
4.863	.6922	.03799	.4168	7.947	.1221
5.268	.7092	.03150	.4307	8.585	.1305

TABLE 11-56 Maximally Flat Delay with Chebyshev Stopband (*Continued*)

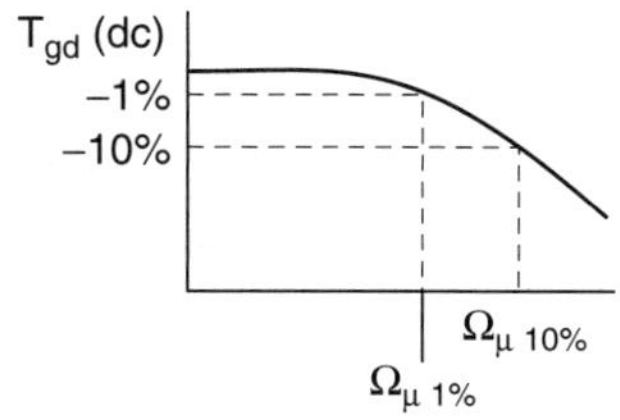

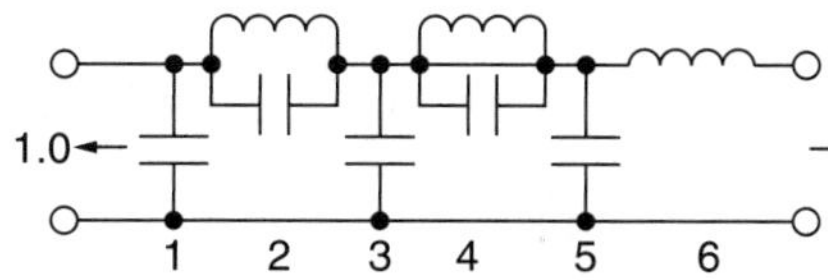

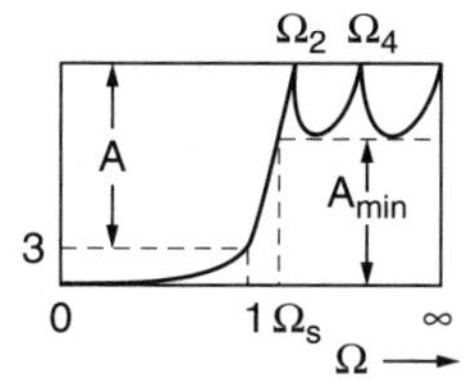

N = 6

A_{min}	Ω_s	T_{gd} (dc)	$\Omega_{\mu\ 1\%}$	$\Omega_{\mu\ 10\%}$	C_1	C_2	L_2
18	2.0530	1.415	2.096	—	.09332	.09400	1.907
26	2.3648	1.631	2.093	—	.1568	.07374	2.047
34	2.6429	1.820	1.910	2.604	2.128	.1820	.6982
42	2.9239	1.985	1.753	2.343	2.150	.1375	.7730
50	3.2353	2.129	1.643	2.190	2.172	.1047	.8371
58	3.6033	2.251	1.549	2.066	2.190	.07957	.8927
66	4.0446	2.352	1.485	1.982	2.206	.06014	.9396
70	4.2965	2.395	1.451	1.939	2.213	.05216	.9602

Ω_2	C_3	C_4	L_4	Ω_4	C_5	L_6
2.362	.4290	.2681	.2033	4.283	.08674	.1107
2.574	.5121	.2078	.2763	4.174	.1506	.1193
2.805	.4206	.2526	.2164	4.277	.08769	.09013
3.067	.4943	.1647	.2960	4.529	.1591	.09865
3.378	.5627	.1134	.3668	4.903	.2144	.1061
3.752	.6231	.08054	.4266	5.395	.2577	.1125
4.207	.6738	.05814	.4751	6.017	.2916	.1178
4.468	.6957	.04967	.4958	6.377	.3056	.1201

TABLE 11-56 Maximally Flat Delay with Chebyshev Stopband (*Continued*)

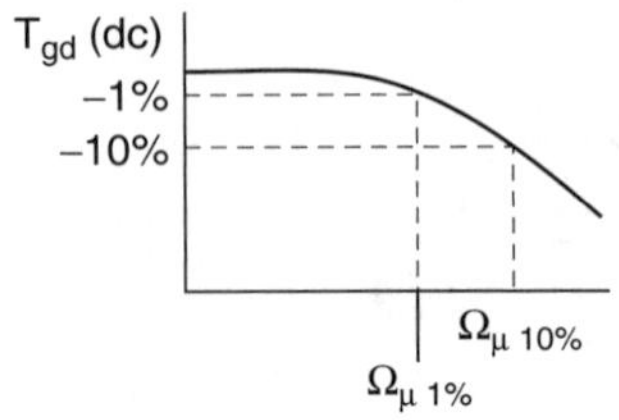

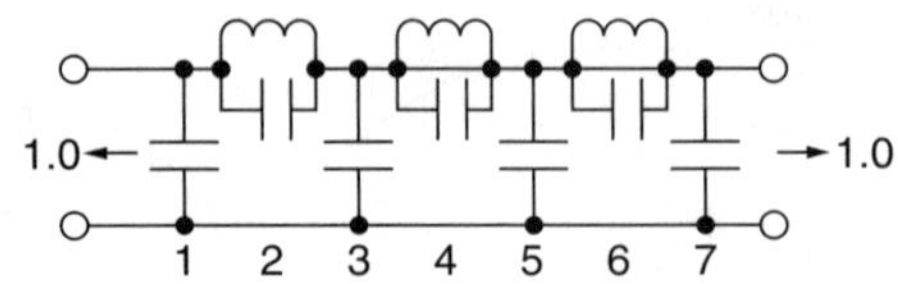

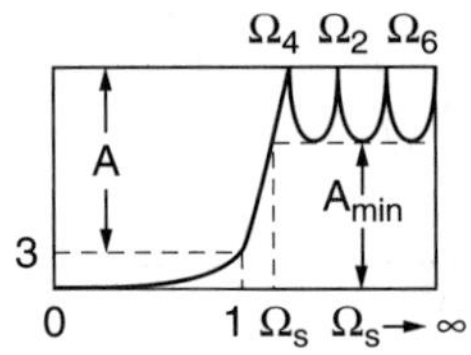

N = 7

A_{min}	Ω_s	T_{gd} (dc)	$\Omega_{\mu\ 1\%}$	$\Omega_{\mu\ 10\%}$	C_1	C_2	L_2	Ω_2
38	2.7259	1.839	2.357	—	.06065	.1728	.3777	3.919
42	2.8814	1.923	2.244	—	.07500	.1560	.4038	3.984
50	3.1288	2.080	2.087	2.704	.1064	.1259	.4543	4.181
58	3.3944	2.221	1.940	2.532	.1367	.1004	.5013	4.457
66	3.6959	2.348	1.839	2.395	.2557	.1169	.3703	4.806
70	3.8657	2.405	1.808	2.330	2.221	.04079	.9757	5.013

C_3	C_4	L_4	Ω_4	C_5	C_6	L_6	Ω_6	L_8
2.367	.2635	.4537	2.891	.2176	.09914	.1905	7.276	.01189
2.375	.2285	.4857	3.002	.2648	.08555	.2152	7.369	.02704
2.396	.1751	.5460	3.234	.3447	.06519	.2596	7.687	.05283
2.421	.1361	.6008	3.497	.4102	.05039	.2981	8.159	.07411
2.343	.07932	.8728	3.800	.5051	.04595	.2832	8.766	.06607
.6640	.1576	.4019	3.973	.3035	.05842	.2056	9.125	.03636

TABLE 11-56 Maximally Flat Delay with Chebyshev Stopband (*Continued*)

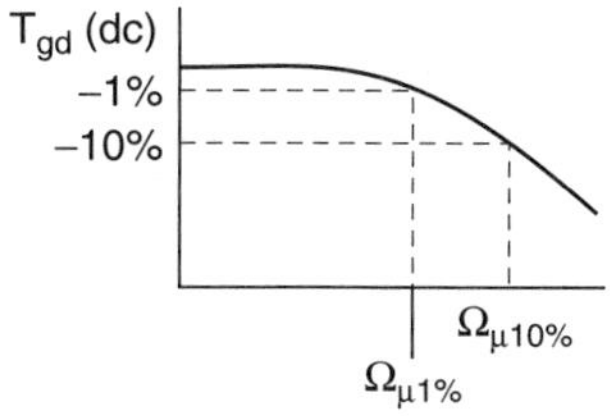

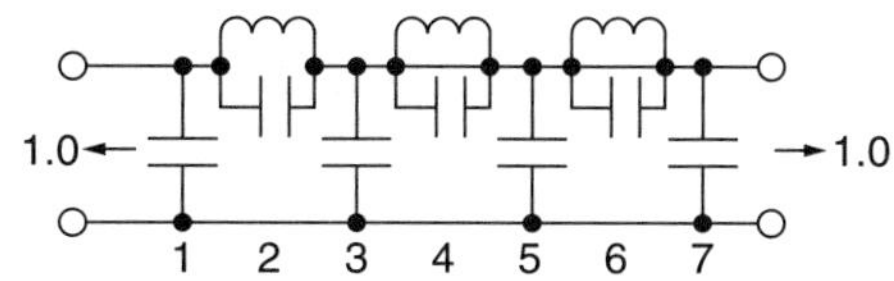

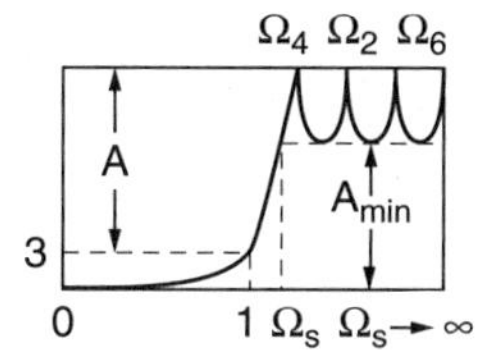

N = 8

A_{min}	Ω_s	T_{gd} (dc)	$\Omega_{\mu\ 1\%}$	$\Omega_{\mu\ 10\%}$	C_1	C_2	L_2	Ω_2
34	2.6318	1.798	2.660	—	.05238	.2052	.3371	3.802
42	2.8660	1.971	2.643	—	.06575	.1851	.3635	3.855
50	3.1053	2.128	2.434	—	2.198	.06844	.9184	3.987
58	3.3358	2.272	2.264	2.906	2.203	.06164	.9265	4.185
66	3.5839	2.403	2.152	2.738	2.209	.05407	.9386	4.439
70	3.7175	2.464	2.100	2.671	.1545	.09950	.4780	4.585

C_3	C_4	L_4	Ω_4	C_5	C_6	L_6	Ω_6	C_7	C_8
2.354	.3089	.4189	2.781	.1356	.1600	.1637	6.179	.07187	.06295
2.363	.2249	.4973	2.990	.2432	.1184	.2204	6.190	.1178	.07113
.5909	.5620	.1736	3.202	.09376	.1597	.1552	6.352	.06576	.06069
.5878	.3242	.2631	3.424	.2016	.1163	.1959	6.625	.1011	.06493
.6106	.2185	.3400	3.669	.2754	.08764	.2333	6.993	.1305	.06877
2.430	.1005	.6877	3.804	.4684	.05036	.3814	7.216	.2339	.09422

CHAPTER 12
INTRODUCTION TO DIGITAL FILTERS

12.1 INTRODUCTION TO SIGNAL PROCESSING

Signal processing refers to the art and science of creating, modifying, codifying, manipulating, extracting, and displaying signal information and attributes. Since the dawn of time, man has been the quintessential signal processor. For millennia, signal processing has been performed by man using the most powerful signal processing engine ever developed, the 25-watt human brain which dedicates about 10 watts to information processing (comparable to an Intel mobile Pentium III processor). As humans evolved, other elements were added to man's signal processing environment and repertoire, such as information coding in terms of intelligent speech[1] and the written word. In time, communication links expanded from the immediate, to local, to global, and now galactic. It was, however, the introduction of electronics that enabled the modern information revolution. Analog electronics gave rise to such innovations as the plain old telephone system (POTS), radio, television, radar/sonar, and a host of other inventions that have revolutionized man's life. Since the introduction of digital technologies a half century ago, man has witnessed an explosion of innovations that have facilitated the conversion of many previous analog solutions into systems having a high digital content. In other instances, digital technology has enabled solutions that previously never existed, such as digital cameras, MP3 players, digital telephony, and so forth. In some cases, digital technology has been disruptive, giving rise to products that were otherwise unfathomable. An example of this is the now ubiquitous personal digital computer.

12.2 INTRODUCTION TO DIGITAL SIGNAL PROCESSING (DSP)

Regardless of the information source, or the type of machine used to process information, engineers and scientists have habitually attempted to reduce signals to a set of parameters that can be mathematically studied, manipulated, combined, dissected or archived. This obsession has been fully actualized with the advent of the digital computer. One of the consequences of this fusion of man and machine has been the development of a new field of study called *digital signal processing*, or *DSP*. Some scholars trace the origins of DSP to the invention of iterative computing algorithms discovered by the ancient mathematicians.

[1]"Now the whole world had one language and a common speech."—Genesis 11.1

Those promoting these activities as being evidence of DSP are, however, missing the important" D"-word. DSP must, at some level, engage digital technology.

The foundations of DSP were laid in the middle of the 20th century in the form of the celebrated sampling theorem. Some attribute this breakthrough to Claude Shannon (1949), others to Harry Nyquist (1928). Whereas their work provided both the direction and motivation for DSP, they too were missing the important "D"-word. In 1965, Cooley and Tukey introduced an algorithm that is now known as the *fast Fourier transform* (FFT). The FFT was indeed a breakthrough in that it recognized both the strengths and weaknesses of a classic von Neumann general-purpose digital computer, and used this knowledge to craft an efficient computer algorithm for computing Fourier transforms. While to some, the FFT was a DSP facilitator, it wasn't until the advent of the DSP μp that DSP truly became a science with a defining technology. Since the introduction of the first generation of DSP μps (Intel 2920, 1979), and driven by Moore's Law, DSP has emerged as both an industry and an academic field of study. Through the combined efforts of engineers, scientists, and technologists, there is now an abundance of DSP resources available to the professional and neophyte alike, allowing them to craft both superior and revolutionary products.

12.3 THE RELATION TO ANALOG FILTERS

DSP is, in many ways, a gift of the sampling theorem and formidable armada of companion theories, methodologies, and tools, such as the celebrated FFT. Nevertheless, it was a mutated digital microprocessor (which became the first generation DSP μp) that gave rise to a discipline we now call DSP. Initially DSP was envisioned as an analog replacement technology. It has been that and much more. The competition and tension between analog and DSP solutions will undoubtedly continue into the foreseeable future. It is, nevertheless clearly apparent to many that DSP will become the dominant signal processing technology of the early 21st century. This claim can be motivated by comparing the assets and liabilities of the two contending technologies.

Digital Advantages

- Both analog and digital systems can be fabricated as highly integrated semiconductor devices. Compared to analog circuitry, digital devices can take full advantage of submicron technologies which are denser than their analog counterparts, providing both economic and performance advantages.
- As semiconductor technologies shrink (deep submicron) and signal voltage levels continue to decline (1.25 V and lower), the intrinsic signal-to-noise ratio found at the transistor level will suffer to the point where their use as analog devices becomes marginal. Furthermore, digital systems are far more tolerant of internal noise.
- Digital systems can operate at extremely low frequencies, which would require unrealistically large component values if implemented as an analog system.
- Digital systems can be designed to operate with increased precision with only an incremental increase in cost, whereas the precision of an analog system is typically limited (ten-bits ~ 60 dB dynamic range).
- Digital systems can be easily programmed to change their functionality, whereas reprogramming analog systems is extremely difficult.
- Digital signals can be easily delayed and/or compressed. This effect is difficult to achieve with analog systems.

- Digital systems require no external alignment, while analog systems need periodic adjustment (due to temperature-drift, aging, and so on).
- Digital systems do not have impedance-matching requirements, while analog systems do.
- Digital systems, compared to analog devices, are less sensitive to additive noise as a general rule.

Analog Advantages

- Analog systems can operate at extremely high frequencies (for instance, microwave and optical frequencies) which exceed the maximum clock rate of digital devices.
- Analog solutions are sometimes more cost-effective (say, a 1st order RC filter) compared to solutions fashioned with digital components (for example, ADC, digital filter, plus DAC).

What should also be appreciated is that DSP is more than a just replacement technology, it is also an enabling, and in some cases, a disruptive technology. DSP allows engineers and technologists to innovate products and solutions that would otherwise not have existed. Based on what has been witnessed over the brief history of DSP, one can only view in awe the possibilities of the future. The applications of DSP are virtually endless and continue to grow. Examples of areas of DSP penetration are

Entertainment
AV receivers
Computing
Digital radio
Home audio
Flat panel displays
Internet audio
Pro audio
Wireless
Special effects

Transportation
Chassis sensors
Power-train
Driver displays
Security systems
Safety systems

Broadband
Wireless LAN
Cable
DSL
VoIP
Home power LAN

Control
Digital power supplies
Embedded sensors
Industrial drives
Motors
Instrumentation

Medical
AED
Monitoring
Hearing aids
Imaging
Prosthetics

Military
Avionics
Countermeasures
Imaging
Munitions
Navigation
Radar/Sonar

Wireless
Handsets
Infrastructure
RF tagging

Security
Biometrics
Smart sensors
Encryption

Telecom
HF radios
Infrastructure
Navigation
Telecom accessories
Wire systems

Video and Imaging
Still cameras
Digital TV
Digital video
Digital recorders
IP video phones
Media devices
Set-top boxes
Streaming media
Surveillance
Video conferencing
Infrastructure

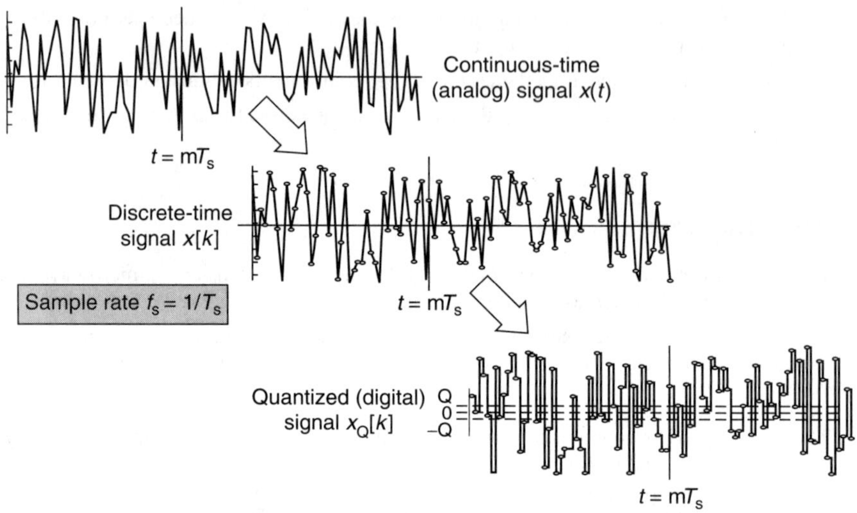

FIGURE 12-1 Signal taxonomy consisting of analog, discrete-time or sampled, and digital or quantized processes.

12.4 SIGNAL REPRESENTATION

In general, three classes of signals exist that attract the attention of signal processing engineers and scientists. They are (see Figure 12-1):

- *continuous-time* (analog) signals (continuously resolved in time and amplitude)
- *discrete-time* or sampled-data signals (continuously resolved in amplitude, discretely resolved in time)
- *digital signals* (discretely resolved in amplitude and time)

Contemporary signal processing systems typically contain a mix of analog, discrete, and digital signals and systems. Digital signals, however, are becoming increasingly dominant in this mix with discrete-time signals becoming rare, normally found in the sample-and-hold circuits of analog to digital converters (ADC).

12.5 DIGITAL DATA REPRESENTATION

Digital signals are represented as digital words normally appearing in *fixed point* (FXP), *floating point* (FLP), or *block floating point* formats. Within the fixed-point family are found sign magnitude (SM), diminished radix (DR), and offset binary (OB) coded numbers. A sign magnitude fixed-point integer X, such that $|X| < 2^{n-1}$, has a unique sign-magnitude representation, given by

$$X = X_{n-1}{:}(2^{n-2}X_{n-2} + \cdots + 2X_1 + X_0) \tag{12-1}$$

where X_i is called the ith-bit, and X_0 is referred to as the least significant bit (LSB). Similarly X_{n-2} is called the most significant bit (MSB) and X_{n-1} is the sign bit. The significance of the least significant bit (LSB) is often set by an ADC. For example, if a signed 12-bit ADC is used to digitize a ±15 volt (V) signal, then the LSB represents a quantization step size of Q = ADC dynamic range/2^n = 30 V (dynamic range)/2^{12} − bits = 7.32 mV/bit. The coding of the resulting data words are motivated in the next example.

Example 12-1 Number Coding

Required:
Convert the integers $X = 10$ and $X = -10$ into five-bit sign-magnitude numbers.

Result:
Using Equation (12-1), obtain

$$X = 10 \rightarrow 0{:}1010$$

or
$$X = (+)(1 \times 2^3 + 0 \times 2^2 + 1 \times 2^1 + 0 \times 2^0) = 10$$

$$X = -10 \rightarrow 1{:}1010$$

$$= (-)(1 \times 2^3 + 0 \times 2^2 + 1 \times 2^1 + 0 \times 2^0) = -10$$

where ":" denotes the sign-bit location.

Diminished radix cases are the 1's (1C) and 2's (2C) complement codes. The latter is by far the most common fixed-point code used in signal processing. In a 2C system, an integer $X \in [-2^{n-1}, 2^{n-1})$ has the unique representation given by

$$X = 2^{n-2}X_{n-2} + \cdots + 2X_1 + X_0 \quad \text{if } X \geq 0$$

and
$$X = -2^{n-1} + 2^{n-2}\overline{X}_{n-2} + \cdots + 2\overline{X}_1 + \overline{X}_0 + 1 \quad \text{if } X < 0 \tag{12-2}$$

where $\overline{X}_i$ denotes the complement of the ith binary digit. The production of positive and negative 2C data words is illustrated in the following example.

Example 12-2 2's Complement Coding

Required:
Convert $X = 10$ and $X = -10$ into a five-bit 2's complement number.

Result:
Using Equation (12-2), obtain

$$X = 10 \rightarrow 0{:}1010$$

or
$$X = 0 \times 2^4 + 1 \times 2^3 + 0 \times 2^2 + 1 \times 2^1 + 0 \times 2^0 = 10$$

$$X = -10; \quad 0{:}1010 = 10$$

$$1{:}0110 \leftarrow \text{complement}$$

$$\underline{+\ 0{:}0001} \leftarrow \text{add 1}$$

$$1{:}0110 \leftarrow -10 \text{ (result)}$$

or
$$X = -1 \times 2^4 + 0 \times 2^3 + 1 \times 2^2 + 1 \times 2^1 + 0 \times 2^0 = -10$$

Fractional numbers are created by scaling X by a power of 2. The value of $X' = X/2^m$ has the same binary representation of X except that the m LSBs are considered to be fractional bits. The interpretation of an integer word as a fractional number is demonstrated below.

Example 12-3 Fractional Number Representation

Required:
Scale $X = 10$ and $X = -10$ by $4 = 2^2$.

Result:
Scaling

$$X = 10 \quad X' = 10/4 = 2.5$$

$$\rightarrow 0{:}1010 \rightarrow 0{:}10.10$$

or $$X' = 0 \times 2^2 + 1 \times 2^1 + 0 \times 2^0 + 1 \times 2^{-1} + 0 \times 2^{-2} = 2.5$$

$$X = -10 \quad X' = -10/4 = -2.5$$

$$\rightarrow 1{:}0110 \quad \rightarrow 1{:}01.10$$

or $$X' = -1 \times 2^2 + 0 \times 2^1 + 1 \times 2^0 + 1 \times 2^{-1} + 0 \times 2^{-2} = -2.5$$

One of the significant attributes of the 2's complement system is its immunity to partial sum overflow provided that the final sum is a valid 2's complement number. This is sometimes referred to as the modulo (2^m) property. A illustration of how overflow can be mitigated using a 2C arithmetic system is shown in the following example.

Example 12-4 Addition Comparison

Required:
Compute the five-bit sum $S = 7 + 5 + 5 - 3 = 14$ using a sign-magnitude and 2's complement adder.

Result:
The sum $S = 7 + 5 + 5 - 3 = 14$ has both a valid five-bit sign magnitude and 2's complement representation. The summation S is computed as shown next:

7	SM 00111	2's 00111
+5	SM + 00101	2's + 00101
12	01100	01100
+5	+ 00101	+ 00101
17	10001 $= -1 \neq 17$	10001 $\neq 17$
−3	+ 10011	+ 11101
14	100100 $= 4 \neq 14$	101110 $= 14$

TABLE 12-1 The CSD Coding Table

x_{i+1}	x_i	c_i	X_i	c_{i+1}
0	0	0	0	0
0	1	0	1	0
1	0	0	0	0
1	1	0	−1	1
0	0	1	1	0
0	1	1	0	1
1	0	1	−1	1
1	1	1	0	1

Observe that an intermediate sign-magnitude overflow is fatal, while the 2's complement accumulator overflow does not affect the final outcome. This is a particularly important observation in the context of digital filter operation. It can provide assurance that a properly scaled digital filter will not introduce large runtime overflow saturation errors during runtime. Whether or not 2's complement arithmetic is used to implement a digital filter, it is normally suggested that saturating arithmetic also be used to mitigate the effect of any possible register overflow. A saturating arithmetic unit will clamp the output value at an extremal value. That is, a saturating arithmetic unit operates linearly over the dynamic range $R \in [-X_{\min}, X_{\max})$, while mapping an input X into an output $Y \in R$, such that

$$Y = \begin{cases} X & \text{if } X \in R \\ -X_{\min} & \text{if } X < -X_{\min} \\ X_{\max} & \text{if } X \geq X_{\max} \end{cases} \tag{12-3}$$

Since digital filters are multiply-accumulate (MAC) intensive, there is an implicit claim that fast arithmetic units are required to perform DSP. Fixed-point solutions have a reputation for providing high arithmetic speeds at low cost. 2C adders can be implemented using shift-add operations, carry-save adders, and carry lookahead adders providing a suite of cost-performance trade-offs. Fast 2C multipliers also span a wide cost-performance range and include multipliers based on fast carry-save or carry-lookahead adders, Booth's algorithm, cellular arrays, and Wallace trees.

Another fixed-point system that is sometimes employed in DSP applications is called the *canonic signed digit* (CSD) system. This system was used with early vacuum tube digital computers because it was maximally dense in zeros which resulted in a high frequency of NO OP operations. The CSD coding algorithm maps a binary-valued 2C number into a ternary value CSD number. The encoding scheme, shown in Table 12-1, shows x_i to be the ith 2C digital, c_i is the ith carry digit, and X_i is the ith canonic digit. The arithmetic processing of CSD coded data is motivated in the next example.

Example 12-5 CSD Arithmetic

Required:

Multiply $X = -5_{10}$ and $Y = -3_{10}$ using a CSD multiplier.

Result:

The product $P = (-5)(-3) = 15$. The 2C values of X and Y are $X = -5_{10} = 1_\Delta 011$ and $Y = -3_{10} = 1_\Delta 101$ respectively. Initializing $c_0 = 0$ (Table 12-1), the canonic coding of $X = -5_{10}$ is shown next.

i	x_{i+1}	x_i	c_i	X_i	c_{i+1}
0	$x_1 = 1$	$x_0 = 1$	$c_0 = 0$	$X_0 = -1$	$c_1 = 1$
1	$x_2 = 0$	$x_1 = 1$	$c_1 = 1$	$X_1 = 0$	$c_1 = 1$
2	$x_3 = 1$	$x_2 = 0$	$c_2 = 1$	$X_2 = -1$	$c_1 = 1$
3	$x_4 = 1\#$	$x_3 = 1$	$c_3 = 1$	$X_3 = 0$	$c_1 = 1$

{# denotes 2C sign extended bit}
Notice that X has been coded as $X = (-4) + (-1) = -5$, and contains two NO-OP operations compared to one in the 2C representation of $X = -5_{10} = 1_{\Delta}011$. The CSD code for the product P can be interpreted as $P = (-4)Y + (-1)Y$ which can be implemented using shift-registers, adders, and assorted logic. Specifically, $P = (-4)(-3) + (-1)(-3) = 15$ as required.

The limited dynamic range of a fixed-point FXP system can be overcome using a floating-point FLP code. A floating-point X is represented as

$$X = \pm m(x)r^{\pm e(x)} \tag{12-4}$$

where $m(x)$ is called the mantissa or significand, $e(x)$ is a biased or offset integer-valued exponent, and r is the radix. The mantissa is usually considered to be a normalized fractional number $1/r < m(x) < 1$. For the case where $r = 2$, the normalized mantissa ranges over [0.5, 1.0] causing the most significant data bit (MSB) to have a value of 1. Since this bit will always be 1 and thereby contributes no new information about the value of $m(x)$, it is often not explicitly displayed. In such cases, it is called a *hidden bit*. In an attempt to provide some degree of structure to this field, the normalized (mantissa) IEEE standard has been proposed and is summarized as follows {() denotes hidden bit}.

Precision	Single	Double	Quad
Word length	32 bits	64 bits	128 bits
Sign	1 bit	1 bit	1 bit
Offset (biased) exponent	8 bits	11 bits	15 bits
Significand	(1) + 23 bits	(1) + 52 bits	112 bits
Max. positive no.	$(2 - 2^{-23})2^{127}$	$(2 - 2^{-52})2^{1023}$	$(2 - 2^{-112})2^{32767}$
Min. positive no.	2^{-126}	2^{-1022}	2^{-32766}

The FLP code can represent numbers over a wide range, but cannot explicitly represent zero or other exceptions such as the result of an invalid operation or *not a number* (NAN), overflow, division by zero, underflow, inexact result, and unnormalized result (QUAD only).

Between the fixed-point and the floating-point numbering systems, the so-called *block floating-point* (BFP) code can be found. A number of DSP application-specific integrated circuits (ASIC), especially FFTs, use block floating points to attempt to expand the dynamic range of a number system which maintain many of the desirable cost-performance

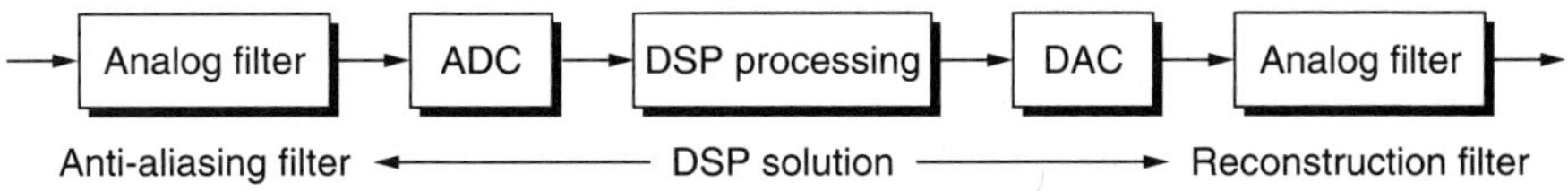

FIGURE 12-2 A DSP system consisting of an input signal conditioner (anti-aliasing filter), ADC, DSP microprocessor, DAC, and output signal conditioner (reconstruction filter).

features of fixed-point. A block floating point representation of an array of numbers $\{x[k]\}$, is given by

$$X = |x[k]_{\max}| = r^E; x[j] = m[j]r^E, 0 \leq m[j] = |x[j]/X| \leq 1 \qquad (12\text{-}5)$$

where the exponent E is set by the largest absolute value in $\{x[k]\}$, and r is called the BFP radix. It can be seen that the BFP system is actually a scaled fixed-point system whose dynamic range is extended by r^E.

12.6 SAMPLING THEOREM

> **Factoid:** Some attribute the sampling theorem to Claude Shannon, and others to Harry Nyquist. Nyquist suggested the sampling theorem in 1928, which was mathematically proven by Shannon in 1949 (MS Thesis). Some books use the term "Nyquist Sampling Theorem," while others use the "Shannon Sampling Theorem" to refer to the underlying theory of sampling.

One of the most important scientific advancements of the 20th century was attributable to Claude Shannon of Bell Laboratories in the form of his celebrated and enduring *Sampling Theorem* (see factoid). Shannon, it should be noted, worked for the telephone company and, as such, was interested in maximizing the number of billable subscribers that could simultaneously use a copper telephone line. His innovation was to sample the individual subscriber's conversations, then interlace the samples together in a process we now call *time-division multiplexing* (TDM). Understanding the Sampling Theorem is core to understanding DSP. The theorem both enables and constrains the performance of the typical DSP system suggested in Figure 12-2. This system consists of an ADC, DAC, a digital or DSP processor, plus analog signal conditioning filters (such as anti-aliasing and reconstruction filters). The Sampling Theorem also motivates the need for these signal conditioning filters.

The *Sampling Theorem* states that if a band-limited signal $x(t)$, whose highest frequency is bounded by some $f_{\max}$, is periodically sampled at some rate f_s, where

$$f_s > 2 \cdot f_{\max}; T_s = 1/f_s \qquad (12\text{-}6)$$

then the original signal $x(t)$ can be reconstructed from the sample values $x(t = kT_s) = x[k]$. It should be strongly noted that the sampling rate must be greater than the Nyquist sample rate[2] ($2f_{\max}$) and <u>not</u> equal to $2f_{\max}$.[2]

[2]Point of terminology: $2 \cdot f_{\max}$ = Nyquist sample rate, $f_s/2$ = Nyquist frequency.

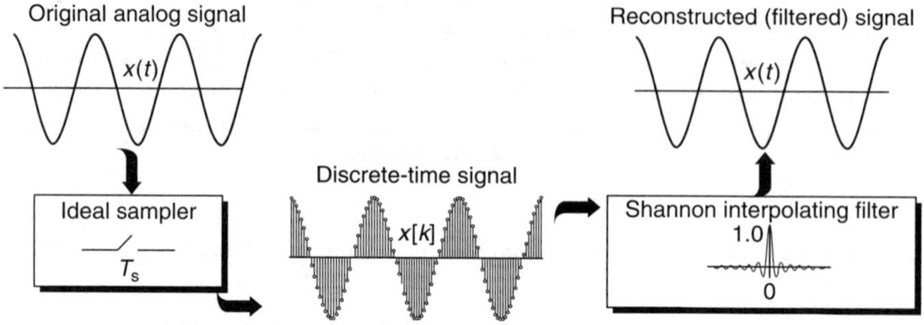

FIGURE 12-3 The Shannon interpolator simulation.

12.7 SIGNAL RECONSTRUCTION

Shannon assumed that the original analog signal $x(t)$ can be replaced with an array of periodically sampled values $x[k]$, called a *time-series*. The reconstruction of $x(t)$ from its sample values is achieved by a process involving an *interpolating filter*. Interpolation can be defined in terms of the interaction of a discrete-time time-series and an analog interpolating filter. Shannon's interpolating filter has an *impulse response* given by

$$h(t) = \sin(\pi t/T_s)/(\pi t/T_s) \equiv \text{sinc}(\pi t/T_s) \tag{12-7}$$

While the input to a Shannon interpolator is a discrete-time time-series, the output is a continuous-time signal process, as motivated in Figure 12-3.

12.8 PRACTICAL INTERPOLATORS

While Shannon's interpolating filter is elegant, it does present a major implementation hazard. The interpolating $h(t) = \text{sinc}(t/T_s)$ is seen to be *non-causal* by virtue of the fact that the filter's response is active for all time $-\infty \leq t \leq \infty$. As a result, Shannon's interpolation filter is not physically realizable. In practice, the interpolation process is described in Figure 12-3 and involves interpolating or smoothing the inter-sample values of a time-series using a realizable analog low-pass filter. When the low-pass filter's passband and cutoff frequencies are known *a priori*, fixed-coefficient op amp–enabled filters can be used. In cases where the sample frequency is variable, then programmable switched capacitor low-pass filters are generally used. The low-pass interpolation op amp or switched capacitor-enabled filter can be used to implement a variety of interpolating filter models (for example, Bessel, Butterworth, Chebyshev, and elliptic). Figure 12-4 compares Bessel and Chebyshev filter

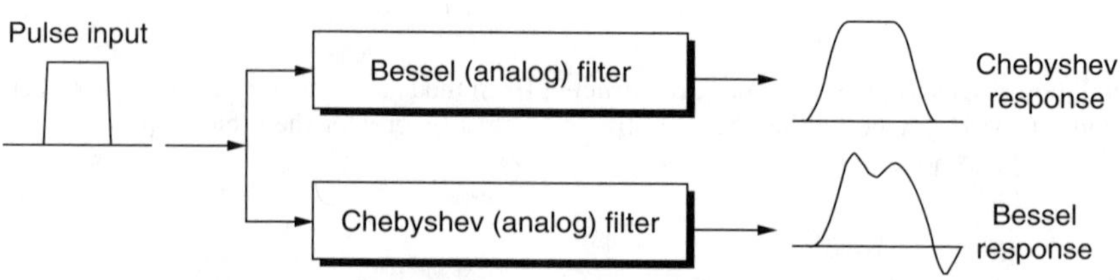

FIGURE 12-4 Processing a piecewise constant signal with a Bessel and Chebyshev analog filter.

responses to a piecewise constant input (for instance, pulse) representing a change in quantized signal level. The Bessel filter response is devoid of the strong overshoot present in the Chebyshev filter's output. As a result, the analog interpolation filters are generally designed using Bessel or Butterworth filter models.

12.9 SAMPLING MODALITIES

In practice, sampling occurs in one of three modes. Assuming that the highest frequency contained in the signal to be sampled is bounded from above by f_{max}, then the sampling modalities are

- *Critically sampled*: $f_s = 2f_{max}$
- *Over sampled*: $f_s > 2f_{max}$
- *Under sampled*: $f_s < 2f_{max}$

Critical sampling presumes that the sample rate is set to the minimum rate that satisfies the sampling theorem. To insure that the analog signal's highest frequency is bounded below the Nyquist frequency, an analog prefilter, called an *anti-aliasing* filter, is often placed between the signal and ADC, as shown in Figure 12-2. A successful anti-aliasing filter will significantly suppress signal energy residing above the Nyquist frequency. However, it is extremely difficult and expensive to construct a "steep-skirt" analog filter that will pass the desired signal components undistorted, and then rapidly transition to a highly attenuated stopband.

Over sampling relaxes the design requirements on the analog anti-aliasing filter. Over sampling occurs when the sample rate is higher than the Nyquist sample rate. To illustrate, consider sampling an audio process consisting of signal components bounded below 20 kHz. The critical sample rate is 40 kHz . Using a 4x over-sampling solution would set the sample rate to 160 kHz, defining the Nyquist frequency to be 80 kHz. Between 20 kHz and 80 kHz, the anti-aliasing filter can achieve an additional 20 to 80 dB of attenuation.

Under sampling can be intentional or unintentional. In either case, the sample rate is chosen to be less than the Nyquist sample frequency. In such instances, a phenomenon called *aliasing* occurs.

12.10 ALIASING

Whenever the Sampling Theorem is violated (for example, $f_s \leq 2f_{max}$), a phenomenon called *aliasing* occurs. Aliasing manifests itself when a reconstructed signal *impersonates* another signal or image. The effects of aliasing can be motivated by sampling a variable frequency sinusoid $x(t) = \cos(2\pi f_0 t)$ at a fixed rate f_s. If the sinusoid's frequency is set to $f_0 = 0.1f_s$ (for instance, 10-percent f_s), the sampled and reconstructed signal, shown in Figure 12-5, is seen to be that of $x(t) = \cos(2\pi f_0 t)$. The experiment is repeated for f_0 equal to 30 percent, and 110 percent of the sampling frequency. It should be recalled that the Sampling Theorem is satisfied whenever f_0 is less than 50 percent of f_s. In such cases, the original signal can be seen to be recovered using Shannon interpolation. However, when the input signal is a frequency equal to 110 percent of the sample frequency, the original signal is no longer evident from viewing the sample values. In fact, a much lower frequency sinusoid appears to be generated. This lower frequency signal is referred to as being aliased since the sample values of the high-frequency signal impersonates those of a lower-frequency signal.

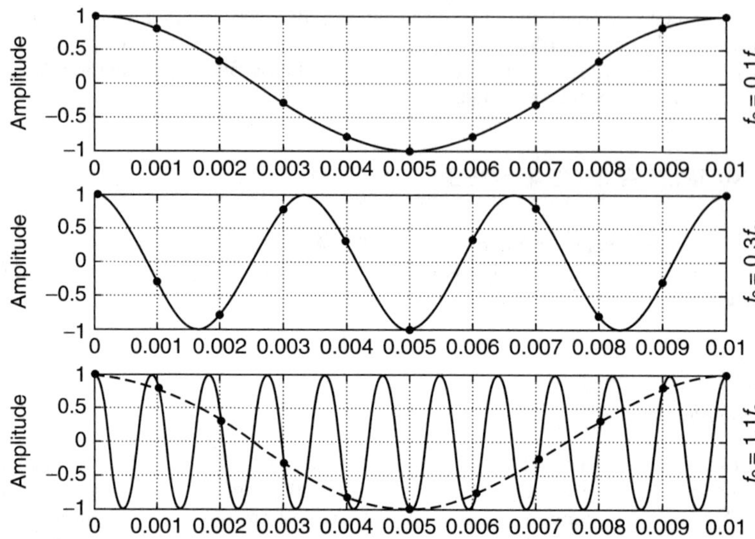

FIGURE 12-5 A signal sampling experiment where the input signal frequency is 10 percent, 30 percent, and 110 percent of the sampling frequency f_s. The individual sample values of the 110 percent case are seen to impersonate the sample values of a sinusoid at 10 percent of the sample rate f_s.

The base-band frequency occupied by aliased reconstructed sinusoidal signals can be mathematically computed. Suppose a signal $x_0(t) = \cos(2\pi f_0 t)$ is sampled at a rate f_s, and then the reconstructed signal, using a Shannon interpolator, produces a base-band signal $x_1(t) = \cos(2\pi f_1 t)$ for $f_1 \in [-f_s/2, f_s/2)$, The relationship between f_0 and f_1 is then defined by

$$f_1 = f_0\text{modulo}(f_s), f_1 \in [-f_s/2, f_s/2) \tag{12-8}$$

If $f_1 = f_0$, then no aliasing has occurred; otherwise, aliasing has taken place. An example of interpreting an aliased signal in the context of Equation (12-8) is provided by the next example.

Example 12-6 Aliasing

Required:

A cosine wave, having a period of 32 μs (for example, $f_0 = (10^6/31) = 31.250$ kHz), establishing a Nyquist rate of $f_N = 62.5$ kSa/s, is *under-sampled* at rate $f_s = (10^6/31) = 32.2581$ kSa/s. Determine the frequency of the reconstructed signal.

Result:

Since $f_s < f_N$, the signal is said to be under-sampled, suggesting that aliasing will occur. From knowledge that $f_0 = 10^6/32 = 10^6/31 - 10^6/(31)(32) = 10^6/31 - 10^6/992 = f_s - 1.0081$ kHz, it then follows that the frequency of the reconstructed signal, as given by Equation (12-8), is $f_1 = (10^6/32) \bmod (10^6/31) = -10^6/992 = -1.0081$ kHz. It can be seen that f_1 is a negative frequency residing within the base band range $f \in [-10^6/62, 10^6/62)$. The reconstructed signal would therefore appear as $y(t) = \cos(2\pi f_0 t) = \cos(-2\pi(1{,}008.06)t) = \cos(2\pi(1{,}008.06)t)$ and has a period of $T = 992$ μs, as shown in Figure 12-6. The analysis can also be performed using the MATLAB script shown next.

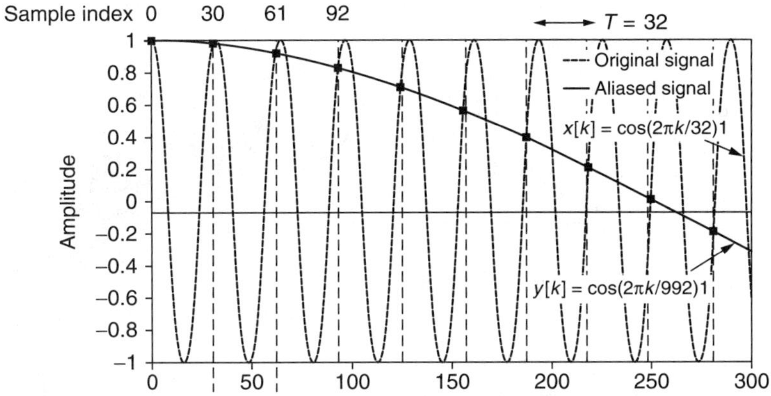

FIGURE 12-6 The original sine wave with a period of 32 μs being sampled at a rate of $10^6/31$ = 32.2581kSa/s. The sample values outline a sinusoid having a period of 992 μs.

```
>> f0 = 10^6/32;  {f0}
>> fs = 10^6/31;  {fs}
>> f1 = mod (f0,fs)   {f1 = f0mod(fs)}
        ans = 31250   {outside of base band, f ∈ [−fs/2, fs/2) = [−16,129, 16,129)}
>> f1 = f1 − fs   correct to base band}
        ans = −1.0081e+003   {f1 = −1,008.1 = −10^6/3}
```

12.11 DATA CONVERSION

The ubiquitous analog to digital converter (ADC) and digital to analog converter (DAC) are core DSP infrastructure technologies. The basic ADC, shown in Figure 12-7, consists of two parts. The first stage performs a *sample-and-hold* operation that physically converts an analog signal $x(t)$ into a discrete-time signal $x[k]$, while the second stage consists of a *quantizer* that maps the discrete-time sample value $x[k]$ into an equivalent digital sample $x_D[k]$. An n-bit ADC quantizes $x[k]$ into one of 2^n possible discrete values. The mechanism by which a digital signal is returned to an analog signal begins with a DAC that translates digital words into an equivalent analog level. This commonly involves a zero-order hold operation that holds an analog level for an entire sample interval. A sequence of discrete-time sample values is then passed to a continuous-time *interpolation* filter that produces a continuous output $x'(t)$.

ADCs and DACs are parameterized in terms of precision (in bits), speed (samples per second or Sa/s), linearity, along with other performance parameters. Some DACs come supplied with embedded analog multipliers that can be used to perform additional waveform shaping using modulation operations. Companding ADCs use either "mu" (μ)-law or

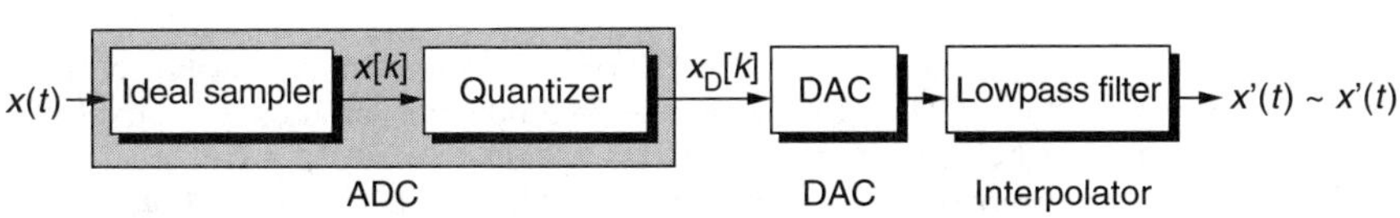

FIGURE 12-7 An ADC-based digital conversion system.

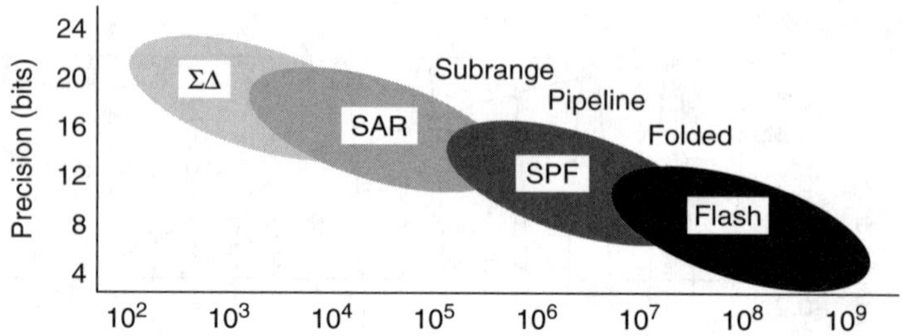

FIGURE 12-8 Various ADC families.

A-law compression algorithms to non-linearly redistribute the converters' dynamic range. For high-end applications, dithering is sometimes used to statistically reduce the effects of quantization. Some ADCs are equipped with high-bandwidth sample and hold circuits for use in under-sampling applications. Both ADC and DAC technologies are abundantly available in commercial products spanning a wide range of cost, precision, and bandwidth choices. ADC architectural choices include sigma-delta ($\Sigma\Delta$), flash, successive approximation (SAR), subrange, pipelined, and folded. Their speed and precision envelopes are compared in Figure 12-8. While the ADC industry is constantly pursuing the goal of high-speed, low-power, and high-precision converters (essentially their "holy grail"), progress remains slow. Whereas Moore's Law for the mainstream digital industry states that digital electronic devices double in density about every 18 months, a constant speed ADC's precision improves at a rate of 1.5-bits every nine years. Based on this model, existing 12-bit ADCs at a rated speed would need 24 years to achieve 16-bit status! Moore's Law would also state that the mainstream digital circuitry attached to the ADC improves by a factor of 64 K over the same time interval!

12.12 FINITE WORDLENGTH EFFECTS

Refer to Figure 12-7 which motivates the analog to digital conversion (ADC) process. The difference between the discrete-time and quantized sample value is called the *quantization error* and is formally defined to be

$$e[k] = x_D[k] - x[k] \tag{12-9}$$

The quantization error, in turn, can be expressed in terms of the *quantization step-size* (denoted by Δ) and is measured in volts or amps per bit. If the input signal $x(t)$ is double-ended (for example, $-A \leq x[k] < A$), then the quantization step-size associated with an n-bit ADC converter is given by

$$\Delta = 2A/2^n \tag{12-10}$$

Statistically, the quantization error is characterized by a uniformly distributed probable density function (PDF). The quantization error production process is motivated in Figure 12-9 and is interpreted for rounding and truncation in Table 12-2.

The mean or average error, along with the maximum error, is predicated on whether rounding or truncation is used. The *error variance* is $\sigma^2 = \Delta^2/12$, regardless of which is

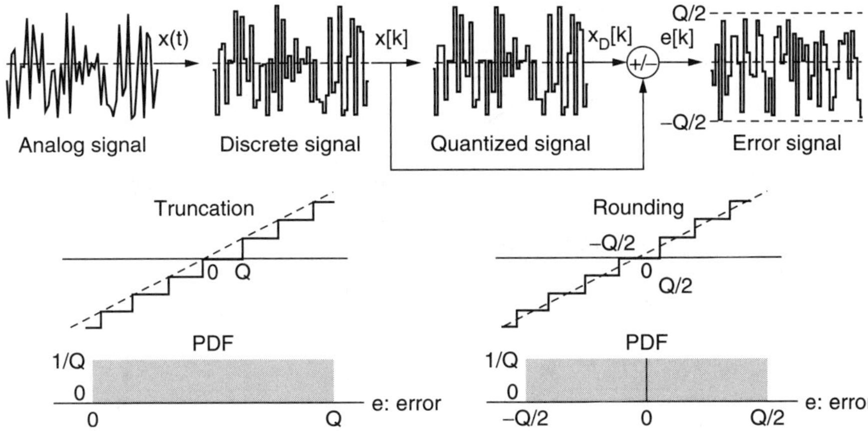

FIGURE 12-9 The quantization error production mechanism (top panel), and the PDF function (bottom panel).

employed. The error *standard deviation* is the square root of the variance, which leads to the error being interpreted in bits as

$$\text{error}_{\text{bits}} = \log_2(\sigma) = \log_2(\Delta) - 1.79 \tag{12-11}$$

and the error in decibels is $\text{error}_{\text{dB}} = \log_2(\sigma) = 20\log_{10}(\Delta) - 10.8$ dB. The production of statistical quantization error bounds and their interpretation is presented in the following example.

Example 12-7 Roundoff Error

Required:

Experimentally verify Equation (12.11).

Result:

Assume that the input to a 16-bit double-ended ADC is assumed to be bounded by $|x(t)| \leq 1$. The quantization step size is $\Delta = 2/2^{16} = 2^{-15}$ and the predicted error variance is $\sigma^2 = \Delta^2/12 = 2^{-30}/12$. Using Equation (12.11), the predicted error in bits is given by $\log_2(\sigma) = \log_2(\Delta) - 1.79 = -15 - 1.79 = -16.79$ bits. The error can also be studied using simulation, a random signal $x(t)$, and an N-bit ADC where n ranges from 2 to 16, bits which corresponds in this case to a fractional range from 1 to 15 bits. The predicted and computed errors are compared in Figure 12-10, where a strong agreement is seen between actual and predicted errors.

TABLE 12-2 Quantization Error Statistics

Policy	Figure	Max. |error|	Error Range	Error Mean	Error Variance
Truncation	12-9 (left)	Δ (LSB)	$e[k] \in [0, \Delta)$	$\Delta/2$ ($\pm\frac{1}{2}$ LSB)	$\Delta^2/12$
Rounding	12-9 (right)	$\Delta/2$ ($\pm\frac{1}{2}$ LSB)	$e[k] \in [-\Delta/2, \Delta/2)$	0	$\Delta^2/12$

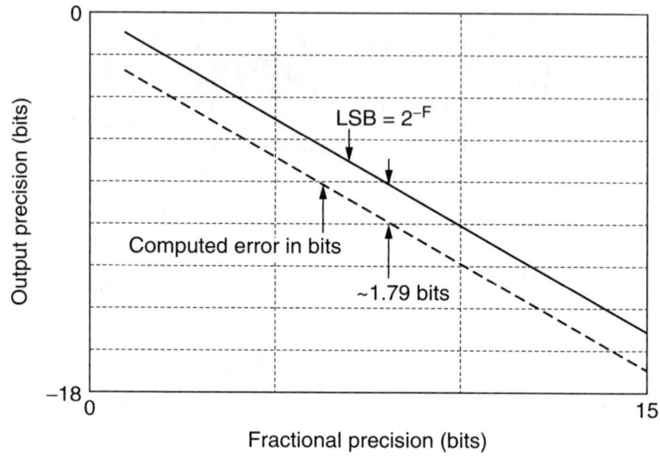

FIGURE 12-10 Results from a computer simulation displaying quantization error (in bits) of a random input for a doubled-ended ADC having 1 to 15 fractional bits of precision. The experimental data clearly illustrates about a 1.79-bit bias in the error statistics.

12.13 MATHEMATICAL SIGNAL AND SYSTEM REPRESENTATION

Discrete-time signals and systems can be modeled in the time domain using discrete difference equations or discrete-time transforms. Transforms, and in particular the standard *z*-transforms, are fundamentally important to the design and analysis of digital filters. The conversion of a continuous-time filter into a discrete-time model can be motivated using a *Laplace transform* and the delay theorem that states that if $x(t) \leftrightarrow X(s)$, then $x(t + \tau) \leftrightarrow e^{-s\tau}X(s)$. A notational simplification can be achieved if the complex exponential term is replaced by $z = e^{s\tau}$, given rise to the so-called *z transform.* The *z* transform of a causal time series $x[k]$ is formally given by

$$X(z) = \sum_{k=0}^{\infty} x[k]z^{-k} \tag{12-12}$$

The *z*-transform is also an important system modeling tool. If the response of a linear constant coefficient system to an input time-series $x[k]$ having a *z*-transform $X(z)$ is the input time-series $y[k]$ having a *z*-transform $Y(z)$, then the system's *transfer function* is given by

$$H(z) = \frac{Y(z)}{X(z)} = \sum_{k=0}^{\infty} h[k]z^{-k} = \frac{\sum_{i=0}^{M} a_i z^{-i}}{\sum_{i=0}^{N} b_i z^{-i}} = \frac{B(z)}{A(z)} \tag{12-13}$$

The *zeros* of $X(z)$ are those values of *z* that set $B(z) = 0$. That is, z_i is a zero of $X(z)$ if

$$B(z) = \prod_{i=1}^{M} (1 - z_i z^{-1}) \tag{12-14}$$

The *poles* of $X(z)$ are those values of z that set $A(z) = 0$. That is, p_i is a poles of $X(z)$ if

$$A(z) = \prod_{i=1}^{N}(1 - p_i z^{-1}) \tag{12-15}$$

The inverse z-transform maps a z-transform $X(z)$ into a discrete-time time series $x[k]$. The standard means of inverting a z-transform is called a *partial fraction* or *Heaviside expansion*.

Another important z-transform tool is called the *bilinear z-transform*. The bilinear z-transform is related to the familiar Laplace transform through the relationship:

$$s = (2/T_s)(z - 1)/(z + 1) \text{ or } z = ((2/T_s) + s)/((2/T_s) - s) \tag{12-16}$$

It is important to realize that the frequency responses of the standard and bilinear z-transform are different. The standard z-transform $z = e^{sT}$ is that of an all-pass filter which repeats itself on $f = f_s$ centers. The bilinear z-transform frequency response is given by

$$e^{j\omega} = ((2/T_s) + j\Omega)/((2/T_s) - j\Omega s) \tag{12-17}$$

where Ω is an analog frequency in radians/second and ω in radians/second is the corresponding digital domain frequency. The frequency relationship established by Equation (12-17) can be reduced to read

$$f_a = (1/\pi T_s)\tan(\pi f_d T_s) \tag{12-18}$$

where f_a is an analog frequency and f_d is the corresponding digital frequency. The nonlinear relationship between the analog and digital frequencies is called *warping*. The warping curve is shown in Figure 12-11. The relationship between analog and digital frequencies associated with a bilinear z transform is illustrated in the next example.

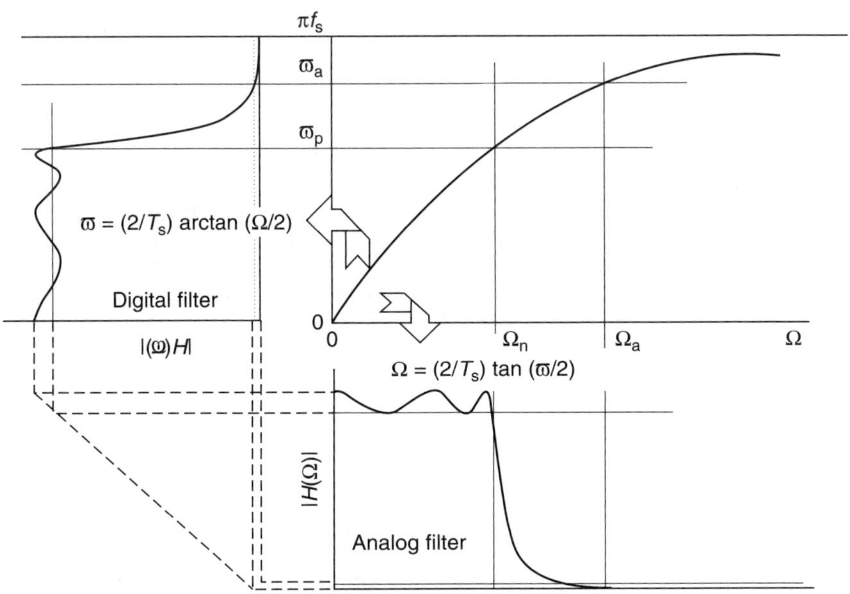

FIGURE 12-11 The relationship between the continuous frequency axis (denoted by (Ω) and the discrete frequency axis (denoted (ω) using the bilinear z-transform.

Example 12-8 Bilinear z transform

Required:
Determine the warped digital domain frequency that corresponds to a 1 kHz analog signal sampled at a rate $f_s = 10$ kHz.

Results:
From $f_a = (1/\pi T_s)\tan(\pi f_d T_s)$, it follows that the inverse relationship is $f_d = (1/\pi T_s)\tan^{-1}(\pi f_a T_s)$. Then

$$f_d = (1/\pi 10^{-4})\tan^{-1}(\pi 10^3 10^{-4}) = (1/\pi 10^{-4})\,2\tan^{-1}(0.1\pi) = 968.9 \text{ Hz} < f_a.$$

12.14 SPECTRAL REPRESENTATION

Besides representing signals in the continuous or discrete-time domain using transforms, signals and systems can also be modeled in the frequency domain. The principal tools used to describe an assumed periodic discrete-time signal or system behavior in the frequency domain is the *discrete Fourier transform* (DFT). A discrete-time time series $x[k]$, consisting of N contiguous samples spaced T_s seconds apart, has a period $T = NT_s$. The DFT of $x[k]$ is defined by

$$X(n) = \sum_{k=0}^{N-1} x[k]\, W_N^{nk}\,;\ (n, k) = 0, 1, \ldots, N-1 \tag{12-19}$$

where $X(k)$ is the kth harmonic of an N-sample DFT, and W_N is the complex exponential; $W_N = \exp(-j2\pi/N) = \cos(2\pi/N) - j\sin(2\pi/N)$. The 0th harmonic, namely $X(0)$, corresponds to the amplitude of the DC signal component, $X(1)$ is the first harmonic located at frequency $f_0 = f_s/N$ Hz, and so forth. The ith harmonic is located at a frequency $f_i = f_0$. The formal frequency spectrum consists of a listing of N complex harmonics with the first $N/2$ harmonics defining the positive base band ($f \in [-0, f_s/2)$ and the next $N/2$ defining the negative base band ($f \in [-f_s/2, 0)$.

The inverse DFT, or IDFT, is given by

$$x[k] = \frac{1}{N}\sum_{n=0}^{N-1} X(n)\, W_N^{-nk}\,;\ (n, k) = 0, 1, \ldots, N-1 \tag{12-20}$$

Because of symmetry properties, the frequency spectrum is symmetrically "folded" about the Nyquist frequency $X(N/2)$. As a result, the Nyquist frequency is also referred to as the *folding* frequency. The first $N/2$ harmonics represent positive frequencies with the remaining $N/2$ harmonics corresponding to negative frequency components.

The relationship between the Fourier transform and the discrete Fourier transform is summarized in Figure 12-12. The Fourier transform accepts a continuous-time (analog)

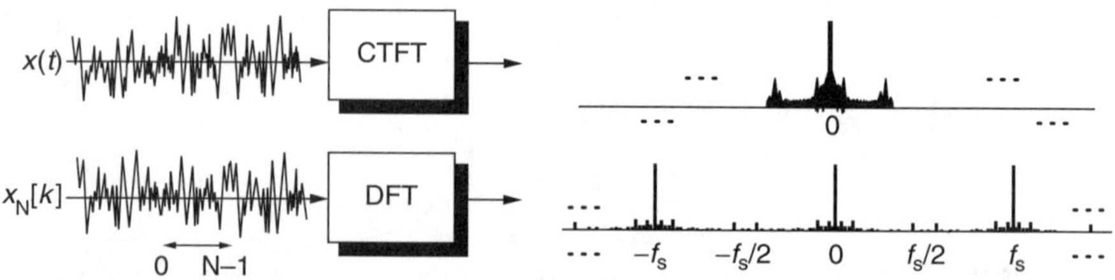

FIGURE 12-12 A comparison of Fourier and discrete Fourier spectra.

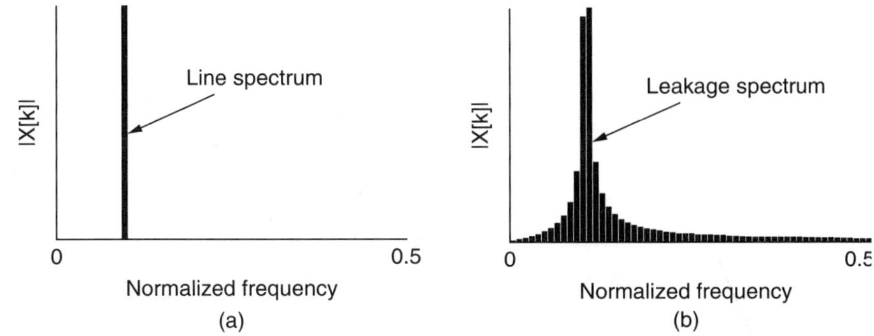

FIGURE 12-13 Spectra representing the DFTs of a sinusoid completing (*a*) 10 (no leakage) cycles and (*b*) 10.5 (leakage) cycles. (The frequency axis was normalized with respect to f_s.)

signal and produces a continuous spectrum defined over all frequencies. An N-point DFT selects N contiguous samples from an infinitely long time series and produces a discretely resolved spectrum. The discrete frequency components residing between $\pm f_s/2$ (the Nyquist frequency) define the *base band*. It is assumed that the signal being transformed is being sampled above the Nyquist rate, resulting in an alias-free base-band spectrum. Referring to Figure 12-12, it can also be noted that the DFT spectrum of the periodically sampled input signal is itself periodically repeated in the frequency domain. These artifacts are due to the sampling process.

Some of the more commonly incurred errors associated with the DFT are called aliasing and leakage. Aliasing occurs when a signal with a frequency above the Nyquist limit is transformed. This effect can be reduced by adding an antialiasing prefilter in front of the DFT. The prefilter is a low-pass, linear phase device whose cutoff frequency is set to the Nyquist frequency, thereby severely attenuating potential aliasing signal components. The second error, leakage, is exemplified in Figure 12-12. This error will occur whenever the periodic assumption is violated. Figure 12-13 reports the N-point DFT of two pure sinusoids, one completing 10 cycles and the other a non-integer 10.5 cycles within the N sample DFT data window. The production of unwanted sideband harmonics in the latter case results in leakage. This condition can be reduced by increasing the window length (for example, increasing N) or using what are called *data windows* (such as the Hamming window).

BIBLIOGRAPHY

Cavicchi, T. *Digital Signal Processing*. New York: John Wiley and Sons, 2000.

Chassaing, R. *Digital Signal Processing and Applications with the C6713 and C6416 DSK*. New York: John Wiley and Sons, 2005.

Ifwachor, E., and B. Jervis. *Digital Signal Processing, 2nd ed.* Reading, Massachusetts: Addison Wesley, 2001.

Mitra, S. *Digital Signal Processing, 3rd ed.* New York: McGraw-Hill, 2006.

Oppenheim, A. V., and R. Schafer. *Digital Signal Processing*. Englewood Cliffs, New Jersey: Prentice-Hall, 1975.

_____. *Digital Signal Processing, 2nd ed.* Englewood Cliffs, New Jersey: Prentice-Hall, 1999.

Taylor, F. J. *Digital Filter Design Handbook*. New York: Marcel Dekker, 1983.

_____ and T. Stouraitis. *Digital Filter Design Using the IBP PC*. New York: Marcel Dekker, 1987.

_____ and J. Mellott. *Hands-On Digital Signal Processing*. New York: McGraw-Hill, 1998.

CHAPTER 13
FINITE IMPULSE-RESPONSE FILTERS

13.1 DIGITAL FILTERS

The digital revolution has created a vast array of new products and solutions along with the field of digital signal processing (DSP) and the study of digital filters. Beginning in the 1970s, engineers began fashioning special-purpose digital filters using discrete components. With the advent of DSP μps and large-scale integrated circuits (LSI), digital filters have become a core DSP technology. They have now matured to the point where they can claim:

- High precision and speed
- Programmability and adaptability
- Robust performance over a wide range of frequencies
- Compact size and interoperability with other digital subsystems
- Low-cost and low-power dissipation
- High-reliability and repeatability

Digital filters have evolved along two major technological paths: fixed coefficient filters and variable coefficient filters. Examples of variable coefficient filters are adaptive systems and neural networks. It is, however, fixed coefficient filters that continue to remain the dominant filter type in commercial applications. Fixed-coefficient filters can also be partitioned into two broad subclasses called *finite impulse response filters* (FIR) and *infinite impulse response* (IIR) filters. Each filter type promotes a unique and often non-complementary set of attributes.

13.2 FIR DIGITAL FILTERS

An Nth-order FIR can be defined in terms of a finite duration *impulse response:*

$$h[k] = \{h_0, h_1, \ldots, h_{N-1}\} \tag{13-1}$$

where the coefficients $\{h_i\}$ are called the filter's *tap weights*. The response of an Nth-order FIR filter (sometimes referred to as a transversal filter) to an input time-series $\{x[k]\}$ can be expressed in terms of the discrete linear convolution sum:

$$y[k] = h_0x[k] + h_1x[k-1] + \cdots + h_{N-1}x[k-N+1] = \sum_{m=0}^{N-1} h_m x[k-m] \tag{13-2}$$

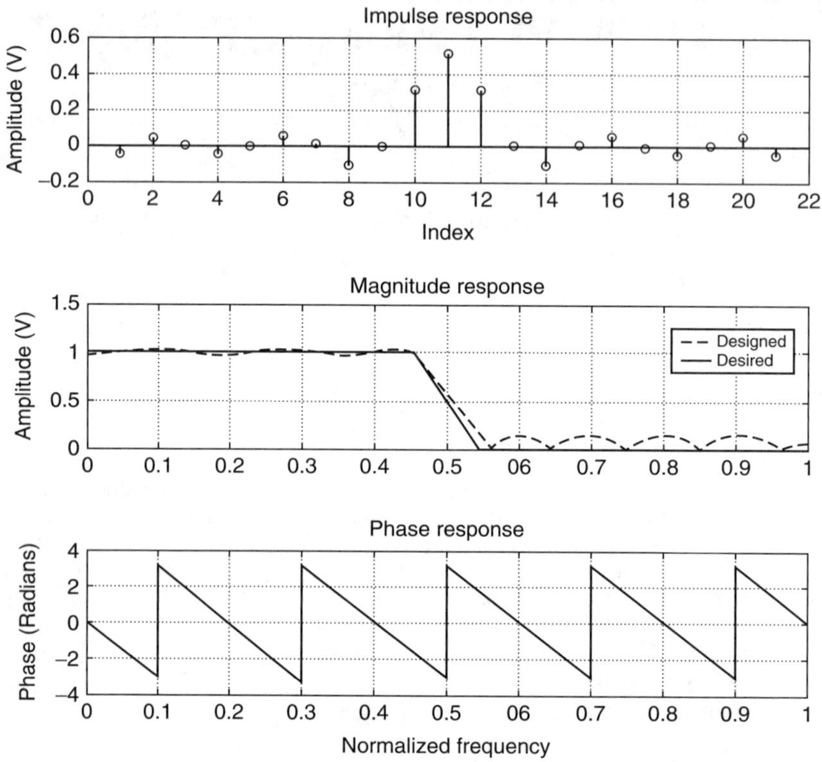

FIGURE 13-1 A typical FIR impulse response (top), a magnitude frequency response (middle), and a phase response (bottom). (Frequency axis normalized with respect to $f_s/2$.)

Equation (13-1) can also be restated in terms of a transfer function:

$$H(z) = \sum_{i=0}^{N-1} h_i z^{-i} \tag{13-3}$$

The typical response of an Nth-order FIR is reported in Figure 13-1. The following example illustrates the behavior of a simple FIR.

Example 13-1 Simple FIR

Required:

A simple 10th-order FIR having a ten-sample decaying exponential impulse response $h[k] = (0.75)^k$, $k \in [0, 9]$, has a transfer function:

$$H(z) = \sum_{k=0}^{9} (0.75)^k z^{-k}$$

Analyze the FIR's impulse, magnitude frequency, and phase responses.

Results:

The FIR's discrete-time and frequency response is reported in Figure 13-2. It can be seen that the magnitude frequency response is that of a low-pass filter with a non–linear-phase response.

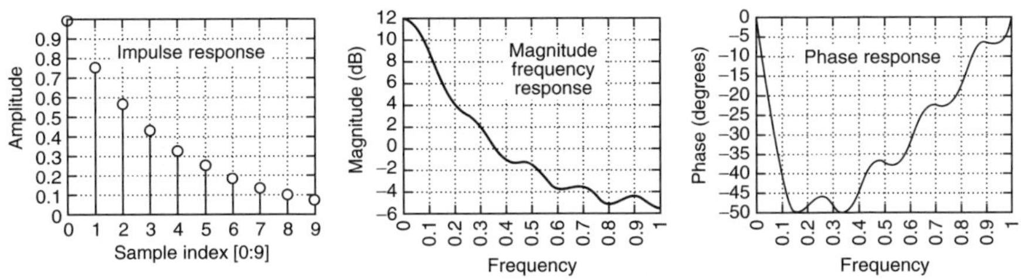

FIGURE 13-2 An impulse response $h_k = h[k]$ (left), a magnitude frequency response (middle), and a phase response (right). (The frequency axis has been normalized with respect to $f_s/2$.)

13.3 STABILITY

It is natural to consider an input time-series $x[k]$ to be bounded on a sample-by-sample basis (for example, $|x[k]| < M_x$). Furthermore, it is expected that the tap weights of a constant coefficient FIR are also bounded on a sample-by-sample basis (for instance, $|h[k]| < M_h$). The partial products terms $h[m]x[k - m]$, found in the linear convolution sum (Equation 13-2), are therefore bounded by $|h[m]x[k - m]| < M_x M_h$. It then follows that the output of an Nth-order FIR is finite and must be bounded by:

$$|y[k]| = \left|\sum_{m=0}^{N-1} h_m x[k - m]\right| \leq NM_x M_h \leq \infty \tag{13-4}$$

Such an FIR is said to be *bounded-input-bounded-output* (BIBO) stable. However, with digital filters (FIR or IIR) the issue is rarely stability, but rather quantifying a filter's maximal runtime dynamic range requirements. If the input time-series is again assumed to be bounded by $|x[k]| \leq M_x$ on a sample-by-sample basis, then the input that maximizes the FIR output is called the *worst case input* and is given by:

$$x_{\text{worst\_case}}[k - m] = M_x \operatorname{sign}(h_m) \tag{13-5}$$

where $\operatorname{sign}(h_m)$ is the sign of h_m (for instance, $\{\pm 1\}$). Therefore, the FIR's response to the worst case input given in Equation (13-5), is

$$y_{\text{worst\_case}}[k] = \sum_{m=0}^{N-1} h_m x_{\text{worst\_case}}[k - m] = \sum_{m=0}^{N-1} h_m M_x \operatorname{sign}(h_m) = M_x \sum_{m=0}^{N-1} |h_m| = M_x G_{\max} \tag{13-6}$$

where the parameter $G_{\max}$, found in Equation (13-6), is called the filter's *worst case gain* and is given by:

$$G_{\max} = \sum_{m=0}^{N-1} |h_m| \leq 2^I \tag{13-7}$$

The worst case gain defines a maximal runtime dynamic range bound on the FIR internal shift registers. Specifically, the worst case gain, defined by Equation (13-7), states that during runtime, I additional bits of dynamic range growth (called "headroom") may be required to protect the system from suffering disabling register overflows. For this reason, FIRs are sometimes configured with extended precision arithmetic units and registers to inhibit runtime dynamic range overflow. This information is of significant importance in implementing fixed-point FIRs in the context of the binary-point assignment. This is motivated by the following example. The experimental exploration of worst case gain is presented in the next example.

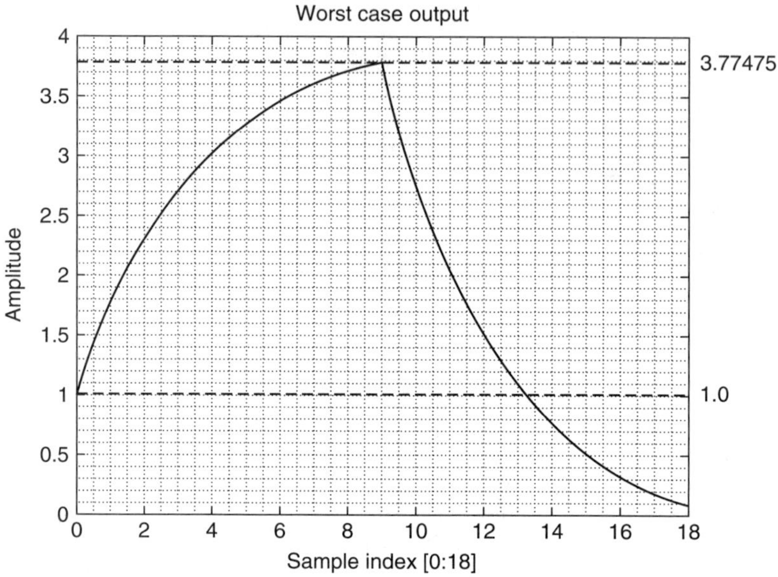

FIGURE 13-3 A worst case filter response.

Example 13-2 Worst case gain

Required:
Consider again the simple $N = 10$th-order FIR studies in Example 13-1 having an impulse response given by $h[k] = (0.75)^k$, $k \in [0, 9]$. Compute and verify the FIR's worst case gain.

Results:
The worst case gain is computed to be

$$G_{\max} = \sum_{k=0}^{9} |h_k| = 3.77475 < 2^2$$

The linear convolution of the FIR's impulse response with a unit bound worst case $M = 10$-sample input ($M_x = 1$) is shown in Figure 13-3. Since the FIR's impulse response is non-negative, the worst case input is $x_{\text{worst\_case}}[k] = 1.0$ for all k. The linear convolution sum produces a measurable response of length $N + M - 1 = 19$ samples. This is consistent with the theory of linear convolution. It can also be noted that even though the input is bounded by unity, the extreme or worst case output has a value $G_{\max} = 3.77475 < 2^2$. This implies that the dynamic range requirement of the FIR can increase by nearly 2-bits during runtime.

13-4 LINEAR-PHASE BEHAVIOR

It can be observed that an FIR is basically a shift-register network. Since digital shift registers can implement precise signal delays, FIRs are tantamount to a linear phase-shifting network. It can therefore be expected that the FIR can possess interesting phase domain

TABLE 13-1 Linear-phase $\phi(\varpi) = \alpha\varpi + \beta$ FIR Relationships

Type	N (order)	Symmetry	α	β
1	N odd	Even	$\alpha = -M; M = (N-1)/2$	$\beta = 0$
2	N odd	Odd	$\alpha = -M; M = (N-1)/2$	$\beta = \pi/2$
3	N even	Even	$\alpha = -M; M = (N-1)/2$	$\beta = 0$
4	N even	Odd	$\alpha = -M; M = (N-1)/2$	$\beta = \pi/2$

attributes. The most important is the FIR's ability to achieve *linear-phase* behavior. Linear-phase performance is critical to many phase-critical applications, such as:

- Digital communication systems
- Phase synchronization systems
- Image processing
- Spectral analysis (for example, Fourier analysis)

An Nth-order FIR filter, having a transfer function like that in Equation (13-3), has a frequency response given by:

$$H(e^{j\varpi}) = \left|H(e^{j\varpi})\right| e^{j\theta(\varpi)} \tag{13-8}$$

where ϖ is a normalized frequency over the range $\varpi \in [-\pi, \pi)$. It's claimed that if an FIR has a symmetric impulse response:

$$h_k = \pm h_{N-1-k} \tag{13-9}$$

then the filter will exhibit linear-phase behavior (for example, $\theta(\varpi) = \alpha\varpi + \beta$). To illustrate, consider the design of a Type 1 FIR (see Table 13-1) having an odd order and even coefficient symmetry ($h_k = h_{N-1-k}$). The center tap coefficient of $H(z)$, shown in Equation (13-1), is located at sample $k = M$ where $N = 2M + 1$. To simplify the analysis of a symmetric FIR, the location of the center tap coefficient can be shifted to $k = 0$ by re-defining the transfer function to be $H(z) = z^{-M}H_0(z)$, $M = (N-1)/2$. It then follows that from Euler's equation:

$$\begin{aligned} H(e^{j\varpi}) &= e^{-jM\varpi}H_0(e^{j\varpi}) = e^{-jM\varpi}\left(h_0 + \sum_{m=1}^{M}(h_m e^{-jm\varpi} + h_{-m}e^{jm\varpi})\right) \\ &= e^{-jM\varpi}\left(h_0 + \sum_{m=1}^{M} 2h_m \cos(m\varpi)\right) = e^{-jM\varpi}C(\varpi) \end{aligned} \tag{13-10}$$

$C(\varpi)$ is a *real* function of ϖ. Therefore, the phase response of a Type 1 FIR is given by:

$$\angle H(e^{j\varpi}) = \phi(\varpi) = -M\varpi. \tag{13-11}$$

In general, the resulting phase response of an FIR having symmetrically distributed coefficients satisfies the linear equation:

$$\theta(\varpi) = -\alpha\varpi + \beta \tag{13-12}$$

The possible values of α and β are summarized in Table 13-1 for various filter orders (N) and symmetry (odd/even) conditions.

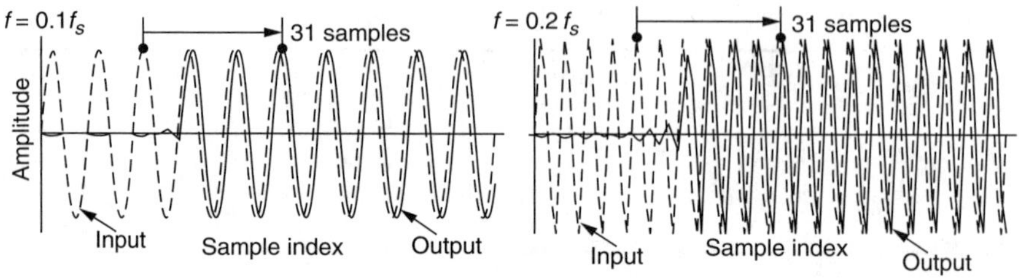

FIGURE 13-4 Input-output response of a 63rd-order Type-1 FIR to sinusoidal inputs at frequencies $f = 0.1f_s$ and $0.2f_s$. The output response is seen to lag behind the input by $\tau = 31$ samples in both cases (the input signal is shown as a dashed line; output response is displayed as a solid line).

Group phase delay, or simply *group delay,* is a standard means of expressing the linearity of a filter's phase response. Group delay, denoted τ, is mathematically defined to be

$$\tau = -d(\theta(\varpi))/d\varpi \tag{13-13}$$

The linear phase of a filter, having a phase response given by Equation (13-12), has a group delay derived from

$$\tau = -d(\theta(\varpi))/d\varpi = -\alpha \tag{13-14}$$

where $-\alpha = M$ as found in Table 13-1. It's measured in samples. An interpretation of group delay is provided by the next example.

Example 13-3 Group delay

Required:
A 63rd-order Type-1 FIR has a theoretical group delay of $\tau = 31$ samples. The physical implication is that a sinusoidal signal's propagation delay through the filter is a constant 31 samples, or $31 \cdot T_s$ seconds regardless of the input signal's frequency. Test this thesis experimentally using test signals at frequencies $f = 0.1f_s$ and $0.2f_s$.

Results:
The convolution results are reported in Figure 13-4. It can be seen that there is a measurable $\tau = 31$ sample difference between common points on the input and output waveforms. This constant propagation delay allows a linear-phase FIR to maintain synchronization between a transmitter and receiver across the frequency spectrum.

13.5 NON–LINEAR-PHASE BEHAVIOR

There are some FIR classes that naturally have a non-symmetric impulse response (for example, lattice, adaptive filters). There are also times when maintaining phase linearity is not an issue, but FIR complexity is an overriding issue. In such cases, non-linear phase can become a design option of choice compared to linear-phase FIRs. It can be noted that only $N/2$ of an Nth-order linear-phase FIR coefficients are actual design parameters since the other $N/2$ is used to impose the symmetry requirement. This condition over-specifies the design constraints imposed on the coefficients. The same magnitude frequency response can be achieved by an $N/2$-order FIR if the symmetry requirements are removed.

13.6 MINIMUM PHASE BEHAVIOR

Minimum-phase filters, also called *minimum delay* filters, have smaller propagation delays than linear-phase filters having similar magnitude frequency response. A minimum-phase filter, however, has a non–linear-phase response. Another point of differentiation is found in the shape of the impulse response. A linear-phase FIR has its largest tap weight coefficients clustered near the center of the impulse response, whereas a minimum phase filter's largest coefficients are located at, or near the beginning of the impulse response. This translates into the fact that a minimum phase filter can respond faster to an external force (for example, step input) than a linear-phase filter.

The zeros of a minimum phase FIR are all interior to, or on the unit circle. That is, the magnitude of the zeros of a minimum phase FIR's filter are all bounded by unity (for instance, $|z_i| \leq 1.0$). A minimum-phase FIR $H_{mp}(z)$ possessing the magnitude frequency response of a non-minimum phase $H(z)$ can be realized by reflecting all the zeros of $H(z)$ that are external to the unit circle to interior locations. If a zero of z_i of $H(z)$ is outside the unit circle (such as $|z_i| > 1$), then it is moved to an interior location using $1/z_i$ (say, $|1/z_i| < 1$). Minimum phase FIRs can be designed using MATLAB's generalized Remez function (*firgr*) contained in Mathwork's Filter Design Toolbox. The design analysis of a minimum phase FIR is presented in the next example.

Example 13-4 Minimum phase FIR

Required:

A 31st-order FIR has an impulse response $h_k = \{1, 1.98, 2.9, 3.725, 4.424, 4.97, 5.349, 5.552, 5.582, 5.449. 5.175, 4.787, 4.322, 3.818, 3.318, 2.867, 2.325, 1.852, 1.448, 1.11, 0.833, 0.61, 0.436, 0.303, 0.203, 0.132, 0.082, 0.048, 0.027, 0.014, 0.006\}$ as shown in Figure 13-5. Classify the FIR as being minimum or non-minimum phase.

Results:

The zeros of the FIR filter are the roots of the polynomial $H(z) = h_0 + h_1 z^{-1} + \cdots + h_{30} z^{-30} = 1 + 1.98z^{-1} + \cdots + 0.006z^{-30}$. The zeros are displayed in Figure 13-5. Since all zeros satisfy the condition $|z_i| \leq 1$, the filter is minimum phase. In addition, the minimum phase FIR has its strongest tap weights in the early stage of the impulse response. This is different from the center clustering associated with the linear-phase filter impulse response exemplified in Figure 13-1.

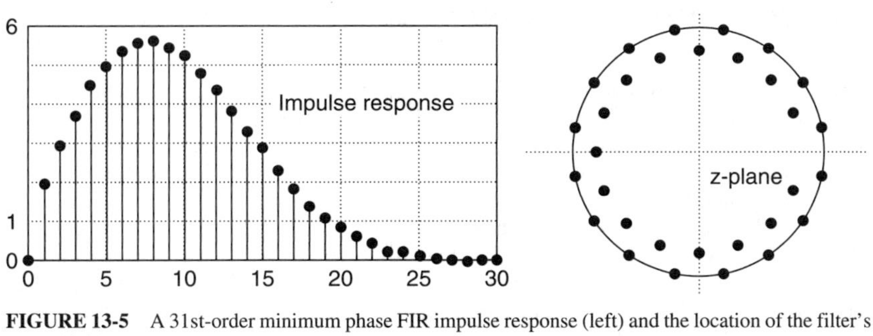

FIGURE 13-5 A 31st-order minimum phase FIR impulse response (left) and the location of the filter's zeros (right).

13.7 FIR DESIGN METHODS

Some basic FIRs have a fixed time or frequency domain function and can be deterministically defined (for example, a raised cosine FIR) once a few key parameters are established. These filters are called *direct design* FIRs. Normally, FIRs are designed to meet a set of pre-specified frequency domain specifications defined with respect to the normalized frequency range $\varpi \in [-\pi, \pi]$, where $\varpi = \pi$ corresponds to the filter's Nyquist frequency. Without constraints, it is assumed that the *desired* FIR's magnitude frequency response is given by $|H_d(e^{j\varpi})|$ and the phase response by $\angle H_d(e^{j\varpi})$. The design objective is to synthesize a physically realizable FIR $H(e^{j\varpi})$ of constrained order (for instance, order $\leq N$) that closely approximates the desired frequency response in some acceptable manner (*i.e.*, $H(e^{j\varpi}) \sim |H_d(e^{j\varpi})|\angle H_d(e^{j\varpi})$). The metrics of comparison are generally defined in terms of an error criterion defined in the frequency-domain, namely:

$$e(\varpi) = H(e^{j\varpi}) - H_d(e^{j\varpi}) \tag{13-15}$$

Classic FIR design strategies are classified as the *window*, *least squares*, and *minimax* design methods. The window and least squares method is based on optimizing an FIR design using a *minimum squared error* (MSE) criterion. The MSE criterion states that the optimum filter minimizes

$$\phi(\varpi) = \sum_{\forall\varpi} e(\varpi)^2 \tag{13-16}$$

The minimax FIR design strategy constructs an FIR whose frequency response satisfies the *minimax error* criterion:

$$\text{minimize}(\text{maximum}(|\varepsilon(\varpi)|)); \quad \forall\varpi \tag{13-17}$$

Both design strategies are shown in Figure 13-6. The advantage of the MSE method is that a solution is generally easy to compute. The shortcoming of the MSE method is that local errors can be large, even though the overall MSE error is small. The minimax error criterion is usually more difficult to satisfy but will guarantee that the worst case error has been reduced to a minimal value. In general, MSE and minimax outcomes are different. In some instances, an MSE design can actually achieve a smaller maximal passband and stopband error than attained by a minimax design. The relative performance of an MSE and minimax FIR can only be compared based on actual design outcomes.

13.8 WINDOW DESIGN METHOD

A straightforward FIR synthesis technique is called the *window method*. It is based on the use of an *inverse discrete Fourier transform* (IDFT) to produce a database that defines a physically realizable Nth-order FIR. In practice, a long *inverse fast Fourier transform* (IFFT) is used to implement the requisite IDFT. It is known that the Fourier transform produces an MSE approximation to the desired FIR response in the frequency domain. The method is called the window method because the M-sample IDFT database is "gated" by an N-sample rectangular data window. Formally, the window design method begins with the computation of an M-sample ($M \gg N$) IDFT of a specified desired magnitude frequency response, $|H_d(e^{j\varpi})|$, and phase response, $\angle H_d(e^{j\varpi})$. The impulse response of the realized N-sample FIR is then extracted from the M-point IDFT. The design of an Nth-order

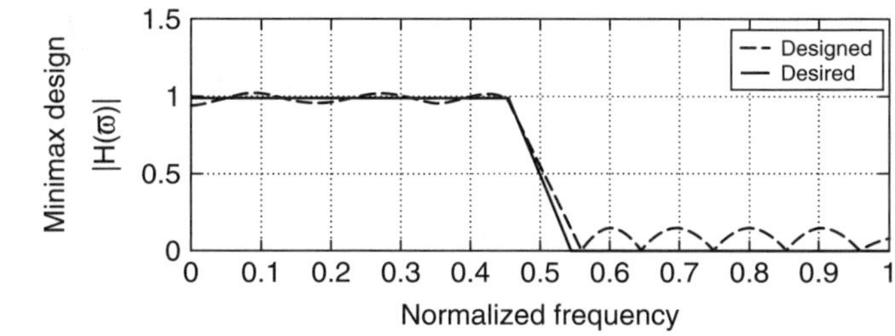

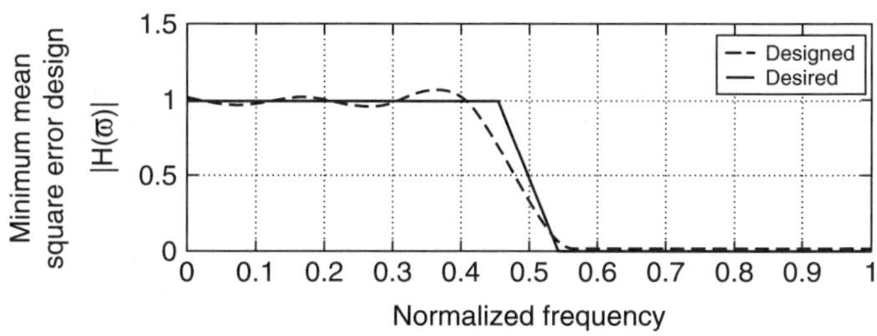

FIGURE 13-6 Common FIR design strategies. Minimax (top), and least mean squares (bottom). The desired response is shown as solid lines, while the realized response is shown as dashed lines. (The frequency axis was normalized with respect to $f_s/2$.)

linear-phase FIR, using the window method, is shown in Figure 13-7. If the desired phase response is linear, then the realized Nth-order FIR is likewise linear.

The design an Nth-order linear-phase FIR Type-1 filter presumes that:

$$\phi(\varpi) = \alpha\varpi + \beta = -\tau\varpi + \beta = -((N-1)/2)\varpi + \beta \qquad (13\text{-}18)$$

where τ is the FIR's group delay. The Type-1 FIR is defined by a complex frequency response that can be expressed as:

$$H(e^{j\varpi}) = |H(e^{j\varpi})|\,(\cos(\phi(\varpi)) + j\sin(\phi(\varpi))) \qquad (13\text{-}19)$$

Equation (13-19) is evaluated at M discrete frequency locations $\varpi_i = i2\pi/M$, $i \in [-M/2, M/2)$ and then presented to an M-harmonic IDFT where $M \gg N$. Upon performing an IDFT, N samples are extracted from the center portion of the M-sample database and used to construct an Nth-order linear-phase FIR. This strategy is shown in the next example. Type 2, 3, and 4 linear-phase FIRs, as well as non–linear-phase FIRs, are simple extensions of the Type-1 design strategy.

Example 13-5 *Window design method*

Required:

Design a Type-1 21st-order linear-phase FIR using the window method. The FIR is to have a desired magnitude frequency response given by:

Desired passband response: $|H_d(e^{j\varpi})| = 1$ for $0 \leq \varpi \leq 0.4\pi$

Desired transition band response: $|H_d(e^{j\varpi})| = 1 - (10/\pi)(\varpi - 0.4\pi)$ for $0.4\pi \leq \varpi \leq 0.5\pi$

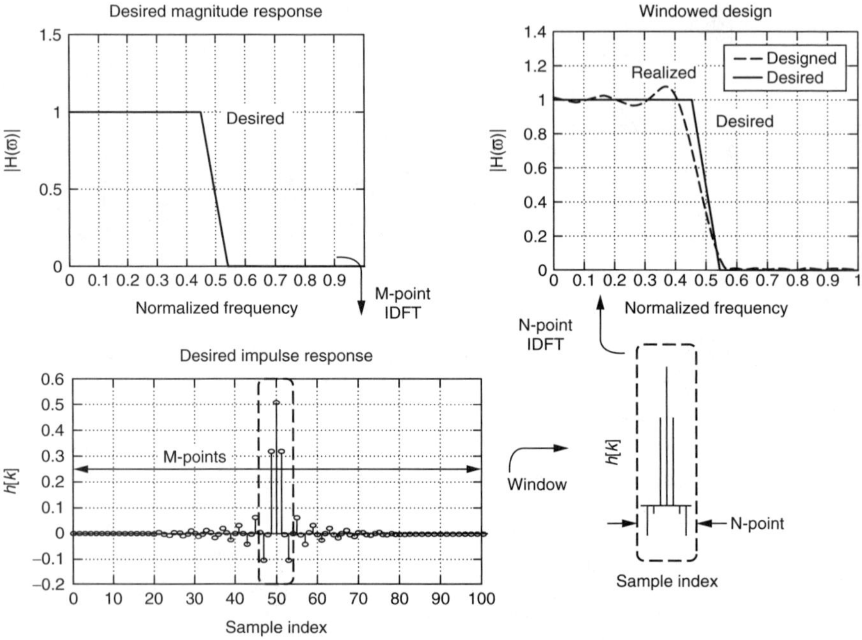

FIGURE 13-7 The direct synthesis method ($M \gg N$). (The frequency axis was normalized with respect to $f_s/2$.)

Desired stopband response: $|H_d(e^{j\varpi})| = 0$ for $0.5\pi \leq \varpi$

Desired phase response: $\phi(\varpi) = -\tau\varpi = -10\varpi$

Results:

The desired piecewise constant magnitude frequency response is first mapped into the time domain using a 1024-point IDFT. The center 21 IDFT sample values, shown in Figure 13-8, are taken for the 1024-point IDFT. The realized magnitude frequency response is seen to be a good approximation of the desired magnitude frequency response. The maximum error occurs at the boundaries of the transition band and is attributable to *Gibbs Phenomenon*. The phase response is linear and the filter's group delay is ten samples.

13.9 NON-RECTANGULAR WINDOW DESIGN METHOD

A rectangular window, also referred to as a rectangular gating function, can introduce a filter error attributable to Gibbs' phenomenon. Data smoothing windows have been developed that can suppress this type of error. Some of the popular fixed coefficient windows that meet this criteria are listed in Table 13-2 (rectangular included for completeness) and are graphically interpreted in the time and frequency domain in Figure 13-9. It should be noted that there are scores of distinct smoothing windows appearing in the literature.

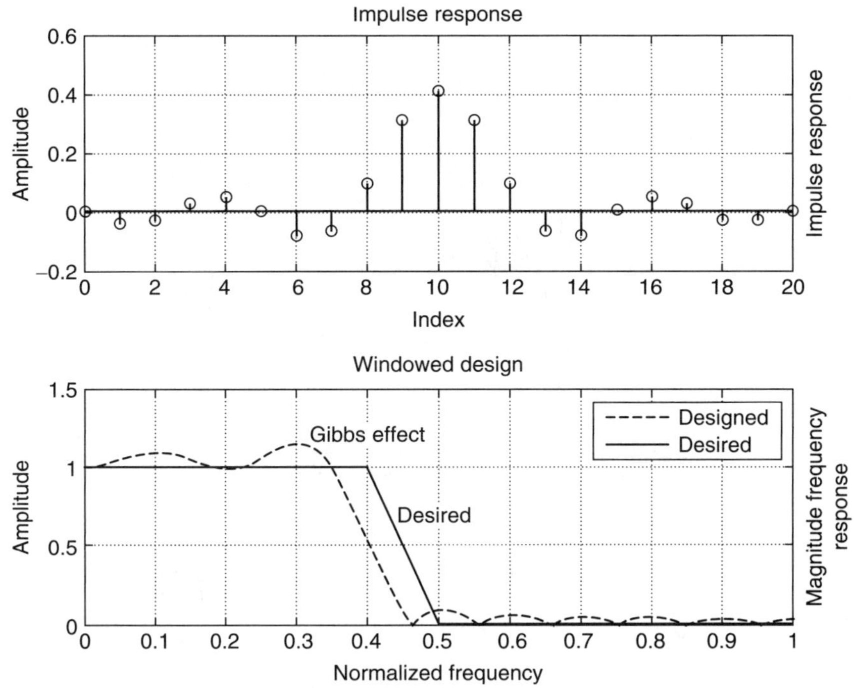

FIGURE 13-8 The window design of a desired Type-I low-pass filter. Shown are the impulse response (top), and the realized magnitude frequency response (bottom). (The frequency axis was normalized with respect to $f_s/2$.)

TABLE 13-2 A Summary of Commonly Used Fixed Window Functions

Window Name	Window Function
Rectangular	$w[k] = 1$ for $k \in [0, N-1]$
Bartlett (triangular)	$w[k] = \begin{cases} \dfrac{2k}{N-1} & \text{for } k \in \left[0, \dfrac{N-1}{2}\right] \\ 2 - \dfrac{2k}{N-1} & \text{for } k \in \left[\dfrac{N-1}{2}, N-1\right]. \end{cases}$
Hann[1]	$w[k] = \dfrac{1}{2}\left(1 - \cos\left(\dfrac{2\pi k}{N-1}\right)\right)$ for $k \in [0, N-1]$
Hamming	$w[k] = 0.54 - 0.46\cos\left(\dfrac{2\pi k}{N-1}\right)$ for $k \in [0, N-1]$
Blackman	$w[k] = 0.42 - 0.5\cos\left(\dfrac{2\pi k}{N-1}\right) + 0.08\cos\left(\dfrac{4\pi k}{N-1}\right)$ for $k \in [0, N-1]$
Flat Top	$w[k] = 0.28106 - 0.520987\cos\left(\dfrac{2\pi k}{N-1}\right) + 0.19804\cos\left(\dfrac{4\pi k}{N-1}\right)$ for $k \in [0, N-1]$

[1]The Hann window is named after Austrian meteorologist Julius Van Hann. It is commonly, and mistakenly referred to as a "Hanning" window.

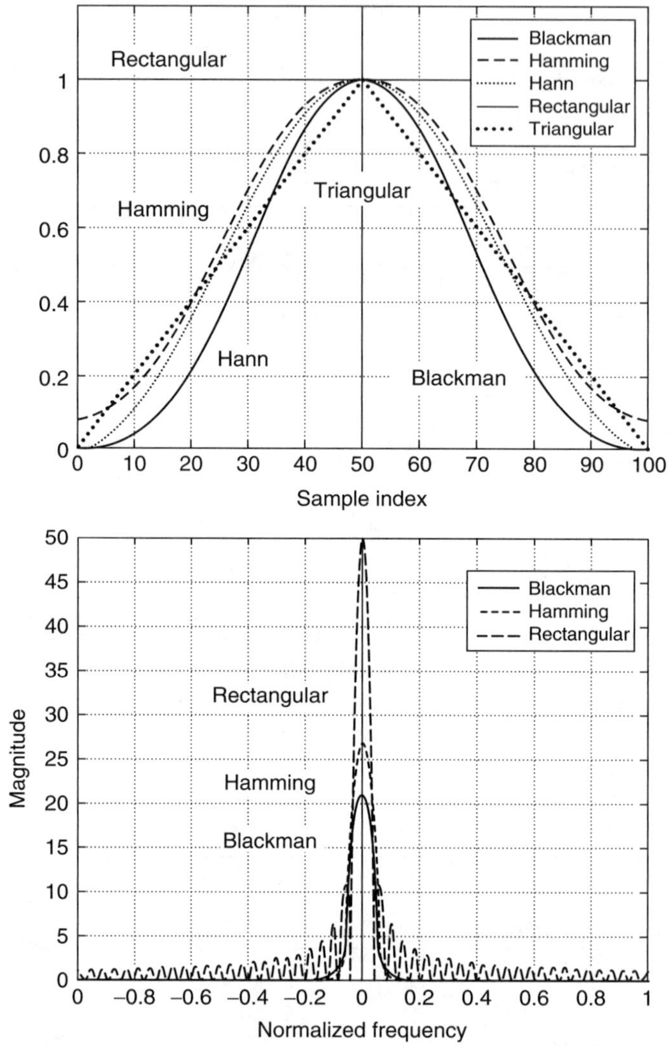

FIGURE 13-9 Examples of selected discrete-time window functions (bottom) and their dB magnitude frequency response (top). (The frequency axis was normalized with respect to $f_s/2$.)

The non-parameterizable windows found in Mathwork's MATLAB Signal Processing Toolbox are

The Modified Bartlett-Hann window
The Bartlett window
The Blackman window
The Four-term Blackman window
The Bohman window
The Flat-Top window
The Gaussian window
The Hanning window
The Hann window
The Nuttall window
The Parzen window
The Rectangular window
The Triangular window
The Tukey window

All of the tabled windows, except the rectangular window, are seen to have sample values at zero or near zero at the window's extremities (for example, sample $k = 0$ and $k = N - 1$).

As a result, any type of discontinuity that may have existed at the ends of a rectangular windowed time-series will be suppressed. The result is a "smoother" impulse response and attendant FIR frequency response.

There are also several useful parameterizable windows that augment the fixed coefficient windows presented in Table 13-2. One is called the *Kaiser window* (*kaiser* in MATLAB), and is given by

$$w[k] = \frac{I_0(\beta\sqrt{1-(k/M)^2})}{I_0(\beta)}; \quad -M \le k \le M \tag{13-20}$$

where $I_0(\beta)$ is the modified 0th-order Bessel function

$$I_0(x) = 1 + \sum_{n=1}^{\infty}\left(\frac{(x/2)^n}{n!}\right)^2 \tag{13-21}$$

The parameter β affects the main lobe bandwidth and stopband attenuation trade-off. A Kaiser window is reported in the time and frequency domain, for $\beta = 1, 3,$ and 10, in Figure 13-10.

Another parameterizable window is the Dolph-Chebyshev (*chebwin* in MATLAB) window, given by

$$w[k] = \frac{1}{2M+1}\left(\frac{1}{\gamma} + 2\sum_{r=1}^{M} T_r\left(\beta\cos\left(\frac{r\pi}{2M+1}\right)\right)\cos\left(\frac{2kr\pi}{2M+1}\right)\right); \quad -M \le k \le M \tag{13-22}$$

where

$$\gamma = \frac{\textit{Amplitude of side lobe}}{\textit{Amplitude of main lobe}}; \quad \beta = \cosh\left(\frac{1}{2M}\cosh^{-1}\left(\frac{1}{\gamma}\right)\right) \tag{13-23}$$

and T_k is a kth-order Chebyshev polynomial satisfying

$$T_r(x) = \begin{cases} \cos(r\cos^{-1}(x)); & |x| \le 1 \\ \cos(r\cosh^{-1}(x)); & |x| > 1 \end{cases} \tag{13-24}$$

The MATLAB defined window functions (Table 13-2, *kaiser,* and *chebwin*) can be applied to any symmetric (linear-phase) FIR impulse response. Since the windows are themselves symmetric, a linear-phase FIR will remain linear phase. An example of a window designed FIR is presented in the following example.

Example 13-6 *Window design method*

Required:

Consider again the rectangular windowed 21st-order linear-phase Type I-FIR previously introduced in Example 13-5 and Figure 13-8. Apply a 21-sample Hamming window to the FIR impulse response and compare the outcome to the rectangular windowed response.

Results:

The outcome is reported in Figure 13-11. Observe that the Hamming window suppresses the large Gibbs overshoot previously located near the transition band edge. The reduction in magnitude frequency response error is seen to be gained at the expense of a widened transition band.

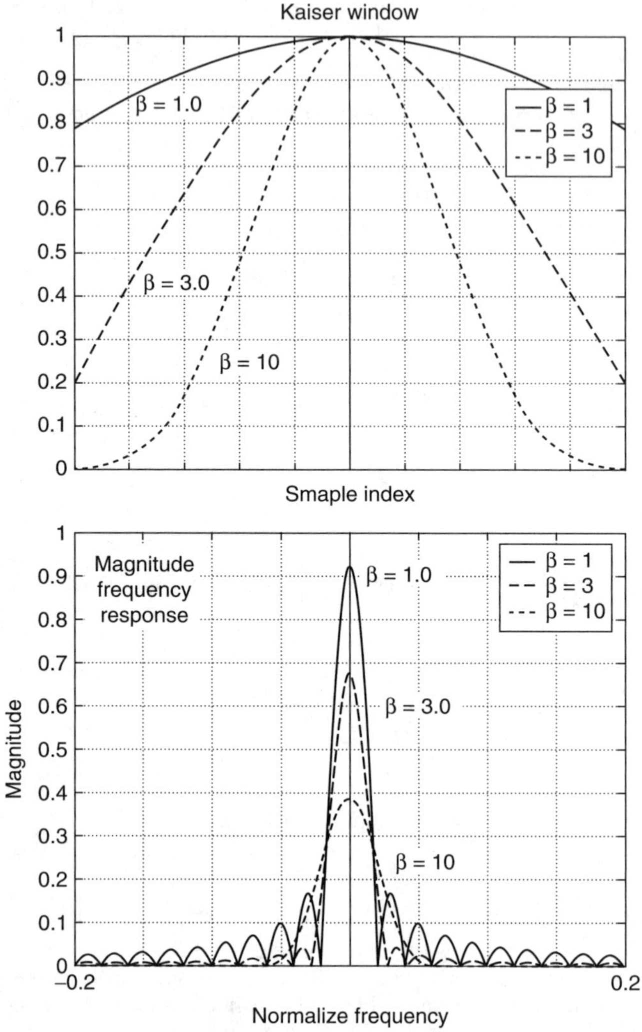

FIGURE 13-10 Kaiser parameterizable window (top) and its magnitude frequency response (bottom). (The frequency axis normalized with respect to $f_s/2$.)

13.10 LEAST SQUARES FIR DESIGN

Classic least squares estimation techniques can also be employed to design an FIR. Consider designing a linear-phase FIR that minimizes the weighted MSE criterion:

$$\sigma = \sum_{i=1}^{K} W(\omega_i)|e(\varpi_i)|^2 = \sum_{i=1}^{K} W(\varpi_i)|H(\varpi_i) - H_d(\varpi_i)|^2 \tag{13-25}$$

where $H(\varpi_i)$ and $H_d(\varpi_i)$ are the realized and desired complex frequency responses, respectively, and $W(\varpi_i) \geq 0$ is a non-negative error weight applied to the ith frequency location

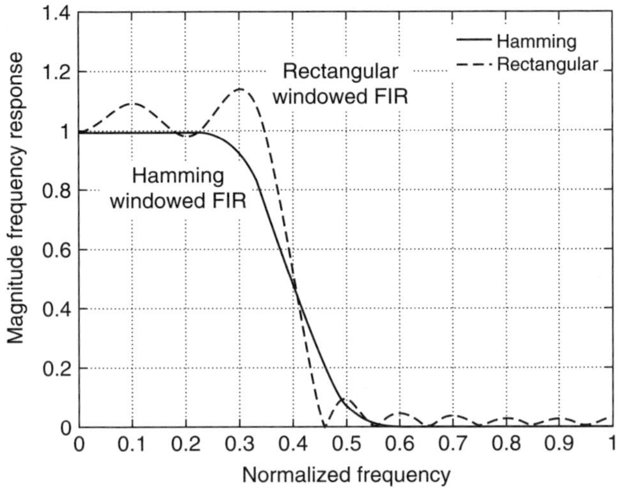

FIGURE 13-11 Comparison of the window FIR design method using the rectangular and Hamming window. (Frequency axis normalized with respect to $f_s/2$).

$\varpi = \varpi_i$. This defines an MSE optimization problem in the frequency-domain. Recall that the (real) magnitude frequency response of a Type-1 linear-phase FIR, modeled after Equation (13-10), is of the form:

$$|H(\varpi)| = \sum_{k=0}^{M} h_k \cos(\varpi k) \tag{13-26}$$

In a matrix-vector framework, the problem can be restated in terms of:

$$\boldsymbol{H} = \begin{bmatrix} W(\varpi_1) & W(\varpi_1)\cos(\varpi_1) & \cdots & W(\varpi_1)\cos(M\varpi_1) \\ W(\varpi_2) & W(\varpi_2)\cos(\varpi_2) & \cdots & W(\varpi_2)\cos(M\varpi_2) \\ \cdots & \cdots & \cdots & \cdots \\ W(\varpi_K) & W(\varpi_K)\cos(\varpi_K) & \cdots & W(\varpi_K)\cos(M\varpi_K) \end{bmatrix};$$
$$\boldsymbol{h} = \begin{bmatrix} h_0 \\ h_1 \\ \cdots \\ h_M \end{bmatrix}; \quad \boldsymbol{d} = \begin{bmatrix} W(\varpi_1)H_d(\varpi_1) \\ W(\varpi_2)H_d(\varpi_2) \\ \cdots \\ W(\varpi_K)H_d(\varpi_K) \end{bmatrix} \tag{13-27}$$

The weighted estimation error would then be

$$\boldsymbol{e} = \boldsymbol{Hh} - \boldsymbol{d}. \tag{13-28}$$

The least squares error estimate is expected to minimize the squared error criteria:

$$\sigma = \boldsymbol{e}^T\boldsymbol{e} = |\boldsymbol{e}|^2 = (\boldsymbol{Hh} - \boldsymbol{d})^T(\boldsymbol{Hh} - \boldsymbol{d}) \geq 0 \tag{13-29}$$

The minimizing coefficient vector h can be obtained by differentiating Equation (13-29) with respect to the parameters of h and setting the derivates to zero, namely:

$$\partial\sigma/\partial\boldsymbol{h} = 0 \tag{13-30}$$

TABLE 13-3 Least Squares FIR Designs

Least Squares FIR #1				Least Squares FIR #2			
K	M	Gain	Coefficients	K	M	Gain	Coefficients
7	5	$H_d(0) = 1$	$h_0 = 0.35$	5	7	$H_d(0) = 1$	$h_0 = 0.49$
		$H_d(2\pi/14) = 1$	$h_1 = 0.569$			$H_d(2\pi/10) = 1$	$h_1 = 0.222$
		$H_d(4\pi/14) = 1$	$h_2 = 0.268$			$H_d(4\pi/10) = 0$	$h_2 = -0.255$
		$H_d(6\pi/14) = 0$	$h_3 = -0.012$			$H_d(6\pi/10) = 0$	$h_3 = -0.348$
		$H_d(8\pi/14) = 0$	$h_4 = -0.167$			$H_d(8\pi/10) = 0$	$h_4 = 0.361$
		$H_d(10\pi/14) = 0$					$h_5 = 0.246$
		$H_d(12\pi/14) = 0$					$h_6 = 0.339$

which results in a solution that is defined by the so-called *normal* equation:

$$\boldsymbol{H}^T\boldsymbol{H}\boldsymbol{h} = \boldsymbol{H}^T\boldsymbol{d} => \boldsymbol{h} = (\boldsymbol{H}^T\boldsymbol{H})^{-1}\,\boldsymbol{H}^T\boldsymbol{d} \tag{13-31}$$

The MATLAB *fircls* function can be used to synthesize a least-squares FIR from a set of discrete frequency domain specifications. The following example illustrates a least squares FIR design.

Example 13-7 Least squares design method

Required:

Design two least squares FIRs having the gain specifications shown in Table 13-3. One design is parameterized by $K = 7$ and $M = 5$, the other by $K = 5$ and $M = 7$. Compare the outcomes.

Results:

The result is reported in Figure 13-12. The system specifications for the case where $K = 7$ and $M = 5$ represents an over-specified solution which lends itself to a successful outcome. The system specifications for the case where $K = 5$ and $M = 7$ represent an under-specified solution which leads to a questionable outcome.

Least squares methods are very robust but, like the window method, can suffer from large localized errors. One of the distinct advantages of the least squares method is that errors in the frequency domain can be selectively weighted by the choice of $W(\varpi_i) \geq 0$. This provides a means of shaping the realized filter's spectral response.

13.11 EQUIRIPPLE FIR DESIGN

The *equiripple* design rule is commonly used to synthesize FIR from a set of frequency domain specifications. The weighted error model for an equiripple design where $\varepsilon(\varpi)$, is

$$\varepsilon(\varpi) = W(\varpi)\,|H_d(e^{j\varpi}) - H(e^{j\varpi})| \tag{13-32}$$

The weighted error is defined in terms of a non-negative *error weight* $W(\varpi) \geq 0$ and the difference between the *desired* ($H_d(e^{j\varpi})$) and *realized* ($H(e^{j\varpi})$) filter's base-band frequency response. Equiripple FIRs meet the *minimax error criteria*, which states that an optimum solution satisfies

$$\delta = \text{minimum}\{\text{maximum}(|\varepsilon(\varpi)|)\}; \quad \varpi \in [0, \varpi_s/2) \tag{13-33}$$

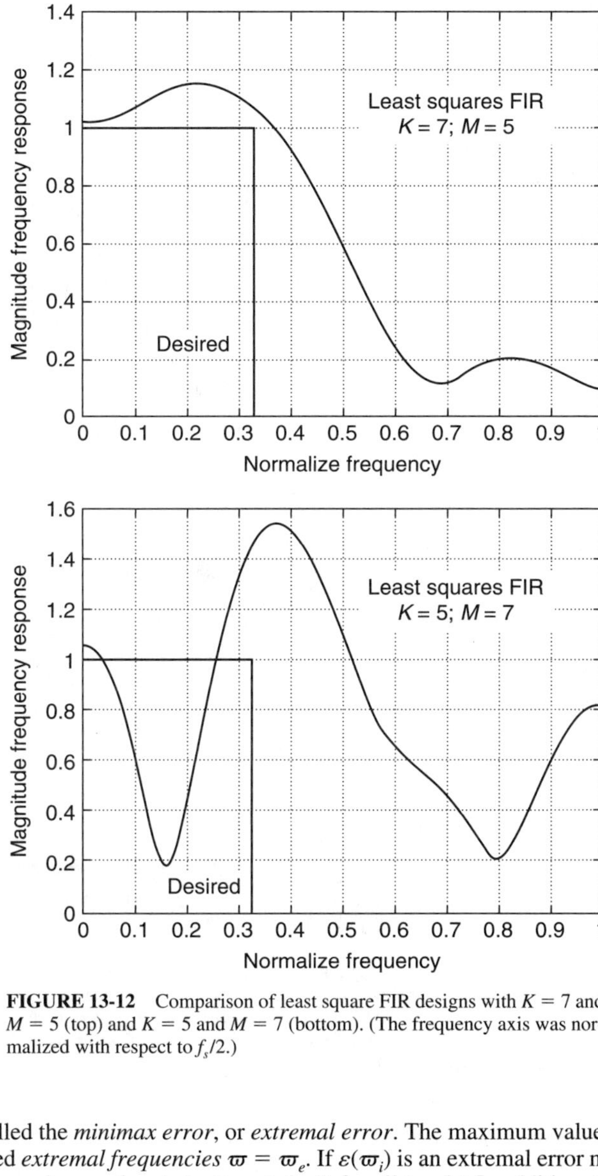

FIGURE 13-12 Comparison of least square FIR designs with $K = 7$ and $M = 5$ (top) and $K = 5$ and $M = 7$ (bottom). (The frequency axis was normalized with respect to $f_s/2$.)

where δ is called the *minimax error*, or *extremal error*. The maximum value of δ occurs at what are called *extremal frequencies* $\varpi = \varpi_e$. If $\varepsilon(\varpi_i)$ is an extremal error measured at the extremal frequencies ϖ_i, then $|\varepsilon(\varpi_i)| = |\varepsilon(\varpi_j)|$ for all i and j. In addition, the signs of the maximal errors are known to alternate from extremal frequency to the next extremal frequency so that $\varepsilon(\varpi_i) = -\varepsilon(\varpi_{i+1})$. Because all the extremal errors are equal in magnitude and alternating in sign, the resulting filter is referred to as an *equiripple FIR*.

For a linear-phase FIR, the location of the maximum errors can be computed using the *alternation theorem* from polynomial approximation theory, implemented using the Remez exchange algorithm. The Remez exchange algorithm iteratively adjusts the iteration's current candidate extremal frequency locations ϖ_i until the minimax criterion is satisfied to

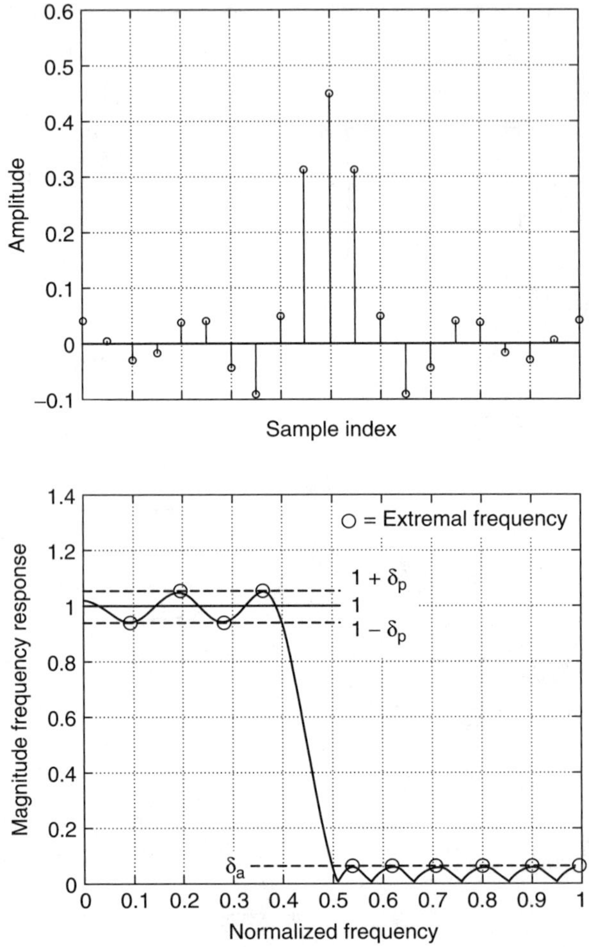

FIGURE 13-13 An equiripple filter approximation to a desired piecewise constant FIR magnitude frequency response showing the impulse response (top) and magnitude frequency response meeting the minimax error criteria at the extremal frequency locations (bottom). (The frequency axis was normalized with respect to $f_s/2$.)

within a fixed tolerance. This method has been used for several decades to design linear-phase FIRs and continues to be reliable.

A typical equiripple FIR approximation of a piecewise constant desired filter model (for example, $H_d(e^{j\varpi})$) is presented in Figure 13-13. The design was conducted using uniform error weights $W(\varpi) = 1$ for all ϖ. The equiripple design objective is shown to be satisfied since the magnitude of the extremal passband and attenuation stopband deviations (errors), denoted δ_p and δ_a respectively, are equal. The filter presented in Figure 13-13, however, demonstrates one of the weaknesses of a uniformly weighted equiripple FIR design. Since the stopband extremal errors (δ_a) are generally small, a uniformly weighted equiripple outcome will also force the passband errors (δ_p) to be likewise small (unrealistically small).

To illustrate, if the equiripple stopband error is -40 dB, then the passband error is likewise -40 dB instead of being a more realistic value of -1 dB or thereabouts. The following example considers the design of a uniformly weighted equiripple FIR.

Finally, the MATLAB *firpm* function can be used to synthesize an equiripple FIR from a set of frequency domain specifications.

Example 13-8 *Uniformly weighted equiripple FIR*

Required:
Design a uniformly weighted 51st-order low-pass equiripple FIR that meets the following specifications

- Sampling frequency $f_s = 100$ kHz:
- Frequency band 1 (passband): $f \in [0.0,10]$ kHz; desired passband gain = 1.0; $W_{\text{passband}} = 1.0$
- Frequency band 2 (stopband): $f \in [15,50]$ kHz; desired stopband gain = 0.0; $W_{\text{stopband}} = 1.0$

Results:
An equiripple FIR was designed and is reported in Figure 13-14. The impulse response has the even symmetry that characterizes a Type-1 linear-phase FIR. The equiripple FIR has an exhibited minimum stopband gain of -47.4-dB passband, and the passband error deviation is likewise -47.4 dB or $|\delta_p| = 0.00428$. Finally, the worst case filter gain (Equation 13-7) is computed to be $G_{\max} = 1.67565 < 2^1$.

The uniform weighting strategy, $W_{\text{passband}}(\varpi) = W_{\text{stopband}}(\varpi)$, results in equal passband and stopband extremal errors (for example, $\delta = \delta_p = \delta_a$). The relationship between δ_p and δ_a can, however, be controlled by choosing dissimilar non-negative *error weights* (for instance, $W_{\text{passband}}(\varpi) \neq W_{\text{stopband}}(\varpi)$). This provides a means of realistically controlling the individual passband and stopband gains. Suppose, for illustrative purposes, that 10 $W_{\text{passband}}(\varpi) = W_{\text{stopband}}(\varpi)$. This artificially makes the stopband extremal error 10 times more serious than the passband error. The resulting solution would exhibit a final error deviation of $\delta_p = 10\delta_a$. In this manner, significant subtle adjustments can be made to the magnitude frequency response envelope of an equiripple design. These relationships can be predicted since the error deviations satisfy

$$\delta = \sqrt{\delta_p \delta_a} \tag{13-34}$$

where δ is the error obtained for a uniformly weighted design (for example, $W(\varpi) = 1$ for all ϖ). The design of a non-uniformly weighted FIR is presented in the following example.

Example 13-9 Non-uniformly weighted equiripple FIR

Required:
Design a non-uniform weighted 51st-order low-pass equiripple FIR that meets the specifications defined in Example 13-8. Example 13-8 considered a uniformly weighted equiripple FIR design for $W_{\text{passband}}(\varpi) = W_{\text{stopband}}(\varpi) = 1.0$, which resulted in a measured stop-band gain that was -47.4 dB and an identical passband error deviation of -47.4 dB, or $\delta = \pm 0.00428$. The passband gain is therefore 1 ± 0.00428, a value that is often considered to be overly restrictive and should therefore be relaxed. A passband deviation on the order of ± 0.1, or -20 dB, may be considered to be more realistic. Explore relaxing the passband gain using error weights.

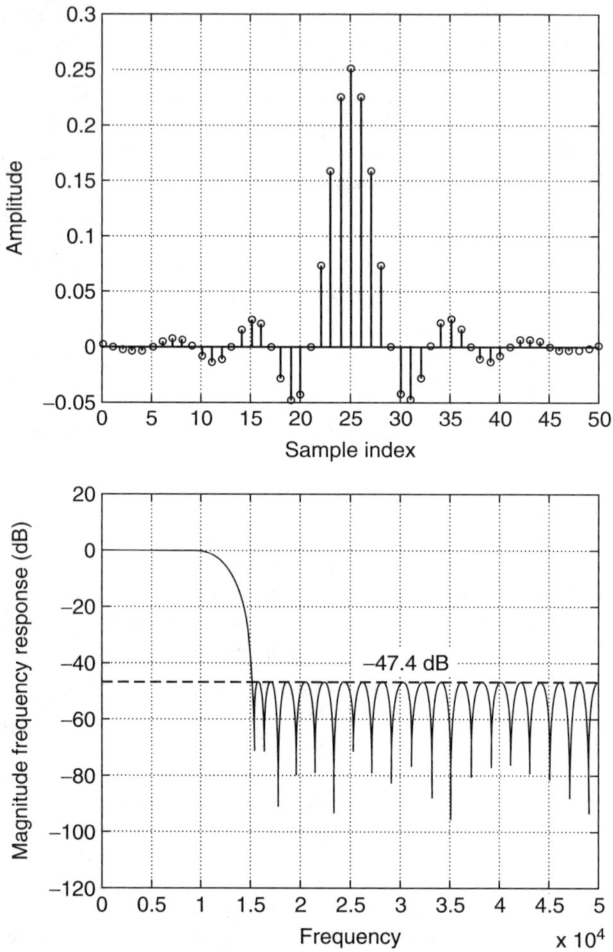

FIGURE 13-14 A uniformly weighted low-pass equiripple Type-1 FIR: impulse response (top), and magnitude frequency response in dB units (bottom).

Results:

Relaxing the passband error deviation and increasing the stopband attenuation can be achieved by assigning the stopband error weight to be more significant than the passband errors. To illustrate, consider setting the stopband error weight to be 1,000 times greater than the passband weight (for example, 1,000 $W_{\text{passband}}(\varpi) = W_{\text{stopband}}(\varpi)$). The equiripple weighted and uniformly weighted FIR design outcomes are reported in Figure 13-15. The weighted design passband deviation is measured to be approximately $\delta_p = \pm 0.0954302 \sim (\pm 0.1)$ and the stopband weight is approximately $\delta_a = -79$ dB. Thus, the relaxation of the passband error deviation facilitated an increase in stopband attenuation by 31.6 dB. That is, $\delta_p \sim 0.1$ and $\delta_a \sim 0.00018$. From this, $\delta = \sqrt{\delta_p \delta_a} = 0.00424264$, which is essentially the value of the uniformly weighted value of δ reported in Example 13-8. The worst case gain of the filter is noted to be $G_{\max} = 1.72627 < 2^1$, which is a slight increase over the uniformly weighted case.

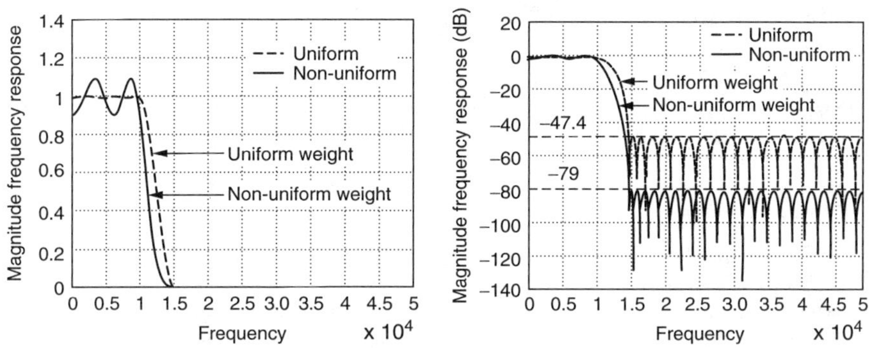

FIGURE 13-15 Uniform and non-uniform weighted low-pass equiripple Type-1 FIRs. Shown are the magnitude frequency response (left), and the magnitude frequency response in dB units (right).

There are several algorithms found in common use that translate the FIR design parameters (δ_p, δ_a, ϖ_p, ϖ_a) into equiripple FIR filter order estimates. The most popular analytic estimation formula is given by

$$N_{\text{FIR}} \approx \frac{-10\log_{10}(\delta_a\delta_p) - 15}{14\Delta\varpi} + 1 \tag{13-35}$$

where $\Delta\varpi$ is the normalized transition bandwidth and is given by $\Delta\varpi = (\varpi_a - \varpi_p)/\varpi_s$. A second estimator is given by

$$N_{\text{FIR}} \approx \frac{-10\log_{10}(\delta_a\delta_p) - 13}{14.6\Delta\varpi} + 1 \tag{13-36}$$

These formulas provide an estimate of an equiripple filter order based on non-uniform error weighting. It should be noted that steep-skirt, or narrow-band equiripple FIRs, having a small $\Delta\varpi$ value, may require a very high-order solution. As a rule-of-thumb, when the normalized transition band $\Delta\varpi$ has a value less than $\Delta\varpi < 0.04$, an equiripple filter is virtually impossible to build. Another problem with high-order FIRs is that the coefficients found out on the tapers (tap weight coefficients found at the extreme ends of the impulse response) have such small values that they cannot be realized within a meaningful fixed-point system. Finally, MATLAB's *firpmord* function can be used to estimate a linear-phase equiripple FIR's order from a listing of frequency domain specifications. An example of order estimation follows.

Example 13-10 Linear phase equiripple FIR order estimation

Required:

Consider again the non-uniformly weighted low-pass equiripple FIR filter studied in Example 13-9, where $\delta_p = 0.1$, $\delta_a = 0.00018$, and $\Delta\varpi = 0.05$. Determine the required Type-1 equiripple filter order.

***Results*:**

An estimate of $N = \lceil 47.3 \rceil = 48$ is obtained from Equation (13-35), and $N = \lceil 48.3 \rceil = 49$ from Equation (13-36). Both estimates are consistent with the measured order of $N = 51$ (Example 13-9).

13.12 EQUIRIPPLE HILBERT FIR DESIGN

Equiripple filters are commonly used in the design of linear-phase low-pass, bandpass, bandstop, and high-pass filters. Another important use of the equiripple paradigm is to construct *Hilbert transforms*. A Hilbert filter has a frequency response given by:

$$H(e^{j\varpi}) = \begin{cases} j: -\pi \leq \varpi < 0 \\ -j: 0 < \varpi \leq \pi \end{cases} \tag{13-37}$$

Observe that a Hilbert filter is essentially an *all-pass filter* in that $|H(e^{j\varpi})| = 1$, for all $\varpi \in [-\pi, \pi)$, but possesses a distinctive *quadrature* phase-shifting property. It is the phase-shifting property that makes the Hilbert filter particularly useful in defining single side-band modulators, quadrature amplitude modulation (QAM) communications systems, and in-phase and quadrature phase (I/Q) channel communication systems. For practical realization reasons, the desired frequency response should be provided guard bands near 0 Hz and the Nyquist frequency. A Hilbert FIR design is presented in the next example.

Example 13-11 *Hilbert FIR design*

Required:
Design a 63rd-order Hilbert FIR having a normalized sample frequency of $f_s = 1$ kHz.

Results:
The machine design of a Hilbert FIR can be improved by adding guard bands at each end of the base band. A design including guard bands is defined in terms of the following specifications:

- Frequency band 1 {guard band}: $f \in [0.0, 0.01]$ kHz; desired gain = 0.0, weight $W(f) = 1$
- Frequency band 2 {passband}: $f \in [0.02, 0.48]$ kHz; desired gain = 1.0, weight $W(f) = 1$
- Frequency band 3 {guard band}: $f \in [0.49, 0.50]$ kHz; desired gain = 0.0, weight $W(f) = 1$

The response of the equiripple Hilbert FIR is reported in Figure 13-16. The FIR passband and stopband deviates from the desired response by an amount $\delta = -15.8039$ dB. The impulse response is also seen to be symmetric, and characteristic of a linear-phase FIR.

13.13 EQUIRIPPLE DIFFERENTIATOR FIR DESIGN

Equiripple filters are also adept at implementing other filter classes, such as a linear-phase *differentiator* of order N. An Nth-order differentiator's frequency response is given by $H(e^{j\varpi}) = (j\varpi)^N$. For practical reasons, the desired frequency response should include a guard band near the Nyquist frequency. That is, $|H(e^{j\varpi})| = 0$ for ϖ near $\varpi = \pi$. The design of a differentiator FIR is motivated in the following example.

Example 13-12 Design of a differentiator FIR

Required:
Design a 51st-order linear-phase 1st-order differentiating filter.

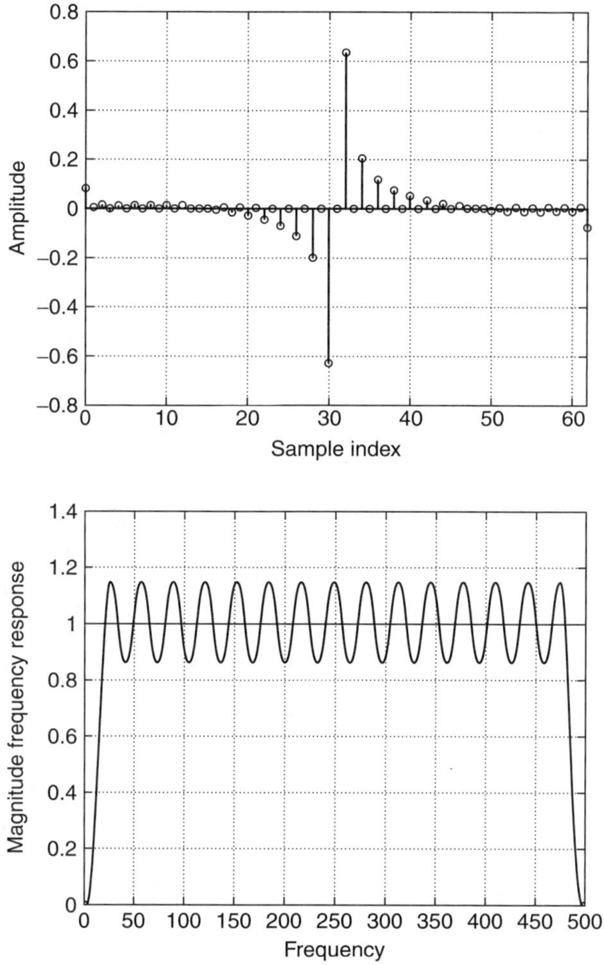

FIGURE 13-16 Hilbert FIR with guard bands: the impulse response (top), and the magnitude frequency response (bottom).

Results:
Create a passband of gain $H(e^{j\varpi}) = (j\varpi)$ out to $0.4f_s$ and a guard band of zero gain over $0.45f_s$ to $0.5f_s$. The results are summarized in Figure 13-17. Notice that the impulse response has an odd symmetry and that the magnitude frequency response supports the differentiation claim out to near $0.4f_s$.

13.14 SPECIAL CASE FIR DIGITAL FILTERS

There are a number of special FIR cases that have high utility over a range of applications. The most important of these are

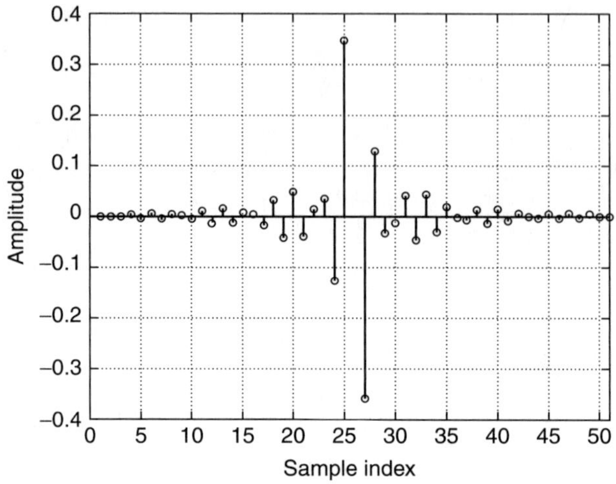

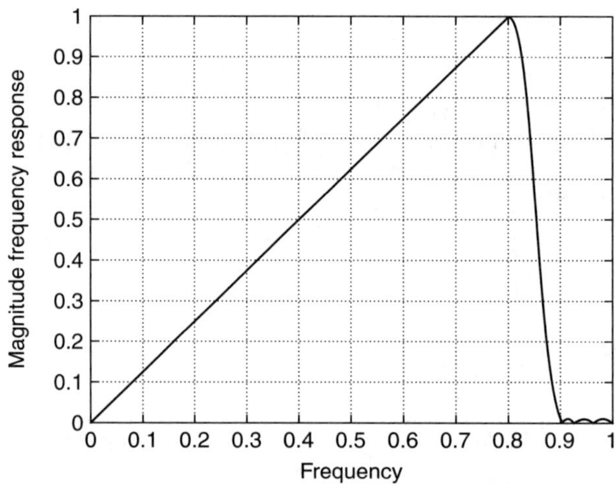

FIGURE 13-17 A differentiator FIR with guard band: An impulse response with odd symmetry (top), and a magnitude frequency response (bottom). (The frequency axis was normalized with respect to $f_s/2$.)

- Multiplier-free moving average and comb filters
- *L*-band filters
- Mirror and complement filters
- Frequency sampling filters
- Stavitzky-Golay filters
- Raised cosine filters

13.15 MULTIPLIER-FREE FIR FILTERS

The linear convolution sum, presented in Equation (13-2), suggests that FIR filtering can be multiply-accumulate intense. A multiplier-free FIR can overcome this obstacle. Such filters would possess only ternary valued coefficients (for instance, $h_i = \{1, 0, -1\}$), which implies that the filter can be implemented using only adders or subtractors. Several important examples of multiplier-free FIRs are the moving average (MA) and comb filters. Their impulse responses are given by:

- An Nth-order *moving average* FIR, having an impulse response:

$$h(n) = \begin{cases} 1 & \text{if } 0 \leq n \leq N-1 \\ 0 & \text{otherwise} \end{cases} \tag{13-38}$$

has a transfer function given by:

$$H_{\text{MA}}(z) = \sum_{i=0}^{N-1} z^{-i} = \frac{(1 - z^{-N})}{(1 - z^{-1})} \tag{13-39}$$

- An Nth-order *comb* FIR, having an impulse response:

$$h(n) = \begin{cases} 1 & \text{if } n = 0 \\ \pm 1 & \text{if } n = N-1 \\ 0 & \text{otherwise} \end{cases} \tag{13-40}$$

has a transfer function given by:

$$H_{\text{comb}}(z) = \left(1 + (\pm 1)z^{-(N-1)}\right) \tag{13-41}$$

The frequency response of a moving average and comb filter are defined by the filter's zeros. For an Nth-order *moving average* FIR, the zeros are defined by the roots of $(1 - z^{-N}) = 0$ and are located at:

$$z_i = e^{j2\pi i/N} \tag{13-42}$$

That is, the N zeros are located on the unit circle separated by $\phi = 2\pi/N$ radians from each other. In addition to these zeros, the moving average transfer function, shown in Equation (13-39), is seen to contain a pole at $z = 1$. However, this pole is canceled by the filter's zero, also located at $z = 1$. The comb filter behaves in a similar manner except for the absence of a canceling pole. The multiplier-free moving average and comb FIRs are illustrated below.

Example 13-13 Moving average and comb FIRs

Required:

Design and simulate a $N = 8$ moving average and comb FIR. Determine the transfer function, zero locations, and frequency response attributes of each FIR case.

Results:

In general, the transfer functions are as follows:

- 8th-order *moving average* FIR:

$$H_{\text{MA}}(z) = \sum_{i=0}^{7} z^{-i} = \frac{(1 - z^{-8})}{(1 - z^{-1})}$$

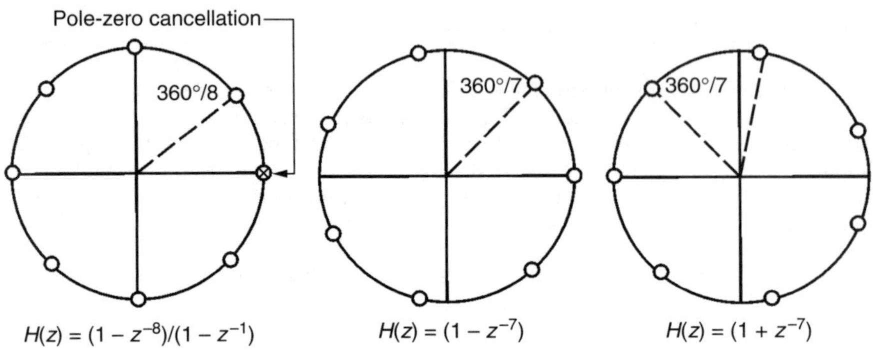

FIGURE 13-18 The pole-zero distribution of an 8th-order moving average (left), an 8th-order comb filter $h_{\text{comb}-}[k]]$ (middle), and an 8th-order comb filter $h_{\text{comb}+}[k]$ (right).

- 8th-order *comb* FIR:

$$H_{\text{comb}+}(z) = (1 + z^{-7});\quad H_{\text{comb}-}(z) = (1 - z^{-7})$$

The pole-zero diagrams for each FIR is displayed in Figure 13-18. The moving average FIR exhibits pole-zero cancellation at $z = 1.0$. The magnitude frequency responses are shown in Figure 13-19. The nulls, or zeros, of the magnitude-frequency response correspond to the location of uncancelled zeros in Figure 13-18. The MA and comb FIR's impulse responses are symmetrically distributed, thus satisfying the linear-phase condition.

13.16 L-BAND FIR FILTERS

An *L*-band *N*th-order FIR is a filter in which every *L*th filter coefficient is zero. An *L*-band FIR, also called a *Nyquist filter*, has a passband that is roughly 1/*L*th the base band frequency range. An *L*-band *N*th-order FIR is therefore less complex than a general *N*th-order FIR due to the increased density of zero coefficients.

A *half-band FIR* is a special FIR *L*-band case which exhibits a magnitude frequency response similar to that shown in Figure 13-20. Observe that frequency response has a point of symmetry in the middle of the base band (in other words, $\varpi = \pi/2$), a point that corresponds to half the Nyquist frequency. The motivation for the half-band FIR is found in the knowledge of the half-wave symmetry property of a Fourier series and duality principle.

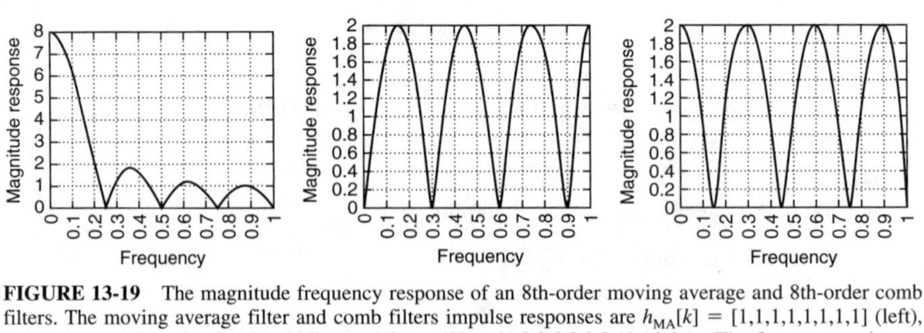

FIGURE 13-19 The magnitude frequency response of an 8th-order moving average and 8th-order comb filters. The moving average filter and comb filters impulse responses are $h_{\text{MA}}[k] = [1,1,1,1,1,1,1,1]$ (left), $h_{\text{comb}-}[k] = [1,0,0,0,0,0,0,-1]$ (middle), and $h_{\text{comb}+}[k] = [1,0,0,0,0,0,0,1]$ (right). (The frequency axis was normalized with respect to $f_s/2$.)

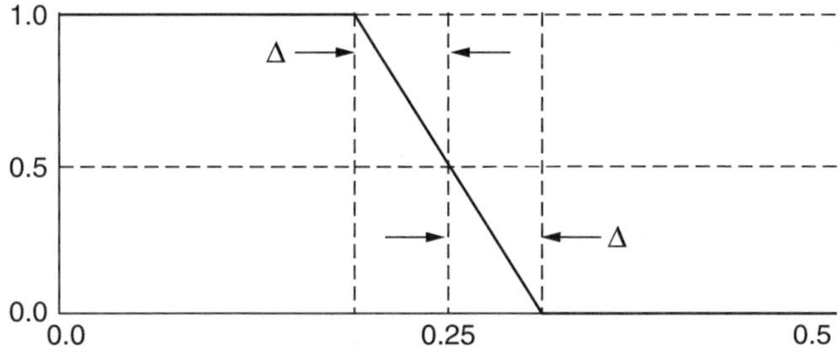

FIGURE 13-20 The desired magnitude frequency response of an Nth-order half-band FIR.

Recall that a periodic signal having half-wave symmetry (say, a periodic pulse-train with a 50-percent duty cycle) possesses a Fourier series consisting only of odd harmonics with the possible exception being the 0th (DC) harmonic. Essentially every other harmonic, or half of all the harmonics, have a value of zero. The duality principal states that the time- and frequency-domain Fourier properties are interchangeable. Symmetry in the frequency domain therefore results in a time-series that has a possible non-zero center tap where every other coefficient has a value of zero, thereby reducing the complexity of a half-band FIR to be essentially half that of a general FIR. The design of a half-band FIR is reported in the next example.

Example 13-14 Half-band FIR

Required:

Consider a 31st-order half-band FIR having the center of the transition band set to $f = 0.25f_s$ with $\pm\Delta f = \pm 0.01f_s$ defining the transition bandwidth. Construct and analyze the resulting half-band FIR.

Results:

An equiripple FIR was designed that meets the stated frequency domain requirements. The FIR has an impulse and magnitude frequency response shown in Figure 13-21. Notice that except for the center tap coefficient, essentially every other tap weight value is zero. As a result, when compared to an arbitrary FIR, the half-band FIR has approximately half the number of multiply-accumulate calls per filter cycle. This translates into a potentially higher real-time filter rate.

$$
\begin{aligned}
h(0) &= -0.1319 = h(30) \\
h(1) &= 0.0000 = h(29) \\
h(2) &= 0.0256 = h(28) \\
h(3) &= 0.0000 = h(27) \\
h(4) &= -0.0289 = h(26) \\
h(5) &= 0.0000 = h(25) \\
h(6) &= 0.0360 = h(24) \\
h(7) &= 0.0000 = h(23) \\
h(8) &= -0.0458 = h(22) \\
h(9) &= 0.0000 = h(21) \\
h(10) &= 0.0640 = h(20) \\
h(11) &= 0.0000 = h(19) \\
h(12) &= -0.1063 = h(18) \\
h(13) &= 0.0000 = h(17) \\
h(14) &= 0.3183 = h(16) \\
h(15) &= 0.5000 = h(15) - \text{center coefficient}
\end{aligned}
$$

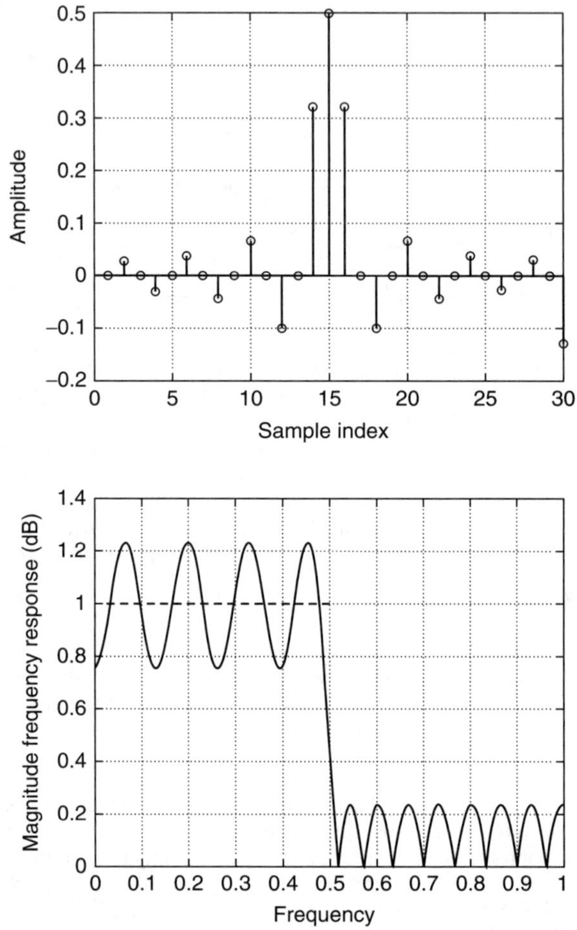

FIGURE 13-21 A 31st-order half-band FIR impulse response (top) and magnitude frequency response (bottom). (The frequency axis was normalized with respect to $f_s/2$.)

13.17 MIRROR AND COMPLEMENT FIR FILTERS

An Nth-order linear-phase FIR can be designed using a host of strategies such as the window, LMS, and equiripple paradigms. In all cases, an original, or parent FIR, has a transfer function $H_{\text{parent}}(z)$ of the form:

$$H_{\text{parent}}(z) = \sum_{k=-L}^{L} h[k]\, z^{-k} \tag{13-43}$$

where $N = 2L + 1$. Based on the parent FIR, other FIRs can be extrapolated. For example, it may be desired to create a new filter that mirrors the frequency response of the parent filter. Assuming that the parent FIR is a low pass filter with a normalized transition bandwidth Δ relative to the sampling frequency, a *mirror* version of the parent would be a high-pass filter having the same normalized transition bandwidth Δ. Mirroring can be achieved by simply modulating the impulse response $h[k]$ by a sinusoid $c[k]$ running at the Nyquist frequency, namely $c[k] = \cos(\pi k) = (-1)^k$. This action will translate (heterodyne) the original frequency response, originally centered about DC, to one centered about the Nyquist frequency. This can be expressed as:

$$h_{\text{mirror}}[k] = (-1)^k h_{\text{parent}}[k] \tag{13-44}$$

That is, the mirror FIR's impulse response is simply that of the parent FIR, but alternating in sign. The spectral relationship between a mirror and its parent filter is graphically interpreted in Figure 13-22.

Another useful filter type is called the *complement filter*. Complement filters can be particularly useful when designing a bank of sub-band filters to cover a range of frequencies where the passband of one does not overlap the passband of the others. A complement FIR is based on an odd order Type-1 unit passband linear-phase parent filter having an impulse

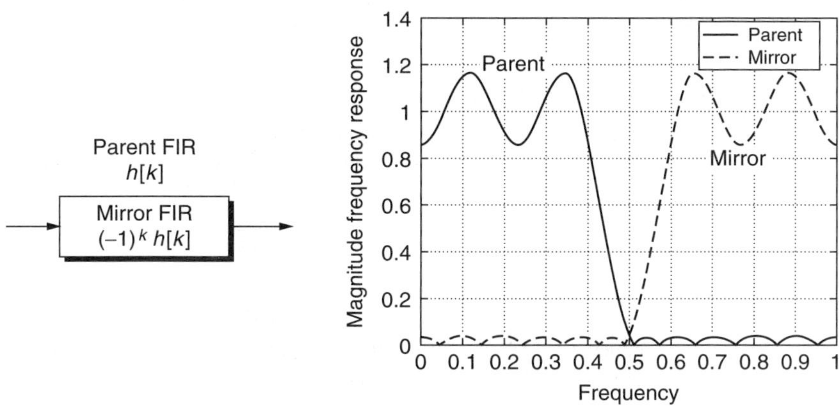

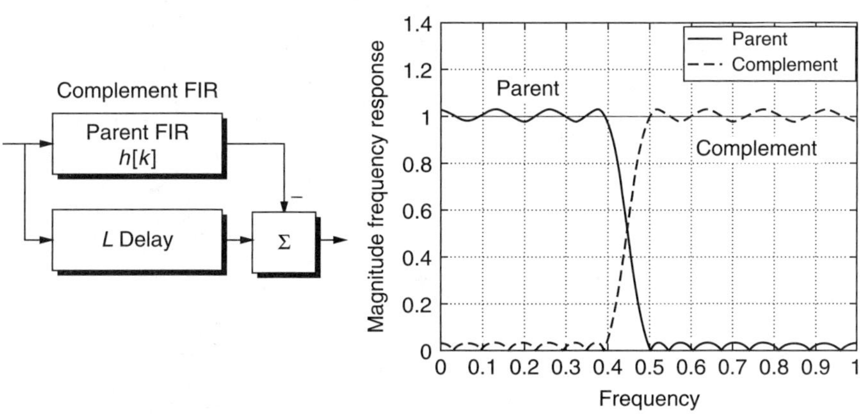

FIGURE 13-22 The relationship of a mirror and complement FIR magnitude frequency response to the parent FIR. (The frequency axis was normalized with respect to $f_s/2$.)

response $h[k]$, defined over $k \in [-L, L]$, and a transfer function given by Equation (13-43). The transfer function of the complement FIR, namely $H_{comp}(e^{j\varpi})$, is defined in terms of that parent filter and satisfies

$$H_{comp}(e^{j\varpi}) + H_{parent}(e^{j\varpi}) = 1 \tag{13-45}$$

That is, when the original and complement filter are combined, a unit gain all-pass filter will result. Equation (13-45) can be rearranged to define the complement filter's transfer function to be

$$H_{comp}(e^{j\varpi}) = 1 - H_{parent}(e^{j\varpi}) \tag{13-46}$$

Equation (13-46) states that the center tap coefficient of the complement FIR equals $h_{comp}[0] = 1 - h_{parent}[0]$. The other complement filter coefficients are the negative of the parent FIR's tap weights. If the parent FIR impulse response begins at $k = 0$, then Equation 13-46 is modified to read

$$H_{comp}(z) = z^{-L} - H(z) \tag{13-47}$$

Equation (13-47) states that the complement filter can be obtained from the parent with the addition of a length L-delay shift register (L = group delay). This action is interpreted in Figure 13-22. The design of a mirror and complement filter is addressed in the following example.

Example 13-15 Mirror and complement FIR

Consider a linear-phase 5th-order moving average (MA) parent FIR having an impulse response $h_{parent}[k] = \{1/5, 1/5, 1/5, 1/5, 1/5\}$. The mirror and complement versions are given by

$$\begin{aligned} h_{mirror}[k] &= \{1/5, -1/5, 1/5, -1/5, 1/5\} \Leftrightarrow H_{mirror}(z) \\ &= 1/5 - 1/5z^{-1} + 1/5z^{-2} - 1/5z^{-3} + 1/5z^{-4} \\ h_{comp}[k] &= \{-1/5, -1/5, 4/5, -1/5, -1/5\} \Leftrightarrow H_{mirror}(z) \\ &= -1/5 - 1/5z^{-1} + 4/5z^{-2} - 1/5z^{-3} - 1/5z^{-4} \end{aligned}$$

The magnitude frequency response of the 5th-order parent MA filter, along with the mirror and complement versions, are displayed in Figure 13-23.

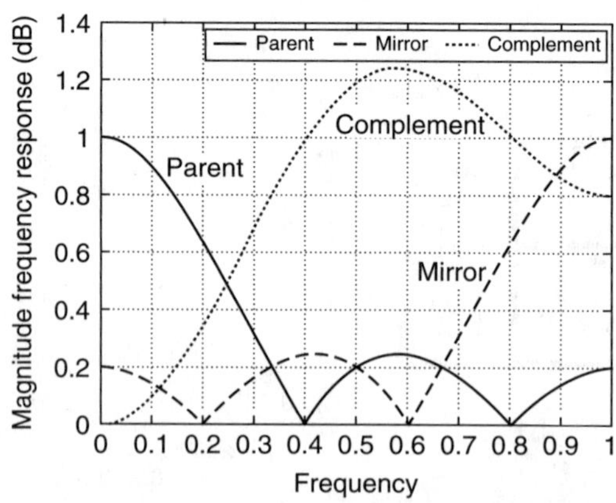

FIGURE 13-23 The magnitude frequency response of the parent FIR, mirror, and complement. (The frequency axis was normalized with respect to $f_s/2$.)

13.18 FREQUENCY SAMPLING FIR FILTERS

An intuitive FIR design strategy, called the *frequency sampling* method, can be used to synthesize an FIR having a frequency response that approximates an arbitrary desired response $H_d(e^{j\varpi})$. An Nth-order frequency sampling filter can be envisioned as a collection of passband filters centered about specific normalized frequencies $\varpi_n = 2\pi n/N$, which collectively has a frequency response:

$$H(e^{j\varpi}) \approx H_d(e^{j\varpi}) = A(\varpi)\angle\phi(\varpi) \tag{13-48}$$

where $|H_d(e^{j\varpi})| = A(\varpi)$ and $\angle H_d(e^{j\varpi}) = \angle\phi(\varpi)$. Furthermore, it is expected that the phase response is linear, satisfying $\phi(\varpi) = -\tau\varpi$ for a Type-1 FIR with τ denoting the FIR's group delay. For notational convenience, let the desired frequency response measured at ϖ_n be denoted $H_d[n] = H_d(e^{j(2\pi n/N)})$. The impulse response of the desired Nth order FIR can be computed using an N-point IDFT of the desired filter specifications, namely:

$$h[k] = \text{IDFT}(H_d(e^{j\varpi})) = \frac{1}{N}\sum_{n=0}^{N-1} H_d(e^{j2\pi n/N})\, e^{j2\pi nk/N} = \frac{1}{N}\sum_{n=0}^{N-1} H_d[n]\, e^{j2\pi nk/N} \tag{13-49}$$

The impulse response is then used to define the FIR's transfer function, specifically:

$$H(z) = \sum_{k=0}^{N-1} h[k]z^{-k} \tag{13-50}$$

The transfer function can be expanded to read

$$H(z) = \frac{1}{N}\sum_{k=0}^{N-1}\left(\sum_{n=0}^{N-1} H_d[n]e^{j2\pi nk/N}\right)z^{-k} \tag{13-51}$$

which can be simplified as shown next:

$$H(z) = \frac{1}{N}\sum_{k=0}^{N-1}\sum_{n=0}^{N-1} H_d[n]\,(e^{j2\pi n/N}z^{-1})^k \tag{13-52}$$

After reversing the order of the summation and rearranging the terms, $H(z)$ can be expressed as

$$H(z) = \frac{1}{N}\sum_{n=0}^{N-1} H_d[n]\sum_{k=0}^{N-1}(e^{j2\pi n/N}z^{-1})^k \tag{13-53}$$

Using a standard summation reduction formula, the parenthetical term in Equation (13-53) can be expressed as

$$\sum_{k=0}^{N-1}(e^{j2\pi n/N}z^{-1})^k = \frac{1 - z^{-N}}{1 - e^{j2\pi n/N}z^{-1}} \tag{13-54}$$

which finally allows the FIR transfer function to be formally expressed as

$$H(z) = \frac{(1 - z^{-N})}{N}\sum_{n=0}^{N-1}\left(\frac{H_d[n]}{1 - e^{j2\pi n/N}z^{-1}}\right) \tag{13-55}$$

Equation (13-55) is graphically interpreted in Figure 13-24. The preamble filter, namely $(1 - z^{-N})$, is seen to be a simple Nth-order comb filter which places nulls (zeros) at the explicit reference frequency locations $\varpi_n = 2\pi n/N$. The terms of the summation represent a collection of sub-band filters having poles at the reference frequencies $\varpi_n = 2\pi n/N$. The

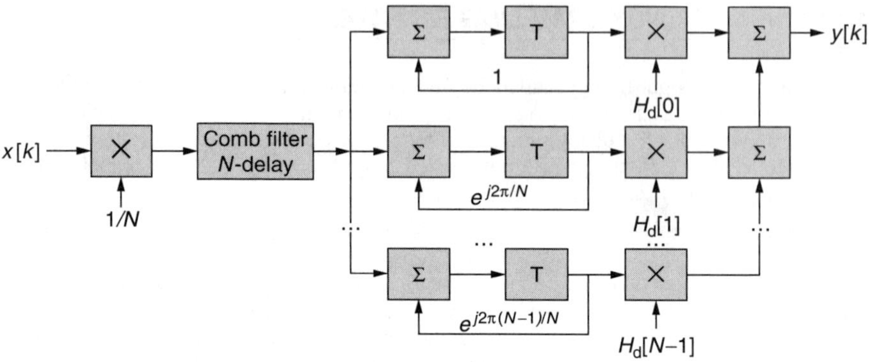

FIGURE 13-24 The basic structure of a frequency sampling FIR.

pole at $\varpi_n = 2\pi n/N$ cancels the comb filter zeros found at the same location. Due to the pole-zero cancellation, the resulting filter is without poles (for example, FIR).

Each sub-band filter, by itself, is a tuned narrowband filter. The nth–sub-band filter, sometimes called the nth resonator, is characterized by

$$H_n(z) = \frac{H_d[n]}{1 - e^{j2n\pi/N}z^{-1}} = \frac{A[n]\angle\phi(n)}{1 - e^{j2n\pi/N}z^{-1}}, \quad n \in [0, N-1] \tag{13-56}$$

The poles of the resonator filter are located along the periphery of the unit circle at locations $z = e^{j2\pi n/N}$. For stability reasons, the filter poles and zeros are often moved slightly interior to the unit circle to reside on an arc of a radius r (for instance, $z = re^{j2\pi n/N}$), where $r < 1$ but is close to unity. Since the complex resonator poles occur in complex conjugate pairs, they can be paired together to define second-order resonator filters having only real coefficients. Collectively, a 2nd-order frequency sampling filter is given by

$$H(z) = \frac{(1 - r^N z^{-N})}{N}\left(\frac{H_d[0]}{1 - rz^{-1}} + \sum_{n=1}^{(N-1)/2} \frac{2A[n](\cos(\phi(n)) - r\cos(\phi(n) - 2\pi n/N)z^{-1})}{1 - 2r\cos(2\pi n/N)z^{-1} + r^2 z^{-2}}\right), \tag{13-57}$$

if N is odd. If N is even, then

$$H(z) = \frac{(1 - r^N z^{-N})}{N}\left(\frac{H_d[0]}{1 - rz^{-1}} + \frac{H_d[N-1]}{1 + rz^{-1}} + \sum_{n=1}^{N/2-1} \frac{2A[n](\cos(\phi(n)) - r\cos(\phi(n) - 2\pi n/N)z^{-1})}{1 - 2r\cos(2\pi n/N)z^{-1} + r^2 z^{-2}}\right) \tag{13-58}$$

The frequency sampling method has been successfully used for some time to design FIR filters having an arbitrary magnitude frequency response envelope. Such filter designs are resource-efficient when the width of the active passband is small compared to the width of the stopband. In this instance, most of the sub-band filters will have zero gain (such as, $A[n] = 0$) and can therefore be left unimplemented. Frequency sampling filters have, however, acknowledged limitations. Design problems can occur when a frequency sampling filter is required to model an ideal piecewise constant filter response characterized by $|H(n)| = 1$ and $|H(n+1)| = 0$ (in other words, an abrupt change) leading to spectral distortions due to

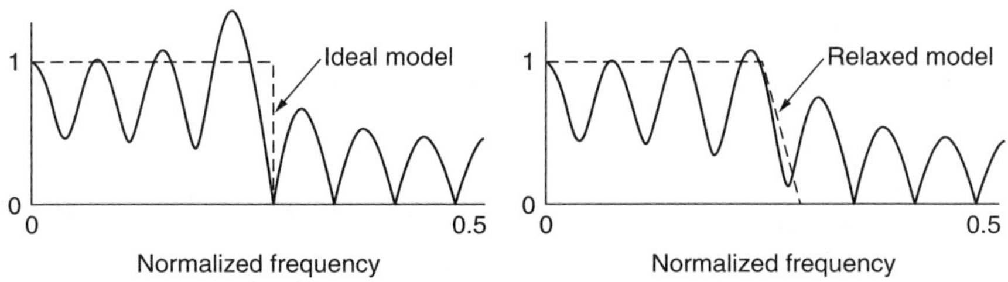

FIGURE 13-25 The frequency response of a frequency sampling filter modeling a zero-phase ideal low-pass (left) and relaxed zero-phase low-pass filter model (right). (The frequency axis was normalized with respect to f_s.)

Gibb's phenomenon. The problem can be mitigated, to a degree, by relaxing the transition band condition so that $|H(n)| = 1 - \alpha$ and $|H(n+1)| = \alpha$ for $0 \angle \alpha \angle 1.0$. This will reduce the slope of the skirt of the filter so it can be more easily modeled by an IDFT. The frequency sampling design strategies are illustrated in following example.

Example 13-16 Frequency sampling FIR

Required:

Design a 15th-order frequency sampling low-pass FIR that emulates an ideal low-pass filter having a normalized transition band centered about 0.5π. The phase response is assumed to be zero (in other words, $\phi(n) = 0$) which actually means that the phase shift at any harmonic frequency can be an integer multiple of 2π. Compare the response to that of a frequency sampling filter having a linear transition band. The target magnitude frequency responses are graphically interpreted in Figure 13-25.

Results:

The frequency sampling filter's center frequencies are located at $\varpi_n = 2\pi n/15 = 0.4188n$, $n \in [0, 14]$. The (desired) filter response establishes that

$$H_d(n) = \begin{cases} 1 + j0, & n \in [0, 3], n \in [12, 14] \\ 0 + j0, & n \in [4, 11] \end{cases}$$

where $n \in [0, 7]$ corresponds to *positive frequencies* and $n \in [9, 14]$ refers to *negative frequencies*. Notice that only 7 of 15 possible filters need to be implemented. The frequency sampling filter, based on the desired filter gain weights, is

$$H(z) = \frac{(1 - z^{-15})}{15}\left(\frac{1}{1 - z^{-1}} + \sum_{n=1}^{3} \frac{2(1 - \cos(2\pi n/15)z^{-1})}{1 - 2\cos(2\pi n/15)z^{-1} + z^{-2}}\right)$$

The comparison filter is defined in terms of relaxed weights, which should create a smoother transition band. The weights of the relaxed filter (for example, $\alpha = 0.23$) are $H(3) = H(12) = 0.707$ and $H(4) = H(11) = 0.23$.

The frequency response of the two frequency sampling filter models are shown in Figure 13-25. Observe that there is significant amplitude ripple and overshoot in the vicinity of the transition band of the sharp transition band design. The relaxed filter design is seen to lessen the overshoot (ripple) locally about the transition band. Also note that neither design produces responses that are truly close replicas of the desired response. To achieve an adequate fit, a filter order much greater than 15 would be required. However, it should be remembered that both filters are specified to be zero-phase designs. Replacing this condition with a linear-phase requirement can significantly improve the quality of the response, as the next example illustrates.

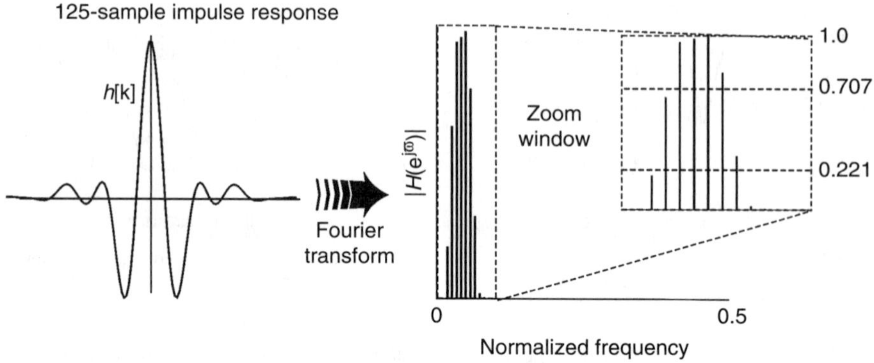

FIGURE 13-26 A frequency sampling FIR impulse response (left) and a magnitude frequency response with zoom expansion (right). (The frequency axis was normalized with respect to f_s.)

Example 13-17 Frequency sampling FIR

Required:

A linear-phase frequency sampling bandpass filter must be designed. The passband ranges over 300 to 700 Hz and the sample rate is 12.5 kHz. Assume that the resonant narrow-band filters are to be set on 100 Hz centers, which defines the filter order to be $N = 125$. The filter is to have a Type-I linear-phase response (odd order, even symmetry) with the phase shift set to π radians per 100 Hz. This condition states that $H_d(n) = (-1)^n A(n)$, as defined next.

Gain and Phase Assignments

$f = 200$ Hz	$A(2) = 0.221$	$\phi(2)\text{mod}(2\pi) = 0$	$H(2) = 0.221$
$f = 300$ Hz	$A(3) = 0.707$	$\phi(3)\text{mod}(2\pi) = \pi$	$H(3) = -0.707$
$f = 400$ Hz	$A(4) = 1$	$\phi(4)\text{mod}(2\pi) = 0$	$H(4) = 1$
$f = 500$ Hz	$A(5) = 1$	$\phi(5)\text{mod}(2\pi) = \pi$	$H(5) = -1$
$f = 600$ Hz	$A(6) = 1$	$\phi(6)\text{mod}(2\pi) = 0$	$H(6) = 1$
$f = 700$ Hz	$A(7) = 0.707$	$\phi(7)\text{mod}(2\pi) = \pi$	$H(7) = -0.707$
$f = 800$ Hz	$A(8) = 0.221$	$\phi(8)\text{mod}(2\pi) = 0$	$H(8) = 0.221$

Results:

The resulting impulse and magnitude frequency response of the frequency sampling filter is shown in Figure 13-26. Notice that the impulse response has a symmetric shape around the center tap coefficient, suggestive of a linear-phase filter. The magnitude frequency response is also seen to follow the general shape of the desired filter response.

13.19 SAVITZKY-GOLAY FIR FILTERS

Savitzky-Golay (SG) smoothing filters (also called digital smoothing polynomial filters or least squares smoothing filters) are used to reduce the effects of broadband noise. In such applications, SG smoothing filters are claimed to outperform a standard FIR moving averaging filter, which tends to filter out a significant portion of a high-frequency signal as well as noise. While SG filters may be more effective in preserving pertinent high-frequency signal components, they are less successful than standard averaging FIR filters in rejecting noise.

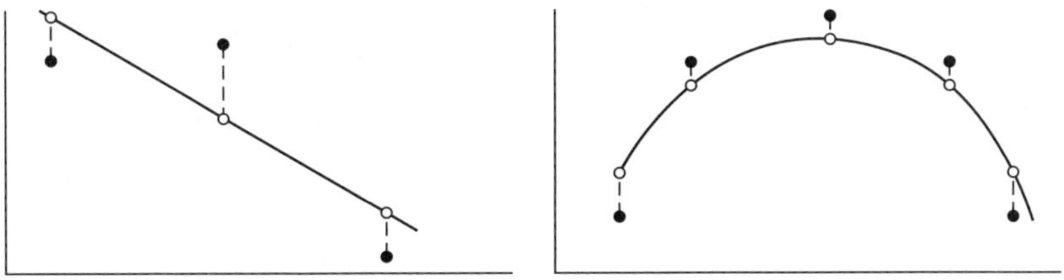

FIGURE 13-27 A linear interpolation (left) and a cubic polynomial interpolation (right).

SG filters are actually a family of filters which are used to smooth data using polynomial interpolation. They process data in an L sample block. The SG smoothing strategy is designed to preserve higher-order statistical moments and are based on least squares techniques that fit a low-order polynomial (typically quadratic or quadric) to a collection of data samples, as shown in Figure 13-27. An example of an SG filter is presented in the next example.

Example 13-18 Savitzky-Golay filter

Required:
Use a 4th-order SG interpolating filter to suppress (smooth) the noise in a chirp signal process for a data frame of length $L = 33$ (16 samples in the past, 16 samples in the future, and the current sample). Compare the results to a 33rd-order moving average FIR.

Results:
The responses of a standard 33-sample moving average and SG smoothing filter based on quadric polynomial interpolation are shown in Figure 13-28. The standard moving average FIR is seen to perform well as a de-noising agent when the signal frequency is low, but becomes problematic at high frequencies due to amplitude roll-off. The SG filter maintains essentially a constant amplitude capability across the entire spectrum. At low frequencies, however, the moving average FIR outperforms the SG filter.

13.20 RAISED FIR FILTERS

The raised cosine and *root raised cosine* filters are common to digital data communications applications. Their use is motivated by their ability to limit inter-symbol interference (ISI). The frequency response of a raised cosine filter is given by

$$H(\varpi) = \begin{cases} 1; \varpi < \varpi_c(1 - \alpha) \\ 0; \varpi > \varpi_c(1 + \alpha) \\ (1 + \cos(2\alpha\varpi_c)/2); \varpi_c(1 - \alpha) < \varpi < \varpi_c(1 + \alpha) \end{cases} \tag{13-59}$$

where α is called the "*roll-off*" parameter. The impulse response of the raised cosine filter is given by

$$h_k = \frac{1}{f_s}\text{sinc}(2kf_c/f_s)\frac{\cos(2\alpha kf_c/f_s)}{1 - (4\alpha kf_c/f_s)^2} \tag{13-60}$$

The magnitude frequency response of an ideal raised cosine low-pass filter is unity across the passband with a cosine transition into a zero-gain stopband. The magnitude frequency response and impulse response are graphically interpreted in Figure 13-29.

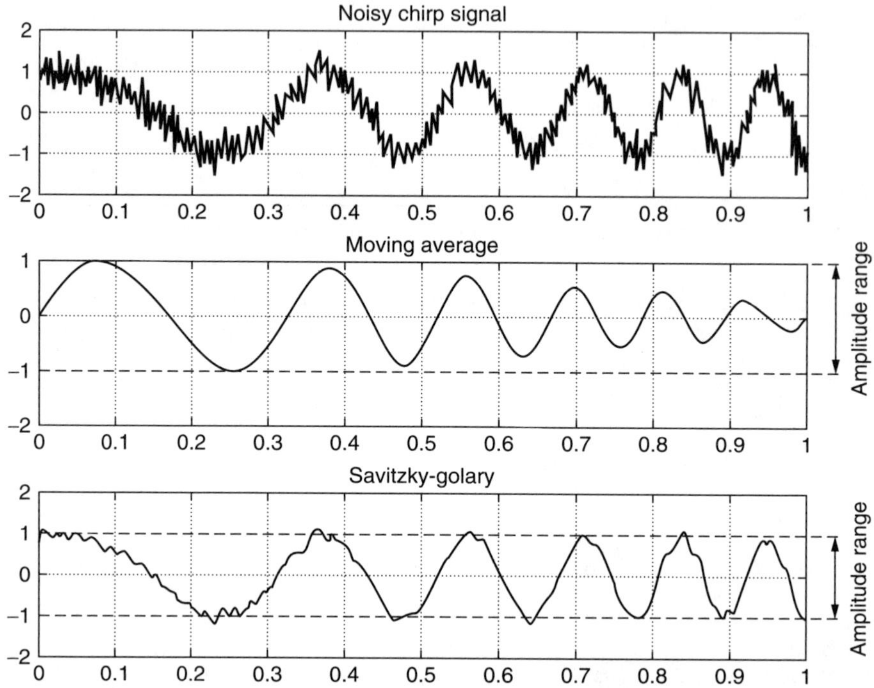

FIGURE 13-28 The responses of a moving average and SG filter to a noise added up-chirp signal.

A root raised cosine filter is used when the raised cosine filtering load is to be spread equally between the transmitter and receiver. The root raised cosine filter's frequency response satisfies

$$H(\varpi) = \begin{cases} 1; \ \varpi < \varpi_c(1 - \alpha) \\ 0; \ \varpi > \varpi_c(1 + \alpha) \\ \sqrt{(1 + \cos(/2\alpha\varpi_c)/2)}; \ \varpi_c(1 - \alpha) < \varpi < \varpi_c(1 + \alpha) \end{cases} \tag{13-61}$$

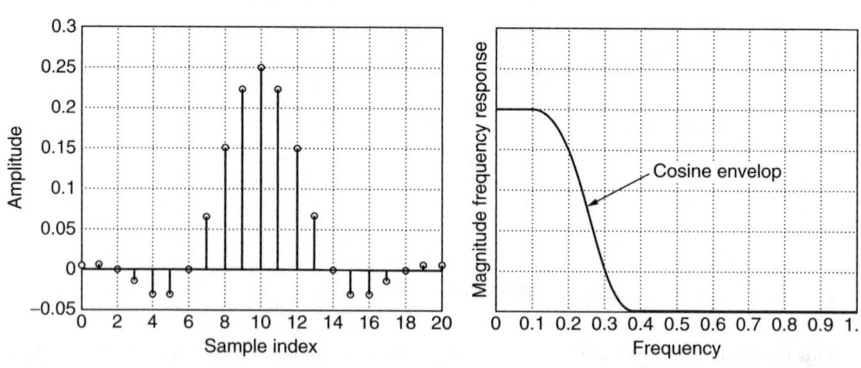

FIGURE 13-29 A raised cosine impulse response (left) and a magnitude frequency response (right). (The frequency axis was normalized with respect to $f_s/2$.)

and the root raised cosine filter's impulse response satisfies

$$h_k = \frac{1}{\pi f_s} \frac{4\alpha \cos((1 + \alpha)2\pi k f_c/f_s) + \sin((1 - \alpha)2\pi k f_c/f_s)/(8\alpha k f_c/f_s)}{\left(\sqrt{\frac{1}{2f_c}}\right)((8\alpha k f_c/f_s)^2 - 1)} \tag{13-62}$$

13.21 MATLAB FIR SUPPORT

> **Caveat:** MATLAB assumes that the order of an FIR is equal to the order of the polynomial $H(z)$ rather than the number of filter coefficient slots in $H(z)$, which is the custom. As a result, the usual Nth-order FIR filter is represented as an $(N - 1)$st-order FIR in MATLAB.

Mathwork's MATLAB, as well as a number of other commercial signal processing software packages, can greatly simplify the FIR design process. It should be expected, however, that individual design routines will be in some state of revision, inclusion, or retirement. As a result, the software reviewed in this section may or may not be available to the reader in the future. Furthermore, software functions routinely may appear repackaged under various names in different software bundles or toolkits. At present, the primary sources of MATLAB FIR design software support are

- SPT = Signal Processing Toolbox
- FT = Filter Design Toolbox
- FT2 = Filter Design Toolbox 2

The basic software options are listed in Table 13-4. Details regarding their functionality and syntax can be found online or by using the product's "help" apparatus.

Traditional FIR modules, appearing in the Signal Processing Toolbox (SPT), are motivated using examples found in Figure 13-30 and Table 13-5.

13.22 FIR ARCHITECTURES

FIRs can be physically implemented in software and/or hardware. Software solutions are realized using programmable general-purpose microprocessors (μp) or digital signal processing microprocessors (DSP μp). Hardware-based solutions are packaged in field-programmable logic arrays (FPGA), or application-specific integrated circuits (ASIC). In all cases, the FIRs are defined in terms of adder, multiplier, shift-register, memory, and data input/output building blocks. How these elements are interconnected establishes the filter's *architecture*. There are several basic FIR-specific architectures that are singled out for special attention.

13.23 DIRECT FORM FIR

An Nth-order causal FIR, having an impulse response $h[k]$, can be expressed in transfer function form as

$$H(z) = \sum_{k=0}^{N-1} h_k z^{-k} \tag{13-63}$$

TABLE 13-4 MATLAB FIR Support

Function	Source	Description	Example
cfirpm	SPT	Design a linear, complex, and nonlinear-phase equiripple FIR.	
fir1	SPT	Design a window-based finite impulse response filter.	Figure 13-30*a*
fir2	SPT	Design a frequency sampling-based finite impulse response filter.	Figure 13-30*b*
fircband	FT	Design an FIR with constrains on the stopbands.	
fircegrip	FT	Design constrained, equiripple, and finite impulse response (FIR) filters.	
fircls	SPT	Design a constrained least square multiband filter.	Figure 13-30*c*
fircls1	SPT	Design a constrained least square low-pass and high-pass linear-phase FIR filter.	Figure 13-30*d*
firgr	FT	Design, using the Parks-McClellan method, an equiripple, Hilbert, and differentiator FIR.	
firhalfband	FT2	Design a half-band FIR filter.	
firlpnorm	FT	Design an FIR with respect to an L_p norm.	
firls	SPT	Design a least square linear-phase FIR filter.	Figure 13-30*e*
firnyquist	FT2	Design an equiripple Nyquist FIR.	
firpm	SPT	Design, using Parks-McClellan, an optimal FIR filter design (a.k.a., Remez).	
firpmord	SPT	Design a Parks-McClellan optimal FIR filter order estimation.	
firpr2chfb	FT2	Design an FIR perfect reconstruction two-channel filter bank.	
firrcos	SPT	Design a raised cosine FIR.	Figure 13-30*f*
gaussfir	SPT	Design a gaussian FIR.	
gremez	FT2	Design a generalized FIR using the Remez method.	
ifir	FT2	Design an interpolating FIR filter.	
intfilt	SPT	Design an interpolating FIR.	
kaiserord	SPT	Estimate parameters for an FIR filter design with a Kaiser window.	
sgolay	SPT	Design a Savitzky-Golay filter.	

The most common implementation of this FIR is called the *direct form* FIR architecture, which is interpreted in Figure 13-31. The FIR's impulse response can be directly inferred from the architecture to be $h[k]$. A direct form FIR consists of a collection of $N - 1$ shift-registers and N tap-weight coefficients h_k with attendant multipliers and adders. For each input sample $x[k]$, a direct FIR would implement the following set of arithmetic operations:

$$\begin{aligned}&\textit{For each input sample } x[k], \textit{ do} \\ &x_0 = x[k] \\ &y[k] = h_0x_0 + h_1x_1 + \cdots + h_{N-1}x_{N-1} \\ &\{\textit{update FIFO stack}\} \\ &x_{N-1} = x_{N-2} \\ &\cdots = \cdots \\ &x_2 = x_1 \\ &x_1 = x_0\end{aligned} \tag{13-64}$$

TABLE 13-5 Traditional FIR Signal Processing Toolbox Support—LP = low pass, HP = high pass, BP = band pass, BS = band stop, Hil = Hilbert, and Diff = differentiator

Function: fir1

LP ✓	HP ✓	BP ✓
BS ✓	Hil	Diff

Example:
Design a 49-tap FIR window method bandpass FIR (order 48 in MATLAB). The normalized passband range is $0.35 \leq \omega \leq 0.65$:

MATLAB:
```
b = fir1(48,[0.35 0.65]);
```

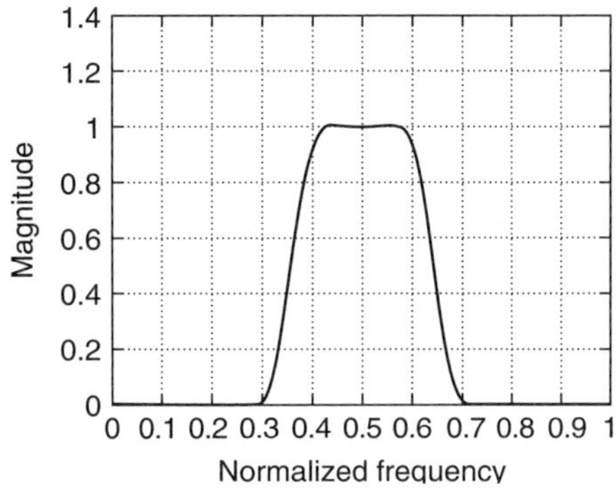

FIGURE 13-30a

Function: fir2

LP ✓	HP ✓	BP ✓
BS ✓	Hil	Diff

Example:
Design a 31-tap low-pass filter (order 30 in MATLAB) having a unity gain to 0.6 of the Nyquist frequency, 0 elsewhere.

MATLAB:
```
f = [0 0.6 0.6 1];
m = [1 1 0 0];
b = fir2(30,f,m);
```

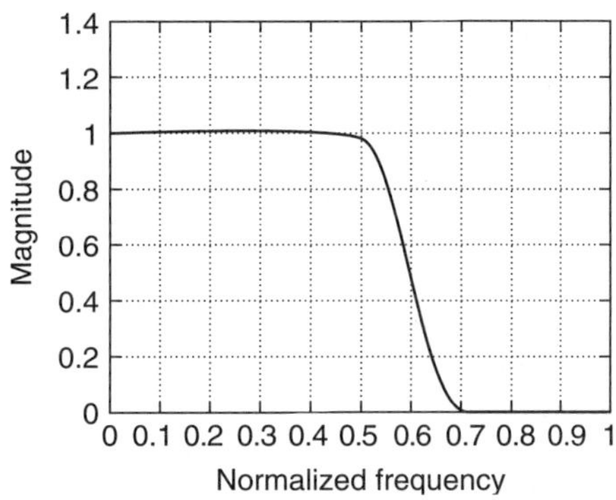

FIGURE 13-30b

Function: fircls

LP ✓	HP ✓	BP ✓
BS ✓	Hil	Diff

Example:
Design an order 151 (order 150 MATLAB) bandpass filter having a 0.4 passband with ±0.02 bounds and ±0.01 stopband bounds.

MATLAB:
```
n = 150; f = [0 0.2 0.6 1];
a = [0 1 0];
up = [.01 1.02 .01];
lo = [-.01 .98 -.01];
b = fircls(n,f,a,up,lo);
```

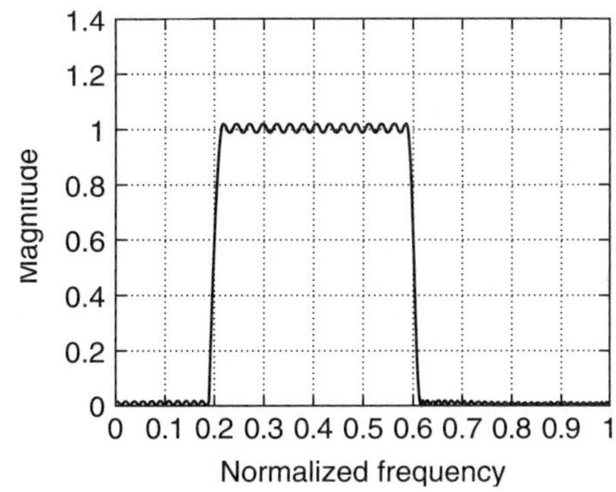

FIGURE 13-30c

(Continued)

TABLE 13-5 Traditional FIR Signal Processing Toolbox Support—LP = low pass, HP = high pass, BP = band pass, BS = band stop, Hil = Hilbert, and Diff = differentiator (*Continued*)

Function: fircls1

LP ✓	HP ✓	BP
BS	Hil	Diff

Example:
Design an order 54 (order 55 MATLAB) low- pass filter with a cutoff frequency located at 0.3:

MATLAB:
```
n = 55;   wo = 0.3;
dp = 0.02; ds = 0.008;
b = fircls1(n,wo,dp,ds,'both');
```

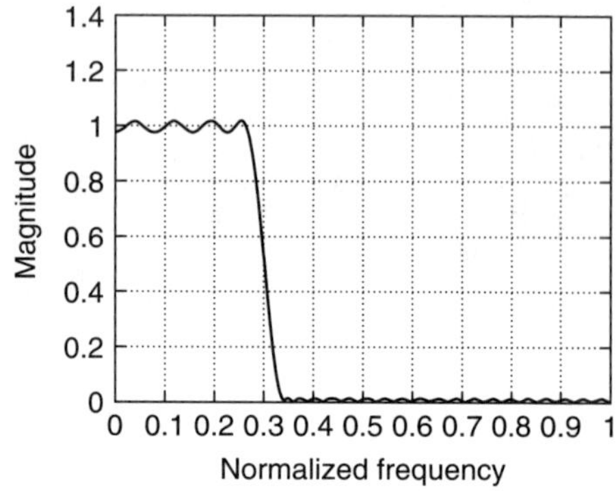

FIGURE 13-30d

Function: firls

LP ✓	HP ✓	BP ✓
BS ✓	Hil ✓	Diff ✓

Example:
Design a 25-tap anti-symmetric Hilbert filter (order 24 in MATLAB) with piecewise linear passbands, and plot the desired and actual frequency response:

MATLAB:
```
F = [0 0.3 0.4 0.6 0.7 0.9];
A = [0 1 0 0 0.5 0.5];
b = firls(24,F,A, 'hilbert');
```

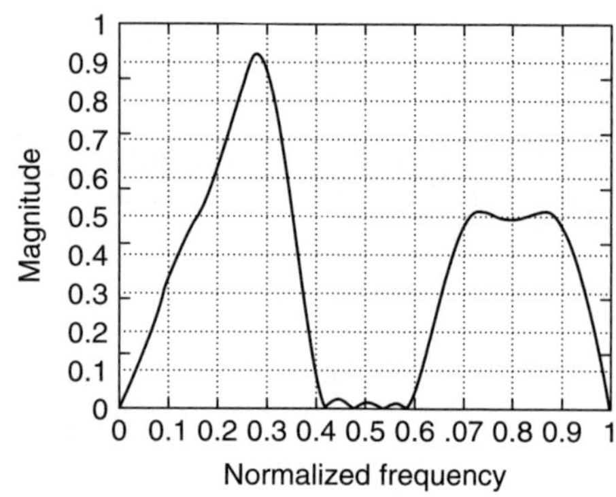

FIGURE 13-30e

Function: fircos

LP ✓	HP	BP
BS	Hil	Diff

Example:
Design a 21-tap raised cosine FIR filter (order 20 in MATLAB) with a cutoff frequency 0.25 of the Nyquist frequency and a transition bandwidth of 0.25:

MATLAB:
```
h = firrcos(20,0.25,0.25);
```

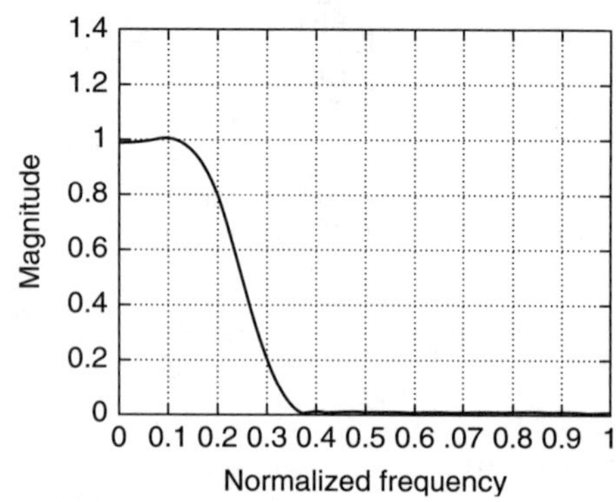

FIGURE 13-30f

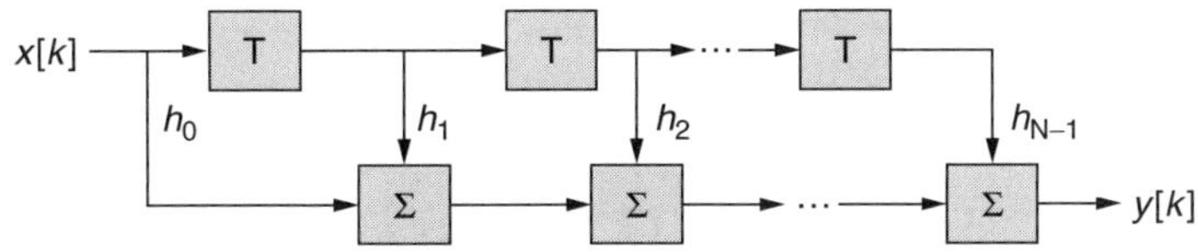

FIGURE 13-31 A direct *N*th-order FIR architecture.

FIRs are well suited for implementation using this sample-by-sample computational process. Once initiated, the routine would be continually repeated over the filter's life cycle. The designers of modern DSP microprocessors have learned how to efficiently implement an array of multiply-accumulate (MAC) calls associated with a sum-of-products or S = AX + Y (SAXPY) algorithms. Many existing DSP microprocessors continue to contain a single MAC unit and data memory, configured as RAM and coefficients memory, fashioned using either RAM or ROM. In many cases, dual-port memory is used to simultaneously supply two operands (x_j, h_i) to the MAC per memory cycle. Coefficients can also be read to and from memory in a circular modulo(N) manner, repeating a fixed sequence each filter cycle. Since many DSP microprocessors can execute an instruction in a single cycle, the computational latency of a single filter cycle can be estimated to be

$$T_{\text{FIR\_cycle}} \cong (N + 1)T_{\text{inst-cycle}} \tag{13-65}$$

The emulation of a direct FIR in a high-level language, such as MATLAB, is a straightforward process. For example, MATLAB's *dfilt.dffir* function can be used to convert a transfer function $H(z)$ into a direct form FIR. The direct FIR architecture is illustrated in the following example.

Example 13-19 Direct architecture FIR

Required:
Implement an FIR with a transfer function $H(z) = 1 + (13/24)z^{-1} + (5/8)z^{-2} + (1/3)z^{-3}$ as a direct FIR.

Results:
The direct form FIR is shown in Figure 13-32.

13.24 TRANSPOSE FORM FIR

Another baseline FIR form is the *transpose FIR* architecture, which is a variation on the direct form. The transpose architecture is shown in Figure 13-33. The FIR's impulse response can be directly inferred from the architecture to be $h[k]$. For implementations involving

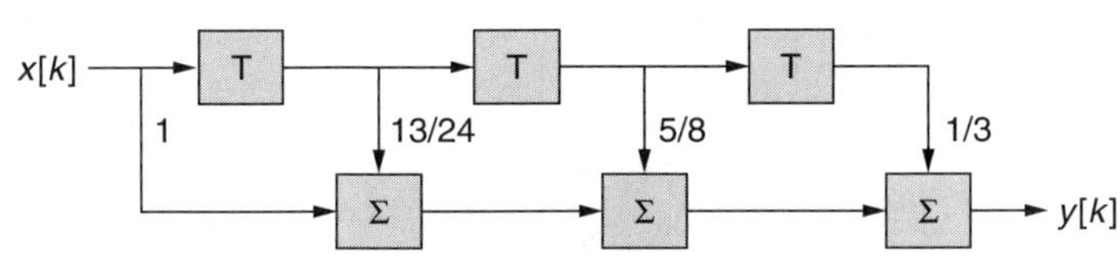

FIGURE 13-32 A direct 3rd-order FIR architecture.

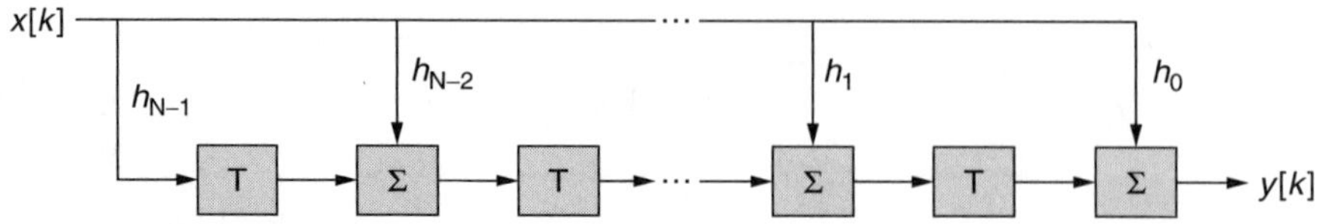

FIGURE 13-33 A transpose Nth-order FIR architecture.

MATLAB, *dfilt.dffirt* can be used to convert a transfer function $H(z)$ into a transpose form FIR. An example of a transpose FIR architecture can be found in the next example.

Example 13-20 Transpose architecture FIR

Required:
Implement an FIR with a transfer function $H(z) = 1 + (13/24)z^{-1} + (5/8)z^{-2} + (1/3)z^{-3}$ as a transpose FIR.

Results:
The transpose architecture is shown in Figure 13-34.

13.25 SYMMETRIC FORM FIR

Many baseline FIRs are linear-phase filters and, as a result, possess either even or odd coefficient symmetry. Coefficient symmetry permits the direct form FIR architecture to be modified as shown in Figure 13-35. The result is called a *symmetric* FIR architecture. The advantage of this architecture is a reduced multiplier budget. If the implementation technology allows additions to be performed at a faster rater than multiplication (such as with ASICs), then a symmetric FIR can have a higher real-time bandwidth (filter cycles per unit time) compared to an equivalent direct form FIR. In some pipelined hardware systems, multiply and add rates are identical, which places the architecture shown in Figure 13-35 at a disadvantage due to its increased adder count. Conventional DSP microprocessors are also designed to optimize multiply-accumulate data flow while other operations, such as add-multiply-add, are not. As a result, symmetric FIR architectures are not necessarily synergistic with general purpose DSP microprocessors. For implementations involving MATLAB, *dfilt.dfsymfir* can be used to convert a transfer function $H(z)$ of even symmetry into a symmetric form FIR. If odd symmetry is present, then *dfilt.dfasymfir* is used.

13.26 LATTICE FORM FIR

The direct, transpose, and symmetric form FIR architecture is often used to implement linear-phase FIRs. Another important class of FIR is called the *lattice* architecture. Lattice architectures are normally associated with non–linear-phase filters. An Nth-order lattice FIR

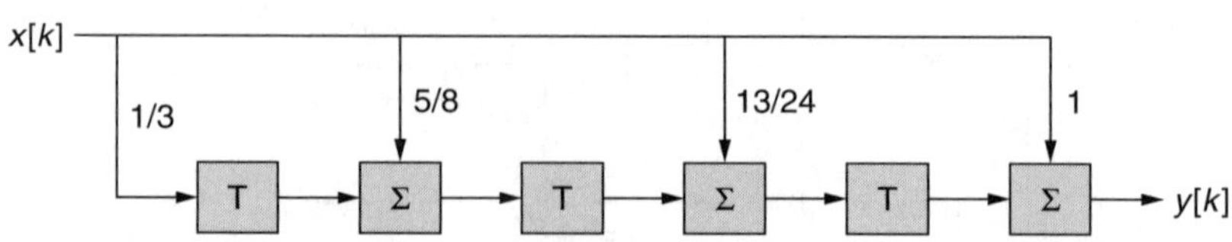

FIGURE 13-34 A transpose 3rd-order FIR architecture.

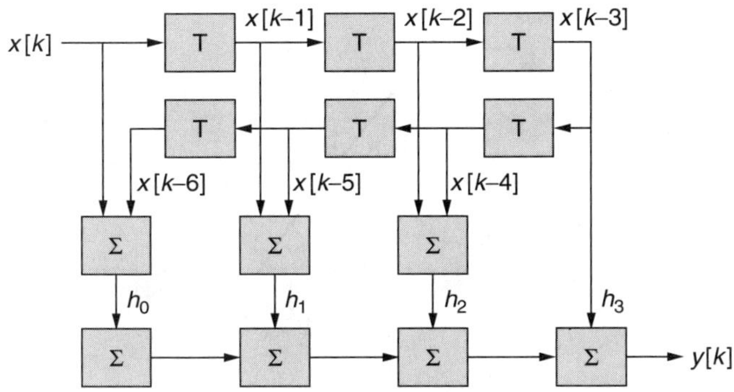

FIGURE 13-35 A symmetric 7th-order FIR architecture.

is shown in Figure 13-36. Compared to an Nth-order direct and transpose FIR, it can be seen that a lattice filter requires twice the number of multiplies per filter cycle (N versus $2N$). Therefore, a lattice filter is physically more complex than a baseline direct, transpose, and symmetric FIR. Nevertheless, a lattice FIR is often preferred over the direct FIR due to the architecture's known ability to suppress coefficient roundoff and other finite wordlength arithmetic effects. In addition, lattice structures are important to the design of adaptive filters (for example, LPC) and applications where linear-phase is not a requirement.

The lattice architecture shown in Figure 13-36 can implement a *monic* FIR filter having a transfer function

$$A(z) = \left[1.0 + \sum_{j=1}^{N-1} a_j z^{-j}\right] \tag{13-66}$$

The transfer function is defined in terms of coefficients a_j in order to differentiate Equation (13-66) from Equation (13-63). In lattice form, the transfer function given in Equation (13-66) is expressed in terms of so-called *PARCOR* (partial correlation) coefficients k_i, as shown in Figure 13-36. The PARCOR coefficients can be reconciled with those of the transfer function $H(z)$ found in Equation (13-63) in an iterative manner using the signal taps, $A_i(z)$ and $B_i(z)$ (see Figure 13-37) as intermediate rubrics. Specifically, the conversion of PARCOR coefficients to transfer function coefficients, for an iteration index $i = 1, 2, \ldots, N$, begins with the assignments

$$A_0(z) = B_0(z) = 1.0 \tag{13-67}$$

$$\begin{aligned} A_m(z) &= A_{m-1}(z) + k_m z^{-1} B_{m-1}(z); \quad m \in [1, N-1] \\ B_m(z) &= z^{-m} A_m(z^{-1}); \quad m \in [1, N-1] \end{aligned} \tag{13-68}$$

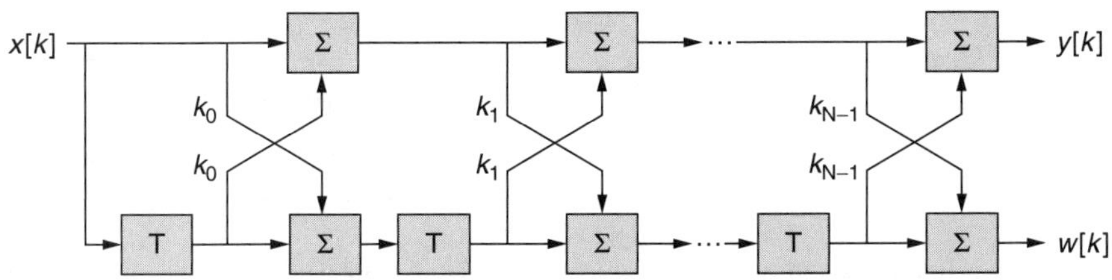

FIGURE 13-36 A lattice FIR architecture.

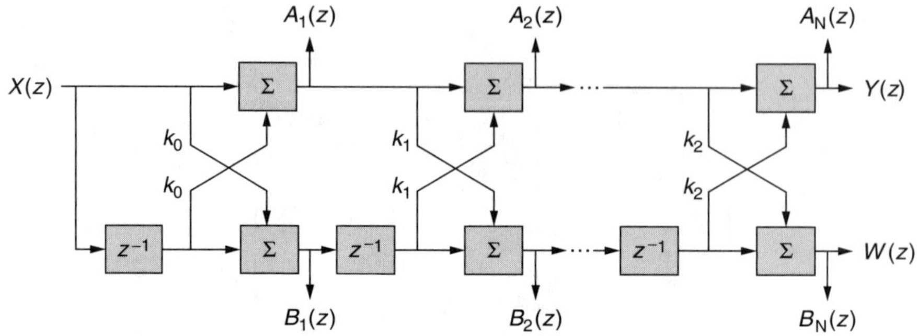

FIGURE 13-37 A lattice FIR interpreted in the *z*-domain.

where the last relation states that $B_m(z)$ is simply $A_m(z)$ written in reverse order. Finally, it can be shown that $A_N(z) = H(z)$. To reverse the conversion process, begin with the transfer function and end with the PARCOR coefficients. The iterative process is defined in terms of

$$A_{m-1}(z) = \frac{A_m(z) - k_m B_m(z)}{1 - k_m^2}; \quad m \in [1, N-1] \tag{13-69}$$

The intermediate transfer functions $A_m(z)$ and $B_m(z)$ are shown in Figure 13-37.

For implementations involving MATLAB, *dfilt.latticeamax* can be used to convert a transfer function $H(z)$ into a maximum phase lattice FIR. The MATLAB function *dfilt. latticeamin* is used to convert a transfer function $H(z)$ into a minimum phase lattice FIR. The design of a lattice form FIR is illustrated in following example.

Example 13-21 Lattice/direct architecture conversion

Required:

Suppose a 3rd-order lattice filter has PARCOR coefficients $k_1 = 1/4$, $k_2 = 1/2$, and $k_3 = 1/3$. Determine the coefficients of the equivalent direct FIR.

Results:

The direct filter coefficients can be computed as follows:

m	$A_m(z)$	$B_m(z)$
1	$A_1(z) = A_0(z) + k_1 z^{-1} B_0(z) = 1 + k_1 z^{-1} = 1 + \frac{1}{4} z^{-1}$	$B_1(z) = z^{-1} A_1(z^{-1}) = \frac{1}{4} + z^{-1}$
2	$A_2(z) = A_1(z) + k_2 z^{-1} B_1(z) = 1 + \frac{1}{4} z^{-1} + \frac{1}{2} z^{-1}\left(\frac{1}{4} + z^{-1}\right) =$ $1 + \frac{3}{8} z^{-1} + \frac{1}{2} z^{-2}$	$B_2(z) = z^{-2} A_2(z^{-1}) =$ $\frac{1}{2} + \frac{3}{8} z^{-1} + z^{-1}$
3	$A_3(z) = A_2(z) + k_3 z^{-1} B_2(z) = 1 + \frac{3}{8} z^{-1} + \frac{1}{2} z^{-2} +$ $\frac{1}{3} z^{-2}\left(\frac{1}{2} + \frac{3}{8} z^{-1} + z^{-1}\right) = 1 + \frac{13}{24} z^{-1} + \frac{5}{8} z^{-2} + \frac{1}{3} z^{-3} = A(z)$	$B_3(z) =$ *not required* $B_3(z) = z^{-3} + \frac{13}{24} z^{-2} + \frac{5}{8} z^{-1} + \frac{1}{3}$

The FIR's transfer function is realized as $A(z) = A_3(z)$. As a side note, the polynomials $A_1(z)$, $B_1(z)$, $A_2(z)$, $B_2(z)$, $A_3(z)$, and $B_4(z)$ correspond to the transfer functions located in

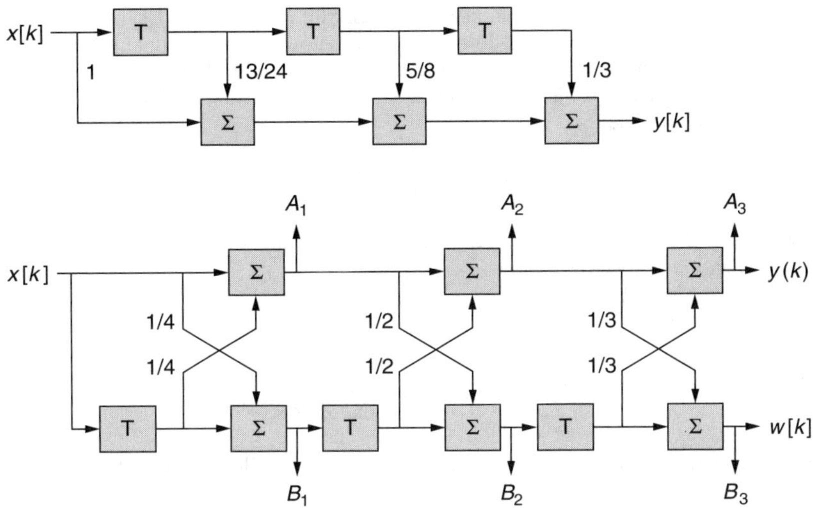

FIGURE 13-38 Direct (top) and lattice (bottom) FIR filters in the z-domain.

the positions shown in Figure 13-38. The process is reversible in that the direct FIR coefficients {1, 13/24, 5/8, 1/3} can be converted into the lattice filter PARCOR coefficients as shown next.

m	$A_m(z)$	$B_m(z)$	
3	$A(z) = A_3(z) = 1 + \frac{13}{24}z^{-1} + \frac{5}{8}z^{-2} + \frac{1}{3}z^{-2}$	$k_3 = a_3 = \frac{1}{3}$	$B_3(z) = z^{-3}A_3(z^{-1}) = \frac{1}{3} + \frac{5}{8}z^{-1} + \frac{13}{24}z^{-2} + z^{-2}$
2	$A_2(z) = \dfrac{A_3(z) - k_3z^{-1}B_3(z)}{1 - k_3^2} = 1 + \frac{3}{8}z^{-1} + \frac{1}{2}z^{-2}$	$k_2 = a_2 = \frac{1}{2}$	$B_2(z) = z^{-2}A_2(z^{-1}) = \frac{1}{2} + \frac{3}{8}z^{-1} + z^{-1}$
1	$A_1(z) = \dfrac{A_2(z) + k_3z^{-1}B_2(z)}{1 - k_2^2} = 1 + \frac{1}{4}z^{-1}$	$k_1 = a_1 = \frac{1}{4}$	$B_1(z) =$ *not required*

These results are summarized in Figure 13-38.

13.27 DISTRIBUTED ARITHMETIC

Upon close inspection of a typical FIR modeled after Equation (13-3), one notes that the FIR coefficients h_i are generally real and known a-priori. As a result, the partial convolution product terms $h_i x[i - k]$ are technically defined by a process known as *scaling* rather then multiplication. One of the most attractive alternatives to implementing fixed-coefficient FIRs is to replace a traditional general-purpose multiplier with *lookup table* (LUT) calls to semiconductor memory. The LUT method simply emulates how children were taught to multiply by memorizing the multiplication tables. A popular LUT-based technology is called *distributed arithmetic* (DA). A DA filter assumes that data is coded as an M-bit 2's-complement data word. Specifically defined

$$x[k] = -x[k:0] + \sum_{i=1}^{M-1} x[k:i]2^{-i} \tag{13-70}$$

where $x[k:i]$ is the ith-bit of sample $x[k]$. Substituting Equation (13-70) into the linear convolution Equation (13-2), we obtain

$$\begin{aligned} y[k] &= \sum_{r=0}^{N-1} h_r\left(-x[k-r:0] + \sum_{i=1}^{N-1} x[k-r:i]2^{-i}\right) \\ &= -\sum_{r=0}^{N-1} h_r x[k-r:0] + \sum_{r=0}^{N-1}\sum_{i=1}^{M-1} h_r x[k-r:i]2^{-i} \end{aligned} \tag{13-71}$$

Upon reversing the order of the double summation, the following results:

$$y[k] = -\sum_{r=0}^{N-1} h_r x[k-r:0] + \sum_{i=1}^{M-1} 2^{-i} \sum_{r=0}^{N-1} h_r x[k-r:i] \tag{13-72}$$

Suppose that a 2^N-word memory lookup table, denoted $\theta[\underline{x}[k]:i]$ contains the preprogrammed mapping shown in Equation (13-72). The LUT is addressable by an N-bit address vector $\underline{\boldsymbol{x}}[k:i] = \{x[k:i], x[k-1:i], \ldots, x[k-N-1:i]\}$, where $x[r:i]$ is a binary-value digit $\{0,1\}$. Furthermore, the output word width of the memory table $\theta[x[k]:i]$ is assumed to be a P-bit word. Collectively:

$$\theta[\underline{x}[k]:i] = \sum_{r=0}^{N-1} h_r x[k-r:i]; \quad x[s:i] \in [0, 1] \tag{13-73}$$

The convolution sum, defined by Equation (13-72), then becomes

$$y[k] = -\theta(\underline{x}[k:0]) + \sum_{i=1}^{M-1} 2^{-i}\theta(\underline{x}[k:i]) \tag{13-74}$$

and consists of a collection of LUT calls, a 2^{-i} scaling, and an accumulation, as shown in Figure 13-39. Note that $[\underline{\boldsymbol{x}}[k]:i]$ is a vector of binary values which is presented to the table θ as a memory address vector. Initially, the address vector is populated by all the bits found in the common least significant-bit (LSB) location of $\underline{\boldsymbol{x}}[k]$. Each iteration, thereafter, moves the address vector $[\underline{\boldsymbol{x}}[k]:i]$ to the next significant-bit location. The first $(M-1)$ lookups produce $\theta(\underline{\boldsymbol{x}}[k:i])$, which have positive weights. It is important to note that the algebraic weight associated with the lookup $\theta(\underline{\boldsymbol{x}}[k:i+1])$ is twice that of $\theta(\underline{\boldsymbol{x}}[k:i])$. This explains the scaling factor 2^{-i} found in Equation (13-74) which implements the scaling process with a dedicated 2's-complement shift-adder, as shown in Figure 13-39. The last lookup, $\theta(\underline{\boldsymbol{x}}[k:0])$, has a negative weight (that is, sign bit location) which is controlled by the accumulator's add/subtract select.

In some instances, the real-time bandwidth of a DA FIR can be significantly higher than that of a filter implemented with a general-purpose multiplier. Suppose $M = 16$-bits and an order $N = 12$ FIR is implemented using a DSP μp and a DA filter. The DSP μp is assumed to have a 100 ns MAC-cycle time and the distributed filter, based on a $2^{12} \times 16$-bit memory table, has a lookup cycle time of 10 ns. A filter cycle for a DSP μp implementation would be on the order of

$$T_{\text{FIR\_cycle}} = N \cdot T_{\text{MAC\_cycle}} \tag{13-75}$$

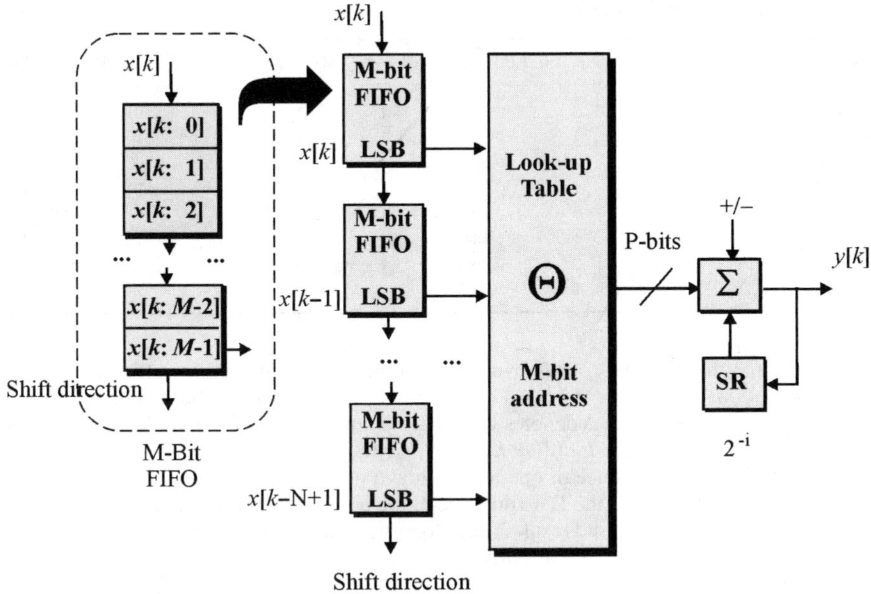

FIGURE 13-39 A distributed arithmetic architecture.

This results in a 1,200 ns filter cycle time, or a real-time rate of 833 kHz. A DA filter would execute the same filter with a cycle time of

$$T_{\text{FIR\_cycle}} = M \cdot T_{\text{Memory\_cycle}} \tag{13-76}$$

This results in a filter cycle time of 160 ns, or a real-time rate of 6.250 MHz (a 750-percent improvement). This is accomplished without any appreciable increase in hardware complexity and is, in many cases, actually less complex than a design based on DSP μp. It is because of the compactness of a DA solution, and its LUT dependency, that this class of filter is popular with FPGA-based designs. A distributed arithmetic design example is presented below.

Example 13-22 Distributed arithmetic FIR

Required:

Implement the simple 4th-order FIR given by $H(z) = 1.0 - 0.9z^{-1} + 0.64z^{-2} - 0.575z^{-3}$ as a distributed arithmetic FIR.

Results:

The worst case gain of the FIR is $G_{\max} = |1.0| + |-0.9| + |0.64| + |-0.575| = 3.115 < 2^2$. This gain requires that at least an additional two bits be assigned to the accumulator to manage the output integer field's dynamic range growth. The largest table lookup value is $\theta[\underline{x}[k]:i]_{\max} = 1.64 < 2^1$ (see Table 13-6) and means that the LUT data format must contain at least one integer bit. Assume that the input samples are coded as signed four-bit 2's complement words with three fractional bits of precision. Finally, assume that the lookup memory table stores data as an eight-bit word that consists of a sign bit, one integer bit (as previously argued), and six fractional bits. The $2^4 = 16$ eight-bit LUT values are pre-computed, as summarized in Table 13-6.

TABLE 13-6 DA Table Contents

$x[k-0{:}i]$	$x[k-1{:}i]$	$x[k-2{:}i]$	$x[k-3{:}i]$	LUT value	$x[k-0{:}i]$	$x[k-1{:}i]$	$x[k-2{:}i]$	$x[k-3{:}i]$	LUT value
0	0	0	0	0	1	0	0	0	0.9843
0	0	0	1	−0.5781	1	0	0	1	0.4062
0	0	1	0	0.6092	1	0	1	0	1.6093
0	0	1	1	0.0468	1	0	1	1	1.0468
0	1	0	0	−0.9062	1	1	0	0	0.0781
0	1	0	1	−1.4843	1	1	0	1	0.4843
0	1	1	0	−0.2656	1	1	1	0	0.7187
0	1	1	1	−0.8437	1	1	1	1	0.1406

Assume that the first four four-bit 2's complement input sample values are $x[3] = 0 \leftrightarrow [0_\Delta 000] = 0, x[2] = -1 \leftrightarrow [1_\Delta 000] = -1, x[1] = 0 \leftrightarrow [0_\Delta 000] = 0$, and $x[0] = 1$ ($[0_\Delta 111] = 7/8$, where Δ denotes the binary point location. The real valued output is given by $y[3] = h_0x[3] + h_1x[2] + h_2x[1] + h_3x[0] = h_1x[2] + h_3x[0] = 0.425$. The DA filter executes the sequence of operations shown in Table 13-7.

At the conclusion of the DA filter cycle, the accumulator holds the value 0.40036, which is close to the desired result. The DA error, which is –0.0246, is generally smaller than if obtained using a general-purpose ALU.

High-order designs are those whose order exceeds the address space of a single lookup table. Higher-order DA filters can be constructed from lower-order distributed filters using the tree architecture, as suggested in Figure 13-40. In such cases, an FIR of an order N is spread across L tables having an address space of n-bits each. In particular, $L = \lceil N/n \rceil$ where $\lceil \circ \rceil$ denotes the ceiling function. This particular design paradigm is also gaining popularity in implementing 2-D FIRs and transforms having coefficients that are known a-priori.

13.28 CANONIC SIGNED DIGIT (CSD)

The canonic signed digit (CSD) system is a ternary-valued numbering system whose digits can take on only the values {1, 0, –1}. CSD codes are dense in zeros which means that their use in multiply-intensive applications increases the no-op occurrence rate. The CSD was employed by a first-generation DSP μp (Intel 2920) in an attempt to accelerate multiplication by maximally populating the data with 0s. Since FIRs are MAC-intensive, CSD can serve as an arithmetic acceleration technology. The most common CSD manifestation represents an FIR coefficient h_i in terms of a limited set of digits, such as the three-term expansion shown next:

$$h_i = H_j2^j + H_k2^k + H_m2^k \tag{13-77}$$

TABLE 13-7 A DA Execution Table

i	Address Vector	Table $\langle\phi\rangle_6$	ACC	ACC = ACC/2 ± ϕ
3	$[x[k{:}0\{\text{LSB}\}] \to [0001]$	−0.5781	0	0+(−0.5781) = −0.5781
2	$[x[k{:}1] \to [0001]$	−0.5781	−0.07226	−0.2890+(−0.5781) = −0.8671
1	$[x[k{:}2] \to [0001]$	−0.5781	−0.21679	−0.4335+(−0.5781) = −1.0116
0	$[x[k{:}3] \to [0100]$	−0.9062	−0.50585	−0.5058+(−0.9062) = 0.40036

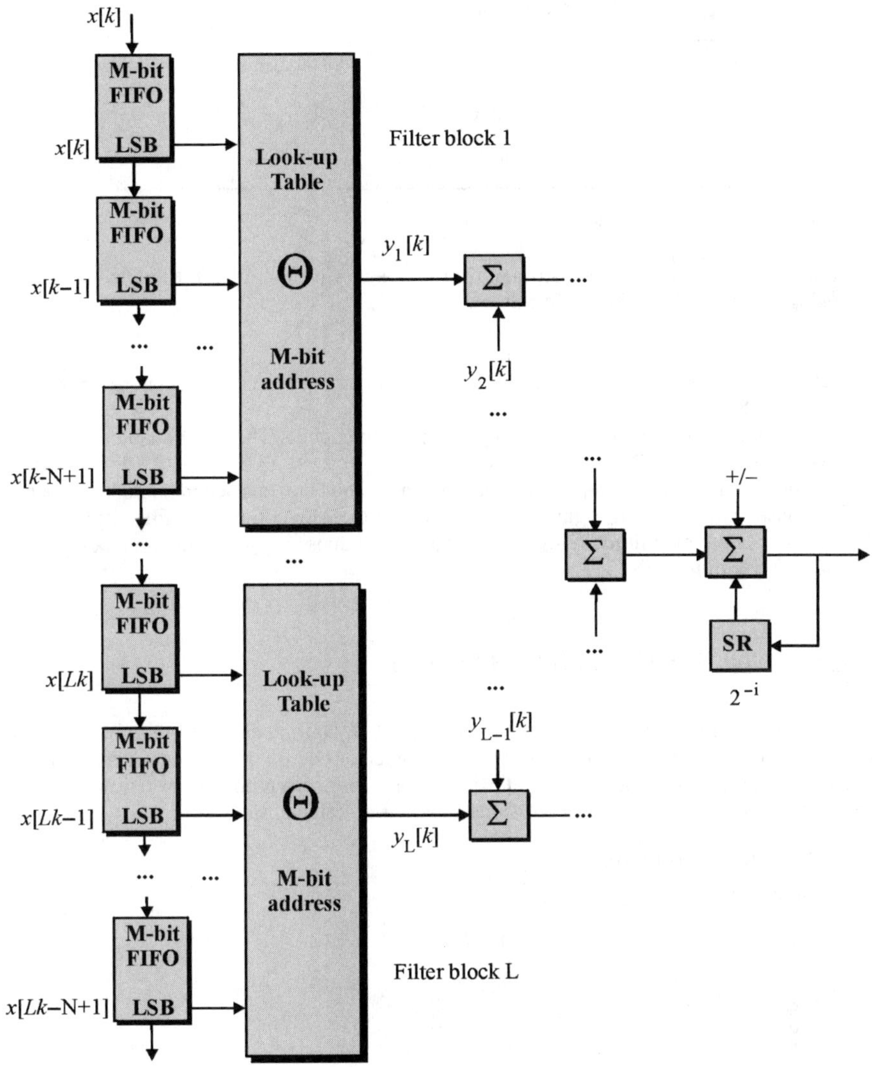

FIGURE 13-40 A high-order DA architecture.

where $H_i = \{1, 0, -1\}$. This means that a conventional N-bit data word can be replaced by a one-, two-, or three-digit number. The product $x[k]c_i$ can therefore be performed using only a few weighted shift-adds in comparison to up to N shift adds for a conventionally coded system. A CSD example follows.

Example 13-23 Canonic signed digit FIR

Required:

The 15th-order FIR, having an impulse response $h[k] = [-1, 4, -16, 32, -64, 136, -352, 1312, -352, 136, -64, 32, -16, 4, -1]$, has a magnitude frequency response

TABLE 13-8 CSD Coefficients

Coefficient	CSD	Number of Binary Digits
1312	1024 + 256 + 32	3
−352	−256 − 64 − 32	3
136	128 + 8	2

that can compensate for an upstream filter having a sin(*x*)/*x* roll-off. Implement the FIR using a canonic signed digit form.

Results:
Using a computer, search the CSD codes for the FIR coefficients, as shown in Table 13-8.

All other coefficients require only one digit to encode. The CSD-enabled FIR filter cycle would require $2 \cdot 5$ (one digit coefficients) $+ 2 \cdot 1 + 3 \cdot 1 + 3 \cdot 1 = 18$ shift-add operations. The uncompensated, compensator, and compensated magnitude frequency responses are shown in Figure 13-41. Note that the low-complexity FIR improves the overall system magnitude frequency response and does so without the high complexity of a conventional FIR.

13.29 FINITE WORDLENGTH EFFECT

The worst case FIR gain is given by $G_{max} = \sum|h_i|$. It shall be assumed that the input has been properly scaled so that no runtime overflow error occurs or that the register's width has been extended by $\lceil \log_2(G_{max}) \rceil$-bits. The errors that occur within an overflow-free system are attributed to what are called finite wordlength effects, such as:

- Coefficient roundoff errors
- Arithmetic errors

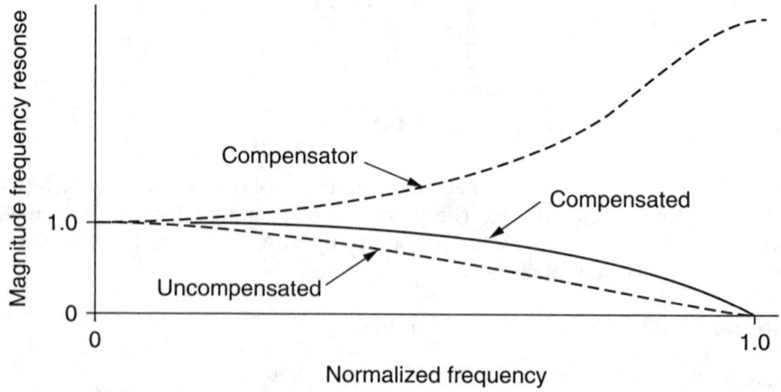

FIGURE 13-41 A CSD implementation of a 15th-order FIR. (The frequency axis was normalized with respect to $f_s/2$.)

13.30 COEFFICIENT ROUNDING

Coefficient roundoff errors correspond to errors attributable to coding real coefficients into data words containing F-bits of fractional precision. It is assumed that the rounding of a real coefficient h_i results in $h_i' = h_i + \varepsilon_i$, where $|\varepsilon_i| \leq 2^{-(F-1)}$ (for instance, ±LSB/2). The size of the error is seen to be a direct function of F, the number of fractional bits used in the finite wordlength representation of h_i. Based on this model, the linear convolution outcome of an ideal and fixed-point system can be compared. Specifically, the linear convolution of an FIR having an impulse response h_k with an input $x[k]$, produces an outcome:

$$y[k] = \sum_{m=0}^{N-1} h_m x[k-m] \tag{13-78}$$

The convolution sum (Equation 13-2), performed using a fixed-point FIR impulse response, results in

$$y'[k] = \sum_{m=0}^{N-1} (h_m + \Delta_m) x[k-m] = y[k] + \sum_{m=0}^{N-1} \Delta_m x[k-m] \tag{13-79}$$

The error, due to coefficient rounding, is

$$e[k] = (y'[k] - y[k]) = \sum_{m=0}^{N-1} \Delta_m x[k-m] \tag{13-80}$$

The error is scaled by the individual values of $x[k]$. If the input time-series has a mean value of zero ($\boldsymbol{E}(x[k]) = 0$), and the FIR coefficients are assumed to be random, then the roundoff error will have a mean value of zero ($\boldsymbol{E}(y[k]) = 0$). Computing the error variance, however, is more challenging since it is a function of the input signal power. If the input is assumed to be an *impulse*, that is $x[k] = \delta[k]$, then Equation (13-80) simplifies to

$$e[k] = (y'[k] - y[k]) = \sum_{m=0}^{N-1} \Delta_m \delta[k-m] = \Delta_k \tag{13-81}$$

where Δk is a uniformly distributed random number defined over $[-2^{-F}/2, 2^{-F}/2)$. The error is seen, in this case, to have a mean value of zero ($\boldsymbol{E}(e[k]) = 0$) and variance $\sigma^2 = 2^{-2F}/12$ (for instance, $Q^2/12$). This is of little practical value, however, since the input signal is rarely just an impulse. If the input is *random* or *arbitrary*, then the error can be modeled as

$$\sigma_e^2 = \frac{N 2^{-2F} \sigma_x^2}{12} \tag{13-82}$$

where σ_x^2 is the signal variance (statistical power). Therefore, the coefficient roundoff error variance associated with an Nth-order FIR is essentially the roundoff error power associated with each rounding (such as, $Q^2/12$), scaled by the signal power and filter order. For example, the error variance of a 64th-order FIR is expected to be twice that of a 32nd-order filter. A standard means of analyzing this data is to interpret the error in bits [for instance, $\log_2(\sigma_e)$]. Therefore going from a 32nd-order to a 64th-order FIR costs only one additional bit of precision due to coefficient rounding. The coefficient roundoff error associated with DA or CSD filters are application-dependent. DA filters are less susceptible to coefficient rounding. The coefficient roundoff errors introduced by CSD encoding is dependent on the original coefficient distribution. Some coefficients lend themselves to exact CSD coding, while others resist. A illustration of the effects of coefficient roundoff is presented in the following example.

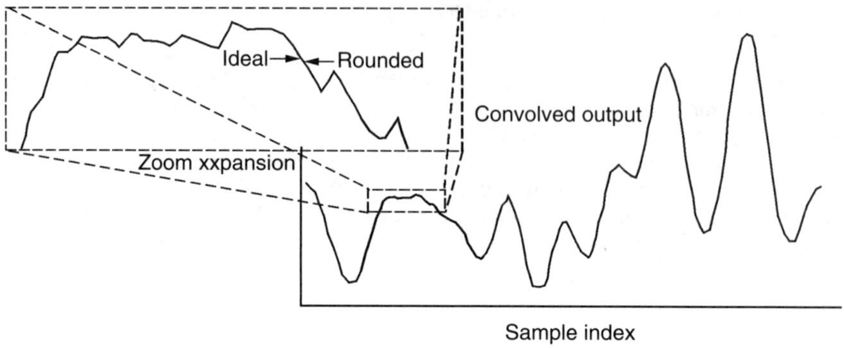

FIGURE 13-42 Coefficient roundoff error effects.

Example 13-24 Coefficient roundoff error

Required:
Experimentally examine the effects of coefficient roundoff errors in a simple FIR.

Results:
Consider a 31st-order FIR defined by a Hamming window. The coefficients are implemented using a floating point and [16:10] fixed-point format, reserving five bits for integer dynamic range growth, and can be used to suppress the possibility of worst-case register overflow. The results are shown in Figure 13-42. The test shows that the errors due to coefficient roundoff are negligible.

13.31 ARITHMETIC ERROR

Arithmetic errors can occur when partial products, of the form $x[k - m] \cdot h_m$, are computed using finite word-length multiply-accumulators (MAC). For an Nth-order FIR, there are N multiplications of two M-bit words having F fractional bits of precision, along with their attendant accumulation. The multiplication outcomes are assumed to be a $2M$-bit full precision word having $2F$ fractional bits of precision. For F sufficiently large ($F > 6$ bits), the rounding of a full precision fixed-point product back to F fractional bits is accomplished with a mean error of zero mean and variance $Q^2/12$, where $Q = 2^{-F}$. The end-to-end FIR error is therefore of zero mean and variance $\sigma^2 = NQ^2/12$. If the full-precision outcomes of each MAC call are presented to an extended precision accumulator (typically $2M + \lceil \log_2(G_{\max}) \rceil$-bits), then the act of multiply-accumulation will introduce no additional error. Returning the final outcome held in the extended precision accumulator register to F bits defines the output error variance to be $\sigma^2 = Q^2/12$.

The arithmetic errors associated with DA or CSD filters represents a set of special cases. A DA FIR is known to enjoy a precision advantage over direct or transpose implementations. An Nth-order DA FIR is assumed to have a statistical error budget of $\sigma^2 = Q^2/9$ versus $\sigma^2 = NQ^2/12$ for traditional arithmetic models. The errors introduced by a CSD implementation are due entirely to coefficient rounding since arithmetic errors are zero if extended precision adders are used.

13.32 SCALING

The worst case gain of an FIR filter gain is denoted $G_{\max}$. Intuitively, it would make sense to actually have the worst case gain fall on a power-of-two value, say 2^g. This can be achieved by scaling the FIR by a factor k, where

$$k = \frac{2^g}{G_{\max}} \tag{13-83}$$

By multiplying each FIR coefficient h_i by k, a new FIR would be created having a maximum gain of $G'_{\max} = 2^g$. This action would be performed offline and not affect the filter's runtime bandwidth. It should be appreciated, however, that the scaled filter's magnitude frequency response will be increased by a factor k. Scaling can take place in a downward direction as well. Logically, a scale factor $k = 2^{g-1}/G_{\max}$ can also be considered. This scale factor k can be used to adjust the gains downward to the nearest power-of-two value. While the absolute magnitude frequency response will be changed, the relative gain will be left unaffected.

13.33 MULTIPLE MAC ARCHITECTURE

As technology provides more powerful and compact arithmetic devices, multiple MACs' architectures are becoming a viable option. An FIR typically requires multiple MAC calls be performed per filter cycle. This provides motivation for implementing multiple MAC FIR architectures. Consider an Nth-order FIR convolution sum given by

$$y[k] = \sum_{m=0}^{N-1} h_m x[k - m] \tag{13-84}$$

where $x[k]$ is a bounded input time-series. The filter tap-weight coefficients are denoted h_k and are assumed to be real. Upon presenting an input to the FIR, the output time-series is given by

$$\begin{aligned} y[0] &= h[0]x[0] \\ y[1] &= h[1]x[0] + h[0]x[1] \\ \cdots &= \cdots \\ y[N-1] &= h[N-1]x[0] + h[N-2]x[1] + \cdots + h[0]x[N-1] \end{aligned} \tag{13-85}$$

For sample indices $k \geq N - 1$, N physical multiplications must be performed per filter cycle. The multiplies can be performed sequentially or concurrently. If M MACs are available for concurrent use, and N is divisible by M, then a $K = N/M$-fold speed-up can be realized. Suppose, however, that $N = KM + K_0$, $K_0 \in [1, N - 1]$, then $K + 1$ MAC cycles would be required where the first K cycles would use all M MACs concurrently and the last cycle would use only K_0 of the available MACs. The efficiency of this action can be mathematically represented by Δ:

$$\Delta = \frac{\text{actual speed} - \text{up}}{\text{ideal speed} - \text{up}} = \frac{N/M}{\lceil N/M \rceil} \tag{13-86}$$

where $\lceil \circ \rceil$ again denotes the ceiling function. The value of Δ is interpreted in Figure 13-43 for other cases. Notice that the efficiency improves as L increases, which simply reflects the fact that the overhead is reduced when a large number of MACs need to be performed.

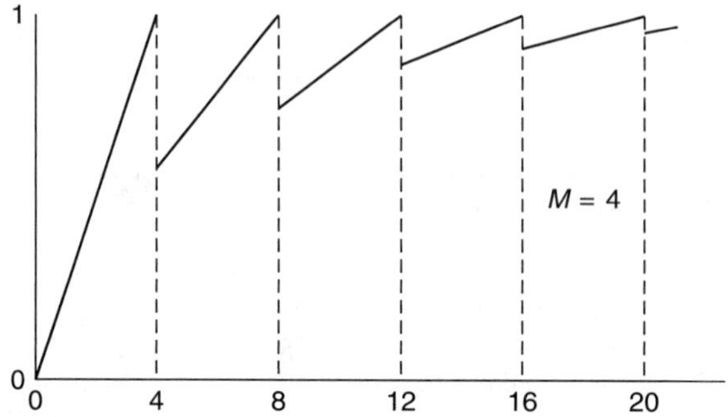

FIGURE 13-43 The speed-up potential of a four-multiplier FIR as a function of filter order.

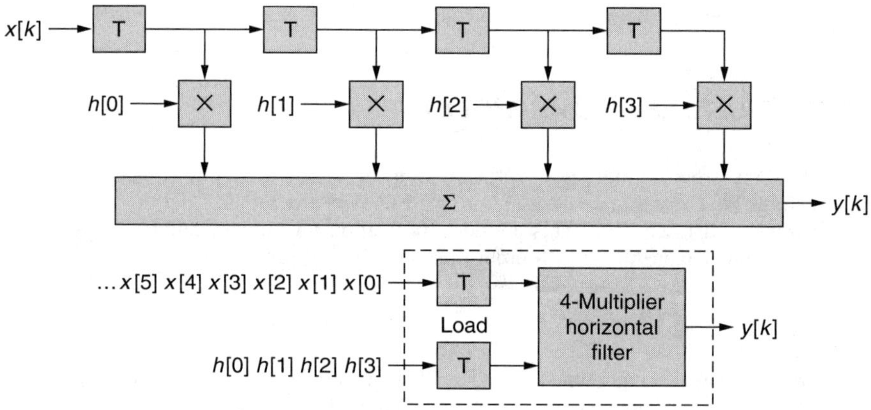

FIGURE 13-44 An example of a 4th-order four-multiplier horizontally (direct)-architected FIR.

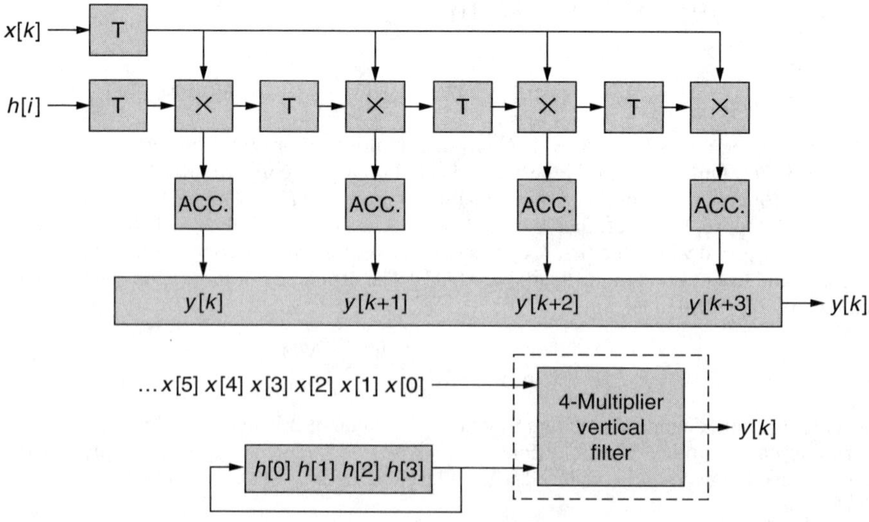

FIGURE 13-45 An example of a 4th-order four-multiplier vertically (transpose)-architected FIR.

TABLE 13-9 Four Multiplier Vertically (Transpose)-Architected FIR Execution

Clock	Cell 0	Cell 1	Cell 2	Cell 3	Sum/Clear
0	$h_3x[0]$	0	0	0	0
1	$h_2x[1]$	$h_3x[1]$	0	0	0
2	$h_1x[2]$	$h_2x[2]$	$h_3x[2]$	0	0
3	$h_0x[3]$	$h_1x[3]$	$h_2x[3]$	$h_3x[3]$	Cell 0 = $y[3]$
4	$h_3x[4]$	$h_0x[4]$	$h_1x[4]$	$h_2x[4]$	Cell 1 = $y[4]$
5	$h_2x[5]$	$h_3x[5]$	$h_0x[5]$	$h_1x[5]$	Cell 2 = $y[5]$
6	$h_1x[6]$	$h_2x[5]$	$h_3x[5]$	$h_0x[5]$	Cell 3 = $y[6]$
7	$h_0x[7]$	$h_3x[6]$	$h_2x[6]$	$h_3x[6]$	Cell 0 = $y[7]$

The multiple MAC filter shown in Figure 13-44 is called a *horizontal architecture* since the MACs are spread horizontally across the convolution space. It is assumed that the coefficients are loaded into registers and physically attached to each MAC.

A *vertical architecture* (same as transpose) is graphically interpreted in Figure 13-45. A sample execution of the first several clock instances of a four-multiplier vertically-architected FIR is shown in Table 13-9.

BIBLIOGRAPHY

Harris, J. F. "On the Use of Windows for Harmonic Analysis with Discrete Fourier Transforms." *Proceedings of the IEEE* (January, 1987).

McClellan, J. H., Parks, T. W., and L. R. Rabiner. "A Computer Program for Designing Optimum FIR Linear Phase Filters." *IEEE Trans. On Audio and Electroacoustics* (December, 1973).

Losada, R., "Practical FIR Filter Design in MATLAB." *The Mathworks Inc.* (January 12, 2004).

Taylor, F. J. *Digital Filter Design Handbook.* New York: Marcel Dekker, 1983.

——— and Stouraitis, T. *Digital Filter Design Software for the IBM PC.* New York: Marcel Dekker, 1987.

Zelniker, G. and Taylor, F. *Advanced Digital Signal Processing: Theory Applicants.* New York: Marcel Dekker, 1994.

CHAPTER 14
INFINITE IMPULSE-RESPONSE FILTERS

14.1 INTRODUCTION

Infinite impulse-response (IIR) filters contain feedback paths that can sustain an impulse response indefinitely. IIRs differentiate themselves from *finite impulse response* (FIR) filters at a number of levels. For a given set of frequency domain specifications:

- IIR solutions are lower order than an FIR, and
- IIR phase response is non-linear.

In general, FIRs are used in phase-sensitive applications and IIRs in phase-insensitive cases. To illustrate their differences, a 6th order IIR and 63rd order FIR meeting the same frequency domain requirements are compared side-by-side in Figure 14-1. Because the IIR is 1/10th the order of the FIR, it can potentially run 10 times faster or use 1/10th the resources. It can be seen that both filters have similar magnitude frequency responses but differ significantly in phase performance. The FIR's phase response, for example, is linear while the IIR's phase response is highly non-linear, as evidenced by their group delay performance.

An IIR can be modeled in terms of a transfer function $H(z)$ expressed in terms of rational polynomials $N(z)$ and $D(z)$, or a collection of poles (β_i) and zeros (α_i), as utilized in Equation (14-1):

$$H(z) = \frac{N(z)}{D(z)} = \sum_{n=0}^{\infty} h(n)z^{-n} = k\frac{\sum_{i=0}^{M} b_i z^{-i}}{1 + \sum_{i=1}^{N} a_i z^{-i}} = K \times \frac{\prod_{i=1}^{M}(z - \alpha_i)}{\prod_{i=1}^{N}(z - \beta_i)} \tag{14-1}$$

The IIR's frequency response is, therefore, given by:

$$H(e^{j\varpi}) = \frac{N(e^{j\varpi})}{D(e^{j\varpi})} = k\frac{\sum_{i=0}^{M} b_i e^{-ji\varpi}}{1 + \sum_{i=1}^{N} a_i e^{-ji\varpi}} = K \times \frac{\prod_{i=1}^{M}(e^{j\varpi} - \alpha_i)}{\prod_{i=1}^{N}(e^{j\varpi} - \beta_i)} \tag{14-2}$$

where $\varpi \in [-\pi, \pi)$. The objective of an IIR design effort is to produce the filter's transfer function. The design process normally begins with the specification of the desired magnitude frequency-response $H(e^{j\varpi})$. The filter's magnitude frequency response is specified in terms of passband and stopband requirements, as shown in Figure 14-2. Specifically, the

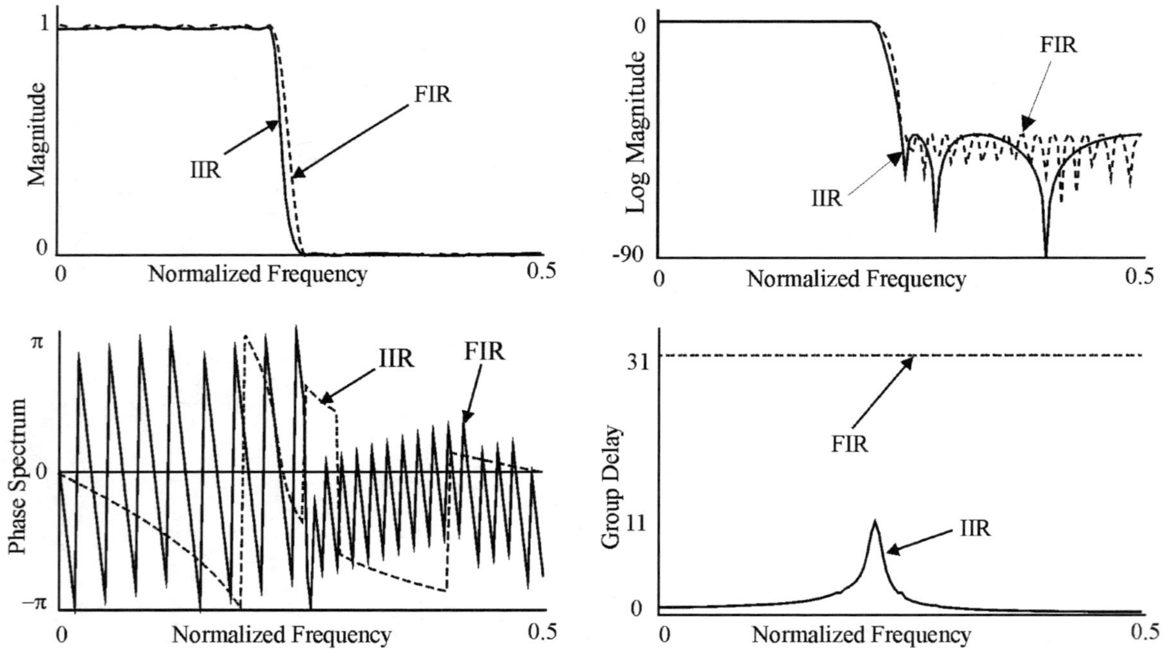

FIGURE 14-1 Comparison of a 63rd-order FIR and a 6th-order IIR low-pass filter. Displayed are the magnitude frequency responses (top left), log magnitude frequency responses (top right), phase responses (bottom left), and group delays (bottom right). (The frequencies have been normalized with respect to f_s.)

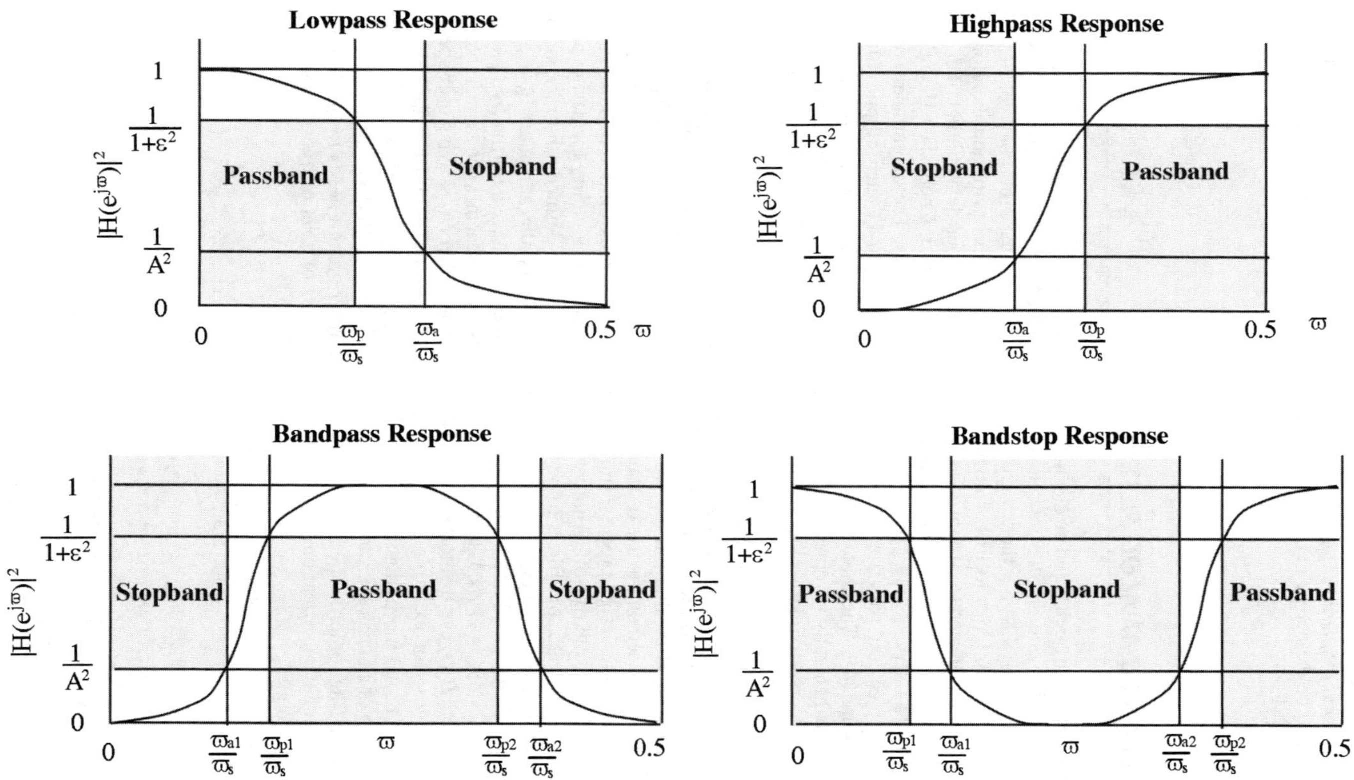

FIGURE 14-2 Typical magnitude-squared design objectives for low-pass, high-pass, bandpass, and bandstop IIR filters.

magnitude frequency response is defined in terms of a passband *critical frequency* (Ω_p), *stopband critical frequency* (Ω_s), and pass- and stopband gain delimiters ε and A. The band gains can also be expressed in decibel units. For example, a −3-dB pass band gain corresponds to $\varepsilon = 1$.

14.2 CLASSIC ANALOG FILTERS

Digital IIR filters are often based on classical Bessel, Butterworth, Chebyshev I, Chebyshev II, and Cauer (elliptic) *analog* filter models having a transfer function $H(s)$. Analog Bessel filters have historically had the flattest group delay behavior of the group of five. Bessel filters, however, are inferior to FIRs in achieving linear phase behavior and therefore have generally been eliminated from consideration as a viable flat group delay digital filter model. The remaining four analog models—Butterworth, Chebyshev I, Chebyshev II, and Cauer (elliptic) filters—serve as the principal enabler of IIR designs. The magnitude frequency response of typical low-pass Butterworth, Chebyshev I and II, and Cauer (elliptic) filters are shown in Figure 14-3. The choice of which model to use is the responsibility of the designer. However, it is generally accepted that for a given filter specification, the elliptic filter will result in the lowest order solution, followed by Chebyshev I and II, and finally the Butterworth option.

In general, a classic analog filter attempts to emulate the response of an ideal low-pass, high-pass, bandpass, bandstop, or all-pass filter. Much is known about these filter classes whose history can be traced to the early days of radio. Radio engineers routinely needed to design band selectable filters in the pre-digital era. As a result, they developed a design strategy that was based on the use of standardized filter models using tables and graphs. In order to expedite the design process, filter options were typically limited to those having a −1 or −3 dB passband roll-off. Another compromise made to simplify the design process was to define all realizable filters to be an extension of a pre-defined nth-order low-pass *analog prototype filter* $H_p(s)$. These prototype filters carry the attributes of a general Butterworth, Chebyshev I and II, or Cauer (elliptic) filter, but have a normalized passband cutoff frequency of $\Omega = 1$ r/s. Using the frequency translations schemes shown in Table 14-1, prototype filters $H_p(s)$ can be morphed into a desired analog filter $H(s)$ having user-specified design attributes (for example, $H_p(s) \rightarrow H(s)$). Included in Table 14-1 are the corresponding MATLAB morphing functions.

The design of an analog prototype filter $H_p(s)$, and therefore $H(s)$, can be finalized once the prototype filter order is specified. The minimum filter order n needs to satisfy a set of frequency domain specifications that are defined in terms of a set of hybrid design parameters including:

$$\text{Gain Transition Ratio: } \eta = \frac{\varepsilon}{\sqrt{A^2 - 1}} \tag{14-3}$$

and application-specific *transition frequency* ratios:

$$\text{Low-pass IIR: } k_d = \Omega_p/\Omega_a \tag{14-4}$$

$$\text{High-pass IIR: } k_d = k_1 \text{ if } \Omega_u^2 \geq \Omega_v^2;\ k_1 \text{ if } \Omega_u^2 < \Omega_v^2 \tag{14-5}$$

$$\text{Bandpass IIR: } k_d = 1/k_1 \text{ if } \Omega_u^2 \geq \Omega_v^2;\ 1/k_1 \text{ if } \Omega_u^2 < \Omega_v^2 \tag{14-6}$$

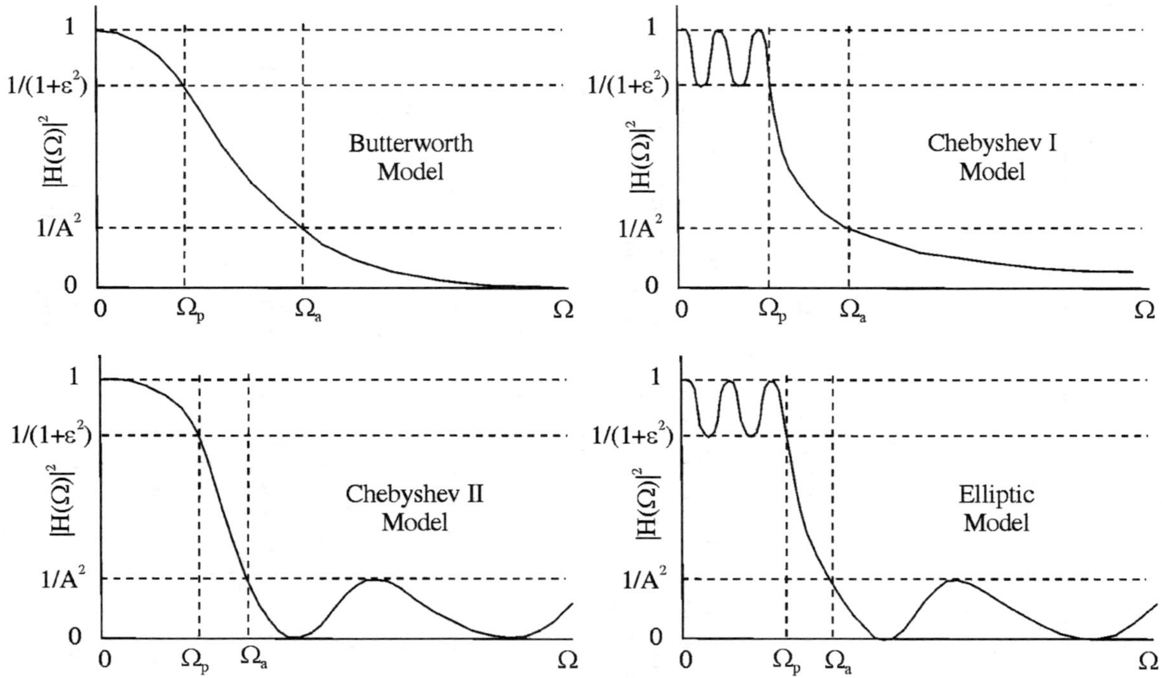

FIGURE 14-3 Magnitude-squared frequency-response profiles for a low pass Butterworth (top left), Chebyshev I (top right), Chebyshev II (bottom left), and elliptic analog filter (bottom right). The Butterworth filter has a smooth passband and stopband, the Chebyshev I has a rippling passband, the Chebyshev II has a rippling stopband, and the elliptic exhibits ripple in both the passband and stopband.

TABLE 14-1 Frequency-Frequency Transforms

Target	Transformation	Final Order	MATLAB
Low-pass	$j\Omega \leftarrow j\Omega/\Omega_p$	n	**lp2lp**
High-pass	$j\Omega \leftarrow \Omega_p/j\Omega$	n	**lp2hp**
Bandpass	$j\Omega \leftarrow ((j\Omega^2 + \Omega_{p1}\Omega_{p2})/(j\Omega + (\Omega_{p2} - \Omega_{p1}))$	$2n$	**lp2bp**
Bandstop	$j\Omega \leftarrow (j\Omega + (\Omega_{p2} - \Omega_{p1}))/((j\Omega^2 + \Omega_{p1}\Omega_{p2})$	$2n$	**lp2bs**

where

$$\Omega_c = \left(\Omega_{p_1}\Omega_{p_2}\right)/\left(\Omega_{p_2} - \Omega_{p_1}\right); \tag{14-7}$$

$$k_1 = \left(\Omega_{a_1}/\Omega_c\right)/\left(1/\left(1 - \Omega_{a1}^2/\Omega_v^2\right)\right); \tag{14-8}$$

$$\Omega_u^2 = \Omega_{a_1}\Omega_{a_2};\ \Omega_v^2 = \Omega_{p_1}\Omega_{p_2}; \tag{14-9}$$

$$k_2 = \left(-\Omega_{a_1}/\Omega_c\right)/\left(1/\left(1 - \Omega_{a2}^2/\Omega_v^2\right)\right) \tag{14-10}$$

Collectively, the parameters η and k_d define the steepness of the filter's skirt. Other hybrid parameters found in common use are $\delta^2 = A^2 - 1$, $d = \delta/\varepsilon$, and $D = \delta^2$. The parameters δ, d, and D are a measure of the depth of the transition region. From this database, the order of the low-pass analog prototype filter can be computed as follows:

$$\text{Butterworth: } n = \log(d)/\log(1/k_d) \tag{14-11}$$

$$\text{Chebyshev I and II: } n = \log((1 + \sqrt{1 - \eta^2})/\eta)/\log(1/k_d + \sqrt{(1/k_d)^2 - 1}) \tag{14-12}$$

$$\text{Cauer (elliptic): } \begin{cases} n \geq \log(16D)/\log(1/q);\ D = d^2;\ d = 1/\eta;\ k' = \sqrt{1 - k_d^2}: \\ q_0 = (1 - \sqrt{k'})/(2(1 + \sqrt{k'}));\ q = q_0 + 2q_0^5 + 15q_0^9 + 15q_0^{13}; \\ D = d^2;\ d = 1/\eta \end{cases} \tag{14-13}$$

In the following example, the order of a -3 dB analog Butterworth, Chebyshev I and II, and elliptic filter are computed.

Example 14-1 Classic Analog Filter

Required:
The critical design parameters for a -3-dB low-pass Butterworth IIR, having a 500 Hz passband and a -40-dB attenuation band beginning at 1000 Hz, are

$\Omega_p = 2\pi \times 500 = 1000\pi$ r/s
$\Omega_a = 2\pi \times 1000 = 2000\pi$ r/s
$k_d = \Omega_p/\omega_a = 0.5$
$\varepsilon = 1$ (-3-dB roll-off)

$A_a = 10^2$ (40-dB attenuation)

$\eta = \varepsilon/\sqrt{A_a^2 - 1} = 1/\sqrt{10^4 - 1} = 0.0100005$

Compute the minimal filter order required of a Butterworth, Chebyshev, and elliptic design.

Results:

The computed filter orders are reported next. The resulting prototype filter provides the basis for the creation of their digital IIR filter counterparts.

Butterworth (Equation 14-11): $n = 6.64378 < 7$

Chebyshev I and II (Equation 14-12): $n = 4.02 < 5$

Elliptic (Equation 14-13): $n = 3.75 < 4$

14.3 MATLAB ANALOG FILTER PRODUCTION

The production of analog filters and prototypes, once an exacting manual process, has today been reduced to executing computer programs. A number of software packages are now available that provide a seamless means of converting user-frequency domain specifications into a desired filter model $H_p(s)$ or $H(s)$. MathWorks Signal Processing Toolbox (SPT) contains many of the program objects needed to design a classic analog filter. Table 14-2 contains the design support needed to create Butterworth, Chebyshev I and II, and Cauer (elliptic) analog prototype filters (for instance, $H_p(s)$). The prototypes $H_p(s)$ can be mapped into a final analog filter $H(s)$ using a frequency-frequency transform, as presented in Table 14-1. Table 14-3 provides the tools needed to implement an analog filter having user-specified frequency domain attributes [such as $H(s)$] and includes an embedded frequency-frequency transform.

TABLE 14-2 MATLAB Analog Prototype Filter Options ($\omega_p = 1$ r/s)

Program	Syntax	Description
buttap: Butterworth low-pass prototype filter	$[z, p, k]$ = buttap(n)	Returns the z = zeros, p = poles, and k = gain of an nth order prototype Butterworth analog low-pass filter
cheb1ap: Chebyshev I low-pass prototype filter	$[z, p, k]$ = cheb1ap(n, R)	Returns the z = zeros, p = poles, and k = gain of an nth-order Chebyshev I analog low-pass prototype filter with an R-dB ripple passband
cheb2ap: Chebyshev II low-pass prototype filter	$[z, p, k]$ = cheb2ap(n, S)	Returns the z = zeros, p = poles, and k = gain of an nth order Chebyshev Type II analog low-pass prototype filter with an S-dB stopband
ellipap: Elliptic low pass prototype filter	$[z, p, k]$ = ellipap(n, R, S)	Returns the z = zeros, p = poles, and k = gain of an nth order elliptic analog low-pass prototype filter with an R-dB passband ripple and S-dB stopband ripple

TABLE 14-3 MATLAB Analog Filter Options

Program	Syntax	Description
butter: Butterworth analog and digital filter design	$[b, a]$ = butter(n, W, 's') $[b, a]$ = butter(n, W, '*ftype*', 's') $[z, p, k]$ = butter(n, W, 's') $[z, p, k]$ = butter(n,W, '*ftype*', 's')	Designs an nth-order analog Butterworth filter with angular cutoff frequency W r/s. Returns the transfer function $H(s) = B(s)/A(s)$ coefficients or z = zeros, p = poles, and k = gain. If $W = [w1\ w2]$, $w1 < w2$, *butter* returns a bandpass filter. '*ftype*' denotes a high-pass, low-pass, or bandstop filter.
cheby1: Chebyshev I filter design	$[b, a]$ = cheby1(n, R, W, 's') $[b, a]$ = cheby1(n, R, W, '*ftype*', 's') $[z, p, k]$ = cheby1(n, R, W, 's') $[z, p, k]$ = cheby1(n, R, W, '*ftype*', 's')	Designs an nth-order analog Chebyshev I filter with angular cutoff frequency W r/s and passband ripple R. Returns the transfer function $H(s) = B(s)/A(s)$ coefficients or z = zeros, p = poles, and k = gain. If $W = [w1\ w2]$, $w1 < w2$, *cheby1* returns a bandpass filter. '*ftype*' denotes a high-pass, low-pass, or bandstop filter.
cheby2: Chebyshev II filter design.	$[b, a]$ = cheby2(n, R, W, 's') $[b, a]$ = cheby2(n, R, W, '*ftype*', 's') $[z, p, k]$ = cheby2(n, R, W, 's') $[z, p, k]$ = cheby2(n, R, W, '*ftype*','s')	Designs an nth-order analog Chebyshev II filter with angular cutoff frequency W r/s and passband ripple R. Returns the transfer function $H(s) = B(s)/A(s)$ coefficients or z = zeros, p = poles, and k = gain. If $W = [w1\ w2]$, $w1 < w2$, *cheby2* returns a bandpass filter. '*ftype*' denotes a high-pass, low-pass, or bandstop filter.
ellip: Elliptic (Cauer) filter design	$[b, a]$ = ellip(n, R, S, W,'s') $[b, a]$ = ellip(n, R, S, W, '*ftype*','s') $[z, p, k]$ = ellip(n, R, S, W, 's') $[z, p, k]$ = ellip(n, R, S, W, '*ftype*', 's')	Designs an nth-order analog Chebyshev II filter with angular cutoff frequency W r/s, passband ripple R, and stopband ripple S. Returns the transfer function $H(s) = B(s)/A(s)$ coefficients or z = zeros, p = poles, and k = gain. If $W = [w1\ w2]$, $w1 < w2$, *cheby2* returns a bandpass filter. '*ftype*' denotes a high-pass, low-pass, or bandstop filter.

14.4 IMPULSE INVARIANT IIR

One obvious means of converting a classical analog filter $H(s)$ into a digital filter $H(z)$ is through the application of the standard z-transform. The standard z-transform is also referred to as an *impulse invariant transform* because the mapping $H(s) \rightarrow H(z)$ preserves the fidelity of the impulse response. Specifically, the standard z-transform produces a digital filter having an impulse response that agrees with the analog filter's impulse response at the sample instances. If the digital filter $H(z)$ is to be used as a replacement for an analog filter with specific time domain attributes, then an impulse invariant digital filter is recommended. If the analog filter to be replaced is defined in terms of frequency domain attributes and requirements, then an impulse invariant design is generally avoided. The reason for this claim can be found in the Sampling Theorem.

The standard z-transform converts an analog filter $H(s)$, with frequency response $H(j\Omega)$, into a digital filter $H(z)$, with frequency response $H(e^{j\varpi})$. Furthermore, the digital filter's base-band frequency response is repeated on $\varpi = \pm 2\pi n$ centers, as shown in Figure 14-4. The frequency limits of a digital filter's baseband extend out the Nyquist frequencies $\pm\varpi_s/2$. Any filter energy existing beyond baseband will be *aliased* back into the impulse invariant digital filter's baseband. Since most analog filters will have a finite gain at any reasonable Nyquist frequency value, aliasing can generally be expected. This condition is graphically interpreted in Figure 14-4 which illustrates how the frequency response

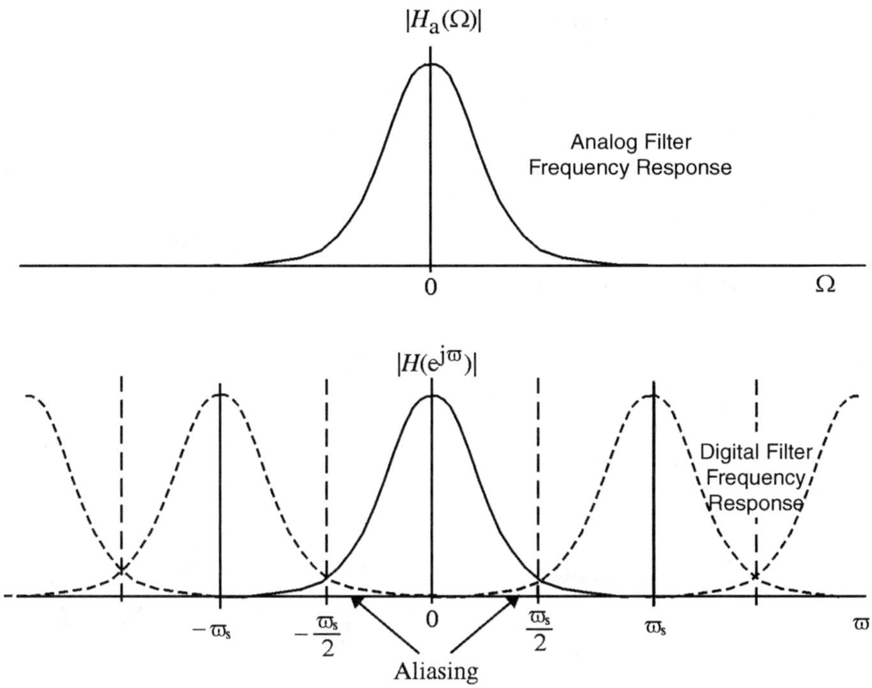

FIGURE 14-4 The spectrum of a typical analog filter and equivalent impulse-invariant digital filter.

of a typical analog filter extends beyond the baseband of the derivative digital filter. This fact points to a serious design limitation that restricts the ability of impulse-invariant filters to meet specific frequency domain requirements. In the next example, an impulse invariant Butterworth filter is designed.

Example 14-2 Impulse Invariant Filter

Required:
A 2nd-order $\varepsilon^2 = 0.2589$ (for example, -1-dB) Butterworth lowpass digital filter, having a 1 kHz passband, has a transfer function given by

$$H(s) = \frac{7.757 \times 10^7}{s^2 - 1.2454 \times 10^4 s + 7.757 \times 10^7}$$

Compare the magnitude frequency response of the analog and derivative digital filter for sample rates of $f_s = 5$ kHz, 10 kHz, and 50 kHz.

Results:
The filter's transfer function can be expressed as a Heaviside expansion, containing terms of the form

$$\frac{\omega_0}{(s+a)^2+\omega_0^2} = \frac{-0.5j}{s+(a-j\omega_0)} + \frac{0.5j}{s+(a+j\omega_0)}$$

$$\overset{L}{\longleftrightarrow} \left(-0.5je^{(-a+j\omega_0)t} + 0.5je^{(-a-j\omega_0)t}\right)u(t) = e^{-at}\sin(\omega_0 t)u(t)$$

The individual exponential terms can then be converted into the z-transform form using the standard z-transform, resulting in

$$\alpha^k \sin(\beta k T_s) \overset{L}{\longleftrightarrow} \frac{2\alpha z \sin(\beta T_s)}{T_s(z^2 - 2\alpha z \cos(\beta T_s) + \alpha^2)}$$

for a sample frequency of $f_s = 5$ kHz and $T_s = 1/f_s$, $\alpha = e^{-6.228\times10^3 T_s}$, and $\beta = 6.228 \times 10^3$. The poles of the impulse invariant digital filter are located at

$$z = e^{-sT_s} = e^{-(6.228\times10^3(1\pm j1)/5\times10^3)} = 0.0919 \pm j0.272686$$

This results in the digital filter model

$$H(z) = \frac{Kz}{(z - 0.0919)^2 + (0.272686)^2}$$

where K can be used to adjust the DC gain to unity. The process can be repeated for $f_s = 10$ kHz and 50 kHz. The analog and digital filter comparisons are displayed in Figure 14-5. Note that the impulse-invariant magnitude frequency response is very

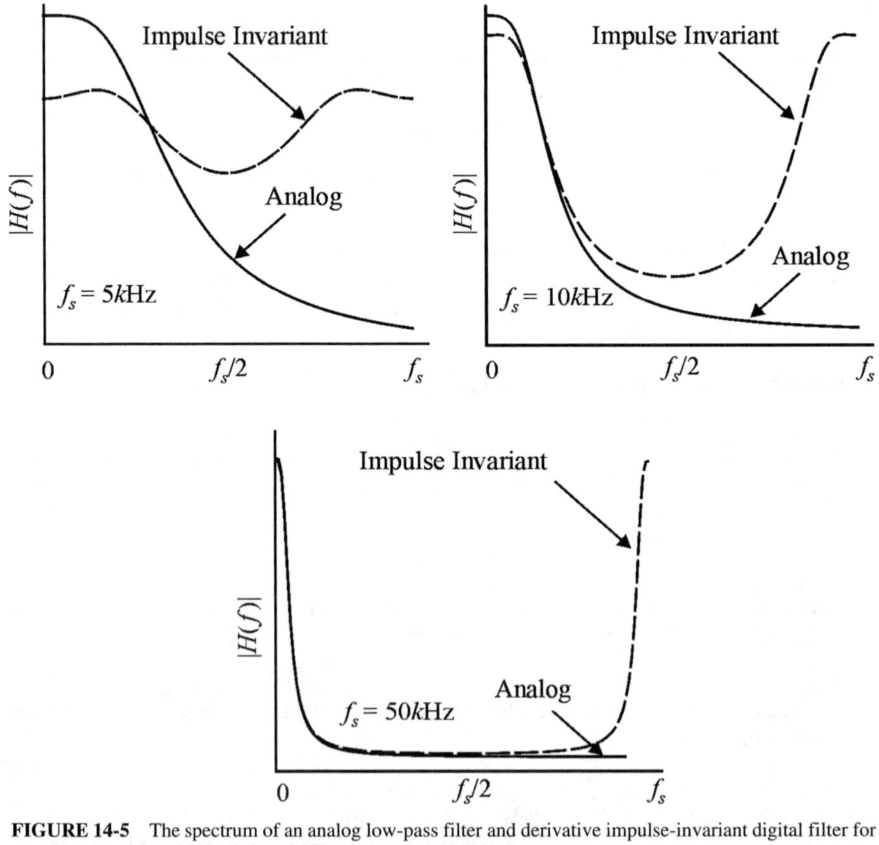

FIGURE 14-5 The spectrum of an analog low-pass filter and derivative impulse-invariant digital filter for sample rates of $f_s = 5$ kHz (left), 10 kHz (right), and 50 kHz (bottom).

sensitive to the choice of sample frequency. The effect of aliasing is especially apparent for the low sample rate case. This design would, in general, be considered unacceptable in most instances that are based solely on meeting posted frequency domain specifications. Obviously, an alternative design strategy is required.

14.5 BILINEAR z-TRANSFORM IIR

The common method used to develop a digital IIR filter version of a classic analog Butterworth, Chebyshev, or Elliptic filter model is the bilinear z-transform (see Section 12.13). The bilinear z-transform is given by

$$s = \frac{2}{T_s}\frac{(z-1)}{(z+1)} \quad \text{or} \quad z = \frac{((2/T_s)+s)}{((2/T_s)-s)} \tag{14-14}$$

The bilinear z-transform is also characterized by a non-linear mapping between analog and digital frequencies. The non-linear frequency distortion is referred to as *warping* and is defined by (see Figure 12-11)

$$\Omega = \left(\frac{2}{T_s}\right)\tan\left(\frac{\varpi}{2}\right) \quad \text{or} \quad \varpi = 2\tan^{-1}\left(\frac{\Omega T_s}{2}\right) \tag{14-15}$$

where Ω is the analog frequency and $\varpi \in [-\pi, \pi)$ is the normalized digital frequency in radians per second (r/s). The bilinear z-transform filter design paradigm is not a straightforward process due to the fact that it involves the use of nonlinear frequency mapping rules. A step-by-step design strategy is outlined next and illustrated in Figure 14-6:

1. Specify the desired digital filter frequency domain requirements and attributes. Prewarp the digital critical digital frequencies ϖ (Equation 14-15) into corresponding critical analog frequencies Ω.
2. From the prewarped analog frequencies Ω, and passband and stopband gain specifications, determine the required order of the analog prototype filter $H_p(s)$.
3. Using frequency-frequency transforms (Table 14-1), convert the analog prototype filter $H_p(s)$ into an analog filter $H(s)$ having the desired prewarped analog frequencies Ω.
4. Convert the pre-warped analog filter $H(s)$ into a digital filter $H(z)$ using a bilinear z-transform (Equation 14-14) which automatically warps the prewarped frequencies back to their original values ϖ.

The next example investigates the computation of an IIR filter order.

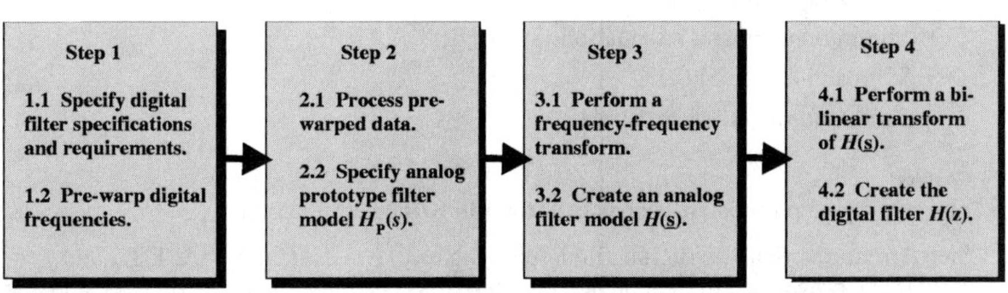

FIGURE 14-6 The IIR design paradigm based on the bilinear z-transform.

Example 14-3 IIR Order Determination

Required:
The determination of the analog filter order is required to complete a bilinear z-transform design process. The order estimation, required in Step 2 of Figure 14-6 can be automated using a set of pre-defined MATLAB functions *buttord, cheb1ord, cheb2ord,* and *ellipord.* Using these functions, determine the required order of the classic low-pass IIRs if sampled at 1000 Hz, having a maximum 3-dB fall-off or ripple in a passband ranging over 0 to 40 Hz, and at least 60 dB of attenuation in the stopband ranging from 150 Hz to the Nyquist frequency 500 Hz.

Results:
The filter orders can be computed as follows:

```
Wp = 40/500; Ws = 150/500;
[n,Wn] = buttord(Wp,Ws,3,60)
n =
        5  {order}
[n,Wn] = cheb1ord(Wp,Ws,3,60)
n =
        4 {order}
[n,Wn] = cheb2ord(Wp,Ws,3,60)
n =
        4 {order}

[n,Wn] = ellipord(Wp,Ws,3,60)
n =
        4 {order}
```

The next example illustrates the step-by-step bilinear z-transform design process.

Example 14-4 Bilinear z-Transform Filter

Required:
Synthesize a digital Butterworth filter using the bilinear z-transform that meets or exceeds the following specifications:

- Maximum band attenuation $= -3$ dB
- Passband $f \in [0, 1]$ kHz
- Minimum stopband attenuation $= -10$ dB
- Stopband $f \in [2, 5]$ kHz
- Sample frequency $f_s = 10$ kHz

Results:
The step-by-step procedure consists of the following set of actions:

Step 1 requires that the digital filter frequencies ($f_p = 1$ kHz, $f_a = 2$ kHz, and $f_s =$ 10 kHz) be prewarped to their analog counterparts. The prewarped analog passband and stopband critical frequencies are

$$\varpi_p = 2\pi f_p/f_s = 2\pi(0.1) = 0.2\pi$$

$$\Omega_p = \frac{2}{T_s}\tan(\varpi_p/2) = 20 \times 10^3 \tan(0.1\pi) = 6498 \text{ r/s} \to 1.0345 \text{ kHz} \qquad \{\text{passband}\}$$

$$\varpi_a = 2\pi f_a/f_s = 2\pi(0.2) = 0.4\pi$$

$$\Omega_a = \frac{2}{T_s}\tan(\varpi_a/2) = 20 \times 10^3 \tan(0.2\pi) = 14531 \text{ r/s} \to 2.312 \text{ kHz} \qquad \{\text{stopband}\}$$

Step 2 requires that an analog prototype filter $H_p(s)$ be constructed. From the given and derived data, the need for a 2nd-order -3-dB Butterworth prototype filter model can be established. The filter has a passband cutoff frequency of $\Omega = 1$, a normalized transition bandwidth ratio of $k_d = \Omega_p/\Omega_a = 6498/14531 = 0.447$, and a stopband gain bounded by -10 dB. The required 2nd-order analog prototype filter is given by

$$H_p(s) = \frac{1}{s^2 + 1.414s + 1}$$

Step 3: The prototype is converted into a prewarped analog filter $H(s)$ using a frequency-frequency transform. In this case, $H_p(s)$ is mapped into a final analog filter using a low-pass to low-pass frequency-frequency transform $s = s/\Omega_p$. The resulting analog filter becomes

$$H_a(s) = \frac{4.3 \times 10^7}{s^2 + 9.2 \times 10^3 s + 4.3 \times 10^7}$$

Step 4: The final step is to apply the bilinear z-transform to $H(s)$ to form $H(z)$. This process warps the critical prewarped frequencies of $H_a(j\varpi)$, namely Ω_p and Ω_a, into the desired digital-critical frequencies ϖ_p and ϖ_a, resulting in

$$H(z) = \frac{0.0676(z + 1)^2}{z^2 - 1.142z + 0.412}$$

The next example reports a Butterworth, Chebyshev I and II, and elliptic IIR design outcome using the bilinear z-transform.

Example 14-5 Classic Bilinear z-Transform Filter

Required:
Synthesize a lowpass Butterworth, Chebyshev I and II, and elliptic IIR filter that meets or exceeds the following specifications using the bilinear z-transform. The digital filter design specifications are

- Maximum passband attenuation = 1.0 dB
- Minimum stopband attenuation = 40.0 dB
- Sampling frequency = 10000.0 Hz
- Passband edge = 1500.0 Hz
- Stopband edge = 2500.0 Hz

Results:

Following the four-step process shown in Figure 14-6, the following outcomes result.
Butterworth, order $n = 8$; actual stopband attenuation = 40.9855 dB

Digital Filter
Scale Factor $k = 0.0005705414218556547$

Numerator Coefficients		Denominator Coefficients	
b 0:	1.000000000000000	a 0:	1.000000000000000
b 1:	8.000000000000000	a 1:	−2.829054555430326
b 2:	28.00000000000000	a 2:	4.309045376441891
b 3:	56.00000000000000	a 3:	−4.092118976022213
b 4:	70.00000000000000	a 4:	2.619609039717636
b 5:	56.00000000000000	a 5:	−1.132089946694619
b 6:	28.00000000000000	a 6:	0.3203684065766790
b 7:	8.000000000000000	a 7:	−0.05378515271965319
b 8:	1.000000000000000	a 8:	0.004084412125652116

Chebyshev I, order $n = 5$; actual stopband attenuation = 44.35666 dB

Digital Filter
Scale Factor $k = 0.002020169397617657$

Numerator Coefficients		Denominator Coefficients	
b 0:	1.000000000000000	a 0:	1.000000000000000
b 1:	5.000000000000000	a 1:	−3.162364647736195
b 2:	10.00000000000000	a 2:	4.760700364549023
b 3:	10.00000000000000	a 3:	−4.052794082948058
b 4:	5.000000000000000	a 4:	1.934390525887203
b 5:	1.000000000000000	a 5:	−0.4152867390282085

Chebyshev II, order $n = 5$; actual stopband attenuation = 44.3566 dB

Digital Filter
Scale Factor $k = 0.05239987594174734$

Numerator Coefficients		Denominator Coefficients	
b 0:	1.000000000000000	a 0:	1.000000000000000
b 1:	2.073170731707318	a 1:	−1.205636106513167
b 2:	3.170731707317074	a 2:	1.213622639361258
b 3:	3.170731707317074	a 3:	−0.4983685124449977
b 4:	2.073170731707318	a 4:	0.1583280351022219
b 5:	1.000000000000000	a 5:	−0.01358662911081143

Elliptic, order $n = 4$; actual stopband attenuation = 51.1447 dB

Digital Filter
Scale Factor $k = 0.02099872314742810$

Numerator Coefficients		Denominator Coefficients	
b 0:	1.000000000000000	a 0:	1.000000000000000
b 1:	1.567841933416378	a 1:	–2.346138658587813
b 2:	2.196319333663452	a 2:	2.690894615786962
b 3:	1.567841933416378	a 3:	–1.585086766991654
b 4:	1.000000000000000	a 4:	0.4130798064394767

The frequency domain behavior of each IIR filter is shown in Figure 14-7. It can be seen that in all cases the synthesized IIR, using the bilinear *z*-transform, is in close agreement with the magnitude frequency domain specifications. Notice that there is a non-linear frequency-dependent group delay ranging from near 0 to 16 sample delays. Furthermore, the most egregious phase or group delay differences are found in the filter's transition band. This non-linear phase distortion is, in general, difficult to correct with an all-pass phase-compensating filter. As a general rule-of-thumb, the non-linear group-delay equalization of an *N*th-order IIR would require another IIR of an order of at least 2*N*.

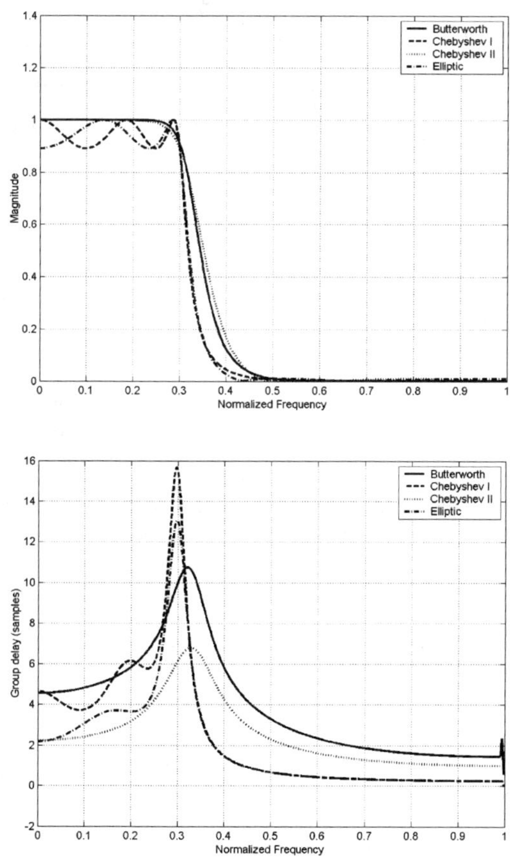

FIGURE 14-7 The IIR performance for the four classic filters in terms of magnitude frequency (top), and group delay (bottom) responses. (The frequency was normalized with respect to $f_s/2$.)

The following example performs a side-by-side time-domain comparison of an IIR designed using an impulse invariant transform and the bilinear z-transform.

Example 14-6 Impulse Invariant vs. Bilinear IIR

Required:
A system having a ramp impulse response $h(t) = tu(t)$ can be periodically sampled at a rate $f_s = 1$ Hz and then transformed into a digital filter using the standard and bilinear z-transform. While the system is unstable, it can nevertheless be used to explore the difference between the two design paradigms.

Results:
The impulse-invariant model $H(z)$ is defined in terms of the standard z-transform of $H(s) = 1/s^2$. The standard and bilinear z-transform of $H(s)$ is, in either case, a straightforward mapping process.

$$H_1(z)|_{\text{standard}-z} = \frac{z}{z^2 - 2z + 1} = \frac{z^{-1}}{1 - 2z^{-1} + z^{-2}}$$

$$H_2(z)|_{\text{bilinear}-z} = 0.25\frac{z^2 + 2z + 1}{z^2 - 2z + 1} = 0.25\frac{1 + 2z^{-1} + z^{-2}}{1 - 2z^{-1} + z^{-2}}$$

The short-term impulse responses can be computed using long division. Specifically:

$$\begin{array}{rl} & 1z^{-1} + 2z^{-2} + 3z^{-3} + \text{ramp time} - \text{series} \\ \text{Standard } z\text{-transform: } z^{-2} - 2z^{-1} + 1 & \overline{)\,z^{-2}} \\ & \underline{z^{-2} - 2z^{-1} + 1} \\ & \text{continue} \end{array}$$

or $h_1[k] = \{0, 1, 2, 3, \ldots\}$, which is recognized to be the sample values of the ramp, thereby reinforcing the impulse invariant claim of the standard z-transform. Next, consider

$$\begin{array}{rl} & 0.25 + 1z^{-1} + 2z^{-2} + \text{others} \\ \text{Bilinear } z\text{-transform: } 1 - 2z^{-1} + z^{-2} & \overline{)\,0.25 + 0.5z^{-1} + 0.25z^{-2}} \\ & \underline{0.25 - 0.5z^{-1} + 0.25z^{-2}} \\ & \text{continue} \end{array}$$

or $h_2[k] = \{0.25, 1, 2, \ldots\}$, which is not a ramp function. That is, the bilinear z-transform produces an impulse response which does not agree with the value of $h(t)$ at the sample instances and, as a result, is not an impulse-invariant mapping.

14.6 MATLAB CLASSIC IIR SUPPORT

At present, the primary sources of MATLAB IIR design software support are

- SPT = Signal Processing Toolbox
- FT = Filter Design Toolbox
- FT2 = Filter Design Toolbox 2

The basic software SPT options are listed in Table 14-4. Details regarding their functionality and syntax and theory of operation can be found online.

Traditional FIR modules, appearing in the SPT, are shown in Figure 14-8 and Table 14-5.

TABLE 14-4 MATLAB IIR Support

Function	Source	Description	Example
butter	SPT	Design a Butterworth analog and digital filter.	Figure 14-8*a*
cheby1	SPT	Design a Chebyshev Type-I filter (passband ripple).	Figure 14-8*b*
cheby2	SPT	Design a Chebyshev Type-II filter (stopband ripple).	Figure 14-8*c*
ellip	SPT	Design an elliptic (Cauer) filter.	Figure 14-8*d*
buttord	SPT	Calculate the order and cutoff frequency for a Butterworth filter.	
cheb1ord	SPT	Calculate the order for a Chebyshev Type-I filter	
cheb2ord	SPT	Calculate the order for a Chebyshev Type-II filter.	
ellipord	SPT	Calculate the minimum order for elliptic filters.	
maxflat	SPT	Design a generalized digital Butterworth filter.	
yulewalk	SPT	Design a recursive digital filter.	
bilinear	SPT	Bilinear transformation method for analog-to-digital filter conversion.	
impinvar	SPT	Impulse-invariance method for analog-to-digital filter conversion.	

TABLE 14-5 Classic IIR Signal Processing Toolbox Support

***Function:* butter**

Example:
Design a 9th-order high-pass Butterworth filter with a cutoff frequency of 300 Hz, where f_s = 1000 Hz ($f_s/2$ = 500).

MATLAB:
```
[b,a] = butter
(9,300/500,'high');
```

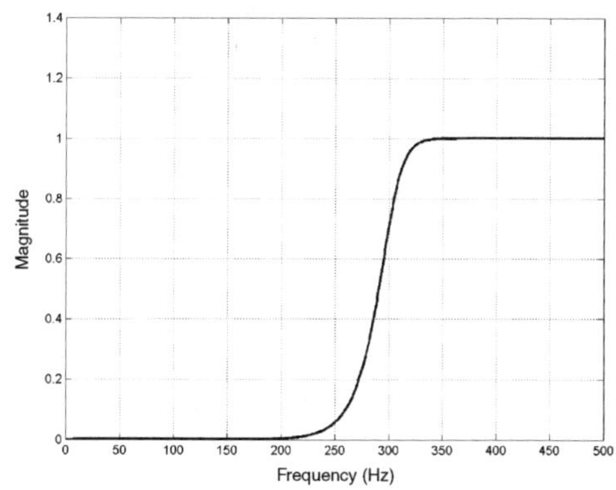

FIGURE 14-8a

***Function:* cheby1**

Example:
Design a 9th-order low-pass Chebyshev Type-I filter with 0.5 dB of ripple in the passband, a cutoff frequency of 300 Hz, and where f_s = 1000 Hz ($f_s/2$ = 500).

MATLAB:
```
[b,a] = cheby1
(9,0.5,300/500);
```

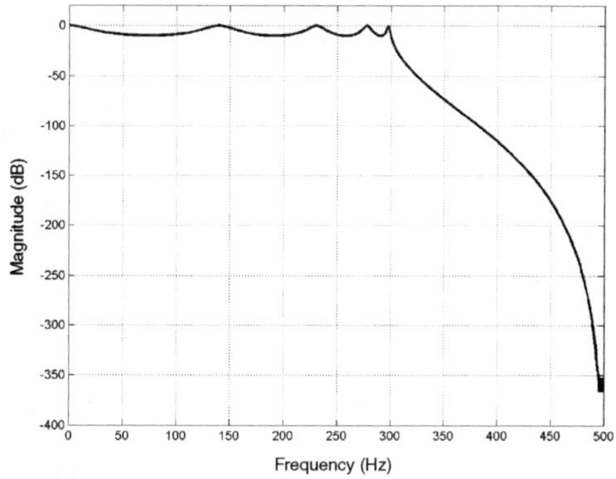

FIGURE 14-8b

(*Continued*)

TABLE 14-5 Classic IIR Signal Processing Toolbox Support (*Continued*)

***Function:* cheby2**

Example:
Design a 9th-order low-pass Chebyshev Type-II filter with stopband attenuation 20-dB down from the passband, with a cutoff frequency of 300 Hz, and where $f_s = 1000$ Hz ($f_s/2 = 500$).

MATLAB:
```
[b,a] = cheby2
(9,20,300/500);
```

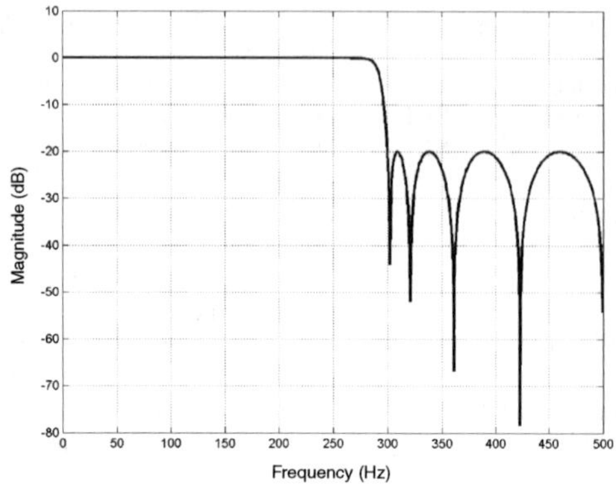

Figure 14-8c

***Function:* ellip**

Example:
Design a 9th-order low-pass elliptic filter with a cutoff frequency of 300 Hz, where $f_s = 1000$ Hz ($f_s/2 = 500$), with 3 dB of ripple in the passband, and 50 dB of attenuation in the stopband.

MATLAB:
```
[b,a] = ellip
(6,3,50,300/500);
```

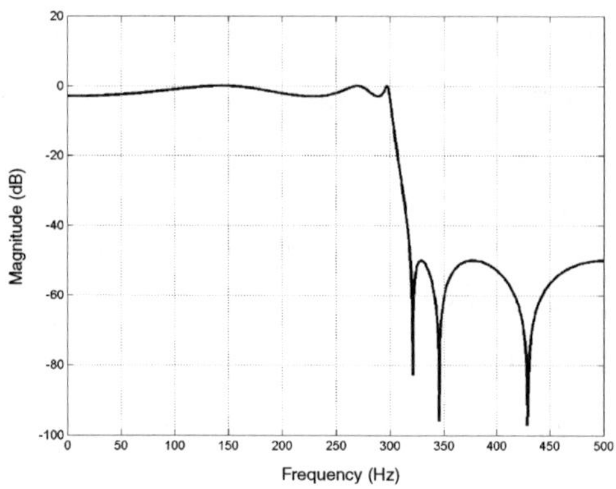

Figure 14-8d

14.7 OTHER IIR MODELS

The characteristics of an IIR can sometimes be deduced from measured system input-output behavior. The challenge is to mathematically translate these measurements into a filter model. An important filter modeling process is based on interpreting a system's input-output power spectrum in the context of an IIR filter model. This leads to a class of filter modeling strategies that are often called *parametric* spectral estimators. Parametric estimators produce a transfer function for a linear system that most likely produces the observed power spectra in some acceptable mathematical manner. While the details of the science of spectral estimation is elegant and important in a number of application areas, for purposes of filter design it is simply viewed as a transfer function production tool.

There are basically three parametric estimation schemes in common use. They are called the *auto-regressive* (AR), *moving-average* (MA), and *auto-regressive moving-average* (ARMA) methods. An *n*th-order auto-regressive model, denoted, AR[n], is given by

$$H(z) = \frac{1}{1 + a[1]z^{-1} + \cdots + a[n]z^{-n}} \tag{14-16}$$

An mth-order moving-average model, denoted MA[m], is given by

$$H(z) = b[0] + b[1]z^{-1} + \cdots + b[m]z^{-m} \tag{14-17}$$

An n, m order auto-regressive moving-average model, denoted ARMA[m, n], is given by:

$$H(z) = \frac{b[0] + b[1]z^{-1} + \cdots + b[m]z^{-m}}{1 + a[1]z^{-1} + \cdots + a[n]z^{-n}} \tag{14-18}$$

The quality of the estimates are case-dependent and exhibit differing spectral responses consisting of spectral peaks and valleys depending upon which modeling technique is employed. Supporting these model-building activities are a number of basic MATLAB objects that are motivated in the following example.

Example 14-7 Parametric IIR Filters

Required:

Design and compare the collection of MATLAB-enabled AR modeling techniques using a common test signal. Specifically, conduct a study of the following AR objectives:

- **Burg Method:** MATLAB's burg function
- **Covariance Method:** MATLAB's pcov function
- **Modified Covariance Method:** MATLAB's pmcov function
- **Yule-Walker Method:** MATLAB's pyulear function

Results:

Define an AR signal generator using the AR model:

$$H(z) = \frac{1}{z^{-4} - 2.2137z^{-3} + 2.9403z^{-2} - 2.1697z^{-1} + 0.9606}$$

If the input is a white random noise process, then the output $x[k]$ is called a random AR sequence. The spectral shape of the output response, $X(e^{j\varpi})$, is shown in Figure 14-9.

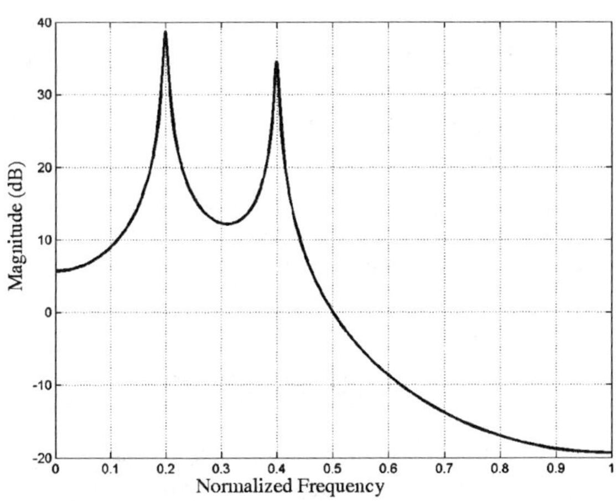

FIGURE 14-9 The spectral power density of a test signal. (The frequencies were normalized with respect to $f_s/2$.)

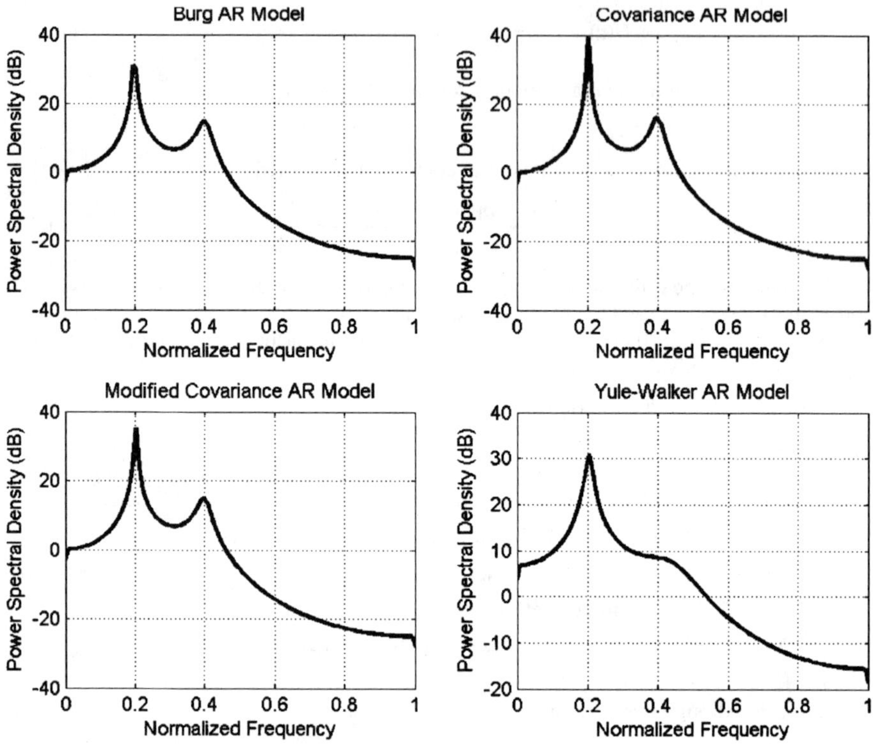

FIGURE 14-10 Parametric estimation performance: Burg (upper left), Covariance (upper right), Modified Covariance (lower left), and Yule Walker (lower right). (The frequencies were normalized with respect to $f_s/2$.)

Burg AR Model: pburg(x, 4) designs a 4th-order Burg AR model based on the test signal $x[k]$.

Covarariance AR Model: pcov(x, 4) designs a 4th-order Covariance AR model based on the test signal $x[k]$.

Modified Covariance Model: pmcov(x, 4) designs a 4th-order Modified Covariance AR model based on the test signal $x[k]$.

Yule-Walker AR Model: pyulear(x, 4) designs a 4th-order Yule Walker AR model based on the test signal $x[k]$.

The outcomes are reported in Figure 14-10. Notice that the results are able to successfully identify the stronger of the two peaks shown in Figure 14-9 and, to a lesser degree, the weaker peak. The quality of the estimate would degrade, however, if the spectral peaks become increasingly close or measurement noise becomes a factor.

14.8 COMPARISON OF FIR AND IIR FILTERS

The principal attributes of FIR are its simplicity and phase linearity. The strength of the IIR is its ability to achieve frequency selective filtering with a low-order non-linear phase design. The difference between the order of an FIR and IIR that's needed to meet a magnitude frequency response objective can be significant. This difference can be equated to differing complexity, which defines the maximum speed, cost, and power consumption of the final design. The following example compares FIRs to IIRs.

Example 14-8 IIR/FIR Comparison

Required:

Compare the estimated order of a low-pass equiripple FIR and elliptic IIR filter having a −0.0873-dB passband, −80-dB stopband, an f_s equal to 8000 Hz, a passband range of $f \in [0, 480]$ Hz, and a stopband range of $f \in [520, 4000]$ Hz.

Result:

The FIR design parameters are given by:

$$\delta_1 = 0.99 \text{ or } (1 - \delta_1) = 10^{-2}$$

$$\delta_2 = 10^{-4}$$

$$\Delta f = 40/8000 = 0.005$$

Based on Equation (13-35), an FIR of order

$$n_{\text{FIR}} = \frac{-10\log(10^{-6}) - 15}{14(0.005)} + 1 = 644$$

would be required to meet the design objectives. Further analysis will reveal that a 12th-order elliptic IIR will satisfy the design statement. If these filters are compared on the basis of the number of multiplications required to mechanize a filter, the nth-order FIR requires n_{FIR} multiplies per filter cycle. The nth-order elliptic filter could require up to $n_{\text{IIR}} = 2n + 1$ multiplies per filter cycle. This results in a complexity ratio of $r \geq 644/25 \sim 26$. If multiplication is the design's principal temporal bottleneck, then the IIR could run up to 26 times faster than the FIR.

14.9 STATE VARIABLE FILTER MODEL

Classic and parametric IIR design strategies result in transfer function $H(z)$. A transfer function, however, only quantifies the filter's input-output filter behavior and does not describe the filter's internal workings. The internal structure of the filter is defined by the system's *architecture*. What is needed to completely quantify the runtime behavior of an IIR is a filter representation methodology that accurately reflects the details of the filter's architecture. This is the role of state *variables*. State variables and the attendant state variable model audit the information entering, exiting, and residing within a digital filter. This information can be monitored by observing the condition, or state, of the filter's registers. This is the role of a state space filter model. A state variable model of a single-input single-output nth-order IIR, illustrated in Figure 14-11, is given by

$$\text{State model: } \mathbf{x}[k+1] = \mathbf{A}\mathbf{x}[k] + b\mathbf{u}[k]$$
$$\text{Output model: } y[k] = c^T\mathbf{x}[k] + d\mathbf{u}[k] \qquad (14\text{-}19)$$

where $\mathbf{x}[k]$ is an n-dimension state vector of state variables, $u[k]$ is the input, and $y[k]$ is the output. The other elements of the state variable model are an $n \times n$ feedback matrix $\mathbf{A}$,

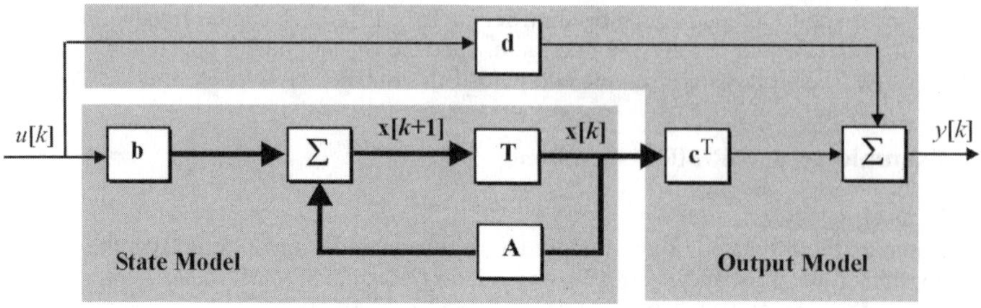

FIGURE 14-11 A state variable IIR filter model.

$1 \times n$ input vector b, $n \times 1$ output vector c, and scalar direct input-output path gain d. The next example interprets an IIR as a state variable model.

Example 14-9 State Simulation

Required:

A 2nd-order system is given by the state space model:

$$A = \begin{bmatrix} 0 & 1 \\ 0.11 & 1 \end{bmatrix}; \quad b = \begin{bmatrix} 0 \\ 1 \end{bmatrix}; \quad c = \begin{bmatrix} 0.11 \\ 0 \end{bmatrix}; \quad d = [1]$$

Sketch the architecture defined by the state model and simulate the system's response to an input $u[k] = \delta[k] - \delta[k-1] - 0.11\delta[k-2]$.

Result:

The state model is graphically interpreted in Figure 14-12.

The response of the system to $u[k]$ is reported next on a sample-by-sample basis for each state. The sample index k corresponds to the time prior to the clock being incremented, and k^+ the instant the clock is incremented.

k	$u[k]$	$x_2[k+1]$	$x_2[k]$	$x_1[k+1]$	$x_1[k]$	$y[k]$
0	1		0		0	
0^+		$1+0+0=1$		0		$0+1=1$
1	−1		1		0	
1^+		$1-1+0=0$		1		$1-1=0$
2	−0.11		0		1	
2^+		$-0.11+0+0.11=0$		0		$0.11-0.11=0$
3	0		0		0	
3^+		0		0		0

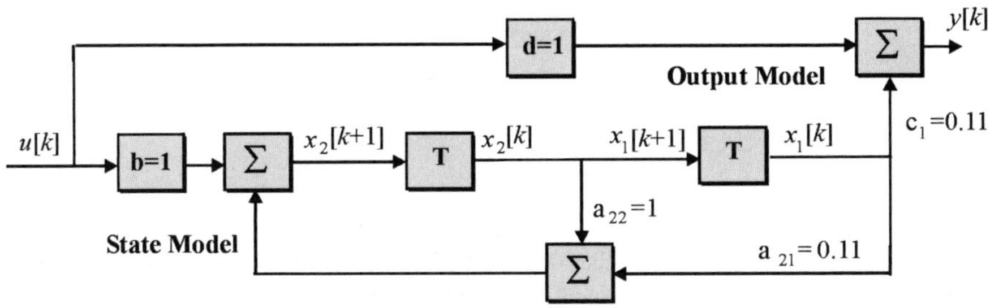

FIGURE 14-12 A second-order system architecture.

A state-determined system is defined by the state four-tuple [**A**, *b*, *c*, *d*], which characterize all the feedback and feedforward paths found within a filter. The state four-tuple [**A**, *b*, *c*, *d*] is also known to be related to the input-output transfer function using the formula

$$H(z) = \mathrm{c}^T(z\mathbf{I} - \mathbf{A})^{-1}\mathrm{b} + d \tag{14-20}$$

The next examples illustrates the conversion of a state variable model into a transfer function using Equation (14-20).

Example 14-10 State to Transfer Function Conversion

Required:
Suppose an IIR is represented by the state four-tuple [**A**, *b*, *c*, *d*], where

$$\mathbf{A} = \begin{bmatrix} 0 & 1 \\ 1 & 1 \end{bmatrix}; \quad \mathrm{b} = \begin{bmatrix} 0 \\ 1 \end{bmatrix}; \quad \mathrm{c} = \begin{bmatrix} 1 \\ 1 \end{bmatrix}; \quad d = 1$$

What is the transfer function $H(z)$ of the corresponding IIR?

Result:
The transfer function associated with the state model is given by

$$H(z) = \mathrm{c}^T(z\mathbf{I} - \mathbf{A})^{-1}\mathrm{b} + d = \frac{1}{z^2 - z - 1}[1 \quad 1]\begin{bmatrix} z-1 & 1 \\ 1 & z \end{bmatrix}\begin{bmatrix} 0 \\ 1 \end{bmatrix} + 1$$

$$= \frac{z+1}{z^2 - z - 1} + 1 = \frac{z^2}{z^2 - z - 1}.$$

The mapping from the state four-tuple [**A**, *b*, *c*, *d*] to $H(z)$ given by Equation (14-20) is unique. However, the mapping from $H(z)$ to a state four-tuple [**A**, *b*, *c*, *d*] is not unique. This explains why there are a number of architectural representations for a common filter transfer function.

14.10 ARCHITECTURE

Initially, an IIR is expressed as a transfer function $H(z)$. The transfer function $H(z)$ can be factored and partitioned in many ways, each resulting in a different architectural expression. A common representation for a transfer function $H(z)$ is as the ratio of rational polynomials, namely

$$H(z) = \frac{N(z)}{D(z)} = k\frac{\sum_{m=0}^{M} b_m z^{-m}}{a_0 + \sum_{m=1}^{N} a_m z^{-m}} \tag{14-21}$$

The transfer function is said to be *monic* if $a_0 = 1.0$. The roots of the polynomial $N(z)$ found in Equation (14-21) defines the location of the filter's *zeros* to be z_i, $i \in [1, M]$. The roots of the polynomial $D(z)$ define the location of the filter's *poles* to be λ_i, $i \in [1, N]$[1]. Collectively, the filter's poles and zeros satisfy

$$N(z) = \prod_{i=0}^{N-1}(z - z_i); \quad D(z) = \prod_{i=0}^{N-1}(z - \lambda_i) \tag{14-22}$$

This gives rise to an alternative factorization of $H(z)$ as a product form of low-order polynomials of the form

$$H(z) = \frac{N(z)}{D(z)} = K\frac{\prod_{i=0}^{N-1}(z - z_i)}{\prod_{i=0}^{N-1}(z - \lambda_i)} \tag{14-23}$$

Furthermore, the poles and zeros can be grouped together to create a collection of low-order subfilters $H_i(z)$ that can be chained together to form

$$H(z) = K\prod_{i=1}^{Q} H_i(z) \tag{14-24}$$

where $Q \le N$. The list of possibilities is virtually endless. There are, however, sets of preferred architectures that have evolved over time which have a proven utility. They represent various trade-offs between maximum real-time speed, complexity, and precision.

14.11 DIRECT II ARCHITECTURE

The Direct II architecture is a common IIR form. It's based on a specific interpretation of Equation (14-21) which expresses an Nth-order IIR, having a transfer function $H(z)$, where

$$\begin{aligned} H(z) &= \frac{N(z)}{D(z)} = \frac{b_0 + b_1 z^{-1} + \cdots + b_N z^{-N}}{a_0 + a_1 z^{-1} + \cdots + a_N z^{-N}} \\ &= \frac{b_0}{a_0} + \frac{(b_1 - b_0 a_1/a_0) z^{-1} + \cdots + (b_N - b_0 a_N/a_0) z^{-N}}{a_0 + a_1 z^{-1} + \cdots + a_N z^{-N}} \\ &= \frac{b_0}{a_0} + \frac{c_1 z^{-1} + \cdots + c_N z^{-N}}{a_0 + a_1 z^{-1} + \cdots + a_N z^{-N}} = d_0 + C(z)\left(\frac{1}{D(z)}\right). \end{aligned} \tag{14-25}$$

[1]For purposes of notational convenience, poles will temporarily be denoted λ_i instead of p_i as is the common case.

and the accompanying state assignment rule

$$\mathbf{x}[k] = \begin{bmatrix} x_1[k] \\ x_2[k] \\ \vdots \\ x_N[k] \end{bmatrix} = \begin{bmatrix} x[k-N] \\ x[k-N+1] \\ \vdots \\ x[k] \end{bmatrix} \tag{14-26}$$

where $\mathbf{x}[k]$ is the N-dimension state vector. From the state assignment, the next state value, namely $\mathbf{x}[k+1]$, can be computed to be

$$\mathbf{x}[k+1] = \begin{bmatrix} x_1[k+1] \\ x_2[k+1] \\ \vdots \\ x_N[k+1] \end{bmatrix} = \begin{bmatrix} x_2[k] \\ x_3[k] \\ \vdots \\ -a_N x_1[k] - a_{N-1}x_2[k] - \cdots - a_2 x_{N-1}[k] - a_1 x_N[k] + u[k] \end{bmatrix} \tag{14-27}$$

which results in both the state and the output equation

$$\begin{aligned} \mathbf{x}[k+1] &= \mathbf{A}\mathbf{x}[k] + \mathrm{b}u[k] \\ y[k] &= \mathrm{c}^T\mathbf{x}[k] + d_0 u[k] \end{aligned} \tag{14-28}$$

where the Direct II state four-tuple $[\mathbf{A}, b, c, d]$ is given by

$$\mathbf{A} = \begin{bmatrix} 0 & 1 & 0 & \cdots & 0 \\ 0 & 0 & 1 & \cdots & 0 \\ \vdots & \vdots & \vdots & \ddots & \vdots \\ 0 & 0 & 0 & \cdots & 1 \\ -a_N & -a_{N-1} & -a_{N-2} & \cdots & -a_1 \end{bmatrix}$$

$$b = \begin{bmatrix} 0 \\ \vdots \\ 0 \\ 1 \end{bmatrix}; \quad \begin{aligned} &\mathrm{c}^T = (\mathrm{c}_N = b_N - b_0 a_N \quad \mathrm{c}_{N-1} = b_{N-1} - b_0 a_{N-1} \quad \cdots \quad c_1 = b_1 - b_0 a_1) \\ &d_0 = b_0 \end{aligned} \tag{14-29}$$

The (i, j) element of $\mathbf{A}$ defines the path gain between state $x_j[k]$ and $x_i[k+1]$, b_i is the path gain between input and $x_i[k+1]$, c_i is the path gain between $x_i[k]$ and the output, and d_0 is the gain of the direct input-output path. These relationships are summarized in Figure 14-13.

The next example illustrates how a transfer function can be converted into a Direct II state variable filter model.

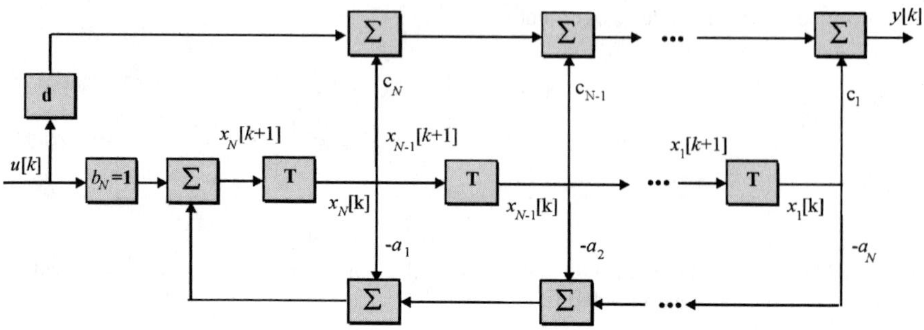

FIGURE 14-13 A Direct II filter architecture.

Example 14-11 Transfer Function to Direct II Conversion

Required:
Consider a 3rd-order digital filter having a transfer function given by

$$H(z) = \frac{z^3 - 0.5z^2 - 0.315z - 0.185}{z^3 - 0.5z^2 + 0.5z - 0.25}.$$

Produce the Direct II state four-tuple, $S = (\mathbf{A}, b, c, d)$, and interpret the results.

Result:
The transfer function can be directly converted into a Direct II state model in accordance with Equation (14-25). In particular, Equation (14-25) states that

$$H(z) = \frac{1 - 0.5z^{-1} - 0.315z^{-2} - 0.185z^{-3}}{1 - 0.5z^{-1} + 0.5z^{-2} - 0.25z^{-3}}$$

$$= 1 + \frac{-0.815z^{-2} + 0.065z^{-3}}{1 - 0.5z^{-1} + 0.5z^{-2} - 0.25z^{-3}}$$

The production of the state four-tuple, $S = [\mathbf{A}, b, c, d]$, follows and is given by

$$\mathbf{A} = \begin{pmatrix} 0 & 1 & 0 \\ 0 & 0 & 1 \\ 0.25 & -0.5 & 0.5 \end{pmatrix}; \quad b = \begin{pmatrix} 0 \\ 0 \\ 1 \end{pmatrix}; \quad c = \begin{pmatrix} 0.065 \\ -0.815 \\ 0.0 \end{pmatrix}; \quad d = 1.$$

which is graphically interpreted in Figure 14-14. Notice that the matrix coefficient a_{31} defines the path gain between $x_1[k]$ and $x_3[k+1]$—that is, $a_{31} = -a_3 = 0.25$.

14.12 MATLAB DIRECT II ARCHITECTURE

Caveat: MATLAB state variable programs make state assignments in a reverse order to that generally found in signal processing literature and textbooks (such as Equation 14-26). Specifically, the usual state x_i appears as state x_{N-i} in MATLAB.

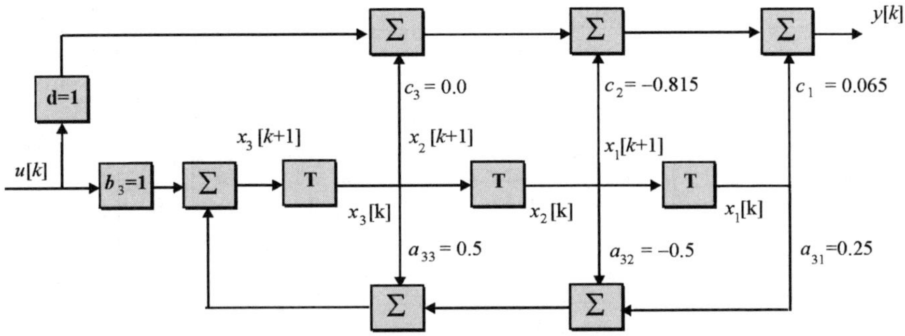

FIGURE 14-14 A Direct II filter architecture.

The conversion of a transfer function $H(z)$ into a Direct II architecture can be performed using the MATLAB programs TF2SS or ZP2SS. Function TF2SS converts a transfer function $H(z) = N(z)/D(z)$ into a Direct-II state form using the syntax $[A,B,C,D]$ = TF2SS(NUM,DEN). Function ZP2SS converts a transfer function $H(z)$, having zeros (Z), poles (P), and input scale factor K into a Direct II state form using the syntax $[A,B,C,D]$ = ZP2SS(Z, P, K). It should be noted that the state assignments are in reverse order to those shown in Figure 14-14. The conversion between the common 3rd-order Direct II indexing scheme used in Figure 14-14, and MATLAB is shown next.

Normal	MATLAB
$x_1[k]$	$x_3[k]$
$x_2[k]$	$x_2[k]$
$x_3[k]$	$x_1[k]$

The next example illustrates the conversion of a transfer function to a Direct II state variable model using MATLAB.

Example 14-12 Matlab Transfer Function to Direct II Conversion

Required:
Consider again Example 14-11 defined in terms of a 3rd-order digital filter:

$$H(z) = \frac{z^3 - 0.5z^2 - 0.315z - 0.185}{z^3 - 0.5z^2 + 0.5z - 0.25}.$$

Produce the state four-tuple, $S = [\mathbf{A}, b, c, d]$ and interpret the results using MATLAB.

Result:
The production of the state four-tuple, $S = (\mathbf{A}, b, c, d)$, using MATLAB proceeds as follows:

```
n = [1 -.5 -.315 -.0185];
d = [1 -.5 .5 -.25];
 [A,B,C,D] = TF2SS(n,d)
```

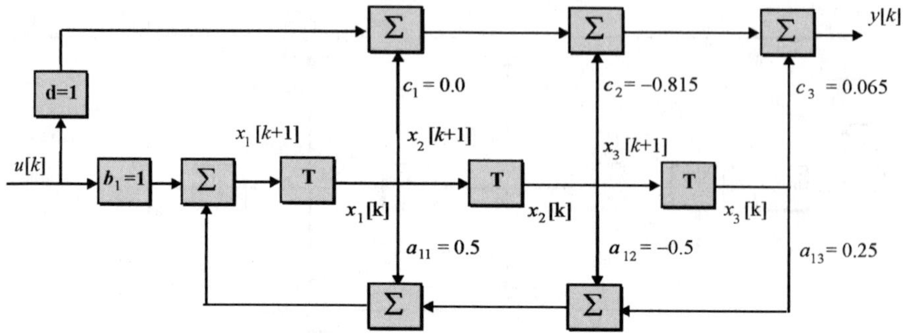

FIGURE 14-15 A Direct II filter architecture using MATLAB.

```
A =
      0.5000    -0.5000      0.2500      {x3 in Figure 14-15}
      1.0000          0           0      {x2 in Figure 14-15}
           0     1.0000           0      {x1 in Figure 14-15}
B =
       1        {b3 in Figure 14-15}
       0
       0
C =
           0    -0.8150      0.0650  {c3, c2, c1 Figure 14-15}
D =
       1
```

and is graphically interpreted in Figure 14-15. It can be seen that the system described in Figure 14-15 is equivalent to that shown in Figure 14-14 except for a different state indexing assignment.

14.13 CASCADE ARCHITECTURE

One of the most popular IIR architectures found in common use today is called the *cascade architecture*. The basic cascade architecture, shown in Figure 14-16, is used to implement a transfer function of the form:

$$H(z) = K\prod_{i=1}^{Q} H_i(z) \tag{14-30}$$

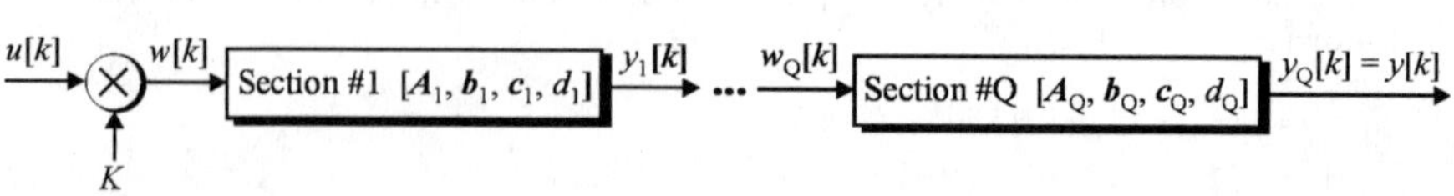

FIGURE 14-16 A cascade filter architecture.

The ith subsystem, denoted $H_i(z)$, is either a 1st- or 2nd-order filter having real coefficients. Each subfilter can be represented in terms of a state model $S_i = [\mathbf{A}_i, b_i, c_i, d_i]$ of order N_i, where $N_i \in \{1, 2\}$, where

$$\sum_{i=1}^{Q} order(H_i(z)) = \sum_{i=1}^{Q} N_i = N \tag{14-31}$$

A cascade architecture links the ith subfilter $S_i = [\mathbf{A}_i, b_i, c_i, d_i]$ with $i + 1$st subfilter $S_{i+1} = [\mathbf{A}_{i+1}, b_{i+1}, c_{i+1}, d_{i+1}]$. Specifically, the output of subsystem S_i, namely $y_i[k]$, is sent to the input of subsystem S_{i+1}, namely $w_{i+1}[k]$. Furthermore, each 1st- or 2nd-order subfilter defined in terms of real coefficients is derived from the poles (λ_i) and zeros (z_i) of $H(z) = N(z)/D(z)$. In particular, 1st-order subfilters are defined in terms of real poles and zeros of $H(z)$. A 2nd-order subfilter is defined by combining a complex pole/zero with their complex conjugate counterparts resulting in a filter section having only real coefficients. The basic first and second order sections can be defined in terms of biquadratic (*biquad*) or Direct II structures. If λ_I is a real pole of $H(z)$, then the ith first-order subsystem $H_i(z)$ is assumed to satisfy one of the two following forms

$$H_i(z) = \frac{q_{i0} + q_{i1}z^{-1}}{1 - \lambda_i z^{-1}} \text{ (biquad)} \tag{14-32}$$

$$H_i(z) = q_{i0} + \frac{r_i z^{-1}}{1 - \lambda_i z^{-1}}; \quad r_i = q_{i1} + \lambda_i q_{i0} \text{ (Direct II)} \tag{14-33}$$

Both architectures are interpreted in Figure 14-17. The principal difference being that the Direct II possesses a state variable model, while the biquad does not.

If λ_i appears in its complex form, then it is combined with its complex conjugate pair λ_i^* to create a 2nd-order section of the form

$$H_i(z) = d_{i0} + \frac{n_i(z)}{1 - \lambda_i z^{-1}} + \frac{n_i^*(z)}{1 - \lambda_i^* z^{-1}} \tag{14-34}$$

Equation (14-34) can be realized in two forms. The first is as a 2nd-order biquad section given by

$$H_i(z) = \frac{w_{i0} + w_{i1}z^{-1} + w_{i2}z^{-2}}{1 - p_{i1}z^{-1} - p_{i2}z^{-2}} \tag{14-35}$$

The second is a Direct II section given by

$$H_i(z) = d_{i0} + \frac{r_{i1}z^{-1} + r_{i2}z^{-2}}{1 - p_{i1}z^{-1} - p_{i2}z^{-2}} \tag{14-36}$$

where the coefficients are real and are given by

$$(z - \lambda_i)(z - \lambda_i^*) = z^2 - p_{i1}z - p_{i2}; \quad d_{i0} = w_{io}; r_{i1} = w_{i1} + p_{i1}w_{i0}; \quad r_{i2} = w_{i2} + p_{i2}w_{i0} \tag{14-37}$$

Both architectures are compared in Figure 14-17. The principal difference is again that the Direct II possesses a state variable model, the biquad does not. Because of this, Direct II implementations have been increasingly gaining favor.

As a general rule, zeros are paired with the closest poles. This *proximity pairing* strategy will generally result in filter design which distributes the subfilter gains uniformly across all

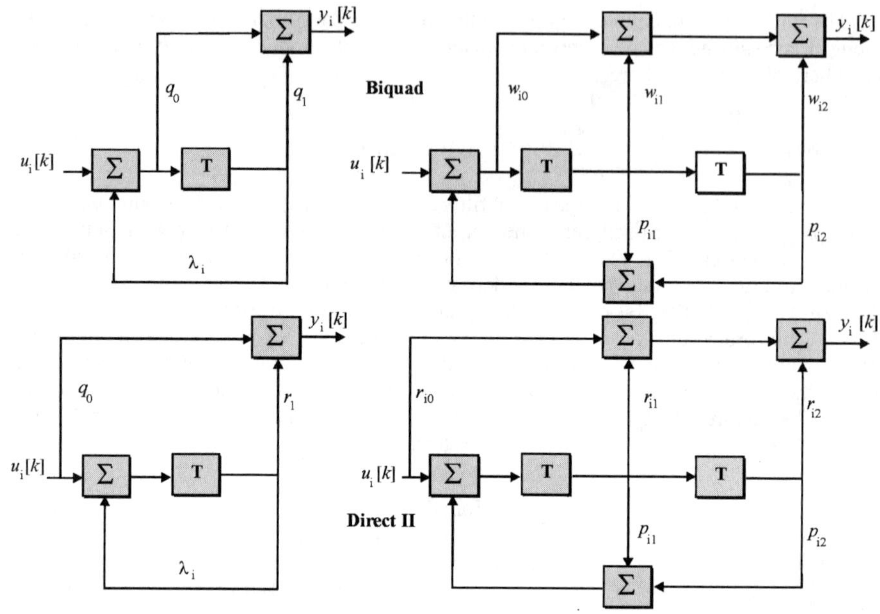

FIGURE 14-17 Biquad (top) and Direct II (bottom) 1st- (left) and 2nd- (right) order subfilter sections.

filter sections. This is considered desirable. Other pairing strategies can result in a few subsystems having excessively large dynamic range requirements while others have small gains. This creates a precision allocation problem that can compromise overall system performance.

14-14 THE MATLAB CASCADE ARCHITECTURE

MATLAB contains a collection of programs that relate to cascade filter implementation. Function *tf2sos* converts a digital filter transfer function data to a set of second-order sections having the form

$$sos = \begin{bmatrix} b_{01} & b_{11} & b_{21} & 1 & a_{11} & a_{21} \\ b_{02} & b_{12} & b_{22} & 1 & a_{12} & a_{22} \\ \cdots & \cdots & \cdots & \cdots & \cdots & \cdots \\ b_{0L} & b_{1L} & b_{2L} & 1 & a_{1L} & a_{2L} \end{bmatrix}$$

where the *i*th row of the array ***sos*** specifies the coefficients of the *i*th subfilter

$$H_i(z) = \frac{B_i(z)}{A_i(z)} = \frac{b_{0i} + b_{1i}z^{-1} + b_{2i}z^{-2}}{a_{0i} + a_{1i}z^{-1} + a_{2i}z^{-2}}$$

Program *zp2sos* converts a transfer function $H(z)$, specified in terms of zero-pole-gain parameters, into an $L \times 6$ *sos* array. Program *sos2ss* maps second-order filter sections into a Direct II state space form, while program *sos2tf* converts a collection of 2nd-order filter sections into a transfer function. For instances where the filter sections are 1st-order, coefficients b_{2i} and a_{2i} are set to zero. Program *sos2zp* converts 2nd-order filter sections into a zero-pole-gain form. The next example converts a transfer function into a Direct II IIR.

Example 14-13 Matlab Cascade Transformations

Required:

Consider a transfer function $H(z) = kH_1(z)H_2(z)$, where

$$H(z) = \frac{-2 + z^{-1} + 2z^{-2} + 4z^{-3} + z^{-4}}{1 + 10z^{-1} + 0z^{-2} - 10z^{-3} - z^{-4}}; \quad H_1(z) = \frac{1 - 1.5z^{-1} - 0.5z^{-2}}{1 + 10z^{-1} + z^{-2}};$$

$$H_2(z) = \frac{1 + z^{-1} + z^{-2}}{1 - z^{-2}}; \quad k = -2$$

Given $H_1(z)$ and $H_2(z)$, compute the Direct II state-space representation of each cascaded filter section. In addition, compute the pole-zero distribution of $H(z)$.

Result:

The two cascade filters $H_1(z) = (1 - 1.5z^{-1} - 0.5z^{-2})/(1 + 10z^{-1} + z^{-2})$ and $H_2(z) = (1 + z^{-1} + z^{-2})/(1 - z^{-2})$ have a MATLAB representation $sos = [1\ -1.5\ -0.5\ 1\ 10\ 1;\ 1\ 1\ 1\ 1\ 0\ -1]$; $\{H_1(z), H_2(z)\}$. The sos array can also be machine generated given the original transfer function as shown below:

```
n=[-2 1 2 4 1]; d=[1 10 0 -10 -1];
[sos,g]=tf2sos(n,d)
sos =
   1.0000 -1.5000 -0.5000  1.0000  10.0000  1.0000
   1.0000  1.0000  1.0000  1.0000   0.0000 -1.0000
g =
    -2
```

From the *sos* database, the cascaded filter sections can be implemented as Direct II filters as follows:

```
sos1=sos(1,:);
[A,B,C,D]=sos2ss(sos1)
A =
   -10.0000    -1.0000
     1.0000          0
B =
     1
     0
C =
   -11.5000    -1.5000
D =
     1
```

```
sos2=sos(2,:);
[A,B,C,D]=sos2ss(sos2)
A =
    -0.0000    1.0000
     1.0000         0
B =
      1
      0
C =
     1.0000    2.0000
D =
      1
```

Finally, using *sos2pz*, the filter's pole-zero distribution can be determined as follows:

```
 [z,p,k] = sos2zp(sos)
z =
   -0.5000 + 0.8660i
   -0.5000 - 0.8660i
    1.7808
   -0.2808
p =
    -1.0000
     1.0000
    -9.8990
    -0.1010
k =
     -2    {k=d}
```

which are the four zeros and four poles of $H(z)$.

14.15 PARALLEL ARCHITECTURE

A parallel IIR architecture is shown in Figure 14-18, and implements the transfer function

$$H(z) = K\left(d_0 + \sum_{i=1}^{Q} H_i(z)\right) \tag{14-38}$$

where $H_i(z)$ is a 1st- or 2nd-order subsystem defined with respect to real coefficients. Each subsystem is represented by a state-determined model, $S_i = [\mathbf{A}_i, b_i, c_i, d_i]$, where

$$\sum_{i=1}^{Q} order(H_i(z)) = \sum_{i=1}^{Q} N_i = N \tag{14-39}$$

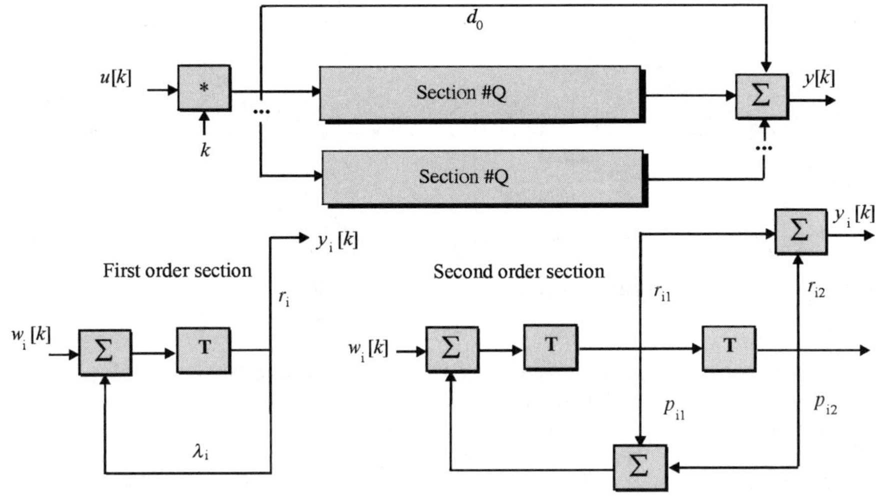

FIGURE 14-18 A parallel architecture.

A parallel architecture, as the name implies, consists of Q subfilters operating in parallel with their outputs sent to a common adder. Specifically, a possibly scaled common input is presented concurrently to all subfilters $S_i = [\mathbf{A}_i, b_i, c_i, d_i]$, producing independent individual output responses $y_i[k]$ which are then combined (added) to form $y[k]$. The state variable model for a parallel system is given by $S = [\mathbf{A}, b, c, d]$, where

$$\mathbf{A} = \begin{pmatrix} A_1 & 0 & 0 & \cdots & 0 \\ 0 & A_2 & 0 & \cdots & 0 \\ 0 & 0 & A_3 & \cdots & 0 \\ \vdots & \vdots & \vdots & \ddots & \vdots \\ 0 & 0 & 0 & \cdots & A_Q \end{pmatrix}; \quad b = \begin{pmatrix} b_1 \\ b_2 \\ \vdots \\ b_Q \end{pmatrix}; \quad c^T = (c_1 \quad c_2 \quad \cdots \quad c_Q); \quad d = \sum_{i=0}^{Q} d_i \tag{14-40}$$

Each subfilter $H_i(z)$ is implemented as a 1st- or 2nd-order biquad or Direct II filter.

14.16 LATTICE / LADDER ARCHITECTURE

In general, an Mth-order linear time invariant IIR can be represented as the *monic* transfer function

$$H(z) = \frac{B(z)}{A(z)} = \frac{\sum_{i=0}^{M} b_i z^{-i}}{1 + \sum_{i=1}^{M} a_i z^{-i}} \tag{14-41}$$

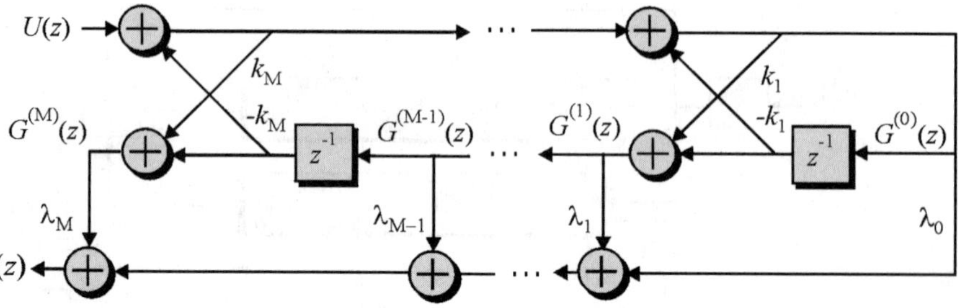

FIGURE 14-19 A lattice-ladder architecture.

The Mth-order system can be realized as the lattice-ladder IIR filter, as shown in Figure 14-19. The filter structure consists of two major subsystems, one providing feedback, the other feedforward data paths. Attached to feedback taps are a collection of tap weights λ_i, called *reflection coefficients*, and are used to synthesize a unique input-output relationship. One of the known properties of a lattice IIR filter is that the reflection coefficients of a stable lattice filter are bounded by unity (for instance, $|k_i| < 1.0$). The IIR lattice filter design procedure is similar in form to that developed for a lattice FIR and is based on iteratively computing the reflection coefficients λ_i.

The derivation of the lattice-ladder parameters begins with restructuring $H(z)$ to read

$$H(z) = \sum_{i=1}^{M} \lambda_i G^i(z) = \sum_{i=1}^{M} \lambda_i B^i(z)/A^{(M)}(z)$$

$$= \sum_{i=1}^{M} \lambda_i b_i z^{-i}/A^{(M)}(z) \hat{=} C^{(M)}(z)/A^{(M)}(z); \quad A^{(M)}(z) = \Delta(z) \tag{14-42}$$

where $B^{(m)}(z)$ was defined in an FIR lattice-ladder discussion to be the reverse copy of $A^{(m)}(z)$ and finally $A^{(M)}(z) = A(z)$. Define

$$C^{(m)}(z) = \sum_{i=1}^{m} c_i^{(m)}(z) = \sum_{i=1}^{m} \lambda_i B^{(m)}(z) = C^{(m-1)}(z) + \lambda_m B^{(m)}(z) \tag{14-43}$$

where $C^{(m-1)}(z)$ is of the degree $m - 1$, the degree of the monic polynomial $B^{(m)}(z)$ is m, where the mth coefficient of $B^{(m)}(z)$ is unity, and the reflection coefficient is

$$\lambda^m = c_m^{(m)} \tag{14-44}$$

The $C^{(m)}(z)$ equation can be rearranged to read

$$C^{(m-1)}(z) = C^{(m)}(z) - \lambda_m B^{(m)}(z) \tag{14-45}$$

which can be used to iteratively compute the coefficients λ_m. The design of a lattice IIR is illustrated in the following example.

Example 14-14 Lattice-Ladder Architecture

Required:

Suppose $H(z) = B(z)/A(z) = C^{(M)}(z)/A^{(M)}(z) = (1 + 2z^{-1} + 3z^{-2} + 4z^{-3})/(1 + (21/32)z^{-1} + (21/64)z^{-2} + (1/8)z^{-3})$ (see Figure 14-20). Derive the lattice-ladder filter parameters.

Result:

The coefficient production process is illustrated next.

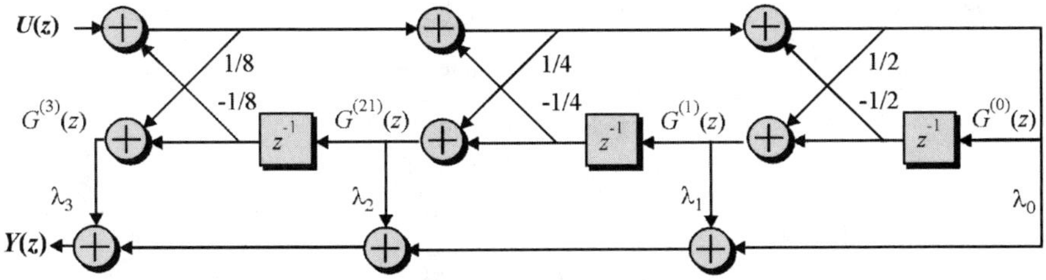

FIGURE 14-20 A 3rd-order lattice-ladder architecture.

$C^{(3)}(z) = 1 + 2z^{-1} + 3z^{-2} + 4z^{-3} \Rightarrow \boldsymbol{\lambda_3} = \boldsymbol{c_3^{(3)}} = 4$

$A^{(3)}(z) = 1 + (21/32)z^{-1} + (21/64)z^{-2} + (1/8)z^{-3}$

Therefore, $B^{(3)}(z) = (1/8) + (21/64)z^{-1} + (21/32)z^{-2} + z^{-3}$ (reverse order)

$C^{(2)}(z) = C^{(3)}(z) - \lambda_3 B^{(3)}(z) = (1 + 2z^{-1} + 3z^{-2} + 4z^{-3}) - (4)\,((1/8) + (21/64)z^{-1} + (21/32)z^{-2} + z^{-3})$

$= (1 + 2z^{-1} + 3z^{-2} + 4z^{-3}) - ((1/2) + (21/16)z^{-1} + (21/8)z^{-2} + 4z^{-3}) = (1/2 + (11/16)\,z^{-1} + (3/8)\,z^{-2})$

$\Rightarrow \boldsymbol{\lambda_2} = \boldsymbol{c_2^{(2)}} = 3/8;$

$A^{(2)}(z) = 1 + (5/8)z^{-1} + (1/4)z^{-2}$

Therefore, $B^{(2)}(z) = (1/4) + (5/8)z^{-1} + z^{-2}$ (reverse order)

$C^{(1)}(z) = C^{(2)}(z) - \lambda_2 B^{(2)}(z) = (1/2 + (11/16)\,z^{-1} + (3/8)\,z^{-2}) - (3/8)((1/4) + (5/8)z^{-1} + z^{-2}) =$

$(1/2 + (11/16)z^{-1} + (3/8)z^{-2}) - (3/32 + (15/64)z^{-1} + (1/8)z^{-2}) = (13/32 + (29/64)z^{-1})$

$\Rightarrow \boldsymbol{\lambda_1} = \boldsymbol{c_1^{(1)}} = 29/64$

$A^{(1)}(z) = 1 + (1/2)z^{-1}$

Therefore, $B^{(1)}(z) = (1/2) + z^{-1}$ (reverse order)

$C^{(0)}(z) = C^{(1)}(z) - \lambda_1 B^{(1)}(z) = (13/32 + (29/64)z^{-1}) - (29/64)\,(1/2 + z^{-1})$

$= (13/32) - (29/128) = 23/128$

$\Rightarrow \boldsymbol{\lambda_0} = \boldsymbol{c_0^{(0)}} = 23/128$

The numerator of the filter's transfer function can be isolated and synthesized by noting that $G^i(z) = B^{(i)}(z)/A^{(M)}(z)$ (Equation 14-42) and that:

$\lambda B^{(i)}(z)$ (numerator terms)	z^{-0}	z^{-1}	z^{-2}	z^{-3}
$\lambda_0 B^{(0)}(z)$: $\lambda_0 = 23/128$	23/128			
$\lambda_1 B^{(1)}(z)$; $\lambda_1 = 29/64$	29/128	29/64		
$\lambda_2 B^{(2)}(z)$; $\lambda_2 = 3/8$	3/32	15/64	3/8	
$\lambda_3 B^{(3)}(z)$; $\lambda_3 = 4$	1/2	21/16	21/8	4
$\Sigma \lambda_i B^{(i)}(z)$	128/128 = 1	128/64 = 2	24/8 = 3	4

The presented design technique is sometimes referred to as the *Gray-Markel* method.

Lattice-ladder filters can be represented in a state variable form, as well as

$$x[k+1] = Ax[k] + bu[k]$$

$$y[k] = c^T x[k] + du[k]$$

$$A = \begin{bmatrix} -k_1 & -k_2 & -k_3 & -k_4 & \cdots \\ 1-k_1^2 & -k_1k_2 & -k_1k_3 & -k_1k_4 & \cdots \\ 0 & 1-k_2^2 & -k_2k_3 & -k_2k_4 & \cdots \\ 0 & 0 & 1-k_3^2 & -k_3k_4 & \cdots \\ \cdots & \cdots & \cdots & \cdots & \cdots \end{bmatrix}; \quad b = \begin{bmatrix} 1 \\ k_1 \\ k_2 \\ k_3 \\ \cdots \end{bmatrix}$$

$$c = \begin{bmatrix} \lambda_1(1-k_1^2) + \lambda_0(k_1) \\ \lambda_2(1-k_2^2) + \lambda_1(-k_2k_1) + \lambda_0(k_2) \\ \lambda_3(1-k_3^2) + \lambda_2(-k_3k_2) + \lambda_1(-k_3k_1) + \lambda_0(k_3) \\ \lambda_4(1-k_4^2) + \lambda_3(-k_4k_3) + \lambda_2(-k_4k_2) + \lambda_1(-k_4k_1) + \lambda_0(k_4) \\ \cdots \end{bmatrix}; \tag{14-46}$$

$$d = \lambda_0 + \sum_{i=1}^{M}(-k_i)\lambda_i$$

14.17 MATLAB LADDER/LATTICE SUPPORT

MATLAB's *latcfilt*, *latc2tf*, and *tf2latc* functions have a lattice-ladder purpose. The function *latcfilt*, for instance, is used to filter a time-series using a lattice-ladder filter, while the function *latc2tf* translates a lattice-ladder filter into a transfer function. Lastly, the function *tf2latc* translates a transfer function into a lattice-ladder filter. The next example demonstrates the use of MATLAB in the design of a lattice IIR.

Example 14-15 Lattice-Ladder Design

Required:
Consider an IIR ARMA(5, 2) filter model given by

$$H(z) = \frac{B(z)}{A(z)} = \frac{1 + 0.6149z^{-1} + 0.9899z^{-2} + 0.0031z^{-4} - 0.0082z^{-5}}{1 + (1/2)z^{-1} + (1/3)z^{-2}}$$

Derive the coefficients of a lattice-ladder implementation of $H(z)$.

Result:
The Signal Processing Toolbox function *tf2latc* accepts an IIR filter in polynomial form and returns the corresponding lattice-ladder coefficients as shown next.

```
[k,v] = tf2latc(b,a)
k =          {k parameters}
```

```
      0.3750
      0.3333
           0
           0
           0
v =              {λ parameters}
      0.6252
      0.1212
      0.9879
     -0.0009
      0.0072
     -0.0082
```

Since $|k_i| < 1$ for all reflection coefficients, the filter is stable.

14.18 NORMAL ARCHITECTURE

Besides Direct II, Cascade, Parallel, and Lattice/Ladder, other architectures are found in common use. Among these are normal, section optimal, and wave architectures of which normal is the most commonly encountered. Normal filters, like many other forms, are based on 2nd-order sections having a transfer function of the form:

$$H(z) = \frac{z^2\beta_0 + \beta_1 z + \beta_2}{z^2 + \alpha_1 z + \alpha_2} = \frac{\beta_0 + \beta_1 z^{-1} + \beta_2 z^{-1}}{1 + \alpha_1 z^{-1} + \alpha_2 z^{-2}}$$

$$= d_0 + \frac{\gamma_1 z^{-1} + \gamma_2 z^{-2}}{1 + \alpha_1 z^{-1} + \alpha_2 z^{-2}} = d_0 + \frac{mz^{-1}}{1 - pz^{-1}} + \frac{m^* z^{-1}}{1 - p^* z^{-1}} \tag{14-47}$$

where m is a Heaviside coefficient and p is one of the poles of the system. If the poles are complex, they appear as complex conjugate pairs. The 2nd-order section, shown in Figure 14-21, is said to be a fully *connected architecture* in that all possible inter-state connections are established. The 2nd-order system is said to be *normal* if the state feedback matrix $\mathbf{A}_N$ satisfies the condition $\mathbf{A}_N^T\mathbf{A}_N = \mathbf{A}_N\mathbf{A}_N^T$ where "T" denotes the matrix transposition. The following example presents a study of a normal filter architecture.

Example 14-16 Normal Matrix

Required:
Show that the **A** matrix, illustrated next, is normal.

$$A = \begin{bmatrix} a_1 & a_2 \\ -a_2 & a_1 \end{bmatrix}$$

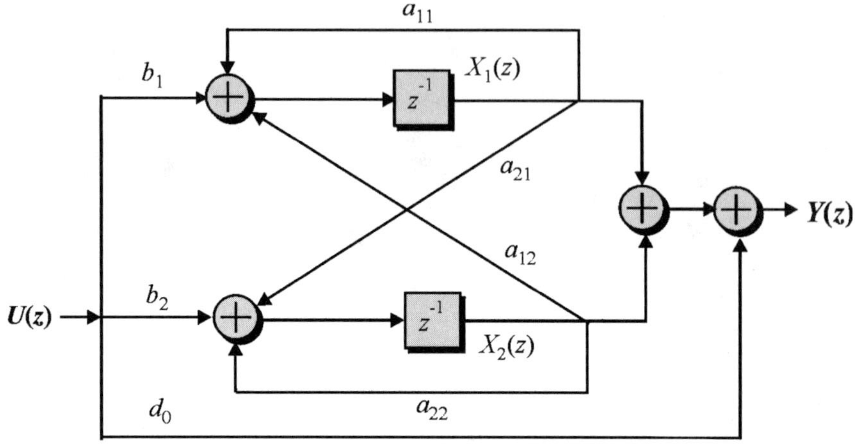

FIGURE 14-21 A fully connected architecture.

Result:

Direct computation results in

$$A^T A = \begin{bmatrix} a_1 & -a_2 \\ a_2 & a_1 \end{bmatrix}\begin{bmatrix} a_1 & a_2 \\ -a_2 & a_1 \end{bmatrix} = \begin{bmatrix} a_1^2 + a_2^2 & 0 \\ 0 & a_1^2 + a_2^2 \end{bmatrix}$$

$$= \begin{bmatrix} a_1 & a_2 \\ -a_2 & a_1 \end{bmatrix}\begin{bmatrix} a_1 & -a_2 \\ a_2 & a_1 \end{bmatrix} = AA^T$$

which demonstrates the normality of **A**.

The connected system shown in Figure 14-21 can be defined in terms of a state variable four-tuple [**A**, b, c, d]. This system, in turn, can be mapped into another 2nd-order system architecture [$\mathbf{A}_N$, b_N, c_N, d_N] using a non-singular *similarity transform* **T**. The new architecture is specified in terms

$$\begin{aligned} \mathbf{A}_N &= \mathbf{T}^{-1}\mathbf{A}\mathbf{T}, \\ b_N &= \mathbf{T}^{-1}b, \\ c_N &= c^T\mathbf{T}^{-1}, \\ d_N &= d. \end{aligned} \tag{14-48}$$

This process is called a *change of basis*. For specific transforms **T**, $\mathbf{A}_N$ can be created as a normal matrix. In fact, the original filter (say, [**A**, ***b***, ***c***, *d*]) can be completely arbitrary. The use of state variable transformation is demonstrated in the next example.

Example 14-17 Normal Matrix

Required:

Convert the state-determined system

$$\mathbf{A} = \begin{bmatrix} -\alpha_1 & 1 \\ -\alpha_2 & 0 \end{bmatrix}; \quad b = \begin{bmatrix} b_1 \\ b_2 \end{bmatrix}; \quad c = \begin{bmatrix} c_1 \\ c_2 \end{bmatrix}; \quad d = d_0$$

into a fully connected normal system.

Result:

The poles of the original system satisfy $\det(z\mathbf{I} - \mathbf{A}) = z^2 + \alpha_1 z + \alpha_2 = (z - p)(z - p^*)$ where $p = \sigma + j\omega$. Define the similarity transform $\mathbf{T}$ to be

$$T = \frac{1}{\omega}\begin{bmatrix} 0 & -1 \\ \omega & \sigma \end{bmatrix}$$

Upon computing $\mathbf{A}_N = \mathbf{T}^{-1}\mathbf{AT}$, $\mathbf{b}_N = \mathbf{T}^{-1}\mathbf{b}$, and $\mathbf{c}_N^T = \mathbf{c}^T\,\mathbf{T}$, it follows that

$$A_N = T^{-1}AT = \left\{\begin{bmatrix} \sigma & 1 \\ -\omega & 0 \end{bmatrix}\right\}\begin{bmatrix} -\alpha_1 & 1 \\ -\alpha_2 & 0 \end{bmatrix}\left\{\frac{1}{\omega}\begin{bmatrix} 0 & -1 \\ \omega & \sigma \end{bmatrix}\right\}$$

$$= \begin{bmatrix} \sigma & (\sigma^2 + \alpha_1\sigma + \alpha_2)/\omega \\ -\omega & -\alpha_1 - \sigma \end{bmatrix} = \begin{bmatrix} \sigma & \omega \\ -\omega & \sigma \end{bmatrix}$$

which is a normal matrix, and b_N and c_N follow.

14.19 STABILITY

The stability of an IIR linear filter can be defined in the transform domain in terms of eigenvalues or poles of the filter's transfer function $H(z) = N(z)/D(z)$. The stability classifications based on knowledge of the location of a filter's pole values, λ_i are summarized in Table 14-6.

A state-determined filter system can be expressed in terms of the state four-tuple $[\mathbf{A}, b, c, d]$. The system poles can be extracted for the $\det(\lambda\mathbf{I} - \mathbf{A})$ and interpreted in accordance with the cases exhibited in Table 14-6. A more sophisticated stability guarantor is called *Lyapunov's Stability Criterion*, which states that if A is an $N \times N$ matrix defining a stable Nth-order IIR, then for any $N \times N$ matrix W, the matrix series

$$\sum_{m=0}^{\infty} \mathbf{A}^m \mathbf{W}(\mathbf{A}^T)^m \tag{14-49}$$

TABLE 14-6 Stability Cases

Pole	Magnitude	Stability Classification	Comments
λ_i	$\lvert\lambda_i\rvert < 1$	Asymptotically stable	If the poles are interior to the unit circle in the z-domain, the filter is stable for all bounded inputs.
λ_i	$\lvert\lambda_i\rvert = 1$	Conditionally stable	For a pole on the unit circle at $z = \lambda_i$, the system is stable provided that the input does not also have a pole at $z = \lambda_i$.
λ_i	$\lvert\lambda_i\rvert = 1$	Unstable	If a pole of multiplicity greater than one resides on the unit circle at $z = \lambda_i$, the system is unstable.
λ_i	$\lvert\lambda_i\rvert > 1$	Unstable	If any of the filter poles are exterior to the unit circle in the z-domain, the filter is unstable.

converges. In addition, there exists a unique positive semi-definite matrix K, and a positive semi-definite matrix Q, such that

$$\mathbf{K} = \mathbf{AKA}^T + \mathbf{bQb}^T \tag{14-50}$$

if and only if **A** is a stability matrix. As stated in the study of FIRs, stability is rarely the issue. What is important is the quantification of the filter's dynamic range requirements. This is a more challenging question than simply determining the stability class of an IIR. The next example performs a stability analysis on a low-order IIR system. Dynamic range requirements are discussed in the next section.

Example 14-18 Stability

Required:
Suppose that a 1st-order (such as, $N = 1$) filter is given by the state four-tuple $[\boldsymbol{a}, 1, c, d]$ where $\boldsymbol{a}$ is an arbitrary constant. Determine the stability classification of the filter.

Result:
The eigenvalue method states that for the system to be stable, the roots to $\det(\lambda\mathbf{I} - \mathbf{A}) = \det(\lambda - a) < 1$ requires $|a| < 1$. Now, let the positive definite matrix $\mathbf{Q}$ be set to $\mathbf{Q} = \mathbf{I}$. If the 1×1 matrix $\mathbf{K} = [\boldsymbol{k}]$ matrix, satisfying $\boldsymbol{k} = \boldsymbol{aka} + bb = \boldsymbol{a}^2\boldsymbol{k} + 1$ is positive semi-definite, then the filter is stable. The 1×1 matrix $\boldsymbol{k}$ can be obtained by solving $\boldsymbol{k} = \boldsymbol{a}^2\boldsymbol{k} + 1$, which results in $\boldsymbol{k} = 1/(1 - \boldsymbol{a}^2)$. The 1×1 martrix $\boldsymbol{k}$ is positive semi-definite if $\boldsymbol{k} \geq 0$, which requires that $|a| < 1$. This is the same conclusion obtained using the eigenvalue method.

14.20 FINITE WORDLENGTH EFFECTS

Digital filters can be studied in the context of their type (such as Butterworth), architecture (say, Cascade), and arithmetic structure. A filter will work as predicted provided the system remains linear. This is almost assured for floating-point implementations. Fixed-point systems, however, are a distinctly different case.

The mechanisms by which a theoretically linear system becomes non-linear are called *finite wordlength effects*. The finite wordlength effects can impact a filter in the manner listed next in descending order of severity:

- Overflow saturation
- Arithmetic rounding
- Coefficient rounding
- Data scaling
- Zero-input limit cycling

Any of these error processes can compromise the performance of a digital filter to the point that the filter is no longer viable. Fortunately, corrective actions are available and can take the form of

- Scaling the input or coefficients
- Increasing the filter's data wordlength
- Selecting an alternative architecture or filter model

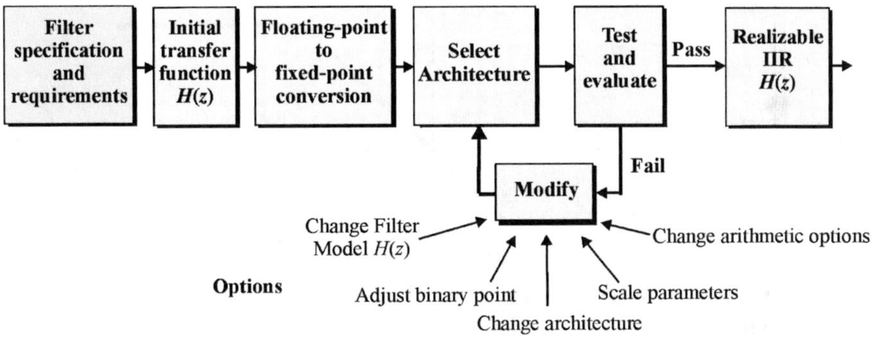

FIGURE 14-22 A fixed-point filter design procedure.

This general IIR design and analysis process is suggested in Figure 14-22. It begins by specifying a filter's transfer function or filter coefficients. The filter is then mapped into a specific architecture. Implicit in this process is a definition of an arithmetic unit which can be classified as floating point or fixed point. For cost and speed reasons, fixed point designs are often preferred to as floating-point instantiation. An N-bit fixed-point number system will be assumed to be represented by a single sign-bit, I-integer bits and F-fractional bits. This defines a $[N{:}F]$ data format where $N = I + F + 1$. The dynamic range of an $[N{:}F]$ system is bounded by $\pm 2^I$ and the value of the least significant bit (LSB) is 2^{-F}. The error variance associated with rounding or truncating a real number to an $[N{:}F]$ formatted word is given by $\sigma^2 = Q^2/12$, where $Q = 2^{-F}$. This is called the quantization step size. It should be noted that a fixed-point system possesses both limited dynamic range and precision. Unless these weaknesses are properly managed, they can compromise system performance. Upon implementation, the performance of the candidate fixed-point filter needs to be analyzed mathematically and/or experimentally to determine if it still meets the design specifications. If the candidate design fails this test, then the design strategy will need to be altered by redefining the data format, architecture, or filter type. Afterward, the test will be repeated.

14.21 OVERFLOW ARITHMETIC

The most serious finite wordlength effect is runtime *register overflow*. Register overflow can introduce significant non-linear distortion into a system's output which can render a filter useless. It is therefore essential that the filter designer eliminate or control the effects of runtime register overflow. There are some standard techniques that can be used to control this problem. One effective means of controlling fixed-point overflow errors is to perform arithmetic using 2's complement (2C) arithmetic unit. Two's complement possesses an important modulo(2^n) property that insures the sum of 2C numbers is a valid 2C outcome, then a 2C accumulator is tolerant to intermediate runtime register overflow.

Another common method used to suppress the potential effects of register overflow is to use saturating arithmetic. A saturating arithmetic unit, upon detecting an overflow condition, clamps the accumulator output to the numbering system's most positive or negative value. That is, the output of an N-bit 2C saturating arithmetic unit is defined to be

$$ACC = \begin{cases} (2^{N-1} - 1) & \text{if } ACC \geq 2^{N-1} \\ ACC & \text{if } -2^{N-1} < ACC < 2^{N-1} - 1 \\ -2^{N-1} & \text{if } ACC \leq -2^{N-1} \end{cases} \tag{14-51}$$

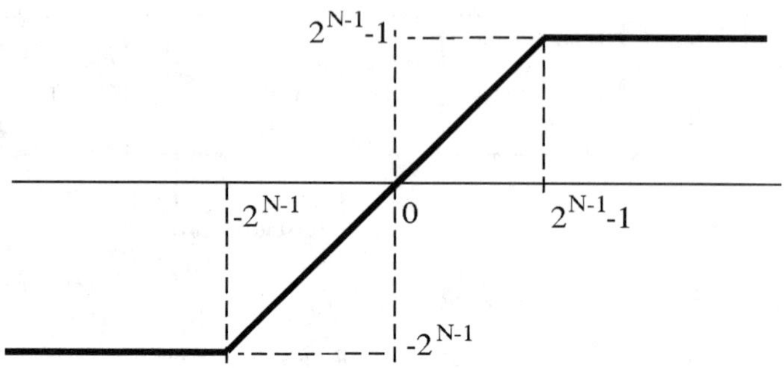

FIGURE 14-23 A saturating arithmetic operational behavior.

The operating characteristics of a saturating arithmetic unit are graphically interpreted in Figure 14-23.

14.22 REGISTER OVERFLOW

Register overflow conditions can be mitigated by scaling the input to a level which will suppress or eliminate the possibility of a register to overflow. This method has a weakness in that determining the required scale-factor experimentally can be tenuous. The tested inputs may not represent an input worst case event and might therefore underestimate scaling needs. Furthermore, scaling reduces the precision of the input which, in turn, reduces output precision. Another means of eliminating runtime overflow is to use extended precision arithmetic and registers. Extended precision registers provide additional "head room" that will allow the filter, even under worst case conditions, to store and preserve a system's states without introducing saturation errors.

The preferred method of overcoming the threat of runtime overflow is to determine the *worst case* filter gain at each register or state location. This is an extension of the worst case design analysis introduced for the study of FIRs. The translation of an FIR into an IIR can be accomplished by assuming the FIR's impulse response is infinitely long. Since a worst case analysis for an FIR was defined in terms of the filter's impulse response, a similar strategy must be employed in the case of IIRs. If we assume that the input is bounded by $|x[k]| \leq M_x$ on a sample-by-sample basis, then the *worst case input* would be given by

$$x_{\text{worst\_case}}[k - m] = M_x \,\text{sign}(h_m) \tag{14-52}$$

where $\text{sign}(h_m)$ is the sign of h_m (i.e., $\{\pm 1\}$). For an IIR, the output response to the worst case input is given by

$$y_{\text{worst\_case}}[k] = \sum_{m=0}^{\infty} h_m x_{\text{worst\_case}}[k - m] \leq \sum_{m=0}^{\infty} h_m M_x \,\text{sign}(h_m) = M_x \sum_{m=0}^{\infty} |h_m| \leq M_x G_{\max} \tag{14-53}$$

where the parameter $G_{\max}$, found in Equation (14-53), is called the filter's *worst case gain* and is given by

$$G_{\max} = \sum_{m=0}^{\infty} |h_m| \leq 2^I \tag{14-54}$$

Technically, G_{max} is the l_1-norm of a filter's input-output response. In order to adapt this FIR-based methodology for use with an nth-order IIR, the l_1-norm of all the states of the system needs to be computed. This requires that n impulse responses, residing between input and state shift registers, needs to be determined. This can be facilitated using the filter's state variable description of the selected architecture. From the elements of [$\mathbf{A}$, b] of the state four-tuple [$\mathbf{A}$, b, c, d], n impulse responses can be computed from input to state n location using a number of mathematical techniques, and can be simulated as well by using

$$\mathbf{h}[k + 1] = \mathbf{A}\mathbf{h}[k] + b\delta[k] \tag{14-55}$$

where $\mathbf{h}[k]$ is an n-vector of impulse responses. Technically, the l_1-norm on all the individual states would be defined by the infinite vector-valued sum

$$\|h_i[k]\|_1 = \sum_{m=0}^{\infty} |h_i[m]| \tag{14-56}$$

where $h_i[k]$ is defined at the output of the ith shift register (for instance, state $x_i[k]$ location). Unfortunately, the production of Equation (14-56) requires the evaluation of an infinite sum.

Assuming that the system is asymptotically stable, the impulse response will eventually converge to zero resulting in a finite steady-state value for $\|\boldsymbol{h}[k]\|_1$. The number of samples needed to insure convergence can be determined using a principle eigenvalue analysis. Another, simpler method is to determine the l_1-norm experimentally. In this mode, Equation (14-56) is approximated by a (possibly long) finite sum of sufficient length. MATLAB's *NORM* function can be used as the l_1 norm of the ith state's impulse response to a given error tolerance. Once computed or estimated, the l_1-norm $\|\boldsymbol{h}[k]\|_1$ can then be employed to establish the dynamic range requirements of the state registers. This concept is motivated in the following example.

Example 14-19 Dynamic Range Determination

Required:

An 8th-order Chebyshev-II IIR low-pass filter is implemented as a Direct II and cascade architecture based on the following specifications:

- Sampling frequency: 100 kHz
- Attenuation frequency: 20 kHz
- Stopband attenuation: 30 dB

The following filter was designed using MATLAB's FDA tool. The Direct II state two-tuple [$\mathbf{A}$, b] is given by

$$\mathbf{A} = \begin{pmatrix} 1.063 & -1.853 & 0.841 & -0.802 & 0.083 & -0.116 & -0.016 & -0.007 \\ 1.0000 & 0 & 0 & 0 & 0 & 0 & 0 & 0 \\ 0 & 1.0000 & 0 & 0 & 0 & 0 & 0 & 0 \\ 0 & 0 & 1.0000 & 0 & 0 & 0 & 0 & 0 \\ 0 & 0 & 0 & 1.0000 & 0 & 0 & 0 & 0 \\ 0 & 0 & 0 & 0 & 1.0000 & 0 & 0 & 0 \\ 0 & 0 & 0 & 0 & 0 & 1.0000 & 0 & 0 \\ 0 & 0 & 0 & 0 & 0 & 0 & 1.0000 & 0 \end{pmatrix};$$

$$b = \begin{bmatrix} 1 \\ 0 \\ 0 \\ 0 \\ 0 \\ 0 \\ 0 \\ 0 \end{bmatrix}$$

The cascade filter consists of the following 2nd-order sections:

$$H_1(z) = \frac{1 + 1.731z^{-1} + z^{-2}}{1 + 0.240z^{-1} + 0.053z^{-2}}$$

$$H_2(z) = \frac{1 + 0.524z^{-1} + z^{-2}}{1 - 0.083z^{-1} + 0.266z^{-2}}$$

$$H_3(z) = \frac{1 - 0.268z^{-1} + z^{-2}}{1 - 0.469z^{-1} + 0.551z^{-2}}$$

$$H_4(z) = \frac{1 - 0.583z^{-1} + z^{-2}}{1 - 0.751z^{-1} + 0.841z^{-2}}$$

such that $H(z) = 0.078\, H_1(z)\, H_2(z)\, H_3(z)\, H_4(z)$. Each 2nd-order Direct II section can be mapped to a state variable model using MATLAB's *SOS2SS* function. The individual 2nd-order Direct II subfilters are shown next:

$$A_1 = \begin{bmatrix} -0.2404 & -0.0535 \\ 1 & 0 \end{bmatrix} \quad c_1 = [1.4905 \quad 0.946]$$

$$A_2 = \begin{bmatrix} 0.0834 & -0.2664 \\ 1 & 0 \end{bmatrix} \quad c_2 = [0.6075 \quad 0.7336]$$

$$A_3 = \begin{bmatrix} 0.4688 & -0.5514 \\ 1 & 0 \end{bmatrix} \quad c_3 = [0.2006 \quad 0.4486]$$

$$A_4 = \begin{bmatrix} 0.7514 & -0.8414 \\ 1 & 0 \end{bmatrix} \quad c_4 = [0.1686 \quad 0.1586]$$

Determine the l_1 state norms for both architectures.

Results:

From the state-determined impulse response vector, the value of $\|\boldsymbol{h}[k]\|_1$ can be estimated using a finite sum approximation to Equation (14-56). The production of the incremented value of $\|h_1[k]\|_1$ is shown in Figure 14-24. Since all Direct II state-determined impulse responses have the same shape, differing only by simple delays, their l_1 norms are identical. Upon computing $\|h_1[k]\|_1$, using a finite sum approximation, the worst case gain is computed to be approximately $\boldsymbol{G}_1 = 2.25$. This indicates that the filter's shift registers need an additional $\log_2(2.25) \sim 1.17$-bits of headroom. Since 1.17-bits is very close to one bit, a one-bit head room would generally be recommended.

The l_1 norms of the state-determined impulse responses of the cascade IIR are shown in Figure 14-25. The largest l_1 norm is found in the fourth subfilter and is about 1.8 ($\log_2(1.8) \sim$ 0.85-bits), which is less than the maximal l_1 norm of the Direct II filter model.

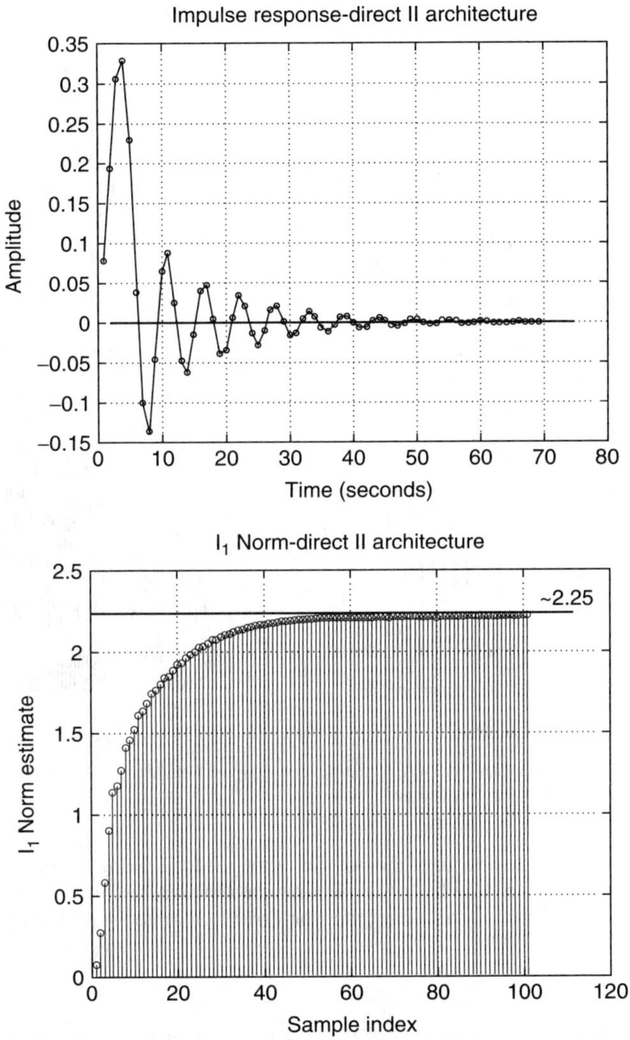

FIGURE 14-24 A Direct II impulse response measured at the $x_1[k]$ shift register location (top) and l_1 norm estimate (bottom).

14.23 ARITHMETIC ERRORS

The next most serious finite wordlength effect is called *arithmetic error*. To illustrate the source of arithmetic errors, consider the fixed-point multiply-accumulate (MAC) or SAXPY ($S = AX + Y$) unit shown in Figure 14-26. It is assumed that A and X are N-bit digital words that are presented to a full-precision multiplier, producing a $2N$-bit product. The $2N$-bit full precision product may be reduced to an S-bit word, $S \leq 2N$, before being presented to a T-bit accumulator ($T \geq S$). The presence of the round by S operation varies from MAC-unit

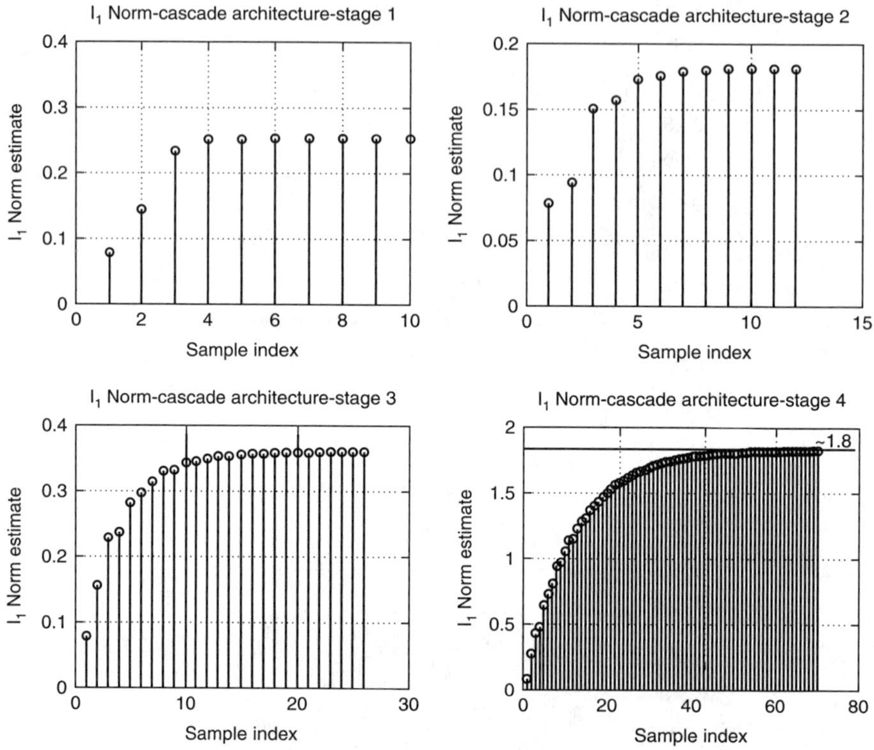

FIGURE 14-25 A cascade with an l_1 norm estimate section by section. (From the upper-left corner, clockwise: Section 1, Section 2, Section 4, and Section 3.)

to MAC-unit. If $T > S$, the accumulator is often referred to as being an extended precision accumulator. The T-bit accumulator is finally reduced to an M-bit final SAXPY outcome, where $M \leq T$. It is generally assumed that the error associated with the case $M \ll 2N$ is given by

$$e = (y[k] - Q_F(y[k])) \tag{14-57}$$

where $Q_F(q)$ denotes the quantization of a real number q to a digital word having F-bits of fractional precision. The error variance is given by

$$\sigma^2 = \frac{Q^2}{12} \tag{14-58}$$

where $Q = 2^{-F}$. The problem is not in modeling the production of the roundoff error itself, but determining what happens to the error after it is produced. This determination is important because errors generated internally to an IIR can be recirculated within the IIR indefinitely.

Mathematical modeling of the arithmetic error process relies on the system being represented in state variable form. Recall that the state model equation is given by $\mathbf{x}[k+1] = \mathbf{A}\mathbf{x}[k] + b\mathbf{u}[k]$, where $\mathbf{x}[k]$ is an N-dimensional state vector. The output model is $y[k] = c^T\mathbf{x}[k] + d\mathbf{u}[k]$. Noise analysis is performed by studying the state model using roundoff error

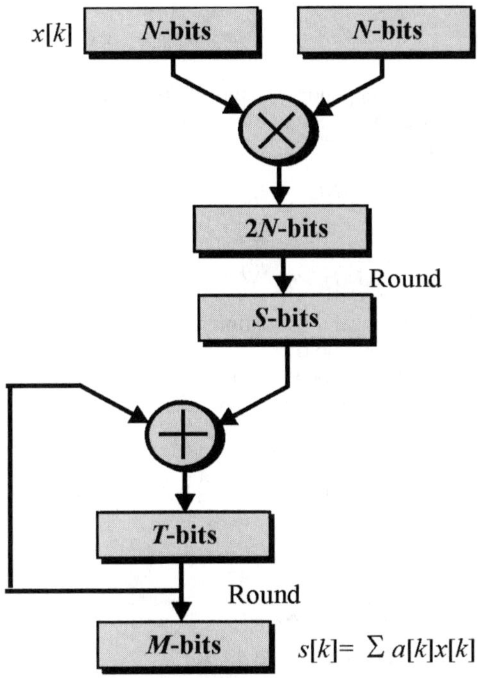

FIGURE 14-26 A fixed-point MAC unit.

sources to drive the filter rather than using the traditional input $\mathbf{u}[k]$. To illustrate, notice that the equation (Equation 14-19) describing the ith next state $x_i[k + 1]$ is found in the ith row of the state equation $\mathbf{x}[k + 1] = \mathbf{A}\mathbf{x}[k] + b\mathbf{u}[k]$. The production of $x_i[k + 1]$ may therefore require up to $N + 1$ multiply-accumulates. Let $m_i \leq N + 1$ denote the number of rounding operations associated with the production of $x_i[k + 1]$. Each distinct roundoff operation contributes a random error of variance $\sigma^2 = Q^2/12$ to the production of $x_i[k + 1]$ (Equation 14-58). If each full-precision product is rounded, then a roundoff error is injected into the ith state shift register for each a_{ij}, $j \in [1, N]$, and d_i having a value other than $\{-1, 0, 1\}$ (in other words, it requires no multiplication). This is called a multiply-round-accumulate architecture. Referring to Example 14-16, $m_1 = 8$, and $m_2 = m_3 = \cdots = m_8 = 0$. If all the full precision products are first accumulated, and then rounded, at most one roundoff error is injected into a state register. For those cases where no multiplication is required to produce $x_i[k + 1]$, $m_i = 0$. Such a system is called a multiply-accumulate-round architecture. Referring to Example 14-19, $m_1 = 1$, and $m_2 = m_3 = \cdots = m_8 = 0$. It should be appreciated that the accumulators, in this case, will probably have to have extended precision in order to accumulate up to $N + 1$ worst case products without overflowing.

It can also be seen that the state $x_i[k]$ is sent to the output with a gain c_i (ignoring the contribution of d). For this interpretation, the predicted noise power gain (NPG_i) defined from the input of the ith state to the output can be computed. Define $NPG_i = W_{ii}$ for $i \in [1, N]$ to be the noise power-gain that defines the noise power amplification of roundoff noise injected into the ith state and appearing at the IIR's output. Collectively, the output noise error variance becomes

$$\sigma^2 = \frac{Q^2}{12}\sum_{k=1}^{N} m_i W_{ii} = \frac{Q^2}{12}\sum_{k=1}^{N} m_i NPG_i = \frac{Q^2}{12}(NPG) \tag{14-59}$$

where m_i defines the number of independent roundoff error sources attached to the input of the ith shift register. The filter's noise power power-gain, denoted NPG in Equation (14-59), can also be interpreted in bits using

$$NG = \sqrt{NPG} \tag{14-60}$$

and then computing

$$NG_2 = \log_2(NG) \quad - \quad \text{bits} \tag{14-61}$$

NG_2 is an estimate of the statistical degradation of IIR's output, in bits, due to accumulated roundoff errors residing within the filter.

The calculation of NPG_i or W_{ii} can be accomplished using a formula obtained from the classic Lyapunov stability theory, which states

$$\mathbf{W} = \mathbf{A}^T\mathbf{W}\mathbf{A} + \mathbf{c}\mathbf{c}^T \tag{14-62}$$

where W_{ii} is the ith on-diagonal term of $\mathbf{W}$, for $i \in [1, N]$. $\mathbf{W}$ can also be computed using the iterative algorithm

Initialize: $\mathbf{W}_0 = \mathbf{c}\mathbf{c}^T$ (14-63)
Loop on i

$\quad\quad \mathbf{W}[i + 1] = \mathbf{A}^T\mathbf{W}[i]\mathbf{A} + \mathbf{W}_0$

$\quad\quad$ Continue until $\mathbf{W}[i + 1] - \mathbf{W}[i] \sim [\mathbf{0}]$

$\mathbf{W}[i + 1] \Rightarrow \mathbf{W}$

The production of the matrix **W** can be bypassed using simulation. Both the analytic and experimental study of noise power gain is illustrated in the following example.

Example 14-20 Finite Wordlength Effects

Required:

The 8th-order Chebyshev-II IIR low-pass filter studied in Example 14-19 was implemented as a Direct II and Cascade architecture. Determine the noise power gain for each filter if extended precision registers are used along with a multiply-accumulate-round architecture.

Results:

The statistical noise power gain of each design can be predicted by calculating the elements of the Lyapunov matrix **W**. For the Direct II design, analysis produces $\mathbf{W}_{11} = \cdots = \mathbf{W}_{88} = 0.35$. The design assumes that extended precision accumulators are used, resulting in $m_1 = 1$ and $m_2 = m_3 = \cdots = m_8 = 0$. Correcting for the scale factor $k = 0.078$, the noise power gain is predicted to be $NPG = \mathbf{W}_{11}/k_1^{\,2} = 58.5$ or $NG_2 = 7.63$, which corresponds to approximately three bits of lost precision in the IIR's output.

The noise power gain for the four-stage cascaded subfilters was determined to be

Stage i	NPG_i
1	3.848
2	1.04
3	0.47
4	0.307

The cascaded noise power gain is given by $NPG = NPG_1 \times NPG_2 \times NPG_3 \times NPG_4/k^2 = 94.91$, or $NG = 9.74$, which corresponds to about a 3.3-bit loss in output precision at the IIR's output (approximately 0.5-bit more than the Direct II model).

The Lyapunov-based method, it should be noted, is based on a number of assumptions that may not completely characterize a filter. It can, however, provide an analytical means of mathematically comparing two or more filters and architectures. Another means of quantifying the statistical behavior of an IIR is using fixed-point simulations. This can be facilitated using a number of contemporary block diagram simulation languages. The following example demonstrates the use of simulation in the study of fixed-point filters.

Example 14-21 Fixed Point Simulations

Required:
Assume the IIRs from Example 14-19 are implemented using an $N = 16$-bit processor possessing full precision multipliers, extended precision accumulators, and a multiply-accumulate-round architecture.

Results:
Using MathWorks' Simulink, an end-to-end fixed-simulation of the 8th-order filter implemented as a Direct II and Cascade filter architectures in Example 14-19, was performed. The key architectural choices defining the simulation:

Data wordlength: $N = 16$-bits

Fractional precision: $F \in [0{:}15]$-bits.

Input data format: $x[k] \in [N{:}F]$-bits

Output data format: $y[k] \in [N{:}F]$-bits

Coefficient data format: $c_k \in [N{:}F]$-bits

Multiplier data paths: $16 \times 16 \rightarrow 32$-bits

Multiplier data paths: $16 \times 16 \rightarrow 32$-bits

Direct II accumulator data paths: $32 + (32 + N_{DII}) \rightarrow 32 + N_{DII}$-bits ($N_{DII} \geq \log_2(2.25) \sim 1.17$-bits)

Cascade accumulator data paths: $32 + (32 + N_C) \rightarrow 32 + N_C$-bits ($N_C \geq \log_2(1.8) \sim 0.8$-bits)

The numerical results of the fixed-point simulation are presented in Figure 14-27 for both architectures. The input forcing function was a long unit-bound uniformly distributed random signal that emulated, at some point in its history, a worst case input. Reviewing the simulated outcome from Figure 14-27, it can be noted that the result can be partitioned into three regimes, namely:

- Too little precision—caused by too few fractional bits of accuracy.
- Linear regime—sufficient dynamic range to inhibit runtime overflow and sufficient fractional precision to eliminate traumatic roundoff errors.
- Too little dynamic range—caused by too few integer bits resulting in a too small dynamic range that results in a plethora of runtime overflow errors.

It can be seen that the Direct II architecture exhibits overflow contamination beginning at two integer bits as predicted by the l_1 norm analysis. The Cascade filter began exhibiting register overflow beginning at 1 integer bit as predicted. Moreover, the

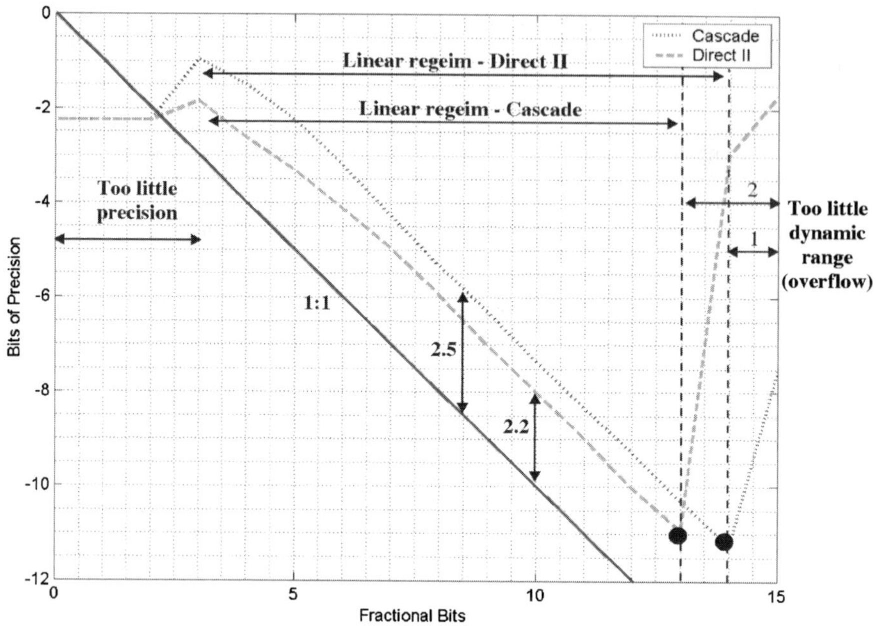

FIGURE 14-27 Statistical filter performance versus the number of fractional bits.

Direct II has slightly better statistical precision over the linear input-output operating range, which was again predicted by the analytical study. In the linear region, the analysis predicted that the Cascade architecture is about 0.3-bit inferior to a Direct II. This too was predicted. The simulation would suggest that the optimum Cascade filter carry a [16:14] format, resulting in a solution having (statistically) about 11.5 fractional bits of precision. The simulation also suggests that the Direct II filter carry a [16:13] format, resulting in a solution having (statistically) about 11 fractional bits of precision.

The conclusion is that the optimum performance point of a fixed-point IIR can be predicted using analytical techniques from a simulation environment. The advantage of the simulation approach is manifold. First, it is easier and faster to construct than creating the tools and environment needed to complete an alternative analytical filter analysis. Second, it is more robust, capable of handling any arbitrary architecture. Third, it is extensible and can be easily modified (for example, a change of design specification). Fourth, it provides open access to the entire signal-processing elements used to construct the simulation. What must be stressed is that the system was studied under worst case conditions. Analyzing the system using impulse or sinusoidal test signals will produce different and erroneous results.

14.24 COEFFICIENT ROUNDING ERRORS

The third most serious source of finite wordlength error is attributed to coefficient rounding. A general IIR model is based on factoring the filter's transfer function $H(z)$. Analytical methods have been developed that are analogous to sensitivity analysis techniques used to study analog filters. These methods are based on computing partial derivatives but can only claim to explain the effects of infinitesimal localized coefficient errors. Coefficient rounding errors typically need to be significant if they are to have any affect on filter performance. With the advent of 16-bit DSP μps and beyond, coefficient rounding errors have become less and less a factor in a successful IIR design. The preferred means of quantifying the effects of coefficient rounding is simulation. The filter coefficients can be quantized to a selected precision, and a simulation be performed using floating-point arithmetic. The difference between the ideal and simulated forced responses is assumed to be attributed to coefficient rounding. Coefficient rounding error is the subject of the following example.

Example 14-22 IIR Coefficient Rounding Sensitivity

Required:

A 6th-order Chebyshev I low-pass IIR has a transfer function

$$H(z) = 0.0096311 \times \frac{1 + 6z^{-1} + 15z^{-2} + 20z^{-3} + 15z^{-4} + 6z^{-5} + z^{-6}}{1 - 1.577z^{-1} + 2.4674z^{-2} - 2.2984z^{-3} + 1.6613z^{-4} - 0.7977z^{-5} + 0.2354z^{-6}}$$

The filter is implemented as a Direct II and Cascade IIR. In both cases, the filters are implemented using the input scale factor $k = 0.0096311 < 2^{-6.698}$. Rounding k to less than six bits will result in the zeroing of the input scale factor (that is, $k_{\text{rounded}} = 0$) completely nulling the input. Determine the coefficient rounding effects for the Direct II and Cascade filters using a [16:8] format.

Results:

Using simulation, the IIR's frequency response and group delay are reported in Figure 14-28. The primary difference between the responses can be attributed to rounding the input scale factor. The eight-bit rounding of k to 0.0078125 introduces a 20-percent downward change in value. The remaining physical effects of coefficient rounding are quantifiable, but seen to be relatively benign. You should know that Cascade architecture has a slight advantage over the Direct II, but represents a slightly more complex implementation.

14.25 SCALING

Register overflow can be inhibited by reducing the dynamic range of the input. Scaling the input by a positive constant, $k < 1$, reduces the chance of runtime overflow. Unfortunately, scaling also reduces the output precision. In particular, the precision of a scaled data word will be reduced by $\log_2(k)$-bits. Equivalently, the output precision will be reduced by a like amount. Therefore, the error variance of the scaled system, compared to the unscaled system, can be expressed as

$$\sigma_k^2 = \frac{(kQ)^2}{12} = k^2\sigma^2 \tag{14-64}$$

As a result, scaling (if required) should only be used minimally and only to the point where overflow is controlled. Using extended precision registers can preserve precision, but does so with an added hardware penalty.

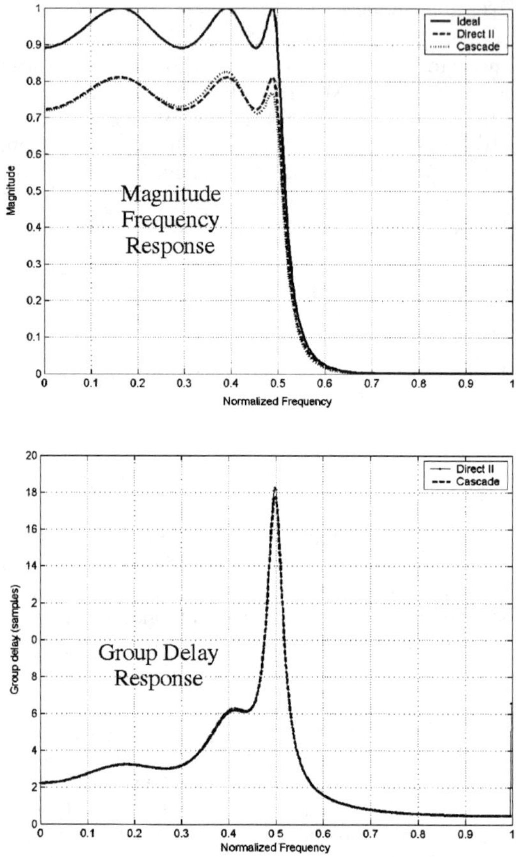

FIGURE 14-28 The magnitude frequency response of a Chebyshev I, Direct II, and Cascade implementation using a [16:8] format (top) and the group delay for a [16:8] coefficient format (bottom). (The frequency was normalized with respect to $f_s/2$.)

14.26 ZERO INPUT LIMIT CYCLING

Another finite wordlength effect is called *zero input limit cycling*, or simply *limit cycling*. Limit cycling causes a digital filter to produce small amplitude changes to appear at the system's output during periods when the input is zero. In voice communication applications, limit cycling can manifest itself as undesirable audible "clicking" sounds that can be heard during quiet (unvoiced) periods. The first generation DSP microprocessors were imprecise (for example, eight bits), and as a result, limit cycling was an annoying problem that could only be reduced through serious engineering labor. With the advent of 16-bit, 24-bit, and floating-point processors, limit cycling has become a secondary problem.

Limit cycling is caused when the response of an unforced stable filter does not successfully decay to zero due to finite wordlength effects. Consider, for illustrative purposes, the simple 1st-order system

$$y[k] = ay[k-1] + x[k] \tag{14-65}$$

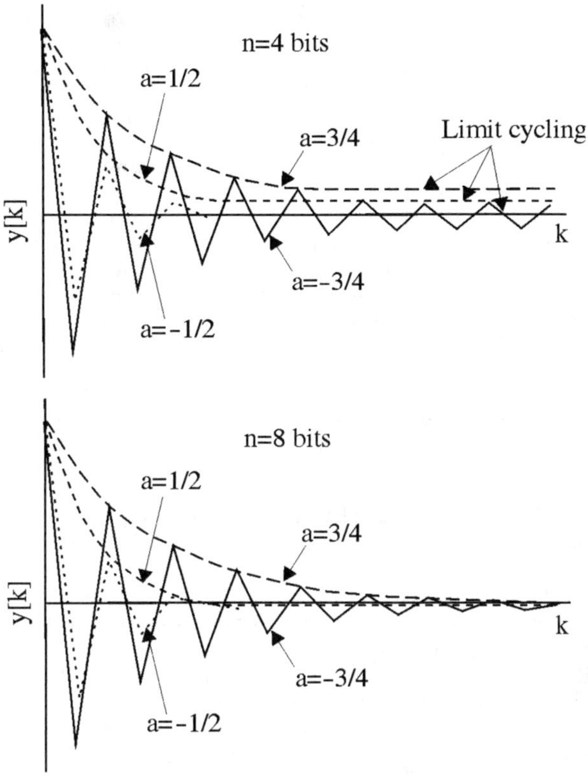

FIGURE 14-29 Limit cycling for the unforced system $y[k] = ay[k-1] + x[k]$, $x[k] = 0$, using a four- and eight-bit fractional precision. Limit cycling is associated with the low-bit resolution response.

If $|a| < 1$, then $y[k] \to 0$ when $x[k] = 0$. If the filter is implemented in fixed-point, then

$$y[k] = Q_F[ay[k-1]] + x[k] \tag{14-66}$$

where $Q_F[q[k]]$ denotes the quantized value of $q[k]$ to F fractional bits of accuracy. Refer to Figure 14-29 and suppose that at the sample instance k, the 1st-order response is $y[k]$ when the input has been $x[k] = 0$ for all $k > K$. If the decay rate of the quantized system is slow (too slow), then the output will decay to an amount less than half a quantization interval (such as $\Delta[k] = (y[k] - y[k-1]) < Q/2$), then upon rounding, $y[k]$ would be returned to its previous value $y[k-1]$. As such, the output would never be able to decay to zero and would have some constant off-set (possibly oscillating), which is called limit cycling. The effects of limit cycling are studied in the following example.

Example 14-23 Limit Cycling Restore Title

Required:

Suppose a system $y[k] = ay[k-1] + x[k]$ has a $[N:4]$ format and $a = 1/2, -1/2, 3/4$, and $-3/4$. Investigate the zero input limit cycling properties of the system. The implementation of these four filters would yield the results shown next. The data is interpreted in Table 14-7

TABLE 14-7 A Zero-Input Limit Cycling Example

k	$a = 1/2 = 0_{\diamond}1000$	$a = -1/2 = 1_{\diamond}1000$	$a = 3/4 = 0_{\diamond}1100$	$a = -3/4 = 1_{\diamond}1100$
0	$0_{\diamond}1111$ (15/16)	$0_{\diamond}1111$ (15/16)	$0_{\diamond}1111$ (15/16)	$1_{\diamond}1111$ (15/16)
1	$0_{\diamond}1000$ (8/16)	$1_{\diamond}1000$ (–8/16)	$0_{\diamond}1011$ (11/16)	$1_{\diamond}0101$ (−11/16)
2	$0_{\diamond}0100$ (4/16)	$0_{\diamond}0100$ (4/16)	$0_{\diamond}0110$ (6/16)	$0_{\diamond}0110$ (6/16)
3	$0_{\diamond}0010$ (2/16)	$1_{\diamond}1110$ (–2/16)	$0_{\diamond}0101$ (5/16)	$1_{\diamond}1100$ (−4/16)
4	$0_{\diamond}0001$ (1/16)	$0_{\diamond}0001$ (1/16)	$0_{\diamond}0100$ (4/16)	$0_{\diamond}0011$ (3/16)
5	$0_{\diamond}0001$ (1/16)	$1_{\diamond}1111$ (–1/16)	$0_{\diamond}0011$ (3/16)	$1_{\diamond}1101$ (−3/16)
6	$0_{\diamond}0001$ (1/16)	$0_{\diamond}0000$ (0/16)	$0_{\diamond}0010$ (2/16)	$0_{\diamond}0010$ (2/16)
7	$0_{\diamond}0001$ (1/16)	$0_{\diamond}0000$ (0/16)	$0_{\diamond}0010$ (2/16)	$1_{\diamond}1111$ (−1/16)
8	$0_{\diamond}0001$ (1/16)	$0_{\diamond}0000$ (0/16)	$0_{\diamond}0010$ (2/16)	$0_{\diamond}0001$ (1/16)
9	$0_{\diamond}0001$ (1/16)	$0_{\diamond}0000$ (0/16)	$0_{\diamond}0010$ (2/16)	$1_{\diamond}1111$ (−1/16)
k	limit cycling at $y[k] = 1/16$	no limit cycling	limit cycling at $y[k] = 2/16$	limit cycling at $y[k] = (\pm 1/16)^k$

for a five-bit 2's complement implementation using as an initial condition $x[0] = 1 \rightarrow 0_{\diamond}1111$ where $\diamond$ denotes the binary point. Determine whether limit cycling can occur.

Results:

The output response is computed for each choice of a and is reported in Table 14-7.

The data indicates limit cycling for a = $\{1/2, -1/2, 3/4, -3/4\}$.

The 4 IIR zero input responses are simulated for four- and eight-bit solutions in Figure 14-29. The filter is initialized to $x[0] = 1.0$ and then run. Note that limit cycling can take place under certain circumstances and that the severity of limit cycling is a function of the decay rate of the response (for example, "a") and the number of bits of arithmetic precision maintained after multiplication. The threat of limit cycling for a 16- or 24-bit design, especially those containing extended precision arithmetic, is very small and negligible.

BIBLIOGRAPHY

Antoniou, A. *Digital Filters: Analysis and Design.* New York: McGraw-Hill, 1979.

Ifwachor, E., and B. Jervis. *Digital Signal Processing.* 2nd ed. Reading, Massachusetts: Addison-Wesley, 2001.

Mitra, S. *Digital Signal Processing.* New York: McGraw-Hill, 2001.

McClellan, J. H., Parks, T. W., and L. R. Rabiner. "A Computer Program for Designing Optimum FIR Linear Phase Filters." *IEEE Trans. On Audio and Electroacoustics* (December, 1973).

Oppenheim, A. V. and R. S. Schafer. *Digital Signal Processing.* Englewood Cliffs, New Jersey: Prentice-Hall, 1975.

Proakis, J., and D. Manaolakis. *Digital Signal Processing.* 3rd ed. Englewood Cliffs, New Jersey: Prentice-Hall, 1996.

Rabiner, L. A., Kaiser, J. F., Herrmann, J., and M. T. Dolan. "Some Comparisons Between FIR IIR Digital Filters." *Bell System Tech. Journal* (February, 1974).

Taylor, F. J. *Digital Filter Design Handbook.* New York: Marcel Dekker, 1983.

——— and T. Stouraitis. *Digital Filter Design Software for the IBM PC.* New York: Marcel Dekker, 1987.

Zelniker, G. and F. Taylor. *Advanced Digital Signal Processing: Theory Applicants.* New York: Marcel Dekker, 1994.

CHAPTER 15
MULTIRATE DIGITAL FILTERS

15.1 INTRODUCTION TO MULTI-RATE SIGNAL PROCESSING

A digital filter accepts a time-series input, produces a time-series output, and in-between modifies the signal in terms of its time and/or frequency domain attributes. Digital filters are normally assumed to operate, end-to-end, at a constant sample-rate f_s where the sample-rate bounds on the fixed sample rates are established by Shannon's Sampling Theorem. Some systems are designed to operate at, or near, the Nyquist sample rate and are said to be *critically sampled*. Other systems operate well above the minimum sample rate and are called *over sampled*. Over sampling can require the use of high speed arithmetic units. In some instances, however, over sampling can reduce design complexity of other parts of the system. This is illustrated in the following example.

Example 15-1 Audio over Sampling

Required:
It is normally assumed that the audio spectrum is band-limited to 20 kHz. Therefore, a standard multimedia 44.1 kHz ADC can provide alias-free data conversion. The input to the ADC, however, should be passed through an analog anti-aliasing filter having a 20 kHz passband and a stopband beginning at 22.05 kHz, or earlier. The problem is that the anti-aliasing filter's transition band is only $\Delta f = 4.1$ kHz wide, which is far too narrow to be realized by any practical analog filter (see Figure 15-1). It is claimed that the design requirements on the analog anti-aliasing filter can be relaxed by over sampling the audio signal by a factor of four (that is, 4x). Analyze the consequences of over sampling on the design of the analog anti-aliasing filter.

Results:
The 4x audio system is assumed to be sampled at $f_s = 4 \times 44.1$ kHz $= 176.4$ kHz. The new Nyquist frequency is therefore 88.2 kHz. Referring to Figure 15-1, it can be seen that the end of the new analog anti-aliasing filter's transition band can, in fact, range out to 156.4 kHz. Designing the required shallow skirt analog anti-aliasing with respect to the new specifications is very manageable.

There are instances when it is preferred, or required, to operate different parts of a filter at different sample rates. Filters operating with multiple sample rates are naturally called *multirate* digital filters. Multirate digital filters can perform a number of tasks, including *sample rate conversion* and signal *bandwidth compression*. A fixed-rate sample rate filter, for example, may require K arithmetic operations per filter cycle. Reducing the sample rate of a signal by a factor M, reduces the arithmetic bandwidth requirements by a like amount.

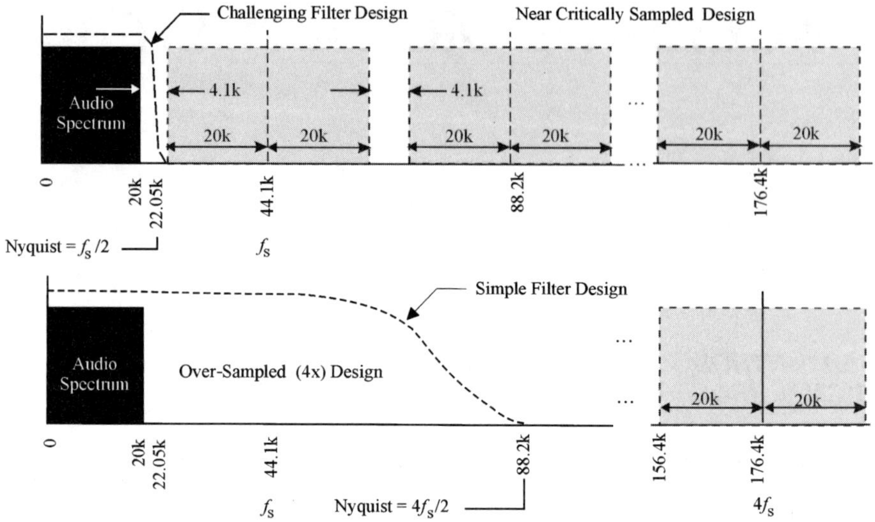

FIGURE 15-1 An over-sampled (4x) audio system with an analog anti-aliasing filter's stopband set to 88.2 kHz.

In other applications, multirate systems are used to achieve performance levels beyond that obtainable with a fixed-rate system.

15.2 DECIMATION

Filter sample rates can be altered using an operation called *decimation* or down sampling. If a time-series $x[k]$ is imported at a sample rate f_s, and exported at a rate f_d, such that $f_s > f_d$, then the signal is said to be *decimated*[1] by a factor M, where

$$M = f_s/f_d \tag{15-1}$$

For a given integer M, the decimated time-series is mathematically given by

$$x_d[k] = x[Mk] \tag{15-2}$$

indicating that only every Mth sample of the original time-series is retained and all others ignored. The decimated time-series also operates at a reduced speed $f_d = f_s/M$ Sa/s. This process is shown in Figure 15-2 for $M = 2$ case.

A decimated by M time-series can be formally modeled as

$$x_d[k] = \sum_{n=-\infty}^{\infty} x[n]\,\delta[k - nM] \tag{15-3}$$

[1]Historically, decimation originally referred to a disciplinary method employed by the Romans in dealing with mutinous soldiers. The mutineers were forced to select balls from an urn containing nine times more white balls than black balls. The holders of black balls would be put to the sword. The result was that every tenth soldier was slain.

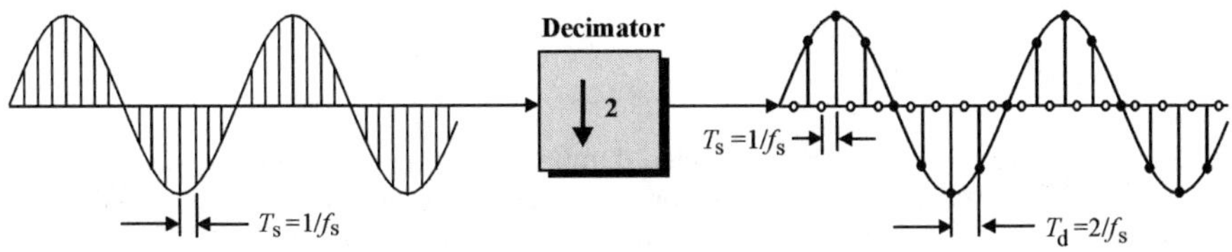

FIGURE 15-2 An illustrated decimate by two processes.

results in a z-transform given by

$$X_d(z) = \sum_{n=-\infty}^{\infty} x_d[n]z^{-n} = \sum_{n=-\infty}^{\infty} z^{-n}\left(\sum_{m=-\infty}^{\infty} x[n]\,\delta(m - nM)\right)$$
$$= \sum_{n=-\infty}^{\infty} x[n]\left(\sum_{m=-\infty}^{\infty} z^{-n}\,\delta(m - nM)\right) = \sum_{n=-\infty}^{\infty} x[n]z^{-nM} = \sum_{n=-\infty}^{\infty} x[n](z^{-n})^M = X(z^M) \tag{15-4}$$

In the frequency domain, Equation (15-4) can be expressed as

$$X_d(e^{j\varpi}) = X(e^{jM\varpi}) \tag{15-5}$$

and is graphically interpreted in Figure 15-3. The input signal's spectrum is assumed to be limited to B Hz and the base band is constrained by Nyquist frequency $f_s/2$ for the undecimated signal. The decimated signal's spectrum is also base band limited to B Hz with respect to a new Nyquist frequency of $f_s/2M$. A decimation by two case is studied in the next example.

Example 15-2 Decimation

Required:

In Example 15-1, a 4x over-sampled solution was investigated. The output sample rate for the over-sampled system was $f_s = 176.4$ kHz. The sample rate can be returned to the

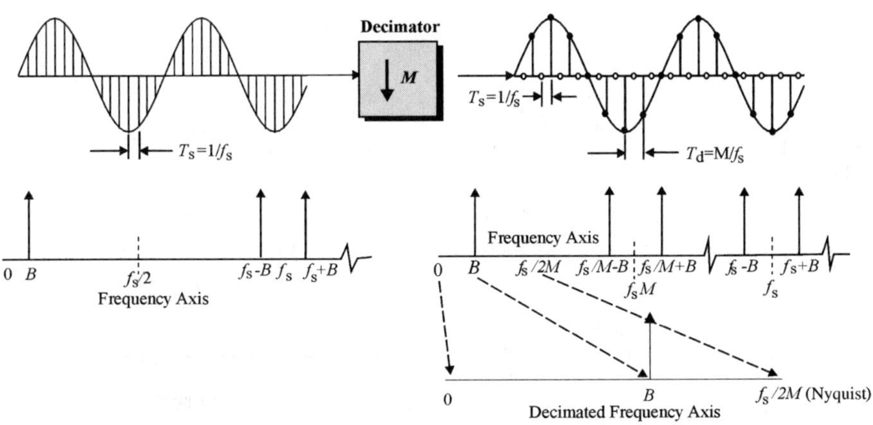

FIGURE 15-3 The magnitude spectrum of a signal $x[k]$ and decimated by M signal, denoted $x_d[k]$, and plotted on a common frequency axis.

multimedia rate $f_d = 44.1$ kHz by decimating the over-sampled signal by four. Analyze the consequence of this action.

Results:

Suppose the audio spectrum is considered to be represented by a single tone located at $f = (1000/16)$ Hz. Using Euler's equation, the signal's spectrum is mathematically defined to be $X(e^{j\varpi}) = 0.5 \times \delta(\varpi - 2\pi 10^3/16) + 0.5 \times \delta(\varpi + 2\pi 10^3/16)$ with respect to the 4x sampled rate f_s. The time-series and spectrum are shown in Figure 15-4. Upon decimating by $M = 4$, Equation (15-5) defines the decimated spectrum to be $X_d(e^{j\varpi}) = X(e^{j4\varpi}) = 0.5 \times \delta((\varpi - 2\pi 10^3)/4) + 0.5 \times \delta((\varpi + 2\pi 10^3)/4)$ with respect to the decimated sample rate of $f_s/4 = 44.1$ kHz. The decimated time-series and spectrum are also displayed in Figure 15-4. It can be seen that the tone is present at its original frequency location, but the width of the output base band has decreased by a factor of four.

In Example 15-1, a relaxed analog anti-aliasing filter was facilitated by over-sampling. What should be appreciated is that any spectral energy found at the analog anti-aliasing filter's input residing beyond $f = 22.05$ kHz will be aliased into the decimated base band. In most audio applications, however, the analog recording spectrum is usually clean beyond 20 kHz and is also free of system-generated noise out to the onset of digital clock noise. Under these assumptions, the choice of a relaxed anti-aliasing filter is valid.

Shannon's sampling theorem also applies to multirate signals and systems. Suppose the highest frequency found in the time-series $x[k]$ is B Hz, as suggested in Figure 15-5. Aliasing can be avoided if the decimated sample rate exceeds $f_d = 2 \times B$ Hz. This means that there is a practical upper limit to decimation. Referring to Figure 15-5, it can be seen that for alias-free decimation to take place:

$$\frac{f_s}{M} - B > B \quad \text{or} \quad M < \frac{f_s}{2B} \tag{15-6}$$

Increasing the decimation rate beyond this value will potentially produce an aliased decimated time-series. In practice, the maximal decimation rate is rarely used. Instead, a more conservative value is generally used, which will allow for a well-defined guard band to be established, as suggested in Figure 15-5. The next example studies the question of maximum decimation rate.

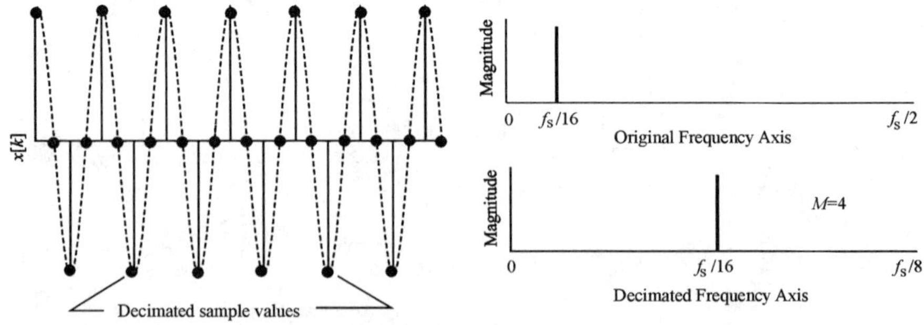

FIGURE 15-4 Shown are an original time-series $x[k]$ and decimated by $M = 4$ signal, along with their corresponding magnitude frequency responses.

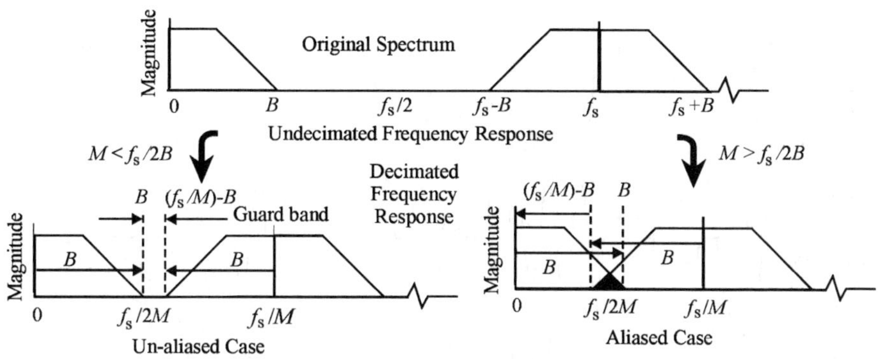

FIGURE 15-5 An unaliased and aliased decimated signal spectra.

Example 15-3 Maximum Decimation Rate

Required:
Suppose that the signal $x(t) = \cos(2 \times \pi \times 10^3 t)$, having a frequency $B = 10^3$ Hz, is over-sampled at a rate $f_s = 10^5$ Hz to form $x[k]$. Determine the maximum decimation rate that can be applied to $x[k]$ that will insure alias-free performance. Analyze the decimated signal spectrum for decimation rates of $M = 16$ and 64.

Results:
The minimum lower bound on the sampling rate (in other words, the Nyquist frequency) is 2×10^3 Hz. Therefore, the maximum decimation rate is bounded by $M < 10^5/(2 \times 10^3) = 50$. The spectrum of the undecimated and decimated signal by a factor 16 signals are reported in Figure 15-6 along a common base band frequency range $f \in [0, 10^5/2)$. The corresponding time-series are also shown in Figure 15-6. Note that the decimated spectrum contains copies (artifacts) of the base band signal on $f_s/16$ Hz centers. The predicted maximum decimation factor was 50 which implies that decimating by 64 should result in aliasing. The resulting decimated by 64 time-series is also shown in Figure 15-6 and is seen to impersonate (alias) a signal having a base-band frequency equal to $f_1 = 10^3 \bmod(10^5/64) = -562.4$ Hz (Equation 12-8). The aliased signal, namely $x_d[k] = \cos(2\pi f_1 k)$, is superimposed over the decimated time-series sample values in Figure 15-5.

Band-limited signals can also be decimated. Decimating such signals results in both a rearranged as well as a translated spectrum. The signal shown in Figure 15-7 has passband activity residing between $m\pi/M \leq \varpi \leq (m + 1)\, \pi/M$, where m is a positive integer. Decimating the band-limited signal by a factor M results in the spectrum shown in Figure 15-7. The specifics of the decimated spectrum are predicated on whether m is an even or odd integer. If m is odd, the spectrum is a reflection of the original spectrum which can be compensated for (that is, reversed) by multiplying the decimated time series, $x_d[k]$, by $(-1)^k$. The following example illustrates how decimation can effect the spectrum of a band-limited signal.

Example 15-4 Decimated Band-Limited Signal

Required:
Using computer simulation, analyze the spectral behavior of a decimated band-limited signal.

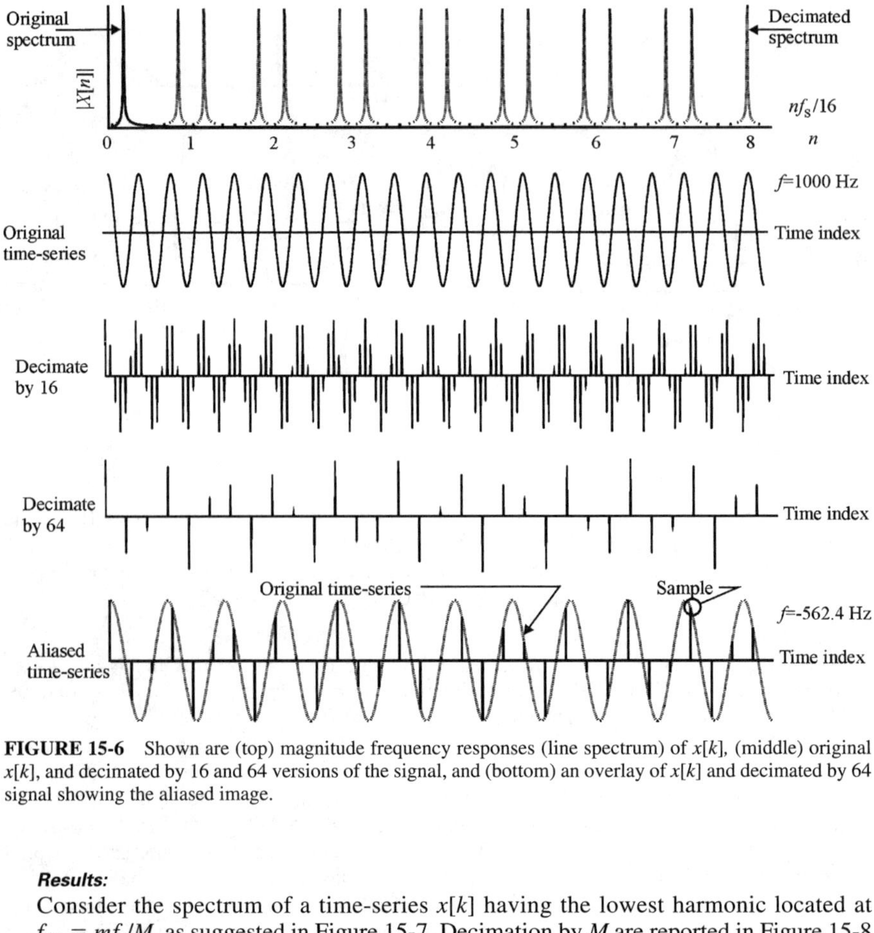

FIGURE 15-6 Shown are (top) magnitude frequency responses (line spectrum) of $x[k]$, (middle) original $x[k]$, and decimated by 16 and 64 versions of the signal, and (bottom) an overlay of $x[k]$ and decimated by 64 signal showing the aliased image.

Results:

Consider the spectrum of a time-series $x[k]$ having the lowest harmonic located at $f_{LO} = mf_s/M$, as suggested in Figure 15-7. Decimation by M are reported in Figure 15-8 for $m = 1$(odd) and $m = 2$(even). Note that for m odd ($m = 1$) and a decimation factor of $M = 8$, the base-band spectrum shows a reflection about $f = 0$ (DC). The reflection distortion can be corrected by modulating the decimated signal by $(-1)^k$. For m even ($m = 2$), the decimated spectrum is seen to be nonreflected.

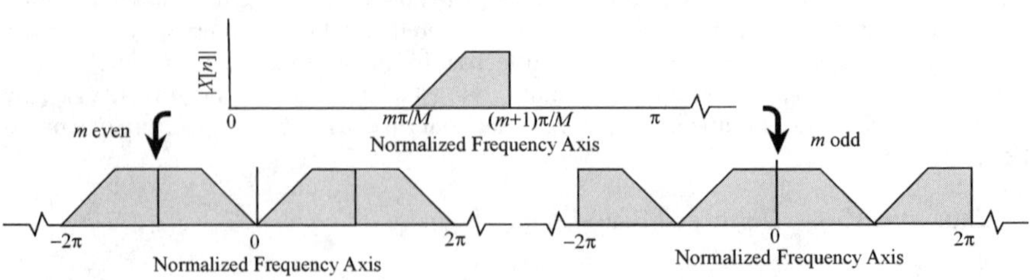

FIGURE 15-7 The spectra of decimation band-limited signals for m even and odd, showing different baseband symmetries of the aliased spectra.

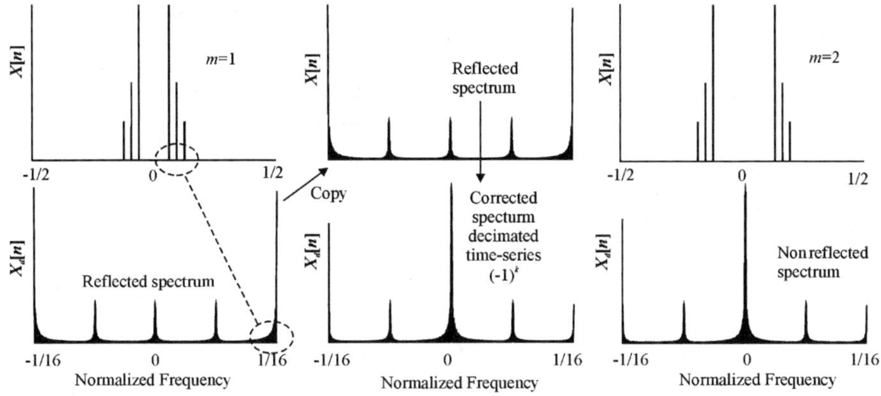

FIGURE 15-8 Shown are (left) a band-limited spectrum for $m = 1$ and decimation factor of $M = 8$. The base-band spectrum is shown to be reflected about $f = 0$ (DC). The spectrum shown in the middle of the figure is the same as that displayed on the left with a corrected reflection using a modulation of the decimated time series by $(-1)^k$. The right panel displays the spectrum of a band-limited signal for $m = 2$ (top) and after decimation by $M = 8$ (bottom). The base-band spectrum is shown correct in its decimated form.

15.3 INTERPOLATION

The antithesis of decimation is called *interpolation* or up-sampling. The use of the verb interpolation is somewhat unfortunate since interpolation also defines a class of methods used to reconstruct a facsimile of an analog signal $x(t)$ from a sparse set of samples $x[k]$. In the context of decimation and interpolation, interpolation simply refers to a mechanism that increases the effective sample rate. Interpolating a time-series $x[k]$, sampled at a rate f_s by a factor N, creates a new time-series $x_i[k]$ given by

$$x_i[k] = \begin{cases} x[k] & \text{if } k = 0 \bmod N \\ 0 & \text{otherwise} \end{cases} \tag{15-7}$$

or

$$x_i[k] = \sum_{m=-\infty}^{\infty} x[k]\,\delta[k - mN] \tag{15-8}$$

Interpolation by N is seen to be equivalent to inserting $N - 1$ zeros in between the samples of the time-series being interpolated. This action is sometimes referred to as *zero-padding*. The result is a new time-series sampled at a rate $f_i = Nf_s$, as shown in Figure 15-9.

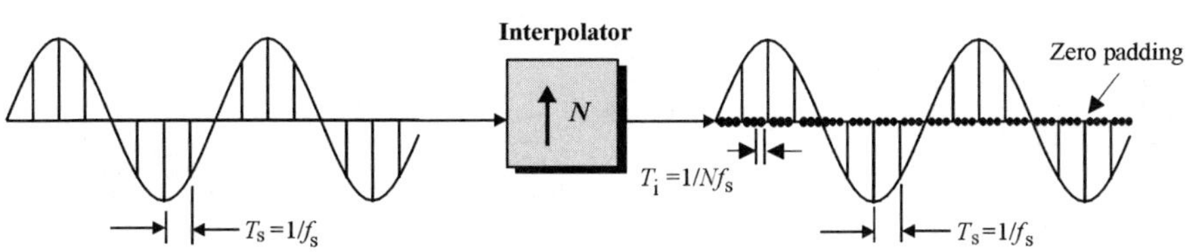

FIGURE 15-9 An illustration of an interpolation by N process.

Interpolation is often directly linked to decimation. To illustrate, suppose $x_d[k]$ is a decimated by M version of the time-series $x[k]$ which was sampled at a rate f_s. The decimated signal is therefore sampled at a rate $f_d = f_s/M$. Interpolating $x_d[k]$ by a factor N would result in a new time-series $x_i[k]$. The sample rate of the interpolated signal is increased from f_d to $f_i = Nf_d = Nf_s/M$. If $N = M$, the output sample rate would be restored to that of f_s.

Relative to the decimated signal $x_d[k]$, the frequency-domain signature of an interpolated by N signal $x_i[k]$ can be defined in terms of the z-transform of Equation (15-8). Specifically:

$$X_i(z) = \sum_{m=-\infty}^{\infty} x_d[m] \sum_{k=-\infty}^{\infty} z^{-k}\delta[k - mN] = \sum_{m=-\infty}^{\infty} x_d[m]z^{-mN} = X_d(z^N) \tag{15-9}$$

In the frequency-domain, Equation (15-9) states that

$$X_i(e^{j\omega}) = X_d(e^{jN\omega}) \tag{15-10}$$

To illustrate, the frequency-domain representation of an interpolated by 2 signal is illustrated in Figure 15-10. It can be noted that the interpolated and original base-band spectra are identical out to $f_s/2$. Thereafter, the interpolated base-band spectrum continues out to $Nf_s/2$, carrying with it a number of artifacts on f_s centers. An illustrated example of an interpolation by two case is shown in the next example.

Example 15-5 Interpolation

Required:

Examine an interpolated by 16 signal by applying a gating function to a pre-decimated signal, as suggested in Figure 15-11. The gating function is periodically "on" for one sample out of N, and off for $N - 1$ samples. Analyze the spectral properties of the interpolated signal.

Results:

A low-frequency multitone process is sampled at a rate f_s/N and displayed in Figure 15-11. The signal is then interpolated by 16 and its spectrum is analyzed. Observe that the resulting interpolated signal's spectrum contains copies of the base-band spectrum located on f_s/N ($N = 16$) centers.

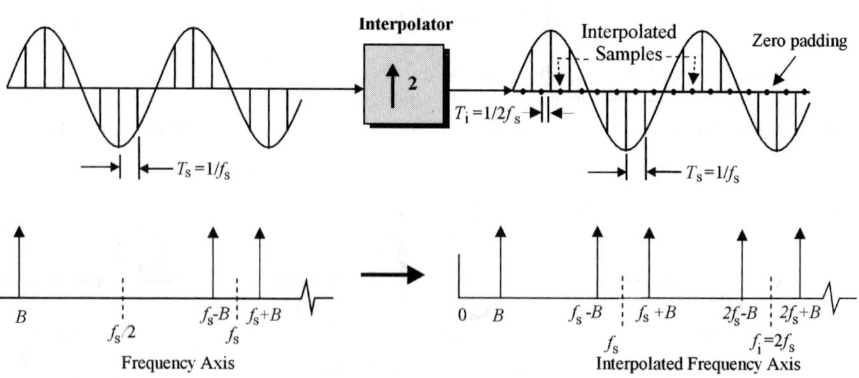

FIGURE 15-10 Frequency response of a signal interpolated $N = 2$.

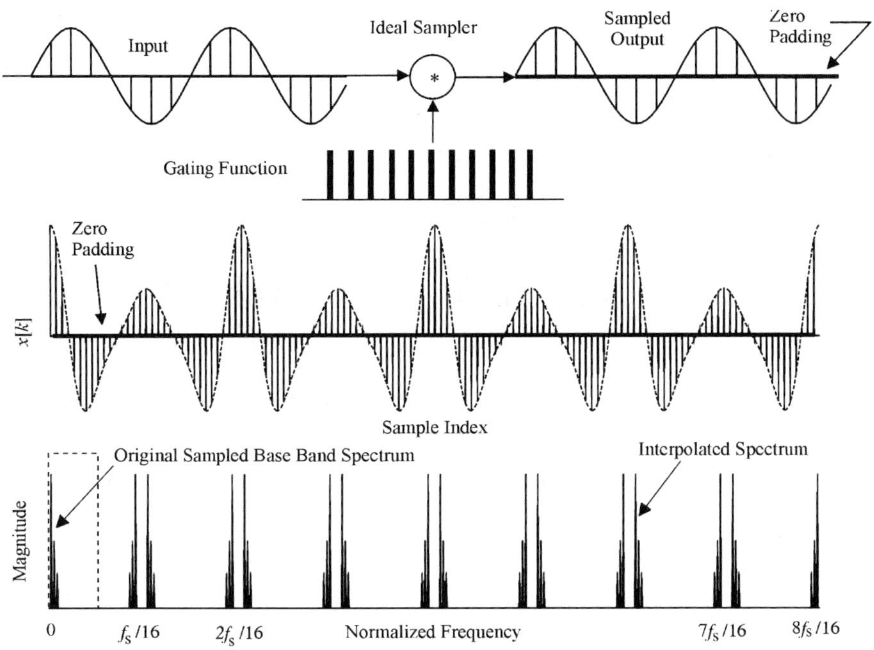

FIGURE 15-11 The gating function model of the interpolation process (top), interpolated time-series (middle), and the spectrum of the interpolated time-series (bottom) showing interpolated images (artifacts) from the original base-band spectrum.

The signal spectrum found at the output of an interpolator, shown in Figure 15-12, preserves the original base band along with periodically spaced artifacts. The unwanted copies, or artifacts, generally need to be removed, using a digital filter, before the signal can be made useful. An ideal Shannon interpolating filter, while being optimal, is not physically

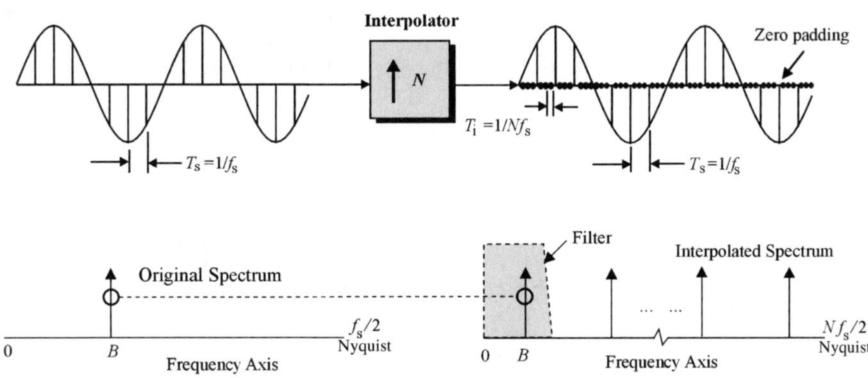

FIGURE 15-12 The spectra of an original signal, zero-padded interpolated signal, and a filtered base-band signal. Also shown is the magnitude response of an artifact removal filter.

realizable. A low-pass filter having a pass band defined over $f \in (-f_s/2N, f_s/2N)$ is generally used as an interpolating filter.

15.4 SAMPLE RATE CONVERSION

A commonly encountered signal processing problem is interfacing two systems having dissimilar sample rates, say f_1 and f_2. Such a system was earlier called a *sample rate converter*. If the ratio of f_1 to f_2 is a *rational* fraction, then direct decimation of interpolation can be used to achieve a sample rate conversion. To illustrate, suppose

$$k = N/M \tag{15-11}$$

where N and M are integers and $f_2 = kf_1$. The system described in Figure 15-13 is called a non-integer sample-rate converter. The indicated low-pass digital filters preserve specific functions. The low-pass filter having a normalized cutoff frequency of π/L removes interpolated artifacts and is called an anti-aliasing filter. The two filters can also be combined into a single filter, as shown in Figure 15-13.An example of sample rate conversion is presented next.

Example 15-6 Sample Rate Conversion

Required:
Two audio subsystems are to be connected. One has a sample rate of 44.1 kHz (multimedia) and the other is sampled at 48 kHz (audio tape). Design a solution.

Results:
Equation (15-11) states that $k = 48{,}000/44{,}100 = 160/147$. Unfortunately, k can be factored no further. The implication is that multimedia systems running at a rate of 44.1 kHz will need to be interpolated by a factor of 160 out to a frequency of 7.056 MHz. This requires that the interpolation filter processes data at a high speed and has a passband that is only 22.05 kHz wide (3 percent of the sample frequency). Such filters are extremely difficult to design. The anti-aliasing low-pass filter, located before the decimator, must have a bandwidth no greater than 24 kHz, based on a 48 k Sa/s output. This filter is likewise difficult to realize. The two digital low-pass interpolating and aliasing filters, if cascaded (see Figure 15.14) would encounter the same problem. An alternative solution would be to convert the 44.1 kHz signal into analog form using a common DAC with appropriate output smoothing, and then resample the analog signal at 40 kHz.

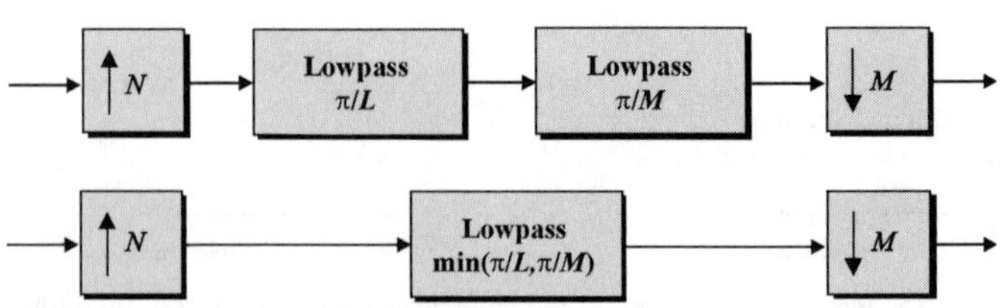

FIGURE 15-13 An equivalent sample rate conversion system.

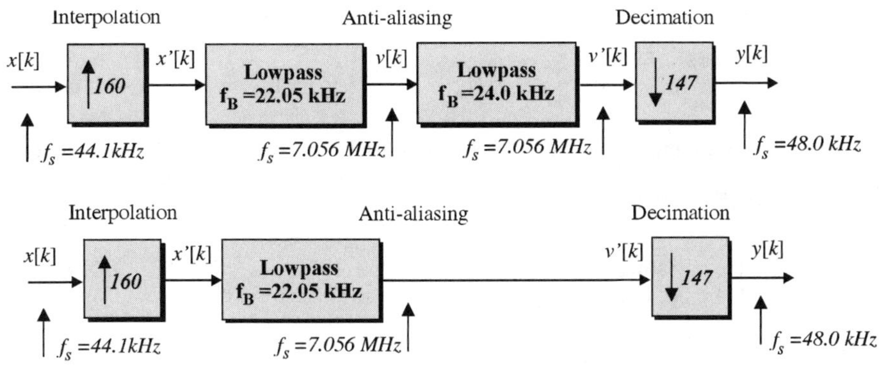

FIGURE 15-14 Two equivalent sample-rate converter architectures.

15.5 POLYPHASE REPRESENTATION

The study of interpolated and decimated signals can be unified using multirate *polyphase* modeling techniques. The polyphase modeling process begins with knowledge of a time-series $x[k]$ and that it can be partitioned into the M distinct data sequences shown in Equation (15-12). This process is called *block decomposition*.

$$\mathbf{BD} = \begin{bmatrix} x[0] & x[M] & x[2M] & \cdots & \Leftarrow & x_0[k] \\ x[1] & x[M+1] & x[2M+1] & \cdots & \Leftarrow & x_1[k] \\ x[2] & x[M+2] & x[2M+2] & \cdots & \Leftarrow & x_2[k] \\ \vdots & \vdots & \vdots & \ddots & & \\ x[M-1] & x[2M-1] & x[3M-1] & \cdots & \Leftarrow & x_{M-1}[k] \end{bmatrix} \tag{15-12}$$

The ith block in the block decomposition data array is given by

$$x_i[k] = \{x[i], x[M+i], x[2M+i], x[3M+i], \ldots\} \tag{15-13}$$

The ith block can be recognized to be equivalent to decimating the original time-series $x[k]$ by M, beginning at sample index $k = i$. In terms of a z-transform, the block decomposed data can be expressed as

$$\mathbf{BD}(z) = \left\{ \begin{array}{l} (x[0] + z^{-M}x[M] + z^{-2M}x[2M] + \cdots) \\ z^{-1}(x[1] + z^{-M}x[M+1] + z^{-2M}x[2M+1] + \cdots) \\ \cdots \quad \cdots \quad \cdots \quad \cdots \\ z^{-(M-1)}(x[M-1] + z^{-M}x[2M-1] + z^{-2M}x[3M-1] + \cdots) \end{array} \right\} \tag{15-14}$$

The ith row of Equation (15-14) defines the ith *polyphase* term $P_i(z)$. Specifically:

$$P_i(z) = \sum_{k=-\infty}^{\infty} x(kM + i)z^{-k} \tag{15-15}$$

The polyphase terms can then be used to synthesize the z-transform of $x[k]$ as

$$X(z) = \sum_{k=0}^{\infty} x[k]\, z^{-k} = \sum_{i=0}^{M-1} z^{-i} P_i(z^M) \tag{15-16}$$

The polyphase representation of a time-series is developed in the following example.

Example 15-7 Polyhase Representation

Required:
Represent the repeating time-series $x[k] = \{\ldots, 0, 1, 2, 3, 4, 3, 2, 1, 0, 1, 2, 3, 4, 3, 2, 1, 0, 1, 2, 3, 4, 3, 2, 1, \ldots\}$, where $x[0] = 0$, in polyphase form for $M = 4$.

Results:
From Equation (15-15), it follows that $P_0(z) = \{\cdots + 0z^0 + 4z^{-1} + 0z^{-2} + 4z^{-3} + \cdots\}$, $P_1(z) = \{\cdots + 1z^0 + 3z^{-1} + 1z^{-2} + 3z^{-3} + \cdots\}$, $P_2(z) = \{\cdots + 2z^0 + 2z^{-1} + 2z^{-2} + 2z^{-3} + \cdots\}$, and $P_3(z) = \{\cdots + 3z^0 + 1z^{-1} + 3z^{-2} + 1z^{-3} + \cdots\}$. It therefore follows from Equation (15-16) that $X(z) = P_0(z^4)\, z^0 + P_1(z^4)\, z^{-1} + P_2(z^4)\, z^{-2} + P_3(z^4)\, z^{-3}$.

A multirate system can also be described in *transposed polyphase* form defined in terms of $Q_i(z)$, where

$$Q_i(z) = P_{M-1-i}(z) \tag{15-17}$$

This results in an equivalent polyphase signal representation are given by

$$X(z) = \sum_{i=0}^{M-1} z^{-(M-1-i)} Q_i(z^M) \tag{15-18}$$

A polyphase signal representation can be used to examine the physical act of decimation. Consider a time-series $x[k]$ which is decimated by a factor M to produce a new time-series $x_d[k]$. According to Equation (15-15), $P_0(z)$ is recognized to be the time-series $x[k]$ decimate by M time-series beginning at index $k = 0$. The decimation equation representing $P_0(z)$ can be expressed in a more complicated form as

$$P_0(z) = \frac{1}{M}\sum_{k=0}^{M-1} X\left(W_M^k z^{1/M}\right);\; W_M = e^{-j2\pi/M} \tag{15-19}$$

The added complexity is needed to provide a mathematical connection between the sparse (decimated) sample set found in $P_0(z)$ and the densely sampled data set represented by $X(z)$. The efficacy of Equation (15-19) can be explored by performing a term-by-term analysis to achieve

$$\begin{aligned}
P_0(z) &= \frac{1}{M}\sum_{k=0}^{M-1} X\left(W_M^k z^{1/M}\right) \\
&= \frac{1}{M}\left(X\left[W_M^0 z^{1/M}\right] + X\left[W_M^1 z^{1/M}\right] + X\left[W_M^2 z^{1/M}\right] + \cdots + X\left[W_M^{M-1} z^{1/M}\right]\right) \\
&= \frac{1}{M}\left\{\begin{aligned}
&x[0] + x[1]z^{-1/M} + x[2]z^{-2/M} + \cdots + x[M]z^{-M/M} + \cdots \\
&+x[0] + x[1]W_M^1 z^{-1/M} + x[2]W_M^2 z^{-2/M} + \cdots + x[M]W_M^M z^{-M/M} + \cdots \\
&\qquad \cdots\cdots \qquad\qquad \cdots\cdots \\
&+x[0] + x[1]W_M^{M-1} z^{-1/M} + x[2]W_M^{2(M-1)} z^{-2/M} + \cdots + x[M]W_M^{M(M-1)} z^{-M/M} + \cdots
\end{aligned}\right\}
\end{aligned}$$

$$= \frac{1}{M}\{Mx[0] + 0x[1]z^{-1/M} + 0x[1]z^{-2/M} + \cdots + Mx[M]z^{-M/M} + \cdots\}$$

$$= \frac{1}{M}(Mx[0] + Mx[M]z^{-M/M} + \cdots) = (x[0] + x[M]z^{-1} + \cdots) \tag{15-20}$$

where z^{-1} represents a single clock delay at the decimated clock rate (in other words, $T_d = MT_s = M/f_s$) and $z^{-1/M}$ corresponds to a $1/M$th clock delay at the undecimated clock rate. In this case, the temporal value of a single $1/M$th clock delay is T_s. The polyphase decomposition formula is studied to decimation rates of two and four in the next example.

Example 15-8 Polyphase Decomposition

Required:
The time-series $x[k] = a^k u[k]$, $|a| < 1$, is sampled at a rate of f_s and represents a decaying exponential. Interpret $x[k]$ and $P_0(z)$ in the context of Equation (15-20) for $M = 4$.

Results:
For $M = 4$, $x_d[k] = x[4k] = \{1, a^4, a^8, \ldots\}$ and $P_0(z) = 1 + a^4z^{-1} + a^8z^{-2}, \ldots$, measured at the decimated sample rate of $f_d = f_s/4$ (note: $P_0(z^4) = 1 + a^4z^{-4} + a^8z^{-8}, \ldots$), the individual terms found in Equation (15-20) are

$$X\left(W_4^0 z^{1/4}\right) = X(z^{1/4}) = \sum_{k=0}^{\infty} a^k z^{-k/4}$$
$$= \{1 + az^{-1/4} + a^2z^{-2/4} + a^3z^{-3/4} + a^4z^{-1}, \ldots\}$$

$$X\left(W_4^1 z^{1/4}\right) = X(-jz^{1/4}) = \sum_{k=0}^{\infty} (-j)^{k/4} a^k z^{-k/4}$$
$$= \{1 - jaz^{-1/4} - a^2z^{-2/4} + ja^3z^{-3/4} + a^4z^{-1}, \ldots\}$$

$$X\left(W_4^2 z^{1/4}\right) = X(-z^{1/4}) = \sum_{k=0}^{\infty} (-a)^k z^{k/4}$$
$$= \{1 - az^{-1/4} + a^2z^{-2/4} - a^3z^{-3/4} + a^4z^{-1}, \ldots\}$$

$$X\left(W_4^3 z^{1/4}\right) = X(jz^{1/4}) = \sum_{k=0}^{\infty} (j)^{k/4} a^k z^{-k/4}$$
$$= \{1 + jaz^{-1/4} - a^2z^{-2/4} - ja^3z^{-3/4} + a^4z^{-1}, \ldots\}$$

And when combined under Equation (15-19), produce

$$P_0(z) = \frac{1}{4}\left(X\left(W_4^0 z^{1/4}\right) + X\left(W_4^1 z^{1/4}\right) + X\left(W_4^1 z^{1/4}\right) + X\left(W_4^1 z^{1/4}\right)\right)$$
$$= 1 + a^4z^{-1} + a^8z^{-2} + \cdots$$

as required.

A logical question can be raised and relates to the preferred location of the decimator in a signal processing stream. The two systems shown in Figure 15-15 are claimed to be functionally equivalent. This relationship is sometimes referred to as the *noble identity*. The topmost path consists of a decimator with the filter running at a decimated speed of $f_d = f_s/M$. The bottom path consists of a filter, running at a speed of f_s, followed by a decimator. Both filters have the same number of coefficients and therefore the same arithmetic complexity. The major difference between the circuits is found in the rate at which the filter arithmetic

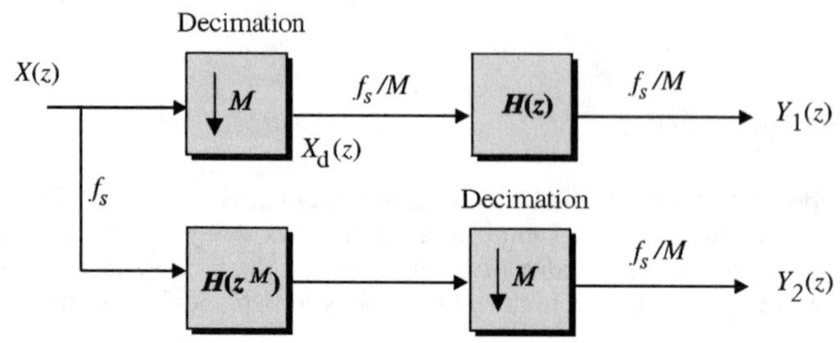

FIGURE 15-15 Equivalent decimated systems ($Y_1(z) = Y_2(z)$).

must be performed. The topmost filter has a real-time arithmetic rate requirement that is $1/M$th that of the bottom filter. Therefore, the top architecture shown in Figure 15-15 is generally preferred due to its reduced computational requirement. A 4th-order polyphase filter is demonstrated in the next example.

Example 15-9 Polyphase Filter

Required:

An FIR filter has a given transfer function $H(z) = 2 + 3z^{-1} + 3z^{-2} + 2z^{-3}$. Analyze the complexity and performance of the two filter instantiations shown in Figure 15-15 for $M = 2$.

Results:

For $M = 2$, it follows that $H(z) = P_0(z^2) + z^{-1}P_1(z^2) = [2 + 3z^{-2}] + z^{-1}[3 + 2z^{-2}]$. The polyphase implementations of the two filter options, presented in Figure 15-15, are detailed in Figure 15-16. Both designs implement a 4th-order FIR using two interleaved 2nd-order polyphase FIRs. The data being processed by Filter A arrives at a rate that's half of that seen by Filter B. Compared to Filter B, Filter A would therefore have the lowest arithmetic demand (MAC/s) of the two choices.

A sample-by-sample analysis of the filtering process displayed in Figure 15-16 and reported in Table 15-1 in terms of the time-series appearing at the output of each 2nd-order filter stage. Specifically:

Both filters produce identical outputs that appear at the decimated rate $f_s/2$. It can also be seen that the polyphase Filter A is economical from an arithmetic resource need.

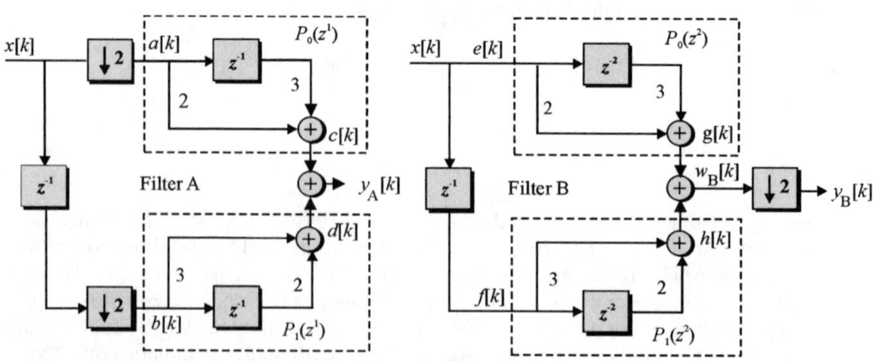

FIGURE 15-16 Equivalent 4th-order polyphase filters.

TABLE 15-1 Polyphase Filter Response

k	⋯	0	1	2	3	4	5
$x[k]$	⋯	$x[0]$	$x[1]$	$x[2]$	$x[3]$	$x[4]$	$x[5]$
$a[k]$	⋯	$x[0]$		$x[2]$		$x[4]$	
$b[k]$	⋯		$x[1]$		$x[3]$		$x[5]$
$c[k]$	⋯	$2x[0] + 3x[-2]$		$2x[2] + 3x[0]$		$2x[4] + 3x[2]$	
$d[k]$	⋯		$3x[1] + 2x[-1]$		$3x[3] + 2x[1]$		$3x[5] + \cdots$
$y_A[k]$	⋯	$\mathbf{2x[0] + 3x[-1] + 3x[-2] + 2x[-3]}$		$\mathbf{2x[2] + 3x[1] + 3x[0] + 2x[-1]}$		$\mathbf{2x[4] + 3x[3] + 3x[2] + 2x[1]}$	
$e[k]$	⋯	$x[0]$	$x[1]$	$x[2]$	$x[3]$	$x[4]$	$x[5]$
$f[k]$	⋯	$x[-1]$	$x[0]$	$x[1]$	$x[2]$	$x[3]$	$x[4]$
$g[k]$	⋯	$2x[0] + 3x[-2]$	$2x[1] + 3x[-1]$	$2x[2] + 3x[0]$	$2x[3] + 3x[1]$	$2x[4] + 3x[2]$	$2x[5] + 3x[3]$
$h[k]$	⋯	$2x[-1] + 3x[-3]$	$2x[0] + 3x[-2]$	$2x[1] + 3x[-1]$	$2x[2] + 3x[0]$	$2x[3] + 3x[1]$	$2x[4] + 3x[2]$
$w_B[k]$	⋯	$2x[0] + 3x[-1]$ $+ 3x[-2] + 2x[-3]$	$2x[1] + 3x[0]$ $+ 3x[-1] + 2x[-2]$	$2x[2] + 3x[1]$ $+ 3x[0] + 2x[-1]$	$2x[3] + 3x[2]$ $+ 3x[1] + 2x[0]$	$2x[4] + 3x[3]$ $+ 3x[2] + 2x[1]$	$2x[5] + 3x[4]$ $+ 3x[3] + 2x[2]$
$y_B[k]$	⋯	$\mathbf{2x[0] + 3x[-1] + 3x[-2] + 2x[-3]}$		$\mathbf{2x[2] + 3x[1] + 3x[0] + 2x[-1]}$		$\mathbf{2x[4] + 3x[3] + 3x[2] + 2x[1]}$	

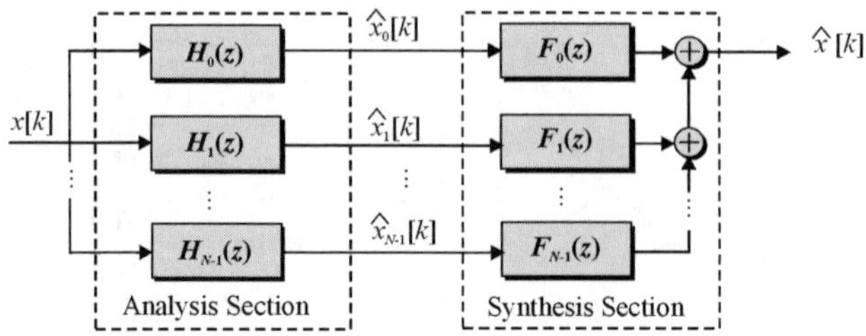

FIGURE 15-17 A typical filter bank system showing the analysis (left), and synthesis (right) filters.

15.6 FILTER BANKS

Multirate systems often appear in the form of *filter banks*. A filter bank maps an input time-series to a collection of *sub-band* filters, denoted $H_i(z)$ and $F_i(z)$ in Figure 15-17. The sub-band filters define what are called the *analysis* and *synthesis* sections of a filter bank. The individual sub-band analysis filters pass their outputs to individual synthesis sub-band filters. Depending on the choice of $H_i(z)$ and $F_i(z)$, the output can be a copy of the input $x[k]$, which is possibly scaled and/or delayed. Other filter banks will produce only an approximation of the input $x[k,]$, while additional strategies will produce an output that is unrestricted. Many filter bank solutions are justified on the basis of the signal properties existing at the analysis-synthesis section interface. Under certain conditions the data rate across this boundary can be made a fraction of the data rate found at the system's input or output. If the data rate at the interface level is low, compared to the filter bank's input and output data rates, then each sub-band filter pair can communicate across a low bandwidth physical channel. This concept is at the core of many bandwidth compression schemes.

Quadrature mirror filters (QMFs) filter banks are a popular means of performing sub-band signal decomposition. The objective of a QMF is often one of compressing the bandwidth requirements for an individual sub band to the point that information can flow through the filter bank across multiple physically band-limited channels. The basic architecture of a two-channel QMF system is shown in Figure 15-18. The two-channel QMF system presents two input-output paths, each having a bandwidth requirement that is half the original bandwidth specification. The top path shown in Figure 15-1 behaves as a low-pass filter, while the bottom path acts as a high-pass filter. The signals found along the top and bottom paths, after decimation, can be expressed in terms of Equation (15-19). Observe that the signals $x_0[k]$ (*viz*: $X_0[z]$) and $x_1[k]$ (*viz*: $X_1[z]$) are decimated by two after being filtered. The signals entering the decimator are $H_0(z)X(z)$ and $H_1(z)X(z)$, respectively, in

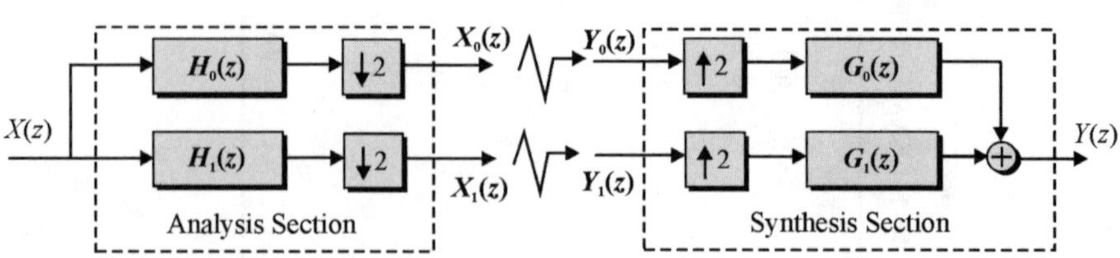

FIGURE 15-18 The basic two-channel QMF filter architecture.

the z-transform domain. For $M = 2$, the signals leaving the decimator, based on Equation (15-19), are

$$X_0(z) = \frac{1}{2}\{X(z^{1/2})H_0(z^{1/2}) + X(-z^{1/2})H_0(-z^{1/2})\}$$
$$X_1(z) = \frac{1}{2}\{X(z^{1/2})H_1(z^{1/2}) + X(-z^{1/2})H_1(-z^{1/2})\} \tag{15-21}$$

The analysis signals $x_0[k]$ and $x_1[k]$ are transmitted to the synthesis filter section along two distinct reduced bandwidth channels. The signals are recovered by the synthesis filter section and restored to the original sample rate f_s. Upon filtering and interpreting $Y_0(z^2)$ and $Y_1(z^2)$, shown in Figure 15-18, it follows that $Y(z)$ can be expressed as

$$Y(z) = G_0(z)Y_0(z^2) + G_1(z)Y_1(z^2) \tag{15-22}$$

The process is graphically interpreted in the frequency domain in Figure 15-19. It can be noted that the output spectrum contains some aliasing contamination which, if left uncorrected, would preclude the reconstruction of an error-free image of the input signal. The source of the aliasing can be traced to the terms shown in Equation (15-23) which is obtained from an expansion of Equation (15-22).

$$Y(z) = \underbrace{\frac{1}{2}\{(H_0(z)G_0(z) + H_1(z)G_1(z))X(z)\}}_{\{alias\ -\ free\ terms\}} + \underbrace{\frac{1}{2}\{(H_0(-z)G_0(z) + H_1(-z)G_1(z))X(-z)\}}_{\{aliased\ terms\}} \tag{15-23}$$

What is needed is a process that can suppress the effects of aliasing. Since an ideal (boxcar) filter is physically impractical, other means must be considered. Alias-free performance can be guaranteed if the alias term in Equation (15-23) is set to zero. This can be trivially achieved if $G_0(z) = H_1(-z)$ and $G_1(z) = -H_0(-z)$. However, interpreting the consequence of such an action is challenging.

Consider the special alias-free case where $H_0(z)$ and $H_1(z)$ are sub-band filters satisfying the mirror relationship $H_1(z) = H_0(-z)$. In the z-domain, this assignment results in the QMF filter condition

$$Y(z) = k\lfloor H_0^2(z) - H_0^2(-z)\rfloor X(z) = T(z)X(z) \tag{15-24}$$

where k is a real scale factor. The filter function $T(z)$ can possess several personalities. The most common persona results in possible distortion. The distortion possibilities are classified as ALD (alias distortion), AMD (amplitude distortion), and/or PHD (phase distortion filters). If a filter is ALD, AMD, and PHD-free, the filter is said to possess the *perfect reconstruction* (PR) property. A PR filter has a transfer function $T(z) = k\,z^{-d}$, which establishes an input-output time-domain relationship given by $y[k] = k\,x[k-d]$. That is, a PR QMF filter uses sub-channel signal processing to reconstruct an output that is a simple scaled and delayed version of the input.

The study of FIRs established the fact that linear phase behavior is often a desired attribute. Suppose it is required to design an Nth–order QMF FIR that is also a linear-phase filter. If N is odd, it can be shown that a null (point of zero gain) will be placed in the output spectrum at the normalized frequency $\varpi = \pi/2$. As a result, an odd order linear-phase FIR can suppress aliasing errors but does not have a flat magnitude frequency response (in

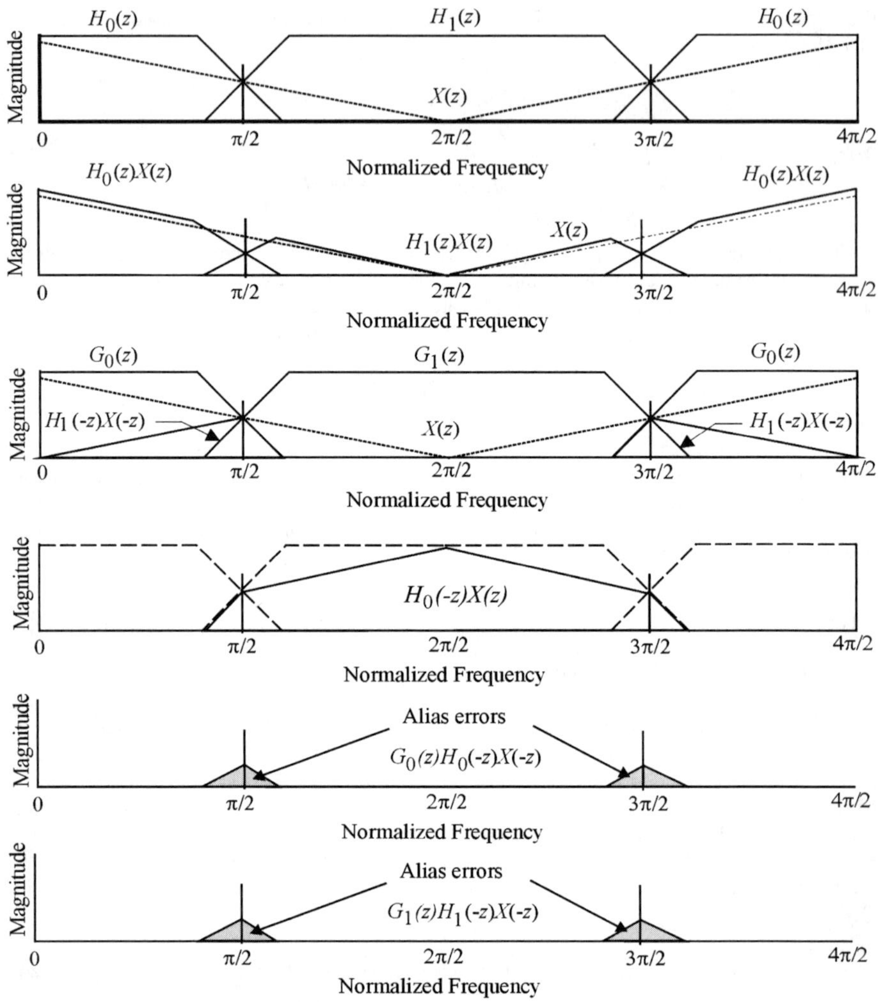

FIGURE 15-19 A graphical interpretation of a QMF filter's spectrum at points along the signal stream.

other words, AMD). If N is even, then the response that is both linear phase and flat is produced only by a trivial two-coefficient FIR having the form

$$H_0(z) = c_0 z^{-2n_0} + c_1 z^{-2(n_1+1)}; \quad H_1(z) = c_0 z^{-2n_0} - c_1 z^{-2(n_1+1)} \tag{15-25}$$

for some integer n_0 and n_1. Unfortunately, this filter has little value in practice. Any other even order linear phase choice of $H_0(z)$ will introduce some distortion. The design of a low-order QMF filter is presented in the next example.

Example 15-10 QMF Filter

Required:

Show that the trivial even order QMF filter, defined in terms of the Harr basis functions $h_0[k] = [1, 1]$ and $h_1[k] = [1, -1]$, results in a linear phase QMF filter.

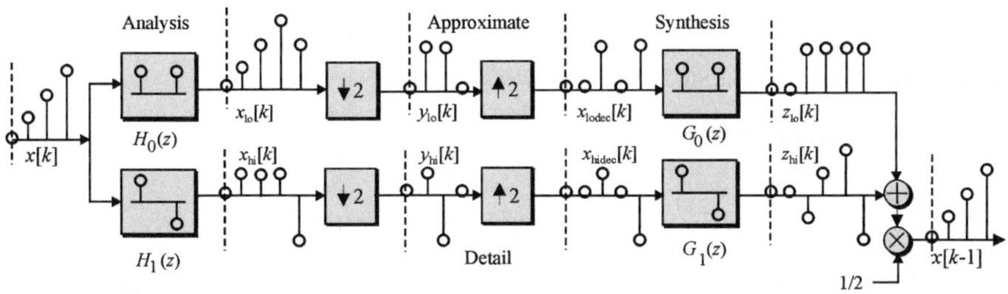

FIGURE 15-20 A multirate perfect reconstruction example.

Results:

The basic functions $h_0 = [1, 1]$ and $h_1 = [1, -1]$ represent 2nd-order FIRs which satisfy the mirror condition $H_1(z) = H_0(-z)$. The filter h_0 is sometimes called the *approximation* filter since it smoothes two successive sample values. The filter h_1 is called the *detail filter* since it responds to changes in $x[k]$. The resulting synthesis filters are given by $G_0(z) = H_1(-z) = H_0(z)$ and $G_1(z) = -H_0(-z) = -H_1(z)$. From Equation (15-24), it follows that

$$T(z) = \left[H_0^2(z) - H_0^2(-z)\right] = 4z^{-1},$$

which defines a perfect reconstructed filter whose output is a scaled and delayed version of the input.

To illustrate the perfect reconstruction capabilities of the filter bank, refer to Figure 15-20 where the multirate system is partitioned into an *analysis* section and a *synthesis* section. Notice that the filter bank consists of decimators, approximation, and detail filters. The output can also be seen to be a perfect reconstruction of the input.

Designing a practical high-order QMF can be a challenging process. It is known that there does not exist any nontrivial, or physically meaningful filters having both a flat response and linear-phase performance. As a result, most practical QMF designs represent some sort of compromise. If the linear phase requirement is relaxed, then a magnitude and phase distortion-less alias-free QMF system can be realized. A popular manifestation of this compromise is called the *perfect reconstruction QMF* (PRQMF) filter. The output of a PRQMF system is equal to the input with a known delay. The design of a PRQMF can follow the recipe shown next.

1. Design a linear phase FIR $F(z)$ as a $(2N - 1)$-order half-band FIR having a ripple deviation δ.
2. Classify the zeros of $F(z)$ as being either interior or exterior to the unit circle. Since many of the zeros of $F(z)$ lie on the unit circle, discriminating between an interior or exterior location can become difficult. To mitigate this problem, add a small offset (bias) to the center tap weight of $F(z)$ to form $F_+(z) = F(z) + q\delta$ where $q > 1.0$, but close to unity. This action adds a small bias $q\delta$ to the frequency response of $F(z)$ across the entire base band. The modified FIR $F_+(z)$ has zeros that are moved slightly off the unit circle. Biasing $F(z)$ in this manner lifts the zeros off the unit circle and forces them to be either interior of exterior.
3. Define an FIR $H(z)$ formed by all the interior zeros of $F_+(z)$.
4. Define $H_0(z) = H(z)$ and $H_1(z) = (-1)^{N-1}z^{-(N-1)}H(-z^{-1})$.
5. Define $G_0(z) = H_1(-z)$ and $G_1(z) = -H(-z)$

By relaxing the linear phase constraint, an all-pass PFQMF system is obtained having an input-output transfer function $T(z) = kz^{-(N-1)}$, where k is a constant of proportionality. These principles are illustrated in the following example.

Example 15-11 PRQMF Design

Required:
The design of a PRQMF system based on a 15th-order (that is, order $(2N - 1)$) linear-phase half-band filter having a transfer function $F(z) = -0.02648z^7 + 0.0441z^5 - 0.0934z^3 - 0.3139z^1 + 0.5 - 0.3139z^{-1} - 0.0934z^{-3} + 0.0441z^{-5} - 0.02648z^{-7}$.

Results:
The magnitude frequency response of $F_+(z) \sim F(z)$ is shown in Figure 15-21. From the half-band FIR, a PRQMF system can be defined using a step-by-step design process, beginning with:

Step 1: $F(z)$ is given and $\delta = 0.0238$.

Step 2: Let $q = 1.01$ and produce $F_+(z) = F(z) + q\delta$.

Step 3: The factors of $F_+(z)$ are (up to the precision of the computing routine) shown next:

z_i	$\lvert z_i \rvert$	Interior/Exterior
$-0.939 \pm j0.398$	1.02	Exterior
$-0.903 \pm j0.382$	0.98	Interior
$-0.451 \pm j0.907$	1.013	Exterior
$-0.439 \pm j0.884$	0.987	Interior
$-0.394 \pm j0.427$	0.581	Interior
0.561	0.561	Interior
$-1.167 \pm j1.264$	1.72	Exterior
1.782	1.782	Exterior

The location of the 14 zeros of $F_+(z)$ are shown in Figure 15-21. Collecting all the zeros residing interior to the unit circle together results in the creation of $H(z)$.

$$H(z) = 1.0 + 1.34z^{-1} + 0.68z^{-2} - 0.24z^{-3} - 0.34z^{-4} + 0.099z^{-5} + 0.239z^{-6} - 0.17z^{-7}$$

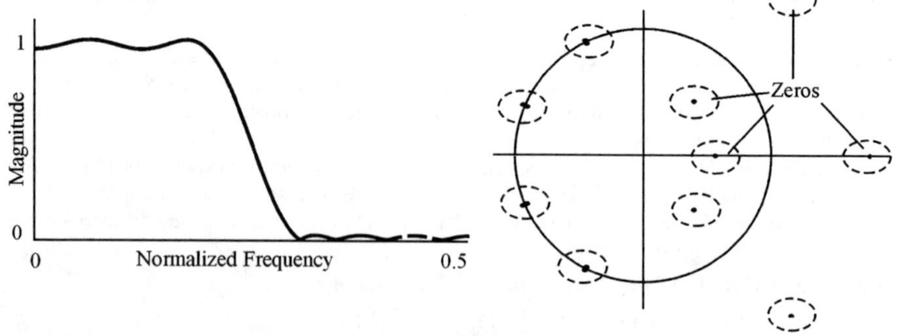

FIGURE 15-21 The magnitude frequency response of $F_+(z)$ and its zero distribution.

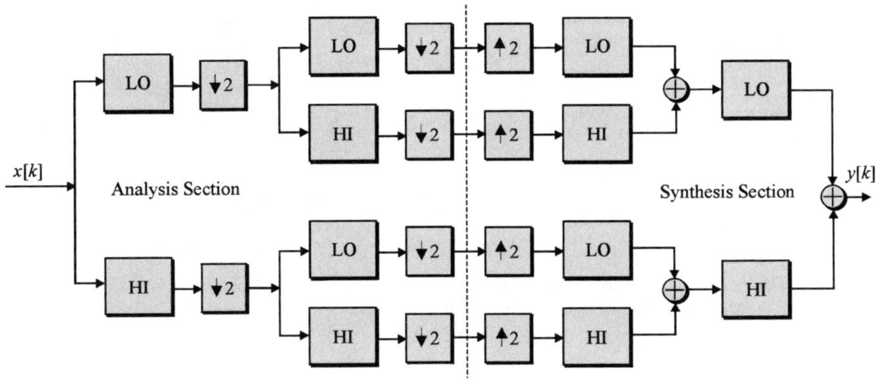

FIGURE 15-22 Dyadic filter bank architecture.

Step 4: Construct $H_0(z)$ and $H_1(z)$.

$H_0(z) = H(z)$

$H_1(z) = -0.17 - 0.24z^{-1} + 0.099z^{-2} + 0.34z^{-3} - 0.24z^{-4} - 0.68z^{-5} + 1.34z^{-6} - 1.0z^{-7}$

Step 5: Construct $G_0(z)$ and $G_1(z)$.

$G_0(z) = -0.17 + 0.24z^{-1} + 0.099z^{-2} - 0.34z^{-3} - 0.24z^{-4} + 0.68z^{-5} + 1.34z^{-6} + 1.0z^{-7}$

$G_1(z) = -1.0 + 1.34z^{-1} - 0.68z^{-2} - 0.24z^{-3} + 0.34z^{-4} + 0.099z^{-5} - 0.24z^{-6} - 0.17z^{-7}$

(Note that individually the filters are non-linear phase.)

In practice, the two-channel QMF filter displayed in Figure 15-18 can be used to motivate the design of an $N = 2^n$ channel filter bank having n-levels. The structure of such a filter bank is suggested in Figure 15-22. The architecture is called a dyadic filter bank and the analysis stage filters are $H_0(z)$ (Lo) and $H_1(z)$ (Hi), while the synthesis stage filters are denoted by $G_0(z)$ (Lo) and $G_1(z)$ (Hi).

15.7 DFT FILTER BANKS

An interesting manifestation of an analysis filter section is called a *uniform DFT filter bank*, or simply *DFT filter bank.* A DFT filter bank has a magnitude frequency response suggested in Figure 15-23 which is effectively that of a bank of identically shaped filters that are uniformly distributed across the base band, and located at distinct center frequencies. The nth filter's response, $H_n(z)$, is defined in terms of a low-pass prototype filter (model) denoted H_0, where

$$H_n(z) = H_0\left(W_M^n z\right) \tag{15-26}$$

for $n \in [0, M-1]$. The complex exponential term W_M performs a modulation service required to translate the envelope of the prototype low-pass filter out to a normalized frequency $\varpi = 2\pi n/M$. The frequency response of the nth filter, centered about $\varpi = 2n\pi/M$, is given by

$$H_n(e^{j\varpi}) = H_0\left(e^{j(\varpi - 2n\pi/M)}\right) \tag{15-27}$$

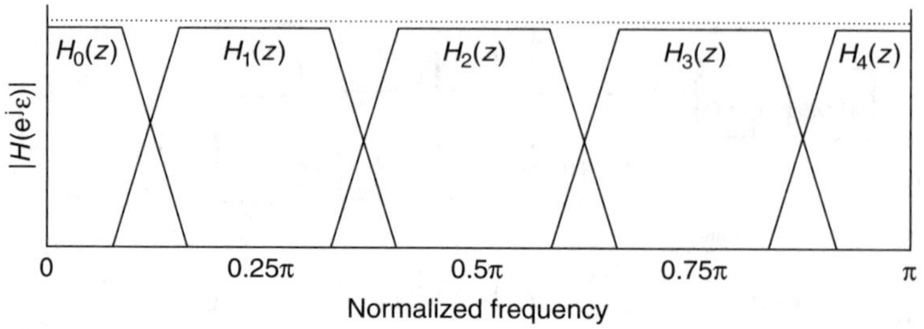

FIGURE 15-23 A uniform DFT filter bank magnitude frequency response for $M = 8$. (The frequency axis was normalized with respect to the Nyquist frequency.)

The 0th filter, or prototype low-pass filter $H_0(z)$, is assumed to be an N_0-order FIR that can be expressed in polyphase form as

$$H_0(z) = \sum_{i=0}^{M-1} z^{-i} P_{0i}(z^M) \tag{15-28}$$

From Equation (15-27), it follows that the nth filter in the DFT bank filter satisfies

$$\begin{aligned} H_n(z) &= \sum_{i=0}^{M-1} \left(W_M^n z\right)^{-i} P_{0i}\left(W_M^n z\right)^M = \sum_{i=0}^{M-1} W_M^{-in} z^{-i} P_{0i}\left(W_M^{Mn} z^M\right) \\ &= \sum_{i=0}^{M-1} W_M^{-in} z^{-i} P_{0i}(z^M) \end{aligned} \tag{15-29}$$

Upon close inspection of Equation (15-29), it can be noted that the equation's structure is that of a DFT formula (for instance, $X[n] = \sum W^{nk} x[k]$). What is intriguing is that an M-point DFT can efficiently perform all the modulation (that is, multiplications by W_M^m) required in the implementation of Equation (15-24). At a system level, the polyphase filter outputs are modulated using an M-sample DFT, as shown in Figure 15-24*a*. The M-point DFT outputs collectively define a filter bank having the frequency domain response shown in Figure 15-18. The complexity of the DFT filter bank can be analyzed in the context of the complexity of the prototype filter and DFT. Consider that the

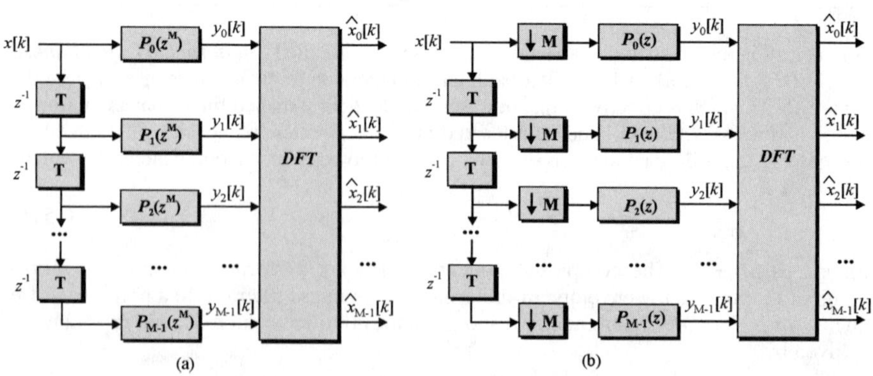

FIGURE 15-24 A DFT filter bank and DFT filter bank with decimators.

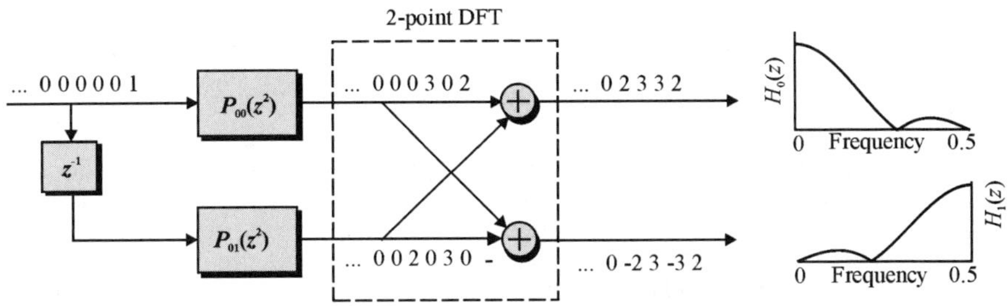

FIGURE 15-25 A DFT filter bank frequency and impulse responses for $M = 2$.

prototype FIR filter, $H_0(z)$, is defined to be of order $N_0 = NM$ filter. Referring to Figure 15-24*a*, it can be seen that there are M polyphase FIRs, each polyphase filter being of order N. The polyphase filter portion of a DFT filter bank solution would therefore require $N_0 = MN$ multiplies per filter cycle distributed across M filters. By adding decimation by M circuits, shown in Figure 15.24*b*, the real-time complexity can be further reduced by a factor of $1/M$. The multiplicative complexity of an M-point DFT can, in practice, be made small and often on the order of $M \log_2(M)$ if an FFT is used to perform the modulation. As a result, a DFT filter bank can be computationally efficient. Furthermore, a high-quality filter bank can be created if the design is based on a well-defined prototype FIR filter $H_0(z)$. The design of a low order DFT filter bank is considered in the next example.

Example 15-12 DFT Filter Bank

Required:

Design and analyze a DFT filter bank for $M = 2$ using a prototype filter $H_0(z) = 2 + 3z^{-1} + 3z^{-2} + 2z^{-3}$.

Results:

The polyphase representation of $H_0(z)$ is $H_0(z) = P_{00}(z^2) + z^{-1}P_{01}(z^2)$, where $P_{00}(z) = 2 + 3z^{-1}$ and $P_{00}(z) = 3 + 2z^{-1}$. Equations (15-29) states that for $W_2^0 = 1$ and $W_2^{-1} = -1$:

$$H_0(z) = \sum_{i=0}^{1} z^{-i}P_{0i}(z^2) = P_{00}(z^2) + z^{-1}P_{01}(z^2) = 2 + 3z^{-1} + 3z^{-2} + 2z^{-3}$$

$$H_1(z) = \sum_{i=0}^{1} W_2^{-i}z^{-i}P_{0i}(z^2) = P_{00}(z^2) - z^{-1}P_{01}(z^2) = 2 - 3z^{-1} + 3z^{-2} - 2z^{-3}$$

The DFT filter bank is presented in Figure 15-25 and is seen to consist of a pair of polyphase filters and a two-point DFT. The impulse response of the DFT filter bank, measured along the top path is {2, 3, 3, 2}, which corresponds to $h_0[k]$. The impulse response measured along the bottom path is {2, −3, 3, −2}, which corresponds to $h_1[k]$.

15.8 CASCADE INTEGRATOR COMB (CIC) FILTER

Wireless and LAN communication systems transmit signals, information rates, and frequencies often beyond those which admit digital processing (for example, 5 GHz). Intermediate frequency (IF) stages are used to reduce signal frequencies to a point that can be accepted by an ADC (for example, 100 MSa/s). The information spectra, however, can be far less than the ADC bandwidth (for example, 100 kHz). For narrow-band communication applications, the channel bandwidth is much lower than the first ADC rate ($f_{\text{bandwidth}}/f_s \ll 11$). For broadband applications, the difference in bandwidth requirements can be less than 10. Nevertheless, the signal sampled by the ADC must be down-converted to a base-band signal before it can be analyzed by a "back-end" processor. This is normally accomplished by mixing the digitized signal with a synthesized sinusoid (sine and cosine) obtained from a direct digital synthesizer or DDS. Using this process, the desired information channel is heterodyned down to DC. Once at DC, a frequency selective filter is used to isolate the information channel of interest from a broadband spectrum. The required down conversion must, however, be performed at high real-time data rates which will normally preclude the use of a common FIR to extract the desired information process from the ADC output. A fast simple solution is needed. Such a solution is called a *digital down converter*, or *channelizer*. The preferred channelizer architecture is called a cascaded integrator-comb (CIC), or Hogenauer filter.

In order to be able to sustain high real-time speeds, multiplier-free filter structures are needed. A moving average (MA) filter, having an impulse response and transfer function:

$$h(n) = \begin{cases} 1 & \text{if } 0 \le n \le N-1 \\ 0 & \text{otherwise} \end{cases}$$
$$H_{\text{MA}}(z) = \sum_{i=0}^{N-1} z^{-i} = \frac{(1 - z^{-N})}{(1 - z^{-1})} \tag{15-30}$$

is a frequency selective multiplier-free FIR. The magnitude frequency response of a MA FIR has a $\sin(x)/x$ envelope. The MA's bandwidth is established by the filter order. In concept, an MA filter can be used to extract a narrow band of information heterodyned down to DC by a DDS. For narrow-band applications, the information bandwidth can be 1/1000th the sample frequency, requiring an MA filter having a very narrow passband and therefore a high order. The problem with a high-order MA filter design is the high shift register count. A CIC filter, abstracted in Figure 15-26, provides an efficient means of implementing a high order MA filter. Each of the N comb filters is clocked at a decimated rate of $f_c = f_s/R$. This effectively defines each comb register delay to be R times longer than an integrator's register delay. Therefore, the N cascaded comb filters have a transfer function given by $H_C(z) = (1 - z^{-R})^N$ which exhibits zeros distributed along the unit circle at locations $z = e^{j2\pi k/R}$, $k \in [0, R)$. When the Nth-order comb is cascaded with an Nth-order integrator, a CIC filter results, having a transfer function:

$$H_{\text{CIC}}(z) = (1 - z^{R})^N/(1 - z^{-1})^N. \tag{15-31}$$

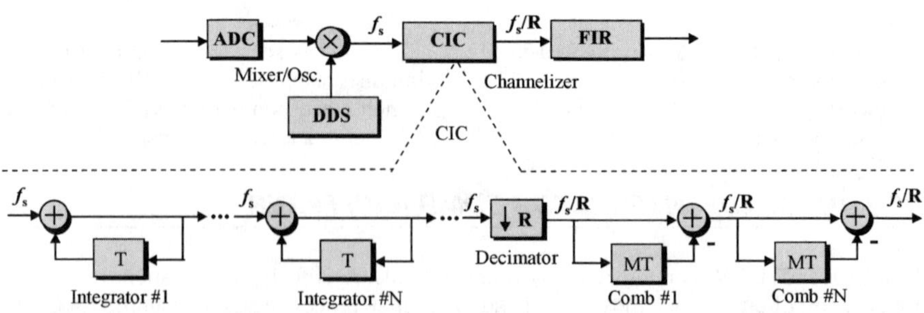

FIGURE 15-26 A CIC filter architecture.

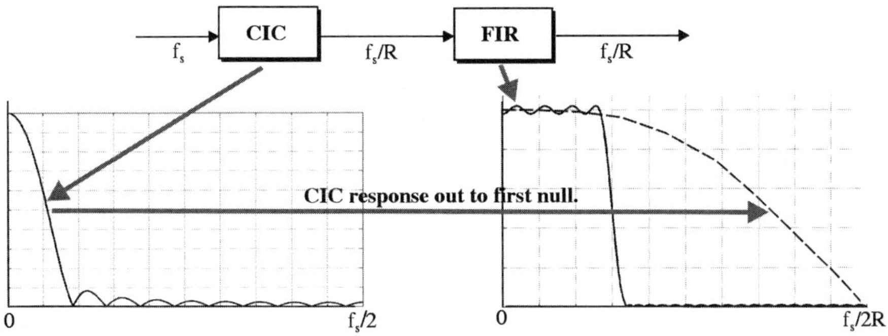

FIGURE 15-27 Basic CIC filter performance along with a post-processing frequency shaping FIR filter.

The CIC filter defined in Equation (15-31) is equivalent to an RN-order filter. The N poles of the integrator at $z = 1$ are cancelled by the N zeros of the comb filter, also at $z = 1$, resulting in high DC gain and a $\sin(x)/x$ roll-off in the side bands. The filter has a maximum gain defined at DC, and is equal to $H_{CIC}(z)_{max} = G_{max} = R^N$. For typical narrow-band applications, having $R = 1024$ and $N = 5$, $G_{max} = (2^{10})^5 = 2^{50}$. This suggests that CIC filters have potentially high internal gains. Finally, the $\sin(x)/x$ CIC response is often post-processed using a low-pass shaping FIR to define the final output spectrum, as suggested in Figure 15-27.

15.9 FREQUENCY MASKING FILTERS

There are instances when a steep skirt (in other words, a narrow transition band) filter is required. Unfortunately, steep skirt fixed-sample rate filters are historically very complex and of high order. Such filters can, however, be designed using multirate techniques based on the *frequency masking* method. The frequency masking method uses what are called *compressed filters*. A compressed by M version of a prototype FIR $H(z)$ is denoted $\mathbf{H}(z) = H(z^M)$ and can be realized by replacing each single clock delay in $H(z)$ with an M sample delay. The compressed filter $\mathbf{H}(z)$ continues to be clocked at the original sample rate f_s.[2] Referring to Figure 15-28, observe how compression scales the frequency axis by a factor $1/M$ and, as a consequence, compresses the FIR's original transition bandwidth by a like amount. It can also be observed that the act of compression populates the base-band spectrum with multiple copies, or artifacts, of the compressed prototype filter's frequency response. The center frequencies for these artifacts are located at $\varpi_k/\varpi_s = k/M$. It is through the intelligent use of compression that steep-skirt filters can be realized.

The frequency-masked FIR architecture is presented in Figure 15-28 and consists of the following definable subsystems:

- $\mathbf{H}_1(z)$, a compressed by M_1 version of an N_1-order FIR $H_1(z)$
- $\mathbf{H}_2(z)$, the compressed by M_2 version of the complement of $H_1(z)$
- $\mathbf{H}_3(z)$, the compressed by M_3 version of an N_3-order FIR $H_3(z)$
- $\mathbf{H}_4(z)$, the compressed by M_4 version of an N_4-order FIR $H_4(z)$
- $\mathbf{H}_5(z)$, an N_5-order FIR

[2]The compressed filter is not an interpolation FIR. An interpolated FIR operates at an elevated sample rate of $f_M = M f_s$, a compressed filter operates at a rate f_s.

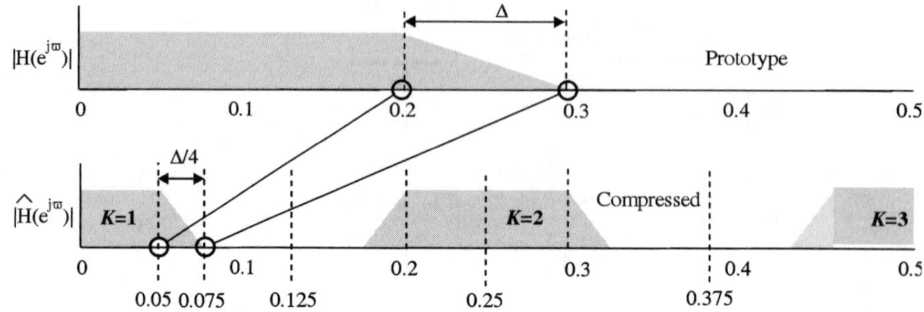

FIGURE 15-28 The magnitude frequency response of compressed by $M = 4$ FIRs showing the critical frequencies of $\varpi_p = 0.2$ and $\varpi_{a1} = 0.3$ being mapped (compressed) to $\varpi = 0.2/4 = 0.05$ and $\varpi = 0.3/4 = 0.075$. The transition bandwidth is scaled from $\Delta = 0.1$ to $\Delta/M = 0.025$. In addition, multiple copies of the compressed spectra are distributed uniformly along the base-band frequency axis.

The compression factor M_1 is chosen in order to map the transition bandwidth of Δ_1 to the final transition bandwidth of $\Delta = \Delta_1/M_1 \ll \Delta_1$. The target filter's low-pass cutoff frequency ϖ_p needs to be made coincident with one of the critical frequencies of a compressed filter $\mathbf{H}_1(z)$ (for example, $K_1\varpi_{p1}/M_1$, as shown in Figure 15-28) or the compressed complement filter $\mathbf{H}_2(z)$. Notice that the passband trailing edge for the first ($K = 0$) compressed image is located at $\varpi_{p1}/4 = 0.05$ and for the second image ($K = 1$), $4\varpi_{p1}/4 = 0.3$. These relationships for the compressed and compressed complement filters are summarized next:

$$\varpi/\varpi_s = \begin{cases} (K\varpi_s + \varpi_{p1})/\varpi_s M_1; & \mathbf{H}_1(z)\text{–based} \\ ((K+1)\varpi(\varpi_s/2) - \omega_p 1)/\varpi_s M_1; & \mathbf{H}_2(z)\text{–based} \end{cases} \tag{15-32}$$

The stopband critical frequencies can be likewise determined, and are a function of M_1 and the original stopband frequency of $H_1(z)$ or a passband of the complement filter $H_2(z)$. Once the compressed or complement compressed critical frequency is chosen, a housekeeping need appears. The compressed artifacts generated by the compressed prototype and compressed complement prototype filters extending beyond the target filter's passband frequency need to be eliminated. This is the role of the *frequency-masking* filters $\mathbf{H}_3(z)$ and $\mathbf{H}_4(z)$. The optional last-stage shaping FIR $\mathbf{H}_5(z)$, shown in Figure 15-29, provides a final level of artifact suppression. These rules that codified the design of a frequency masking filter are shown next.

- The component FIR filters $\mathbf{H}_1(z)$, $\mathbf{H}_2(z)$, $\mathbf{H}_3(z)$, $\mathbf{H}_4(z)$, and $\mathbf{H}_3(z)$ should be designed to have their transition bands somewhere in the middle of the base-band range $f \in [0, f_s/2)$. This will ensure that no unusual passband or stopband widths are imposed on the component filters.
- The filter $\mathbf{H}_1(z)$, or its complement, must have a critical passband frequency ω_p that maps to the target passband frequency for a compression of M_1 and a copy index K.

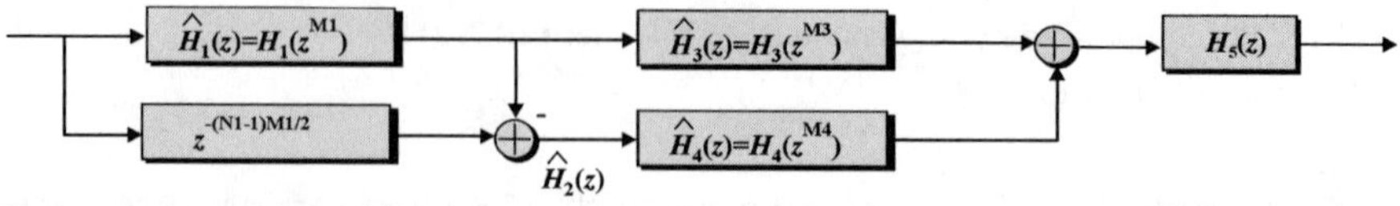

FIGURE 15-29 The anatomy of a frequency-masked (steep skirt) FIR.

- The design should minimize the solution's transition bandwidth, which is given by

$$\gamma = \min\left\{\frac{1}{\Delta_{\hat{H}_1(z)}} + \frac{1}{\Delta_{\hat{H}_3(z)}} + \frac{1}{\Delta_{\hat{H}_5(z)}}, \frac{1}{\Delta_{\hat{H}_4(z)}} + \frac{1}{\Delta_{\hat{H}_4(z)}} + \frac{1}{\Delta_{\hat{H}_5(z)}}\right\} \tag{15-33}$$

which corresponds to the estimated bandwidths of the upper and lower paths. The value of γ can be reduced if all the compressed filters have similar transition bandwidths. For the case where the component filters are of differing orders (N_i) and transition bandwidths (Δ_i), then it should be designed so that all the values of $N_i\Delta_i$ are similar.

As a rule, the highest order FIR section in a frequency-masked system is generally $\mathbf{H}_1(z)$ (therefore $\mathbf{H}_2(z)$), followed by $\mathbf{H}_3(z)$, $\mathbf{H}_4(z)$, and finally $\mathbf{H}_5(z)$. This suggests that their individual uncompressed transition bandwidths should appear in the reverse order. For linear phase solutions, the group delay of the upper and lower paths need to be the same. If $N_4 < N_3$, then filter $\mathbf{H}_4(z)$ will need to be equipped with an additional $(N_4 - N_3)/2$ shift register delay in order to equalize the group delays of the upper and lower paths. Finally, as a general rule, the passband deviation of each filter can be chosen to be 25–33 percent of the target deviation, in order to account for the degradation (increase) in passband ripple due to cascading. An example of the design process associated with a steep-skirt FIR is presented in the next example.

Example 15-13 Steep-skirt FIR Design

Required:

Design a steep-skirt FIR low-pass filter having the following specifications:

- A passband defined over $f \in [0.0, 0.1]f_s$ (in other words, $\varpi_p = 0.1$) with a maximum deviation of -0.175 dB from 0 dB
- Stopband defined over $f \in [0.1025, 0.5]f_s$ (or $\varpi_a = 0.1025$) with a gain of -40 dB or less
- A transition bandwidth of $(0.1025 - 0.1)f_s = (0.0025)f_s$.

Results:

It is worth noting that satisfying the specifications would require a linear-phase equiripple filter having an order in excess of 700. This is, in most instances, unacceptable. The design of a steep-skirt filter begins with a definition of the prototype $\mathbf{H}_1(z)$ in terms of the critical design parameters (ϖ_{p1}, ϖ_{a1}, Δ_1), compression ratio (M_1), and replication constant (K_1). Since the target normalized transition bandwidth satisfies $0.0025 = \Delta_1/M_1$, a list of acceptable Δ_1 and M_1 pairs can be assembled using a direct computer search. A reasonable, but by no means unique choice is $M_1 = 17$, resulting in $\Delta_1 = 0.0428$. Next, for $M_1 = 17$, the targeted passband cutoff frequency needs to be expressed in terms of the compression filter parameters ϖ_{ps} or ϖ_{a1}, and K_1 (these parameters also apply to the compressed complement filter as well.) Again, a direct computer search can be used to sort out the parametric options, as illustrated in Figure 15-30. Figure 15-30 reports the outcome if $\varpi_p = 0.1$, $\varpi_{p1} = 0.2573$, $\varpi_{a1} = 0.3$, $K = 2$, and $M = 17$. That is, the target passband frequency is obtained by compressing the complement filter $\mathbf{H}_2(z)$ and $K = 2$.

Finally, the passband gains of the component filters need to be specified. Suppose the minimum passband gain for the upper path is $G_{\text{upper}} = G_1G_3G_5$, and $G_{\text{lower}} = G_2G_4G_5$ for the lower path. Assume, for the purpose of discussion, that all individual gains are comparable such that $G_{\text{upper}} \sim G_1^3$ and $G_{\text{lower}} \sim G_1^3$. Since the specified minimum pass band gain is on the order of -0.175 dB, it follows that $G_{\text{upper}} = G_{\text{lower}} = 0.98$, or the gain deviation is $(1 - 0.98) = 0.02 \sim -34$ dB. For design consistency, let the passband deviation of all the filters (in other words, $G_1, \ldots, G_5$) be essentially the same. Then if $G_{\text{upper}} = G_{\text{lower}} = 0.98 = G_1^3$, $G_1 = 0.9933$ or the passband deviation is on the order of

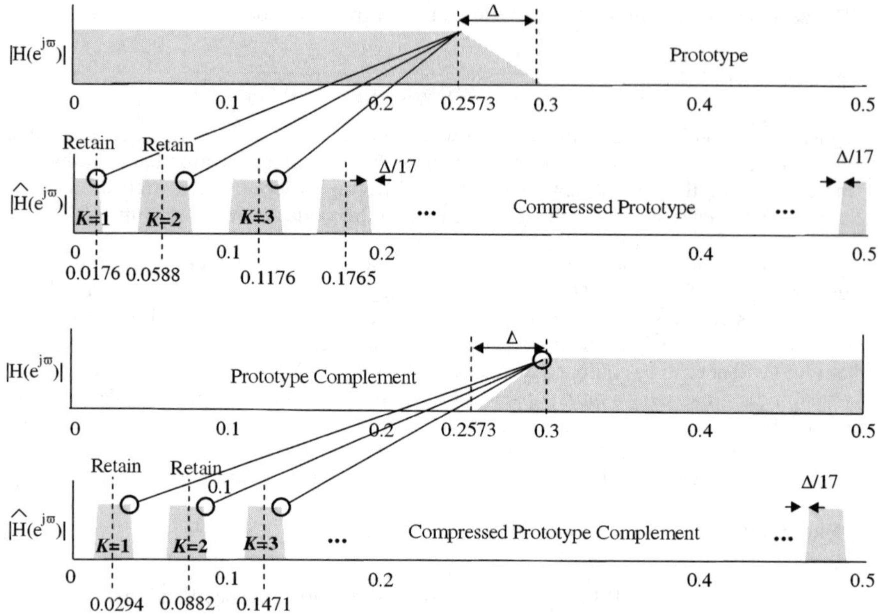

FIGURE 15-30 Frequency-masked design example for $\varpi_p = 0.1$, $M = 17$, $K = 2$, $\varpi_{p1} = 0.2573$, and $\varpi_{a1} = 0.3$. The final design retains two copies ($K = 2$) of the compressed prototype spectrum and two copies ($K = 2$) of the compressed complement response.

$1 - 0.9933 = 0.0067 \sim -43$ dB. The minimum stopband attenuation is essentially set by the stopband attenuation of filter $\mathbf{H}_5(z)$ if K_3 and/or K_4 is greater than unity. In such instances, the filters $\mathbf{H}_3(z)$ and/or $\mathbf{H}_4(z)$ will spawn spectral artifacts that are outside the final solution's passband, which are suppressed by $\mathbf{H}_5(z)$.

The *frequency-masking* filter process is summarized in Table 15-2. All component filters are equiripple FIRs with critical frequencies ϖ_{pi} and ϖ_{ai}, as described next. The behavior of the filter $\mathbf{H}_2(z)$ is established by $\mathbf{H}_1(z)$. The filters are generally designed to have a passband deviation on the order of -40 dB and a value of $N_i\Delta_i \sim 2.8$. The

TABLE 15-2 Frequency Masked Filter

Item	$\mathbf{H}_1(z)$	$\mathbf{H}_2(z)$	$\mathbf{H}_3(z)$	$\mathbf{H}_4(z)$	$\mathbf{H}_5(z)$
Passband edge (ϖ_{pi})	0.257200	0.300000	0.229412	0.300000	0.100000
Stopband edge (ϖ_{ai})	0.300000	0.257200	0.307500	0.398382	0.200599
Passband ripple (δ_{pi})	−43 dB	−56 dB	−45 dB	−58 dB	−42 dB
Stopband ripple (δ_{ai})	−56 dB	−43 dB	−58 dB	−71 dB	−42 dB
Filter order N_i	63	63	37	37	23
Transition bandwidth Δ_i	0.0428	0.0428	0.0781	0.0984	0.1006
$N_i\Delta_i$	2.8676	2.8676	2.8897	3.64	2.438
Compression factor M_i	17	17	3	3	1

exception is the 37th-order $\mathbf{H}_4(z)$ (same order as $\mathbf{H}_3(z)$), where $N_4\Delta_4 \sim 3.6$. A 29th-order $\mathbf{H}_4(z)$ could have been used, resulting in a $N_4\Delta_4 \sim 2.85$ if eight additional delays (four pre-delays, four post-delays) are added to equalize the group delays. Choosing a 37th-order FIR over a 29th-order FIR will simply result in the lower path having a slightly different gain deviation.

Figure 15-31 reports the spectral response of the complete 160th-order solution (note: $\mathbf{H}_2(z)$ is assumed to be implemented as a delay-enabled complement FIR requiring no additional coefficient multipliers). The design is based on a compressed critical frequency obtained from $\mathbf{H}_2(z)$ ($\varpi_{a1} = 0.3$ is compressed to $\varpi_p = 0.1$). It can also be seen that the filters $\mathbf{H}_3(z)$ and $\mathbf{H}_4(z)$ pass the first $K_1 = 2$ copies of $\mathbf{H}_1(z)$ and

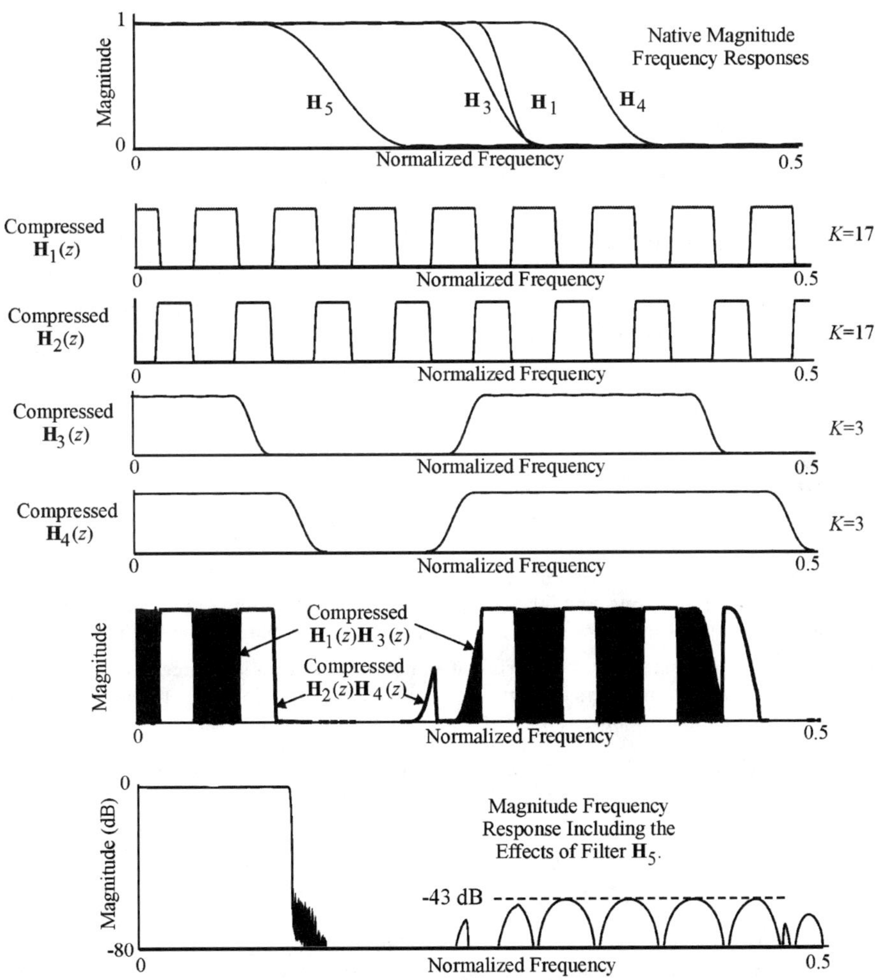

FIGURE 15-31 From top to bottom are shown the uncompressed component filter responses, compressed component filter responses, the composite filter response before the application of the housekeeping filter $\mathbf{H}_5$, and the final filter response including the effect of filter $\mathbf{H}_5$.

$\mathbf{H}_2(z)$, respectively. The estimated transition bandwidths of the upper and lower paths are

$$\frac{1}{\Delta_{\text{upper}}} = \frac{1}{\Delta_{\hat{H}_1}} + \frac{1}{\Delta_{\hat{H}_3}} + \frac{1}{\Delta_{\hat{H}_5}} = \frac{1}{0.0428/17} + \frac{1}{0.0781/3} + \frac{1}{0.1006} = \frac{1}{0.0216}$$

$$\frac{1}{\Delta_{\text{lower}}} = \frac{1}{\Delta_{\hat{H}_2}} + \frac{1}{\Delta_{\hat{H}_4}} + \frac{1}{\Delta_{\hat{H}_5}} = \frac{1}{0.0428/17} + \frac{1}{0.0984/3} + \frac{1}{0.1006} = \frac{1}{0.0230}$$

which results in a value of $\Delta_{\text{steep-skirt}} = 0.0023 < 0.0025$.

The resulting steep-skirt linear phase FIR is analyzed in Figure 15-31, and quantitatively as

- Passband edge: $0.1f_s$
- Stopband edge: $0.1025f_s$
- Passband ripple: < 0.1 dB
- Stopband ripple: > -42 dB

which are seen to meet or exceed the design specifications.

15.10 MATLAB MULTIRATE SUPPORT

Mathwork's Signal Processing Toolbox contains a set of basic tools that support multirate system simulation and analysis. They include

- *decimate*
- *downsample*
- *interp*
- *resample*
- *upfirdn*
- *upsample*

Some of these functions contain embedded filters; others simply implement basic decimation and interpolation operations. These methods are abstracted in the following:

decimate: Decimation reduces the original sampling rate for a sequence to a lower rate. The decimation program filters the input data with an *n*th-order low-pass filter and then resamples the resulting smoothed signal at a lower rate (see Figure 15-32).

```
Example: Decimate a signal by a factor of four.
t = 0:.00025:1;
x = sin(2*pi*30*t) + sin(2*pi*60*t);
y = decimate(x,4);
```

downsample: Downsample decreases the sampling rate of a signal $x[k]$ using an integer factor by keeping every *n*th sample starting with the first sample.

interp: Interp increases the original sampling rate by an integer factor. Program interp performs low-pass interpolation by inserting zeros into the original sequence and then applying a low-pass filter to smooth the data (see Figure 15-33).

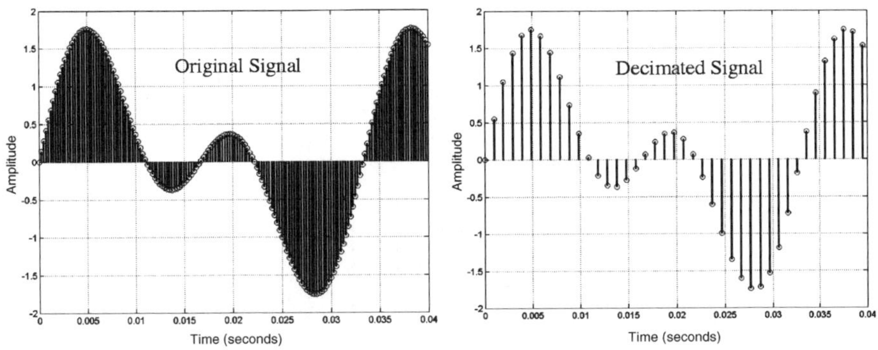

FIGURE 15-32 MATLAB decimation.

```
Example: Interpolate a signal by a factor of four:
t = 0:0.001:1; % Time vector
x = sin(2*pi*30*t) + sin(2*pi*60*t);
y = interp(x,4);
```

resample: Resample changes a signal's $x[k]$ sample rate by a rational factor p/q using a polyphase filter implementation. The parameters p and q must be positive integers. The length of the re-sampled signal is equal to [length(x) $\times$ p/q)]. Resample applies an anti-aliasing (low-pass) FIR filter to the signal during the re-sampling process (see Figure 15-34).

```
Example: Resample a simple linear sequence at 3/2 the
original rate:
fs1 = 10;               % original sampling frequency in Hz
t1 = 0:1/fs1:1;         % time vector
x = t1;                 % define a linear sequence
y = resample(x,3,2);    % resample
```

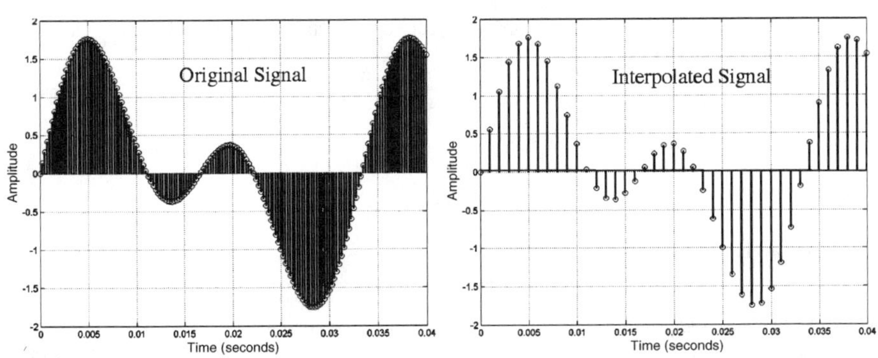

FIGURE 15-33 The MATLAB interpolation.

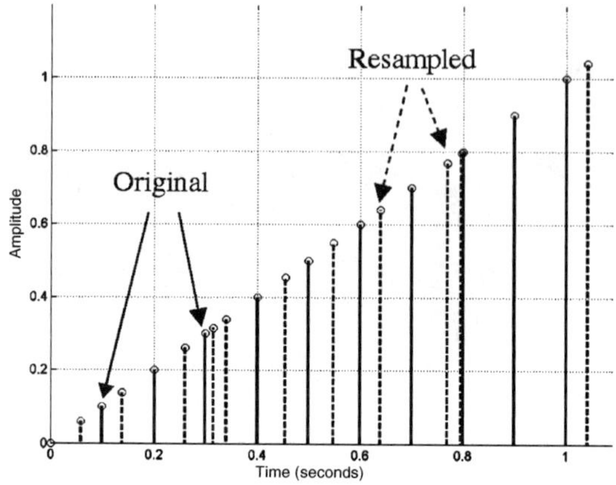

FIGURE 15-34 A MATLAB re-sampling (sample rate conversion).

upfirdn: Program upfirdn performs a cascade of three sequential operations beginning with up-sampling the input signal by a factor p (inserting zeros), then FIR filtering the up-sampled signal, and finally down-sampling the result by a factor of the integer q.

Example: Change the sampling rate by a factor of 147/160 (corresponds to a 48 kHz (DAT rate) to 44.1 kHz (CD rate))

```
L = 147; M = 160;                  % Interpolation/decimation factors.
N = 24*M;
h = fir1(N,1/M,kaiser(N + 1,7.8562));
h = L*h; % Passband gain = L
Fs = 48e3;                         % Original sampling frequency: 48 kHz
n = 0:10239;                       % 10240 samples, 0.213 seconds long
x = sin(2*pi*1e3/Fs*n);            % Original signal, sinusoid at 1 kHz
y = upfirdn(x,h,L,M);              % 9408 samples, still 0.213 seconds
```

upsample: Upsample increases the sampling rate of a signal by an integer factor via inserting $n - 1$ zeros between samples.

Example: Increase the sampling rate of a sequence by 3:

```
x = [1 2 3 4];
y = upsample(x,3);
```

BIBLIOGRAPHY

Cavicchi, T. *Digital Signal Processing*. New York: John Wiley and Sons, 2000.

Chassaing, R. *Digital Signal Processing and Applications with the C6713 and C6416 DSK*. New York: John Wiley and Sons, 2005.

Harris, F. *Multirate Signal Processing for Communications Systems*. Englewood Cliffs, New York: Prentice-Hall, 2004.

Ifwachor, E., and B. Jervis. *Digital Signal Processing 2nd ed.*, San Francisco: Addison Wesley, 2001.

Mitra, S. *Digital Signal Processing, 3rd ed.* New York: McGraw-Hill, 2006.

Oppenheim, A. V., and R. Schafer. *Digital Signal Processing*. Englewood Cliffs, New York: Prentice-Hall, 1975.

———. *Digital Signal Processing 2 ed.* Englewood Cliffs, New York: Prentice-Hall, 1999.

Suter, B. *Multirate and Wavelet Signal Processing*. Lighting Source, 1998.

Taylor, F. J. *Digital Filter Design Handbook.* New York: Marcel Dekker, 1983.

——— and T. Stouraitis. *Digital Filter Design Using the IBP PC*. New York: Marcel Dekker, 1987.

——— and Mellott, J. *Hands-On Digital Signal Processing*. New York: McGraw-Hill, 1998.

Vaidyanathan, P. P. *Multirate Systems and Filter Banks*. Englewood Cliffs, New York: Prentice Hall, 1993.

CHAPTER 16
DIGITAL FILTER TECHNOLOGY

16.1 INTRODUCTION TO SIGNAL PROCESSING

DSP, at one level or another, is technology-driven. A DSP technology roadmap was established in the early 1970s using the definition of what was to become the DSP microprocessor or DSP μp.[1] Beginning in the late 1970s, the first generation of DSP μp (that is, Intel 2920) made a commercial appearance. Since then, DSP has become a unique and valuable fusion of science, mathematics, and technology. While there have been significant theoretical and algorithmic advancements made to the field over the past three decades (for example, Cooley-Tukey FFT, 1975), technology has been the real engine for DSP growth. Today, the principal beneficiary of the DSP revolution is communications (for example, cellular telephony) followed by consumer electronics and computers. Communications applications are generally well articulated (such as a digital down converter), leading to the creation of high-performance compact wireless and network systems. DSP is also leading the march towards software-defined radio (SDR), enables new services such as satellite radio (for example, XM Radio and Sirius Radio), and a host of terrestrial innovations (high definition or HD Radio). The impact of DSP on consumer electronics is case-dependent. The value of DSP in such applications as MP3 for audio players, JPEG for digital cameras, and MPEG for DVD players cannot be challenged. DSP is also a critical computer technology in the form of disk controllers, multimedia, and image and video processing. Other areas having an obvious DSP content are biomedicine, transportation, and defense, just to name a few. It is apparent that the future of this field is intimately tied to semiconductor innovations and man's ingenuity. Gordon E. Moore, co-founder of Intel, has provided a means of predicting the future in terms of *Moore's Law*, which was expressed as:

$$(Nt/A)(t_1) = (Nt/A)(t_0) \times 1.58^{(t_1-t_0)} \tag{16.1}$$

where $(Nt/A)(t)$ denotes the number of transistors normalized by area at some point in time. Moore's empirical formula predicts the doubling of transistor density every 18 months. For decades, Moore's Law has been on target, in lock step with improvements in semiconductor lithography that enabled smaller scale sizes, which in turn resulted in

- Decreased power dissipated per transistor ($P \sim CV^2f$ + leakage)
- Lower cost, increasingly complex chips

[1]U.S. Patent 3,812,470; Programmable Digital Signal Processor, John C. Murtha et al. (Westinghouse), 1974.

TABLE 16-1 Interpretation of Moore's Law

Item	Circa 1972	Circa 2002	Factor
CPU speed	6 MHz	600 MHz	100
Memory	128 KB	256 MB	2,000
Memory access time	850 ns	100 ns	8.5
Network	3 Mbps (Ethernet)	100 Mbps (Ethernet)	30
Disk capacity	2.5 MB	100 GB	2,400

Moore's Law can be illustrated in terms of semiconductor memory, noting that

- Memory capacity has improved 1,000,000-fold since 1970
- Memory bandwidth has improved 100-fold since 1970
- Memory latency had been reduced 10-fold since 1970

While transistor density has been exponentially increasing, bandwidth and latency are seen to lag. Moore's Law applied to processors suggests that more complex multiprocessor chips, using one or more processors to perform a single function (for example, in graphics), will be commonplace. Moving data to and from the processor will, unfortunately, remain problematic. Further insights can be drawn from a side-by-side comparison of two general-purpose digital machines configured in Table 16-1. The first system is defined at the time the Intel 8080 microprocessor was introduced. The second machine is more recent. Based on this model, the future can be seen to be potentially amazing. There are apparently only two limits on where future digital technologies will take man. The first is the loss of man's creative instincts. The second is that man is approaching the physical technological limits predicted by semiconductor physics. It has turned out that atoms are too big and that light is too slow to meet our future needs.

16.2 PROCESSOR FORMS

One of the principle driving forces of the digital revolution has been processor technology. The form and function of digital processors can vary widely and are often differentiated by their semiconductor technology expression and intended application. Specifically, they can be classified as

- General-purpose μps chips or cores
- DSP μp floating-point chips or cores
- DSP μp fixed-point chips or cores
- Field programmable gate arrays (FPGAs)
- Function-specific integrated circuits
- Application-specific integrated circuits (ASICs)

Chips are self-contained entities that are placed onto printed circuit boards (PCBs) and used to define system-level solutions. Cores are semiconductor designs supplied as software statements (RTL, VHDL, targeted net lists) that can be embedded into a chip design, enhancing the outcome's functionality.

General-purpose processors have a long history of use in implementing DSP algorithms. Some of the processors have evolved DSP-friendly instruction sets and are often released as embeddable processor cores. Due to their high internal clock rate and superscalar architecture, general-purpose processors can often become credible DSP engines (for example, Intel Pentium with MMX extensions). Their disadvantage is cost, power dissipation, execution time variability, and a lack of DSP-intensive instructions.

DSP μps have been continuously refined since their initial entry into the marketplace. It is often assumed DSP μps are the majority technology in DSP applications, but this would be false. Nevertheless, DSP μps are powerful agents, often executing instructions in parallel using an optimized DSP instruction set. They can exhibit superscalar capabilities and some are based on a very long instruction word (VLIW) architecture that issues multiple instructions per instruction cycle.

Function-specific integrated circuits, sometimes called application-specific standard parts (ASSP), are single-chip devices that serve high-valued, high-volume needs and are intended for mass resale. Examples are digital down converters, JPEG decoders, and so forth. These devices are often implemented using what is called standard cell and cell-based design strategies. Application-specific integrated-circuits (ASIC) often provide the services associated with an ASSP, but are technically intended for internal product use. The boundary between ASSPs and ASICs has blurred over time, resulting in both classes often being referred to as ASICs. Structured ASICs, also called platform ASICs, are based on a mask layer that is partially predefined by the ASIC provider. Structured ASICs can provide high-performance approaching that of a custom chip without the excessive high nonrecurring expenditures (NRE) associated with a custom design. Nevertheless, the NRE costs can be considerable. Compared to custom designs, a structured/platform ASIC can reduce the time-to-market by a significant amount. Together, ASIC/ASSPs are the dominant DSP technology. The future of this class of technology is strongly tied to the utilization of reusable intellectual property (IP) to define important core functions (for instance, FFT).

FPGAs are members of a class of devices called programmable logic. The major advantage of FPGAs is their design flexibility and ability to prototype complex solutions. Design changes can be made quickly and often remotely. However, FPGA flexibility is purchased at the expense of the device's footprint, power consumption, and unit cost. FPGAs have a growing presence in mainstream DSP as a rapid prototyping technology and in developing few-of-a-kind solutions. FPGAs have been shown to be a viable digital filter technology, especially in those cases where a filter can be implemented using distributed arithmetic. More recently, FPGA architectures have been developed and marketed that contain arrays of embedded multiply-accumulators, providing a more general-purpose capability at an elevated cost per part.

In the life cycle of virtually every hardware-orchestrated DSP design there is a point where the implementation technology needs to be critically accessed. The choice is influenced by many factors beyond cost and performance. Comparing an ASIC to an FPGA design, for example, is heavily predicated on market size as suggested in Figure 16-1 where the cross-over point is assumed to be 1,000 units.

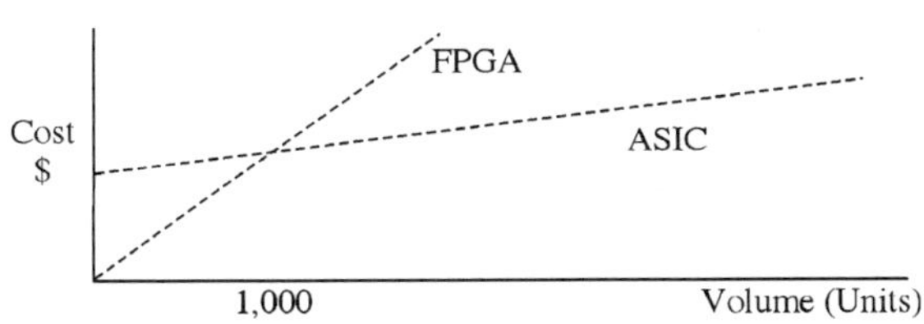

FIGURE 16-1 The economics of FPGA versus ASIC designs.

Another concept gaining increased attention is the notion of a system-on-a-chip (SOC) solution. The ability to concentrate the majority of processor elements and communication functions on a single chip provides a wealth of operational and economic advantages. DSP solutions are also making increased use of intellectual property (IP) in the form of predefined partial solutions. IP providers create partial solutions that can be used to assemble more complex solutions (for example, ARM10, FFT). IP provides a pathway that can mitigate time-to-market pressures and facilitate the realization of solutions of increasing complexity. IP is generally delivered as licensed RTL or VHDL/Verilog code, or net lists, that define solution elements. IP can be classified as hard cores (being fixed optimized designs), soft cores that provide only a device's functional description, or firm cores that are a blend of hard and soft cores.

16.3 GENERAL-PURPOSE MICROPROCESSORS (μps)

To more fully understand the deviating paths taken by DSP technologists in the quest of the perfect processor, their underlying technical philosophies and strategies need to be understood. These issues were presented by Taylor and Mellott[2] in a general context which motivates the discussion found in the next few sections. The study begins with stating the expected attributes and liabilities of a general-purpose processor and DSP processor.

Features found on most general-purpose processors (not inclusive) include

- Multiple data types supported by the processor hardware
- Multilevel cache memories
- Paged virtual memory management in hardware
- Support for hardware context management including supervisor and user modes
- Unpredictable instruction execution timing
- Large general-purpose register files
- Orthogonal instruction sets
- Simple or complex memory addressing, depending upon whether the processor is RISC or CISC

The most important data types employed by general-purpose processors are character types, followed by the integer types. From the viewpoint of market share, the majority of general-purpose processors are employed in business applications that involve text and database processing. Floating-point arithmetic is generally not crucial in most applications run on general-purpose computers, although there are niche markets where this is not true (for example, technical and scientific workstations).

Cache memories have been demonstrated to be a powerful enhancement agent for many general-purpose processors. The inclusion of sometimes substantial-sized cache memories in general-purpose computers are based on the assumption that application programs demonstrate instruction or data locality, sometimes called the "cache assumption." The cache assumption is frequently used to justify the design of shared memory multiprocessing general-purpose computers where the main memory is connected to the processors via a shared bus. If the cache assumption is violated, the performance of single and multiprocessing general-purpose computers is generally degraded.

[2]F. Taylor and J. Mellott. *Hands-On Digital Signal Processing*. New York: McGraw-Hill, 1998.

Large register files are included in many general-purpose architectures, although there are exceptions (for example, Intel x86). Since most general-purpose machines operate on scalar data, and if the cache assumption holds, large register files are generally beneficial. General-purpose registers and orthogonal instruction sets tend to enable compilers to produce efficient object and assembly language code. Since external memory can only be accessed by load and store operations, it is desirable to keep more operands on hand in the register file to enhance performance. As a result, the load-store architectures used in many RISC processors are enhanced through the use of large register files.

Hardware support for the management of virtual memory and multiple process contexts is desirable in general-purpose computers. Most general-purpose processors support the timeshared execution of multiple processes; even single-user desktop computers are generally running many processes in the background. Virtual memory allows programs to run in a degraded manner if their primary memory requirements exceed available resources. The penalty for virtual memory is increased data access latency due to address translation penalties and long page-fault latencies. The latter is generally managed by switching the processor context to another process so that the processor does not idle while a page fault is being serviced. Support for multiple process contexts by a general-purpose computer is therefore crucial for optimal utilization of the processor resource among multiple tasks.

Instruction execution timing on general-purpose processors is generally unpredictable. This is the result of myriad features designed to enhance the general performance of the processor. Cache memory and virtual memory introduce a substantial amount of uncertainty in instruction execution timing. The amount of time required to read or write a particular location in memory will depend upon whether or not a cache "hit" occurs, at which level of the cache it hits, and whether or not that virtual address resides in the table look-aside buffer (TLB). Main memory latency can be affected when a cache fault occurs and the cache needs to be refreshed. Also, variable latencies occur when other processors, like direct memory access (DMA), cause access contention problems. Various architectural enhancements, such as superscalar execution, speculative execution, out-of-order execution, and branch target caches, may further confound any attempt to measure the execution time of an instruction.

Another class of general-purpose processor commonly encountered is the microcontroller. Most microcontrollers are derived from successful general-purpose microprocessor designs, although some are original designs. Microcontrollers are typically targeted at embedded applications which typically do not require significant arithmetic performance. Microcontrollers usually eliminate features such as large cache memories and virtual memory, and instead, add integrated peripheral interfaces to support the intended embedded applications.

16.4 DSP PROCESSOR (DSP μp)

Since the 1970s, the semiconductor industry has experienced geometric growth in the number of transistors that can be placed on a chip. One of the principal beneficiaries of this high-density, deep submicron capability is DSP. For example, the Texas Instruments TMS320 family of DSP μps has logically migrated from its original fixed-point versions (TMS320C10) to today's vast collection of floating-point and VLIW multiprocessor devices. All these solutions are based on the use of dominating MAC operations used to form sums of products over the real or complex field (for example, convolution, correlation, and transforms). A variation on the basic MAC call is the vector *SAXPY* (Scalar = *AX* Plus *Y*) operation. SAXPY algorithms range from linear to non-linear, in 1-D to *M*-D configurations.

General-purpose processors have often exhibited a Von Neumann architecture. Von Neumann architectures are defined in terms of a single block of memory, as illustrated in Figure 16-2. Von Neumann machines were extremely popular, enabling many important general-purpose computing innovations and commercially successful systems. The early days of computing, however, were dominated by machines having a so-called Harvard architecture. DSP processor designers have been rediscovering this design paradigm in increasing numbers. Machines employing Harvard architectures have dedicated buses serving various parts of the system (for example, memory bus, instruction bus, I/O bus, and so on), as shown in Figure 16-2. The designers of DSP processors realized that a Harvard architecture was synergistic to highly repetitive multiply-accumulate (MAC) DSP algorithms. It was found that such operations could be accelerated using Harvard architectures along with direct memory access (DMA) support. More recent interpretations of the Harvard architecture introduced additional layers of instruction cache and high-bandwidth I/O ports to data memory paths, as suggested in Figure 16-2.

While DSP architectures are varied, more often than not they are differentiated by the market they attempt to serve. Regardless of the details, DSP processors are expected to exhibit the following characteristics:

- High-arithmetic performance often involving extended precision multipliers and arithmetic units
- One or two data types being supported by the processor hardware
- Limited data cache memory
- No memory management hardware
- No support for hardware context management (task switching)
- Exposed pipelines
- Predictable instruction execution timing
- Limited register files, instead providing special purpose registers

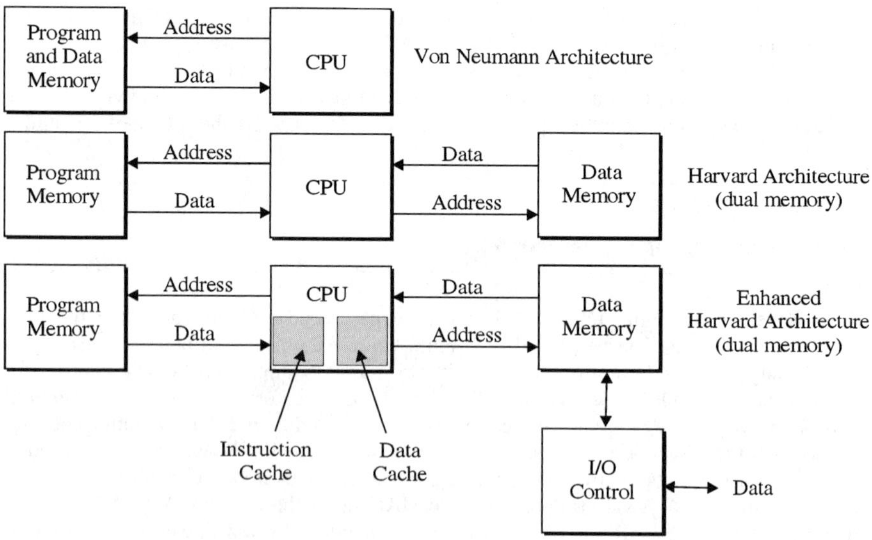

FIGURE 16-2 Architectural choices.

- Non-orthogonal instruction sets
- Enhanced memory addressing modes
- On-board fast RAM, ROM, and DMA

DSP signal processors are designed around a different set of assumptions than those which drive the design of general-purpose processors. First, digital signal processors generally operate on arrays of data rather than scalars; therefore, the scalar load-store architectures found in general-purpose RISCs are not essential. The economics of software development for digital signal processors is also different from that for general-purpose applications. Digital signal processing problems tend to be algorithmically smaller than, for example, a word processor. In many cases, the ability to use a slower and less expensive digital signal processor by expending some additional software engineering effort is economically attractive: a good return-on-investment may be achieved even if a few dollars per unit of manufacturing cost can be saved in a product that will ship a million units. This fact is often reflected in serious design optimization efforts associated with DSP solutions. As a consequence, the most serious programming of digital signal processors is done in assembly language rather than high-level languages (HLL), although working with high-level languages is becoming increasingly popular.

16.5 DSP ADDRESSING MODES

Modern general-purpose processors can have many addressing modes (such as CISC processors) or few addressing modes (for instance, RISC processors). CISC processors can support addressing modes such as direct, register or memory indirect, indirect indexed, indirect with displacement, and indirect indexed with displacement. Furthermore, the indexed modes may support pre- and post-increment or decrement of the indices. Historically, complex addressing modes have resulted in higher code entropy which has two consequences. First, the productivity of the assembly language programmer is enhanced, and second, the resulting object code is more compact. A number of factors have contributed to the disappearance of complex addressing modes characteristic of CISC processors. The first is the change in the economics of hardware costs versus software development costs. Decades ago, software development was less costly and required investment in hardware. Handcrafted assembly-level language code was commonly used to create application programs. Today, hardware is inexpensive relative to software development costs. Furthermore, contemporary time-to-market pressures are altering this landscape. As a result, most applications are coded using high-level languages.

Another issue is related to the first. It has proven difficult to get compilers to take full advantage of complicated addressing modes and nonorthogonal instruction sets. Another strike against complex addressing modes in general-purpose computers is that the complex addressing modes tend to cause pipeline stalls, due to the complicated data interdependencies produced by the complex addressing modes. Even modern CISC implementations have been optimized so that better performance results when complex addressing modes are avoided. Eschewing complex addressing modes has led to the adoption of a load-store philosophy that allows functional units to accept issues without stalling due to data dependencies associated with data stored in memory. By moving to register indirect load-store architecture, all of the more complex addressing operations are performed in software, thus allowing greater flexibility in scheduling instruction issues. A register indirect load-store architecture synthesizes more complicated "addressing modes" with several simple instructions. The compiler is free to statically arrange these instructions with an awareness of the impact of adjacent instructions on the scheduling of processor resources. The processor may also elect to rearrange the execution of these simple instructions within the constraints of available resources and data dependencies. In contrast, the classic CISC has the micro-operations of each instruction statically scheduled in the micro-program for each instruction.

DSP applications frequently require non-sequential access to data arrays using modular or bit-reversed addressing. These addressing modes are not easily supported in general-purpose RISC or CISC processors. For maximum performance in digital signal processing applications, it is sensible to add dedicated hardware support for these addressing modes. To summarize, the addressing modes required include

- Address register indirect
- Address register indirect with unit stride and non-unit stride modular indexing
- Address register indirect with bit-reversed indexing

Existing DSP architectures are single-issue so, with the exception of the special modes indicated, the address register file and arithmetic unit would be similar to that found in general-purpose architectures. In order to support multiple issues, it will be necessary to define either a hardware or software mechanism to support concurrent address generation for multiple function units.

16.6 CIRCULAR BUFFERING

The nature of DSP algorithms suggests that DSP processors are expected to perform SAXPY-type operations frequently and efficiently. These operations often involve the pair-wise multiplicative combination of a variable and filter coefficient. Both sample data and coefficients can be viewed as being read from a circular buffer, as shown in Figure 16-3. The data is accessed by a pointer that indicates which data value is to be read from the buffer at a specific point in the program's execution. Modern DSP μps efficiently implement these buffers as well as provide a means to read and write data to the buffer and expel data that is no longer needed. The use of a circular buffer in implementing an *L*th-order FIR is illustrated using the following pseudo-code.

```
MOVE {current sample to circular data buffer}
UPDATE {update pointers to the circular data buffer}
CLEAR {zero accumulator}
BEGIN {begin FIR loop}
READ {read FIR coefficient from circular coefficient
buffer}
UPDATE {update pointers to the circular coefficient buffer}
READ {read data from circular data buffer}
```

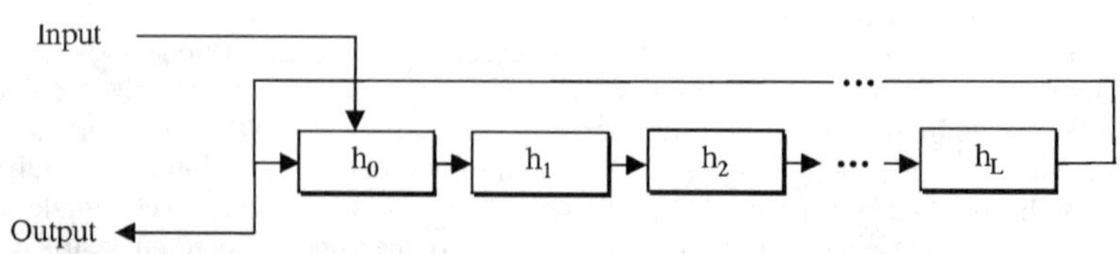

FIGURE 16-3 A circular buffer.

```
UPDATE {update pointers to the circular data buffer}
MULTIPLY {multiply data and coefficient}
ADD {send result to the accumulator}
END {end loop}
MOVE {move output result to destination}
REPEAT {repeat the process for the next input sample}
```

16.7 DSP PROCESSOR FEATURES

Execution time in DSP applications is normally dominated by sum-of-product (SOP) operations. Typical examples are computing inner products and linear convolutions. These operations are based on a sequence of multiply-accumulates that can be implemented using a simple computing loop. It is therefore reasonable to provide instruction set support for executing a loop for a fixed number of times. In fact, looping based upon the value of a counter is the most common branching operation in digital signal processors. As a result, many DSP μps have dedicated instructions to reduce any computational penalty in managing a loop count. For example, DSP processors normally support an instruction that causes the next machine instruction to be repeated a fixed number of times. As a consequence of this, a justification for dedicating substantial resources to a branch-target cache cannot be found. Branch-target caching makes more sense in general-purpose applications since many of these applications have branching patterns that are difficult to predict at compile-time.

Most integer digital signal processors are actually fixed-point arithmetic machines. The fixed-point format is achieved by integrating shifters with the multiplier-accumulator so as to allow pipelined adjustment of operands and results. The multipliers and accumulators included in most fixed-point digital signal processors are oversized or extended precision to allow transient computations to exceed the normal word-width of the processor. Exposed pipelines are usually avoided, however, in the quest for higher performance at lower unit cost. However, some processors (such as the TMS320C50) have switched to exposed pipelines. Exposed pipelines present some challenging programming issues that can be exasperating given how applications may require the programmer to reach more deeply into pipelines. Since digital signal processors are designed to support the real-time processing of large quantities of sampled data, they generally have support for enhanced dataflow. Modified or enhanced Harvard architectures are often needed to interface with on-board memories. Some digital signal processors include DMA controllers that are capable of performing memory-memory and memory-I/O move operations concurrent with computational tasks. An independent DMA controller would typically be used to load new data into the on-chip memory while some computation is performed. This allows an internal Harvard architecture to be better exploited by keeping the processor busy with computation rather than programmed data I/O. Currently, these DMA resources are managed explicitly by the programmer. In order to support rapid code development and portability, it is important that the management of the DMA resources be simplified, at least, if not moved completely into the programming tools.

16.8 DSP PROCESSOR PARALLELISM

As VLSI technology has improved, it has become possible to include additional hardware resources to enhance the performance of general-purpose and application-specific processors. In order to increase throughput in a traditional Von Neumann machine, additional hardware resources are added to exploit opportunities for instruction-level parallelism. The techniques that have been developed to exploit opportunities for instruction-level parallelism are super-pipelining, superscalar architecture, dataflow processors, and very long instruction word architecture. Since software development costs have spiraled upwards, a significant effort has been focused in the area of automatic compiler-based optimization of high-level language code.

The technique of super-pipelining has been exploited by some processors, such as the Intel Pentium Pro, in order to achieve high throughput. Super-pipelining is attained by adding pipeline stages so as to achieve a very short machine cycle, thus allowing a high issue rate. While instructions are issued sequentially at a high rate, they take many cycles to complete. So while one instruction is started, several or many previous instructions may be in various stages of completion. The disadvantage of super-pipelining is that it increases latency (the time from when an instruction is issued to when it is completed) and makes pipeline flushes more expensive. From a hardware perspective, the addition of pipeline registers requires significant extra hardware resources. In order to hide the pipeline from the programmer and/or compiler, the processor must keep track of resources that have been committed to instructions that are in progress in the pipeline. If resource conflicts occur, the pipeline is stalled, and "bubbles" are introduced into the pipeline. Instructions are generally ordered by the compiler or programmer to avoid pipeline stalling whenever possible.

From a commercial viewpoint, attempts at parallelism for digital signal processing have relied upon expensive multiprocessor communications or alternatively multiple independently programmed ALUs. Unfortunately, these solutions carry a high price tag and can become I/O pin-bound. The newer high-end DSP μps combine an RISC processor with multiple ALUs under independent program control. The ALUs are optimized for signal- or image-processing operations, and the device is optimized for specific market segments quantified in terms of speed, cost, and power dissipation. One must also be mindful that DSP algorithms developed for uniprocessors do not necessarily translate well to a multiprocessor environment.

16.9 FIXED-POINT VERSUS FLOATING-POINT

Choosing whether to implement a design in fixed-point, block floating-point, or floating-point can lead to tension between the various stakeholders in the design outcome. Consider the comparison presented in Table 16-2 that compares fixed- and floating-point solutions. Block floating-point solutions fall in between fixed- and floating-point. While strong arguments can be made to support either path, it nevertheless remains a fact that fixed-point solutions are, by far, the predominant DSP technology of the day.

16.10 DSP BENCHMARKS

Technologists often presume that MIPS and MFLOPS are the standard measure of a DSP processor's capability. A more important predictor is the device's millions of multiply-accumulates per second (MAC/s) rate. Potentially the most important tool used to classify and differentiate DSP technologies is the benchmark. A good benchmark is intended to isolate a particular device feature or function, and exercise the feature rigorously. There are, however, no standard sets of benchmarks that are universally accepted. A valid benchmark can, however, provide insights into a technology's capabilities measured in terms of cycle

TABLE 16-2 Floating-Point Fixed-Point Comparison

Item	Floating-point (e.g., TI TMS320C67)	Fixed-Point (e.g., TI TMS320C64)
Precision	Best—Has virtually unlimited precision. Eliminates the possibility of runtime overflow or other finite wordlength effects.	Worst—16-bit or higher precision, both limited in dynamic range and computational precision giving rise to finite wordlength effects.
Speed	Worst—Due to the complexity of floating-point, and possibly long pipeline delays.	Best—Simple arithmetic data flow and potentially high clock rates.
Cost	Worst—Potentially high hardware cost.	Best—Available in a wide range of cost-performance modalities.
Power dissipation	Worst—Can exhibit high power consumption.	Best—Range of power choices. Some have been specifically designed for low-power mobile and handheld operation.
Arithmetic Execution Latency	Worst—Floating point add has data-dependent delays which can be managed, to a degree, using a highly pipelined arithmetic system architecture.	Best—Simple arithmetic system design providing opportunities to develop efficient real-time execution code.
Time-to-market	Best—Faster due primarily to side-stepping finite wordlength concerns.	Worst—Managing finite wordlength effects can require additional analysis and testing.

count, memory use, execution time, energy consumption, and cost performance. Elements of common DSP benchmarks are

- Real FIR
- Complex FIR
- Single step real FIR
- Circular block FIR
- Single step adaptive FIR
- Single step real IIR
- Block IIR (biquad)
- Cascaded block IIRs
- Linear convolution
- Autocorrelation
- Vector add
- Vector product
- Maximum element in an array
- Complex FFT radix-4
- Complex FFT radix-2

- Inverse complex FFT radix-2
- Viterbi decoder
- Bit manipulation

16.11 ADC/DAC OPERATION

Analog to digital (ADC) and digital to analog (DAC) conversion are core DSP technologies (see Section 12.11). They often set the performance and precision bounds on a DSP system design. Figure 12-7 establishes that an ADC operation is actually performed by two distinct operations, namely an ideal sampler (sample-and-hold) and a quantizer. An n-bit quantizer converts the sample and hold circuit's output value into one of 2^n discrete values. In reality, the internal details of an ADC can differ. Each instance represents a different ADC architecture. The more common architectures are summarized in this section.

Delta-Sigma ADC

A *delta-sigma* ($\Delta\Sigma$) converter combines over-sampling with subtractive feedback control, as shown in Figure 16-4. A $\Delta\Sigma$ converter can be of high-precision if the signal to be converted has a low bandwidth relative to the sample rate. As the "delta" (Δ) implies, a $\Delta\Sigma$ ADC relies on representing a signal with small differential amplitude changes. Using over-sampling, the difference between adjacent samples of a band-limited signal can be guaranteed to be small. The analog front-end of a $\Delta\Sigma$ ADC contains a number of low-resolution components that can quantify these small changes. The back-end processor includes a feedback path that is used to synthesize small differential corrections $\pm\Delta V$ to the previous estimate $x'(t)$ of value of $x(t)$. Since the $\Delta\Sigma$ ADC is operating at a high over-sample rate, it is assumed that only small changes in $x(t)$ could occur on an inter-sample basis. The ADC will continue to make small adjustments in the estimate of $x'(t)$ based on the measure error $\varepsilon(t) = x(t) - x'(t)$. There is, however, a fundamental caveat associated with $\Delta\Sigma$ ADCs. Initially, the error $\varepsilon(t)$ is normally large, causing the $\Delta\Sigma$ ADC to take maximum corrective action (that is, administer a long series of small incremental changes). This is called the adaptation phase. Once adaptation takes place, the error approaches zero and the small corrective changes are applied selectively. In this context, is should be evident that the signal being converted should be smooth (in other words, small differential changes between samples). If noise or high-frequency signal components are present, an $\Delta\Sigma$ ADC can exhibit erratic behavior.

Flash ADC. A flash converter consists of a collection of dedicated analog comparators and multilevel threshold detectors, as suggested in Figure 16-5. An n-bit flash converter

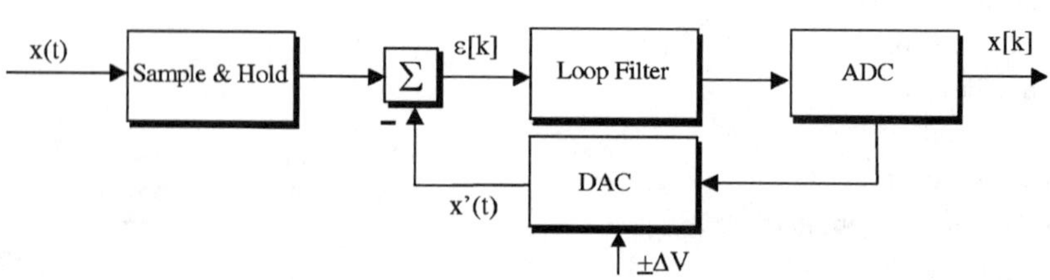

FIGURE 16-4 A $\Delta\Sigma$ ADC architecture.

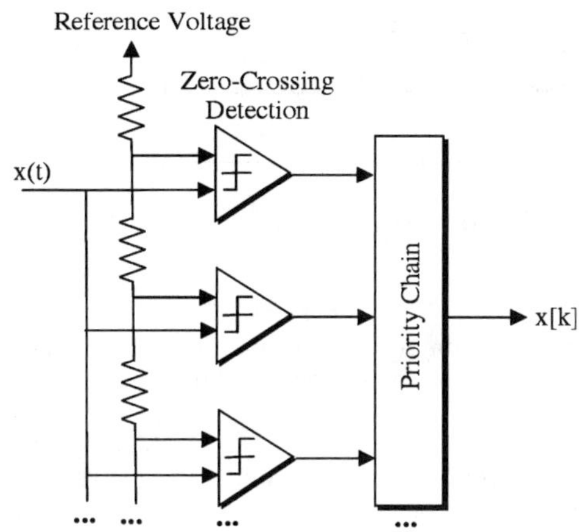

FIGURE 16-5 Flash ADC architecture.

incorporates 2^n dedicated analog comparators. While being the fastest ADC architecture, flash converters consume a great deal of power and are of low word width.

Successive Approximation ADC. Successive approximation converters are popular for mid-range applications. A successive approximation ADC is shown in Figure 16-6 and consists of a successive-approximation-register (SAR), comparators, and DACs. The SAR produces a monotonically increasing binary valued sequence (ramp). The digital ramp is converted into a discrete-time (analog) ramp using a DAC. The monotonic sequence continues to increase until a threshold is passed, based on the zero-crossing outcome. The time required to detect this sign-change is directly correlated to the ADC digital output. The speed of a successive approximation ADC is directly linked to the converter's word width.

Subrange ADC. Subrange converters subdivide a signal's dynamic range into smaller units as suggested in Figure 16-7. The M most-significant-bits (MSBs) are obtained by a direct conversion by an M-bit ADC, with the others following in successive order. For a two-stage N-bit subrange ADC, the M most-significant-bits would be followed by the final (N-M) bits. The advantage of this method is that compared to a flash converter, a subrange

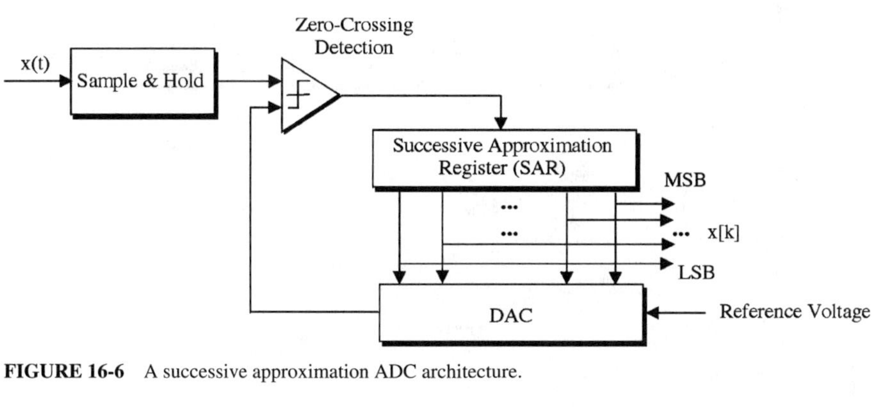

FIGURE 16-6 A successive approximation ADC architecture.

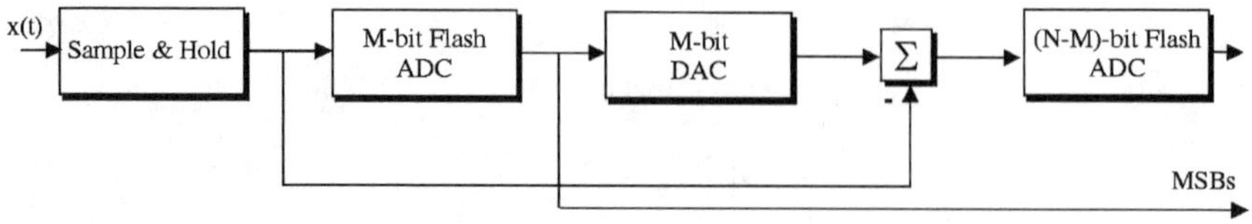

FIGURE 16-7 Subrange ADC architecture.

converter enjoys a significant complexity reduction. However, multiple stages increase the conversion latency and compound the sample-and-hold circuit requirements.

Pipelined ADC. Pipelined ADCs distribute the conversion process over multiple stages using collections of short wordlength ADCs and DACs, as shown in Figure 16-8. Each stage consists of one ADC synchronized to the succeeding ADC. A pipelined ADC requires sample and hold circuits, quantizers, and DACs for all but the last stage.

Folded ADC. A folding or interpolating ADC employs a number of "folding blocks," as suggested in Figure 16-9. The coarse converters produce the most significant bits of the folded circuit output. The least significant bits are defined by a modulo(V) rule. Interpolation is used to reduce the number of folding blocks required to achieve a given precision. A folding block contains N-coupled differential pairs whose output is directly converted by an N-bit ADC. Without interpolation, the folded ADC has the same complexity as a flash ADC.

The following example illustrates the behavior of a typical ΔΣ converter.

Example 16-1 Delta-Sigma (ΔΣ) ADC

Required:
A delta-sigma (ΔΣ) ADC operates at a sample rate f_s and consists of the elements shown in Figure 16-10. They are

- Differencer: $\varepsilon(t) = x(t) - x'(t)$
- Integrator: $i(t) = \int (\varepsilon(t))\, dt$
- 1-bit ADC: $x_s[k] = \text{sign}(i(t))$

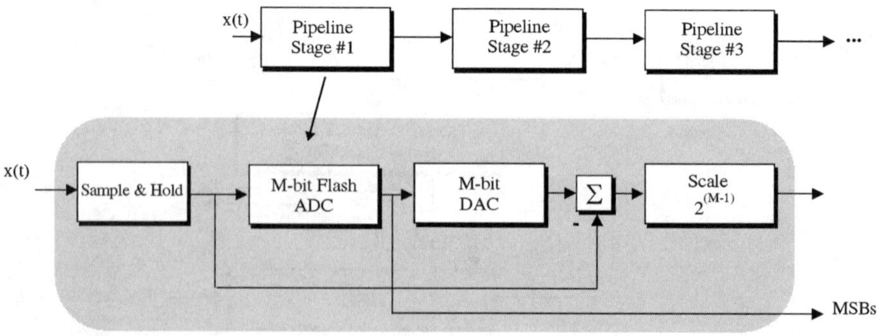

FIGURE 16-8 Pipelined ADC architecture.

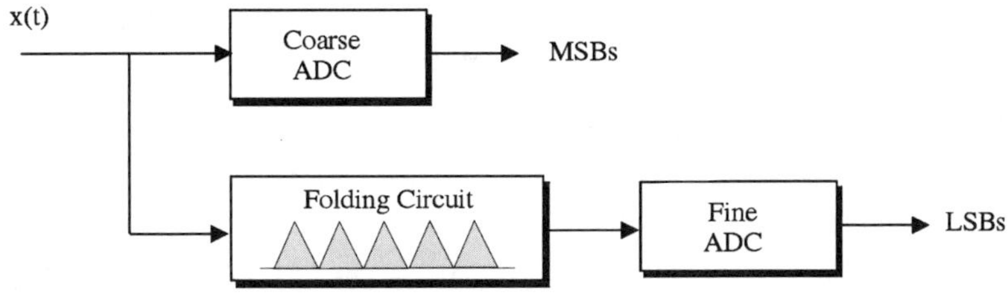

FIGURE 16-9 Folded ADC architecture.

- 1-bit DAC: $x'(t) = \Delta V_{\mathrm{ref}}\, x_s[k] = \pm\Delta V_{\mathrm{ref}}$
- Digital low-pass filter: $x_s[k] \rightarrow x_r[k]$
- Decimator: Down-sample by M [decimation]

The input signal $x(t)$ arrives and is differenced with a synthesized signal $x'(t)$ to form an error signal $\varepsilon(t) = x(t) - x'(t) = x(t)$. The difference signal $\varepsilon(t)$ is then passed through a finite aperture analog integrator that smoothes the error signal, producing what represents the average value of ε(t) over a sliding interval of time. In practice, the smoothing is performed by a "leaky" analog integrator. A so-called 1-bit ADC, running at an accelerated rate of Mf_s, produces an output that can only change $\pm\Delta V$ (that is, the least significant bit (LSB) value) between samples. In other words, the estimate $x'(t)$ is allowed to change only by an amount $\pm\Delta V$ from sample to sample. The averaged error signal provides the feedback information that will cause the error to converge towards zero. Once convergence has occurred, the system will simply "toggle" the LSB to maintain convergence. Finally, it should be noted that the over-sample rate, Mf_s, is returned to its original value f_s using a process called *decimation*.

Suppose, for illustrative purposes, that the $\Delta\Sigma$ ADC input is a constant $x(t) = 8$ and $\Delta V = 1$. Assume that the integrator is modeled as a moving average FIR having an impulse response $h[k] = (1/4)[1, 1, 1, 1]$. Assume further that the output digital low-pass filter is an accumulator. Determine the response of the $\Delta\Sigma$ ADC starting from an original at-rest state.

Results:

The response of the $\Delta\Sigma$ ADC is traced in Table 16-3. It can be noted that initially a large error is present (acquisition phase), which is reduced by the feedback loop. When converged, only small errors are present, requiring simple $\pm\Delta V$ corrections, as shown in Figure 16-11.

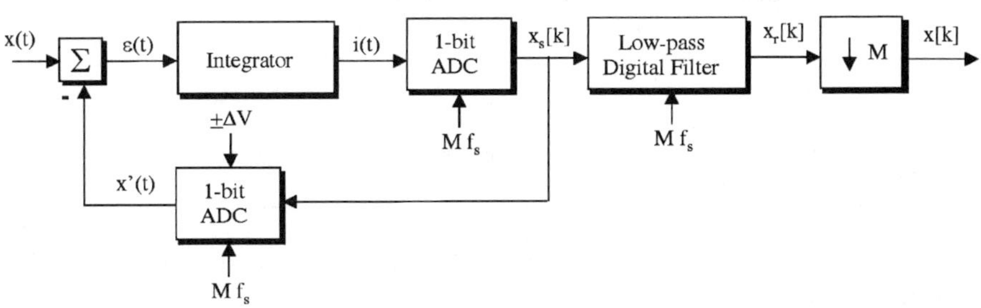

FIGURE 16-10 ΔΣ ADC architecture.

TABLE 16-3 A ΔΣ ADC Example

Sample k	Input $x(t)$	$\varepsilon(t) = x(t) - x'(t^-)$	$i(t)$	$x_s[k]$	$x'(t)$
0	8	8 (initial)	0 (initial)	0 (initial)	**0 (initial)**
1	8	8	8/4 = 2	1	**1**
2	8	7	15/4 = 3.75	1	**2**
3	8	6	21/4 = 5.25	1	**3**
4	8	5	26/4 = 6.5	1	**4**
5	8	4	22/4 = 5.5	1	**5**
6	8	3	18/4 = 4.5	1	**6**
7	8	2	14/4 = 3.5	1	**7**
8	8	1	10/4 = 2.5	1	**8**
9	8	0	6/4 = 1.5	1	**9**
10	8	−1	2/4 = 0.5	1	**10**
11	8	−2	−2/4 = –0.5	0	**9**
12	8	−1	−4/4 = –1	0	**8**
13	8	0	−4/4 = –1	0	**7**
14	8	1	−2/4 = –0.5	0	**6**
15	8	2	2/4 = 0.5	1	**7**
16	8	1	4/4 = 1	1	**8**
17	8	0	4/4 = 1	1	**9**
...	...	...	...	...	**...**

The next example illustrates issues associated with designing a multiplexed ADC system.

Example 16-2 Multiplexed ADC System

Required:

Consider the application illustrated in Figure 16-12. The ADC solution requires 16-bits of resolution over four independent signals, each signal having bandwidths bounded by 15 kHz, 15 kHz, 15 kHz, and 45 kHz, respectively. What is the minimum sample rate that will meet Shannon's sampling criterion.

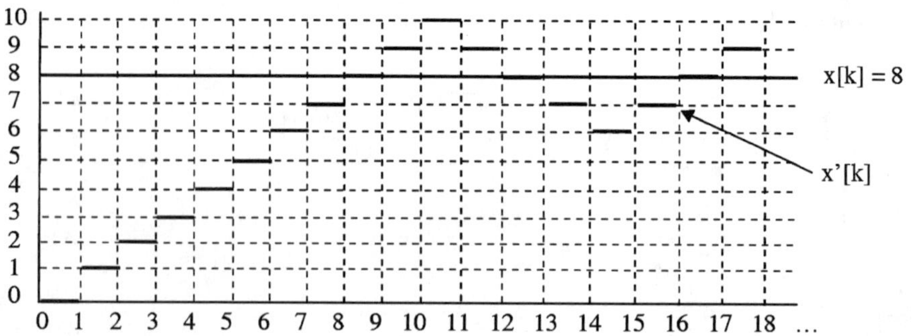

FIGURE 16-11 ΔΣ ADC acquisition and convergence.

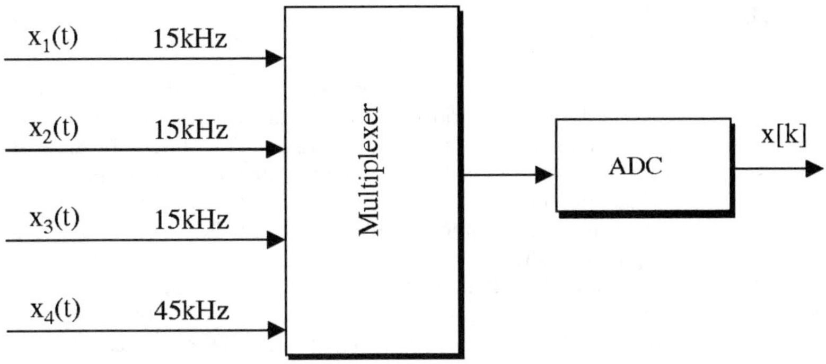

FIGURE 16-12 A multiplexed multichannel ADC system.

Results:
Consider sequencing the multiplexer to output an interleaved time-series { . . . , $x_4[kT_s]$, $x_1[(k+1)T_s]$, $x_4[(k+2)T_s]$, $x_2[(k+3)T_s]$, $x_4[(k+4)T_s]$, $x_3[(k+5)T_s]$. The signal $x_4(t)$ is polled three times more often than $x_1(t)$, $x_2(t)$, and $x_3(t)$. The signal $x_4(t)$ needs to be sampled at a Nyquist frequency of 90 kHz or higher. Since only $\frac{1}{2}$ the samples are of $x_4(t)$, the multiplexed sample rate needs to be 180 kSa/s, or higher.

The next example compares ADC performance based on the type of converted employed.

Example 16-3 ADC Type Selection

Required:
A multichannel problem is considered in Example 16-2. The system is to operate at a 180 k Sa/s rate, and is to be implemented using a successive approximation Analog Device AD974 ADC and/or a delta-sigma ($\Delta\Sigma$) Analog Device AD7722. Both ADCs can operate at sample rates above 180 k Sa/s. Discuss the design options and choices.

Results:
The ADC requirements are summarized in Table 16-4. The SAR ADC accepts each sample as a discrete event, immediately converting the sample into a digital word. Unfortunately, the $\Delta\Sigma$ converter needs to acquire and track an analog signal (acquisition), and produces a reliable output only after convergence. When multiple signals are presented to a $\Delta\Sigma$ ADC, the converter is perpetually in an acquisition mode. As a result, four dedicated $\Delta\Sigma$ ADCs are needed to meet the data conversion requirement which can be satisfied with one SAR converter.

TABLE 16-4 Analog Device's Candidate ADCs

Converter	Architecture	Rated Throughput	Required # of Converters
AD974	Successive Approximation	200 ksps	1
AD7722	Delta-Sigma ($\Delta\Sigma$)	195 ksps	4

16.12 ADC METRICS

Digital data acquisition and conversion systems are found in virtually every modern communication system, DSP solution, electronic instrument, and micro-controller application. As a technology, data conversion systems are evolving at a slow rate compared to mainstream semiconductors. Two data acquisition parameters that are considered key to many applications are speed and precision. For mobile and untethered applications, a third parameter, power dissipation, is equally important. An n-bit ADC is assumed to map an analog sample into an equivalent digital word having a sign bit ($\pm$), I integer bits, and F fractional bits, where $n = I + F + 1$. The output of an ADC is quantized into 2^n possible levels, with each level separated by a quantization step-size $Q = 2^{-F}$. Roundoff errors are assumed to be a uniformly distributed random process having zero mean and variance $\sigma^2 = Q^2/12$. The error variance can also be expressed in bits as:

$$\ln_2(\sigma) = \ln_2(Q/\sqrt{12}) = \ln_2(2^{-F}) - \ln_2(\sqrt{12}) = -F - 1.79 \text{ bits} \qquad (16\text{-}2)$$

The next example demonstrates the process of quantifying fixed point wordlength effects and errors.

Example 16-4 ADC Parameters

Required:

Assume that an analog signal $x(t)$, bounded between $\pm A$ v, is mapped to a digital word with an n-bit ADC. The quantized signal is given by $x_D[k] = Q\lceil x(kT_s)/\Delta\rceil \cdot \Delta] = Q\lceil x_s[k]/\Delta\rceil \cdot \Delta$, where Δ is the quantization step size given by $\Delta = 2A/2^n$, $Q\lceil \bullet \rceil$ denotes quantization (truncation or rounding), and n is the precision of the quantizer in bits. Consider a ± 10 v ($A = 10$) 10-bit ($n = 10$) signed ADC. Determine the quantization step size, integer and fractional field widths, and quantization error statistics.

Results:

From the given data, the quantization step size is $\Delta = 20/1024{\sim}20$ mV/bit. The ADC output corresponds to a 10-bit data word having one sign bit and nine-data bits, where $2^9 \cdot \Delta = 512 \cdot 0.02 \sim 10$ (single-ended range limit). The LSB of the ADC has a weight of $\Delta = 20$ mV. Assuming rounding and a quantization error ($e = x - Q[x]$) that is uniformly distributed over $[-\Delta/2, \Delta/2)$, the error statistics are

$$E(e) = 0 \text{ (mean)}$$

$$\sigma^2 = \Delta^2/12 \text{ (variance)} \qquad (16\text{-}3)$$

where $\log_2(\sigma) = 7.43$-bits. What does -7.43 bits of precision actually mean?

First, the quantization error, in bits, is given by $\log_2(\sigma) = \log_2(\Delta) - 1.79$ bits (Equation (12-10)). From the computed value of $\log_2(\sigma) = -7.43$, first consider removing the statistical bias of 1.7-bits from -7.43. This results in the actual fractional ADC precision being about $(7.43 - 1.79) = 5.64$ bits. Notice also that 5.64 fractional bits covers a dynamic range [0, 1] (i.e., $2^{5.64} \times \Delta = 2^{5.64} \times 0.02 = 0.990442{\sim}1.0$), the range of the fractional bits of the ADC. The output of the $n = 10$-bit ADC can, therefore, be interpreted as having one sign bit, 5.6 fractional bits, leaving about 3.4 integer bits of precision. Observe that $2^{3.4} = 9.849{\sim}10$, the limits of the single-ended ADC's dynamic range. If the data format can be formally expressed as $[\pm{:}\ I \diamond F]$, where $\diamond$ denotes the binary point, I is the number of integer bits, then F is the fractional word width in bits. The system under study would technically have the data format [±: 3.4 ▵ 5.6]. In "*DSP-speak*," the format is represented as $Q(10{:}5.6)$ which is of the form $[N{:}F]$, where n is the word width in bits. The values of n, I, and F are normally integers having values of 16, 24, or 32 bits (for example, TI C2*x*, C5*x*, [16:*x*], Motorola 5600, [24:*x*]). In "TI-speak," the format is also referred to as $Q(5.6)$, where $n = 16$ is implicit in the line of TI DSP μps. For unsigned systems, $n = I + F$ and for signed systems, $n = I + F - 1$ (accounting for the sign bit).

ADC performance and quality can be quantified in terms of:

- Sample rate (Sa/s)
- Analog input (sample and hold) bandwidth
- Power and packaging
- Statistical metric.

The basic quantization error model gives rise to a set of statistical performance metrics shown next, along with a few deterministic measures.

Static Measures	**Dynamic Measures**
Differential non-linearity (DNL)	Signal-to-noise ratio (SNR) $SNR_{max} = 6.02n + 1.76$ dB
Integral non-linearity (INL)	Signal-to-noise + Distortion ratio (SNDR)
Monotonic	Dynamic range
Offset	Effective number of bits (ENOB) $ENOB = (SNDR_{max} - 1.76)/6.02$
Gain error	Spurious-free-dynamic range (SFDR) $SFDR_{max} = 9N–6$ dBc
Latency	(with respect to the fundamental frequency)

In some instances, DACs are integrated into an ADC design. For DACs, the quality metrics include

Static Measures	**Dynamic Measures**
Input/output characteristics	Settling time
Differential nonlinearity (DNL)	Glitch area
Integral nonlinearity (INL)	Signal-to-noise ratio (SNR) $SNR_{max} = 6.02\,N + 1.76$ dB
Offset	Signal-to-noise + Distortion ratio (SNDR)
Gain error	
Latency	

The next example illustrates the SFDR analysis process.

Example 16-5 Spurious-Free-Dynamic-Range (SFDR)

Required:

Spurious-free-dynamic range (SFDR) is shown in Figure 16-13. It is assumed that a pure sinusoid is presented to a converter. The amplitude difference between the ***rms*** value of the fundamental and that of the largest harmonic spur defines the spurious free dynamic range. A 10-bit ADC digitizes a sinusoid whose frequency is chosen to produce a leakage-free FFT spectrum, which is interpreted in Figure 16-13. The non-fundamental harmonics are due to the quantization of the sine wave. Analyze the result.

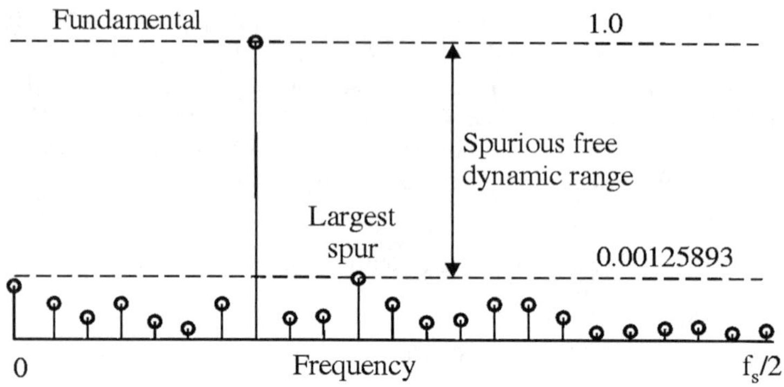

FIGURE 16-13 SFDR interpretation.

Results:
An ideal unsigned 10-bit converter would have, for example, a $20 \times \log(2^{-10}) = -62$ dB signal-to-noise ratio (SNR). The SDR is given by $20 \times \log(0.00125893) = -58$ dB. An SFDR of -58 dB for a 10-bit ADC is considered typical and acceptable.

16.13 ADC TECHNOLOGY ISSUES

One of the problems encountered by electronic device designers when attempting to integrate analog and digital circuitry onto the same substrate is the degradation of analog circuit performance resulting from digital clock current draw and attendant device noise. That is, most of the negative effects that digital circuits have had on analog circuits are caused by clocked Boolean logic (CBL). The circuits result in strong clock-synchronized current draws, glitches, and spikes. This noise is, by nature, impulsive and highly correlated. A standard mitigation technique is called *big A little D* which dictated the minimum use of digital circuitry when being integrated with sensitive analog circuits. Many other techniques have been studied to suppress the undesirable properties of CBL, with the most widely applied being clock gating and the use of multiphase clocks. Another technique used in sampled data applications (for example, data converters) is to perform analog operations during digital "quiet time." This technique becomes less practical at high clock rates. Clock dithering has been applied to continuous time systems with limited success. The dither process provides a limited degree of de-correlation while introducing modulation artifacts, and impacting system performance by increasing constraints on circuit timing. The problem with these techniques is that they attempt to suppress the effects of phenomena that are inherent to CBL, as opposed to eliminating the effects in first place. Economic forces (for example, wireless, entertainment, and the military), are nevertheless requiring engineers to develop truly mixed-mode integrated devices with larger amounts of digital circuits on the same substrate. Properly architected delay-insensitive logic (DIL) can randomize switching noise as well as reduce total noise power, enhancing the performance of mixed-mode devices. Essentially some of the problems of digital switching-induced noise can be mitigated by careful device floor planning and best standard practice layout techniques; however, those techniques have met with only limited success.

16.14 ADC APPLICATIONS

ADCs are now commodities with multiple suppliers vending hundreds of devices with speed ranging from a few hundred hertz to multiple gigahertz. Precision selection also varies widely, ranging from a few bits to over 20 bits. Inexpensive ADCs are facilitating technologies for a wide range of applications, including

- Wireless and cellular systems
- Sensor signal processing and instrumentation
- Beam forming and agile antennas
- Intelligent interfaces (speech, object recognition)
- Entertainment systems (including video and audio)
- Machine and platform control (including military systems)

What's important to understand is that most DSP applications today are ADC limited and not processor bound! This limitation is not always speed. Designing a very low speed ADC (for example, 10 Sa/s for biological signals) can be a very challenging problem. It is generally considered physically unrealistic to design an ADC that operates at very slow conversion rates (<1 k Sa/s) since its implementation would require prohibitively expensive analog sample-and-hold circuits. If sampling at low-frequencies is required, then it is normally recommended that signals be *over-sampled* at a rate consistent with a commercially available ADC. The unwanted samples are then simply discarded. For example, decimating a 10 k Sa/s ADC output by a factor of 1000 would result in an effective 10 Sa/s sample rate. An ADC also naturally defines a trade-off between conversion speed, power consumption, and precision. Implementing ADC conversions at high RF rates is also considered to be impractical to impossible at this time. While 20 G Sa/s ADCs have been developed for use with digital instrumentation, it comes with high power and packaging requirements.[3] In order to operate at low-power, as required in mobile applications, lower than desired conversion rates must often be accepted. An ADC can also be easily overcome with a narrowband jamming signal by using up all the allowable dynamic ADC range. In adjusting the ADC dynamic range to respond to the strong narrowband jammer, the percentage of dynamic range occupied by the actual signal can be insignificant.

All these problems would vanish if high-speed, low-power, and high-precision converter ADCs were available. Such a technology would revolutionize the communications and the electronics industry. High speed is needed to gain closure with high-frequency signals found at the antenna level. Low-power is needed in battery-powered mobile applications. High precision will make these systems more capable and more resilient to jamming, fading, and multipath corruption. The problem is that it may take some time before these promises are fulfilled. It is also argued that a modified Moore's Law exists for ADCs. Whereas Moore's Law for the mainstream digital industry states that digital density will double every 18 months, the precision of a constant speed ADC improves only at a rate of 1.5-bits per nine years. Based on this model, an existing 12-bit ADC with a rated speed would need 24 years to achieve 16-bit status! Moore's Law would also state that the mainstream digital circuitry attached to the ADC improved by a factor of 64k over the same time interval!

The next example explores the problem of undersampling.

[3]Agilent Labs (www.agilent.com)

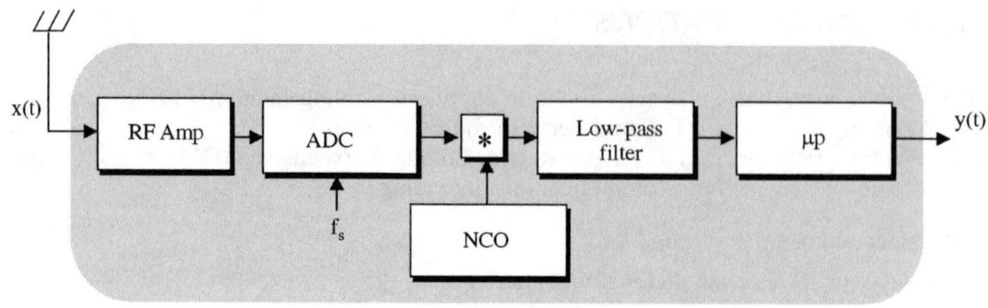

FIGURE 16-14 A typical RF receiver.

Example 16-6 ADC Under sampling RF receiver

Required:

An RF receiver is proposed in Figure 16-14. It consists of an analog RF/IF section that translates an RF signal down to an IF center frequency of 70 MHz with 1 MHz-wide information sidebands. The IF output is presented to a 25.344 Sa/s ADC, which under-samples the analog signal by a factor of 6. The digitized signal is then translated down to DC by mixing the ADC output with the output of a digital numerically controlled oscillator (NCO). The information is then isolated by a digital low-pass filter, and then processed by a back-end microprocessor. Analyze the spectra at various parts of the system and analyze the feasibility of the under-sampled receiver.

Results:

The spectrum found at the output of the ADC and beyond is shown in Figure 16-15. The base band extends over the frequency range $f \in [-12.667 \text{ kHz}, 12.667 \text{ kHz})$. It can be observed that information sidebands are aliased (folded) back to base band. If the NOC is set to 6.032 MHz, the information spectrum is translated down to 0 Hz and $\pm$12.064 MHz. The low-pass digital filter passes the $\pm$1 MHz spectrum centered about 0 MHz. There is, however, a caveat that goes along with this method called *IF sampling*. In order for the samples produced by the ADC's sample-and hold circuit, the analog circuit needs to

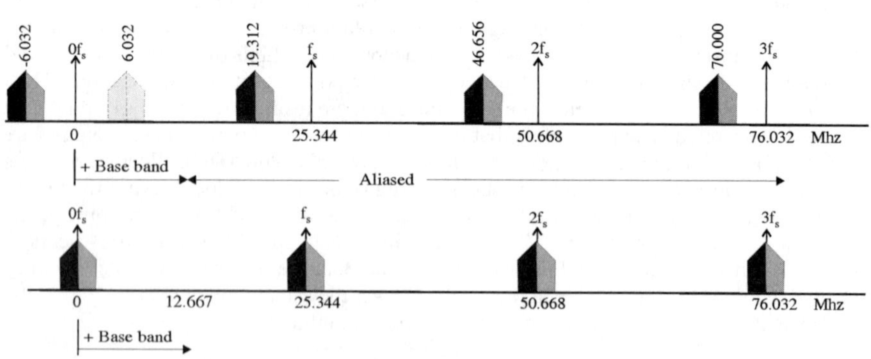

FIGURE 16-15 An illustration of an under-sampled RF system. Notice that in the top figure, the IF band is aliased down to $f_0 = f \bmod(f_s) = 70 \times 10^6 \bmod (76.032 \times 10^6) = -6.032 \times 10^6$ Hz. The base-band information spectrum centered about $+6.032 \times 10^6$ Hz is actually due to the aliasing of the -70×10^6 Hz IF spectrum. Upon modulating the under-sampled IF signal with a 6.032 MHz sinusoid produced by the NCO, the spectrum shown in the bottom panel of the figure results.

have at least a 70 MHz bandwidth. If the sample-and-hold circuit had a bandwidth consistent with a 25.344 MHz ADC, the IF signals would be distorted because of excessive smoothing. IF-sampling ADC manufacturers provide a product that has a high analog bandwidth (far higher than is required by a slower quantizer). The solution could also have been achieved with a 70 M Sa/a class ADC, whose output is reduced to a sample rate of 25.344 MHz using decimation.

16.15 ADC ENHANCEMENTS

Dithering is a means of extending the precision of an ADC device. The object of dithering is to keep the least significant bits of an ADC in a constant state of change. To illustrate, consider the case where the input signal is $x(t) = K\,\Delta + \frac{1}{2}\,\Delta$ where Δ is the value of the ADC's LSB (volts/bit), for K an integer. The ADC output is the binary-coded version of K and the error would be a constant $\frac{1}{2}$ LSB or $\Delta/2$. Next, assume that a small amount of noise $n(t)$ is added to the input signal, producing a new signal $x'(t) = K\Delta + \frac{1}{2}\,\Delta + n(t)$ where $n(t)$ is normally distributed with a standard deviation of $\sigma = 2\Delta/3$. The LSB of the converter's output will now be continuously perturbed. Half the outputs are now K and the other half, $K + 1$, with the average error converging to zero. This is a $\frac{1}{2}$ LSB change in favor of what is referred to as an *additive dither solution*. Figure 16-16 describes this process in a more general context.

Figure 16-16 presents a case where a signal is first shown being quantized with the output generally falling at, or near, an integer value of 1024. Next, a small amount of normally distributed noise is added to the signal before conversion ($n(t) = N(0, (2\Delta/3)^2)$). The quantized output is now seen to be in a dynamic state of change. The result is a slight increase in the noise floor (3 dB typical) and a proportionally larger improvement in the total harmonic distortion (THD). A variation on this theme is to high-pass filter the dither noise before adding it to the input signal, or use a triangularly distributed noise instead of a normally distributed dither. These actions assist in decoupling the dither from the input signal.

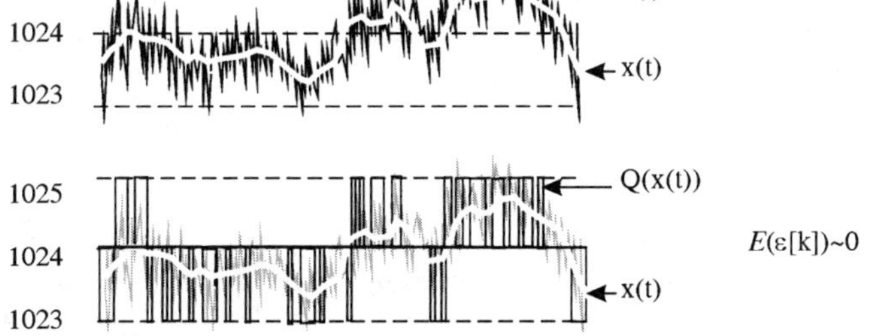

FIGURE 16-16 ADC dithering beginning with the original signal (top), dithered signal (middle), and quantized signal (bottom).

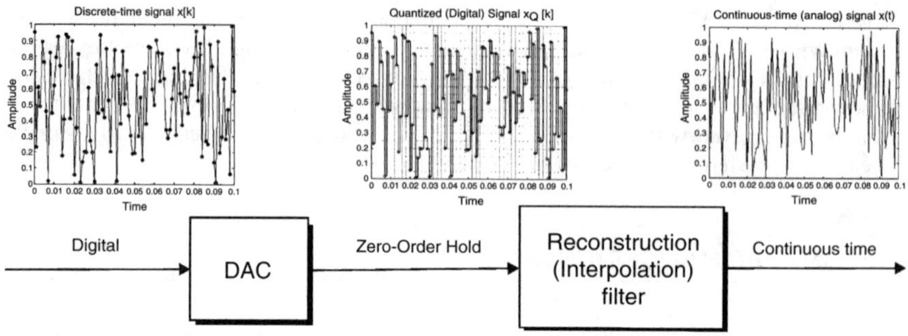

FIGURE 16-17 A typical DAC interpolation system.

Another version is called the *subtractive dithering*. In the subtractive method, dither is added to the signal before quantization, the dithered signal is then quantized, and the additive noise is subtracted from the quantized signal (regeneration). The benefits are improved linearity and noise floor over the additive scheme, but at the high cost of added complexity. Whether additive or subtractive, dithering is generally used only in high-end (typically audio) applications.

16.16 DAC TECHNOLOGY

In concept, a string of quantized samples can be read from memory and directly converted into a discrete-time signal. The quantized values can then be electronically translated into discrete-time analog levels using a simple resistor-ladder circuit. The resulting discrete-time signal, however, does not in general qualify as an acceptable continuous-time signal. A better translation can be achieved by having the sample value's (signal level) held constant for an entire sample period. This operation is called a *zero-order hold* and produces a piecewise constant analog signal. In fact, most digital-to-analog converters (DAC) operate as a zero-order hold device, as shown in Figure 16-17. The zero-order hold output lacks the smoothness required of many applications. Additional smoothing can be obtained by presenting the DAC's zero-order hold output to a simple analog low-pass interpolating filter, as displayed in Figure 16-17. The low-pass *interpolation* or *reconstruction* filter "smoothes" the DAC output and removes irregularities. To illustrate, consider Figure 16-17 where the DAC output samples are then sent to an analog low-pass filter that "smoothly" connects together a string of sample values. For over-sampled signals, the smoothing filter can have strong low-pass tendencies. As the sample rate decreases, the passband of the smoothing filter needs to be proportionally increased. A filtered signal reconstruction is illustrated in Figure 16-18.

16.17 DSP SOFTWARE

As programmable DSP hardware becomes increasingly capable and robust, the rate at which they are being embedded into both high- and low-end products is accelerating (for example, software-defined radio). Today, and into the foreseeable future, these DSP solutions are

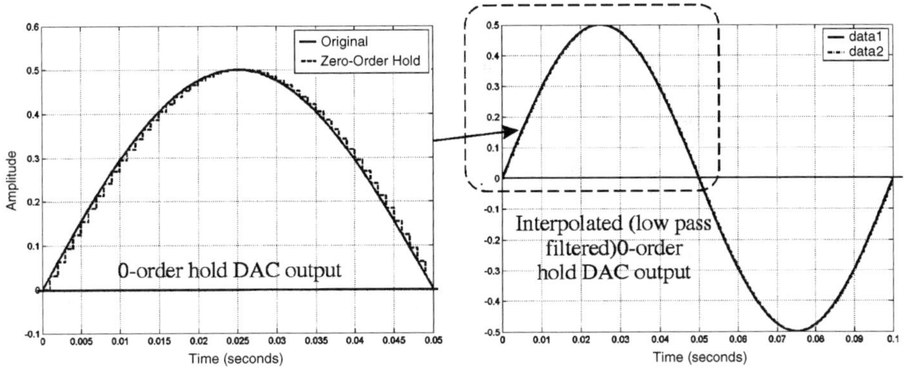

FIGURE 16-18 An interpolation demonstration for a DAC output reconstruction. Shown are a sine wave $x(t)$, zero-order hold values (left), and its reconstructed signal (right).

being orchestrated by software. DSP software solutions generally differ from general-purpose solutions. For instance, software-enabled DSP solutions must be able to operate at real-time data rates and are SAXPY-intensive. The need for specialized and efficient software development tools and environments are self-evident. These programming environments are reviewed in this section.

Assembly Language

Historically, it was assumed that hand-crafted optimized assembly code was the primary means of maximizing the performance of a DSP processor. The control provided by assembly code can, in concept, maximize the capabilities of a processor and often allows a lower-cost slower processor or part to out-perform a more costly device. The historical problem with this approach is that extensive prerequisite programming skills and experience are required to successfully complete an assembly-level solution. Many of these skills can only be acquired by years of in-field experience. Compounding these problems are barriers that limit code migration and portability from one processor to another, or one programmer to another. Maintaining assembly level code, the lack of prototyping tools, and limited debugging support exasperate the problem. As a result, most modern DSP code is developed at a higher level, using assembly-level programming to refine and optimize those parts of the code which continue to be problematic.

C Language. Today, there are several factors that draw DSP programmers from assembly-level activities to high-level languages (HHL). First, programmable DSP processors are becoming increasingly capable of implementing complicated solutions that raise serious development and support questions if implemented using low-level languages. Employing high-level languages and development environments can often overcome this problem. Secondly, time-to-market pressures force programmers to sacrifice performance in order to rapidly implement a design. The logical high-level language choice has become C, which now serves as a de facto standard. The original motivation to adopt C was due to the fact that the language was mature, well known to many programmers, and in the public domain. C can, however, obscure optimization opportunities making it difficult for compilers to target specialized processors. In addition, standard C did not support fixed-point arithmetic, divided memory, or circular buffering. Fortunately, these early shortcomings were fixed. Compiler developers have added "intrinsic" functions to the compiler that converts specific operations directly into efficient assembly code. The C language has also been extended with the addition of specialized data types and constructs that closely match the demands of DSP processors. An example is Embedded C. Embedded C adds fractional data types, saturation arithmetic, and multiple memory spaces to C. Still another innovation is C++ which, compared to C, is a more complex and powerful language with strong object-oriented programming support. Embedded C++ is an attempt to standardize C++ for embedded DSP applications. In the process, it removed some features from C++, such as multiple inheritance, exception handling, templates, and namespaces. Texas Instrument's TMS320C6x C/C++ compiler, for example, supports Embedded C++ but also retains support for C++ namespaces. The programming environment is illustrated in the following example.

Example 16-7 Code Composer

Required:

The primary programming environment for high-end Texas Instruments DSP μps is called Code Composer. It is an integrated environment and is shown in Figure 16-19. It consists of editing, compiling, assembly, and linking products that can export code which can be used to simulate DSP μp performance, develop and test prototype designs, or be targeted to products under development or that are acquired. Evaluate the software system's ability to develop efficient code.

Results:

While compiled code may not run as efficiently as expertly crafted assembly code, it is generally good enough. The veracity of this claim is based on the quality and features of the resident compiler, and whether or not it is targeted for use with a processor of interest (see Table 16-5). Examples, based on a TI TMS320C6713, are shown in Table 16-5. While compiled C code may not run as fast or as compactly as assembly code, C continues to narrow the difference.

In another Texas Instruments test of C versus assembly-level programming, the results shown in Table 16-6 were obtained. The data indicates that runtime performance need not be sacrificed by developing solutions in a high-level language, especially in the case where the programmed objects are mainstream DSP operations.

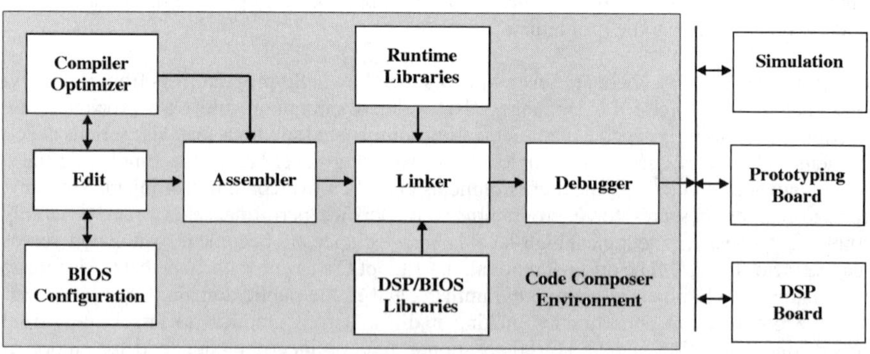

FIGURE 16-19 The Code Composer software development environment.

TABLE 16-5 TMS320C62 compiler benchmarks

Algorithm	Application	ASM Cycles	ASM Time	C Cycles	C Time	% Efficiency
Mean squared error (image)	Motion compensation	348	1.16 μs	402	1.3 μs	87%
Codebook search	CELP voice coders	977	3.26 μs	961	3.20 μs	100%
Vector max. (40 elements)	Search algorithms	61	0.20 μs	59	0.20 μs	100%
10th-order FIR 40 input samples	VSELP voice coders	238	0.79 μs	280	0.93 μs	85%
2304 array search	Search algorithms	1185	3.95 μs	1318	4.39 μs	90%
16th-order IIR	Filter	43	0.14 μs	38	0.13 μs	100%
Five-stage cascaded IIR	Filter	70	0.23 μs	78	0.25 μs	93%
MAC dual 40 sample arrays	VSELP voice coders	61	0.20 μs	58	0.10 μs	100%
Vector sum dual 40 sample arrays	General	51	0.17 μs	47	0.16 μs	100%

Source: Texas Instruments.

It is generally accepted that code can be developed in a high-level language faster than it can at the assembly level. Another advantage gained by using high-level languages is access to powerful debugging and maintenance tools. Errors found by a compiled code are more easily isolated, identified, and repaired, while errors in assembly code can sometimes elude detection, appearing at the most inopportune times. C-based code is also highly profitable and reusable. Integrating software objects in the form of 3rd-party IP is more easily facilitated within a high-level language-defined environment such as C. Libraries are also abundant, simplifying the programming process. This process and collection of resources are ultimately leading to the implementation of DSP objects, such as digital filters.

A core DSP filter operation is the implementation of a linear convolution call of the form $y[k] = h[k]*x[k]$. For the case of an FIR filter, the linear convolution sum can be expressed as

$$y[k] = h[k] * x[k] = \sum_{i=0}^{N-1} h[i]x[k-i] \tag{16-4}$$

TABLE 16-6 TMS320C6000 compiler benchmarks

Algorithm	Assembly Cycles	C cycles	% Efficiency
Codebook search (CELP voice coder)	977	961	102%
FIR (L = 40)	238	280	85%
IIR (L = 16)	43	38	113%
IIR (cascade, ten sections)	70	75	93%
MAC (L = 40)	61	58	105%
Vector sum (L = 44)	51	47	109%

Source: Texas Instruments.

which can be expressed in C-language as

```
y[k] = 0.0;
for (i = 0; i < N; i++)
y[k] = y[k] + h[i]*x[i-k];
```

It is claimed that this common C code is inefficient. The reason for this claim is

- The variable *y*[*k*] is accessed multiple times during a filter cycle.
- Accessing the array element *h*[*i*] with the index *i* was inefficient.
- The production of the index [*i*–*k*] requires arithmetic intervention.

Direct array accessing a sequentially stored data from an array is, in general, sub-optimal. Since the compiler only has knowledge of the address of the leading element array, additional arithmetic labor is required to generate the actual desired memory address. Specifically, the following operations are required to generate an address index:

- Load the array's leading address
- Load the value of counter index *i*
- Load the value of the sample index *k*
- Compute the offset *i*–*k*
- Add the offset to the starting address

A more efficient method of accessing data sequentially stored in an array is through the use of a pointer. Pointers provide the C language programmer with the ability to efficiently access array elements.

In C, the syntax **ptr* indicates that *ptr* is a pointer. This in turn means that

- The variable *ptr* is declared to contain an address.
- The "*" operator means the data is to be read from that address.

Pointers can be modified after the data has been accessed. The syntax **ptr++* means

- The variable *ptr* is declared to contain an address.
- The "*" instructs that the data be read from that address.
- The "++" states that, after having read the data, the pointer ptr is auto-incremented to point to the next sequential data element.

To illustrate, consider the code segment:

```
float *y_ptr, *h_ptr, *x_ptr;   {q_ptr is to be
treated as address pointer}
y_ptr = &y[0];                  {one time initiation of
the point}
for (i = 0; i < N; i++)
  *y_ptr = *y_ptr + *h_ptr++ * *x_ptr--;
```

Each pointer needs to be initialized only once, eliminating the need to perform offset arithmetic within the loop. The pointers are then automatically incremented or decremented.

DSP operations can become memory-intensive resulting in possible bottlenecks. Many modern DSP processors make multiple memory accesses a single instruction cycle. The inner loop of the FIR program requires two data memory reads and one data write to memory per cycle. Such activity can exacerbate memory contention and bandwidth problems. To respond to the problem, DSP processor architects have added arrays of registers that are closely coupled to the processing units. As a result, data supporting the inner loop calculations can be stored locally. This can be accomplished using the following code:

```
register float temp;
temp = 0.0;
for (k = 0; k < N; k++)
temp = temp + *c_ptr++ * *x_ptr--;
```

The C declaration "register float temp" establishes that the variable temp is to be held in a processor register. The inner loop therefore requires only two memory accesses to read the two operands **c_ptr* and **x_ptr*. The formal initialization of temp to zero can be bypassed by implementing the following code:

```
register float temp;
temp = *c_ptr++ * *x_ptr--;
for (k = 1; k < N; k++)
temp = temp + *c_ptr++ * *x_ptr--;
```

The inner loop can then be used to develop an efficient C implementation of an FIR filter:

```
float y[N], c[N], x[N];
float *y_ptr, *c_ptr, *x_ptr;
register float temp;
int n, k;
y_ptr = &y[0];
for (n = 0; n < N-1; n++) {
c_ptr = &c[0]; x_ptr = &x[N-1];
temp = *c_ptr++ * *x_ptr--;
for (k = 1; k < N; k++)
temp = temp + *c_ptr++ * *x_ptr--;
*y_ptr++ = temp;
}
}
```

The execution of a single cycle of this program is seen to involve a multiply and accumulate. The remainder of the program involves loop control, initialization, and data I/O operations. These are operations that are intrinsic to DSP and DSP processors. Over time, DSP hardware and software technologists have optimized these operations. Pipeline operations are illustrated in the following example.

Example 16-8 Software Pipelining

Required:
The core DSP operation is an arrayed multiply-accumulate that can be modeled as an inner product:

$$y = \sum_{i=1}^{L} a_i x_i \tag{16-5}$$

Any machine capable of implementing Equation (16-5) can implement Equation (16-3). High-level software can interpret such operations as pipelining opportunities. Specifically, software pipelining refers to a process by which multiple data streams are interlaced to make maximum use of available hardware resources. Compare a classical Von Neumann implementation of a five-term inner product executed on a TI TMS320C6000-class machine having an architecture described in Figure 16-20. The 6000 family includes DiVinci media processors (for example, 320DM64), fixed-point DSP μps (such as 320C64), and floating-point DSP μps (say, 320C67). The processor is linked to external memory and an assortment of signals through buses. The processor also contains a block on internal memory that is organized along functional lines. The processor shown contains two complete processors (multiprocessors) consisting of registers, and the following function units.

- D = Data Unit
- M = Multiplier unit
- L = ALU (accumulator) unit
- S = Shift unit

The compiler can orchestrate how these units function, as well as encourage data locality during runtime.

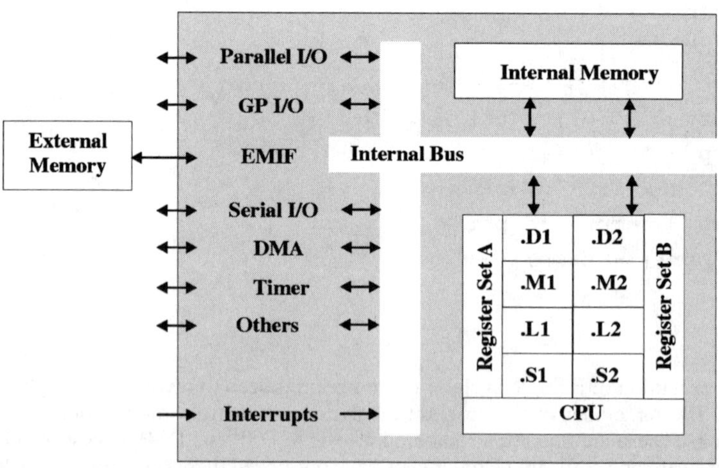

FIGURE 16-20 The TI TMS320C6000-class processor architecture.

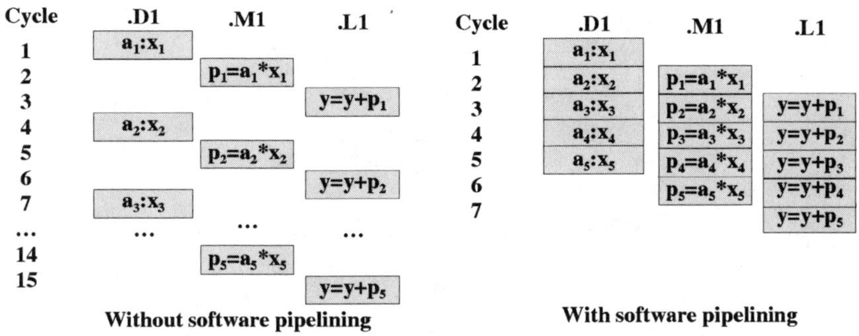

FIGURE 16-21 An illustration of software pipelining.

Results:
Without software pipelining, the operational behavior of the processor is shown in Figure 16-21 and is seen to have an execution latency of 15 cycles. With pipelining, also shown in Figure 16-21, the execution time is reduced to seven cycles.

The study of software pipelining continues in the next example.

Example 16-9 Software Pipelining

Required:
Develop a code that will implement the inner product described by Equation (16-5).

Results:
In the context of the C6000 series processor, the relevant commands are

- MPY (multiply using the .M unit)
- ADD (add using the .L unit)
- SUB (subtract using the .L unit)
- B (program looping control using the .S unit)
- MVK (initialize a counting loop using the .S unit)
- LDH (data load using the .D unit)
- STW (data store using the .D unit)

The data load options are

- LDB—load byte (8 bits)
- LDH—load half-word (16 bits)
- LDW—load word (32 bits)
- LDDW—load double word (64 bits)

The required code for the case where the data are half words is shown in Figure 16-22.

Prototyping Languages. Historically, interpretive languages like MathWorks' MATLAB have been used as a rapid prototyping tool, producing high-level simulation, and used in algorithm development. These tools can provide an environment that can be used to

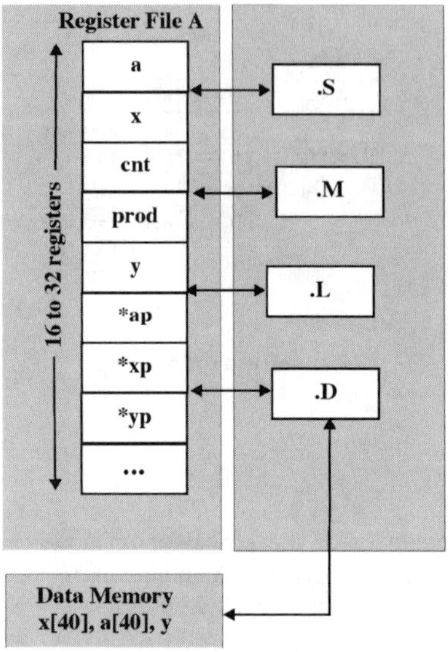


```
        MVK   .S   40,cnt
loop:
        LDH   .D   *ap++, a
        LDH   .D   *xp++, x
        MPY   .M   a, x, prod
        ADD   .L   y, prod, y
        SUB   .L   cnt, 1, cnt
[cnt]   B     .S   loop
        STW   .D   y, *yp
```

FIGURE 16-22 An illustration of a 40th-order FIR implement.

synthesize, analyze, and visualize a candidate solution. Many of these software systems can support fixed-point studies and can be integrated with a user interface having a data-flow or block-diagram motif (for example, Simulink). In some cases, the prototype code can be automatically translated into C code that, in turn, can be targeted for a particular DSP processor (for example, Simulink and Real-Time Workshop). Solutions can be tested using simulation, as well as studied discretely in a single-step clock mode. The efficiency of the code produced using this strategy is reported to be highly variable and is application-dependent.

Libraries. Family-specific function libraries (algorithm libraries) are collections of software expressions of a common DSP algorithm. Technology providers generally offer free libraries to customers that support basic signal-processing functions (for example, FFT) with some license restrictions as a marketing device. Third-party IP developers market function-specific libraries targeting a particular market sector or application domain (such as JPEG compression). Their use typically requires a license. As a general rule, it should be assumed that 90 percent of an original product solution should be derived from the use of pre-defined code or legacy code. Product differentiating features therefore become the remaining 10 percent of the code development.

Reference Designs. There is an old adage that "a good engineer is a lazy engineer because he/she will always find a simple way to solve a complicated problem." Indeed, many designers would like to create a solution with a minimum need to develop code. This is the role of a reference design, which is a package that integrates the key software and hardware components required to complete a solution. Reference designs can serve as starting points, or provide an entire solution with possibly minor modifications. The result is serious cost and time-to-market savings. There are several caveats, however. First, the final solution is predicated on the quality of software and design decisions supplied by others, resulting in a loss of intellectual control. Second, the reference design usually represents a particular solution strategy which may, in the context of the customer, be sub-optimal or self-limiting. Thirdly, vendor-supplied reference designs maximize the solution's use only if the design contains vendor-supplied components.

Profiler. High-end DSP processors are becoming increasingly complex, exacerbating optimization attempts. Multiple processor and data paths can easily become over-subscribed, producing what are called "hot spots." Profilers are tools that can display numerically and/or graphically the internal system load. This information can be used to "tune" the code in an attempt to balance and optimize data flow. This is typically an iterative process which can involve manipulating compiler options, selecting different algorithm forms, plus other details.

16.18 DIGITAL FILTER IMPLEMENTATION

The design of digital filters has been greatly simplified through the release of hundreds of software design packages. Some are highly integrated, while others focus on a single class of filter. These tools accept design requirements in text or graphic form, process the data, and produce a set of filter coefficients. This process is generally straightforward, but does not completely define a filter solution. The synthesized filter needs to be implemented, which requires an architecture assignment followed by the filter's physical implementation in hardware, firmware, or software. Besides selecting an architecture and target implementation technology (for example, microprocessor, DSP mp, ASIC, FPGA), the designer needs to insure that the final outcome meets all posted requirements in terms of

- Filter frequency and time-domain behavior
- Real-time throughput
- Power dissipation
- Packaging
- Cost

While synthesizing the filter coefficients may take seconds, insuring that the overall design requirements are met can be far more demanding. Through the intelligent use of analysis software and analysis techniques, filters can be successfully architected so as to maximize runtime speed and precision, and minimize the expectation of register overflow.

Other decisions that affect this process are

- Developing the solution in-house
- Developing the solution based on external IP
- Developing the solution based on custom design and IP integration

Once a filter has been synthesized, typically using high-level DSP design software (for example, MATLAB), then an architecture needs to be specified and the filter implemented. For the case of an FIR, the architectural choices are normally:

- Direct FIR
- Transpose FIR
- Symmetric FIR
- Distributed Arithmetic FIR

Direct and transpose forms are the most area efficient, but performance degrades with filter order. A symmetric FIR can reduce the physical complexity of the filter for those cases where the filter coefficients are symmetrically distributed (for instance, a linear-phase FIR). The use of distributed arithmetic can result in the highest performance design, but performance rapidly degrades with word width. In practice, the implementation of such basic DSP operations, like an FIR filter, is based on tested intellectual property (IP) code obtained from the technology supplier or third parties. Using IP software, designs can be rapidly compiled and tested. The problem with this paradigm is that hand-crafted code can often reuse on-chip resources that solutions based on a set of independent IP code packages can not.

Digital filters, whether FIR, IIR, or multirate, can be implemented in software or hardware. Maximum performance is gained when the software solution is executed on DSP-capable hardware. An example is a Texas Instruments (TI) TMS320C6x DSP μp, a VLIW fixed-point processor having multiple processor cores. The internal structure of the processor is motivated in Figure 16-20. The dual multiply-accumulate units (MAC) can accelerate SAXPY execution. The implementation of a typical FIR digital filter is presented in the next example.

Example 16-10 DSP μp FIR Implementation

Required:
Design and implement an FIR using a DSP μp.

Results:
MathWorks' MATLAB provides two filter design tools for the creation of both FIR and IIR filters. One tool is called the Signal Processing Toolkit, or SPTool, and the other is called the Filter Design and Analysis Toolkit, or FDATool (see Figure 16-23). Using the FDA Toolkit, FIRs can design a filter and export the coefficients as a C header file (see Figure 16-24).

The coefficients, saved in file *lp1000.h* can be integrated into the FIR program shown next.

```
/* FIR Implementation */

#include "dsk6713.h"
#include "dsk6713_aic23.h"
#include "lp1000.h"
```

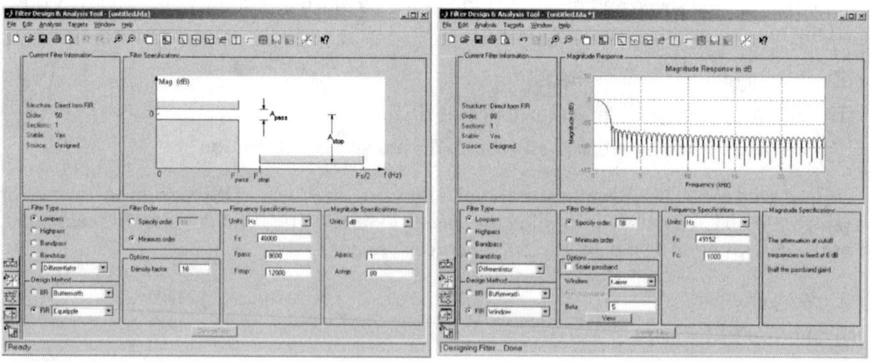

FIGURE 16-23 The FDA Toolkit environment showing the design screen (left) and a partially completed design (right).

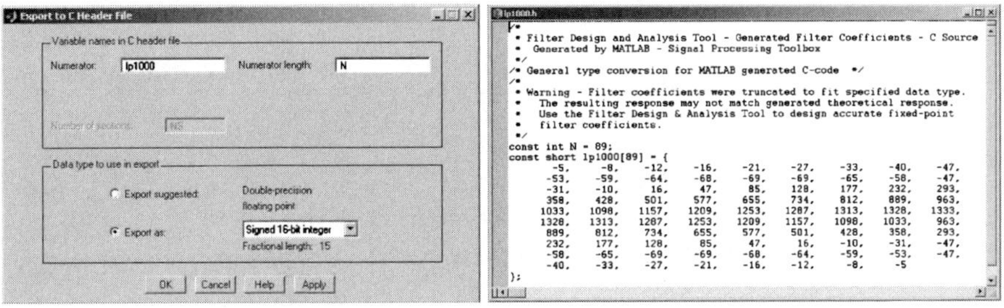

FIGURE 16-24 The FDA Toolkit environment coefficient exportation screen (left) and some sample output (right).

```
// Globals

static int leftSample, rightSample;
int leftChanOut = 1;
int leftChanIn = 1;
int yn = 0;          // initialize filters output
short dly[89];       // delay samples
int START = 0;

/*=============Config Settings and Handle
Declerations==============*/
// Codec configuration settings - DEFAULT
DSK6713_AIC23_Config config = {
    0x0017,  /* 0 DSK6713_AIC23_LEFTINVOL  Left line input
channel volume */
    0x0017,  /* 1 DSK6713_AIC23_RIGHTINVOL Right line input
channel volume */
    0x01f9,  /* 2 DSK6713_AIC23_LEFTHPVOL  Left channel
headphone volume */
    0x01f9,  /* 3 DSK6713_AIC23_RIGHTHPVOL Right channel
headphone volume */
    0x0011,  /* 4 DSK6713_AIC23_ANAPATH    Analog audio path
control */
    0x0004,  /* 5 DSK6713_AIC23_DIGPATH    Digital audio
path control */
    0x0000,  /* 6 DSK6713_AIC23_POWERDOWN  Power down
control */
    0x0043,  /* 7 DSK6713_AIC23_DIGIF      Digital audio
interface format */
    0x0081,  /* 8 DSK6713_AIC23_SAMPLERATE Sample rate
control */
    0x0001   /* 9 DSK6713_AIC23_DIGACT     Digital
interface activation */
};
```

```
// Declare BSL Handle for AIC23 Codec
DSK6713_AIC23_CodecHandle hCodec;
/*=============================================================
============*/

/*==============main()
Routine=========================================*/
//  Initializes BSL and hardware interrupts

void main()
{
     // Initialize the BSL
     DSK6713_init();
     // Open the Codec
     hCodec = DSK6713_AIC23_openCodec(0, &config);
     // Set the sampling frequency
     DSK6713_AIC23_setFreq(hCodec,
DSK6713_AIC23_FREQ_48KHZ);
     // Enable the McBSP interrupt for IRQ_EVT_XINT1
     IRQ_enable(IRQ_EVT_XINT1);
     // Enable the McBSP interrupt for IRQ_EVT_RINT1
     IRQ_enable(IRQ_EVT_RINT1);
     return;
}
/*=============================================================
============*/
/*==============HWI
Routines=========================================*/

// outHWI - The Interrupt Service Routine called when the
McBSP wants more data
void outHWI(void)
{
     if (leftChanOut)
     {
            short i;
            leftChanOut = 0;

            if (START == 1)
            {
                 dly[0] = leftSample;  // newest sample @
top of buffer
                 yn = 0;          // initialize filter
output
```

```
                    for (i = 0; i < 89; i++)
                          yn += (lp1000[i] * dly[i]);
     //y[n] += h[i]*x[n-i]
                    for (i = 88; i >0; i--)     //
starting @ bottom of the buffer
                    dly[i] = dly[i-1];          // update
delays with data move
              while(!DSK6713_AIC23_write(hCodec, yn>>15));
// output shift by 16
         }
         else
              while(!DSK6713_AIC23_write(hCodec,
leftSample));
     }
     else
     {
          leftChanOut = 1;
          rightSample = 0;
          while(!DSK6713_AIC23_write(hCodec, rightSample));
     }
}


// inHWI - The Interrupt Service Routine called when the
A/D wants more data
void inHWI(void)
{
     if(leftChanIn)
     {
           leftChanIn = 0;
           while(!DSK6713_AIC23_read(hCodec, &leftSample));
     }
     else
     {
           leftChanIn = 1;
           while(!DSK6713_AIC23_read(hCodec,
&rightSample));
     }
}
/*=====================================================
============*/
```

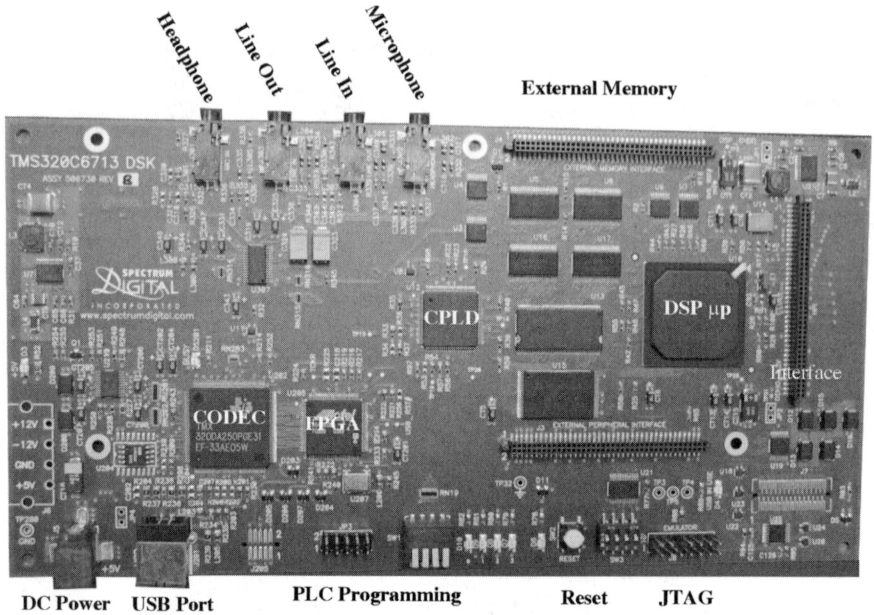

FIGURE 16-25 A C6713 DSK board.

The code can be executed in an emulation mode, or to a targeted processor, development system, or board. For example, the FIR filter can be downloaded and executed immediately on a Texas Instruments C6713 DSK board (see Figure 16-25).

The C6713 DSK environment includes the following assets.

- C Compiler
  - Code Composer is a C-to-ASM compiler designed for use with the TI chipsets.
  - The Code Composer compiler also includes optimization tools to balance code size and performance to meet the application needs.
- C6000 Signal Processing Libraries and Peripheral Drivers
  - Chip Support Libraries
  - TI includes CSLs to function all of the on-board peripherals of the chip.
- Board Support Libraries
  - The DSK includes drivers and functions to run all of the on-board hardware easily. If a custom board was designed, TI has specific hardware that the drivers are already written.
- DSP Libraries
  - TI has available for FREE download DSP functions that are pre-optimized for each chipset. These functions include FFTs, FIR, IIR, matrix math, and so on.

In addition, the DSK also provides access to the following resources.

- Timer/Counters
  - Two 32-bit timer/counters
  - Can generate interrupts to the CPU, events to DMA/EDMA, and can pulse or toggle-value on output pins

- Hardware Interrupts (HWI)
  - Four configurable external interrupt pins
  - One non-maskable interrupt
  - Reset pin
- General-Purpose Input/Output (GPIO)
  - Observe or control the signal of a single pin
  - Dedicated pins on the '6713
- External Memory Interface (EMIF)
  - Access to synchronous or asynchronous memory
  - Works with PC100 SDRAM
  - Byte-wide data access
- Multi-Channel Buffered Serial Port (McBSP)
  - Used to connect serial codecs (such as combined A/D and D/A converters)
  - Two synchronous serial ports
  - Full duplex
  - Up to 100-Mbps performance
- TMS320C6713 DSP
  - 225 MHz, floating point, 256 kB internal RAM
- External SDRAM
  - 16 MB with a 32-bit interface
- External Flash
  - 512 kB, 8-bit interface
- AIC23 Codec
  - Stereo, 8 kHz–96 kHz sampling rate, 16- to 24-bit samples, mic, line-in, line-out, and speaker jacks
- CPLD
  - Used for "glue" logic
- Four User LEDs
  - Writable/readable through the CPLD
- Four User DIP Switches
  - Readable through the CPLD

The implementation of a FIR digital filter with an FPGA is considered in the next example.

Example 16-11 FPGA FIR Implementation

Required:

Design and implement an FIR using an FPGA.

Only some of today's FPGAs come equipped with embedded general-purpose multipliers. The multiplier-less FPGA, if they can be employed, offer the most cost-effective FPGA design opportunity. They can accept FIR designs in several modes. One common method is called distributed arithmetic which performs multiplication using a set of table lookups. Another approach is based on synthesizing a multiplier (scaling) using binary-valued (2^i) multiplication. Illustrate this method by implementing an FIR having an impulse response $h = \{3.5, -2.25, 1.125, -1\}$.

Result:

The implementation of a basic FIR using a multiplier-less FPGA can be achieved with a set of binary shift operations at the register level. The following pseudo-code illustrates this type of FIR realization.

$$H = \{h[0] = 3.5, h[1] = -2.25, h[2] = 1.125, h[3] = -1\}$$

$$X = \{x[k-3], x[k-2], x[\mathrm{k}-1], x[k]\}$$

$$y[k] = HX^T = h[0]\cdot x[k-3] + h[1]\cdot x[k-2] + h[2]\cdot x[k-1] + h[3]\cdot x[k]$$

$$y \leftarrow 2^1\cdot x[k-3] - 2^1\cdot x[k-2] + 2^0\cdot x[k-3] + 2^0\cdot x[k-1] - 2^0\cdot x[k]$$
$$+ 2^{-1}\cdot x[k-3] - 2^{-2}\cdot x[k-2] + 2^{-3}\cdot x[k-1]$$

BIBLIOGRAPHY

Cavicchi, T. *Digital Signal Processing*. New York: John Wiley and Sons, 2000.

Chassaing, R. *Digital Signal Processing and Applications with the C6713 and C6416 DSK*. New York: John Wiley and Sons, 2005.

Kester, W. "Undersampling Applications." *Analog Devices* (June, 2002).

Maloberi, F. "High Speed Data Converters for Communication Systems." *IEEE Circuits and Systems Magazine* Q1 (2001).

Mitra, S. *Digital Signal Processing*. 3rd ed. New York: McGraw-Hill, 2006.

Harris, F. *Multirate Signal Processing for Communications Systems*. Englewood Cliffs, New Jersey: Prentice-Hall, 2004.

Ifwachor, E., and B. Jervis. *Digital Signal Processing*. 2nd ed. Reading, Massachusetts: Addison-Wesley, 2001.

Oppenheim, A. V., and R. Schafer. *Digital Signal Processing*. Englewood Cliffs, New Jersey: Prentice-Hall, 1975.

Oppenheim, A., and R. Schafer. *Digital Signal Processing*. 2nd ed. Englewood Cliffs, New Jersey: Prentice-Hall, 1999.

Suter, B. *Multirate and Wavelet Signal Processing*. Lighting Source, 1998.

Taylor, F. J. *Digital Filter Design Handbook*. New York: Marcel Dekker, 1983.

——— and T. Stouraitis. *Digital Filter Design Using the IBP PC*. New York: Marcel Dekker, 1987.

——— and J. Mellott. *Hands-On Digital Signal Processing*. New York: McGraw-Hill, 1998.

Texas Instruments. "Code Composer Studio White Paper." *TI White Paper SPRA520*. www.ti.com.

Williston, K. "Inside DSP on Tools: Software Building Blocks for Signal Processing Applications." *Inside DSP* (June, 2005).

CHAPTER 17
SWITCHED-CAPACITOR FILTERS

17.1 INTRODUCTION

Switched-capacitor filters have gained widespread acceptance because they require no external reactive components, are precisely controlled by a clock derived from a crystal or other precise oscillator, and can be programmed easily. They come close to truly digital filters without requiring the mathematical complexity of computation and the sophistication necessary for implementation. Miniaturization is achieved by implementing these devices on a chip, while stability is a function of the external components which are typically low-temperature–coefficient resistors and a crystal-derived clock. They can be used for frequencies ranging from a small fraction of a hertz to a few hundred kilohertz.

17.2 THE THEORY OF SWITCHED-CAPACITOR FILTERS

The theory of switched-capacitor filters is based on the ability to very closely match capacitor ratios on an MOS integrated circuit. The actual magnitudes of the capacitors can vary significantly from device to device due to differences in the dielectric constant of the MOS process. The ratios between capacitors on the same device, however, are extremely precise since the areas allocated to each capacitor are very closely controlled by the manufacturing process.

The Switched Resistor

Figure 17-1 illustrates a capacitor C connected to ground, two switches S_1 and S_2, and an applied voltage E. The charge on a capacitor is given by

$$Q = CE \tag{17-1}$$

where Q is the charge in coulombs and E is the voltage applied to the capacitor.

The current through a capacitor is Q/T since an ampere is defined as a coulomb per second. The AC current through the capacitor of Figure 17-1 is

$$I_{ac} = \frac{\Delta Q}{\Delta T} = \frac{C \Delta E}{\Delta T} \tag{17-2}$$

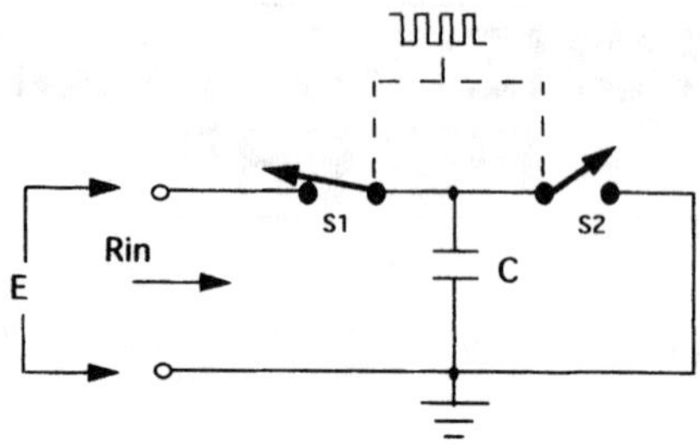

FIGURE 17-1 A resistor simulated by a switched capacitor.

where ΔT is the time between the closing of S_1 and S_2, which are operated alternately, and where ΔE is the variation in voltage across the capacitor. The capacitor is thus switched between voltage E and ground during alternate parts of the switching cycle, making ΔE equal to E.

Since ΔT is determined by a clock frequency F_{clock}, we can say

$$I_{\text{ac}} = C\Delta E\, F_{\text{clock}} \tag{17-3}$$

If we simply transpose and solve for the equivalent resistance R_{in}, in terms of voltage divided by current, we obtain

$$R_{\text{in}} = \frac{1}{CF_{\text{clock}}} \tag{17-4}$$

Equation (17-4) indicates that a capacitor can be made to behave like a resistor whose value is inversely proportional to the capacitance and to the clock-switching frequency. The switching rate should be much higher than any short-term variation in E.

The Basic Integrator as a Building Block. Figure 17-1 illustrates how a resistor can be implemented using clock-controlled switches and a capacitor. This is exactly the technique used in monolithic switched-capacitor filters. The switches are actually MOSFETs and have relatively low resistance compared with the resistor values implemented. The capacitors are controlled by the geometry of the surface of the substrate.

The circuit of Figure 17-2*a* shows a conventional inverting integrator. The transfer function of this circuit is

$$T(s) = -\frac{1}{sRC} \tag{17-5}$$

If we use a switched-capacitor resistor for R, the circuit of Figure 17-2*b* is obtained. This is an inverting integrator implemented using a switched-capacitor approach. We can substitute Equation (17-4) for R in Equation (17-5) and obtain the following transfer function:

$$T(s) = -\frac{1}{SC_b\dfrac{1}{C_aF_{\text{clock}}}} \tag{17-5}$$

Simplifying Equation (17-5) results in

$$T(s) = -\frac{F_{\text{clock}}C_a}{SC_b} \tag{17-6}$$

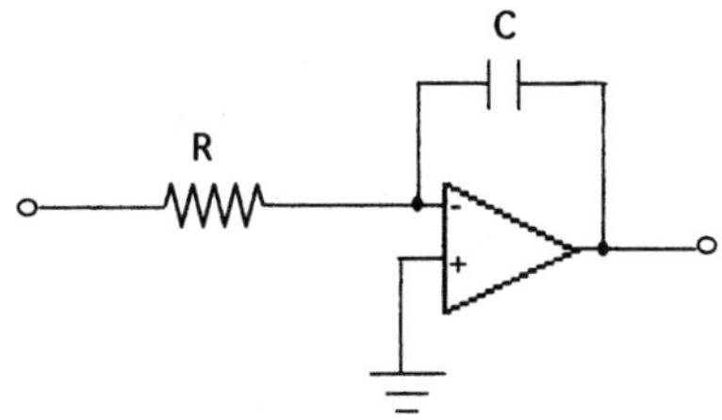

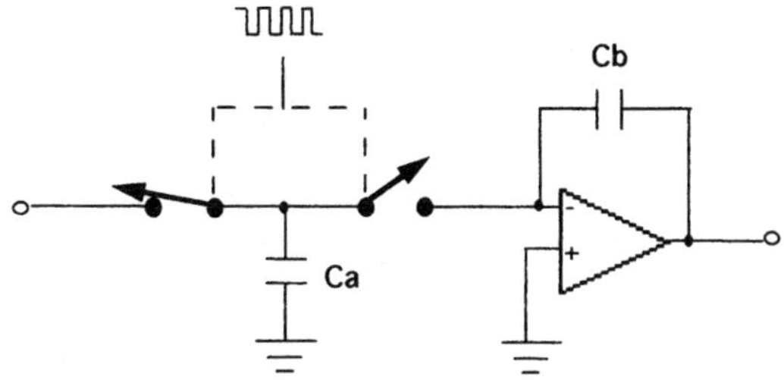

FIGURE 17-2 (*a*) A basic integrator; (*b*) a switched-capacitor integrator.

Equation (17-6) indicates that the transfer function of a switched-capacitor integrator can be scaled proportionally by a clock frequency F_{clock} and is determined by the constant C_a/C_b, which is a *ratio* of two capacitors. Since this ratio can be very precisely determined by chip geometries, we can build a filter having properties that are directly proportional to an externally supplied clock and independent of absolute magnitudes of chip capacitors. Variations of the circuit of Figure 17-2*b* are used as basic building blocks for switched-capacitor filters.

The Limitations of Switched-Capacitor Filters

Noise. A complex switched-capacitor filter chip contains many capacitors within the device. In order to minimize the absolute magnitude of these capacitors to keep the devices small, internal resistor values are relatively higher than in most CMOS ICs. As a result, the output noise of switched-capacitor filters is higher than that produced by a conventional active filter. The noise generated by switched-capacitor filters consists of two types: clock feedthrough and thermal noise. The clock of a switched-capacitor filter is normally 50 or 100 times the operating frequency range, so it can be easily filtered by a simple RC at the output. Thermal noise is in-band and is typically 80 dB below operating signal levels. It can be minimized by optimizing gain distribution for maximum signal-to-noise ratio. Careful PC layout and decoupling should occur as well.

Charge injection of the CMOS switches into the integrating capacitors and high internal impedance levels result in larger DC offsets than with active filters. However, some capacitive coupled architectures are available which can eliminate this problem.

Frequency Limitations. Switched-capacitor filters have a limiting figure of merit which is the product of Q *times* the center frequency. In other words, the higher the center frequency, the lower the realizable Q. For relatively low Qs, traditional switched-capacitor filters are realizable for frequencies up to 200 kHz or so. With newer submicron technologies, the range has been extended to a few MHz.

Since we are dealing with a sampled system, aliasing effects can occur. If input signals occur above $F_S/2$, where F_S is the internal sampling frequency (typically 50 or 100 times the cutoff), we get foldover, which could make the signals of interest unusable if the foldover components are significant in amplitude. Sometimes an input RC is useful to prevent this phenomenon, especially if the input signal consists of rectangular pulses rich in harmonic content.

17.3 UNIVERSAL SWITCHED-CAPACITOR SECOND-ORDER FILTERS

Figure 17-3 illustrates the basic architecture for a universal switched-capacitor building block using external resistors for determining characteristics in conjunction with a clock. This structure was first used in a device called an MF10 developed by National, which contains two such circuits. It was followed by the National LMF100 and has become the fundamental architecture for most complex switched-capacitor filters. Maxim also makes an MF10 and a next generation dual-filter building block with improved performance and single supply operation as well as many other features—what they call their MAX7490/7491 series.

Linear Technology has a device called an LTC1067, which is a next generation version of the MF10 but with significantly new features such as single supply operation, low noise, and wide dynamic range. Both Maxim and Linear Technology have quad versions of this architecture for more complex filters. Linear Technology has a single and triple version as well.

These devices, as well as others, are listed in the selection guide of section 17.4, and more details are provided in the manufacturers' data sheets, which are available at their web sites. In addition, software is available from the manufacturers' web sites to perform design computations. A program called FilterCAD is also included on the CD-ROM. This program (from Linear Technology) can help to quickly design a switched-capacitor filter from a set of input parameters and will then provide a schematic of the filter along with the predicted frequency response and time response.

Although the following discussion and example is based on the MF10, the general design equations and methodology still apply to the other versions of universal switched-capacitor filters as they all maintain essentially the same architecture.

This circuit realizes a second-order transfer function and can provide a bandpass, low-pass, high-pass, notch, or all-pass output by adding external resistors. The cascading of multiple sections results in switched-capacitor filters of very high complexity.

The basic configuration consists of an operational amplifier, a summing node, and two integrators. A MOS switch under the control of input *S* connects one of the summing node inputs either to the second integrator or to ground. The output of the operational amplifier can either be a notch, all-pass, or high-pass response depending on how the device is

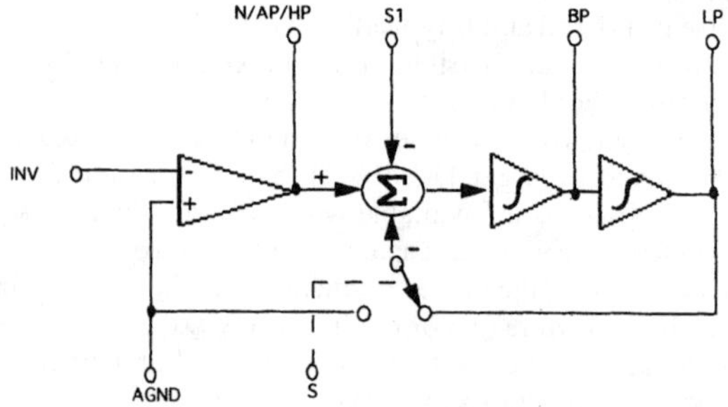

FIGURE 17-3 A block diagram of a universal second-order filter.

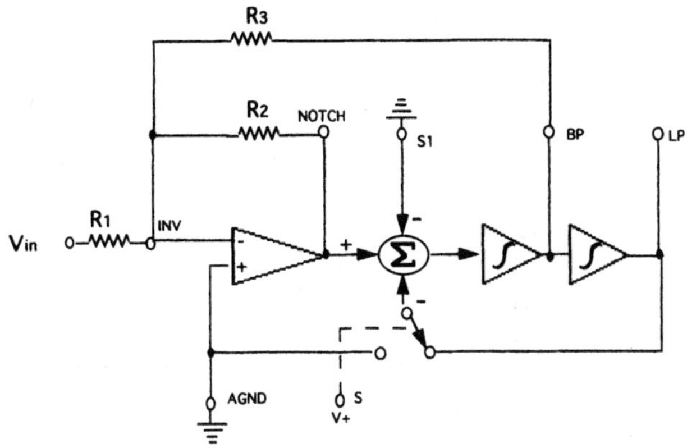

FIGURE 17-4 Mode 1: Bandpass, low-pass, $F_{notch} = F_0$.

configured. The two integrators, on the other hand, can provide band-pass and low-pass outputs, and in most cases, these outputs can be provided simultaneously.

Although not shown, a clock at either 50 or 100 times the center (pole) frequency must be provided for the National MF10 and LMF100. A logic input to the chip determines whether the clock is 50 or 100 times F_0. Some newer devices can also internally generate clock frequencies of a few MHz proportional to a single external passive component.

Modes of Operation. The universal filter structure of Figure 17-3 is quite flexible and can be reconfigured into various forms by adding external resistors and an amplifier. Each form or *mode* has unique properties which determine the filters' characteristics. Figures 17-4 through 17-10 show the most popular modes of operation. The design equations for each mode are given in Table 17-1.

Operating Mode Features

Mode 1. The circuit of mode 1 provides simultaneous bandpass and notch outputs at center frequencies that are equal to each other and directly determined by $F_{clock} \div 50$ or $F_{clock} \div 100$. As a result, we can design clock-tunable bandpass/bandstop networks with

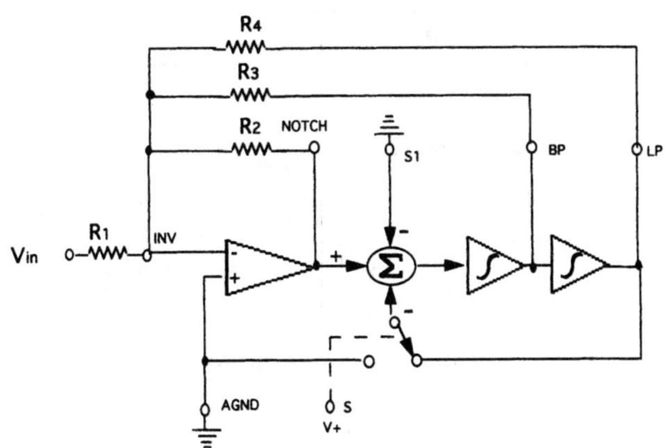

FIGURE 17-5 Mode 2: Bandpass, low-pass, $F_{notch} \le F_0$.

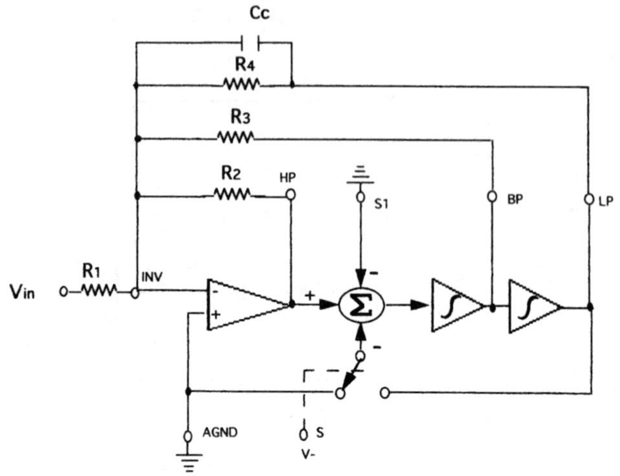

FIGURE 17-6 Mode 3: Bandpass, low-pass, high-pass.

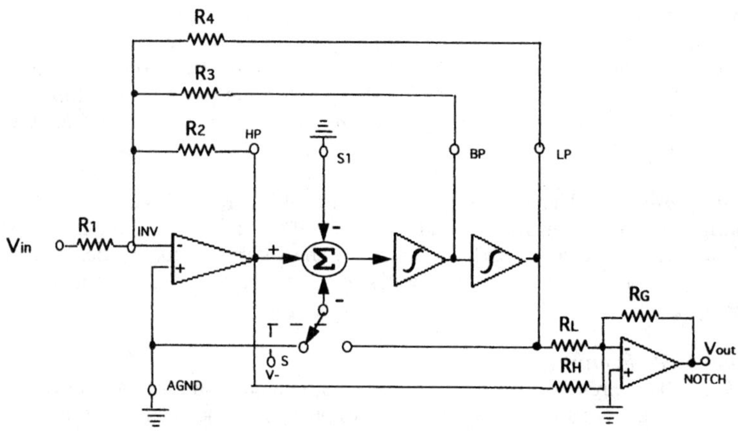

FIGURE 17-7 Mode 3a: Bandpass, low-pass, F_{notch} using an external operational amplifier.

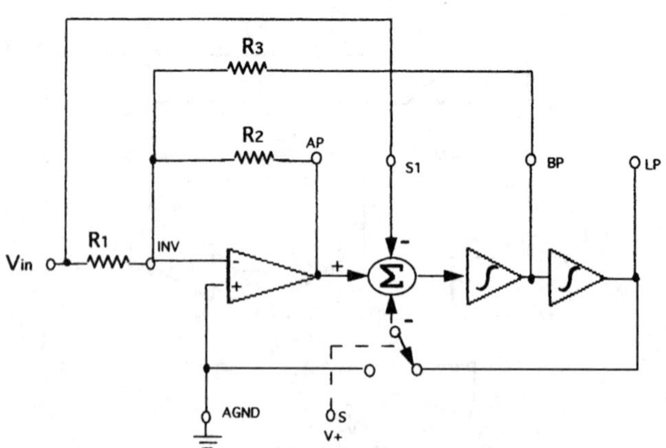

FIGURE 17-8 Mode 4: All-pass, bandpass, low-pass.

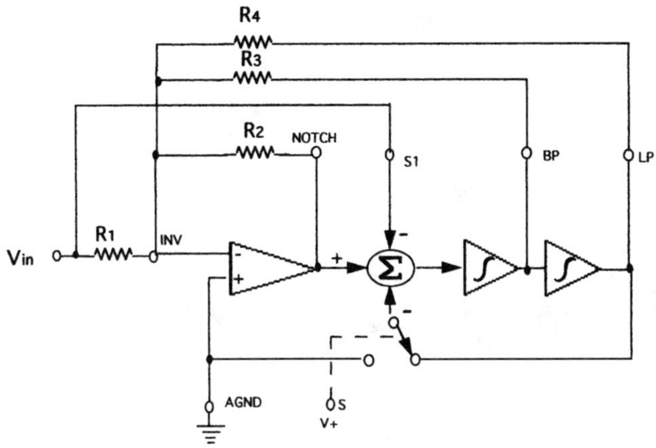

FIGURE 17-9 Mode 5: Numerator complex zeros, bandpass, low-pass.

frequency stability and accuracy as a direct function of the clock. Circuit Q is determined by the ratio R_3/R_2. A low-pass output is available as well.

Mode 2. The mode 2 configuration can provide bandpass and low-pass outputs as well as a notch output, providing that $F_{notch} \leq F_0$. Unlike the mode 1 configuration, center frequency F_0 can be different than F_{notch}. This circuit is useful for cascading second-order sections to create an elliptic-function high-pass, band-reject, or band-pass response requiring transmission zeros below the passband poles.

Mode 3. The circuit of mode 3 allows us to use a single resistor ratio (R_2/R_4) to tune F_0. Because the feedback loop is closed around the input summing amplifier, a slight Q enhancement occurs where the circuit Q is slightly above the actual design Q. A small capacitor across R_4 (10–100 pF) will result in a compensation phase lead which eliminates the Q enhancement.

Mode 3a. Mode 3a is probably the most versatile of all modes. An external operational amplifier sums together the low-pass and bandpass outputs of mode 3 to form a notch output which can be on either side of center frequency F_0. This is especially useful for an elliptic-function low-pass, high-pass, band-reject, or bandpass response requiring transmission zeros both above and below the passband poles. Gains can be distributed between the internal circuitry and the external operational amplifier to achieve maximum dynamic range.

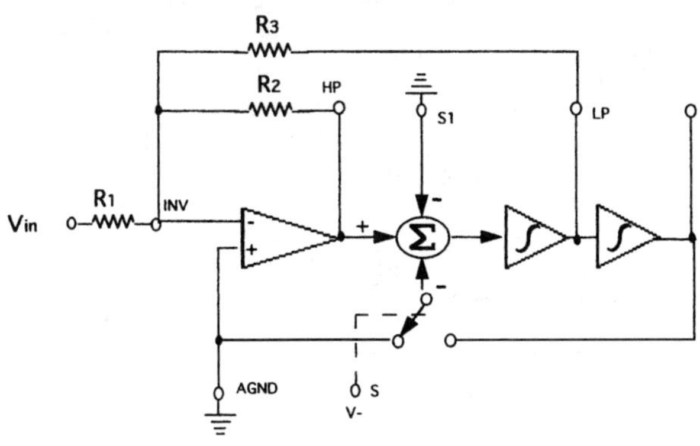

FIGURE 17-10 Mode 6a: Single-pole low-pass, high-pass.

TABLE 17-1 Design Equations for Modes of Operation

Mode	F_0	Q	F_{notch}	LP Gain	BP Gain	HP Gain	Notch Gain	All-Pass Gain
1	$\frac{F_{clock}}{100(50)}$	$\frac{R_3}{R_2}$	F_0	$-\frac{R_2}{R_1}$	$-\frac{R_3}{R_1}$		$-\frac{R_2}{R_1}$ (as $F \to 0$)	
2	$\frac{F_{clock}}{100(50)} \times \sqrt{1+\frac{R_2}{R_4}}$	$\frac{\sqrt{\frac{R_2}{R_4}+1}}{R_2/R_3}$	$\frac{F_{clock}}{100(50)}$	$-\frac{R_2/R_1}{R_2/R_4+1}$	$-\frac{R_3}{R_1}$		$-\frac{R_2/R_1}{R_2/R_4+1}$ (as $F \to 0$)	
3	$\frac{F_{clock}}{100(50)} \times \sqrt{\frac{R_2}{R_4}}$	$\sqrt{\frac{R_2}{R_4}} \times \frac{R_3}{R_2}$		$-\frac{R_4}{R_1}$	$-\frac{R_3}{R_1}$	$-\frac{R_2}{R_1}$		
3a	$\frac{F_{clock}}{100(50)} \times \sqrt{\frac{R_2}{R_4}}$	$\sqrt{\frac{R_2}{R_4}} \times \frac{R_3}{R_2}$	$\frac{F_{clock}}{100(50)} \times \sqrt{\frac{R_H}{R_L}}$	$-\frac{R_4}{R_1}$	$-\frac{R_3}{R_1}$	$-\frac{R_2}{R_1}$	$Q\left[\frac{R_G R_2}{R_H R_1} - \frac{R_G R_4}{R_L R_1}\right]$ (at F_0)	
4	$\frac{F_{clock}}{100(50)}$	Q_{pole} $\frac{R_3}{R_2}$ Q_{zero} $\frac{R_3}{R_1}$		$-\left[\frac{R_2}{R_1}+1\right]$	$-\frac{R_3}{R_2}\left(1+\frac{R_2}{R_1}\right)$			$-\frac{R_2}{R_1}$ ($R_2 = R_1$ for all-pass)
5	$\frac{F_{clock}}{100(50)} \times \sqrt{1+\frac{R_2}{R_4}}$	Q_{pole} $\frac{\sqrt{1+\frac{R_2}{R_4}}}{R_2/R_3}$ Q_{zero} $\frac{\sqrt{1-\frac{R_1}{R_4}}}{R_1/R_3}$	$\frac{F_{clock}}{100(50)} \times \sqrt{1-\frac{R_1}{R_4}}$	$-\left(\frac{R_2+R_1}{R_2+R_4}\right) \times \frac{R_4}{R_1}$	$-\left(\frac{R_2}{R_1}+1\right) \times \frac{R_3}{R_2}$			
6a	$F_c = \frac{F_{clock}}{100(50)} \frac{R_2}{R_3}$			$-\frac{R_3}{R_1}$		$-\frac{R_2}{R_1}$		

Mode 4. An all-pass output can be provided by using mode 4 in addition to low-pass and bandpass outputs. As in the case of mode 1, the center frequency F_0 is directly determined by $F_{clock} \div 50$ or $F_{clock} \div 100$. Resistor R_1 must be made equal to R_2 for all-pass characteristics.

Mode 5. This mode can be used for implementing complex numerator zeros for a transfer function and offers low-pass and high-pass outputs as well.

Mode 6a. A single low-pass or high-pass real pole can be implemented using the mode 6a configuration. The major benefit of the added complexity of this implementation over a single RC is that the 3-dB cutoff F_c is clock-dependent. This makes this approach very useful for filters requiring variable characteristics controlled by an external clock frequency. If the cutoff is fixed, an ordinary RC implementation of the real pole is more cost-effective.

Using the MF10 and LMF100 Dual Universal Second-Order Filter. Figure 17-11 is a block diagram of an MF10 in a 20-pin package available as either a DIP (dual inline package) or a surface mount version. In addition to the pins that make up the basic switched-capacitor filter, other pin functions are provided as follows:

Power Supply Connections. Typically, the MF10 operates from ±5-V supplies. Separate digital and analog positive and negative power supply pins are provided, which are VD+ (pin 8), VA+ (pin 7), VD− (pin 13), and VA− (pin 14). Pins 7 and 8 can be tied together, provided that adequate bypassing exists. The same applies to pins 13 and 14. Analog ground (pin 15) is normally connected to ground, and the level shift function (pin 9) is also connected to ground.

The MF10 is also capable of operating from a single positive supply. In that case, the VD– and VA– pins, as well as level shift, are connected to system ground. The analog ground (pin 15) should be biased at half of the positive rail and adequately decoupled to system ground.

50/100 Clock Select. The 50/100 function (pin 17) determines whether the clock is either in the 50:1 mode or the 100:1 mode. When pin 17 is connected to VD+ and VA+ the mode is 50:1. If pin 17 is connected to analog ground, the mode is 100:1. Some devices also allow pin 17 to provide a sample-and-hold capability to freeze the outputs when connecting pin 17 to VD− and VA−.

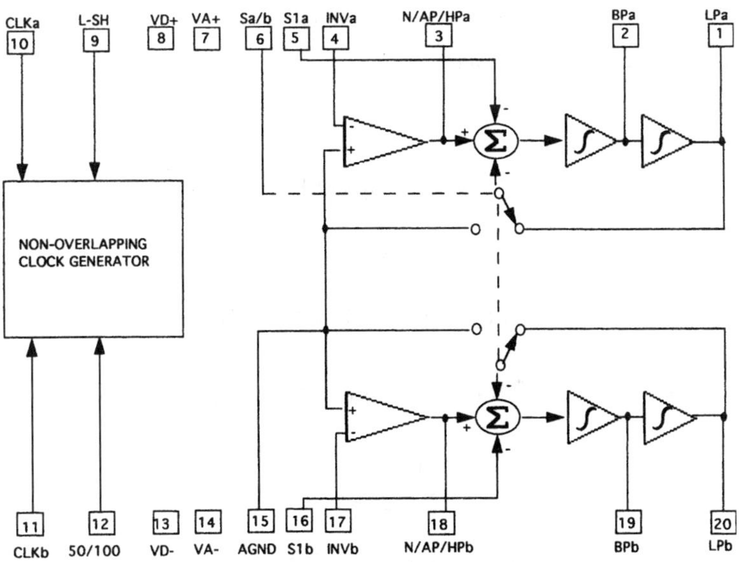

FIGURE 17-11 The block diagram of the MF10 dual universal second-order filter.

Clock Input Pins. Two separate TTL-compatible clock input pins for the A section and B section filters are provided on pins 10 and 11, respectively. The clock duty cycle should be close to 50 percent.

$S_{a/b}$ *Control.* This function controls the two CMOS switches that connect the summer input to either analog ground or the low-pass output function. When $S_{a/b}$ is connected to the negative rail, both switches go to analog ground. If $S_{a/b}$ is connected to the positive rail, both switches are connected to the two low-pass outputs. The switch positions are determined by the requirements of the operating mode.

Example 17-1 Design of a Switched Capacitor Elliptic-Function Low-Pass Filter

Required:

Design a switched-capacitor low-pass filter that has a 0.5-dB maximum ripple below 100 Hz and greater than 18-dB rejection above 155.6 Hz. In addition, the filter should be programmable for a cutoff at any frequency in the range of 100 Hz through 10 kHz, while maintaining the frequency ratio of 1.6 for greater than 18-dB attenuation.

Results:

(*a*) Open ***Filter Solutions***

Check the ***Stop Band Freq*** box.

Enter **.177** in the ***Pass Band Ripple (dB)*** box.

Enter **100** in the ***Pass Band Freq*** box.

Enter **155.6** in ***Stop Band Freq*** box.

Check the ***Frequency Scale Hertz*** box.

(*b*) Click the ***Set Order*** control button to open the second panel.

Enter **18** for the ***Stop band Attenuation (dB).***

Click the ***Set Minimum Order*** button and then click ***Close.***

3 Order is displayed on the main control panel.

(*c*) Click the ***Transfer Function*** button.

Check the ***Casc*** box.

The following is then displayed:

Continuous Transfer Function

$$Wn = 1095$$

$$\frac{306.6 \quad (S^2 + 1.199e{+}06)}{(S^2 + 358.2{*}S + 5.529e{+}05) \quad (S + 664.8)}$$

$$Wo = 743.6$$
$$Q = 2.076$$

3rd Order Low Pass Elliptic

Pass Band Frequency = 100.0 Hz	Stop Band Ratio = 1.556
Pass Band Ripple = 177.0 mdB	Stop Band Frequency = 155.6 Hz
	Stop Band Attenuation = 18.56 dB

(*d*) The design parameters are summarized as follows:

$$\text{Section } Q = 2.076$$

$$\text{Section } \omega_0 = 743.6$$

$$\text{Section } \omega_\infty = 1095$$

$$\alpha_0 = 664.8 \text{ (from the denominator)}$$

(*e*) Convert rad/sec to Hz (divide by 2π) and restate design parameters.

$$\text{Section } Q = 2.076$$

$$\text{Section } f_0 = 118.3 \text{ Hz}$$

$$\text{Section } f_\infty = 174.1 \text{ Hz}$$

$$\text{Real Pole } f_c = 105.7 \text{ Hz}$$

(*f*) Design a two-section switched-capacitor low-pass filter as follows:

Section 1:

$$f_0 = 118.3 \text{ Hz}$$

$$Q = 2.076$$

$$f_\infty = 174.1 \text{ Hz}$$

Section 2:

Real pole where $f_c = 105.7$ Hz

Use a single MF10 dual second-order filter IC. The A section can be used to implement section 1 and the B section for section 2. The design proceeds as follows:

Section 1 (A Section):

Use mode 3a

Let $$F_{\text{clock}} = 5000 \text{ Hz}$$

Select 50:1 clock mode

Let $$R_1 = R_4 = 10 \text{ k}\Omega$$

Using the f_0 equation for mode 3a from Table 17-1 (where $f_0 = 118.3$ Hz):

$$R_2 = 13.99 \text{ k}\Omega$$

Using the Q equation for mode 3a from Table 17-1 (where $Q = 2.076$):

$$R_3 = 24.57 \text{ k}\Omega$$

Using the f_{notch} equation for mode 3a from Table 17-1 (where $f_{\text{notch}} = f_\infty = 174.1$ Hz):

Let $$R_L = R_G = 10 \text{ k}\Omega$$

then $$R_H = 30.31 \text{ k}\Omega$$

Section 2 (B Section):

Use mode 6a

Let $$F_{\text{clock}} = 5000 \text{ Hz}$$

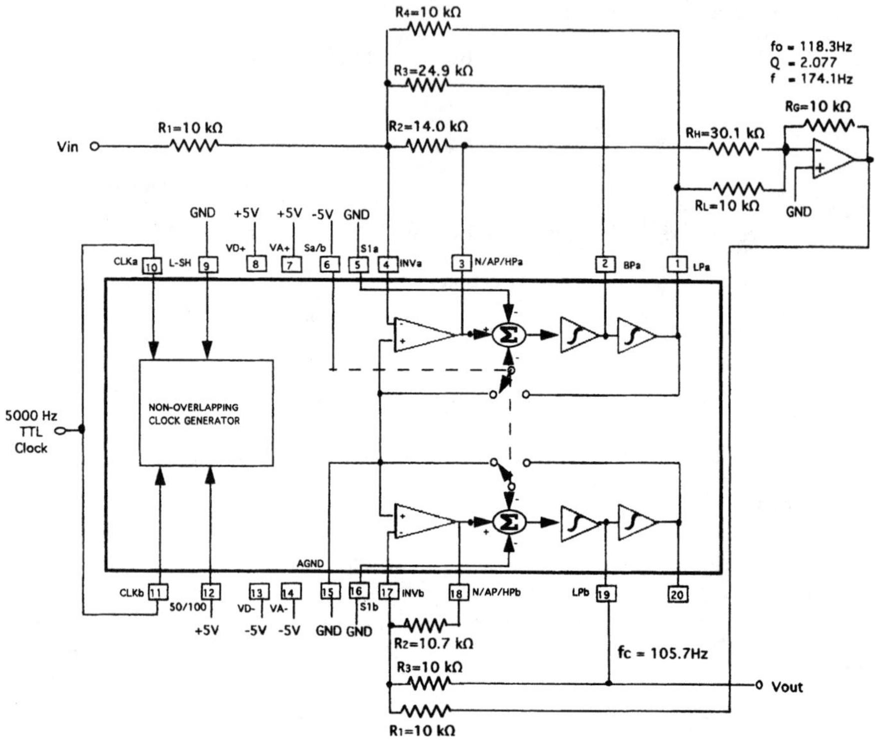

FIGURE 17-12 The low-pass filter of Example 17-1.

Using the f_c equation for mode 6a from Table 17-1 (where $f_c = 105.7$ Hz):

Let $$R_1 = R_3 = 10\ \text{k}\Omega$$

then $$R_2 = 10.57\ \text{k}\Omega$$

(*g*) The schematic of the resulting filter is shown in Figure 17-12. The response curves for clock frequencies of 5000 Hz and 500 kHz are given in Figure 17-13, where the cutoffs are 100 Hz and 10 kHz. For any clock frequency between 5000 Hz and 500 kHz, the filter cutoff will be exactly one-fiftieth of the clock frequency. All resistor values computed have been rounded off to their nearest 1-percent standard value.

17.4 TYPES OF SWITCHED-CAPACITOR FILTERS

Universal

Universal switched-capacitor building blocks consist of independent universal second-order filters using the configuration of Figure 17-3. Up to four sections can be contained within a single package. This architecture allows adjustment of filter parameters such as center frequency, Q, notch location, and gain. These parameters are also all controlled by relationships between resistors and the applied clock frequency. Figure 17-14 shows a

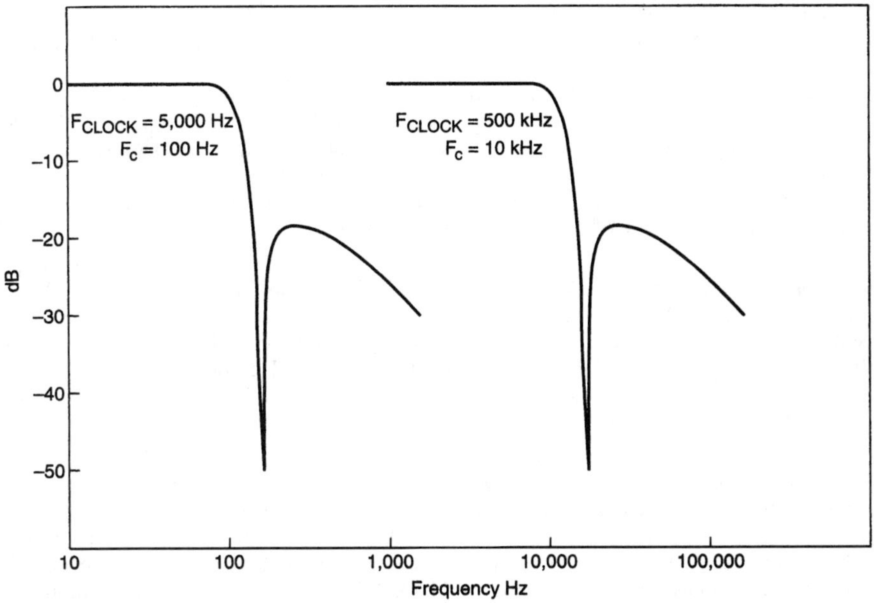

FIGURE 17-13 The frequency response curve for Example 17-1.

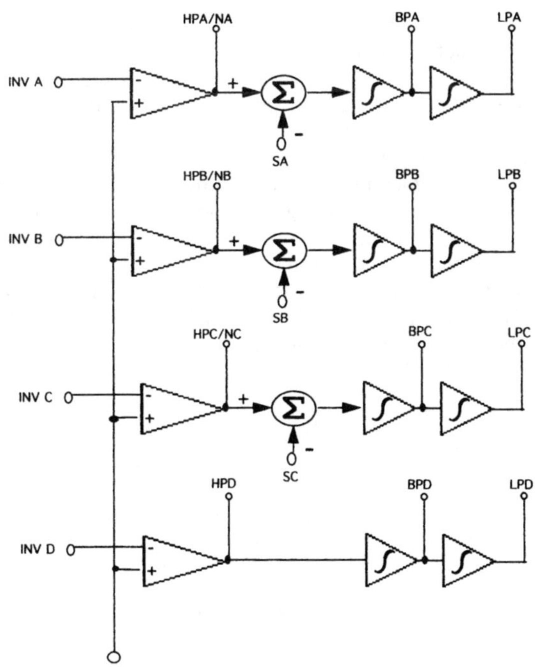

FIGURE 17-14 A block diagram of the Linear Technology LTC1064 and LTC1068.

block diagram of the Linear Technology LTC1064 and LTC1068, containing four second-order sections. Refer to the references and the manufacturers' web sites for detailed application information on these devices.

Filters of varying complexity can be designed using these building blocks. The design procedure consists of transforming a complex filter transfer function into individual first- or second-order sections as required. Each section is defined in terms of parameters, such as center or cutoff frequency, Q, and notch frequency, and then implemented using cascaded ICs. This was illustrated in Example 17-1 for a third-order elliptic function low-pass filter using the MF10 type of dual switched capacitor filter.

Microprocessor Programmable Universal Switched Capacitor Filters. This family of switched capacitor filters uses an external program to load in the filter mode of operation, center/corner frequency f_0 as a function of a clock division ratio, and Q for each section in the form of coefficients in the ASCII format. No external components are required. MAXIM MAX260/261/262 devices fall into this family and are dual second-order universal switched capacitor filters.

Pin Programmable Universal Switched Capacitor Filters. Unlike the microprocessor versions, these filters are defined by hard-wiring dedicated pins to either positive or negative voltages to define f_0 as a function of a clock division ratio and Q, as well as the mode of operation. This would be the MAXIM MAX263/264 product. The MAX267/268 is dedicated to the bandpass mode only, while the MAX 265/266 combines pin programming for f_0 and the use of resistors to set Q.

Dedicated Switched Capacitor Filters. Dedicated switched-capacitor filters are already configured to provide a defined transfer function, and as a result do not require external resistors. Typically, they provide low-pass responses and are available with standard shapes such as Butterworth, Chebyshev, Bessel, and elliptic-function. Designing with these devices is simply a matter of connecting a clock having a frequency, which is a defined multiple of the cutoff frequency and connecting power and ground to the device.

In some cases, the manufacturer simply customizes one of the universal filters by adding some components internally to the IC. A representative example of this approach is the Linear Technology LTC1064-1, which is shown in Figure 17-15. This device is an eighth-order elliptic-function low-pass filter created from an LTC-1064 and requires no external components. It has a passband ripple of ± 0.15 dB and makes the transition from passband to stopband within a steepness factor of 1.5, achieving over 68 dB of stopband attenuation, where the clock-to-cutoff frequency ratio is 100:1.

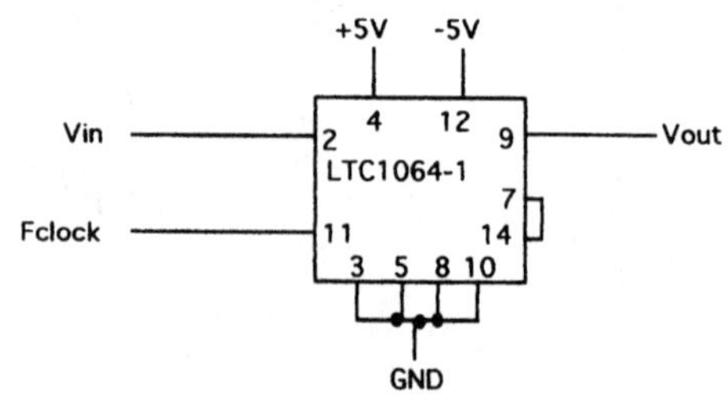

FIGURE 17-15 An eighth-order elliptic-function low-pass filter using the LTC1064-1.

TABLE 17-2 Switched Capacitor Filter Selection Guide: Universal Switched Capacitor Filters

Device	Type	Response	Order	Freq. Accur.	Max. Center Freq.	Ratio Clock: F_0	Special Features
LMF100	Universal	Universal	Dual N = 2	±0.2%	100 kHz ($f_0 \times Q$ up to 1.8 MHz)	50:1, 100:1	MF10 pin-compatible
MF10	Universal	Universal	Dual N = 2	±0.6%	30 kHz ($f_0 \times Q$ up to 200 kHz)	50:1, 100:1	Industry standard
LTC1059	Universal	Universal	Single N = 2	±0.3%	40 kHz	50:1, 100:1	Low noise
LTC1060	Universal	Universal	Dual N = 2	±0.3%	30 kHz ($f_0 \times Q$ up to 1.6 MHz)	50:1, 100:1	MF10 pin-compatible
LTC1061	Universal	Universal	Triple N = 2	±1.2%	35 kHz	50:1, 100:1	Low noise, single supply
LTC1064	Universal	Universal	Quad N = 2	±0.3%	140 kHz	50:1, 100:1	Low noise
LTC1067	Universal	Universal	Dual N = 2	±0.2%	20 kHz	100:1	Low noise, single supply
LTC1067-50	Universal	Universal	Dual N = 2	±0.2%	20 kHz	50:1	Low noise, single supply
LTC1068-200	Universal	Universal	Quad N = 2	±0.3%	25 kHz	200:1	Low noise, single supply
LTC1068	Universal	Universal	Quad N = 2	±0.3%	50 kHz	100:1	Low noise, single supply
LTC1068-50	Universal	Universal	Quad N = 2	±0.3%	50 kHz	50:1	Low noise, single supply
LTC1068-25	Universal	Universal	Quad N = 2	±0.3%	200 kHz	25:1	Low noise, single supply
MAX260	Universal	Universal	Dual N = 2	±1%	7.5 kHz	Programmable	Microprocessor programmable
MAX261	Universal	Universal	Dual N = 2	±1%	57 kHz	Programmable	Microprocessor programmable
MAX262	Universal	Universal	Dual N = 2	±1%	140 kHz	Programmable	Microprocessor programmable
MAX263	Universal	Universal	Dual N = 2	±1%	57 kHz	Programmable	Pin programmable
MAX264	Universal	Universal	Dual N = 2	±1%	75 kHz	Programmable	Pin programmable
MAX265	Universal	Universal	Dual N = 2	±0.2%	40 kHz	Programmable	Pin and resistor programmable
MAX266	Universal	Universal	Dual N = 2	±0.2%	140 kHz	Programmable	Pin and resistor programmable
MAX7490	Universal	Universal	Dual N = 2	±0.2%	40 kHz	100:1	Single supply +5 V
MAX7491	Universal	Universal	Dual N = 2	±0.2%	40 kHz	100:1	Single supply +3 V

TABLE 17-2 Switched Capacitor Filter Selection Guide: Dedicated Switched Capacitor Filters (*Continued*)

Device	Type	Response	Order	Freq. Accur.	Max. Cutoff Freq.	Ratio Clock: F_0	Special Features
LTC1062	Low-pass	Butterworth	5th	±1%	20 kHz	100:1	Zero DC offset
LTC1064-1	Low-pass	Elliptic	8th	±1%	20 kHz	100:1	Low noise
LTC1064-2	Low-pass	Butterworth	8th	±1%	140 kHz	50:1, 100:1	Low noise
LTC1064-3	Low-pass	Bessel	8th	±1%	95 kHz	75:1, 120:1, 150:1	Low noise
LTC1064-4	Low-pass	Elliptic	8th	–	100 kHz	50:1, 100:1	Low noise
LTC1069-1	Low-pass	Elliptic	8th	–	12 kHz	100:1	Single supply
LTC1069-6	Low-pass	Elliptic	8th	–	20 kHz	50:1	Single supply, low power
LTC1069-7	Low-pass	Linear Phase	8th	–	200 kHz	25:1	Single supply
LTC1569-6	Low-pass	Linear Phase	10th	±3.5%	64 kHz	1:1, 4:1, 16:1	Cutoff frequency is resistor programmable
LTC1569-7	Low-pass	Linear Phase	10th	±3.5%	300 kHz	1:1, 4:1, 16:1	Cutoff frequency is resistor programmable
MAX280-281	Low-pass	Butterworth/Bessel	5th	–	20 kHz	100:1/101:1	No DC error, internal or external clock
MAX291-294	Low-pass	Butterworth/Bessel/Elliptic	8th	–	25 KHz	100:1	Internal or external clocks
MAX295-297	Low-pass	Butterworth/Bessel/Elliptic	8th	–	50 KHz	50:1	Internal or external clocks
MAX7400/ 7403/7404/7407	Low-pass	Elliptic	8th	–	10 KHz	100:1	Single supply, low power, internal or external clocks
MAX7401/7405	Low-pass	Bessel	8th	–	5 kHz	100:1	Single supply, low power, internal or external clocks
MAX7418-7425	Low-pass	Butterworth/Bessel/Elliptic	5th	–	45 kHz	100:1	Single supply, low power, internal or external clocks, offset adjustment.

17.5 FILTERCAD 3.0 SOFTWARE

A program called FilterCAD is included on the CD-ROM. This program from Linear Technology can help to quickly design a switched-capacitor filter from a set of input parameters and then provide a schematic of the filter along with the predicted frequency response and time response. Two basic modes of operation are available: *Quick Design* and *Enhanced Design*. *Quick Design* guides you through a design similar in operation to a "wizard" where it asks you questions along the way. A design based on standard responses such as Butterworth, Bessel, Chebyshev, and elliptic polynomials will result. The *Enhanced Design* mode is more interactive, allowing custom designs as well as standard polynomials. The user is encouraged to explore the various Help topics to gain more familiarity with this program.

There is a folder called FCAD on the CD-ROM. Copy this folder to the desktop, open the folder, and then double-click SETUP.EXE. The program will then install. From time to time Linear Technology may update this program, which is available from their web site.

17.6 THE SWITCHED CAPACITOR FILTER SELECTION GUIDE

Switched-capacitor filters offer circuit designers a solution to filtering needs while utilizing a minimal printed circuit board area as compared with passive and active filters. They feature high stability and precision as well as the ability to change frequency characteristics under the control of an external clock. A selection guide to switched-capacitor filters is provided in Table 17-2.

Note: Many of these dedicated devices are simply a customization by a mask of an existing universal product. This table is not all-inclusive since other customized products exist or can be created for a set of requirements.

BIBLIOGRAPHY

Linear Technology. *Monolithic Filter Handbook.* Milpitas, California: Linear Technology, 1990.

————————. *Linear Databook Supplement.* Milpitas, California: Linear Technology, 1990.

Maxim Integrated Circuit Products Inc. *Application Note: The Basics of Anti-Aliasing: Using Switched Capacitor Filters.* Sunnyvale, California: Maxim Integrated Circuit Products Inc., 2005.

National Semiconductor Corporation. *Switched-Capacitor Handbook.* Santa Clara, California: National Semiconductor Corporation, 1985.

————————. *Linear Databook, Rev 1.* Santa Clara, California: National Semiconductor Corporation, 1988.

CHAPTER 18
INTRODUCTION TO MICROWAVE FILTERS

18.1 IMPLEMENTATION OF FILTERS

The passive (*LC*) filters discussed previously work quite well at frequencies up to a few hundred megahertz. Beyond this range, components deviate significantly from anything close to ideal. Parasitics start to dominate, and component values become impractical, while capacitors become inductors and vice versa. Distances between components turn important, and traces on a PC board introduce unwanted capacitance and inductance. The methodology discussed in this section concerns using PC board traces to create transmission lines by controlling their properties, and then configuring these transmission lines into an architecture resulting in filters. The resulting filters are then based on distributed parameters rather than lumped inductors and capacitors. At submicrowave frequencies, this approach is not feasible since the dimensions based on fractions of a wavelength become impractical.

For low-power applications, stripline and microstrip filters are extensively used because of their low cost and repeatability. For high-power requirements, waveguide structures are utilized. Waveguides are covered in great detail in Matthaei (see Bibliography) and will not be covered here.

This chapter is introductory and is by no means intended as a design guide for microwave filters. The reader should instead refer to the multiple references cited in the Bibliography for further information on this topic.

18.2 MICROSTRIP AND STRIPLINE TRANSMISSION LINES

The two basic forms of PC board transmission lines are *microstrip* and *stripline*. The microstrip approach uses a track (or tracks) on one side of a PC board and a ground plane on the other. A stripline is similar but the trace is sandwiched between two ground planes, one on either side of the board. This is shown in Figure 18-1.

One of the advantages of stripline over microstrip is that the "sandwiching" of the signal trace between two ground planes contains the EMI fields. Another advantage is that this form of construction results in a more uniform ground plane surrounding the center conductor than microstrip. The microstrip approach is more difficult to analyze because of leakage fields from the PC track since it is not fully surrounded by a ground plane. These leakage fields reduce the "effective" dielectric constant of the board, which raises the impedance. This is because some of the field lines pass through air, which has a dielectric constant of 1 rather than 4.7, a figure typical of PC board material FR4.

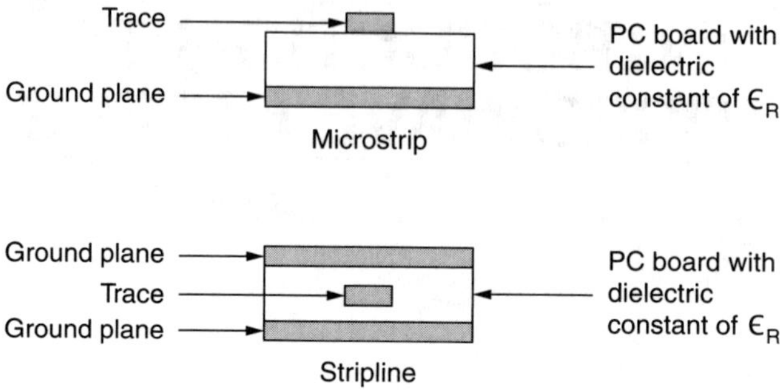

FIGURE 18.1 Microstrip and stripline construction.

Transmission lines have a characteristic impedance Z_O and an electrical length λ[1]. The calculations involved in determining these equivalent electrical properties of microstrip and stripline transmission lines are based on the width and thickness of the traces, the distance from the ground plane and the dielectric constant ε_R (sometimes called relative permittivity), and are quite complex. A number of programs are available that can calculate the resulting electrical properties of microstrip and stripline implementations. An extremely useful one is included on the CD-ROM. It's called *txline,* and is from Applied Wave Research, Inc., a world leader in electronic design automation (EDA) software.

A program called *Filter Solutions* is included on the CD-ROM. This program is limited to elliptic function *LC* filters (up to $N = 10$) and is a subset of the complete program which is available from Nuhertz Technologies® (www.nuhertz.com). Many microwave filter design capabilities are included in the full version, such as microstrip and stripline low-pass, high-pass, bandpass and bandstop filters, diplexers, multi-conductor microstrip, and stripline band-pass filters and lumped tubular filters.

18.3 RICHARDS' TRANSFORMATION

If a transmission line is *shorted* at the opposite end and is $\frac{1}{4}$ of a wavelength in length ($\lambda/4$), an applied sinusoidal signal will be reflected back to the input, *exactly in phase* with the input, preventing any current from flowing. This would be analogous to the infinite impedance of a parallel resonant circuit, at resonance.

If a transmission line is *open* at the opposite end and is $\frac{1}{4}$ of a wavelength in length ($\lambda/4$), an applied sinusoidal signal will be reflected back to the input 180° out of phase with the input and cancel the signal.

Hence an attenuation pole will occur at this frequency analogous to a series resonant circuit, at resonance, in parallel with the source

Below resonance, a parallel resonant circuit appears inductive and a series resonant circuit is capacitive. Transmission lines can then be used to implement inductors and capacitors for microwave filters providing the wavelengths are below $\lambda/4$. Typically, a filters'

[1]The characteristic impedance Z_O of a line can be defined as the impedance seen at the input of that line if it were infinitely long, while the electrical length is expressed as a multiple or submultiple of the wavelength λ of a periodic signal traveling through a medium. The wavelength of a signal is the distance it travels during one cycle. In free space, it is $\lambda = c$ (speed of light) $/f$ (frequency). In another medium, it is reduced by the multiplication factor $1/\sqrt{\varepsilon_R}$.

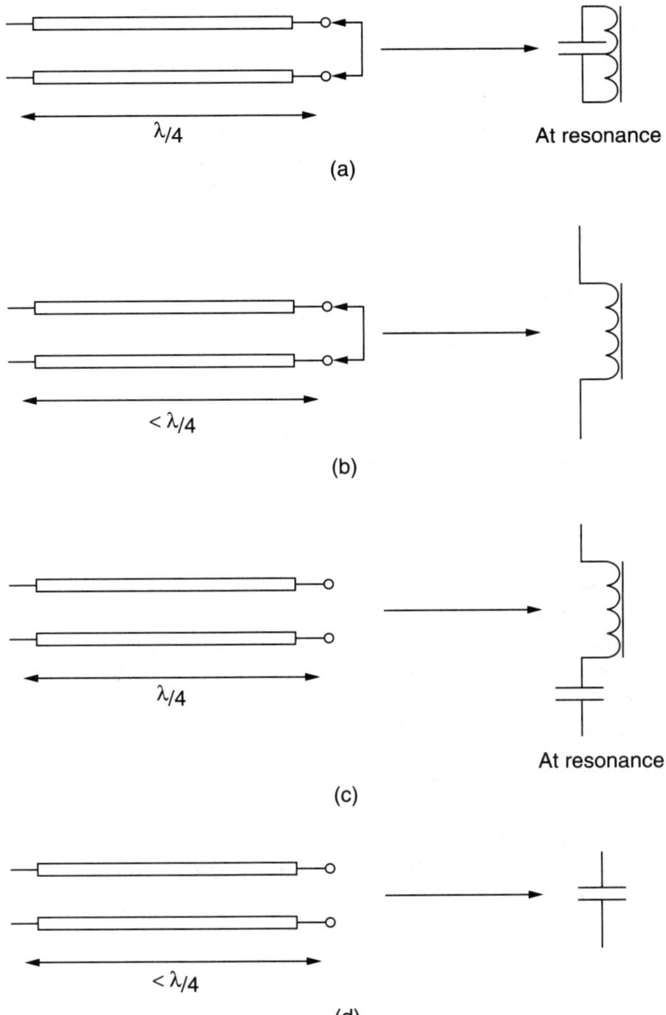

FIGURE 18-2 Shorted and open transmission line stub equivalents: (*a*) shorted; length = $\lambda/4$, (*b*) shorted; length < $\lambda/4$, (*c*) open; length = $\lambda/4$, and (*d*) open; length < $\lambda/4$.

cutoff frequency ω_C is chosen at $\lambda/8$. The attenuation poles then occur at $2\omega_C$ ($\lambda/4$). The behavior of open and shorted transmission line "stubs" is shown in Figure 18-2.

Richards' transformation (see Bibliography) enables the conversion of lumped element filters into distributed filters using PC board traces functioning as transmission lines. Shorted and open transmission line "*stubs*" approximate the behavior of inductors and capacitors over a finite frequency range. Each stub has an electrical length of $\lambda/8$ at a cutoff frequency ω_C. For a low-pass filter, ω_C is the 1 radian/sec cutoff in the normalized case. At $2\omega_C$, attenuation poles will occur where the shorted stub behaves like a parallel resonant circuit at resonance, and the open stub behaves like a series resonant circuit at resonance.

Line with Short Circuit at Output. The impedance seen at the input of a *lossless* transmission line terminated in a short circuit is

$$Z_{SC} = jZ_O \tan(\gamma l) \tag{18-1}$$

where γ is the propagation constant for the lossless line equal to $j\omega\sqrt{(LC)}$ and l is the unit length. (L and C are the distributed inductance and capacitance of the transmission line per unit length.) Therefore, we can express the input impedance with a short circuit at the output as

$$Z_{SC} = jZ_O \tan(\omega\sqrt{LC}\, l) \tag{18-2}$$

Note that the L and C in Equation (18-2) are the distributed parameters of the transmission line per unit length l.

The reactance of an inductor is given by the well-known expression

$$X_L = j\omega L \tag{18-3}$$

So if we can establish an equivalency between Equations (18-2) and (18-3), we can substitute a shorted transmission line stub for an inductor. Richards' transformation results in the following expression:

$$j\omega L = jZ_O \tan\left[\frac{\pi}{2}, \frac{\omega}{\omega_Q}\right] \tag{18-4}$$

Equation (18-4) implies that a shorted transmission line having a characteristic impedance Z_O has an equivalent inductance of Z_O over the frequency range between ω and ω_Q, where ω_Q corresponds to the $\frac{1}{4}$ wavelength frequency. For $\omega = \omega_Q$, the value of $j\omega L$ is infinite (tan $\pi/2$ or tan 90° is infinity). *We have mapped the $S = j\omega$ frequency plane into a new compressed frequency plane bounded by ω_Q, which would correspond to ω = infinity in the original frequency plane.*

For a shorted line: $L = Z_O$ over the frequency domain of $0 \leq \omega \leq \omega_Q$

Line with Open Circuit at Output. In a similar manner, for a lossless transmission line open at the far end, Richards' transformation results in an expression for the susceptance ($1/X_C$) of a capacitor :

$$j\omega C = jY_O \tan\left[\frac{\pi}{2}, \frac{\omega}{\omega_Q}\right] \tag{18-5}$$

Equation (18-5) then implies that an open transmission line having a characteristic admittance $Y_O (1/Z_O)$ has an equivalent capacitance of Y_O over the frequency range between ω and ω_Q, where ω_Q corresponds to the $\frac{1}{4}$ wavelength frequency.

For an open line: $C = Y_O$ over the frequency domain of $0 \leq \omega \leq \omega_Q$

These results imply that we can implement a lumped element low-pass filter by replacing each inductor L by a shorted transmission line having a Z_O equal to L, and each capacitor C by an open transmission line, having a Y_O equal to C. The lines must have a wavelength of $\lambda/4$ (90° or $\pi/2$) at the highest frequency of operation where an attenuation pole will occur. The cutoff frequency ω_C should correspond to $\lambda/8$.

It should be recognized that as we go above ω_C, the impedance of the stubs do not exactly track the original lumped element impedances, so deviations from the response of the theoretical low-pass prototype will occur. Also, it is important to be aware of the fact that the response is *periodic* and repeats every $4\omega_C$. This periodicity is a general characteristic of distributed-element filters and should be anticipated.

Example 18-1 Design of a Microstrip Low-Pass Filter

Required:
Design a 0.5 dB $N = 5$ Chebyshev Low-Pass filter normalized to 1 radian/sec using a microstrip-distributed transmission line.

Result:
Figure 18-3*a* illustrates a $n = 5$ 0.5 dB Chebyshev low-pass filter obtained from Table 11-30 normalized for a 3 dB cutoff of 1 radian/sec and 1 Ω. The circuit of Figure 18-3*b* replaces each capacitor *C* by a parallel stub open-circuited transmission line having a characteristic impedance *Z* equal to the reciprocal of the capacitor *C*. Each inductor *L* is realized by a series stub short-circuited transmission line having a characteristic impedance *Z* equal to *L*.

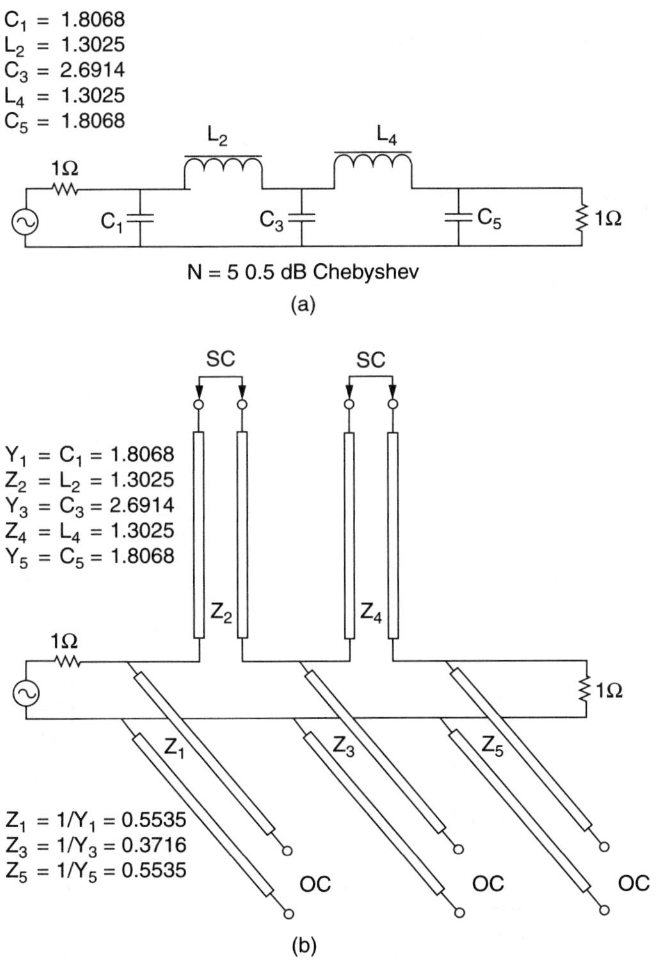

FIGURE 18-3 A normalized $n = 5$ 0.-dB Chebyshev low-pass filter with (*a*) lumped constant values from Table 11-30, and (*b*) a microstrip-distributed transmission line equivalent.

18.4 KURODA'S IDENTITIES

The implementation of Figure 18-3 requires series stubs with a short circuit at the far end. These shorted stubs are extremely difficult to implement on a PC board without using other approaches such as coaxial transmission lines. The physical geometry of the PC board containing these series shorted stubs in addition to the parallel open circuited stubs becomes awkward and impractical.

Kuroda's identities allow the transformation of series stubs into shunt stubs and vice versa. This is an exact transformation and not an approximation. This transformation is accomplished by introducing a building block called a "Unit Element" or UE, which is a transmission line having a length of $\lambda/8$ at ω_C, and a characteristic impedance of 1-Ω in the normalized case.

Figure 18-4 shows the application of Kuroda's identities to make this conversion.

Series to Shunt Stub. To transform a series stub (inductive) to a shunt stub (capacitive), as shown in Figure 18-4*a*, first compute *n* as follows:

$$n = 1 + \frac{Z_1}{Z_O} \tag{18-6}$$

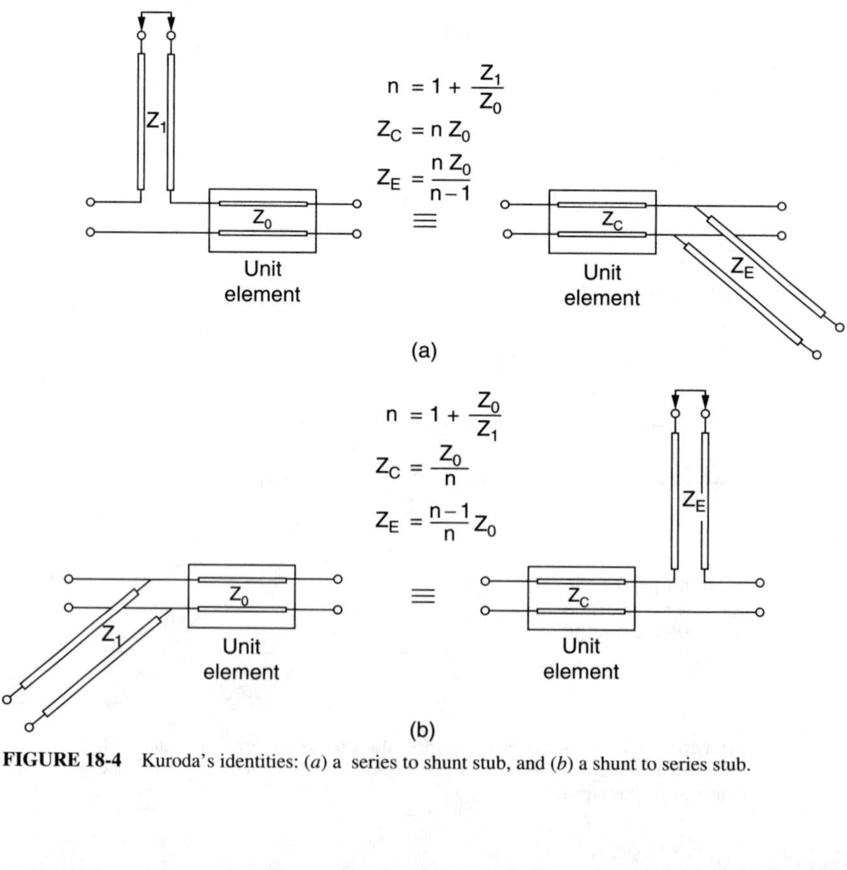

FIGURE 18-4 Kuroda's identities: (*a*) a series to shunt stub, and (*b*) a shunt to series stub.

then

$$Z_C = nZ_O \tag{18-7}$$

$$Z_E = \frac{nZ_O}{n - 1} \tag{18-8}$$

Shunt to Series Stub. To transform a shunt stub (capacitive) to a series stub (inductive), shown in Figure 18-4*b*, first compute *n* as follows.

$$n = 1 + \frac{Z_O}{Z_1} \tag{18-9}$$

then:

$$Z_C = \frac{Z_O}{n} \tag{18-10}$$

$$Z_E = \frac{n - 1}{n} Z_O \tag{18-11}$$

Combining Richards' Transformation and Kuroda's Identities to Design a Low-Pass Filter. The following example is a continuation of Example 18-1 converting all series stubs to parallel stubs using Kuroda's identities.

Example 18-2 Implementing the Microstrip Filter of Example 18-1

Required:
Implement the filter of Example 18-1 using a microstrip approach for a cutoff f_c of 1GHz and a source and load impedance of 50-Ω.

Result:

(a) The normalized low-pass filter is shown in Figure 18-5*a*. Replace each capacitor *C* with a shunt open-circuit stub having a characteristic impedance of 1/*C* and each inductor *L* with a series shorted stub having a characteristic impedance of *L*. This is shown in Figure 18-5*b*.

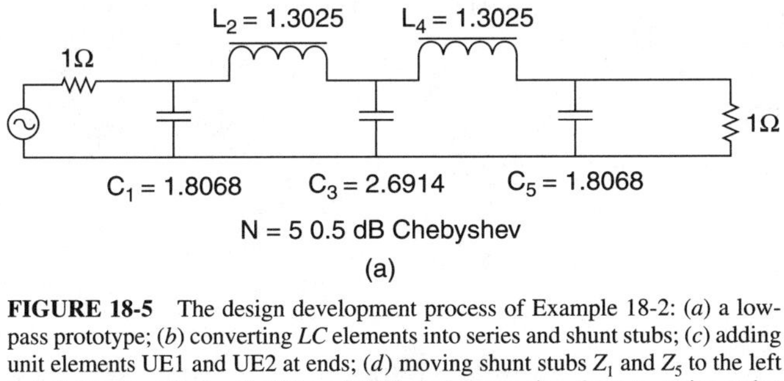

FIGURE 18-5 The design development process of Example 18-2: (*a*) a low-pass prototype; (*b*) converting *LC* elements into series and shunt stubs; (*c*) adding unit elements UE1 and UE2 at ends; (*d*) moving shunt stubs Z_1 and Z_5 to the left and right respectively of UE1 and UE2, and converting them to series stubs; (*e*) adding unit elements EU3 and UE4; (*f*) shifting series stubs Z_1 and Z_2 to the left, and Z_4 and Z_5 to the right, converting them into shunt stubs; and (*g*) a final implementation using Microstrip traces on PC board.

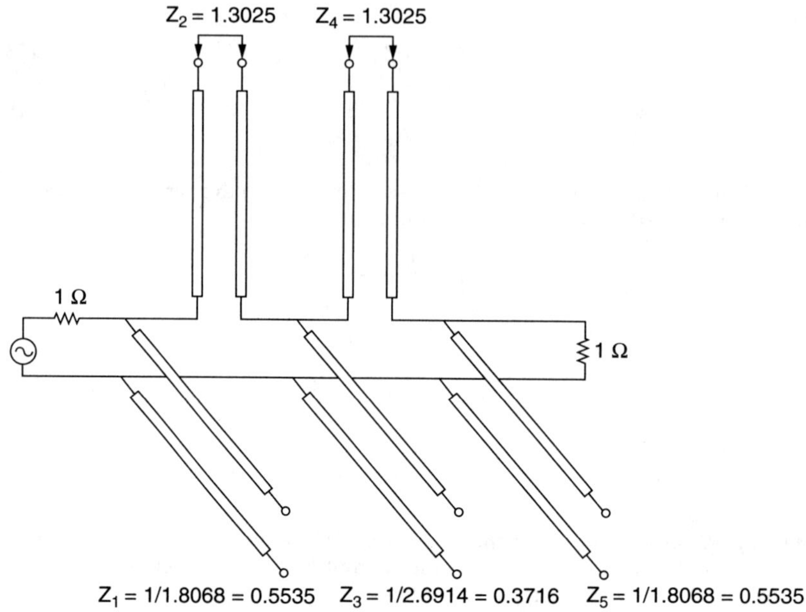

(b)

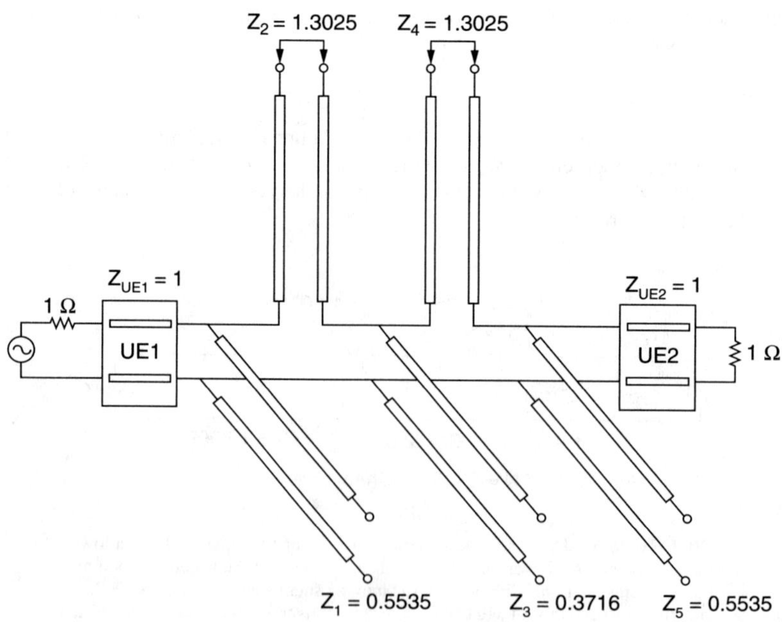

(c)

FIGURE 18-5 (*Continued*)

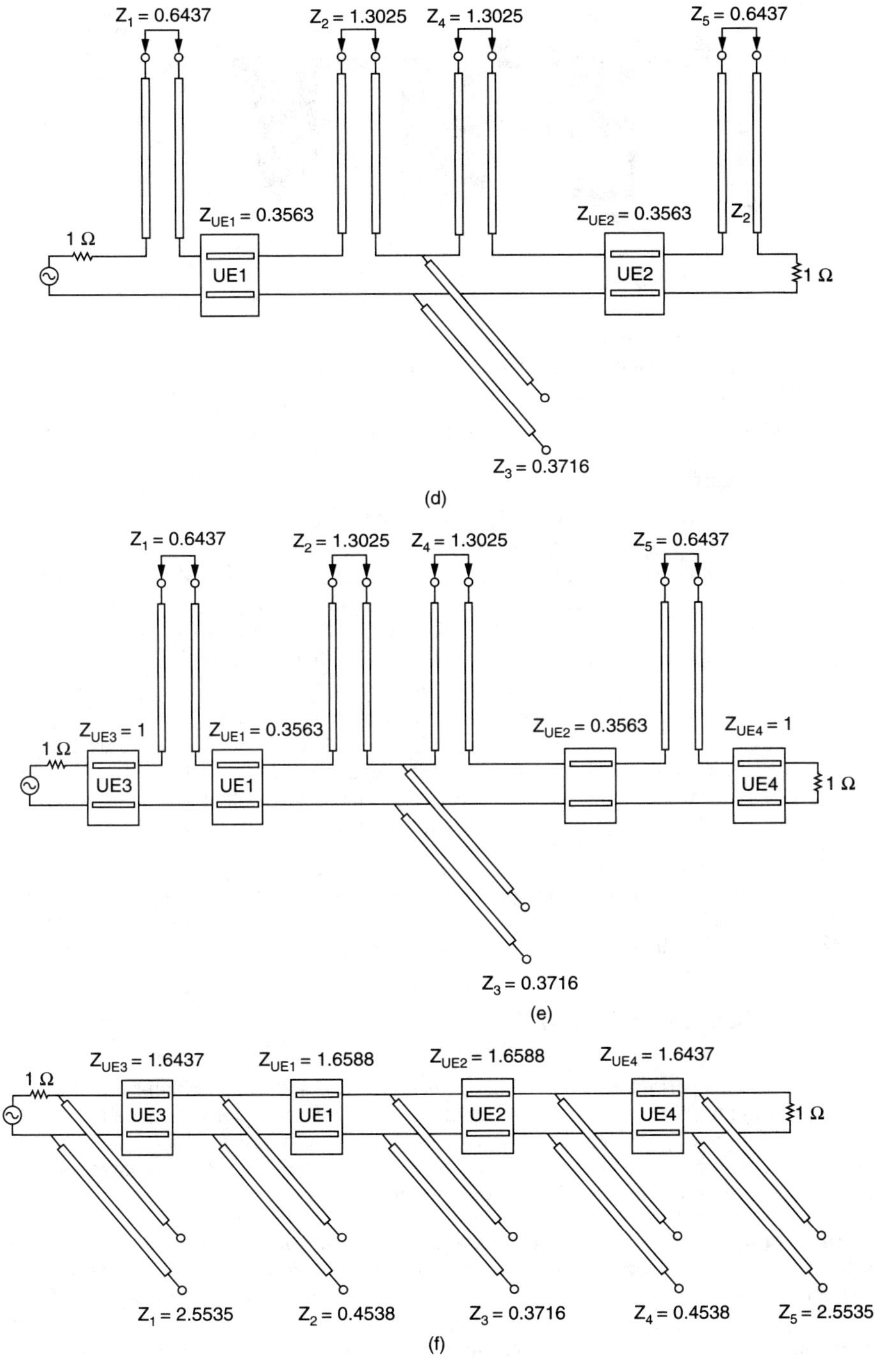

FIGURE 18-5 (*Continued*)

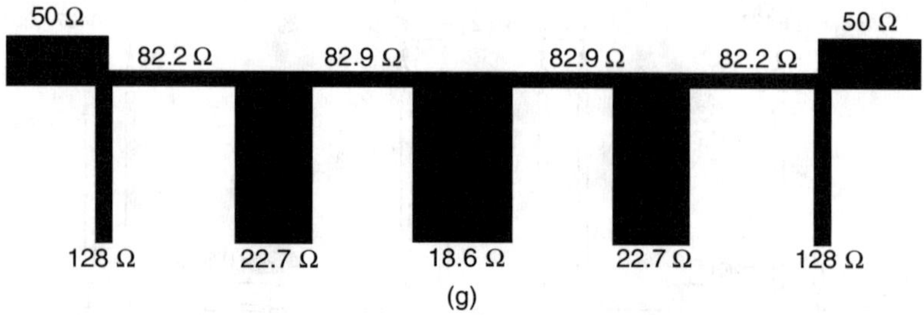

FIGURE 18-5 (*Continued*)

(b) The objective is to convert all series stubs to parallel stubs. First, add unit elements, UE1 and UE2, at each end. This is shown in Figure 18-5*c*.

(c) Move shunt stubs Z_1 and Z_5 to the left and right respectively of UE1 and UE2, converting them to series stubs, as shown in Figure 18-5*d*. The values of Z_1, Z_5, UE1, and UE2 will change in accordance with Equations (18-9) through (18-11) (Figure 18-4*b*) using Kuroda's identities.

(d) Add unit elements EU3 and UE4, as shown in Figure 18-5*e*.

(e) Shift series stubs Z_1 and Z_2 to the left, and Z_4 and Z_5 to the right, converting them into shunt stubs. The values of UE1 through UE4 and Z_1, Z_2, Z_4, and Z_5 will all change as the series stubs are shifted across the unit elements and converted into shunt stubs in accordance with Equations (18-6) through (18-8) (Figure 18-4*a*) using Kuroda's identities. The final circuit is shown in Figure 18-5*f* consisting of all shunt stubs.

(f) The microstrip implementation is illustrated in Figure 18-5*g*, where the design is impedance-scaled to 50-ohms. This figure actually shows the PC board traces representing the characteristic impedances as indicated. Each section should have an electrical wavelength of $\lambda/8$ at an f_c of 1 GHz. At 2 GHz, the wavelength is $\lambda/4$, where maximum attenuation will occur since each stub has an attenuation pole at this frequency. *Keep in mind the presence of a ground plane separated by the PC board thickness.*

The dimensions can be computed using the *txline* program included on the CD-ROM. For example, for a PC board of fiberglass resin (FR4 material), use a dielectric constant of 4.7 and copper as the conductor. Enter the cutoff frequency, impedance, and wavelength ($\lambda/8$ or 45°). The dimensions are indicated on the right after clicking the arrow (→).

18.5 BANDPASS FILTERS

There are various permutations of using open and shorted stubs and transmission line sections for designing filters. Matthaei (see Bibliography) covers numerous design techniques using stubs in great detail, along with a wide range of other microwave filter design methods. The mathematics for the following filters is beyond the intent of this chapter and the reader should refer to Matthaei and the other references listed.

Bandpass Filters Using Shorted Parallel Stubs. An implementation for bandpass filters using shorted $\lambda/4$ stubs connected by transmission lines $\lambda/4$ long at the bandpass center frequency is shown in Figure 18-6*a*. Figure 18-2*a* indicates that an open stub having a wavelength of $\lambda/4$ at a particular frequency is equivalent to a parallel resonant circuit at resonance—that is, exhibiting infinite impedance. The $\lambda/4$ sections connecting the three parallel stubs transform the center parallel stub into a *series* impedance that is a series

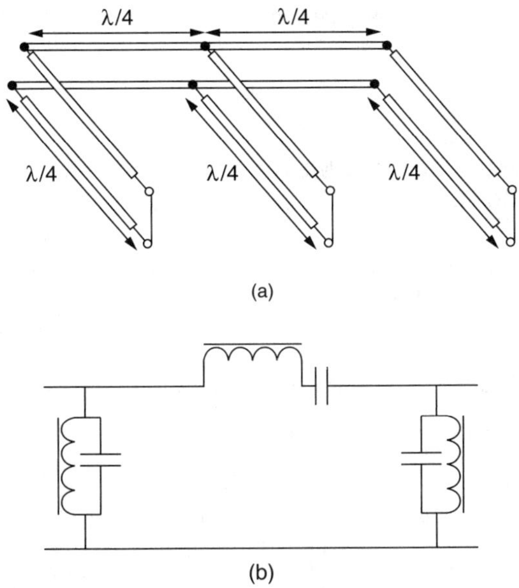

FIGURE 18-6 (*a*) A bandpass filter using λ/4 shorted parallel stubs and interconnecting λ/4 transmission lines, and (*b*) the equivalent bandpass *LC* circuit.

resonant circuit. The equivalent circuit is shown in Figure 18-6*b*. This approach is most effective when applied to narrow-band bandpass filters. If the stubs were of the open-circuit type, a band-stop filter would result.

Bandpass Filters Using Edge-Coupled Half-Wavelength Lines. The filter shown in Figure 18-7*a* is a narrow band edge coupled $n = 5$ bandpass filter. This approach is limited to relatively narrow bandwidths. The analogous *LC* circuit is shown in Figure 18-7*b*

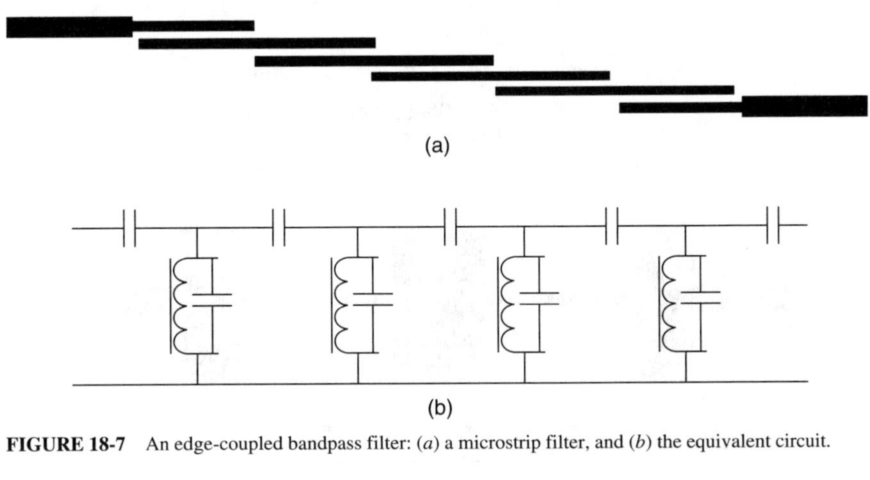

FIGURE 18-7 An edge-coupled bandpass filter: (*a*) a microstrip filter, and (*b*) the equivalent circuit.

and is similar to the narrow-band capacitive coupled resonators covered in Section 5.1. Other versions involve laying the sections end-to-end rather than using parallel coupling.

18.6 ADDITIONAL DESIGN METHODS USING PC BOARD TRACES

The following section discusses how PC board traces can be used as filter elements rather than the use of discrete components.

Using PC Board Traces to Replace Inductors and Capacitors

Rather than using a transmission line approach with open stubs, an alternate method uses traces on the PC board interacting with the ground plane to replace inductors and capacitors. The length of these traces must be much less than $\lambda/4$ in the passband. A long narrow trace can replace an inductor and a wide trace can replace a capacitor. An intuitive explanation for this is that the wide capacitor trace has more surface area and hence more capacitance to the ground plane. The length of the narrow trace is proportional to the inductance. Figure 18-8 illustrates what form this type of implementation would take. The reader should refer to the Winder citation (see the Bibliography) for the computational details.

Inductors can be made more efficiently as far as PC board utilization goes by taking forms other then straight traces. Figure 18-9 shows inductors in a spiral format. Design equations are can be found by researching the Wadell reference (see Bibliography). Removal of ground planes under traces will increase inductance.

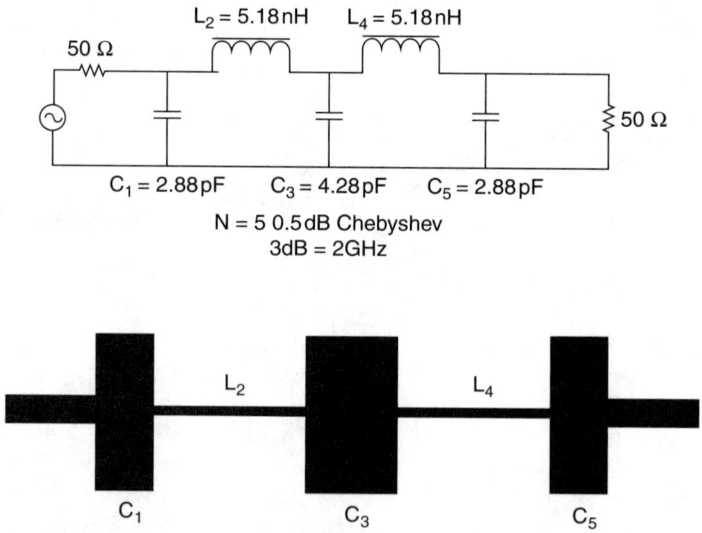

FIGURE 18-8 A microstrip 2 GHz filter using PC board traces.

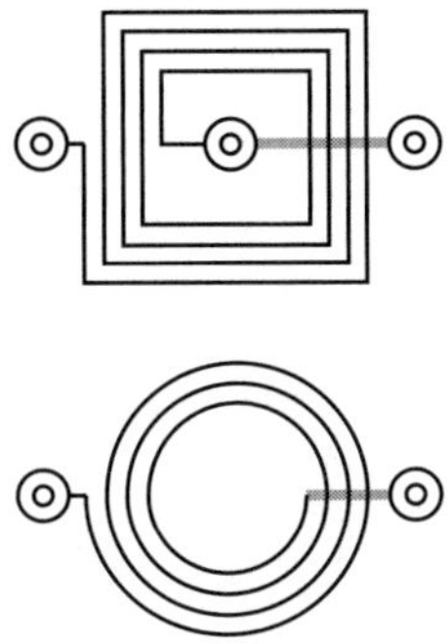

FIGURE 18-9 Spiral inductors.

BIBLIOGRAPHY

Kuroda, K. "Some Equivalent Transformations in Ladder-Type Networks with Distributed Constants." *Inst. Elec. Commun. Engr.* Monograph Series on Circuit Theory (February, 1957) (in Japanese).

Matthaei, G. L., Young, L., and E. M. T. Jones. *Microwave Filters, Impedance-Matching Networks, and Coupling Structures.* Massachusetts: Artech House, 1980.

Rhea, R. W. *HF Filter Design and Computer Simulation.* Georgia: Noble Publishing Company, 1994.

Richards, P. I. "Resistor Transmission Line Circuits." *Proceedings of the IRE* 36 (February, 1948): 217.

Wadell, B. C. *Transmission Line Design Handbook.* Massachusetts: Artech House, 1991.

Winder, S. *Analog and Digital Filter Design.* Massachusetts: Elsevier Science, 2002.

APPENDIX A
DISCRETE SYSTEMS MATHEMATICS

A.1 DIGITAL FILTER MATHEMATICS (THE z-TRANSFORM)

The principal analysis tool for continuous or analog systems has been the Laplace transform. Digital filters are, however, more closely related to discrete-time systems, which are normally modeled using z-transforms. The z-transform, it should be noted, is directly related to the Laplace transform. Specifically, if $x(t)$ has a Laplace transform $X(s)$, such that:

$$x(t) \overset{L}{\leftrightarrow} X(s) \tag{A-1}$$

then the delay theorem states that

$$x(t - T_s) \overset{L}{\leftrightarrow} e^{-sT_s}X(s)$$

where $T_s = 1/f_s$. The term e^{-sT_s} is ubiquitous in discrete-time signal and system representation and analysis. As a result, it has been given a shorthand representation:

$$z = e^{sT_s} \quad \text{or} \quad z^{-1} = e^{-sT_s} \tag{A-3}$$

Specifically, z^{-n} denotes a signal delay of n samples with each sample delay being T_s seconds in duration. It immediately follows when using the z notation, that the z-transform of the time-series $x(nT_s)$ can be expressed as

$$X(z) = X(s)|_{z=\exp(sT_s)} \tag{A-4}$$

The Laplace transform of a time-series $x[k] = x(kT_s)$ is given by

$$X(s) = x[0] + x[1]e^{-sT_s} + x[2]e^{-s2T_s} + \cdots = \sum_{k=0}^{\infty} x[k]e^{-skT_s} \tag{A-5}$$

which has a z-transform representation given by

$$X(z) = x[0] + x[1]z^{-1} + x[2]z^{-2} + \cdots = \sum_{k=0}^{\infty} x[k]\, z^{-k} \tag{A-6}$$

The resulting transform is represented, in this case, as an infinite sum. If the time series, for example, represents the impulse response of an FIR filter, then the sum would be finite

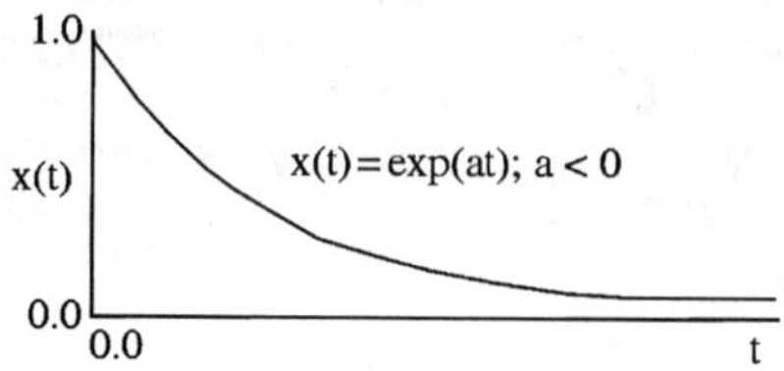

FIGURE A-1 An example exponential signal.

in length. In some cases, infinite sums can be reduced to a rational polynomial description $X(z) = N(z)/D(z)$, where $N(z)$ and $D(z)$ are polynomials of finite order. As an example of the transform a real exponential can be found in the following example.

Example A-1 Real Exponential

Required:

Suppose $x(t)$ is a decaying exponential given by $x(t) = e^{at}$ for $t \geq 0$, 0 for $t < 0$, and $a < 0$. The trajectory of $x(t)$ is shown in Figure A-1. What is $X(z)$ if the sample rate is $f_s = 1/T_s$ samples per second?

Result:

The sampled time-series is given by

$$x(nT_s) = e^{anT_s}$$

It therefore follows that

$$X(s) = \sum_{n=0}^{\infty} e^{(anT_s)} e^{(-nsT_s)}$$

and

$$X(z) = X(s)|_{z=e^{(sT_s)}} = \sum_{n=0}^{\infty} e^{(anT_s)}(z^{-n}) = \sum_{n=0}^{\infty} (e^{aT_s}z^{-1})^n$$

The above equation can be placed in closed form by observing that if $|x| < 1$, then

$$\sum_{i=0}^{\infty} x^i = 1 + x + x^2 + x^3 + \cdots = \frac{1}{1 - x}$$

It therefore follows that:

$$X(z) = \sum_{n=0}^{\infty} (e^{aT_s}z^{-1})^n = \frac{1}{1 - e^{aT_s}z^{-1}} = \frac{z}{z - e^{aT_s}} = \frac{z}{z - \alpha}$$

where $\alpha = e^{aT_s}$.

The methods by which z-transforms are produced and manipulated are very similar to those techniques used to support signal and system analysis using Laplace transforms. Specifically, signals and systems are normally obtained by intelligently using pre-defined information stored in tables of a standard z-transform, such as the entries found in Table A-1. Table A-1 contains the common z-transforms of discrete-time elementary signals.

The study of a real exponential is reported in the following example.

TABLE A-1 Elementary Functions

$x(t)$ or $x(k)$	$X(z)$
δ_K (Kronecker delta)	1
$\delta(k-m)$	z^{-n}
1 (unit step)	$z/(z-1)$
k (ramp)	$z/(z-1)^2$
k^2	$\dfrac{z(z+1)}{(z-1)^3}$
k^3	$\dfrac{z(z^2+4z+1)}{(z-1)^4}$
k^4	$\dfrac{z(z^3+11z^2+11z+1)}{(z-1)^5}$
k^5	$\dfrac{z(z^4+26z^3+66z^2+26z+1)}{(z-1)^6}$
a^k	$\dfrac{z}{z-a}$
ka^k	$\dfrac{az}{(z-a)^2}$
k^2a^k	$\dfrac{az(z+a)}{(z-a)^3}$
k^3a^k	$\dfrac{az(z^2+4az+a^2)}{(z-a)^4}$
k^4a^k	$\dfrac{az(z^3+11az^2+11a^2z+a^3)}{(z-a)^5}$
k^5a^k	$\dfrac{az(z^4+26az^3+66a^2z^2+26a^3z+a^4)}{(z-a)^6}$
$(k+1)a^k$	$\dfrac{z^2}{(z-a)^2}$
$\dfrac{(k+1)(k+2)a^k}{2!}$	$\dfrac{z^3}{(z-a)^3}$
$\dfrac{(k+1)(k+2)(k+3)a^k}{3!}$	$\dfrac{z^4}{(z-a)^4}$
$\dfrac{(k+1)(k+2)\cdots(k+m)a^k}{m!}$	$\dfrac{z^{m+1}}{(z-a)^{m+1}}$
$\sin(k\omega T_s)$	$\dfrac{z\sin(\omega T_s)}{z^2-2z\cos(\omega T_s)+1}$
$\cos(k\omega T_s)$	$\dfrac{z[z-\cos(\omega T_s)]}{z^2-2z\cos(\omega T_s)+1}$
$a^k\sin(k\omega T_s)$	$\dfrac{az\sin(\omega T_s)}{z^2-2az\cos(\omega T_s)+a^2}$
$a^k\cos(k\omega T_s)$	$\dfrac{z[z-a\cos(\omega T_s)]}{z^2-2az\cos(\omega T_s)+a^2}$
$\sin(k\omega T_s+\theta)$	$\dfrac{z[z\sin\theta+\sin(\omega T_s-\theta)]}{z^2-2z\cos(\omega T_s)+1}$
$\cos(k\omega T_s+\theta)$	$\dfrac{z[z\cos\theta+\cos(\omega T_s-\theta)]}{z^2-2z\cos(\omega T_s)+1}$

TABLE A-1 Elementary Functions (*Continued*)

$x(t)$ or $x(k)$	$X(z)$
$ka^k \sin(k\omega T_s)$	$\dfrac{z(z-a)(z+a)a\sin(\omega T_s)}{[z^2 - 2az\cos(\omega T_s) + a^2]^2}$
$ka^k \cos(k\omega T_s)$	$\dfrac{az(z^2\cos(\omega T) - 2az + a^2\cos(\omega T_s)]}{[z^2 - 2az\cos(\omega T_s) + a^2]^2}$
$\sinh(k\omega T_s)$	$\dfrac{z\sinh(\omega T_s)}{z^2 - 2z\cosh(\omega T_s) + 1}$
$\cosh(k\omega T_s)$	$\dfrac{z[z - \cosh(\omega T_s)]}{z^2 - 2z\cosh(\omega T_s) + 1}$
$\dfrac{a^k}{k!}$	$\exp(a/z)$
$\dfrac{(1na)^k}{k!}$	$a^{1/z}$
$\dfrac{n!}{(n-k)!k!}a^k b^{n-k}$	$\dfrac{(bz+a)^n}{z^n}$
$\dfrac{1}{k}$ (for $k = 1, 2, 3, \ldots$)	$ln\left(\dfrac{z}{z-1}\right)$
$\dfrac{k(k-1)}{2!}$	$\dfrac{z}{(z-1)^3}$
$\dfrac{k(k-1)(k-2)}{3!}$	$\dfrac{z}{(z-1)^4}$
$\dfrac{k(k-1)(k-2)\cdots(k-m+1)}{m!}$	$\dfrac{z}{(z-1)^m}$

Example A-2 Real Exponential

Required:

Suppose $a < 0$, $x(t) = e^{at}$ for $t \geq 0$, and 0 otherwise. The signal is sampled at a rate of $f_s = 1/T_s$ samples per second. The resulting time series then becomes

$$x[nT_s] = e^{anT_s} = \alpha^n$$

where $\alpha = e^{aT_s}$. Interpret the z-transform of $x[nT_s]$ in terms of the objects found in Table A-1.

Result:

$X(z)$ was determined in Example A-1 to be

$$X(z) = \frac{1}{1 - e^{aT_s}z^{-1}} = \frac{z}{z - e^{aT_s}} = \frac{z}{z - \alpha}$$

which corresponds to the ninth entry in Table A-1.

The study of a complex exponential is reported in the following example.

Example A-3 Complex Exponential

Required:

Sampling $x(t) = \exp(-j\omega_0 t)$ results in a time-series $x[nT_s] = \exp(-j\omega_0 nT_s)$. Determine the common z-transform of $x(t)$.

Result:
Using Example A-1 as a model, $X(z)$ is given by

$$X(z) = \sum_{n=0}^{\infty} e^{-j\omega_0 nT_s} z^{-n} = \frac{1}{1 - e^{-j\omega_0 T_s} z^{-1}}$$

Using the standard trigonometric identity

$$e^{j\theta} = \cos(\theta) + j\sin(\theta)$$

it follows that

$$X(z) = \frac{1}{[1 - e^{-j\omega_0 T_s} z^{-1}]} = \frac{1}{[1 - \cos(\omega_0 T_s) z^{-1}] + j\sin(\omega_0 T_s) z^{-1}}$$

$$= \frac{1 - \cos(\omega_0 T_s) z^{-1} - j\sin(\omega_0 T_s) z^{-1}}{[1 - \cos(\omega_0 T_s) z^{-1}]^2 + [\sin(\omega_0 T_s) z^{-1}]^2} = \frac{1 - \cos(\omega_0 T_s) z^{-1}}{D(z)} - \frac{j\sin(\omega_0 T_s) z^{-1}}{D(z)}$$

where

$$D(z) = 1 - 2\cos(\omega_0 T_s) z^{-1} + z^{-2}$$

Furthermore, the z-transform of a pure sine and cosine signal can be deduced from the preceding equation. That is

$$Z[\cos(\omega_0 T_s)] = \frac{1 - \cos(\omega_0 T_s) z^{-1}}{D(z)} = \frac{z^2 - \cos(\omega_0 T_s) z}{z^2 D(z)}$$

$$Z[\sin(\omega_0 T_s)] = \frac{\sin(\omega_0 T_s) z^{-1}}{D(z)} = \frac{\sin(\omega_0 T_s) z}{z^2 D(z)}$$

where $z^2 D(z) = z^2 - 2\cos(\omega_0 T_s)\, z + 1$.

The objects found in Table A-1 can be combined together to define the common z-transform of an arbitrary signal $x(t)$. The process by which the individual z-transforms are combined to synthesize more complicated transforms are governed by a set of rules described in Table A-2.

The next example demonstrates how the elements of Table A-2 can facilitate the analysis of a complex signal.

Example A-4 Higher Order Signals

Required:
Suppose $x(t) = t_2$, $x_1(t) = t$, and $x(t) = t x_1(t)$. What is the z-transform of $x(t)$?

Result:
From property 6 of Table A-2, and knowledge that $X_1(z) = T_s z/(z - 1)^2$, it follows that:

$$X(z) = -T_s z dX_1(z)/dz = -T_s z\left[T_s \frac{1}{(z - 1)^2} - 2T_s \frac{z}{(z - 1)^3}\right]$$

$$= -T_s^2 z\left[\frac{(z - 1) - 2z}{(z - 1)^3}\right] = T_s^2 z(z + 1)/(z - 1)^3$$

The next example demonstrates the use of Table A-2 entries for a specific signal.

TABLE A-2 Functional Relationships

Rule	If	Then
Linearly	$x(t) = x_2(t)$	$X(z) = X_1(z) + X_1(z)$
Scaling	$x(t) = ax_1(t)$	$X(z) = aX_1(z)$
Delay (left shift)	$x(t) = x_1(t + kT_s)$	$X(z) = z^k X_1(z) - z^k x_1(0) - z^{k-1} x_1(T_s) - \cdots - zx(kT_s - T_s)$
Advance (right shift)	$x(t) = x_1(t - kT_s)$	$X(z) = z^{-k} X_1(z) + z^{-k+1} x_1(-T_s) + \cdots + z^{-1} x_1(-kT_s + T_s) + x_1(-kT_s)$
Exponential scaling	$x(t) = \exp(-at)x_i(t)$	$X(z) = X_1[\exp(aT_s)z]$
Time bias	$x(t) = tx_1(t)$	$X(z) = -T_s z\, dX_1(z)/dz$
Reciprocal time bias	$x(t) = x_1(t)/t$	$X(z) = (-1/T_s)\int (X_1(z)/z)dz$
Radix "a" bias	$x(t) = a^t x_1(t)$	$X(z) = X_1(z/a^{T_s})$
Summation	$x(t) = \Sigma y(iT), i = 1, \ldots$	$X(z) = zY(z)/(z - 1)$
Periodic sequences	$x(t) = y(t = NT_s)$	$X(z) = z^N Y(z)/(z^N - 1)$
Convolution	$z(t) = \Sigma x(t - nT_s)\, y(t)$	$Z(z) = X(z)\, Y(z)$

Example A-5 Higher Order Signals

Required:
Suppose $x(t) = te^{-at}$ along with the assignments $x_1(t) = t$ and $x(t) = \exp(-at)x_1(t)$. What is the z-transform of $x(t)$?

Result:
From property 5 of Table A-2, and knowledge that $x_1(t) = T_s z/(z - 1)^2$, it follows that:

$$X(z) = X_1[\exp(aT_s)z] = \frac{T_s[\exp(aT_s)z]}{[\exp(aT_s)z - 1]^2} = \frac{T_s \exp(-aT_s)z}{[z - \exp(-aT_s)]^2}$$

An important z-transform relationship is referred to as the *initial value theorem*, which can be used to compute $x(0)$ given knowledge of $X(z)$. Specifically:

$$\lim_{t \to 0} x(t) = x(0) = \lim_{z \to \infty} X(z) \qquad \text{(A-7)}$$

The typical use of the initial value theorem is studied in the following example.

Example A-6 Initial Value Theorem

Required:
From Example A-5 it was shown that if $x(t) = t\exp(at)$, then

$$X(z) = \frac{T_s \exp(-aT_s)z}{[z - \exp(-aT_s)]^2}.$$

Using the initial value theorem, determine that $x(0) = 0$.

Result:
A direct application of the initial value theorem yields

$$\lim_{z \to \infty} X(z) = \frac{T \exp(-aT)z}{[z - \exp(-aT)]^2} = 0$$

which is in agreement $x(t) = t\exp(at)$ evaluated at $t = 0$.

Another important relationship is referred to as the *final value theorem*, which equates $x(t = 0)$ to the z-transform of $x(t)$, namely $X(z)$. The final valued theorem is given by

$$\lim_{t\to\infty} x(t) = \lim_{z\to 1}\left[\frac{(z-1)}{z}X(z)\right] \tag{A-8}$$

if $X(z)$ has no more than one pole on the unit circle.

The typical use of the final value theorem is studied in the following example.

Example A-7 Final Value Theorem

Required:
Determine the final value of $x(t) = \exp(-at)$, having a z-transform $X(z) = z/[z - \exp(-aT_s)]$, using the final value theorem.

Result:
For $aT_s < 1.0$, $X(z)$ has no pole residing on the unit circle. A direct application of the final value theorem states that

$$\lim_{z\to 1}\frac{(z-1)z}{z(z-\exp(-aT_s))} \to 0$$

which is the expected final value of $x(t)$.

There exists a systematic technique of converting a known Laplace transform $X(s)$ into its equivalent z-transform $X(z)$. It makes use of the *residue theory* of calculus and is given by

$$X(z) = \sum_k \text{Res}(\{X(s)/[1 - \exp(sT_s)z^{-1}]\}, p_k) \tag{A-9}$$

where $\{p_k\}$ is a pole of $X(s)$. If $X(s)$ is expanded as a Laurent series, having the form

$$X(s) = \frac{a_{-1}}{s} + \frac{a_{-2}}{s^2} + \cdots \tag{A-10}$$

the coefficient a_{-1} is called the residue of $X(s)$ at $s \to \infty$. The residue can be computed as

$$a_{-1} = \lim_{s\to\infty} sX(s) \tag{A-11}$$

Suppose $X(s) = N(s)/D(s)$ where $D(s)$ is an nth-order polynomial consisting of k distinct poles. Then $D(s)$ can be factored as

$$D(s) = \prod_{i=1}^{k}(s - p_i)^{n(i)} \tag{A-12}$$

where $n(i)$ is called the *multiplicity* of the root located at $s = p_i$ and $n = \Sigma n(i)$, $i = 1, \ldots, k$. The series expansion of $X(s)$ is given by

$$X(s) = \frac{a_{-11}}{(s-p_1)} = \frac{a_{-21}}{(s-p_1)^2} + \cdots + \frac{a_{-n(1)1}}{(s-p_1)^{n(1)}} + \frac{a_{-12}}{(s-p_2)} + \cdots + \frac{a_{-n(k)k}}{(s-p_k)^{n(k)}} \tag{A-13}$$

The residue of $X(s)$ at pole location p_i is the coefficient a_{-1i}. That is, it is the coefficient associated with the simple (first-order) pole $(s - p_i)^1$. In general

$$a_{-1k} = \text{Res}(\{X(s)/[1 - \exp(sT_s)z^{-1}]\}, p_k)$$
$$= \frac{1}{(n(k) - 1)!} \lim_{s \to p_k} \frac{d^{n(k)-1}(s - p_k)^{n(k)} X(s)/(1 - \exp(sT_s)z^{-1})}{d^{n(k)-1} s} \tag{A-14}$$

If all the poles of $X(s)$ (in other words, the roots of $D(s)$) are first order (simple), then the residues are

$$a_{-1k} = \lim_{s \to p_k} \left\{ \frac{(s - p_k)X(s)}{[1 - \exp(sT_s)z^{-1}]} \right\} \tag{A-15}$$

The use of residue calculus is demonstrated in the following examples.

Example A-8 Residue Method

Required:
Suppose $x(t) = \exp(-at)$ for $t \geq 0$, and 0 otherwise. Then, $X(s) = 1/(s + a)$. Compute the residue of $X(s)$.

Result:
From $X(s) = 1/(s + a)$, it follows that

$$a_{-11} = \lim_{s \to -a} \frac{(s + a)[1/(s + a)]}{[1 - \exp(sT_s)z^{-1}]} = \frac{1}{1 - \exp(-aT_s)z^{-1}} = \frac{z}{z - \exp(-aT_s)}$$

which agrees with the result listed as the ninth entry in Table A-1.

Example A-9 Residue Method

Required:
Suppose $x(t) = t$ for $t \geq 0$, and 0 otherwise. Then, $X(s) = 1/s^2$. Compute the residue of $X(s)$.

Result:
From $X(s) = 1/s^2$, $n(1) = 2$, it follows that

$$X(s) = \frac{a_{-11}}{(s - 0)} + \frac{a_{-21}}{(s - 0)^2}$$

and

$$a_{-11} = \lim_{s \to 0} \frac{d\{s^2(1/s^2)/[1 - \exp(sT_s)z^{-1}]\}}{ds} = \lim_{s \to 0} \frac{d\{1/[1 - \exp(sT_s)z^{-1}]\}}{ds}$$
$$= \lim_{s \to 0} \frac{T \exp(sT_s)z^{-1}}{[1 - \exp(sT_s)z^{-1}]^2} = \frac{T_s z^{-1}}{(1 - z^{-1})^2} = \frac{T_s z}{(z - 1)^2} = X(z)$$

Example A-10 Residue Method

Required:
Suppose $x(t) = \cos(\omega_0 t)$ for $t \geq 0$, and 0 otherwise. Then $X(s) = s/(s^2 + \omega_0^2) = s/(s + j\omega_0)(s - j\omega_0)$. Compute the residue of $X(s)$.

Result:
For poles $p_1 = -j\omega_0$ and $p_2 = j\omega_0$, it follows that

$$a_{-11} = \lim_{s \to j\omega_0} \frac{(s + j\omega_0)s}{(s + j\omega_0(s - j\omega_0)[1 - \exp(sT_s)z^{-1}]} = \frac{-j\omega_0}{-2j\omega_0[1 - \exp(-j\omega_0 T_s)z^{-1}]}$$
$$= \frac{1}{2[1 - \exp(-j\omega_0 T_s)z^{-1}]}$$
$$a_{-12} = \lim_{s \to j\omega_0} \frac{(s = j\omega_0)s}{(s - j\omega_0)(s + j\omega_0)[1 - \exp(sT_s)z^{-1}]} = \frac{j\omega_0}{2j\omega_0[1 - \exp(j\omega_0 T_s)z^{-1}]}$$
$$= \frac{1}{2[1 - \exp(j\omega_0 T_s)z^{-1}]}$$

Finally

$$X(z) = \Sigma \mathrm{Res}(\{X(s)/[1 - \exp(sT_s)z^{-10}]\}, p_k) = a_{-11} + a_{-12}$$
$$= \frac{1}{2}\left[\frac{1}{1 - \exp(-j\omega_0 T_s)z^{-1}} + \frac{1}{1 - \exp(j\omega_0 T_s)z^{-1}}\right]$$
$$= \frac{1}{2}\left[\frac{1 - \exp(-j\omega_0 T_s)z^{-1} + 1 - \exp(j\omega_0 T_s)z^{-1}}{1 - \exp(-j\omega_0 T_s) + \exp(j\omega_0 T_s)z^{-1} + z^{-2}}\right]$$
$$= \frac{1 - \cos(\omega_0 T_s)z^{-1}}{1 - 2\cos(\omega_0 T_s)z^{-1} + z^{-2}} = \frac{z(z - \cos(\omega_0 T_s)]}{z^2 - 2\cos(\omega_0 T_s)z + 1}$$

A.2 INVERSE z-TRANSFORM

The inverse z-transform defines a mapping that will convert a given z-transform, say $X(z)$, into its original parent time-series $x[k]$. Formally, the inverse z-transform of $X(z)$ is given by

$$x(n) = Z^{-1}(X(z)) = \frac{1}{2\pi j}\int_C X(z)z^{n-1}\,dz \tag{A-16}$$

where C is a close path found in the z plane. In general, the definition of the inverse z-transform is too difficult and awkward to use on a routine basis. A similar condition exists in the study of Laplace transforms where an integral definition of the inverse Laplace transform can also be found. In the case of Laplace transforms, standard tables of transform pairs are normally used to bypass the formal integration process. In the case of z-transforms, standard tables plus other short-cut techniques are used. The primary z-transform inversion methods are

- Long division
- Partial fraction expansion
- Residue calculus

Long Division. The first inversion method, long division, converts a rational polynomial description of a z-transform signal, namely $X(z) = N(z)/D(z)$ into time-series $x[k]$. More specifically, if

$$X(z) = N(z)/D(z) \tag{A-17}$$

then

$$D(z)\,\overline{\big)\,N(z)}\quad x[0] + x[1]z^{-1} + x[1]z^{-2} + \cdots \tag{A-18}$$

The result is a z-transform $X(z) = x[0] + x[1] + x[2]\,z^{-1} + \cdots$ From the formal definition of the standard z-transform defined by Equation (A-6), it immediately follows that

$$\{x(n)\} = \{a_0, a_1, a_2, \ldots\} \tag{A-19}$$

This method can prove useful if the time domain analysis can be restricted to a few sample values locally clustered around $t = 0$. An example of this is the transient analysis of a highly damped system. The long division method is illustrated in the following example.

Example A-11 Long Division

Required:
Consider again Example A-4 for a given $X(z) = T_s^2 z(z + 1)/(z - 1)^3$. Assume $T_s = 1$ for the purpose of numerical analysis. Then

$$N(z) = z^2 + z;\ D(z) = z^3 + 3z^2 + 3z - 1$$

What are the first four sample values of $x[k]$?

Results:
Evaluating $X(z) = N(z)/D(z)$ using long division will yield

$$\begin{array}{r|l} & z^{-1} + 4z^{-2} + 9z^{-3} \\ \hline z^3 - 3z^2 + 3z - 1 & z^2 + z \\ & \underline{z^2 - 3z + 3z^0 - z^{-1}} \\ & \quad 4z - 3z^0 + z^{-1} \\ & \quad \underline{4z - 12z^0 + 12z^{-1} + 4z^{-2}} \\ & \qquad 9z^0 - 11z^{-1} + 4z^{-2} \end{array}$$

or $X(z) = 0z^0 + 1z^{-1} + 4z^{-2} + 9z^{-3} + \cdots$ and $x[k] = \{0, 1, 4, 9, \ldots\}$. From Example A-4, it was shown that the inverse z-transform of $X(z)$ should be $x[k] = k^2$, which agrees with the first four computed sample values of the long-division expansion of $x[k]$.

Partial Fraction Expansion. When working with Laplace transforms, the inverse of a given nth-order $X(s)$ usually begins with a partial fraction, or Heaviside expansion. The form of the expansion appears as follows:

$$X(s) = \sum \alpha_i X_i(s) \tag{A-20}$$

The arithmetic order of the individual terms $X_i(s)$ (that is, n_i) is less than or equal to the order of $X(s)$ itself. If the order of $X_i(s)$ is sufficiently low (for instance, $n_i = 1$ or 2), it is reasonable to assume that $X_i(s)$ can be matched to the Laplace transforms of the elemental functions. This same technique can also be used to invert a given z-transform $X(z)$. Specifically, assume that an nth-order $X(z)$ can be factored into

$$X(z) = \sum \alpha_i X_i(z) \tag{A-21}$$

If the order of $X_i(z)$ is n_i is sufficiently small to be contained in Table A-1 then the inversion of $X_i(z)$ can be accomplished using the table of elemental functions presented in Table A-1.

For

$$X(z) = \frac{N(z)}{D(z)} \begin{array}{l} \text{order}\{N(z)\} \leq n \\ \text{order}\ \{D(z)\} \leq n \end{array} \tag{A-22}$$

the poles of $X(z)$ are given by those values of z which set $D(z) = 0$, and the zeros are solutions to $N(z) = 0$. The poles, or roots of $D(z)$, can appear in one of two forms:

- Distinct
- Repeated

The distinct, or non-repeated root case, results in a representation $D(z)$ given by

$$D(z) = \prod_{i=1}^{N} (z - p_i) \tag{A-23}$$

where p_i is either a real or complex number. If the poles are complex, they are assumed to appear in complex conjugate pairs. That is, if p_i is a complex root, then so is p_i^*, the complex conjugate of p_i.

Suppose that the n roots of an nth-order polynomial $D(z)$ are

$$\begin{array}{ll} z = r_i & \text{if } r_i \text{ is real and } D(r_i) = 0, \\ & i = 1, 2, \ldots, K \\ z = p_i \quad \text{or} \quad p_i^* & \text{if } p_i \text{ is complex and} \\ & D(p_i) = D(p_i^*) = 0, i = K \\ & +1, \ldots, N \end{array} \tag{A-24}$$

Under such conditions

$$\begin{aligned} X(z) &= a_0 + \sum_{i=1}^{K} \frac{a_i z}{(z - r_i)} + \sum_{i=K}^{(N-K)/2} \frac{b_i z}{(z - p_i)} + \frac{c_i z}{(z - p_i^*)} \\ &= a_0 + \sum_{i=1}^{K} a_i X_i[z : r_i] + \sum_{i=1}^{(N-K)/2} b_i X_i[z : p_i] + c_i X_i\left[z : p_i^*\right] \end{aligned} \tag{A-25}$$

where the factors $X_i[z : v_i]$ are defined in the obvious manner. The preceding expression is called a Heaviside expansion of $X(z)$. The coefficients $\{a_i\}$, $\{b_i\}$, and $\{c_i\}$ satisfy

$$a_0 = \lim_{z \to 0} X(z) \tag{A-26}$$

$$\begin{aligned} a_1 &= \lim_{z \to r_i} \frac{(z - r_i)X(z)}{z} \qquad i = 1, 2, \ldots, K \\ b_i &= \lim_{z \to p_i} \frac{(z - p_i)X(z)}{z} \qquad i = 1, 2, \ldots, (N - K)/2 \\ c_i &= \lim_{z \to p_i^*} \frac{\left(z - p_i^*\right)X(z)}{z} \qquad i = 1, 2, \ldots, (N - K)/2 \end{aligned}$$

The inverse z-transform of $X_i[z : r_i]$ (that is, the real pole) is defined by the elementary function $x(t) = \exp(-a_i t)$ (see Table A-1):

$$Z(x(t)) = \frac{z}{z - \exp(-a_i T_s)} = X_i[z : r_i] \tag{A-27}$$

where $r_i = \exp(-a_i T)$. Similarly, if $x(t) = \exp(-j\omega_i t)$, then

$$Z[x(t)] = \frac{z}{z - \exp(-j\omega_i T_s)} = X_i[z : p_i] \tag{A-28}$$

which can be similarly interpreted if $p_i^* = \exp(j\omega_i T)$. It follows that

$$x(nT) = Z^{-1}(X(z)) = a_0 + \sum_{i=1}^{K} a_i \exp(-r_i T_s) + \sum_{i=1}^{(N-K)/2} b_i \exp(-j\omega_i T_s) + c_i \exp(j\omega_i T_s) \tag{A-29}$$

which is the inverse z-transform of $X(z)$. The next example demonstrates a partial fraction expansion of a function having distinct roots.

Example A-12 Partial Fraction Expansion

Required:

Suppose

$$X(z) = \frac{z^2}{(z - 1)(z - \exp(-T_s)}$$

Perform a Heaviside expansion of $X(z)$ of the form

$$X(z) = a_0 + \frac{a_1 z}{(z - 1)} + \frac{a_2 z}{(z - \exp(-T_s))}$$

Determine the Heaviside coefficients a_0, a_1, and a_2.

Result:

Evaluating

$$a_0 = \lim_{z\to 0} X(z) = 0$$

$$a_1 = \lim_{z\to 1}(z - 1)\frac{X(z)}{z} = \lim_{z\to 1}\frac{z}{[z - \exp(-T_s)]} = \frac{1}{1 - \exp(-T_s)}$$

$$a_2 = \lim_{z\to\exp(-T_s)}[z - \exp(-T_s)]\frac{X(z)}{z}$$

$$= \lim_{z\to\exp(-T_s)}\frac{z}{(z - 1)} = \frac{\exp(-T_s)}{\exp(-T_s) - 1} = \frac{1}{1 - \exp(-T_s)}$$

It then follows that $x(nT_s) = \begin{cases} a_1 + a_2 \exp(-nT_s) & \text{for } n \geq 0, \\ 0 & \text{otherwise} \end{cases}$

For the repeated root case, the nth-order polynomial $D(z)$ can be factored in terms of L distinct roots such that

$$D(z) = \prod_{i=1}^{L}(z - p_i)^{n(i)};\ n = \sum_{i=1}^{L} n(i) \tag{A-30}$$

where $n(i)$ is said to be multiplicity of the root p. If p_i is a complex root, then so is its complex conjugate value p_i^*. If $n(i) = 1$, then the ith pole is non-repeated. Otherwise, for

$$X(z) = N(z)/D(z) \tag{A-31}$$

then

$$X(z) = a_0 + \cdots + \frac{c_{i1}z}{z - p_i} + \frac{c_{i2}z}{(z - p_i)^2} + \cdots + \frac{c_{in(i)}z}{(z - p_i)^{n(i)}} + \cdots \tag{A-32}$$

where

$$\begin{aligned}
a_0 &= \lim_{z\to 0} X(z) \\
c_{in(i)} &= \lim_{z\to p_i}(z - p_i)^{n(i)}\frac{X(z)}{z} \\
c_{i1} &= \lim_{z\to p_i}\frac{d[(z - p_i)^{n(i)}X(z)/z]}{dz} \\
c_{ik} &= \lim_{z\to p_i}\frac{1}{k!}\frac{d^{n(i)-k}[(z - p_i)^{n(i)}X(z)/z]}{dz^{n(i)-k}}
\end{aligned} \tag{A-33}$$

for $k \in [1, n(i)]$. The next few examples illustrate a partial fraction expansion of a function having repeated roots.

Example A-13 Partial Fraction Expansion

Required:
The function $X(z) = z^2/(z - 1)^2$ has a partial fraction expansion given by

$$X(z) = \frac{a_{11}z}{z - 1} + \frac{a_{12}z}{(z - 1)^2} \qquad n(1) = 2$$

Compute the Heaviside coefficients.

Result:
Computing according to Equation (A-33):

$$\begin{aligned}
a_{12} &= \lim_{z\to 1}\frac{(z - 1)^2X(z)}{z} = \lim_{z\to 1} z = 1 \\
a_{11} &= \lim_{z\to 1}\frac{d[(z - 1)^2X(z)/z]}{dz} = \lim_{z\to 1}(dz/dz) = \lim_{z\to 1} 1 = 1
\end{aligned}$$

Therefore

$$X(z) = \frac{z}{z - 1} + \frac{z}{(z - 1)^2}$$

which consists of two elementary terms which can be inverted using standard tables of z-transforms (as shown in Table A-1). That is

$$\begin{aligned}
Z^1[z/(z - 1)] &= 1 && \text{for } t \ge 0\text{; 0 otherwise} \\
Z^1[z/(z - 1)^2] &= nT_s && \text{for } t \ge 0\text{; 0 otherwise}
\end{aligned}$$

Therefore, $x(n) = 1 + nT_s$ for $t \ge 0$, and 0 otherwise.

Example A-14 Partial Fraction Expansion

Required:
Suppose

$$X(z) = \frac{3z^3 - 5z^2 + 3z - 0.5}{(z-1)^2(z-0.5)}$$

The partial fraction expansion of $X(z)$ has the form

$$\begin{array}{cccc} X(z) = a_0 & + \dfrac{a_{11}z}{(z-0.5)} & + \dfrac{a_{12}z}{(z-1)} & + \dfrac{a_{22}z}{(z-1)^2} \\ \uparrow & \uparrow & \uparrow & \uparrow \\ \text{Kronecker} & \text{exponential} & \text{step function} & \text{ramp function} \end{array}$$

Compute the Heaviside coefficients.

Result:
Computing according to Equation (A-33):

$$a_0 = \lim_{z\to 0} X(z) = \frac{0.5}{0.5} = 1$$

$$a_{11} = \lim_{z\to 0.5}\frac{(z-0.5)X(z)}{z} = \lim_{z\to 0.5}\frac{(3z^3 - 5z^2 + 3z - 0.5)}{z(z-1)^2}$$

$$= \frac{(3/8 - 5/4 + 3/2 - 1/2)}{(1/2)(1/4)} = 1$$

$$a_{12} = \lim_{z\to 1}\frac{(z-1)^2X(z)}{z} = \lim_{z\to 1}\frac{(3z^3 - 5z^2 + 3z - 1/2)}{z(z-1/2)} = \frac{(3 - 5 + 3 - 1/2)}{1/2} = 1$$

$$a_{22} = \lim_{z\to 1}\frac{d(z-1)^2X(z)/z}{dz}$$

$$= \lim_{z\to 1}\left\{\frac{9z^2 - 10z + 3}{z(z-1/2)} - \frac{(3z^2 - 5z^2 + 3z - 1/2)(z - 1/2 + z)}{[z(z-1/2)]^2}\right\}$$

$$= \left[\frac{9 - 10 + 3}{1(1/2)}\right] - \left\{\frac{(3 - 5 + 3 - 1/2)(1/2 + 1)}{[1(1/2)]^2}\right\} = 1$$

Upon collecting terms

$$x(nT) = \delta_K + (0.5)^{nT} + (1.0) + (nT)$$

The first few sample values of $x(nT)$ can be seen to be $\{x(n)\} = \{3, 5/2, 13/4, \ldots\}$ for $n = 0, 1$, and 2. As a check, consider the first several terms of $x(nT)$ produced by long division. Here

$$X(z) = \frac{N(z)}{D(z)} = \frac{3z^3 - 5z^2 + 3z - 1/2}{z^3 - 5/2z^2 + 2z - 1/2}$$

or

$$D(z)\,\overline{\big)N(z)}\quad 3 + (5/2)z^{-1} + (13/4)z^{-2} + \cdots$$

which agrees with the derived time series.

If complex roots are encountered, the combining of first-order terms into second-order terms can be achieved using the following relationship:

1. Let $p_i = \alpha_i = j\beta_I$ and $n(i) = 1$ (nonrepeated) and $p_i^* = \alpha_i + j\beta_i$ and $n^*(i) = 1$ (nonrepeated)

$$X(z) = \frac{N(z)}{D(z)} = \frac{N(z)}{D'(z)(z - \alpha - j\beta)(z - \alpha + j\beta)}$$

$$D'(z) = D(z)/(z - \alpha - j\beta)(z - \alpha = j\beta) \tag{A-34}$$

2. Partial fraction expansion

$$X(z) = \cdots + \frac{a_k z}{(z - \alpha - j\beta)} + \frac{a_{k+1} z}{(z - \alpha + j\beta)} \tag{A-35}$$

$$a_k = \lim_{z \to \alpha + j\beta} \frac{(z - \alpha - j\beta)N(z)}{zD(z)} = \lim_{z \to \alpha + j\beta} \frac{N(z)}{zD'(z)(z - \alpha + j\beta)}$$

$$= \frac{N(\alpha + j\beta)}{(\alpha + j\beta)D'(\alpha + j\beta)(2j\beta)} \tag{A-36}$$

$$a_{k+1} = \lim_{z \to \alpha - j\beta} \frac{(z - \alpha + j\beta)N(z)}{zD(z)} = \lim_{z \to \alpha - j\beta} \frac{N(z)}{zD'(z)(z - \alpha - j\beta)}$$

$$= \frac{N(\alpha - j\beta)}{(\alpha - j\beta)D'(\alpha - j\beta)(-2j\beta)}$$

3. Define

$$H_i(z) = \frac{a_k z}{z - p_i};\ H_{i+1}(z) = \frac{a_{k+1} z}{z - p_i^*} \tag{A-37}$$

4. Define

$$H_k(z) \triangleq H_i(z) + H_{i=1}(z) = \frac{a_k z}{z - p_i} + \frac{a_{k+1} z}{z - p_i^*}$$

$$= \frac{(a_k + a_{k+1})z^2 + (-p_i^* a_k - p_i a_{k+1})z}{z^2 - 2\alpha_i z + \left(\alpha_i^2 + \beta_i^2\right)} = \frac{c_i z^2 + d_i z}{z^2 + e_i z + f_i} \tag{A-38}$$

where c_i, d_i, e_i, and f_i are real and defined in the obvious manner.

5. Invert $H_k(z)$ using the table of elementary functions.

The complex root case is illustrated in the following example.

Example A-15 Partial Fraction Expansion

Required:
Suppose

$$X(z) \frac{z^3 - (\sqrt{2}/2)z^2 + z/2}{(z - \sqrt{2}/2)(z^2 - \sqrt{2}z + 1)}$$

The second-order polynomial $(z^2 - \sqrt{2}z + 1)$ can be factored into a pair of terms

$$z^2 - \sqrt{2}z + 1 = \left(z - \frac{\sqrt{2}}{2} - j\frac{\sqrt{2}}{2}\right)\left(z - \frac{\sqrt{2}}{2} + j\frac{\sqrt{2}}{2}\right) \triangleq (z - p_1)(z - p_2)$$

where $p_2 = p_1^*$. The partial fraction expansion of $X(z)$ becomes

$$X(z) = a_0 + \frac{a_i z}{z - \sqrt{2}/2} + \frac{a_2 z}{z - \sqrt{2}/2 - j\sqrt{2}/2} + \frac{a_3 z}{z - \sqrt{2}/2 + j\sqrt{2}/2}$$

Compute the Heaviside coefficients.

Result:
Computing

$$a_0 = \lim_{z \to 0} X(z) = 0$$

$$a_1 = \lim_{z \to \sqrt{2}/2} \frac{(z - \sqrt{2}/2)X(z)}{z} = \frac{(1/2) - (1/2) + (1/2)}{(1/2) - 1 + 1} = 1$$

$$a_2 = \lim_{z \to \sqrt{2}/2 + j\sqrt{2}/2} \frac{(z - \sqrt{2}/2 - \sqrt{2}/2)X(z)}{z}$$

$$= \lim_{z \to \sqrt{2}/2 + \sqrt{2}/2} \frac{z^2 - (\sqrt{2}/2)z^1 + 1/2}{(z - \sqrt{2}/2)(z - \sqrt{2}/2 + j\sqrt{2}/2)}$$

$$= \frac{(1/2 + j1 - 1/2) - (1/2 + j1/2) + 1/2}{(j\sqrt{2}/2)(+j\sqrt{2})} = \frac{j1/2}{-1} = -j/2$$

$$a_3 = \lim_{z \to \sqrt{2}/2 + j\sqrt{2}/2} \frac{(z - \sqrt{2}/2 - \sqrt{2}/2)X(z)}{z}$$

$$= \lim_{z \to \sqrt{2}/2 + \sqrt{2}/2} \frac{z^2 - (\sqrt{2}/2)z^1 + 1/2}{(z - \sqrt{2}/2)(z - \sqrt{2}/2 + j\sqrt{2}/2)}$$

$$= \frac{(1/2 + j1 - 1/2) - (1/2 - j1/2) + 1/2}{(-j\sqrt{2}/2)(-j\sqrt{2})} = \frac{-j1/2}{-1} = j/2 = a_2^*$$

Then, for $p_2 = -\sqrt{2}/2 - j\sqrt{2}/2$, it follows that

$$X(z) = a_0 + \frac{1z}{z - \sqrt{2}/2} + \frac{(-jz/2)}{(z + p_2)} = \frac{(jz/2)}{(z + p_2^*)}\bigg|_{a_0 \to 0}$$

$$= \frac{z}{z - \sqrt{2}/2} + j\frac{z}{2}\left(\frac{-p_2^* + p_2}{z^2 - \sqrt{2}z + 1}\right)$$

$$= \frac{z}{z - \sqrt{2}/2} + j\frac{z}{2}\left(\frac{-j\sqrt{2}}{z^2 - \sqrt{2}z + 1}\right) = \frac{z}{z - \sqrt{2}/2} + \frac{\sqrt{2}z/2}{z^2 - \sqrt{2}z + 1}$$

Upon inspection of Table A-1 of the elementary functions, one finds

$$Z^{-1}(z/(z - \sqrt{2}/2)) = \sqrt{2}/2)^{nT}$$

$$Z^{-1}[(\sqrt{2}z/2)/(z^2 - \sqrt{2}z + 1)] = \sin(\omega_0 nT_s)$$

for $\sin(\omega_0 T_s) = \sqrt{2}/2$.

Residue Calculus. The residue theorem was introduced earlier and was used to map the Laplace transform of $x(t)$, namely $X(s)$, into the z-transform of the time series of $\{x(nT)\}$. The residue theorem can also be used to invert a z-transform.

A function $X(z)$ is said to be analytic at a point $z = z_0$ if and only if it is single valued and uniquely differentiable at z_0. If $X(z)$ is analytic at z_0, it can be expanded as a Taylor series of the form

$$X(z) = \sum_{i=0}^{\infty} a_i (z - z_0)^i$$

$$a_0 = X(z_0) \tag{A-39}$$

$$a_n = \frac{1}{n!} \frac{d^n X(z)}{dz^n}\bigg|_{z=z_0}$$

If $X(z)$ is not analytic at z_0, a Laurent series is required and is given by

$$X(z) = \sum_{i=1}^{\infty} a_{-1}/(z - z_0)^i \tag{A-40}$$

where a_{-1} is called the residue of $X(z)$ at z_0 and is given, in general, by

$$a_{-1} = \frac{1}{2\pi j} \oint X(z)\,dz \tag{A-41}$$

Given knowledge of the residues of $X(z)$, the parent time series can be reconstructed as

$$X(n) = \sum_{K} \text{Res}(X(z)z^{n-1}, p_k) \tag{A-42}$$

where Res $[a, b]$ reads the residue of a at the point b. The following example demonstrates the residue method of inversion.

Example A-16 Residue Inversion

Required:

Suppose $X(z) = z^2/(z - 1)^2$. Using long division, it can be seen that

$$(z - 1)^2 = z^2 - 2z + 1 \;\overline{\big)\, z^2}\quad 1 + 2z^{-1} + 3z^{-2} + \cdots$$

Using a partial fraction expansion

$$X(z) = a_0 + \frac{a_{11}z}{(z - 1)} + \frac{a_{12}z}{(z - 1)^2}$$

where

$$a_0 = \lim_{z \to 0} X(z) = 0$$

$$a_{12} = \lim_{z \to 1} \frac{(z-1)^2 X(z)}{z} = \lim_{z \to 1} z = 1$$

$$a_{11} = \lim_{z \to 1} \frac{d[(z-1)^2 X(z)/z]}{dz} = \lim_{z \to 1} (dz/dz) = 1$$

Therefore

$$X(z) = z_0 + \frac{z}{z-1} + \frac{z}{(z-1)^2}$$

and

$$x(n) = 1 + nT_s \text{ for } \geq 0\text{; 0 otherwise}$$

Use the residue theorem to verify this result.

Result:
The Laurent expansion of $X(z)$ is given by

$$X(z)z^{n-1} = \frac{z^{n+1}}{(z-1)^2} = \frac{a_{-1}}{z-1} + \frac{a_{-2}}{(z-1)^2}$$

Compute a_{-1}, a residue, using the rules for repeating roots as presented in the partial fraction study

$$a_{-1} = \lim_{z \to 1} \frac{d(z-1)^2 X(z) z^{n-1}}{dz} = \lim_{z \to 1} \frac{dz^{n+1}}{dz} = n + 1$$

which is in agreement with the other derivations.

Some care must be taken to ensure that all of the residue of $X(z)z^{n-1}$ are specified for all n. Example A-17 demonstrates this.

Example A-17 Residue Inversion

Required:
Suppose $X(z) = (z = 1)^3/[(z + 1/2)(z - 1/2)^2]$. The term

$$X(z)z^{n-1} = \frac{(z+1)^3 z^{n-1}}{(z+1/2)(z-1/2)^2}$$

has poles at $z = -1/2$ of multiplicity 1, and at $z = 1/2$ or multiplicity 2 for all $n > 1$. For the $n = 0$, an additional pole is created by $z^{n-1} = z^{-1}$ and is seen to reside at the origin (that is, $z = 0$). Therefore, the analysis must attend to these cases. Use the residue theorem to invert $X(z)$.

Result:
Begin with the case $k = 0$.

$$X(z)z^{-1} = \frac{a_{-1}}{z} + \frac{b_{-1}}{z+1/2} + \frac{c_{-1}}{z-1/2} + \text{others}$$

$$\leftarrow \quad \text{residues} \quad \rightarrow$$

$$a_{-1} = \lim_{z\to 0} z[X(z)z^{-1}] = \lim_{z\to 0}\frac{(z+1)^3}{(z+1/2)(z-1/2)} = \frac{1}{(1/2)(-1/2)^2} = 8$$

$$b_{-1} = \lim_{z\to 1/2}(z+1/2)[X(z)z^{-1}] = \lim_{z\to 1/2}\frac{(z+1)^3}{z(z-1/2)^2} = \frac{(1/2)^3}{(-1/2)(-1)^2} = -\frac{1}{4}$$

$$c_{-1} = \lim_{z\to 1/2}\left\{\frac{d[(z-1/2)^2(X(z)z^{-1})]}{dz}\right\} = \lim_{z\to 1/2}\left(\frac{d\{(z+1)^3/[z(z+1/2)]\}}{dz}\right)$$

$$= \lim_{z\to 1/2}\left\{\frac{3(z+1)^2}{z(z+1/2)} - \frac{(z+1)^3(z+z+1/2)}{[z(z+1/2)]^2}\right\} = \frac{3(3/2)^2}{1/2(1)} - \frac{(3/2)^3(3/2)}{[(1/2)(1)]^2}$$

$$= \frac{27}{2} = -3\frac{27}{4} = -6\frac{3}{4}$$

Therefore

$$x(n=0) = \sum_k \text{Res}(X(z)z^{-1}, p_k)$$

$$= a_{-1} + b_{-1} + c_{-1} = 8 - \frac{1}{4} - 6\frac{3}{4} = 1$$

As a check, the first term in long division inversion of $X(z)$ yields

$$z^3 - z/2 - z/4 + 1/8 \,\overline{\big)\, z^3 + 3/z^2 + 2z + 1} \quad 1 + \cdots$$

or $x(k=0) = 1$ as derived.

Continuing, for the second case $k > 0$.

$$X(z)z^{n-1} = \frac{b_{-1}}{z+1/2} + \frac{c_{-1}}{z-1/2} + \text{others}$$

$$b_{-1} = \lim_{z\to -1/2}(z+1/2)[X(z)z^{n-1}] = \lim_{z\to 1/2}\frac{(z+1)^3 z^{n-1}}{(z-1/2)^2} = \frac{(1/2)^3(-1/2)^{n-1}}{-1^2}$$

$$= -(1/2)^2(-1/2)^n = -(-1/2)^n/4$$

$$c_{-1} = \lim_{z\to 1/2}\frac{d(z-1/2)^2X(z)z^{n-1}}{dz} = \lim_{z\to 1/2}\frac{d[(z+1)^3z^{n-1}/(z+1/2)]}{dz}$$

$$= \lim_{z\to 1/2}\left[\frac{(n-1)z^{n-2}(z+1)^3 + 3z^{n-1}(z+1)^2}{(z+1/2)} - \frac{(z+1)^3z^{n-1}}{(z+1/2)^2}\right]$$

$$= (n-1)\left(\frac{1}{2}\right)^{n-2}\left(\frac{3}{2}\right)^3 + 3\left(\frac{1}{2}\right)^{n-1}\left(\frac{3}{2}\right)^2 - \left(\frac{3}{2}\right)^3\left(\frac{1}{2}\right)^{n-1}$$

$$= (n-1)\frac{27}{2}\left(\frac{1}{2}\right)^n + \left(\frac{27}{4} - \frac{27}{4}\right)\left(\frac{1}{2}\right)^n$$

$$= 27\left(\frac{n-1}{2} + \frac{1}{4}\right)\left(\frac{1}{2}\right)^n = \frac{27}{2}\left(n - \frac{1}{2}\right)\left(\frac{1}{2}\right)^n$$

Then $\quad x(n) = \sum \text{Res}[X(z)z^{n-1}, p_k] \qquad n > 0$

$$= b_{-1} + c_{-1} = -\frac{1}{4}\left(-\frac{1}{2}\right)^n - \frac{27}{4}\left(\frac{1}{2}\right)^n + \frac{27}{2}n\left(\frac{1}{2}\right)^n$$

Using a partial fraction expression, as a check, one considers

$$X(z) = a_0 + \frac{a_1 z}{(z + 1/2)} + \frac{a_2 z}{(z - 1/2)} + \frac{a_3 z}{(z - 1/2)^2}$$

$$a_0 = 0$$

$$a_1 = \lim_{z \to 1/2} (z + 1/2)\frac{X(z)}{z} = \lim \frac{(z + 1)^3}{z(z - 1/2)^2} = \frac{(1/2)^3}{(-1/2)(-1)^2} = -\frac{1}{4}$$

$$a_3 = \lim_{z \to 1/2} \frac{(z + 1)^3}{z(z + 1/2)} = \frac{(3/2)^3}{(1/2)(1)} = \frac{27}{4}$$

$$a_2 = \lim_{z \to 1/2} \frac{d[(z - 1/2)^2 X(z)/z]}{dz} = \lim_{z \to 1/2} \frac{d[(z + 1)^3/z(z + 1/2)]}{dz}$$

$$= \lim_{z \to 1/2} \left\{ \frac{3(z + 1)^2}{z(z + 1/2)} - \frac{(z + 1)^3(z + z + 1/2}{[z(z + 1/2)]^2} \right\} = \frac{3(3/2)^2}{(1/2)(1)} - \frac{(3/2)^3(3/2)}{[(1/2)(1)]^2}$$

$$= \frac{27}{2} - \frac{27}{2}\left(\frac{3}{2}\right) = \frac{-27}{4}$$

Therefore

$$X(z) = -\frac{1}{4}\frac{z}{z + 1/2} - \frac{27}{4}\frac{z}{(z - 1/2)} + \frac{27}{4}\frac{z}{(z - 1/2)^2}.$$

Referring to Table A-1, it can be noted that every term except the last $[z/(z - 1/2)^2]$ has a known inverse. The term in question will require the application of property 6 which results in the following analysis (for $T_s = 1$). Specifically, if

$$x(n) \leftrightarrow X(z)$$

then

$$nx(n) \leftrightarrow -z dX(z)/dz$$

Since $z/(z - 1/2) \leftrightarrow (1/2)^n$, it follows that

$$n(1/2)^n \leftrightarrow -z\frac{d[z/(z - 1/2)]}{dz}$$

where

$$-z\frac{d[z/(z - 1/2)]}{dz} = -z\left[\frac{1}{(z - 1/2)} - \frac{z}{(z - 1/2)^2}\right]$$

$$= -z\left[\frac{-1/2}{(z - 1/2)^2}\right] = \frac{1}{2}\frac{z}{(z - 1/2)^2}$$

which is seen to have the form of the unrecognized term. Therefore

$$X(z) = -\frac{1}{4}\left(\frac{z}{z + 1/2}\right) - \frac{27}{4}\left(\frac{z}{z - 1/2}\right) + \frac{27}{4}\left(\frac{z}{(z - 1/2)^2}\right)$$

$$= -\frac{1}{4}\left(\frac{z}{z + 1/2}\right) - \frac{27}{4}\left(\frac{z}{z - 1/2}\right) + \frac{27}{2}\left[\frac{1}{2}\frac{z}{(z - 1/2)^2}\right]$$

and

$$x(n) = -\frac{1}{4}\left(-\frac{1}{2}\right)^n - \frac{27}{4}\left(\frac{1}{2}\right)^n + \frac{27}{2}\left[n\left(\frac{1}{2}\right)^n\right]$$

for $n > 0$. This agrees with the result derived using the residue theorem.

A.3 MATLAB INVERSION

The MATLAB function *residue* is used to compute Heaviside coefficients. The name residue is derived from the residue calculus equation described by Equation 1. The MATLAB function *residue* is summarized next.

```
RESIDUE Partial-fraction expansion (residues).
[R,P,K] = RESIDUE(B,A) finds the residues, poles and direct
term of a partial fraction expansion of the ratio of two poly-
nomials B(s)/A(s).
If there are no multiple roots,

      B(s)       R(1)       R(2)             R(n)
      ----  = -------- + -------- + ··· + -------- + K(s)
      A(s)    s - P(1)   s - P(2)         s - P(n)

Vectors B and A specify the coefficients of the numerator and
denominator polynomials in descending powers of s.  The
residues are returned in the column vector R, the pole loca-
tions in column vector P, and the direct terms in row vector
K.  The number of poles is n = length(A)-1 = length(R) =
length(P). The direct term coefficient vector is empty if
length(B) < length(A), otherwise length(K) = length(B)-
length(A)+1.

If P(j) = ··· = P(j+m-1) is a pole of multiplicity m, then the
expansion includes terms of the form

         R(j)         R(j+1)                R(j+m-1)
       -------- + ------------  + ··· + ------------
       s - P(j)   (s - P(j))^2           (s - P(j))^m

[B,A] = RESIDUE(R,P,K), with three input arguments and two
output arguments, converts the partial fraction expansion
back to the polynomials with coefficients in B and A.
```

The MATLAB function *residue* was developed primarily to invert Laplace transforms $X(s)$ but can be easily adapted for use with z-transforms. Specifically, suppose

$$X(z) = \alpha_0 + \frac{\alpha_{i,n_i} z}{\left(z - \lambda_i\right)^{n_i}} + \cdots + \frac{\alpha_{j,n_j} z}{\left(z - \lambda_j\right)^{n_j}} + \cdots + \frac{\alpha_{j,k} z}{\left(z - \lambda_j\right)^{k}} + \cdots \tag{A-43}$$

where all the terms are in one-to-one correspondence with element in Table A-1. It can be observed that all the numerator terms in Equation (A-43) have a multiplicative factor z. Since function *residue* is intended for use with a Laplace transform, it assumes that the partial fraction expansion can take the form:

$$X(s) = \frac{\alpha_0}{s} + \frac{\alpha_{i,n_i}}{\left(s - \lambda_i\right)^{n_i}} + \cdots + \frac{\alpha_{j,n_j}}{\left(s - \lambda_j\right)^{n_j}} + \cdots + \frac{\alpha_{j,k}}{\left(s - \lambda_j\right)^{k}} + \cdots \tag{A-44}$$

where all the component terms can be found in the standard table of Laplace transforms. This form can be reproduced in the z-domain by expanding $X'(z) = X(z)/z$ instead of $X(z)$. The partial fraction expansion of $X'(z) = X(z)/z$ is simply

$$X'(z) = \frac{X(z)}{z} = \frac{\alpha_0}{z} + \frac{\alpha_{i,n_i}}{(z - \lambda_i)^{n_i}} + \cdots + \frac{\alpha_{j,n_j}}{(z - \lambda_j)^{n_j}} + \cdots + \frac{\alpha_{j,k}}{(z - \lambda_j)^k} + \cdots \tag{A-45}$$

Observe that, in general, α_{j,n_j} can be isolated and analyzed using the Heaviside method. The use of MATLAB to perform the inversion is illustrated in the next set of examples.

Example A-18 MATLAB Inversion

Required:
Suppose

$$X(z) = \frac{z^2}{(z - 1)(z - e^{-T_s})}$$

where $x[k]$ is causal and $T_s > 0$. It is apparent that all the poles of $X(z)$ are distinct and are $z = 1$ and $z = e^{-T_s}$. The general form of the partial fraction expansion of $X(z)$ is

$$X(z) = \alpha_0 + \alpha_1\left(\frac{z}{z - 1}\right) + \alpha_2\left(\frac{z}{z - e^{-T}}\right)$$

and $X'(z) = X(z)/z$ is

$$X'(z) = \frac{X(z)}{z} = \frac{\alpha_0}{z} + \alpha_1\left(\frac{1}{z - 1}\right) + \alpha_2\left(\frac{1}{z - e^{-T}}\right)$$

Manually computing the Heaviside coefficients results in the following outcome:

$$\alpha_0 = \lim_{z\to 0}\left(\frac{zX(z)}{z}\right) = \lim_{z\to 0}\left(\frac{z^2}{(z - 1)(z - e^{-T})}\right) = 0$$

$$\alpha_1 = \lim_{z\to 1}\left(\frac{(z - 1)X(z)}{z}\right) = \lim_{z\to 1}\left(\frac{z}{(z - e^{-T})}\right) = \left(\frac{1}{(1 - e^{-T})}\right)$$

$$\alpha_2 = \lim_{z\to e^{-T}}\left(\frac{(z - e^{-T})X(z)}{z}\right) = \lim_{z\to e^{-T}}\left(\frac{z}{(z - 1)}\right) = \left(\frac{e^{-T}}{(e^{-T} - 1)}\right) = \left(\frac{1}{(1 - e^{T})}\right).$$

Therefore

$$X(z) = 2\left(\frac{z}{z - 1}\right) - 1\left(\frac{z}{z - 0.5}\right) \Rightarrow x[k] = \left(2 - 0.5^k\right)u[k]$$

Using program "*residue*" performs a partial fraction expansion of $X'(z)$ for $\exp(-T_s) = 0.5$. That is

$$X'(z) = \frac{z^2}{z(z - 1)(z - e^{-T_s})} = \frac{z^2}{z^3 - 1.5z^2 + 0.5z}$$

$$= \frac{\alpha_0}{z} + \alpha_1\left(\frac{1}{z - 1}\right) + \alpha_2\left(\frac{1}{z - 0.5}\right)$$

Results:

It follows that

```
» b=[1,0,0];        {=z^2}
» a=[1,-1.5,0.5,0];       {=z^3-1.5z^2+0.5z}
» residue(b,a)
ans =
       2
      -1
       0
```

which agrees with the manually computed residues.

Example A-19 MATLAB Inversion

Required:

Suppose:

$$X(z) = \frac{z^2}{(z-1)^2}$$

and it follows that

$$X'(z) = \frac{X(z)}{z} = \frac{z^2}{z(z-1)^2}$$

$X'(z)$ has poles at $z = 0$, 1, and 1 (repeated). The partial fraction expansion of $X'(z)$ is

$$X'(z) = \frac{X(z)}{z} = \alpha_0 \frac{1}{z} + \alpha_{1,1}\left(\frac{1}{z-1}\right) + \alpha_{1,2}\left(\frac{1}{(z-1)^2}\right)$$

A manual analysis of the problem yields

$$\alpha_0 = \lim_{z\to 0} \frac{zY(z)}{z} = 0$$

$$\alpha_{1,2} = \lim_{z\to 1} \frac{(z-1)^2 Y(z)}{z} = \lim_{z\to 1}(z) = 1$$

$$\alpha_{1,1} = \lim_{z\to 1} \frac{d}{dz}\left(\frac{(z-1)^2 Y(z)}{z}\right) = \lim_{z\to 1}\left(\frac{d(z)}{dz}\right) = 1,$$

resulting in a partial fraction expansion of

$$X'(z) = \frac{X(z)}{z} = \frac{1}{z-1} + \frac{1}{(z-1)^2} \Rightarrow X(z) = \frac{z}{z-1} + \frac{z}{(z-1)^2}$$

where all the terms are found in Table A-1. Perform the partial fraction expansion using MATLAB.

Result:

Using *residue*, the following outcome resulted.

```
EDU» b=[1,0,0]      {=z^2}
b =
        1       0       0
EDU» a=[1,-2,1,0]      {=z^3-2z^2+z}
a =
        1      -2       1       0
```

```
EDU» residue(b,a)
ans =
      1
      1
      0
```

which agrees with the manually computed residues.

Another MATLAB function, called *residuez*, is promoted as a partial fraction expansion tool for use with z-domain polynomials. The MATLAB function *residuez* is summarized next.

```
RESIDUEZ Z-transform partial-fraction expansion.
[R,P,K] = RESIDUEZ(B,A) finds the residues, poles and direct
terms of the partial-fraction expansion of B(z)/A(z),

 B(z)        r(1)                  r(n)
 ---- = ------------ +··· ------------ + k(1) + k(2)z^(-1) ···
 A(z)   1-p(1)z^(-1)       1-p(n)z^(-1)

B and A are the numerator and denominator polynomial coeffi-
cients, respectively, in ascending powers of z^(-1).  R and P
are column vectors containing the residues and poles, respec-
tively.  K contains the direct terms in a row vector.  The
number of poles is n = length(A)-1 = length(R) = length(P)

The direct term coefficient vector is empty if length(B) <
length(A); otherwise, length(K) = length(B)-length(A)+1

If P(j) = ··· = P(j+m-1) is a pole of multiplicity m, then the
expansion includes terms of the form

      R(j)                R(j+1)                   R(j+m-1)
 -------------- + ------------------ + ··· + ------------------
 1 - P(j)z^(-1)   (1 - P(j)z^(-1))^2         (1 - P(j)z^(-1))^m

[B,A] = RESIDUEZ(R,P,K) converts the partial-fraction expan-
sion back to B/A form.
```

The MATLAB *residuez* does, however, have a weakness that is motivated in the next two examples.

Example A-20 MATLAB Inversion

Required:
Suppose

$$X(z) = \frac{z^2}{(z-1)(z-0.5)} = \frac{1}{(1-z^{-1})(1-0.5z^{-1})} = \alpha_1\frac{1}{(1-z^{-1})} + \alpha_2\frac{1}{(1-0.5z^{-1})}$$

Use *RESIDUEZ* to compute α_1 and α_2. Specifically, show

$$X(z) = \alpha_1\frac{1}{(1-z^{-1})} + \alpha_2\frac{1}{(1-0.5z^{-1})}$$

Results:
The program *RESIDUEZ* performs a partial fraction expansion of $X(z) = N(z)/D(z)$ as follows:

```
b=[1];
a=[1,-1.5,0.5,0];
residue(b,a)
ans =
      2
      1
```

which defines $X(z)$ to be

$$X(z) = \frac{2}{(1 - z^{-1})} + \frac{-1}{(1 - 0.5z^{-1})}$$

It therefore follows that

$$x[k] = (2 - 0.5^k)u[k]$$

Example A-21 MATLAB Inversion

Required:
Suppose

$$X(z) = \frac{1}{(1 - z^{-1})^2} = \alpha_1\frac{1}{(1 - z^{-1})} + \alpha_2\frac{1}{(1 - z^{-1})^2}$$

Use *RESIDUEZ* to compute α_1 and α_2.

Results:
Using MATLAB's *RESIDUEZ*, the Heaviside coefficients are computed to be

```
b=[1];
a=[1,2,1];
residuez(b,a)
ans =
      0
      1
```

which correspond to

$$X(z) = \frac{1}{(1 - z^{-1})^2} = 0\frac{1}{(1 - z^{-1})} + 1\frac{1}{(1 - z^{-1})^2}$$

However, this is not a transform form that is recorded in a standard z-transform table (see Table A-1).

From the first principles, it is known that

$$X(z) = \frac{1}{\left(1 - z^{-1}\right)^2} = \frac{z^2}{(z - 1)^2}$$

$$X'(z) = \frac{X(z)}{z} = \frac{z^2}{z(z - 1)^2} = \alpha_0\frac{1}{z} + \alpha_{1,1}\left(\frac{1}{z - 1}\right) + \alpha_{1,2}\left(\frac{1}{(z - 1)^2}\right)$$

Observe that $X(z)$ is now defined in terms of an eigenvalue located at $z = 0$ of multiplicity 1, and $z = 1$ of multiplicity 2. The partial fraction expansion for $X'(z) = X(z)/z$ can be computed using *RESIDUE* as shown next:

```
b=[1,0,0]
b =
        1       0       0
a=[1,-2,1,0]
a =
        1      -2       1       0
residue(b,a)
ans =
        1
        1
        0
```

which results in a partial fraction:

$$X'(z) = \frac{1}{z-1} + \frac{1}{(z-1)^2} \Rightarrow X(z) = \frac{z}{z-1} + \frac{z}{(z-1)^2}$$

All the component terms forming $X(z)$ are found referenced in a table of standard z-transforms (Table A-1). The resulting time-series $x[k]$ is given by

$$x[k] = (1 + k)u[k]$$

The conclusion to be drawn is that *RESIDUE* and *RESIDUEZ* both produce numerically correct results, but *RESIDUEZ* can produce an outcome which is difficult to interpret when the transform being inverted has poles of multiplicity greater than one. The *RESIDUE*-based method will produce an outcome which is in direct correspondence with the entries found in a table of standard z-transforms.

A.4 DISCRETE FOURIER TRANSFORM (DFT)

The representation of a signal by its frequency domain signature is called spectrum analysis. The decomposition of an arbitrary waveform into a set of sinusoidal components was known to Fourier in 1822. During the intervening decades, the techniques of Fourier have been used to develop a new theory, focusing on the design of analog communication systems and engineering analyses. During this time interval, little was accomplished in the area of computing spectra. However, the advent of the digital computer provided the means by which efficient computer algorithms could be brought to bear on the problem. Initially, the availability of large-scale digital computers in the 1960s had little initial impact on spectral analysis. Software-based systems were slow and they consumed large amounts of computer memory. It was not until the Cooley-Tukey algorithm was published and accepted that the spectral analysis harnessed the power of modern digital computers. The Cooley-Tukey algorithm is a representative of a general class of discrete Fourier transforms, called the fast Fourier transforms (FFT).

The continuous Fourier transform (CFT) transforms a class of time-domain signals, denoted $x(t)$, into the frequency domain. The inverse operation is called the inverse continuous Fourier transform (ICFT). These operations are mathematically defined as follows:

$$X(f) = \int_{-\infty}^{\infty} x(t)\exp(-j2\pi ft)\,dt \quad \text{(CFT)} \tag{A-46}$$

$$x(t) = \int_{-\infty}^{\infty} X(f)\exp(j2\pi ft)\,dt \quad \text{(ICFT)} \tag{A-47}$$

Here, $X(f)$ is called the complex spectrum of $x(t)$, consisting of a magnitude and phase signal representation. The spectrum $X(f)$ is a description of $x(t)$ in the frequency domain. It is characterized by a complex number representing magnitude and phase.

A special case of the Fourier transform is the Fourier series (FS). If $x(t)$ is periodic over some interval of time $[-T/2, T/2]$, it is possible to express the frequency-domain behavior of $x(t)$ as a continuous rather than discrete signal process. A Fourier series can appear in either the trigonometric or the exponential form. The familiar trigonometric series is given by

$$x(t) = a_0 + \sum_{i=-\infty}^{\infty} a_i \cos(2\pi it/T) + b_i \sin(2\pi it/T) \quad \text{(FS)} \tag{A-48}$$

where

$$a_0 = \frac{1}{T}\int_{-T/2}^{T/2} x(t)dt \tag{A-49}$$

$$a_n = \frac{1}{T}\int_{-T/2}^{T/2} x(t)\cos(2\pi nt/T)\,dt \tag{A-50}$$

$$b_n = \frac{1}{T}\int_{-T/2}^{T/2} x(t)\sin(2\pi nt/T)\,dt \tag{A-51}$$

For an exponential transform, use

$$\begin{aligned} \sin(a) &= [\exp(ja) - \exp(-ja)]/2j \\ \cos(a) &= [\exp(ja) = \exp(-ja)]/2j \end{aligned} \tag{A-52}$$

The result is an exponential Fourier series given by

$$x(t) = \sum_{i=-\infty}^{\infty} A_i \exp(j2\pi it/T) \tag{A-53}$$

where

$$A_0 = a_0 \tag{A-54}$$

$$A_i = \begin{bmatrix} (a_i - jb_i)/2 \\ (a_i + jb_i)/2 \end{bmatrix} \quad \begin{matrix} \text{for } i > 0 \\ \text{for } i < 0 \end{matrix} \tag{A-55}$$

The notable difference between a general CFT and the FS is spectral resolution.

The CFT spectrum is a continuous image into the frequency domain, ranging from $-\infty$ to ∞ Hz. The spectrum generated by the Fourier series is a discrete cover of the interval. In fact, if the interval of signal repetition (called the fundamental period) is T, thereby defining the fundamental frequency of $x(t)$ to be f_0 where

$$f_0 = \frac{1}{T_s} \tag{A-56}$$

then the Fourier series exhibits spectral information at kf_0 Hz only, where $k = (\ldots, 2, -1, 0, 1, 2, \ldots)$. In terms of harmonics,

$$
\begin{aligned}
|k| &= 0 \text{ implies the 0 harmonic (DC)} \\
|k| &= 1 \text{ implies the first harmonic (fundamental)} \qquad \text{(A-57)} \\
|k| &= 2 \text{ implies the second harmonic}
\end{aligned}
$$

The discrete Fourier transform (DFT) is conceptually similar to both the CFT and the FS. It is like the CFT in form, except integration is replaced by summation. It is also like the Fourier series in that it is a transformation of a time-domain record of length T into the discrete frequency domain. However, it differs from both in that it transforms a discrete time-domain record into the frequency domain. A principal difference is that the frequency spectrum of a DFT does not range from $-\infty$ to ∞ Hz but resides in a finite subinterval of that infinite interval. The hierarchy of Fourier transforms is diagrammed in Figure A-2.

A time-series record of a signal $x(t)$ is generated by sampling. If a sample is taken every T_s seconds for a total of T_0 seconds, the time series record of $x(t)$ would consist of N sample values, where

$$T_0 = NT_s \qquad \text{(A-58)}$$

Consider the time-series record

$$x[k] = x(0), x(1), x(2), \ldots, x(N-1) \qquad \text{(A-59)}$$

Again, let the fundamental frequency (first harmonic) of $x(t)$ be denoted f_0, where

$$f_0 = 1/T_0\,Hz \qquad \text{(A-60)}$$

or in radians per second

$$\omega_0 = 2\pi f_0 \text{ rad/s} \qquad \text{(A-61)}$$

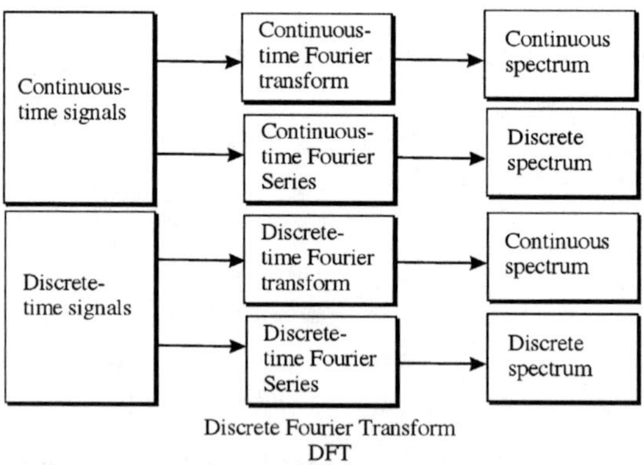

FIGURE A-2 The family of Fourier transforms show signal origin and spectrum destination. The familiar Discrete Fourier Transform (DFT) is functionally equivalent to the DTFS. The difference being that the DTFS requires that the transformed signal be periodic; the DFT only assumes that the transformed signal is periodic.

Then, the discrete Fourier transform can be defined to be

$$X(i) = \sum_{k=0}^{N-1} x(k) W_N^{ik} \qquad (i = 0, 1, \ldots, N-1) \qquad \text{(DFT)} \tag{A-62}$$

and the inverse discrete Fourier transform is given by

$$x(k) = \frac{1}{N} \sum_{k=0}^{N-1} X(i) W_N^{-ik} \qquad (i = 0, 1, \ldots, N-1) \qquad \text{(IDFT)} \tag{A-63}$$

where

$$W_N^{ik} = \exp(-2\pi jik/N) \tag{A-64}$$

The frequency components of the discrete Fourier transform are located at kf_0, $k = 0, 1, 2, \ldots, N-1$. This process is characterized in block diagram form in Figure A-3.

One of the interesting features of the DFT is that both the time-series sample values and Fourier coefficients, namely $x(k)$ and $X(i)$, are periodic. That is, for m an integer

$$x(k) = x(mN + k) \text{ time-series} \tag{A-65}$$

$$X(i) = X(mN + i) \text{ harmonics} \tag{A-66}$$

In addition, if $x(k)$ is a real sequence of sample values, there is symmetry in the transformation of the time-series. This symmetry in the frequency domain is about the so-called folding frequency f_F, where

$$f_F = \frac{Nf_0}{2} = \frac{f_s}{2} \tag{A-67}$$

In addition, the real components of $X(f)$, $\text{Re}[X(f)]$, have even symmetry about f_F, whereas the imaginary components, $\text{Im}[X(f)]$, have odd symmetry (antisymmetry). The symmetry properties of $X(f)$ allow the spectral components of $X(f)$, existing from $N/2$ to $N-1$, to be interpreted as negative frequency locations. The periodic behavior of $X(f)$ gives rise to the relation

$$X(-K) = X(N-K) \quad (k = 0, 1, 2, \ldots, N-1) \tag{A-68}$$

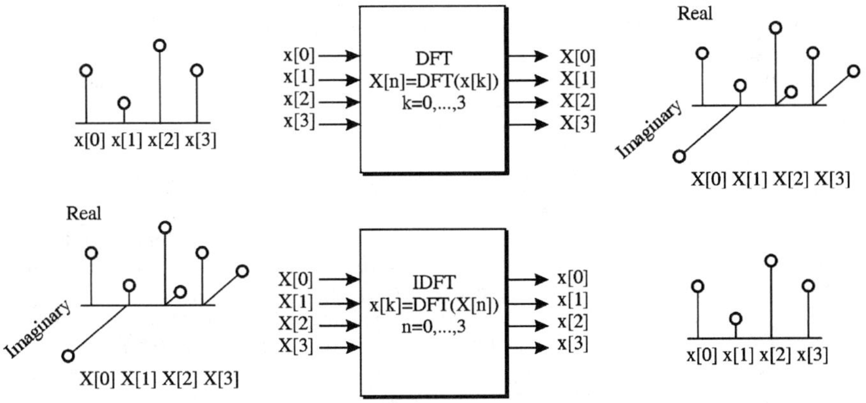

FIGURE A-3 A model of a DFT and IDFT transform for $N = 4$.

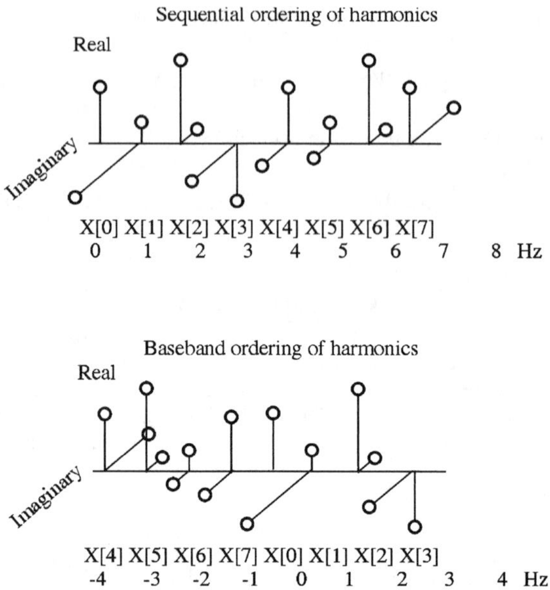

FIGURE A-4 The relationship between sequential ordering of an eight-point DFT output and the positive and negative (base-band) frequencies. The sample rate is assumed to be 8 Hz and the harmonic spacing is 1 Hz.

Then it follows that $X(k)$ and $X(-k)$ form a complex conjugate pair. That is,

$$X(k) = X(-k)^* \tag{A-69}$$

where $*$ denotes complex conjugation. The discussed symmetry properties are interpreted graphically in Figure A-4.

The Cooley-Tukey FFT algorithm is a numerically efficient method of producing discrete transforms. The generation of a spectrum using the direct DFT equation would require approximately N^2 complex multiplications and additions. An FFT-produced spectra would require approximately $2N\log_2(N)$ complex multiply and add operations. The computational savings associated with the FFT method are shown in Figure A-5. The increased efficiency is accomplished through the use of recursive operations on small groupings of data.

A.5 DFT ERROR SOURCES

The utility of an DFT and FFT analysis is great. However, there are latent error sources embedded in this method that the user must be aware of. These error sources fall into three categories: aliasing, leakage, and the picket fence effect.

Aliasing. As the name implies, a signal may be impersonated (an alias) by another. It is the result of choosing a sample rate that is too slow to completely characterize the input signal. As a consequence, a bogus low-frequency signal impersonates the higher-frequency input signal. The aliasing problem can be removed by requiring that the sample rate chosen be at least twice that of the highest frequency residing in the signal to be sampled. This

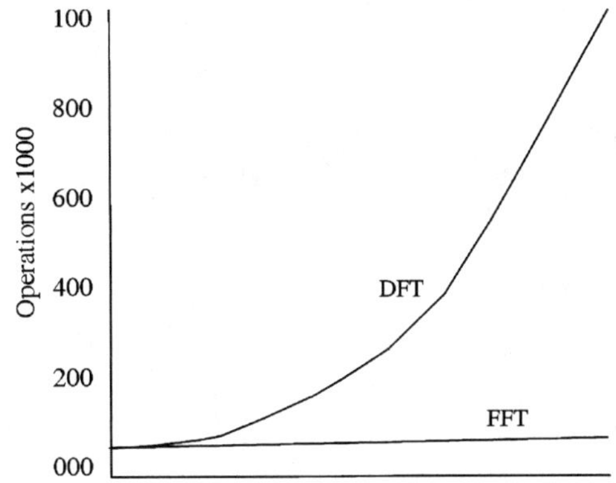

FIGURE A-5 The efficiency of the FFT versus the DFT.

condition is often called the Nyquist condition. For example, suppose a signal $x(t)$ is bandwidth-limited to a frequency $f_{\max}$. That is, there is a bandwidth limitation imposed on $x(t)$ in that all frequency components of $x(t)$ reside at frequencies less than or equal to $f_{\max}$ Hz; then, a non-aliasing sample rate would be f_s where f_s is greater than or equal to $2f_{\max}$ Hz. The effects of aliasing are shown in Figure A-6.

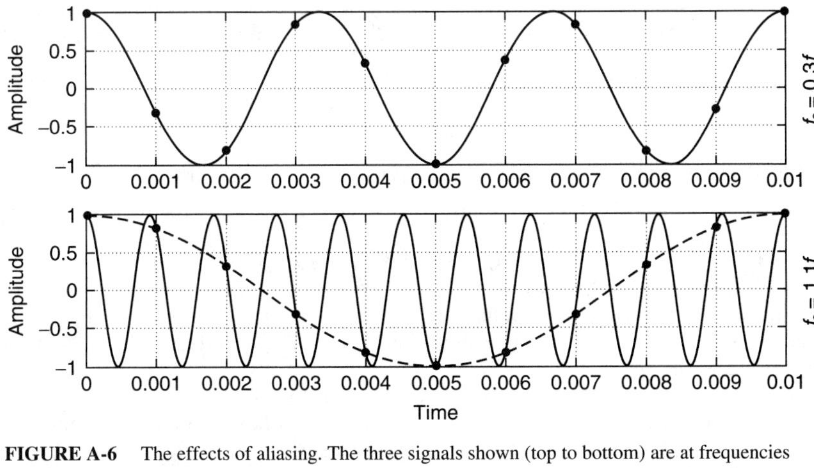

FIGURE A-6 The effects of aliasing. The three signals shown (top to bottom) are at frequencies $f_0 = 0.1f_s$, $f_0 = 0.3f_s$, and $f_0 = 1.1f_s$. The reconstructed signals from the sample values of the first two signals are alias-free. The last signal shows a low frequency reconstructed (aliased) signal.

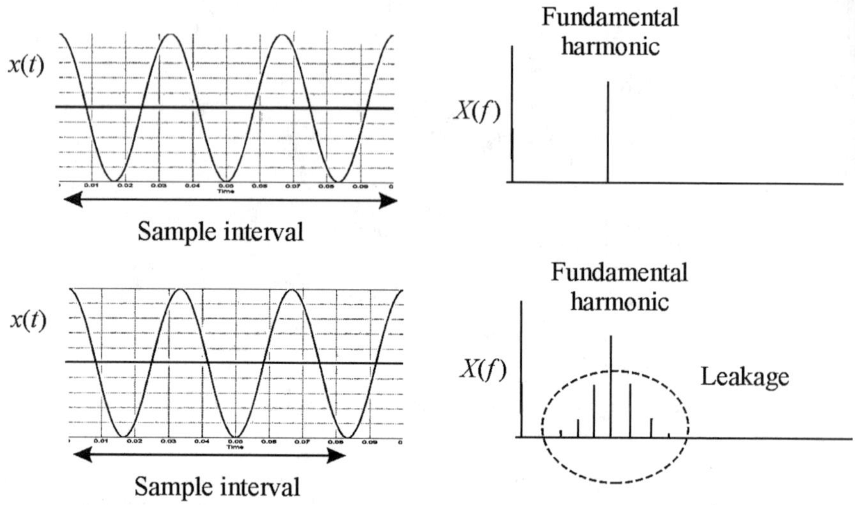

FIGURE A-7 The effects of leakage.

Leakage. Leakage is an undesirable harmonic distortion. The following example dramatizes the problem. Suppose a signal $x(t)$ is given by $x(t) = \sin(2\pi f_0 t)$. Then, only one nontrivial DFT frequency component exists because the time-series is periodic in the sampling interval. However, if $x(t) - \sin[(1.5)\pi f_0 t]$, 1.5 periods are presented to the DFT. The resulting transform consists of a number of components which account for the energy in a signal not exactly at f_0 Hz. The problem of leakage is suggested in Figure A-7.

The leakage problem can be minimized by establishing a sampling interval that is long compared to the fundamental period of the signal to be analyzed. The problem of leakage also may be reduced by a periodicity-inducing filter. Such a filter is used to make an aperiodic time-series appear "nearly" periodic. For example, a popular periodicity-inducing filter is the Hamming window. Over a period T_0, the Hamming window, denoted $H(t)$, satisfies

$$H(t) = \frac{1}{2}[1 - \cos(2\pi f_0 t)] \tag{A-70}$$

The Hamming window will alter an input process, say $x(t)$, as follows:

$$y(t) = H(t) \cdot x(t) \tag{A-71}$$

where $y(t)$ represents the filter output. Suppose the value of $x(0)$ is different from the value of $x(T_0)$. If so, then the process $x(t)$ is obviously aperiodic. However, the windowed output $y(t)$ will have identical end-point values, namely, $y(0) = y(T_0) = 0$. In addition, $y(t)$ will appear to be generally more periodic over the interval 0 to T_0. The effects of windowing are shown in section 13.9.

The Picket Fence Effect. A third source of error is the picket fence effect. The DFT can be envisioned as a set of narrow bandpass filters whose center frequencies are located at k/T Hz, $k = 0, 1, 2, \ldots, N - 1$. The frequency response measured about each center frequency is that of a moving average (MA) filter (that is, $\sin(x)/x$ response), and is detailed in Section 13.15. The FFT base-band response therefore resembles a picket

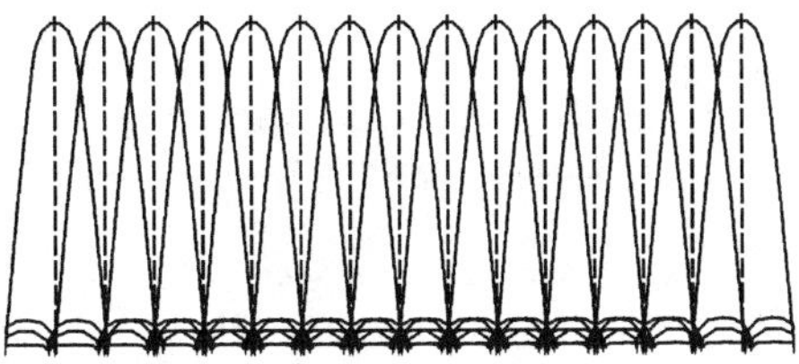

FIGURE A-8 The picket fence effect.

fence, as illustrated in Figure A-8. The frequency components of an input signal $x(t)$ that reside at $k/T = kf_0$ Hz will be transformed without distortion. However, the frequency components of $x(t)$ that reside at noninteger multiples of f_0 are transformed with distortion.

Ensemble-averaging techniques can be used to minimize the effects of the picket fence effect. Interpolation methods, which minimize the effects of this error source, are also available.

BIBLIOGRAPHY

Blahut, R. E. *Fast Algorithms for Digital Signal Processing*. Reading, Massachusetts: Addison-Wesley, 1985.

Bracewell, R. *The Fourier Transform and Its Applications*. New York: McGraw-Hill, 1978.

Brigham, E. O. *The Fast Fourier Transform*. Englewood Cliffs, New Jersey: Prentice-Hall, 1974.

Cadzow, J. A. *Discrete Time Systems*. Englewood Cliffs, New Jersey: Prentice-Hall, 1974.

Cheney, E. W. *Introduction to Approximation Theory*. New York: McGraw-Hill, 1966.

Childers, D. G., ed. *Modern Spectrum Analysis*. New York: IEEE Press, 1978.

Churchill, R. V. *Introduction to Complex Variables and Applications*. New York: McGraw-Hill, 1978.

Digital Signal Processing Committee of IEEE ASSPS. *Programs for Digital Signal Processing*. New York: IEEE Press, 1979.

Fourier, J. *Theorie analytique de la chaleur*, 1822. Reprint of English translation. New York: Dover, 1955.

Guillemin, E. A. *Synthesis of Passive Networks*. New York: John Wiley and Sons, 1957.

Haykin, S. *Communication Systems*. New York: John Wiley and Sons, 1983.

Heideman, M. T., D. Johnson, and C. S. Burrus. "Gauss and the History of the Fast Fourier Transform." *IEEE ASSP Magazine* 1, no. 4 (October, 1984): 14–21.

Jury, E. I. *Theory and Application of the z-Transform Method*. New York: John Wiley and Sons, 1964.

Kaiser, J. F. "Digital Filters." In F. F. Kuo and J. F. Kaiser (eds.), *Systems Analysis by Digital Computer*. New York: John Wiley and Sons, 1966.

Kuo, F. F., and J. F. Kaiser. *Systems Analysis by Digital Computer*. New York: John Wiley and Sons, 1966.

McClellan, J. H., T.W. Parks, and L. R. Rabiner. "A Computer Program for Designing Optimum FIR Linear Phase Digital Filters." *IEEE Trans. Audio Electroacoust* AU-21 (December, 1973): 506–526.

Oppenheim, A. V., and R. W. Schafer. *Digital Signal Processing*. Englewood Cliffs, New Jersey: Prentice-Hall, 1975.

Papoulis, A. *The Fourier Integral and Its Applications*. New York: McGraw-Hill, 1962.

Rabiner, L. R., and B. Gold. *Theory and Application of Digital Signal Processing*. Englewood Cliffs, New Jersey: Prentice-Hall, 1975.

Roberts, A. R., and C. T. Mullis. *Digital Signal Processing*. Reading, Massachusetts: Addison-Wesley, 1987.

Shannon, C. E. "Communicators in the Presence of Noise." *Proc. of the IRE* 30, no. 1, 1949.

———. *Digital Filter Handbook*. New York: Marcel Dekker, 1984.

——— and S. L. Smith. *Digital Signal Processing in Fortran*. Lexington, Massachusetts: Lexington Books, 1965.

Tretter, S. A. *Introduction to Discrete Time Spectral Analysis*. New York: John Wiley and Sons, 1975.

Van Valkenburg, M. E. *Analog Filter Design*. New York: Holt, Rinehart, and Winston, 1982.

Widrow, B., and S. D. Stearns. *Adaptive Signal Processing*. Englewood Cliffs, New Jersey: Prentice-Hall, 1985.

Wylie, C. R., Jr. *Advanced Engineering Mathematics*. New York: McGraw-Hill, 1966.

APPENDIX B
SOFTWARE SUMMARY

B.1 THE FLTRFORM.XLS SPREADSHEET OF FORMULAS

Fltrform.xls is an EXCEL spreadsheet arranged by chapter that contains all the significant formulas in the chapter to simplify some of the calculations. By entering variables, the end result is precisely calculated, in real time, so values can be optimized if desired. Simply copy this spreadsheet from the CD-ROM to your computer.

B.2 FILTER SOLUTIONS (BOOK VERSION) SOFTWARE FOR THE DESIGN OF ELLIPTIC-FUNCTION LOW-PASS FILTERS

Previous editions of this book contained extensive numerical tables of normalized values which had to be scaled to the operating frequencies and impedance levels during the design process. This is no longer the case.

A program called Filter Solutions is included on the CD-ROM. This program is limited to elliptic-function *LC* filters (up to $N = 10$) and is a subset of the complete program, which is available from Nuhertz Technologies (www.nuhertz.com). The reader is encouraged to obtain the full version, which in addition to passive implementations covers many filter polynomial types and includes transmission line, active, switched capacitor, and digital along with many very powerful features. It's been integrated into Applied Wave Research's (AWR) popular Microwave Office software.

The program is quite intuitive and self-explanatory, so the reader should be able to explore its many features on his/her own. In addition, all design examples using this program elaborate on its usage and provide helpful hints. The user can obtain transfer functions, pole-zero plots, reflection coefficients, and step and impulse response, as well as phase, group-delay, and attenuation plots versus frequency for each low-pass design up to a 10th-order elliptic-function low-pass.

Extract and install this program by running FSBook.exe, which is contained on the CD-ROM. You can copy it to the desktop first.

All examples in the book using *Filter Solutions* are based on starting with program default settings. To restore these settings, click the Initialize button, then choose Default, and then Save.

B.3 ELI 1.0 FOR THE DESIGN OF ODD-ORDER ELLIPTIC-FUNCTION LOW-PASS FILTERS UP TO THE 31ST ORDER

This program permits the design of odd-order elliptic-function *LC* low-pass filters up to a complexity of 15 nulls (transmission zeros) or the 31st order. It is based on an algorithm developed by Amstutz[1].

The program inputs are Passband edge (Hz), Stopband edge (Hz), Number of nulls (up to 15), Stopband rejection in dB, and Source and Load terminations (which are always equal). The output parameters are Critical Q (theoretical minimum Q), Passband ripple (dB), Nominal 3 dB cutoff, and a list of component values along with resonant null frequencies.

To install the program, first copy ELI1.zip from the CD-ROM to the desktop and then double-click and extract it to the C:\ root directory, or right-click the folder and choose Extract All, designating the C:\ root directory as the location. A folder, eli1, will be created in the C:\ root directory, and a desktop shortcut called eli1.bat will be placed on the desktop. (If this doesn't happen, go to the C:\eli1 folder and create a shortcut on the desktop using eli1.bat.)

To run, double-click the eli1.bat shortcut and enter inputs as requested. Upon completing the execution, a dataout.text file will open using Notepad, which should contain the resulting circuit description.[2] This text file can be printed out.

B.4 FILTERCAD 3.0 FOR THE DESIGN OF SWITCHED CAPACITOR FILTERS

FilterCAD 3.0 from Linear Technology can help you quickly design a switched-capacitor filter from a set of input parameters, and will then provide a schematic of the filter along with the predicted frequency response and time response. There are two basic modes of operation: *Quick Design* and *Enhanced Design*. *Quick Design* will guide you through a design similar in operation to a "wizard" where it will ask you questions along the way. A design based on standard responses such as Butterworth, Bessel, Chebyshev, and elliptic polynomials will result. The *Enhanced Design* mode is more interactive, allowing custom designs as well as standard polynomials. The user is urged to explore the various Help topics to become more familiar with the program. From time to time Linear Technology may update this program, so it is also available from their web site.

The CD-ROM contains a folder called FCAD. Copy this folder to the desktop, open the folder, and then double-click SETUP.EXE. The program will then install. (See the Start menu to access the program.)

B.5 TX LINE FOR THE DESIGN OF MICROSTRIP, STRIPLINE, AND OTHER STRUCTURES FOR MICROWAVE FILTERS

TX Line is a very useful program from Applied Wave Research, Inc. which can calculate the resulting electrical properties of microstrip and stripline implementations, as well as other structures.

Copy this program from the CD-ROM to the desktop and then double-click and extract it to the C:\ root directory, or right-click the folder, choose Extract All, and select the C:\ root directory as your destination. Create a desktop shortcut icon to txline.exe.

---

[1]P. Amstutz, "Elliptic Approximation and Elliptic Filter Design on Small Computers," *IEEE Transactions on Circuits and Systems* CAS-25, no. 12 (1978).

[2]If the number of nulls is excessive for the response requirements (indicated by zero passband ripple), the final capacitor may have a negative value as a result of the algorithm. Reduce the number of nulls, increase the required attention, or define a steeper filter—or perform a combination of these.

INDEX

LICENSE AGREEMENT

THIS PRODUCT (THE "PRODUCT") CONTAINS PROPRIETARY SOFTWARE, DATA AND INFORMATION (INCLUDING DOCUMENTATION) OWNED BY THE McGRAW-HILL COMPANIES, INC. ("McGRAW-HILL") AND ITS LICENSORS. YOU RIGHT TO USE THE PRODUCT IS GOVERNED BY THE TERMS AND CONDITIONS OF THIS AGREEMENT.

LICENSE: Throughout this License Agreement, "you" shall mean either the individual or the entity whose agent opens this package. are granted a non-exclusive and non-transferable license to use the Product subject to the following terms:
(i) If you have licensed a single user version of the Product, the Product may only be used on a single computer (i.e., a single CPU). If licensed and paid the fee applicable to a local area network or wide area network version of the Product, you are subject to the terms of following subparagraph (ii).
(ii) If you have licensed a local area network version, you may use the Product on unlimited workstations located in one single buildin selected by you that is served by such local area network. If you have licensed a wide area network version, you may use the Product on unlimited workstations located in multiple buildings on the same site selected by you that is served by such wide area network; provide however, that any building will not be considered located in the same site if it is more than five (5) miles away from any building inclu such site. In addition, you may only use a local area or wide area network version of the Product on one single server. If you wish to us Product on more than one server, you must obtain written authorization from McGraw-Hill and pay additional fees.
(iii) You may make one copy of the Product for back-up purposes only and you must maintain an accurate record as to the location of t back-up at all times.

COPYRIGHT; RESTRICTIONS ON USE AND TRANSFER: All rights (including copyright) in and to the Product are owned by McGraw-Hill and its licensors. You are the owner of the enclosed disc on which the Product is recorded. You may not use, copy, decor disassemble, reverse engineer, modify, reproduce, create derivative works, transmit, distribute, sublicense, store in a database or retriev system of any kind, rent or transfer the Product, or any portion thereof, in any form or by any means (including electronically or otherw except as expressly provided for in this License Agreement. You must reproduce the copyright notices, trademark notices, legends and of McGraw-Hill and its licensors that appear on the Product on the back-up copy of the Product which you are permitted to make hereu All rights in the Product not expressly granted herein are reserved by McGraw-Hill and its licensors.

TERM: This License Agreement is effective until terminated. It will terminate if you fail to comply with any term or condition of this License Agreement. Upon termination, you are obligated to return to McGraw-Hill the Product together with all copies thereof and to p all copies of the Product included in any and all servers and computer facilities.

DISCLAIMER OF WARRANTY: THE PRODUCT AND THE BACK-UP COPY ARE LICENSED "AS IS." McGRAW-HILL, ITS LICENSORS AND THE AUTHORS MAKE NO WARRANTIES, EXPRESS OR IMPLIED, AS TO THE RESULTS TO BE OBTAI BY ANY PERSON OR ENTITY FROM USE OF THE PRODUCT, ANY INFORMATION OR DATA INCLUDED THEREIN AND ANY TECHNICAL SUPPORT SERVICES PROVIDED HEREUNDER, IF ANY ("TECHNICAL SUPPORT SERVICES"). McGRAW-HILL, ITS LICENSORS AND THE AUTHORS MAKE NO EXPRESS OR IMPLIED WARRANTIES OF MERCHANTABILITY OR FITNESS FOR A PARTICULAR PURPOSE OR USE WITH RESPECT TO THE PRODUCT. McGRAW-HILL, ITS LICENSORS, AND THE AUTHORS MAKE NO GUARANTEE THAT YOU WILL PASS ANY CERTIFICATION EXAM WHATSOEVER BY USING THIS PRODUCT. NEITHER McGRAW-HILL, ANY OF ITS LICENSORS THE AUTHORS WARRANT THAT THE FUNCTIONS CONTAINED IN THE PRODUCT WILL MEET YOUR REQUIREMENT THAT THE OPERATION OF THE PRODUCT WILL BE UNINTERRUPTED OR ERROR FREE. YOU ASSUME THE ENTIRE R WITH RESPECT TO THE QUALITY AND PERFORMANCE OF THE PRODUCT.

LIMITED WARRANTY FOR DISC: To the original licensee only, McGraw-Hill warrants that the enclosed disc on which the Prod recorded is free from defects in materials and workmanship under normal use and service for a period of ninety (90) days from the dat purchase. In the event of a defect in the disc covered by the foregoing warranty, McGraw-Hill will replace the disc.

LIMITATION OF LIABILITY: NEITHER McGRAW-HILL, ITS LICENSORS NOR THE AUTHORS SHALL BE LIABLE FOR INDIRECT, SPECIAL OR CONSEQUENTIAL DAMAGES, SUCH AS BUT NOT LIMITED TO, LOSS OF ANTICIPATED PROF OR BENEFITS, RESULTING FROM THE USE OR INABILITY TO USE THE PRODUCT EVEN IF ANY OF THEM HAS BEEN ADVISED OF THE POSSIBILITY OF SUCH DAMAGES. THIS LIMITATION OF LIABILITY SHALL APPLY TO ANY CLAIM CAUSE WHATSOEVER WHETHER SUCH CLAIM OR CAUSE ARISES IN CONTRACT, TORT, OR OTHERWISE. Some state not allow the exclusion or limitation of indirect, special or consequential damages, so the above limitation may not apply to you.

U.S. GOVERNMENT RESTRICTED RIGHTS: Any software included in the Product is provided with restricted rights subject to subparagraphs (c), (1) and (2) of the Commercial Computer Software-Restricted Rights clause at 48 C.F.R. 52.227-19. The terms of th Agreement applicable to the use of the data in the Product are those under which the data are generally made available to the general p by McGraw-Hill. Except as provided herein, no reproduction, use, or disclosure rights are granted with respect to the data included in Product and no right to modify or create derivative works from any such data is hereby granted.

GENERAL: This License Agreement constitutes the entire agreement between the parties relating to the Product. The terms of any P Order shall have no effect on the terms of this License Agreement. Failure of McGraw-Hill to insist at any time on strict compliance w this License Agreement shall not constitute a waiver of any rights under this License Agreement. This License Agreement shall be co and governed in accordance with the laws of the State of New York. If any provision of this License Agreement is held to be contrary that provision will be enforced to the maximum extent permissible and the remaining provisions will remain in full force and effect.